Nano-Structured Photovoltaics

Presenting a comprehensive overview of a rapidly burgeoning field blending solar cell technology with nanotechnology, the book covers topics such as solar cell basics, nanotechnology fundamentals, nanocrystalline silicon-based solar cells, nanotextured-surface solar cells, plasmon-enhanced solar cells, optically improved nanoengineered solar cells, dye-sensitized solar cells, 2D perovskite and 2D/3D multidimensional perovskite solar cells, carbonaceous nanomaterial-based solar cells, quantum well solar cells, nanowire solar cells, and quantum dot solar cells. The book provides an in-depth and lucid presentation of the subject matter in an elegant, easy-to-understand writing style, starting from basic knowledge through principles of operation and fabrication of devices to advanced research levels encompassing the recent breakthroughs and cutting-edge innovations. It will be useful for graduate and PhD students, scientists, and engineers.

Nano-Structured Photovoltaics

Solar Cells in the Nanotechnology Era

Vinod Kumar Khanna

CRC Press
Taylor & Francis Group
Boca Raton London New York

CRC Press is an imprint of the
Taylor & Francis Group, an **informa** business

First edition published 2023
by CRC Press
6000 Broken Sound Parkway NW, Suite 300, Boca Raton, FL 33487-2742

and by CRC Press
4 Park Square, Milton Park, Abingdon, Oxon, OX14 4RN

CRC Press is an imprint of Taylor & Francis Group, LLC

© 2023 Vinod Kumar Khanna

ISBN: 978-1-032-07556-3 (hbk)
ISBN: 978-1-032-10403-4 (pbk)
ISBN: 978-1-003-21515-8 (ebk)

DOI: 10.1201/9781003215158

Typeset in Times
by Apex CoVantage, LLC

Dedicated to my beloved parents,
Late Shri Amarnath Khanna and Shrimati Pushpa Khanna,
For my upbringing and education.
My grandson Hansh, daughter Aloka, and wife, Amita,
For their affection and unfailing support.

Contents

Part I   Preliminaries and Nanocrystalline
Silicon Photovoltaics

Part II   Nanotechnological Approaches to Sunlight Harvesting

Part III   *Electrochemical Photovoltaics Using Nanomaterials*

Part IV Photovoltaics with 2D Perovskites and Carbon Nanomaterials

Part V   Quantum Well, Nanowire, and Quantum Dot Photovoltaics

Preface

The sun is a huge celestial body endowed with infinite and inexhaustible energy. Solar cell technology has reached an advanced development level, but still much needs to be accomplished. Nanotechnology can provide solutions to many problems faced by solar cell technology. The union of two technologies, solar cells and nanotechnology, can work wonders. Interesting improvements in solar cell performance facilitated by nanotechnology constitute the subject matter of this book. A general research trend in third-generation solar cells entails the increasing incorporation of nanostructures or nanostructured materials in cells for improving efficiency and reducing manufacturing cost of photovoltaics. In pursuance of this approach, many types of solar cells are being nanoarchitectured and fabricated by economical solution-processable techniques. The present-day nanostructured solar cells still lag behind conventional silicon and GaAs solar cells with regard to efficiency. However, the field is replete with immense opportunities. Novel nanomaterials and associated physical phenomena taking place exclusively at ultrasmall dimensions can usher in a revolution in photovoltaics.

The goal of this book is very straightforward, namely, to describe different types of solar cells in which nanomaterials are used or where nanotechnological principles are applied. The book seeks to weave together the widely scattered knowledge on nanostructured photovoltaics within the pages of a single treatise in a form that is readily accessible to students, engineers, and researchers. Both solar cell and nanotechnologies are progressing rapidly and are regularly flooded with research papers. The available information is scattered in research papers, magazine articles, and webpages. The book seeks to provide a cohesive, accessible framework in which this diversified growth is seamlessly intertwined for easy reading. There are many books on P-N junction solar cells, but very few books on nanotechnological approaches for improving solar cell efficiency and reducing production costs. So there is a scope for a new book to fill this gap. The relevance of this topic in the present scenario of energy crisis and the quest for pollution-free energy sources need hardly be emphasized.

The book is organized into five parts. These five parts contain a total of 15 chapters, as follows:

Part I: Preliminaries and Nanocrystalline Silicon Photovoltaics (Chapters 1–3)
Part II: Nanotechnological Approaches to Sunlight Harvesting (Chapters 4–6)
Part III: Electrochemical Photovoltaics Using Nanomaterials (Chapter 7)
Part IV: Photovoltaics with 2D Perovskites and Carbon Nanomaterials (Chapters 8–9)
Part V: Quantum Well, Nanowire, and Quantum Dot Photovoltaics (Chapters 10–15)

Chapter 1 describes the evolution of generations of solar cells and related solar cell technologies, viz, the first-generation single-crystal or multicrystalline silicon wafer-based solar cells, too expensive to manufacture and having highest efficiency; the second-generation cells, including a-Si:H, CdTe, and CIGS cells, and usually called thin-film solar cells, made from thin layers of semiconductors, consuming less material and therefore available at lower costs; and the third-generation cells, such as mesoscopic dye-sensitized solar cells, quantum well, nanowire, and quantum dot solar cells. Emerging nanostructure devices are in research phase. Many solar cells of the third generation are limited to laboratories and are still away from commercial production.

Chapter 2 demystifies the main terms of nanotechnology, dubbed the technology of matter at nanoscale. Nanotechnology is the buzzword in recent times, and many of its applications are already on the market. The manifold ways in which nanomaterials are used in solar cells or nanotechnology is applied to augment cell performance constitute the focal theme of this book.

Chapter 3 presents solar cells based on nanocrystalline silicon. Hydrogenated nanocrystalline silicon (nc-Si:H), sometimes called microcrystalline silicon, contains crystallites of size 10–75 nm in a matrix of amorphous silicon (a-Si:H). It is more stable towards light than a-Si:H and can absorb the lower-frequency portion of the solar spectrum.

Chapter 4 takes a look at techniques of nanotexturing the surfaces of solar cells to increase absorption of light. Due to the short optical path restriction imposed by thin absorber thickness, an efficient light-trapping strategy is necessary. Significant suppression of photoreflectance of the solar cell surface is achieved by nanotexturing the surface. Nanotextures have proven to be very effective in photon management.

Chapter 5 deals with enhancing solar cell performance by surface plasmon oscillations. Scattering of light by noble metal nanoparticles anchored on the semiconductor surface or inside the absorber layer of the solar and excited at their surface plasmon resonance frequency is utilized to increase optical absorption in solar cells. The antenna-like response of plasmonic nanoparticles caused by dipole oscillation of surface plasmons drastically improves the coupling of sunlight with the semiconductor. The coupling is increased in both dipolar and multipolar scattering regimes by tuning the shapes and sizes of the nanoparticles.

Chapter 6 surveys additional light management nanostructures for solar cells in extension to those presented in Chapter 4. These include the use of dielectric nanoparticles in place of metal nanoparticles, building metallic nanostructures on the rear surface of the solar cell, negative nano-textures, and so forth.

Chapter 7 explores dye-sensitized solar cells. A sensitizer dye attached to the surface of a mesoporous wide bandgap oxide layer (TiO_2, ZnO or Nb_2O_5) film containing nanosize particles absorbs incident sunlight. Photo-induced electron injection from the dye into the conduction band of the semiconductor causes separation of charges. Electron donation from an electrolyte (iodide/triiodide redox couple in an organic solvent) restores the original state of the dye, preventing the recapturing of electron by the oxidized dye, while reduction of triiodide at a counter electrode regenerates the dye. The difference in the Fermi level of the electron in the solid and the redox potential of the electrolyte determines the voltage produced. A large fraction of sunlight can be harvested by using sensitizer dyes with broad absorption bands together with oxide films of nano-crystalline morphology.

Chapter 8 delves into the solar cells made with 2D perovskite nanomaterials, which are layers of perovskites separated by bulky organic cation spacers. The prime motivation for these solar cells is to mitigate the stability and lifetime problems that have stalled the proliferation of 3D perovskite solar cells in the market.

Chapter 9 deals with carbon nanostructures-based solar cells. The use of carbon nanotubes is destined to become pervasive in solar cells because they can be used as a light-sensitive component, for making carrier-selective contacts and for passivation. In silicon solar cells, carbon nanotubes serve as hole-selective contacts, while in organic solar cells, they provide exciton dissociation in the photoactive layer or act as electron acceptors. Laminated CNT networks have been used to make metal electrode-free perovskite solar cells. Apart from CNTs, graphene also has a great potential as a transparent electrode material.

Chapters 10 and 11 overview the progress in quantum well solar cells. The quantum well solar cell is an alternative to tandem solar cell for enhancing efficiency. The cell contains a multiquantum well in the intrinsic region of a P-I-N diode. The bandgap of a quantum well structure can be modulated to match with the solar spectrum to utilize photons from a larger part of this spectrum.

Chapters 12 and 13 outline the developments in nanowire solar cells. Radial as well as axial P-N junctions are fabricated on the nanowires for solar cell realization. Comparatively smaller amount of semiconductor material is consumed in depositing nanowires. The restrictions on material quality are also flexible. Both these factors make nanowire solar cells more cost-effective. Over and above, nanowires display good light-trapping characteristics. So nanowire arrays act as an antireflection layer on the surface of traditional solar cells.

Chapters 14 and 15 are devoted to discussions on quantum dot solar cells. Quantum dots are used in solar cells in a multiplicity of ways. Arrays of quantum dots are used as photoelectrodes. In this arrangement, the array is formed with small inter-QD distances to enable intense electronic coupling for formation of minibands so that long-range electronic transport is possible. An intermediate band solar cell is implemented with a quantum dot superlattice consisting of an array of quantum dots to provide the necessary intermediate band. Quantum dots are used in place of dyes to sensitize nanocrystalline TiO_2 layer to build a variant of the familiar dye-sensitized solar cell, called the quantum dot–sensitized solar cell. Quantum dots are dispersed in a mixture of electron- and hole-conducting polymers. Photovoltaic effects are observed in structures in which QDs form junctions with organic semiconductor polymer. The phenomenon responsible for high conversion efficiency in QD-based solar cells is multiple exciton generation (MEG), whereby more than one, generally two, electron-hole pairs are created by impact with a single incoming photon. The same does not happen in bulk impact ionization.

With the inclusion of a vast canvas of topics spanning over a variety of nanostructured solar cells, the book caters to the needs of a wide-ranging audience, including undergraduate/postgraduate and PhD students in electrical/electronics engineering, nanotechnology, and related disciplines and practicing engineers and scientists engaged in this field. It is hoped that students of electrical and electronics engineering as well as research students working on solar cells will find the book immensely useful. For professionals such as scientists and practicing engineers, the book will be a valuable resource and handy companion helping in updating awareness and keeping pursuit of current research trends.

Let us therefore wear the spectacles of nanotechnology and look through these glasses. We observe solar cell technology in a nanotechnological perspective. Let us input nanotechnology ingredients to make solar cells—these ingredients can be in the form of a nanomaterial, such as carbon nanotubes, or a nanofabrication technique, e.g., nanotexturing—and see how it works wonders for us! There are myriad ways of doing this. That is what this book is all about.

I have an interesting story to tell
There is an electronic device called solar cell
Which generates electricity from sunlight
To light our lamps in the night
There is a technology with the name "nanotechnology"
Dealing with nanoscale objects
Which brings in surface area and quantum mechanical effects
And can revolutionize solar cell prospects
So solar cell meets nanotechnology
And explains its ideology
Nanotechnology says, "All right"
And they work together with all their might
To improve solar cell performance and reduce price
They progress, forge ahead, and rise
Leapfrogging and soaring to reach a new success height
Making our homes more luminous and bright
"Well done!" we write
"Three cheers!" we exclaim with delight
Solar cells look very nice
When seen through nanotechnology eyes
Their beautiful nanostructured appearance
Is an enthralling and exhilarating appearance

Vinod Kumar Khanna
Chandigarh, India

Acknowledgments

I am thankful to God for the invaluable blessings and for giving me the strength and wisdom to complete this work.

I am extremely grateful to the pioneering scientists and engineers for their meticulous, painstaking, and patient work on solar cells and nanotechnology. Their informative research papers and articles are cited in the bibliographies appended at the end of each chapter.

I am indebted to the commissioning editor and editorial assistants at CRC Press for their motivation, kind cooperation, and support. The editorial assistant Dr. Danny Kielty supplied images for Figure 1(a)–(d), Figure 2.5(a)–(b), Figure 2.7(b)–(c), and the book cover.

Last but not the least, I acknowledge the full-fledged support of my family.

Vinod Kumar Khanna
Chandigarh, India

About the Author

INTRODUCTION

Vinod Kumar Khanna is an independent researcher at Chandigarh, India. He is a retired chief scientist from the Council of Scientific and Industrial Research (CSIR)—Central Electronics Engineering Research Institute (CEERI), Pilani, India, and a retired professor from the Academy of Scientific and Innovative Research (AcSIR), Ghaziabad, India. He is a former emeritus scientist, CSIR, and professor emeritus in AcSIR, India. His broad areas of research were the design, fabrication, and characterization of power semiconductor devices and micro- and nanosensors.

ACADEMIC QUALIFICATIONS

He received his MSc degree in Physics with specialization in Electronics from the University of Lucknow in 1975, and PhD degree in Physics from Kurukshetra University in 1988 for the thesis entitled "Development, Characterization and Modeling of the Porous Alumina Humidity Sensor."

WORK EXPERIENCE AND ACCOMPLISHMENTS

His research experience spans over a period of 40 years, from 1977 to 2017. He started his career as a research assistant in the Department of Physics, University of Lucknow, from 1977 to 1980. He joined CSIR–Central Electronics Engineering Research Institute, Pilani (Rajasthan), in April 1980. At CSIR-CEERI, he worked on several CSIR-funded as well as sponsored research and development projects. His major fields of research included power semiconductor devices and microelectronics/MEMS and nanotechnology-based sensors and dosimeters.

In power semiconductor devices area, he worked on the high-voltage and high-current rectifier (600A, 4,300V) for railway traction, high-voltage TV deflection transistor (5A, 1,600V), power Darlington transistor for AC motor drives (5A, 1,600V), fast-switching thyristor (1,300A, 1,700V), power DMOSFET, and IGBT. He contributed towards the development of sealed-tube Ga/Al diffusion for deep junctions, surface electric field control techniques using edge beveling and contouring of large-area devices and floating field limiting ring design, and characterization of minority-carrier lifetime as a function of process steps. He also contributed towards the development of P-I-N diode neutron dosimeter and PMOSFET-based gamma ray dosimeter.

In the area of sensor technology, he worked on the nanoporous aluminum oxide humidity sensor and ion-sensitive field-effect transistor-based microsensors for biomedical, food, and environmental applications; microheater-embedded gas sensor for automotive electronics; MEMS acoustic sensor for launch vehicles; and capacitive MEMS ultrasonic transducer for medical applications.

SEMICONDUCTOR FACILITY CREATION AND MAINTENANCE

He was responsible for setting up and looking after diffusion/oxidation facilities, edge beveling and contouring, reactive sputtering, and carrier lifetime measurement facilities. As the head of MEMS and the Microsensors Group, he looked after the maintenance of six-inch MEMS fabrication facility for R&D projects, as well as the augmentation of processing equipment under this facility at CSIR-CEERI.

SCIENTIFIC POSITIONS HELD

During his tenure of service at CSIR-CEERI from April 1980 till superannuation in November 2014, he was promoted to various positions, including one merit promotion. He retired as Chief Scientist and Professor (AcSIR, Academy of Scientific and Innovative Research) and Head of MEMS and the Microsensors Group. Subsequently, he worked for three years as Emeritus Scientist, CSIR, and Emeritus Professor, AcSIR, from November 2014 to November 2017. After completion of the emeritus scientist scheme, he now lives at Chandigarh. He is a passionate author and enjoys reading and writing.

MEMBERSHIP OF PROFESSIONAL SOCIETIES

He is a fellow and life member of the Institution of Electronics and Telecommunications Engineers (IETE), India. He is a life member of the Indian Physics Association (IPA), Semiconductor Society, India (SSI), and Indo-French Technical Association (IFTA).

FOREIGN TRAVEL

He is widely travelled. He participated and presented research papers in IEEE Industry Application Society's (IEEE-IAS) Annual Meeting at Denver Colorado, USA, in September–October 1986. His short-term research assignments include deputations to Technische Universität Darmstadt, Germany, in 1999; at Kurt-Schwabe-Institut fur Mess-und Sensortechnik e.V., Meinsberg, Germany, in 2008; and at Fondazione Bruno Kessler, Trento, Italy, in 2011, under collaborative programs. He was a member of the Indian delegation to the Institute of Chemical Physics, Novosibirsk, Russia, in 2009.

SCHOLARSHIPS AND AWARDS

Awarded National Scholarship by the Ministry of Education and Social Welfare, Goverment of India, on the basis of Higher Secondary result, 1970; CEERI Foundation Day Merit Team Award, for projects on fast-switching thyristor (1986), power Darlington transistor for transportation (1988), P-I-N diode neutron dosimeter (1992), and high-voltage TV deflection transistor (1994); Dr. N. G. Patel Prize for Best Poster Presentation in the 12th National Seminar on Physics and Technology of Sensors 2007, BARC, Mumbai; and CSIR-DAAD Fellowship in 2008 under Indo-German Bilateral Exchange Programme of Senior Scientists, 2008.

RESEARCH PUBLICATIONS AND BOOKS

He has published 194 research papers in leading peer-reviewed national/international journals and conference proceedings. Prior to the present book, he has authored 18 books and has also contributed 6 chapters in edited books. He has 5 patents to his credit, including 2 US patents.

About the Book

Taking a close look at how nanotechnology can improve performance of solar cells, this book provides an overview of nanomaterials and nanofabrication processes that can be integrated in the solar cell manufacturing lines. It provides a comprehensive, state-of-the-art survey of nanostructured solar cells, combining foundational information with latest research findings in a simple, easily digestible, and lucid format. The book begins by familiarizing the reader with the preliminaries of solar cells and nanotechnology and covers in detail various types of solar cells that have benefitted from application of nanotechnological approaches. The contents include solar cell basics, nanotechnology fundamentals, nanocrystalline silicon-based solar cells, nanotextured-surface solar cells, plasmon-enhanced solar cells, optically improved nanoengineered solar cells, dye-sensitized solar cells, 2D perovskite and 2D/3D perovskite multidimensional solar cells, carbonaceous-nanomaterial-based solar cells, quantum well solar cells, nanowire solar cells, and quantum dot solar cells. It bridges the gap between an elementary treatise and research-level compendium in order that the scientific achievements in the frontline research areas can be assimilated by a wide cross section of readership, including students, researchers, and practicing engineers.

Acronyms and Abbreviations

A	ampere
AAO	anodized aluminum oxide (template)
AC	alternating current
ACN	acrylonitrile, acetonitrile
ALD	atomic layer deposition
AM	air mass coefficient
AM0	air mass zero
AM1.5	air mass 1.5
AM1.5G	air mass 1.5 global
ARC	antireflection coating
AVAI	aminovaleric acid iodide
AZO	aluminum-doped zinc oxide
BAI	n-butylammonium iodide
BARC	bottom antireflective coating
BCB	benzocyclobutene
BCP	bathocuproine
BDT	1,3-benzenedithiol
BHJ	bulk heterojunction
BSF	back surface field
C	coulomb
°C	degree centigrade
CE	counter electrode
CIGS/CIGSe	cadmium indium gallium selenide
cm	centimeter
CNT	carbon nanotube
Cos-	negative directed cosine lattice
Cos+	positive directed cosine lattice
CPS	concentrated power system
CSS	closed space sublimation
CTAB	cetyltrimethylammonium bromide
CVD	chemical vapor deposition
0D	zero-dimensional
1D	one-dimensional
2D	two-dimensional
$2Dg\text{-}C_3N_4$	two-dimensional graphitic carbon nitride
3D	three-dimensional
DBH	depleted bulk heterojunction (solar cell)
DC	direct current
DEA	diethanolamine
DH	depleted heterojunction (solar cell)
DI	deionized (water)
DJ	Dion-Jacobson (perovskite)
DMD-TCE	dielectric-metal-dielectric transparent conducting electrode
DMF	N,N-dimethylformamide
DMII	1,3-dimethylimidazolium iodide
DMPII	1-propyl-2,3-dimethylimidazolium iodide
DMSO	dimethyl sulfoxide

DMZ	dimethyl zinc
DNA	deoxyribonucleic acid
DoS	density of states
DSSC	dye-sensitized solar cell
DSUCNPs	dye-sensitized upconversion nanoparticles
e_0	ground state
e^-	electron
e_1	first excited state
EC	ethylene carbonate, ethyl cellulose
EDT	1,2-ethanedithiol
E-GaIn	gallium-indium eutectic
ETL	electron transport layer
EtOH	ethyl alcohol
eV	electron volt
F	farad
FA^+	formamidinium cation
FF	fill factor
FK 209 Co(III)-TFSI	Tris(2-(1H-pyrazol-1-yl)-4-tert-butylpyridine)-cobalt(III)tris(bis(trifluoromethyl sulfonyl)imide))
4FPEAI	4-fluoro-phenethylammonium iodide
FTO	fluorine-doped tin oxide
GA	guanidinium, $CH_6N_3^+$ ion
GABr	guanidinium bromide
GAI	guanidinium iodide
GO	graphene oxide
GPa	gigapascal
GRIM	Grignard metathesis
GuNCS	guanidinium thiocyanate
h.	hour
h^+	hole
HMDST	hexamethyldisilathiane
HOMO	highest occupied molecular orbital
HPD	high-pressure depletion (regime)
HRT	high-resistivity transparent (film)
HTL	hole transport layer
HTM	hole transport material
Hz	hertz
IPA	isopropyl alcohol
IR	infrared
ITO	indium tin oxide
ITO-IR	indium tin oxide-infrared
J	joule
keV	kiloelectronvolt
kg	kilogram
kV	kilovolt
Li-TFSI	lithium bis(trifluoromethanesulfonyl)imide
LPCVD	low-pressure chemical vapor deposition
LSPR	localized surface plasmon resonance
LUMO	lowest unoccupied molecular orbital
M	molar (concentration)
m	meter

m	metallic (CNTs)
mA	milliampere
MA^+	methylammonium cation
MABr	methylammonium bromide
MACE or MacEtch	metal-assisted chemical etching
MACl	methylammonium chloride
MAI	methylammonium iodide
$MAPbI_3$	methylammonium lead iodide
MBE	molecular beam epitaxy
MEG	multiple exciton generation
MEH-PP	poly[2-methoxy-5-(2'-ethyl-hexyloxy)-1,4-phenylene vinylene]
MeOH	methanol
MeO-2PACz	(2-(3,6-Dimethoxy-9H-carbazol-9-yl)ethyl) phosphonic acid
MePN	3-methoxypropionitrile
meV	millielectronvolt
mg	milligram
MHz	megahertz
min.	minute
mL	milliliter
mM	millimolar
MOCVD	metal-organic chemical vapor deposition
MWCNT	multiwalled carbon nanotube
MOVPE	metal-organic vapor phase epitaxy
m.p.	melting point
MPA	3-mercaptopropionic acid
MPN	3-methoxypropionitrile
MQW	multiple quantum well
mTorr	millitorr
MV	megavolt
mV	millivolt
mW	milliwatt
n	a number
N^+	heavily doped N-region
N^{++}	very heavily doped N-region
NDs	nanodisks
NIL	nanoimprint lithography
NIR	near infrared
nM	nanomolar
nm	nanometer
NMBI	N-methylbenzimidazole
NMP	N-methyl-2-pyrrolidone
NPs	nanoparticles
NRs	nanorods
ns	nanosecond
NSs	nanospheres
NW	nanowire
O_2	oxygen
OA	oleic acid
ODCB	o-dichlorobenzene
ODE	1-octadecene
P^+	heavily doped P-region

P++	very heavily doped P-region
PBDTTT-CF	poly[1-(6-{4,8-bis[(2-ethylhexyl)oxy]-6-methylbenzo[1,2-*b*:4,5-*b*′]dithiophen-2-yl}-3-fluoro-4-methylthieno[3,4-*b*]thiophen-2-yl)-1-octanone]
PC	propylene carbonate
$PC_{61}BM$	[6,6]-phenyl-C_{61}-butyric acid methyl ester
$PC_{71}BM$	[6,6]-phenyl C_{71} butyric acid methyl ester
PD2FCT-29DPP	diketopyrrolopyrrole-based polymer with benzothiadiazole derivatives
PDMS	polydimethylsiloxane
PEAI	phenylethylammonium iodide
PECVD	plasma-enhanced chemical vapor deposition
PEDOT:PSS	poly(3,4-ethylenedioxythiophene) polystyrene sulfonate
PEG	poly(ethylene glycol)
PERC	passivated emitter and rear cell
P3HT	poly(3-hexylthiophene-2,5-diyl) (block copolymer)
P3HT-*b*-PS	poly(3-hexylthiophene)-block-polystyrene
PIE	proton implant exfoliation
pm	pico meter
$PM_{2.5}$	particulate matter (particles < 2.5μm in diameter)
PMMA	polymethyl methacrylate
ps	picosecond
PS-B(OR)$_2$	polystyrene block having the pinacol boronic ester as the end functional group
PSPR	propagating surface plasmon resonance
PTAA	poly[bis(4-phenyl)(2,5,6-trimethylphenyl)amine
PV	photovoltaic
PVD	physical vapor deposition
QCSE	quantum-confined Stark effect
QD	quantum dot
QD-IBSC	quantum dot intermediate band solar cell
QDSSC	quantum dot–sensitized solar cell
QW	quantum well
RCA	Radio Corporation of America (cleaning of silicon wafers)
RF	radio frequency
RGO	reduced graphene oxide
RHEED	reflection high-energy electron diffraction
RPM	revolutions per minute
RTA	rapid thermal annealing
RTIL	room-temperature ionic liquid
S	Siemen
s	second
s	semiconducting (CNTs)
SAG	selective area growth
SAM	self-assembled monolayer
SAQDs	self-assembled quantum dots
SCAPS	solar cell capacitance simulator
sccm	standard cubic centimeters per minute
SCIL	substrate conformal imprint lithography
SDS	sodium dodecyl sulfate
SEM	scanning electron microscope
SHJ	silicon heterojunction
SILAR	successive ionic layer adsorption and reaction
SOI	silicon-on-insulator (wafer)

Spiro-OMeTAD	2,2',7,7'-Tetrakis[N,N-di(4-methoxyphenyl)amino]-9,9'-spirobifluorene
SPP	surface plasmon polariton
SPR	surface plasmon resonance
sq.	square
SS-MBE	solid-source molecular beam epitaxy
SWCNT	single-walled carbon nanotube
TBAC	tetrabutylammonium chloride
TBAI	tetrabutylammonium iodide
TBP,	tBP 4-tert-butylpyridine
TCO	transparent conducting oxide
TE	transverse electric (mode of the electromagnetic field)
TEG	triethylgallium
TFSA	bis(trifluoromethanesulfonyl)amide
Ti(OBu)$_4$	tetrabutyl orthotitanate, titanium(IV) butoxide
TM	transverse magnetic (mode of the electromagnetic field)
TMA	trimethyl aluminum
TMG	trimethyl gallium
TMS	bis(trimethylsilyl)sulfide
TPa	terapascal
TTIP	titanium tetraisopropoxide
UCNPs	upconversion nanoparticles
UV	ultraviolet
UV-NIL	ultraviolet nanoimprint lithography
V	volt, voltage
VHF	very high frequency
VLS	vapor-liquid-solid (method of nanowire growth)
VTD	vapor-transport deposition
v/v	volume/volume
W	watt

Chemical Symbols

Ag	silver (argentum)
AgNDs	silver nanodisks
$AgNO_3$	silver nitrate
Al	aluminum
AlGaAs	aluminum gallium arsenide
AlGaInP	aluminum gallium indium phosphide
AlInN	aluminum indium nitride
AlInP	aluminum indium phosphide
Al_2O_3	aluminum oxide
AlSb	aluminum antimonide
AlZnO (AZO)	aluminum-doped zinc oxide
As	arsenic
As_2H_2	diarsene, an arsenic hydride
AsH_3	arsine
a-Si	amorphous silicon
a-Si:H	amorphous silicon: hydrogenated
Au	gold (Aurum)
$Au@SiO_2$	gold-silicon dioxide (core-shell NPs)
$BaSnO_3$	barium stannate
BBr_3	boron tribromide
Be	beryllium
B_2H_6	diborane
Br	bromine
C_{60}	buckminsterfullerene
CBD	chemical bath deposition
$Cd(NO_3)_2$	cadmium nitrate
CdS	cadmium sulfide
CdSe	cadmium selenide
CdTe	cadmium telluride
CF_4	carbon tetrafluoride
$(CF_3SO_2)_2 NH$	bis(trifluoromethanesulfonyl)amide (TFSA)
CH_4	methane
CHF_3	fluoroform
CH_3NH_3I	methylammonium iodide
$CH_3NH_3PbI_3$	methylammonium lead iodide
$C_6H_8O_6$	ascorbic acid
$C_6H_{12}O_6$	glucose
Cl_2	chlorine
CO_2	carbon dioxide
Cr	chromium
Cs^+	cesium ion
CsBr	cesium bromide
c-Si	crystalline silicon
$c-TiO_2$	compact TiO_2
Cu	copper
CuS	copper (II) sulfide
Cu_2S	copper (I) sulfide

$CuSO_4$	copper (II) sulfate
Er	erbium
F	fluorine
FA	formamidinium ion, $CH_5N_2^+$
FAI	formamidinium iodide
$Fe(NO_3)_3$	iron(III) nitrate, ferric nitrate
FeSe	iron (II) selenide
Ga	gallium
GaAs	gallium arsenide
GaAsP	gallium arsenide phosphide
GaAsSb	gallium arsenide antimonide
$Ga(CH_3)_3$	trimethyl gallium
GaInAs	gallium indium arsenide
GaInNP	gallium indium nitride phosphide
GaInP	gallium indium phosphide
GaN	gallium nitride
$g\text{-}C_3N_4$	graphitic carbon nitride
Ge	germanium
H	hydrogen
$HAuCl_4$	chloroauric acid
HBr	hydrobromic acid
HCl	hydrochloric acid
HF	hydrofluoric acid
HNO_3	nitric acid
H_2O	water
H_2O_2	hydrogen peroxide
H_3PO_4	phosphoric acid
H_2PtCl_6	hexachloroplatinic acid
H_2Se	hydrogen selenide
H_2SiF_6	hexafluorosilicic acid
H_2SO_4	sulfuric acid
HsO_3F	fluorosulfuric acid
I	iodine
(I)a-Si:H	intrinsic amorphous silicon: hydrogenated
In	indium
InAs	indium arsenide
InGaAs	indium gallium arsenide
InGaAsP	indium gallium arsenide phosphide
InGaN	indium gallium nitride
InGaP	indium gallium phosphide
In_2O_3	indium (III) oxide
InP	indium phosphide
KCl	potassium chloride
KOH	potassium hydroxide
KrF	krypton fluoride
LiF	lithium fluoride
LiI	lithium iodide
MA	methylammonium ion, CH_6N^+
MABr	methylammonium bromide
MgF_2	magnesium fluoride
Mn	manganese

Mn^{2+}	manganese cation
Mo	molybdenum
MoO_3	molybdenum trioxide
N	nitrogen
$NaBH_4$	sodium borohydride
$Na_3C_6H_5O_7$	trisodium citrate
NaF	sodium fluoride
NaHS	sodium hydrosulfide
NaOH	sodium hydroxide
Na_2S	sodium sulfide
$Na_2S_2O_8$	sodium persulfate
$NaYF_4$	sodium yttrium fluoride
Nb_2O_5	niobium pentoxide
NCS	thiocyanato group
nc-Si:H	nanocrystalline-silicon: hydrogenated
$nc-SiO_2$	nanocrystalline silicon dioxide
Nd	neodymium
NH_4OH	ammonium hydroxide
Ni	nickel
NiO	nickel (II) oxide
NO	nitric oxide
NO^+	nitrosonium ion
NO_3^-	nitrate anion
$NOBF_4$	nitrosonium tetrafluoroborate
O_2	oxygen
P	phosphorous
$PbAc.3H_2O$	lead (II) acetate trihydrate
$PbBr_2$	lead (II) bromide
$PbCl_2$	lead (II) chloride
PbI_2	lead (II) iodide
$Pb(NO_3)_2$	lead (II) nitrate
PbO	lead (II) oxide
PbS	lead sulfide
PbSe	lead selenide
Pd	palladium
PEN	polyethylene naphthalate
PET	polyethylene terephthalate
PH_3	phosphine
$POCl_3$	phosphorous oxychloride
Poly-Si	polysilicon
Pt	platinum
S	sulfur
S^{2-}	sulfide anion
Sb_2S_3	antimony trisulfide
SCN^-	thiocyanate anion
$(SCN)_2$	thiocyanogen
Se	selenium
Si	silicon
SiF_4	silicon fluoride
SiF_6^{2-}	silicon hexafluoride anion
SiH_4	silane

$Si_xN_y/SiN_x/Si_3N_4$ silicon nitride
SiO_2 silicon dioxide
Sn tin
S_n^{2-} polysulfide dianion
SnO_2 tin (IV) oxide
Ta_2O_5 tantalum pentoxide
TBA^+ tetrabutylammonium ion
Ti titanium
$TiCl_3$ titanium (III) chloride
$TiCl_4$ titanium tetrachloride
TiO titanium (II) oxide
TiO_2 titanium dioxide
Yb ytterbium
Zn zinc
$Zn(CH_3COO)_2 \cdot 2H_2O$ zinc acetate dihydrate
$Zn(NO_3)_2$ zinc nitrate
ZnO zinc oxide
ZnS zinc sulfide

Mathematical Symbols

a	radius of the spherical-shaped metal nanoparticle
a_0	Bohr radius
A	material-independent constant
A, B	constants
a, b	mole fractions of atoms in the alloy
$\left|A_{\text{Junction}}\right|_{\text{Axial cell}}$	cross-sectional area of the junction of axial solar cell
$\left|A_{\text{Junction}}\right|_{\text{Radial cell}}$	cross-sectional area of the junction of radial solar cell
b	bandgap energy bowing parameter
C, D	constants
d	diameter of the nanowire
D_n	diffusion coefficient of electron
E	energy of the electron
e	elementary charge
$\mathbf{E}_0$	static electric field, electric field of the illuminating optical radiation
$\mathbf{E}_1$	electric field inside the spherical nanoparticle
$\mathbf{E}_2$	electric field outside the spherical nanoparticle
$E_1, E_2, E_3, \ldots E_n$	energy levels occupied by electrons
E_{1e}	energy for the ground state of the electron in the quantum well
$\left|E_{1e}\right|_{\text{VBE}=0}$	energy of the electron energy level in the ground state taking the energy of top edge of valence band as zero
E_{1hh}	ground state energy of a heavy hole in the valence band
E_{1lh}	ground state energy of a light hole in the valence band
$\left|E_{1hh}\right|_{\text{VBE}=0}$	energy of heavy hole energy level in the ground state taking the energy of top edge of valence band as zero
$\left|E_{1lh}\right|_{\text{VBE}=0}$	Energy of light hole energy level in the ground state taking the energy of top edge of valence band as zero
E_B	binding energy of the exciton
E_{Bandgap}	bandgap of a semiconductor
E_C	energy of the bottom edge of conduction band
E_{C1}, E_{C2}	energies of bottom edges of the conduction bands of two semiconductors 1 and 2
E_{Exciton}	exciton energy
E_F	Fermi level
$E_{\text{FPbS QD}}$	Fermi level energy in PbS QD
E_G	energy gap (of the semiconductor)
$E_{G,\,\text{Bulk}}$	energy gap of the semiconductor in bulk
$E_{G,}^{\text{Bulk}}$	bandgap energy of the bulk semiconductor (same as $E_{G,\,\text{Bulk}}$)
E_G, E_g	energy gaps of the larger bandgap barrier layer and smaller bandgap quantum well layers in a quantum well solar cell
$E_G^{\text{Effective}}$	effective energy gap
$E_{G-\text{Heavy hole-to-conduction}}$	heavy hole-to-conduction band: energy level difference = Energy gap
$E_{G-\text{Light hole-to-conduction}}$	light hole-to-conduction band: energy level difference = Energy gap
$E_{G,\,\text{NW}}$	effective energy gap in the nanowire
$\left|E_{G,\,\text{NW}}\right|_{\text{Electron-Heavy hole}}$	effective energy gap in the nanowire for the electron-heavy hole case

$\left. E_{G,\,NW} \right\|_{\text{Electron–Light hole}}$	effective energy gap in the nanowire for the electron-light hole case
E_G^{QD}	effective bandgap of the quantum dot
$\left. E_G^{QD,\,CdS} \right\|_{R=5nm}$	energy gap of the CdS quantum dot of radius 5nm
$E_G^{\text{Relaxed material}}$	energy gap of the material in the relaxed state
$E_{G,\,QW}$	energy gap of the quantum well
$\left. E_{G,\,QW} \right\|_{hh}$	energy gap of the quantum well for heavy hole case
$\left. E_{G,\,QW} \right\|_{hh}$	energy gap of the quantum well for light hole case
$E_{G,\,\text{Well material}}$	energy gap of the quantum well material
$E_n(x),\ E_n(y),\ E_n(z)$	quantized energies of the electron as a function of distance in the X-, Y-, and Z-directions; n is an integer
$E_n(x,\ y,\ z)$	quantized energy of the electron as a function of position $(x,\ y,\ z)$; n is an integer
$\left. E_n(x,y,z) \right\|_{n=1,\ \text{Electron}}$	quantized energy of the electron as a function of position (x, y, z) for the ground state $n = 1$ of the electron
E_V	valence band edge
$E_{V1},\ E_{V2}$	energies of top edges of the valence bands of two semiconductors 1 and 2
E_{Vacuum}	vacuum energy level
FF	fill factor
H	height
h	Planck's constant, height of the nanowire
$\hbar$	reduced Planck's constant
I	current
$I(x)$	intensity of light at a depth x below the top surface
$I(0)$	intensity of light on the top surface $(x = 0)$
I_{MPP}	current at maximum power point
I_{SC}	short-circuit current
$J_0(x)$	dark saturation current density at a depth x below the surface
$J_{\text{Light}}(0)$	light-generated current density at the surface $(x = 0)$
$J_{\text{Light}}(x)$	light-generated current density at a depth x below the surface
J_{SC}	short-circuit current density
k	wave vector
k_B	Boltzmann constant
l	absorption depth
L	thickness, distance, a side of the cubical box
L_n	diffusion length of electron
$L_{wx},\ L_{wy},\ L_{wz}$	dimensions of this rectangular well along the X-, Y-, and Z-directions
$m,\ m^*$	effective mass of a particle
m_0	rest mass of electron
m_e	mass of the electron
m_e^*	effective mass of electron
m_h^*	effective mass of hole
m_{hh}^*	effective mass of heavy hole
m_{lh}^*	effective mass of light hole
$(m,\ n)$	chiral vector
n	a number, integer

n, n_1, n_2	refractive indices
$n_{\text{Effective}}$	effective refractive index
n_x, n_y, n_z	quantum numbers in the X-, Y-, and Z-directions
P, p	period
$\mathbf{p}$	dipole moment inside the spherical nanoparticle
p	momentum
$P_{\text{Incident}}(\lambda)$	incident optical power at wavelength (λ)
P_M	maximum power
P_{MPP}	maximum power point
$\left. P \right|_{N \text{ QWs}}$	carrier escape probability for an N quantum well solar cell
$P_{\text{Reflected}}(\lambda)$	reflected optical power at wavelength (λ)
$\left. P \right|_{\text{Single QW}}$	carrier escape probability for a single quantum well solar cell
R	reflectance, radius of the axial junction solar cell, radius of the spherical box, radius of the quantum dot
r	hydrogen dilution ratio, distance of the point at which the field is observed from the center of the sphere, radius of the cylindrical-shaped nanowire
R_1	bulk component (contribution from bulk) towards recombination rate of nanowire
R_2	surface component (contribution from surface) towards recombination rate of nanowire
r_B	exciton Bohr radius
R_{Bulk}	recombination rate per unit volume of the bulk of nanowire
$\text{Re}\{\varepsilon(\omega)\}$	real component of the dielectric function of the metal nanoparticle
$R_{\text{Effective}}$	effective reflectance
R_{NW}	recombination rate per unit volume of the nanowire
R_{Surface}	recombination rate per unit volume of the surface of nanowire
S	surface recombination velocity
T	transmittance, temperature on Kelvin scale
t_w	thickness of the quantum well
V_{MPP}	voltage at maximum power point
V_{NW}	volume of the nanowire
V_{OC}	open-circuit voltage
$\left. V_{\text{oc}} \right|_{\text{Corrected}}$	corrected open-circuit voltage
x	fraction
α	polarizability of the spherical nanoparticle
$\alpha(\lambda)$	absorption coefficient of the light of wavelength λ
ΔE	difference between the conduction and valence band energies
ΔE_C	discontinuity of conduction band (conduction band offset), energy difference between the energies E_{C1}, E_{C2} of the bottom edges of conduction bands of two semiconductors
$\Delta E_G^{\text{Quantum confined Stark effect}}$	increment in energy gap by quantum-confined Stark effect
$\Delta E_G^{\text{Quantum size effect}}$	increment in energy gap by quantum size effect
$\Delta E_G^{\text{Strain effect}}$	increment in energy gap by the effect of strain
ΔE_V	discontinuity of valence band (valence band offset), energy difference between the energies E_{V1}, E_{V2} of the top edges of valence bands of two semiconductors
Δn	excess electron concentration
ε	frequency-dependent complex permittivity of the metal nanoparticle

ε_0	permittivity of free space
ε_{GaAs}	dielectric constant of gallium arsenide
ε_{Medium}	frequency-dependent complex permittivity of the semiconductor medium
ε_{PbS}	dielectric constant of PbS
ε_S	relative permittivity of the semiconductor
$\varepsilon_{Silicon}$	dielectric constant of silicon
ε_{TiO2}	dielectric constant of TiO_2
η	efficiency, electric field enhancement factor
λ	wavelength of light
$\lambda_{de\ Broglie}$	de Broglie wavelength
$\lambda_{Thermal\ de\ Broglie}$	thermal de Broglie wavelength
μ	reduced mass of the charge carriers
$\left|\mu\right|_{Electron-Heavy\ hole}$	reduced mass of electron and heavy hole
$\left|\mu\right|_{Electron-Light\ hole}$	reduced mass of electron and light hole
μm	micrometer (micron)
μs	microsecond
ν	frequency
$\sigma_{Absorption}$	absorption cross section of the spherical nanoparticle
$\sigma_{Scattering}$	scattering cross section of the spherical nanoparticle
τ_{Bulk}	carrier lifetime in the bulk of the nanowire
$\tau_{Effective}$	effective lifetime of charge carrier in the nanowire
τ_{Escape}	escape time
τ_n	lifetime of electron
$\tau_{Recombination}$	lifetime of recombination of carriers
$\tau_{Relaxation}$	relaxation time of electron
$\tau_{Thermionic}$	thermionic emission time
$\tau_{tunneling}$	tunneling time
ϕ	work function
$\Phi_1,\ \Phi_2$	scalar potentials inside and outside the spherical nanoparticle
$\phi_{Incident}(\lambda)$	flux received on the surface at wavelength λ
$\phi_{Reflected}(\lambda)$	flux reflected from the surface at wavelength λ
χ	electron affinity
$\psi(x)$	wave function of the electron as a function of position x
Ω	ohm
ω	frequency of the electric field
ω_{LSPR}	LSPR resonance frequency
ω_{Plasma}	plasma frequency

Part I

Preliminaries and Nanocrystalline
Silicon Photovoltaics

1 Solar Cell Basics

1.1 PROGRESSION FROM FOSSIL FUELS TO RENEWABLE ENERGY SOURCES

1.1.1 FOSSIL FUELS, THE LIFEBLOOD OF MODERN CIVILIZATION

Our early ancestors relied on deadwood as the main fuel ever since the discovery of fire. Wood was available in abundance in the forests. As science and technology evolved, man became increasingly dependent on fossil fuels. Coal, petroleum, natural gas, and orimulsion are the four main fuels formed by decomposition and decay of the buried remains of living organisms (plants and animals) below the Earth's crust over millions of years. Orimulsion is a naturally occurring bitumen (70%) mixed with water (30%). Bitumen is a petroleum-based hydrocarbon found in oil sands.

Coal has been in use since more than 3,000 years ago (Kentucky coal education 2007). The first recorded use of oil is in the year 3000 BC, the birth of modern oil industry took place in 1859, and motorcar fueling with oil was done in 1885 (BBC Teach 2022). Today fossil fuels constitute the principal energy sources being used by mankind. They act as an essential cornerstone of our technological development by providing the energy for our electricity-generating plants and catering to the industrial manufacturing, transportation, and lighting demands. The basis of present civilization is energy, and the sources of energy predominantly utilized currently are the fossil fuels.

1.1.2 EVILS AND LIMITATIONS OF FOSSIL FUELS

1.1.2.1 Land and Habitat Destruction

The landscape of Earth, including forests and mountains and the ecosystem, is devastated by the underground mining, drilling, and processing activities associated with exploitation of fossil fuels. Vast stretches of land are lost to their infrastructure. They vitiate our groundwater and water supply system.

1.1.2.2 Greenhouse Effect

The gases emitted during the burning of fossil fuels contribute profoundly to atmospheric pollution. Carbon dioxide, methane, and nitrous oxide released into the atmosphere during consumption of fossil fuels by cars, trucks, buses, and various industries are all greenhouse gases. They absorb and trap the solar heat reflected from the Earth's surface, which would otherwise escape into space. The greenhouse gases are given this name because they act like the glass walls of a greenhouse. The *greenhouse* is a building with glass roof and walls which remains warm even during night by trapping the sun's heat received during daytime. It is used to grow tender and off-season plants, which do not tolerate extreme warmth or cooling. The activities of a person, organization, or community release a certain amount of greenhouse gases into the atmosphere. This amount is known as its carbon footprint.

1.1.2.3 Global Warming

The trapping of heat by gaseous emissions from fossil fuels makes our planet hotter, leading to global warming, the phenomenon of increase in average temperature near the surface of the Earth over the past few centuries. The average global surface temperature is estimated to have risen by 1.07°C between 1850 and 2019. Climate scientists have warned that global warming must be limited to 1.5°C by 2040 to avert serious devastating climatic changes affecting us in the form of droughts, floods, storms, etc.

1.1.2.4 Depletion of Fossil Fuels

Fossil fuels are nonrenewable energy sources. Although in large-scale use for around 200 years only, oil reserves are expected to last for the ensuing ~ 50 years, and coal reserves for ~ 100 years. It took

DOI: 10.1201/9781003215158-2

millions of years to form the fossil fuels that are used today. These fuels cannot be formed over a short duration of time; hence, they are adjudged as nonrenewable. Mankind must therefore look for alternatives. Solar, wind, hydroelectric, and geothermal energy are invaluable gifts of nature. Sunlight is a prominent renewable energy source, a nondepletable source, which is always replenished by nature.

1.1.3   PROMISES OF SOLAR ENERGY FOR SUSTAINABLE DEVELOPMENT

Sunlight is the most abundant renewable energy source with manifold advantages:

   (i) It provides long-term assurance of energy lasting as long as the sun exists.
  (ii) It is an environment-friendly and safe energy source posing no concerns of global warming and climate changes.
 (iii) It is a clean option for energy, nondamaging to the land and habitat.
  (iv) It supports sustainable development. The paradigm of sustainable development seeks to provide economic and social development with environmental protection. In order that the development is sustainable, the needs of the present generation must be met in such a way that there is no compromise with the ability of future generations to meet their needs, a dream realizable with widespread use of solar energy.

Figure 1.1 shows a solar power station and a few of the myriad application areas of solar panels, ranging from household electricity and street lighting to artificial satellites. This chapter will provide a quick synopsis of the fundamental principles of solar cells and a review of the main types of solar cells.

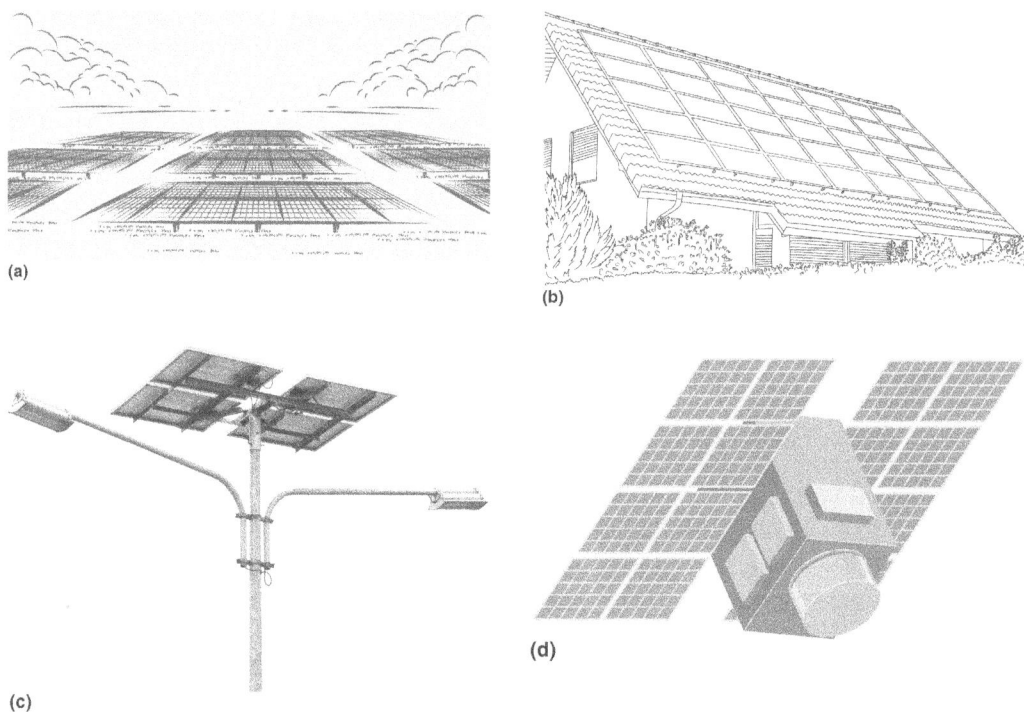

(a)

(b)

(c)

(d)

FIGURE 1.1   Solar energy: (a) solar cell panels and wind turbines as renewable energy sources, (b) a house with solar panels on rooftop, (c) solar street lighting, and (d) telecommunication satellite with solar panels.

1.2 SOLAR POWER GENERATING SYSTEM

There are two mainstream solar power generation technologies: photovoltaic (PV) power systems and concentrated power systems (CPSs).

1.2.1 PHOTOVOLTAIC POWER SYSTEM

In the PV power system, solar energy is converted into electricity by utilizing the photovoltaic effect concerned with the creation of voltage across a semiconductor material and the associated current flow in it when exposed to light. The device used for this energy conversion is known as a photovoltaic cell or solar cell. This device is the topic of discourse in this book. *Photovoltaics* is the technology dealing with the production of current in a material by photovoltaic effect.

1.2.2 CONCENTRATED POWER SYSTEM

In a CPS, the energy from the sun is focused with the help of lenses and mirrors, and the heat concentrated is used to produce steam for driving turbines to generate electricity. It is essentially a thermal-to-work conversion system, like the thermal power system.

1.3 PHOTOVOLTAIC POWER SYSTEM

Two types of systems are employed to tap the electricity generated in the solar cells. These are grid-fallback and grid-tie systems (Burdick and Schmidt 2017).

1.3.1 OFF-GRID (STAND-ALONE OR GRID FALLBACK) SOLAR POWER SYSTEM

The off-grid system (Figure 1.2) is intended for a place with no grid supply. Sunlight is received by a matrix of flat, thin solar cells wired together and sealed into weatherproof solar modules containing 48–72 cells and producing 25–40 V at 150–300W. A group of several solar modules working together forms a solar panel. Assemblies of solar panels are arranged on the roof of a house, constituting a solar array. The panels are fitted on house rooftops with a tilt angle of 10° for self-cleaning by rain. Thus, the solar cell is a subset of the solar module, the solar module is a subset of the solar array, and the solar array contains numerous individual solar cells (Zakery et al 2018).

The electricity produced by the solar array must be stored. Electricity is stored by charging the batteries. For storage, a battery pack is used. The pack is a combination of cells which are properly connected in series or parallel configuration to supply the correct voltage or current.

The battery pack is monitored and controlled by a device called the charge controller. It avoids undercharging or overcharging of batteries, both of which are harmful to batteries. From the charge controller, DC loads can be directly supplied electricity, but for AC loads, the DC obtained from the battery pack is converted by an inverter to the required voltage and power level. Various equipment, such as electric lamps, fridge, TV, washing machines, etc., in a house are fed this AC.

1.3.2 GRID-TIE SOLAR POWER SYSTEM

This system is always connected to the utility grid. The solar panels are directly converted to an inverter, not to a battery pack, as in the off-grid system. The aim is to synchronize the output with the external grid supply. The inverter takes care of the voltage and frequency of the power from the solar system. It maintains the values of electrical parameters at the same level as that of the external grid.

When the electricity required by the house exceeds that generated by the solar power system, power is pulled from the grid. On the other hand, when the electricity requirement of the house is

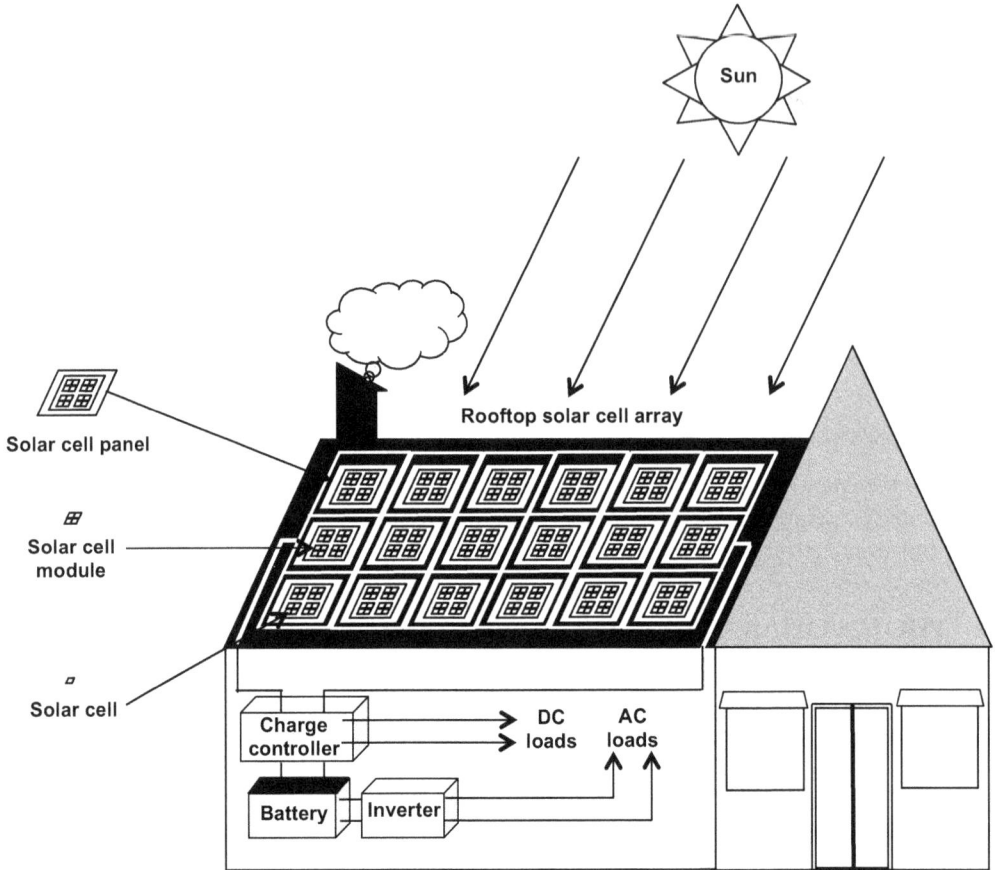

FIGURE 1.2 Components of an off-grid solar photovoltaic power generating system fitted on a house rooftop: the solar cell (a small device), the solar module (a collection of solar cells), the solar panel (several modules working cooperatively), the solar array (assemblies of solar panels), a charge controller (battery-monitoring device), a battery pack (charge storage unit charged by the electricity from solar array), DC loads (those directly supplied DC from the battery), AC loads (those supplied AC from inverter), and an inverter (circuit for converting DC from the battery into AC).

less than that produced by the solar system, meaning, that the solar power is in surplus, the excess solar power is fed back to grid, turning the utility meter in the reverse direction. As there is no means for energy storage, the grid-tie system goes down during grid power outage.

1.4 CONSTRUCTION AND WORKING OF A SOLAR CELL

Figure 1.3(a) illustrates the structure and working principle of a solar cell. The solar cell consists of a semiconductor P-N junction diode, in which there is an N-region containing excess of electrons and a P-type region with a deficiency of electrons or a surplus of holes. In order that the solar photons can reach up to greater depths inside the solar cell, the metallization of the front contact is designed as a rectangular grid or made with finger electrodes, leaving empty regions without metal between metal lines for absorption of sufficient sunlight by the cell. The backside contact is a continuous metal film.

Driven by concentration gradient, the electrons diffuse across the junction from the N-type region to the P-type region. Consequent upon electron diffusion to the P-side, positively charged

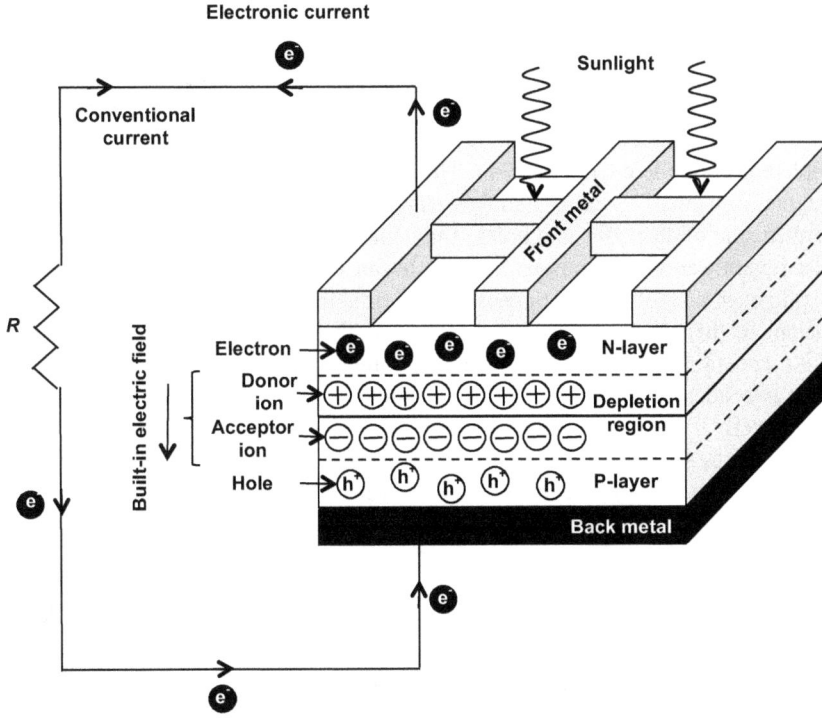

(a)

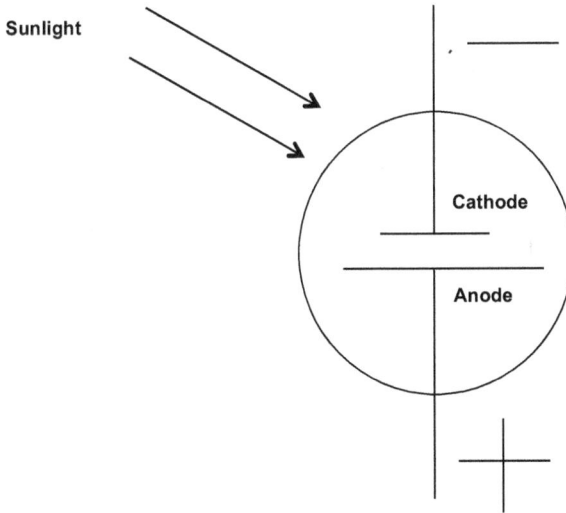

(b)

FIGURE 1.3 Solar cell: (a) Construction and principle of operation: The front-side N-layer is covered with a finger metal electrode, while the backside P-layer is fully metallized. A depletion region is formed at the P-layer/N-layer interface, setting up a built-in electric field. Sunlight falling on the front-side N-layer produces a current flow across the external resistor R. (b) Circuit diagram symbol of a solar cell: The anode and cathode terminals and incident sunlight are indicated.

donor ions are left behind on the N-side. Like electrons, the holes also see a concentration gradient and therefore diffuse from the P-type region to the N-type region. Due to the diffusion of holes to the N-side, the P-side contains negatively charged acceptor ions. This electron and hole migration to opposite sides affects the free-charge carrier density in a thin region at the interface of the N- and P-regions. The exodus of electrons and holes produces a thin layer on both sides of the junction, which is depleted of mobile charge carriers. The region containing no free carriers is known as a depletion region, in view of its exhausted free-charge carrier population.

The interdiffusion of electrons and holes, and hence the widening of the depletion region, continues until a dynamic equilibrium is established, because the concentrations of donor and acceptor ions build up to such a level that it opposes further transference of electrons/holes due to electrostatic repulsion. At this stage, the depletion region width stops increasing. The charge separation at the junction constitutes an electric field, i.e., a potential difference is set up between the ionic charges on the two sides of the P-N junction. This built-in potential plays a central role in the operation of the solar cell.

The exposure of the solar cell to photons having energies greater than the energy gap of the semiconductor causes the creation of electron-hole pairs because the solar radiation dislodges electrons from their bonds, and the electron vacancy is known as a hole. The electrons and holes thus produced by sunlight move under the influence of the built-in electric field in directions determined by the electric field. The electrons travel in a direction opposite to that of the electric field. When the two terminals of the solar cell are connected to a load, a current flows in the external circuit. Electrons/holes produced by bombardment with photons of sunlight inside the depletion region as well as those produced within a diffusion length of the depletion region are able to reach the contacts, while carriers beyond this length are mostly lost by recombination and do not contribute to the current in external circuit. The direction of conventional current is opposite to that of the electronic current.

The circuit diagram symbol of a solar cell is shown in Figure 1.3(b). It consists of a negative electrode or cathode and a positive electrode or anode. Arrows indicate the sunlight falling on the solar cell.

1.5   OPTOELECTRICAL CHARACTERISTICS AND PARAMETERS OF A SOLAR CELL

The electrical characteristics of a solar cell under different conditions are shown in Figure 1.4. The measurement conditions are mentioned in the diagram. Note that the forward characteristics of a solar cell diode in dark shown in Figure 1.4(a) lie in the first quadrant. In opposition, the I-V and P-V characteristics of the solar cell shown in Figure 1.4 (b) lie in the fourth quadrant. The current as well as power are negative for the solar cell (Figure 1.4[b]). Negative values mean that the current and power are generated. For a forward-biased solar cell diode in dark (Figure 1.4[a]), the current is positive because the current is supplied by the biasing battery and power is dissipated, not produced.

The electrical characteristics are measured by the circuit given in Figure 1.5.

1.5.1   SHORT-CIRCUIT CURRENT (I_{SC})

The current I flowing through the solar cell = short-circuit current (I_{SC}) = current flowing in the cell when the two terminals of the device are short-circuited so that there is no load in the circuit, and this will happen when the voltage across the cell is zero, i.e., $V = 0$. So $I = I_{SC}$ when $V = 0$.

1.5.2   OPEN-CIRCUIT VOLTAGE (V_{OC})

The voltage across the cell V = open-circuit voltage (V_{OC}) = voltage across the cell under open-circuit condition. In this condition, the current flowing through the cell $I = 0$. Hence, $V = V_{OC}$ when $I = 0$.

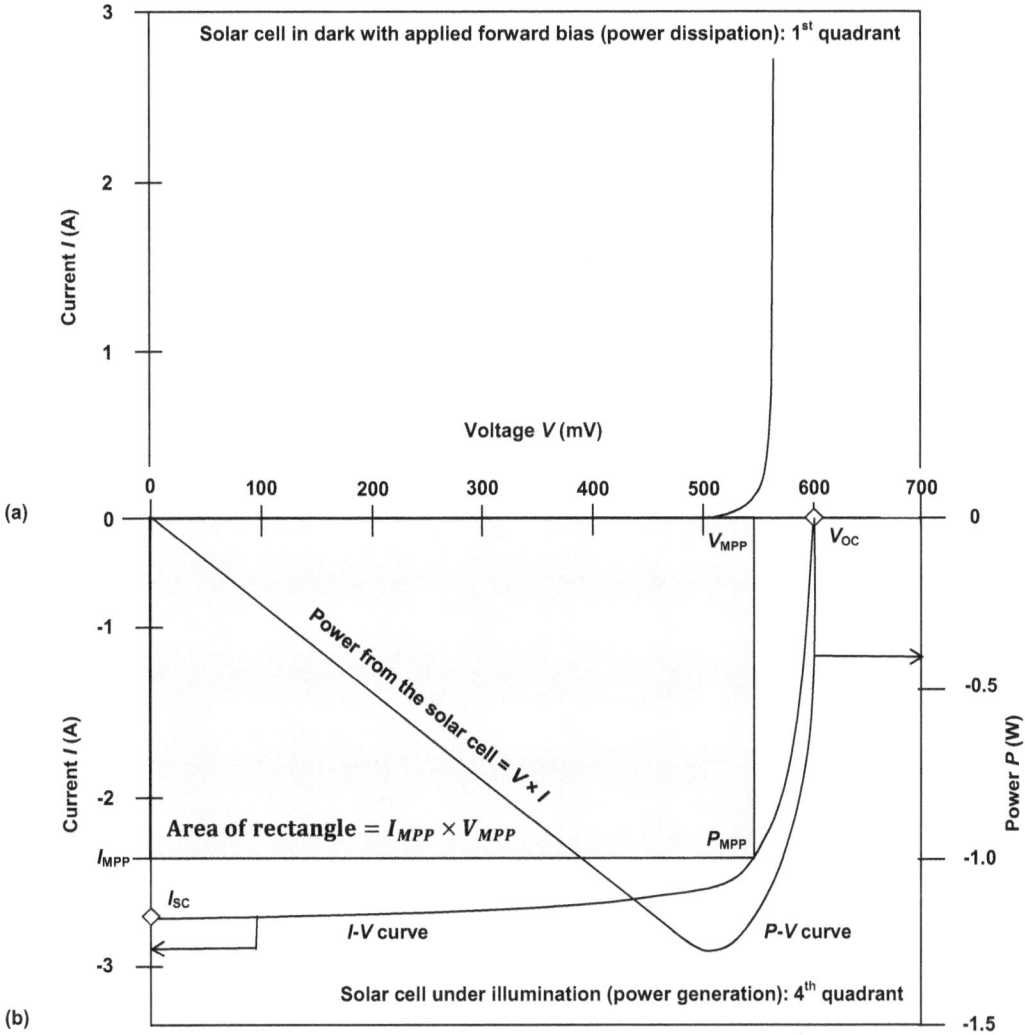

FIGURE 1.4 Electrical characteristics of solar cells: (a) Current-voltage characteristic of a solar cell diode in the dark. The solar cell is enclosed in a darkened box for plotting the characteristic. The forward voltage across the diode is varied by a potentiometer, and the forward current is measured. A range of forward voltages is applied to generate the plot. (b) Current-voltage (I-V) and power-voltage (P-V) characteristics of a solar cell exposed to sunlight. For drawing the characteristics, the solar cell is illuminated with sunlight, the voltage across the solar cell is changed with a variable resistor (potentiometer) connected in the circuit, and the current is measured for each value of voltage.

The voltage V_{OC} = maximum voltage obtained from a solar cell because no voltage is dropped across any load in the circuit in absence of current flow.

1.5.3   MAXIMUM POWER (P_M) AND MAXIMUM POWER POINT (P_{MPP})

With the help of the potentiometer, the voltage across the solar cell is adjusted to different values and the current is measured at each voltage. The product of voltage with the corresponding current gives the power produced by the solar cell. The power increases with increasing voltage, reaching

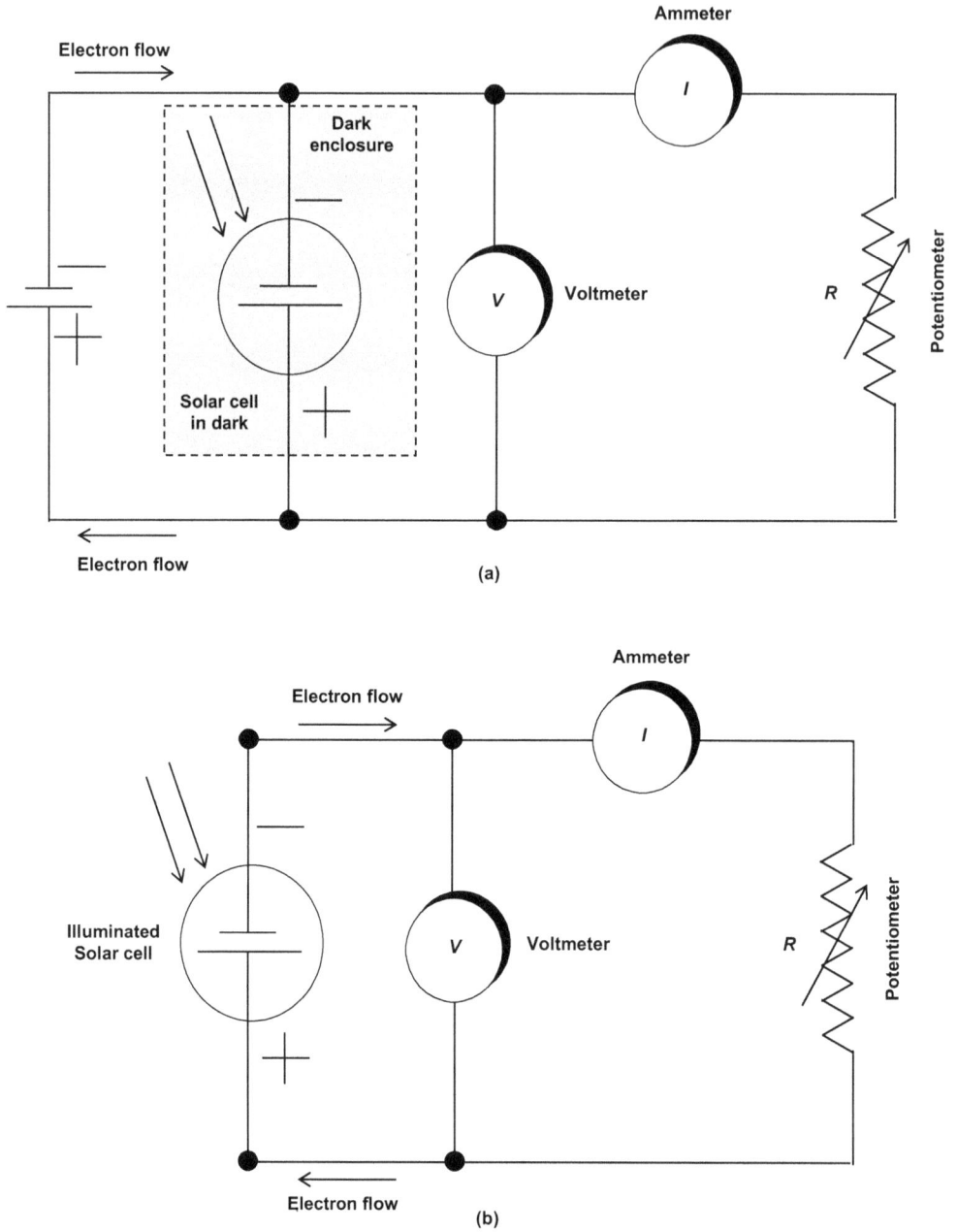

FIGURE 1.5 Circuits for recording current-voltage characteristics of solar cells: (a) solar cell kept inside a dark enclosure and biased in forward mode, and (b) solar cell kept under sunlight without any applied bias. Both circuits contain a voltmeter connected in parallel to the solar cell to measure the potential difference across the cell, an ammeter in series with the cell to measure the current flowing through the cell and a potentiometer to vary the voltage across the cell. In *a*, a battery is connected to forward-bias the solar cell diode.

a peak value, and then declines with further rise of voltage. The point of reaching maximum power P_M is called maximum power point P_{MPP}. The maximum power is given by the area of the largest rectangle that can be fitted in the space between the *I-V* curve and the coordinate axes. The potential at the maximum power point is the voltage at maximum power point (V_{MPP}), and the associated current is the current at maximum power point (I_{MPP}). P_{MPP}, V_{MPP}, and I_{MPP} are marked on the graph. The maximum power is given by

$$P_M = V_{MPP} \times I_{MPP} \tag{1.1}$$

1.5.4 FILL FACTOR

Besides short-circuit current and open-circuit voltage, two vital parameters characterizing the performance of a solar cell are its fill factor and efficiency. The fill factor, abbreviated as *FF*, and a measure of the closeness of the current-voltage characteristics of solar cell to a square shape, is defined as the ratio of maximum power given by the area of the largest area rectangle fitted between the *I-V* curve of solar cell and coordinate axes, and the product of open-circuit voltage and short circuit current of the solar cell.

$$FF = \frac{\text{Maximum power}}{\text{Open} - \text{circuit voltage} \times \text{Short} - \text{circuit current}} = \frac{P_M}{V_{OC} \times I_{SC}} = \frac{V_{MPP} \times I_{MPP}}{V_{OC} \times I_{SC}} \tag{1.2}$$

Also,

$$P_M = V_{OC} \times I_{SC} \times FF \tag{1.3}$$

1.5.5 POWER CONVERSION EFFICIENCY

A vital parameter giving an idea about the performance of a solar cell is its power conversion efficiency η. It is a fraction equal to the output power P_M divided by input power P_{IN}, i.e., P_M/P_{IN}. Generally, it is expressed as a percentage. The intensity of the sunlight, the solar spectrum, and the temperature of the solar cell determine the efficiency. Different manufacturing firms and research laboratories report the efficiency values for solar cells made by them. To assess their relative ranking, these claims need to be compared against a common benchmark. Meaningful comparison fosters competition. Therefore, the efficiencies must be measured and specified under standard conditions laid out for the solar spectra. Also, the measurement conditions for efficiency must be explicitly stated. From equation 1.3,

$$\eta = \frac{P_M}{P_{IN}} = \frac{V_{OC} \times I_{SC} \times FF}{P_{IN}} \tag{1.4}$$

1.5.6 AM0 AND AM1.5 SOLAR SPECTRA

The spectrum of sunlight varies from one place on the Earth to another. At a given place on Earth, the spectra measured at different hours of the day are different. In view of these unavoidable variations, reference spectra are defined for characterization of solar cells. These spectra serve as standards for comparison of performance parameters of solar cells fabricated at different research laboratories and manufacturing plants across the world. Test results reported under standard conditions help in relative evaluation of the solar cell technologies developed with respect to a common reference platform.

The air mass (AM) coefficient is a parameter used to characterize the spectrum of sunlight after its propagation through the Earth's atmosphere. It is defined as the ratio = the optical path length traversed by sunlight through the atmosphere/the path length measured in the vertically upwards direction pointing towards the zenith. Space-qualified solar cells are measured at AM0, or air mass zero. *AM0* means the spectrum outside the atmosphere pertaining to 5,800K black body radiation. Measurements on solar cells for terrestrial applications are done under 1.5 air mass conditions at 25°C. The term "AM1.5G," or "1.5 air mass global," refers to solar spectrum for 1.5 atmosphere thickness. The solar spectrum for 1.5 atmosphere thickness corresponds to spectrum for a solar zenith angle of 48.2°. *Solar zenith angle* is the angle subtended by the sunrays with the vertical direction. The letter *G* is affixed to represent "global," meaning, that this standard includes both direct and diffuse radiation obtained after scattering from the atmosphere.

1.5.7 SHOCKLEY-QUEISSER DETAILED BALANCE LIMIT OF EFFICIENCY OF P-N JUNCTION SOLAR CELL

As the energy bandgap of semiconductor is enlarged, the open-circuit voltage of the solar cell increases. At the same time, the short-circuit current density falls. On the opposite side with the narrowing of the bandgap of the semiconductor, the open-circuit voltage of solar cell decreases, while the short-circuit current density rises. As both open-circuit voltage and short-circuit current density appear in equation 1.4 for efficiency of a solar cell, an optimum value of bandgap exists, which provides maximum efficiency of conversion of power.

Shockley-Queisser limit is the maximum theoretical efficiency calculated for a solar cell with a single ideal P-N junction under the assumption of radiative recombination of electrons and holes as the only loss mechanism in solar cell (Shockley and Queisser 1961). The calculation approximates the solar spectrum with 6,000K black body radiation. It gives a bandgap of 1.1eV and places the upper limit of efficiency at 30%. Later calculations with AM1.5G spectrum as the standard spectrum for nonconcentrated conversion for a solar cell with a highly reflective rear mirror yield the bandgap as 1.34eV and ultimate efficiency limit as 33.7%. The mirror maximizes the light emission from the front surface opposite to the direction of incident rays.

1.6 SOLAR CELL GENERATIONS

The evolution of solar cells is divided into several generations (Figure 1.6). We look at the salient features of solar cells of different generations to know which generation of solar cells we shall be discussing in this book.

1.6.1 FIRST GENERATION

First-generation cells are wafer-based solar cells fabricated on single-crystal to multicrystalline bulk silicon wafers (Roghini and Enrichi 2020). These are the oldest solar cells, hence called traditional cells, and are the most widely used cells owing to their high efficiencies, but they are also the most expensive variety. Their high manufacturing costs have prompted scientists to look for alternative technologies that can make solar cells available to users at affordable budgets.

1.6.2 SECOND GENERATION

While the first generation of cells is made of high-quality, defect-free, thick wafers of semiconductor material, the second-generation cells consist of thin films of nonmonocrystalline materials deposited on low-cost plastic substrates. The material consumption is vastly cut down because of the use of thin films. These solar cells include amorphous silicon, cadmium telluride (CdTe), and copper indium gallium diselenide (CIGS) cells. They are cost-effective relative to the first generation, but at the sacrifice of efficiency, which has comparatively lower values for the second-generation cells. Although efficiency is lower, the cost is also less, enabling lower cost per watt. CdTe is a toxic material and environmentally hazardous, so its disposal is problematic. The indium in CIGS cells is a rare earth metal found in 0.21 parts per million in the Earth's crust; hence, there is a need for a better alternative.

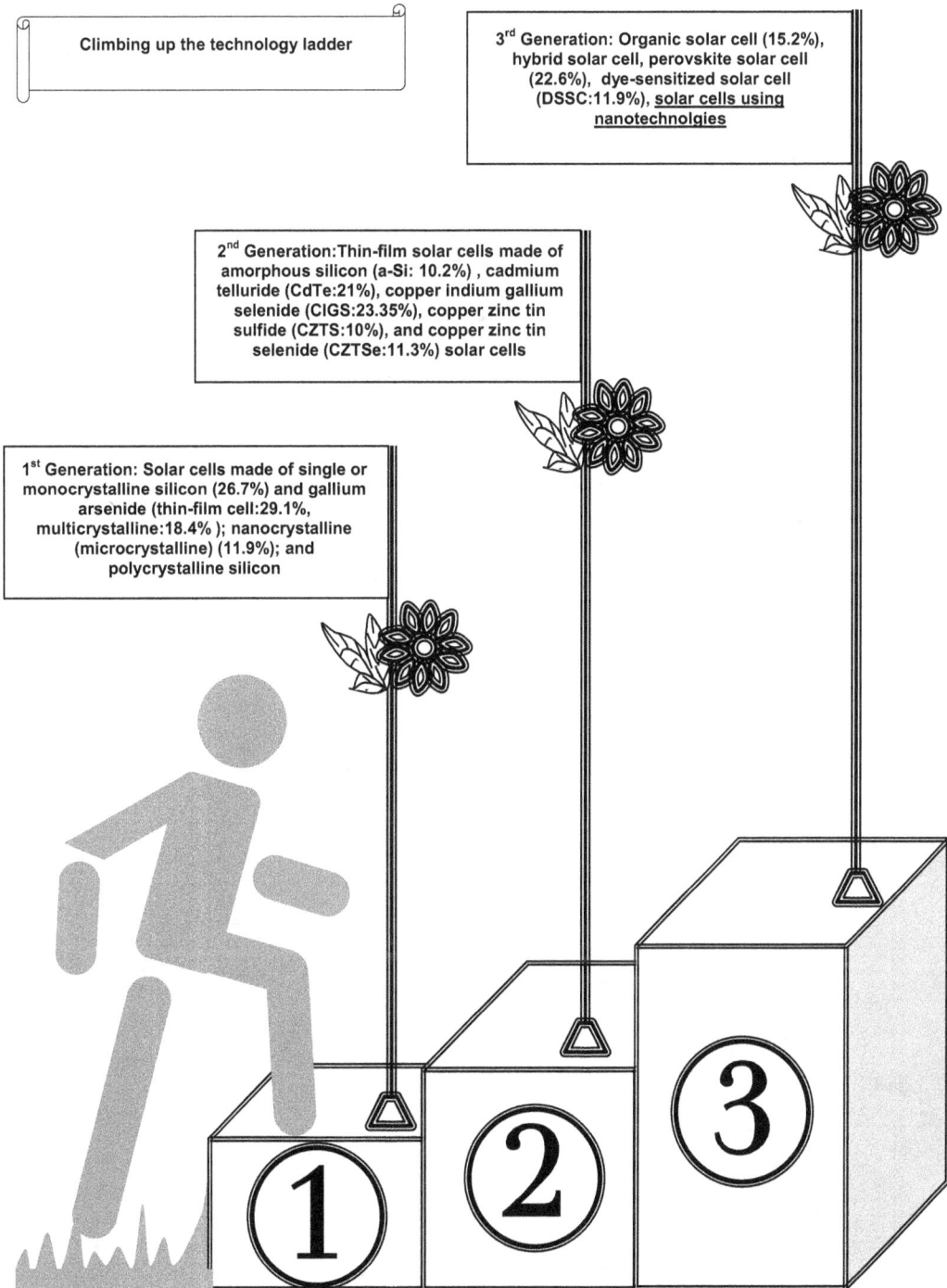

Climbing up the technology ladder

3rd Generation: Organic solar cell (15.2%), hybrid solar cell, perovskite solar cell (22.6%), dye-sensitized solar cell (DSSC:11.9%), solar cells using nanotechnolgies

2nd Generation:Thin-film solar cells made of amorphous silicon (a-Si: 10.2%) , cadmium telluride (CdTe:21%), copper indium gallium selenide (CIGS:23.35%), copper zinc tin sulfide (CZTS:10%), and copper zinc tin selenide (CZTSe:11.3%) solar cells

1st Generation: Solar cells made of single or monocrystalline silicon (26.7%) and gallium arsenide (thin-film cell:29.1%, multicrystalline:18.4%); nanocrystalline (microcrystalline) (11.9%); and polycrystalline silicon

FIGURE 1.6 Solar cell generations showing the three generations of devices: first, which are made of single crystal semiconductors, e.g., silicon or gallium arsenide, or nanocrystalline/polycrystalline silicon or multi-crystalline gallium arsenide; second, made of thin films of materials, such as a-Si, CdTe, CIGS, CIGSe solar cells; and third, including organic, perovskite, dye-sensitized cells, as well as emerging nanotechnology-based cells. Efficiencies of solar cells are mentioned within brackets in accordance with Green et al (2022) under AM1.5 spectrum at 1,000Wm^{-2} power density 25°C and are improving continuously with ongoing research. Due to the unabated progress, the present efficiencies may be higher than shown here.

1.6.3 THIRD GENERATION

Third-generation solar cells extend the use of thin-film technology by employing novel techniques and materials as well as inexpensive solution-based processes for cost reduction. These solar cells aim at achieving a high performance at low costs by using nanomaterials and applying nanotechnologies.

The third-generation cells may not employ traditional P-N junction approach. Many cells are solution-processed cells using organic macromolecules, inorganic nanoparticles, or hybrids. In this generation are polymer: fullerene solar cells, perovskite solar cells, mesoscopic dye-sensitized solar cells, nanocrystal or quantum dot solar cells, and quantum well solar cells. Mention must also be made of the hybrid polymer solar cells with one component an organic semiconducting polymer and the other component an inorganic semiconducting nanoparticle, together constituting a single multispectrum layer. Several upcoming technologies still await commercial acceptability.

These third-generation solar cells, along with nanocrystalline silicon solar cells, will be described in this book.

1.7 SOLAR CELL TECHNOLOGIES

1.7.1 MONOCRYSTALLINE SILICON SOLAR CELL

This is the solar cell which is in most widespread commercial use today. The device shown in Figure 1.7 has the structure of an aluminum back surface field (BSF)/P-type Si/N-type Si/SiO$_2$/

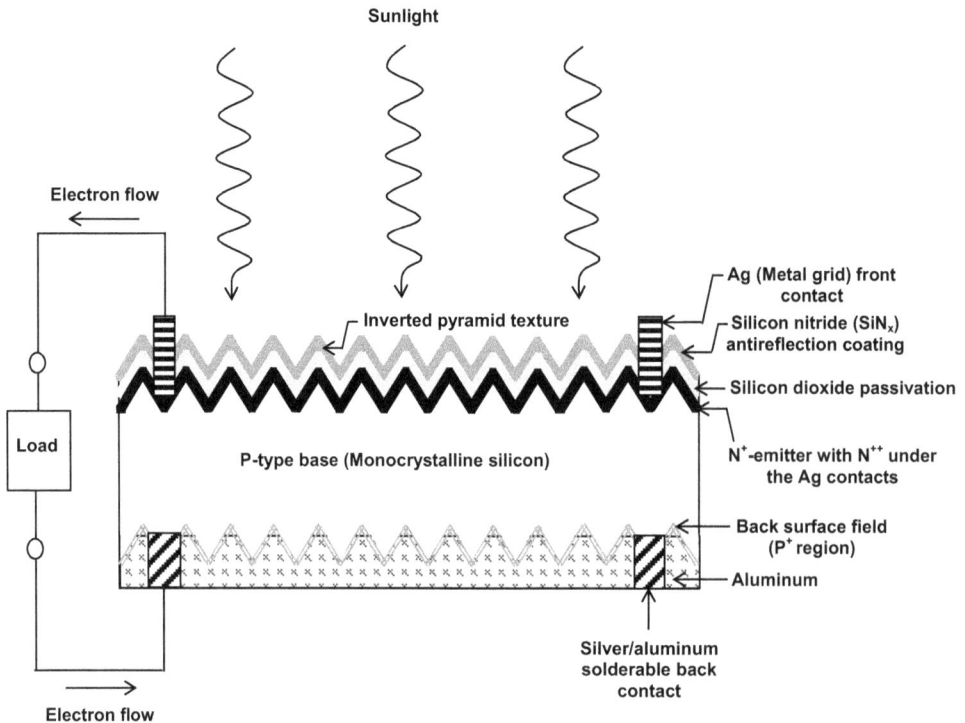

FIGURE 1.7 Monocrystalline silicon solar cell. Both surfaces of P-type silicon wafer are textured. A diode is fabricated with N$^+$ emitter on a P-type monocrystalline base, and P$^+$ back surface field layer is formed at the rear surface. The front surface is passivated with SiO$_2$, and a SiN$_x$ antireflection layer is deposited. Front contact metal is Ag; solderable back contact is made with Ag/Al. Sunlight falls on SiN$_x$ layer, and electrical output is obtained between Ag and Ag/Al contacts.

SiN$_x$/Ag (Liu et al 2014). The back surface field is produced by forming a heavily doped (P$^+$) region at the rear surface, yielding a high-low P$^+$/P-junction. The electric field created across the high-low P$^+$/P-interface acts as a barrier to the flow of minority carriers.

The device is made on a P-type silicon wafer of resistivity 1–3Ω and thickness 200µm. The process involves the following steps: (i) the surface of P-type single crystal wafer is texturized in KOH/isopropanol solution, forming inverted pyramid-shaped protuberances, (ii) phosphorous diffusion is carried out to form N$^+$ emitter, (iii) the phosphosilicate glass is removed by wet-etching, (iv) the SiO$_2$ layer is grown and the SiN$_x$ antireflecting coating is deposited, (v) the Ag front contacts are screen-printed and dried, (vi) the Ag/Al back contacts are screen-printed and dried, and (vii) the Al film is printed, dried, and sintered. The solar cell is illuminated from the side of the Ag grid, and electrical output is collected between the Ag front and the Ag/Al back contacts.

1.7.2 Gallium Arsenide Solar Cell

This solar cell consists of a P-GaAs substrate (350µm, 5–10 $\times$ 10^{18}cm^{-3}) on the lower side of which the back metallization film is deposited to form ohmic contact for P-anode (Figure 1.8). On the upper side, the following layers are deposited sequentially by metal organic vapor phase epitaxy

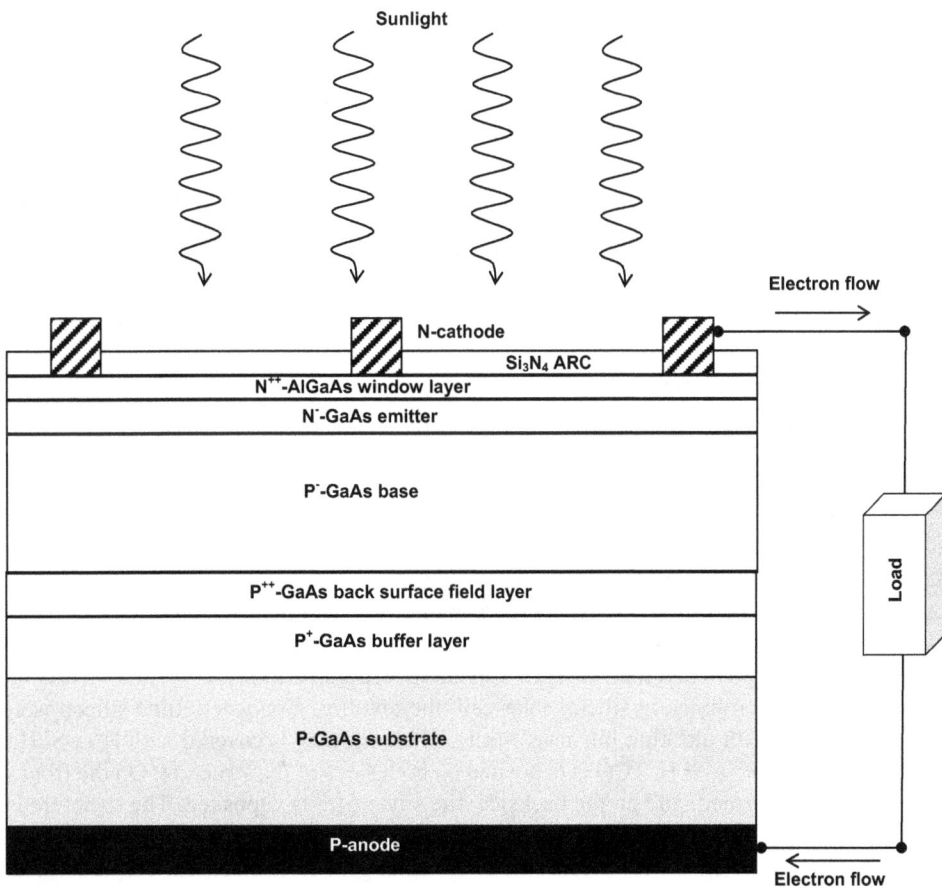

FIGURE 1.8 GaAs solar cell showing the different layers in the structure from top downwards: N-cathode-Si$_3$N$_4$ ARC layer/N^{++}-AlGaAs window layer/N$^+$-GaAs emitter/P-GaAs base/P^{++}AlGAs BSF layer/P$^+$-GaAs buffer layer/P-GaAs substrate/P-anode. Sunlight falls on the N-cathode-Si$_3$N$_4$ ARC layer, and electric current is drawn between the N-cathode and P-anode.

(MOVPE): P^+-type GaAs buffer layer (500nm, $5 \times 10^{18} cm^{-3}$), $P^{++}Al_{0.20}Ga_{0.80}As$ back surface field layer (200nm, $2 \times 10^{19} cm^{-3}$), P-GaAs base (1,000nm, $1 \times 10^{17} cm^{-3}$), N^+-GaAs emitter (50nm, $2 \times 10^{18} cm^{-3}$), $N^{++}-Al_{0.20}Ga_{0.80}As$ window layer (30nm, $5 \times 10^{18} cm^{-3}$), Si_3N_4 antireflection coating, and front ohmic contact (N-cathode). Typical doping concentrations and thicknesses are given in brackets (Selvamanickam 2018).

The window layer is a small-thickness, large-bandgap, heavily doped layer taken sufficiently thin for not absorbing any light and transparent enough to allow easy passage of incoming light to the absorber layer (here P-GaAs base layer); hence, it is called the window layer (Kamdem et al 2019). Its primary use concerns limiting the surface recombination. At the surface, the periodicity of atoms of the crystal lattice is interrupted, leading to a high recombination rate. Surface recombination velocity is a function of surface roughness and contamination from processing gases (Tumpa et al 2015). The window layer minimizes front surface recombination near the window layer/emitter layer interface by achieving a low surface recombination velocity. Thus, it prevents any minority carrier produced by a photon absorbed near the surface of the absorber layer from being lost by recombination. Its other function is to conduct the current to the metal grid; hence, it is heavily doped.

The buffer layer serves as an intermediary between the BSF layer and the substrate to improve interface quality.

1.7.3 AMORPHOUS SILICON SOLAR CELL

An extremely thin layer of a-Si:H is necessary to make an amorphous silicon solar cell. Consumption of a very small amount of material bestows a distinct cost advantage. This happens because of the larger absorption coefficient of a-Si:H than crystalline Si, necessitating a much smaller thickness of a-Si:H than crystalline silicon. Further, inexpensive glass, plastic substrates, or flexible metal foils can be used for making solar cells using low-temperature deposition techniques, such as plasma-enhanced chemical vapor deposition for a-Si:H and physical vapor deposition methods, like thermal evaporation or sputtering for metal contacts, or thick film screen-printing process.

A generic amorphous silicon solar cell structure consists of a P-I-N diode, with the P-, I-, and N-layers made of amorphous silicon (Figure 1.9). The structure has the form Al/N-a-Si:H/I-a-Si:H/P-a-Si:H/ITO/glass substrate with I-a-Si:H photoactive layer, and N-a-Si:H and P-a-Si:H carrier transport layers (Carlson and Wronski 1976; Singh et al 2013; Qarony et al 2017).

1.7.4 SILICON HETEROJUNCTION SOLAR CELL

In a silicon heterojunction device (Figure 1.10), the two semiconductor materials used for forming the heterojunction are amorphous and monocrystalline silicon. While single-crystal silicon is an indirect bandgap semiconductor with an energy gap of 1.1eV, hydrogenated amorphous silicon is a direct bandgap semiconductor with an energy gap 1.5–2.0eV. The energy gap of a-Si:H depends on the conditions of its growth and the hydrogen content incorporated in it.

For making the heterojunction silicon solar cell, the textured monocrystalline silicon wafer is coated on both sides with ultrathin intrinsic a-Si:H. Then one side is covered with N^+-a-Si:H, and the opposite side with P^+-a-Si:H. TCO is deposited on both sides of the structure. On the front side, the Ag metal grid is formed, and on the backside, the silver film is deposited. The structure is Ag grid/TCO/N^+a-Si:H/intrinsic a-Si:H/P-type monocrystalline base/intrinsic a-Si:H/P^+a-Si:H/TCO/Ag. When sunlight falls on the solar cell at the Ag metal grid side, electrical output is obtained between the Ag metal grid and the backside Ag film.

Why do we need a heterojunction solar cell? By making monocrystalline/amorphous silicon heterojunction, the recombination-active contacts in a monocrystalline silicon solar cell (Figure 1.7) are shifted from the surface of monocrystalline silicon, as revealed by comparing Figure 1.10 and Figure 1.7. This happens because the film of the wide bandgap material, here I-a-Si:H, is interposed

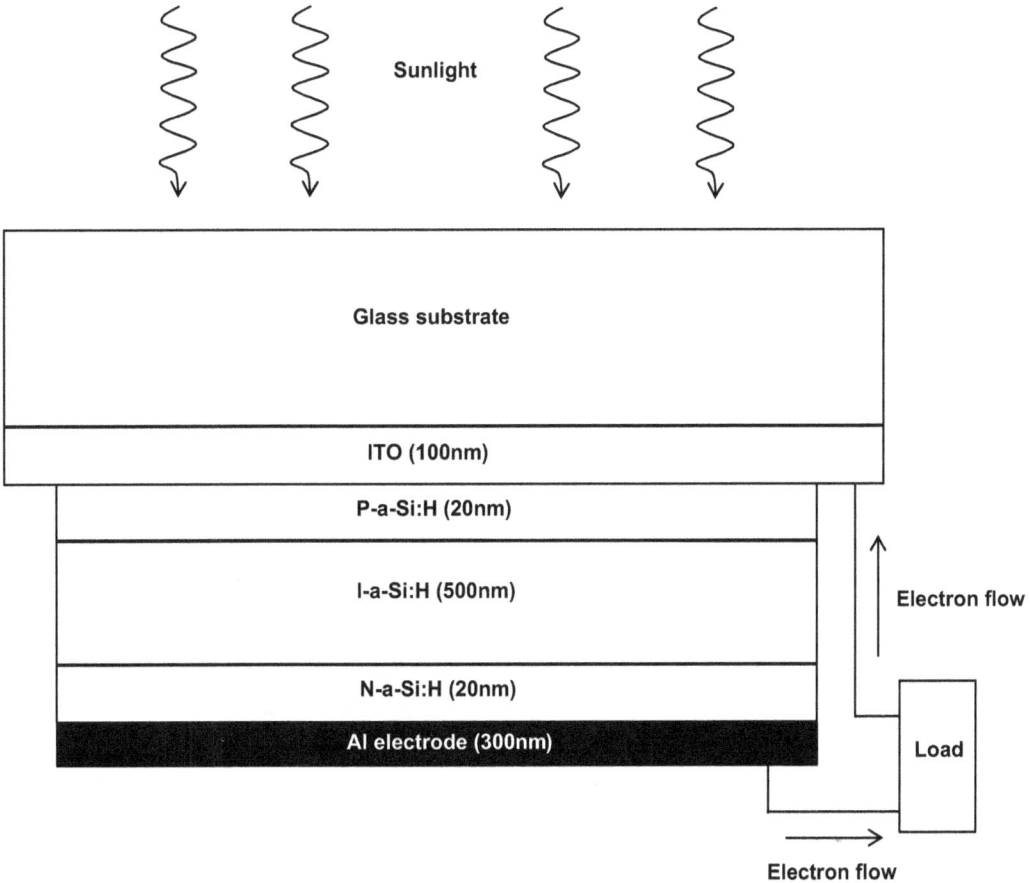

FIGURE 1.9 Amorphous silicon solar cell consisting of a glass substrate on which the front side transparent conducting oxide (indium tin oxide, ITO, In_2O_5Sn, solid solution of In_2O_3 with SnO_2) is deposited, followed by the P-, intrinsic, and N-doped amorphous silicon layers, and then finally the aluminum film. Sunlight falls on the glass substrate. Electrical contacts are made with ITO and Al electrodes. Electric current flows between the contacts when sunlight falls on the glass substrate side of the solar cell.

between the contacts and the monocrystalline silicon. The intrinsic hydrogenated amorphous silicon film reduces the density of surface states at monocrystalline silicon surface by hydrogenation. Also, it can be doped N- or P-type for contact fabrication, enabling fabrication of high-efficiency solar cells (Taguchi et al 2014). Thus, the doped N- or P-type a-Si:H overlayers forming heterojunctions with monocrystalline silicon are used as electron and hole collectors, obtaining nearly recombination-free contacts (Seif et al 2014).

1.7.5 CADMIUM TELLURIDE SOLAR CELL

CdTe solar cell is a low-cost photovoltaic cell produced quickly using high-throughput manufacturing. It not only shows stable performance in terrestrial applications but can also withstand the high-energy proton and electron flux and UV radiation in outer space (Romeo et al 2001). This solar cell (Figure 1.11) is a thin-film heterojunction solar cell using two semiconductors: N-type CdS (window layer) and P-type CdTe (absorber layer). CdTe is an II-VI direct band semiconductor with an ideal bandgap 1.5eV, matching with the solar spectrum. The cell is made on an ITO-coated glass substrate on which the N-type CdS layer is deposited by any one among

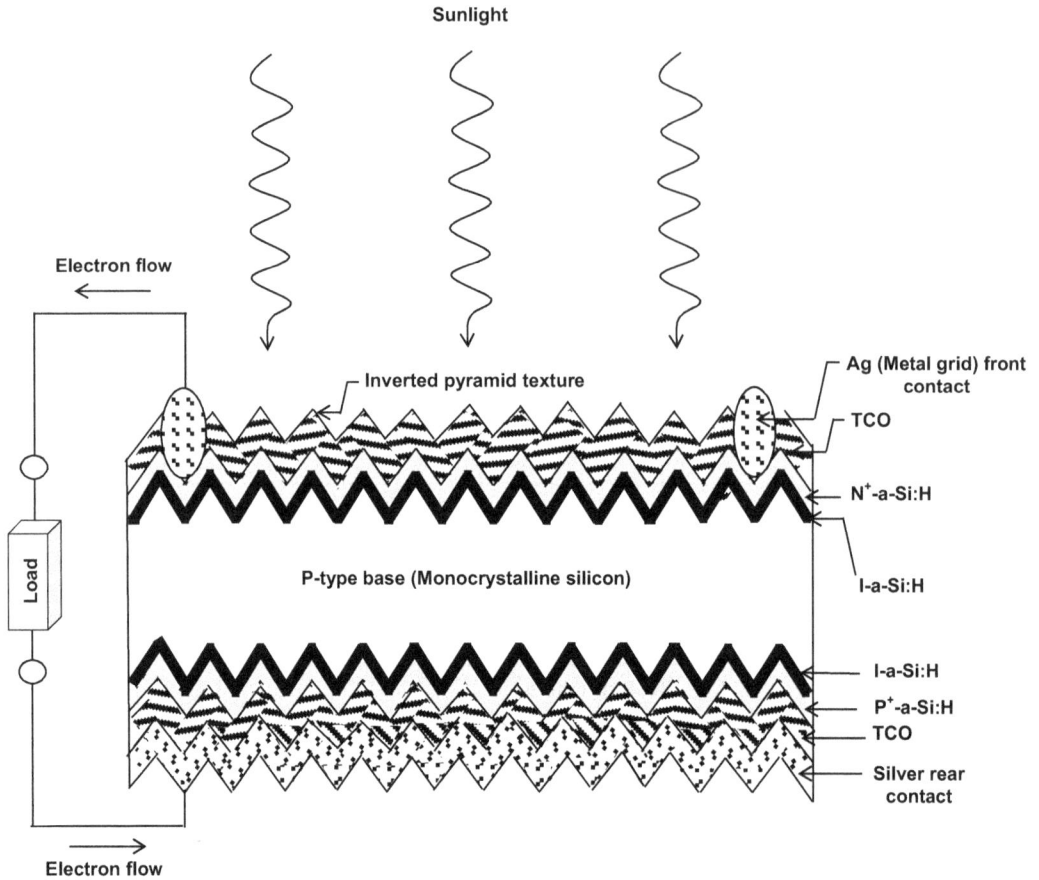

FIGURE 1.10 Silicon heterojunction solar cell in which the heterostructure is formed between the mono-crystalline silicon and the amorphous silicon layers. Above the P-type monocrystalline silicon layer, the layers successively placed are a plasma-enhanced chemical vapor deposition (PECVD)–formed intrinsic a-Si:H layer, an N⁺ PECVD a-Si:H emitter, a physical vapor deposition (PVD) antireflection TCO layer, and a PVD Ag grid. Below the monocrystalline silicon layer, the layers are an intrinsic PECVD a-Si:H layer, a PECVD P⁺-type a-Si:H layer, a PVD TCO layer, and a screen-printed Ag contact layer. The solar cell is illuminated on the Ag metal grid side, and external connections are made to Ag electrodes.

the techniques vapor-transport deposition (VTD), closed-space sublimation (CSS), sputtering or chemical bath deposition. The CdTe film is deposited by CSS, sputtering, or thermal evaporation (Romeo and Artegiani 2021). After CdTe deposition, the film stack is annealed at 400°C in $CdCl_2$ and oxygen ambience for reduction of CdTe defect density, raising P-type conductivity and passivation of grain surfaces, thereby increasing the carrier lifetime above 3ns (Basol and McCandless 2014). A high-resistivity transparent (HRT) film, such as zinc-doped tin oxide having ~ 30nm thickness, is interposed between the TCO and CdS layers in high-efficiency CdTe solar cells to improve the quality of the CdS film and reduce leakage current. The TCO (ITO) is the front contact. The rear contact is a silver film deposited on CdTe.

1.7.6 Cadmium Indium Gallium Selenide (CIGSe) Solar Cell

This solar cell using a thin-film CIGS absorber layer is a high-efficiency, large-scale-manufacturing option to silicon solar cells, allowing easy integration with vehicles, such as airplanes and cars,

Sunlight

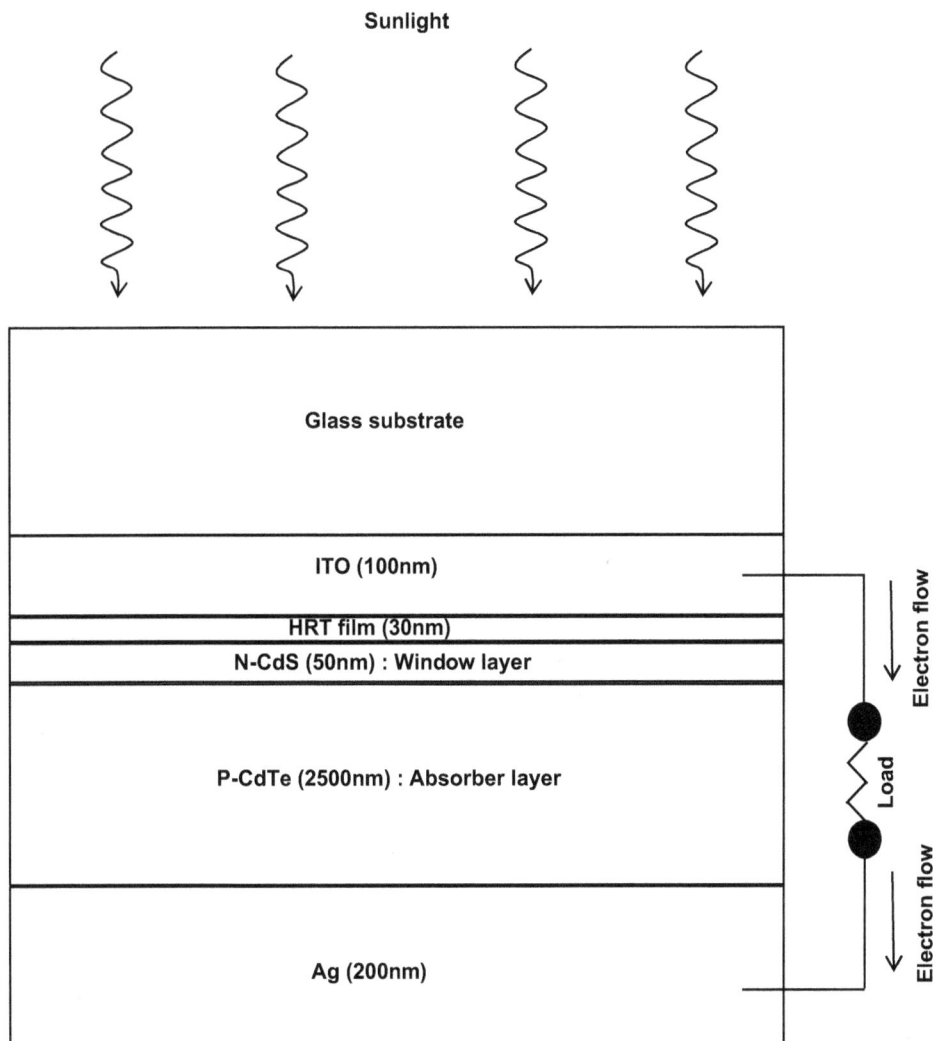

FIGURE 1.11 Cadmium telluride solar cell in which a P-N junction is formed between N-CdS and P-CdTe layers, and contacts are made with transparent conducting oxide on the front surface and silver film on the back surface. Between the TCO and CdS layers, there is a high-resistivity transparent (HRT) film. Electric current flows between the ITO and Ag electrodes when sunlight is incident on the glass surface of the solar cell.

with insignificant addition to weight. CIGSe stands for copper indium gallium (di) selenide, i.e., $Cu(In,Ga)Se_2$, a I-III-VI$_2$ high optical absorption coefficient, direct bandgap semiconductor with energy gap tunable from 1.01 to 1.68eV for best matching with solar spectrum by changing the Ga fraction in $CuIn_xGa_{1-x}Se_2$, $x = 0$ to 1 (Belghachi and Limam 2017).

The cross section of a CIGSe solar cell is shown in Figure 1.12. On a molybdenum-coated soda lime glass substrate or some flexible substrate (polymer or metal foil), the P-type CIGSe layer is formed either by coevaporation or cosputtering and annealing in selenide vapor or by any one of low-cost, nonvacuum deposition processes, such as chemical spray pyrolysis, electrodeposition, or paste coating methods, e.g., screen-printing, doctor blading, or curtain coating (Kaelin et al 2004). Then, an N-type CdS buffer layer is deposited by chemical bath deposition and covered with a bilayer of DC sputter-deposited intrinsic and aluminum-doped zinc oxide TCO (front contact);

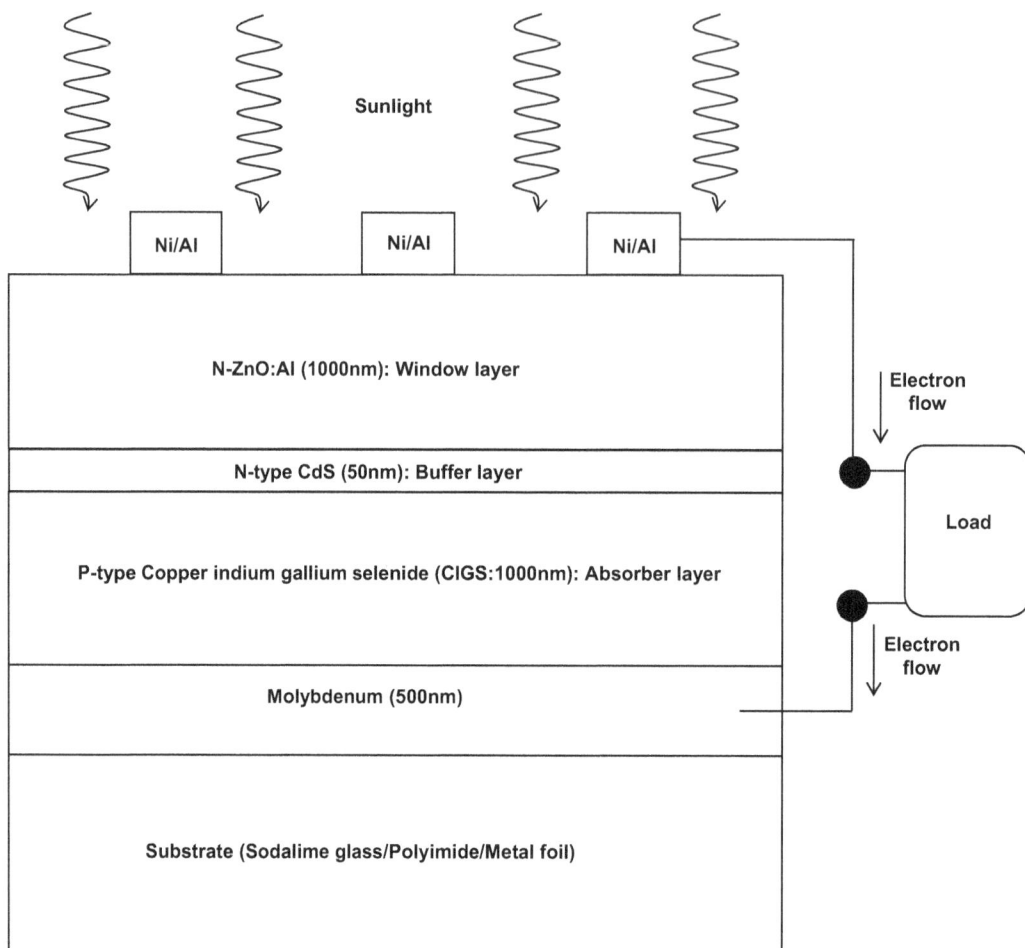

FIGURE 1.12 CIGSe solar cell in which a P-N heterojunction is formed between the N-type CdS and the P-type CIGSe films, and contacts are made with the Ni/Al grid to the ZnO:Al on the front side and molybdenum on the backside. Sunlight falls on the Ni/Al grid-ZnO:Al surface. Electric current flows between the Ni/Al and the Mo contacts.

Mo film is the back contact. The N-type CdS buffer layer fixes the electrostatic conditions in the absorber layer. It also provides structural stability to the solar cell (Mostefaoui et al 2015).

The photons are absorbed by the CIGSe layer, producing electron-hole pairs. The depletion region formed at the heterojunction between the N- and P-type materials separates the charge carriers, leading to a net flow of current, which is obtained across the Ni/Al and the Mo contacts.

1.7.7 PEROVSKITE SOLAR CELL

In this solar cell, a perovskite material is used as the light absorber. *Perovskite* is a name given to any material which has the same crystal structure as naturally occurring mineral calcium titanate $CaTiO_3$. Perovskites are compounds having the chemical formula ABX_3, where A and B are cations, with A larger than B, and X is an anion (Green et al 2014). Like oxide perovskites, there are halide perovskites, which have received attention for solar cells. The commonly used organic-inorganic perovskite is methylammonium lead trihalide containing the methylammonium cation $CH_3NH_3^+$ and a halogen anion, e.g., iodine, bromine, or chlorine. Perovskite materials offer multifarious advantages of low-cost, strong optical absorption, bandgap tuning, and long carrier diffusion lengths together with ease of device fabrication.

The perovskite solar cell (Figure 1.13) is fabricated on an indium-doped tin oxide or fluorine-doped tin oxide-coated glass substrate (Tang et al 2017). On this substrate, an N-type semiconductor is deposited as an electron transport layer (ETL). Compact TiO_2 (c-TiO_2) is usually used as the ETL; other materials are SnO_2, C_{60}, and its derivatives.

Over the ETL, the perovskite layer is formed. The perovskite layer is covered with a P-type semiconductor as a hole transport layer (HTL). Spiro-OMeTAD is the common HTL material. Other examples are poly(3-hexylthiophene) (P3HT), poly(3,4-ethylenedioxythiophene)-poly(styrenesulfonate) (PEDOT:PSS), and poly(triaryl)amine (PTAA).

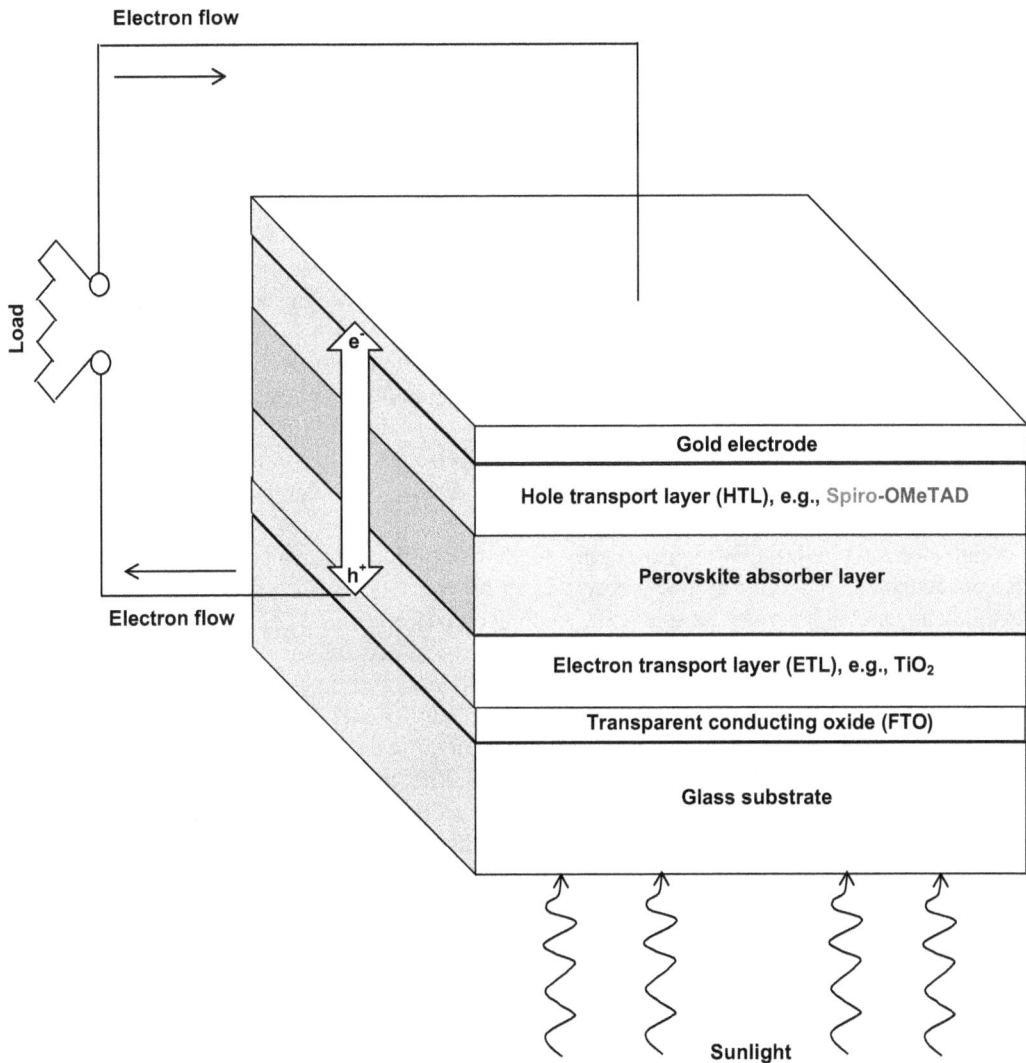

FIGURE 1.13 Perovskite solar cell (PSC), consisting of transparent conducting oxide and gold electrodes on the opposite sides of a trilayer structure: electron transport layer/perovskite absorbing layer/hole transport layer. Upon illumination with sunlight, the electron-hole pairs produced in the perovskite layer are extracted by the electron and hole transport layers and supplied to the respective electrodes, producing a potential difference across them.

The final layer is a metal layer over the HTL. Thus, the general configuration of a perovskite solar cell consists of the layered structure ITO/FTO-coated glass substrate/ETL/perovskite/HTL/metal. The perovskite layer, ETL, and HTL are deposited by spin-coating for small-area solar cells and by screen-printing or spray-coating for larger-size cells.

Upon illumination with sunlight from the glass substrate side, electron-hole pairs are produced in the perovskite layer. The electrons move to the ETL and are collected by the ITO layer. The holes move to the HTL and are received by the metal electrode.

Perovskite solar cells suffer from long-term stability problem arising from degradation of perovskite upon exposure to humid environments or oxygen (Rong et al 2018). UV illumination and thermal stress, too, promote degradation. The TiO_2 ETL induces photocatalytic degradation under UV radiation. UV-stable materials, such as $BaSnO_3$, may be used in place of TiO_2. Stability can also be improved by depositing an interfacial modifier, e.g., CsBr or Sb_2S_3 film, between the ETL and the perovskite layer.

Performance of ZnO as an ETL has been improved by replacing with polydopamine-modified aluminum-doped ZnO (AZO:PDA) (Liu et al 2022). It optimizes interfacial contact with perovskite. By inhibition of the detrimental interfacial reaction, the $FA_{0.9}Cs_{0.1}PbI_3$ perovskite solar cells show an efficiency of 21.36%, with better photostability and thermostability, retaining 90% of their efficiency after 360 h. at 85°C.

Interfaces of multication and halide perovskite have been functionalized with an organometallic compound ferrocenyl-bis-thiophene-2-carboxylate (FcTc) (Li et al 2022). Inverted perovskite solar cells have shown an efficiency of 25%. The efficiency is preserved up to 98% after 1,500 h. of continuous operation at maximum power point. The devices show high stability in damp heat tests carried out at 85% relative humidity and 85°C temperature.

To avoid instability from HTL, the HTL can be substituted with an inorganic material. It can be separated from the perovskite layer by an intervening buffer layer. It can also be protected by suitable encapsulation.

Device instability caused by corrosion of metal electrodes by ions migrating from perovskite layer is prevented by inserting a barrier layer of MoO_x, Cr, or reduced graphene oxide for inhibiting ionic diffusion.

High electron mobility, low defect density single-crystalline TiO_2 nanoparticles having low lattice mismatch and high affinity with perovskite absorber are synthesized by a facile solvothermal method and used in a perovskite solar cell, yielding an efficiency of 24.05% in small-area devices and maintaining 90% of starting performance after nonstop working for 1,400 h. (Ding et al 2022).

Nanomaterials based on two-dimensional organic-inorganic hybrid lead halide perovskites have emerged as the likely candidates to surmount the stability challenges posed by 3D perovskites (Stoumpos et al 2016). The hydrophobicity of bulky organic cations in these nanomaterials provides a strong barrier to moisture ingress in the perovskite, retarding the inception of instability. 2D perovskites are used in two ways: (i) as the primary light-absorbing medium or (ii) as a secondary protective encapsulant, leaving the primary light-absorbing role to 3D perovskite. Both kinds of solar cells, 2D perovskite and 2D/3D perovskite solar cells, have been demonstrated, and stable operation has been shown in unfavorable conditions. These will be discussed at length in Chapter 8.

1.7.8 Organic Solar Cell

Organic solar cells use small organic molecules or polymers for absorption of light and transport of charge carriers. Organic molecules have a high absorption coefficient. Light is absorbed in a small amount of material, requiring a few-hundred-nanometer-thick films.

Organic solar cells can be fabricated by inexpensive solution processing. They are light in weight and easily disposable. They do not impact the environment in an unfriendly manner. Their potential to exhibit transparency can be utilized for applications in windows.

Disadvantages are low efficiency, about two to three times less than inorganic solar cells, and low stability due to photochemical degradation. Durability or lifetime of these solar cells is shorter than those of their inorganic counterparts (Chen 2019).

There are two approaches to organic solar cell realization viz bilayer heterojunction and bulk heterojunction solar cells (Kietzke 2008). In a bilayer heterojunction solar cell (Figure 1.14(a)),

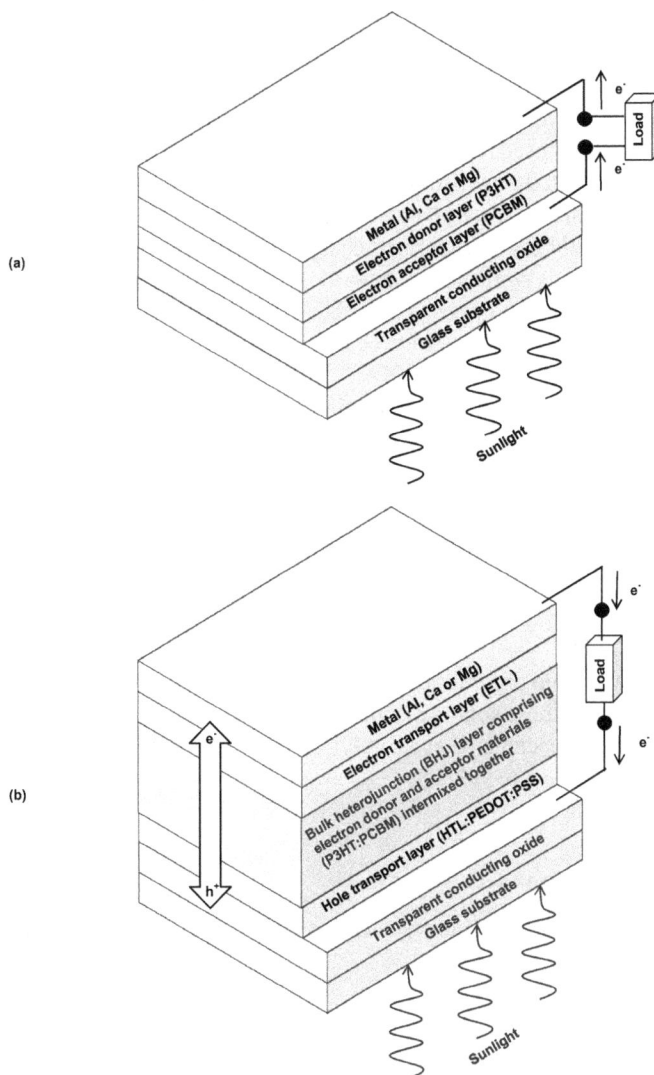

FIGURE 1.14 Organic solar cells: (a) bilayer heterojunction-type cell with separate electron donor and acceptor layers, (b) bulk heterojunction-type cell with donor and acceptor layers blended together to form a single layer, known as the bulk heterojunction (BHJ) layer. Regioregular poly(3-hexylthiophene-2,5-diyl) (P3HT) is a commonly used donor material, while [6,6]-phenyl-C_{61}-butyric acid methyl ester (PCBM), a fullerene derivative, i.e., the solubilized form of buckminsterfullerene C_{60}, is an acceptor material. Layers of electron and hole transport materials in cell (b) facilitate the extraction and transport of the respective photogenerated carriers, electrons for ETL, and holes for HTL. In both cases, a transparent conducting oxide deposited on a glass substrate is used as one electrode, while a metallic film serves as the other electrode. Electric current flows between the electrodes when sunlight strikes the solar cell from the glass substrate side.

a heterojunction between an electron donor material and an electron acceptor material constitutes the active medium. It is formed either by sublimation or spin-coating of the electron donor layer over the electron acceptor material. The active medium is 100nm thick, which is much less than for monocrystalline solar cell (100μm) and thin-film inorganic solar cells (10μm). Incident photons produce excitons. These excitons dissociate into electrons and holes, which diffuse to respective contacts through electron and hole transporting materials. Since the diffusion lengths of charge carriers are ~ 3–10nm, therefore these solar cells must be very thin; otherwise, a large fraction of carriers will be lost by recombination. But reduction of cell thickness also decreases absorption of light. This problem is avoided in a bulk heterojunction (BHJ) solar cell.

In a bulk heterojunction solar cell (Figure 1.14(b)), donor and acceptor materials are blended together and cast as a mixture. If small organic molecules are used, they are deposited by cosublimation. When polymers are used, the mixture of polymers is deposited by spin-coating. The two materials phase separate into domains of two types of materials, which are only a few nanometers apart. Due to the small distance, a large number of carriers are able to diffuse to contacts without recombining. To ensure delivery of separated charge carriers to respective electrodes, an electron transport layer for electrons is deposited on one side of the bulk heterojunction and a hole transport layer for holes is deposited on the opposite side. These transport layers serve as trickling pathways for the carriers. A metal anode is used for collection of electrons, and a TCO cathode for hole collection.

It must be emphasized that satisfactory operation of a bilayer heterojunction solar cell is possible if the thicknesses of individual layers are less than the diffusion length of excitons. Similarly, a bulk heterojunction solar cell will function properly only if the thicknesses of phase-separated domains are smaller than the exciton diffusion length. Noncompliance with these conditions will result in recombination and loss of a large fraction of free carriers.

1.7.9   Hybrid Solar Cell

This is a solar cell in which an organic semiconductor is combined with an inorganic semiconductor in the active medium. In these solar cells, the morphology of the active medium must be tailored at the nanoscale (Müller-Buschbaum et al 2017). These solar cells will be described at several places in the book, e.g., Secs. 6.6, 9.4, 9.5.

1.7.10   Dye-Sensitized Solar Cell

These solar cells use TiO_2 and ZnO nanostructures as photoanodes (Bera et al 2021) and will be described in Chapter 7.

1.8   DISCUSSION AND CONCLUSIONS

Two kinds of power generating systems were explained viz photovoltaic and concentrated power systems. The photovoltaic system can be off-grid or grid-tie type. The significance of solar cell as a component of photovoltaic power system must be appreciated. The main characterization parameters of a solar cell are short-circuit current, open-circuit voltage, maximum power and maximum power point, fill factor, and power conversion efficiency. Measurements performed with respect to reference AM0 and AM1.5 solar spectra can be mutually compared. A quick journey was made into the principles of established solar cells from monocrystalline silicon solar cell to organic solar cells (Table 1.1).

TABLE 1.1
Summary of Different Types of Solar Cells

Sl. No.	Type of Solar Cell	Structure
1.	Monocrystalline silicon solar cell	Aluminum back surface field (BSF)/P-type Si/N-type Si/SiO$_2$/SiN$_x$/Ag
2.	Gallium arsenide solar cell	P-GaAs substrate/P$^+$-type GaAs buffer layer/P^{++}Al$_{0.20}$Ga$_{0.80}$As back surface field layer/P$^-$-GaAs base/N$^+$-GaAs emitter/N^{++}-Al$_{0.20}$Ga$_{0.80}$As window layer/Si$_3$N$_4$ antireflection coating/front ohmic contact (N-cathode)
3.	Amorphous Si solar cell	Glass substrate/ITO/P-a-Si:H/I-a-Si:H/N-a-Si:H/Al
4.	Silicon heterojunction solar cell	Ag/TCO/P$^+$a-Si:H/intrinsic a-Si:H/P-type monocrystalline base/intrinsic a-Si:H/N$^+$a-Si:H/TCO/Ag grid
5.	Cadmium telluride solar cell	ITO-coated glass substrate/HRT/N-type CdS layer/P-type CdTe layer/Ag
6.	Cadmium indium gallium selenide (CIGS) solar cell	Glass substrate/P-type CIGS layer/N-type CdS buffer layer/intrinsic and Al-doped ZnO TCO
7.	Perovskite solar cell	FTO-coated glass substrate/ETL/perovskite/HTL/metal
8.	Organic solar cell	Glass/TCO/electron acceptor layer/electron donor layer/metal

Questions and Answers

1.1 Distinguish between (a) photovoltaic and concentrated solar power systems and (b) off-grid and grid-tie solar power systems. Answer: (a) Photovoltaic systems use solar cells, while concentrated power systems use lenses and mirrors to focus heat for driving turbines for electricity generation. (b) An off-grid system is not connected to a grid, while the grid-tie system is connected to a grid.

1.2 Explain the construction and working of a solar cell in simple language. Answer: It is an electronic device consisting of a P-type semiconductor and an N-type semiconductor. At the junction of the two semiconductors, there is a depletion region consisting of negatively charged acceptor ions on the P-side and positively charged donor ions on the N-side, setting up an electric field directed from N- to P-side. Sunlight striking the device produces electron-hole pairs which move in the electric field and cause an electric current flow when the P- and N-sides are connected through a load by a wire.

1.3 Explain the significance of the following parameters for a solar cell: (a) short-circuit current density, (b) open-circuit voltage, (c) fill factor, and (d) power conversion efficiency. Answer: (a) Maximum current delivered by the cell, (b) maximum voltage across the cell, (c) a measure of squareness of current-voltage characteristics (the squarer, the better), and (d) a measure of how much incident solar energy on solar cell is converted into usable electric power.

1.4 What is meant by air-mass coefficient? What do the terms "AM0" and "AM1.5G" mean? Answer: Air mass coefficient = 1/cos (zenith angle). Zenith angle is the angle between sunrays and vertical direction. AM0, or air mass zero, represents extraterrestrial solar spectrum, i.e., solar spectrum outside the atmosphere (zero atmosphere). AM1.5 is the solar spectrum at zenith angle 48.19°. In AM1.5G, the letter G stands for "global," meaning, a spectrum including the direct light from the sun as well as diffused light received after scattering from the atmosphere, representing total or global radiation.

1.5 What do you mean by "1 sun" in the context of a solar cell? What irradiance does "5 suns" signify? Answer: 1 sun refers to a level of solar irradiance (flux received by a surface per unit area) equal to 1kWm^{-2}. Hence, "5 suns" corresponds to an irradiance of 5kWm^{-2}.

1.6 How will you push the performance of a solar cell beyond the Shockley-Queisser limit for the efficiency of a single-junction cell? Answer: (i) By multijunction configuration using P-N junctions made of semiconductors of different bandgaps, (ii) by upconversion of sub-bandgap light, (iii) by multiple exciton generation spawning two or more electron-hole pairs by absorption of a single high-energy photon, and (iv) by concentrator cells using lenses or mirrors to focus sunlight (Guo et al 2015).

1.7　(a) Into how many generations are solar cells divided? (b) Give one example of solar cells of each generation. (c) In which generation are nanotechnology-based solar cells placed? Answer: (a) Three, (b) single-crystal silicon solar cell (first generation), amorphous silicon thin-film solar cell (second generation), dye-sensitized solar cell (third generation); (c) third.

1.8　(a) How is back surface field produced in a monocrystalline silicon solar cell? (b) What is the use of this field? Answer: (a) By making a high low (P^+/P^-) junction, (b) to act as a barrier to minority-carrier flow.

1.9　(a) What is the purpose of the window layer in a GaAs solar cell? (b) Mention its features. Answer: (a) To reduce front surface recombination and conduct current to the metal grid. (b) Thin, transparent, heavily doped layer of large bandgap material.

1.10　A single-crystal silicon solar cell is made from a thick wafer of silicon, while only a thin film of amorphous silicon is sufficient to make a solar cell. Why? Answer: Because of much higher absorption coefficient of amorphous silicon than single-crystal silicon.

1.11　A heterojunction consists of layers of two dissimilar semiconductors. (a) What are the two constituent silicon layers commonly used in a silicon heterojunction solar cell? (b) What are the main differences between these layers? (c) How are the recombination-active contacts shifted by making a heterojunction? Answer: (a) Amorphous and mono-crystalline silicon, (b) amorphous Si: direct bandgap, 1.5–2.0eV, monocrystalline Si: Indirect bandgap, 1.1eV, (c) through the I-a-Si:H layer interposed between the contacts and monocrystalline Si.

1.12　(a) Is CdTe solar cell a thin-film solar cell? (b) Is it a heterojunction solar cell? If so, what semiconductors are used for making the heterojunction? (a) Yes, (b) yes, P-CdTe, and N-CdS.

1.13　(a) How is the bandgap of CIGS matched with the solar spectrum? (b) What material qualities favor its use as a material for solar cells? (c) Mention one process by which the CIGS layer is formed during solar cell fabrication. Answer: (a) By varying x in $CuIn_xGa_{1-x}Se_2$, $x = 0$ to 1, (b) high optical absorption coefficient, direct bandgap semiconductor with bandgap tunable from 1.01–1.68eV, (c) coevaporation or cosputtering of elemental constituents (Cu, In, Ga), and annealing in selenide vapor.

1.14　(a) What are perovskites? (b) Name the most popular perovskite material used in solar cells. (c) Give examples of electron and hole transport materials used in perovskite solar cells. Answer: (a) Ceramic oxide with formula (Cation A) (Cation B smaller than A) (Anion)$_3$, (b) methylammonium lead trihalide, (c) ETL:TiO_2, HTL: Spiro-OMeTAD.

1.15　What conditions create instability issues in perovskite solar cells? Suggest possible solutions to improve stability. Answer: Exposure to moisture, oxygen, UV, or thermal stress causes instabilities. For better stability, the intrinsic stability of the perovskite absorber needs to be improved and a modifier must be included at ETL/perovskite layer interface. Good-quality HTL layer and durable encapsulant materials are necessary. 2D perovskite and 2D/3D perovskite-based solar cells must be fabricated (see Chapter 8).

1.16　Which is the more stable material for solar cells: (a) 3D perovskite and (b) 2D perovskite? Answer: (b) 2D perovskite.

1.17　(a) What are the pros and cons of making an organic solar cell? (b) Which is better: bilayer or bulk heterojunction solar cell? In what respect? Answer: (a) Pros: Lightweight, eco-friendly organic solar cells are realized by easier, cheaper processing. It is possible to make transparent solar cells for window-integrated photovoltaics. Cons: Lower efficiency and stability and shorter lifetime than standard cells. (b) Bulk heterojunction solar cell, because in these cells the charge carriers can easily reach contacts without recombining as they have to travel through the shorter distance between phase-separated materials.

1.18 What is the necessary condition for satisfactory working of a bilayer heterojunction organic solar cell regarding the relative magnitudes of individual layers and diffusion length of excitons? Answer: Thicknesses of individual layers < the diffusion length of excitons.

1.19 What is the requirement for satisfactory operation of a bulk heterojunction organic solar cell regarding the distance between phase-separated domains and diffusion length of excitons? Answer: Thicknesses of phase-separated domains < the exciton diffusion length.

1.20 Why is a hybrid solar cell called by this name? Explain the significance of the word "hybrid." Answer: It is a hybrid cell because it combines an inorganic semiconductor with an organic semiconductor.

REFERENCES

Basol B. M. and B. McCandless 2014 Brief review of cadmium telluride-based photovoltaic technologies, Journal of Photonics for Energy, 4: 040996–1 to 040996–11.

BBC Teach 2022, www.bbc.co.uk/teach/how-did-oil-come-to-run-our-world/zn6gnrd.

Belghachi A. and N. Limam 2017 Effect of the absorber layer band-gap on CIGS solar cell, Chinese Journal of Physics, 55(4): 1127–1134.

Bera S., D. Sengupta, S. Roy and K. Mukherjee 2021 Research into dye-sensitized solar cells: A review highlighting progress in India, Journal of Physics: Energy, 3(032013): 1–30.

Burdick J. and P. Schmidt 2017 Install your Solar Panels: Designing and Installing a Photovoltaic System to Power Your Home, Storey Publishing, LLC, North Adams, MA, p. 6.

Carlson D. E. and C. R. Wronski 1976 Amorphous silicon solar cell, Applied Physics Letters, 28(11): 671–673.

Chen L. X. 2019 Organic solar cells: Recent progress and challenges, ACS Energy Letters, 4(10): 2537–2539.

Ding Y., B. Ding, H. Kanda, O. J. Usiobo, T. Gallet, Z. Yang et al 2022 Single-crystalline TiO$_2$ nanoparticles for stable and efficient perovskite modules, Nature Nanotechnology, 17: 598–605, https://doi.org/10.1038/s41565-022-01108-1.

Green M. A., A. Ho-Baillie and H. J. Snaith 2014 The emergence of perovskite solar cells, Nature Photonics, 8: 506–514.

Green M. A., E. D. Dunlop, J. Hohl-Ebinger, M. Yoshita, N. Gopidakis and X. Hao 2022 Solar cell efficiency tables (version 59), Progress in Photovoltaics: Research and Applications, 30: 3–12.

Guo F., N. Li, F. W. Fecher, N. Gasparini, C. O. R. Quiroz, C. Bronnbauer et al 2015 A generic concept to overcome bandgap limitations for designing highly efficient multi-junction photovoltaic cells, Nature Communications, 6(7730): 1–9.

Kaelin M., D. Rudmann and A. N. Tiwari 2004 Low cost processing of CIGS thin film solar cells, Solar Energy, 77(6): 749–756.

Kamdem C. F., A. T. Ngoupo, F. K. Konan, H. J. T. Nkuissi, B. Hartiti and J.-M. Ndjaka 2019 (October) Study of the role of window layer Al$_{0.8}$Ga$_{0.2}$As on GaAs-based solar cells performance, Indian Journal of Science and Technology, 12(37): 1–9.

Kentucky coal education 2007 Copyright © 1996–2007 Kentucky Foundation, www.coaleducation.org/q&a/who_discovered_coal.htm

Kietzke T. 2008 Recent advances in organic solar cells, Advances in OptoElectronics, 2007(Article ID 40285): 15.

Liu C. P., M. W. Chang and C. L. Chuang 2014 Effect of rapid thermal oxidation on structure and photoelectronic properties of silicon oxide in monocrystalline silicon solar cells, Current Applied Physics, 14: 653–658.

Liu G., Y. Zhong, H. Mao, J. Yang, R. Dai, X. Hu, Z. Xing, W. Sheng, L. Tan and Y. Chen 2022 Highly efficient and stable ZnO-based MA-free perovskite solar cells via overcoming interfacial mismatch and deprotonation reaction, Chemical Engineering Journal, 431(Part 2): 134235.

Li Z., B. Li, X. Wu, S. A. Sheppard, S. Zhang, D. Gao, N. J. Long and Z. Zhu 2022 Organometallic-functionalized interfaces for highly efficient inverted perovskite solar cells, Science, 376(6591): 416.

Mostefaoui M., H. Mazari, S. Khelifi, A. Bouraiou and R. Dabou 2015 Simulation of high efficiency CIGS solar cells with SCAPS-1D software, International Conference on Technologies and Materials for Renewable Energy, Environment and Sustainability, TMREES15, Energy Procedia, 74: 736–744.

Müller-Buschbaum P., M. Thelakkat, T. F. Fässler and M. Stutzmann 2017 Hybrid photovoltaics—from fundamentals towards application, Advanced Energy Materials, 7(1700248): 1–25.

Qarony W., M. I. Hossain, M. K. Hossain, M. J. Uddin, A. Haque, A. R. Saad and Y. H. Tsang 2017 Efficient amorphous silicon solar cells: Characterization, optimization and optical loss analysis, Results in Physics, 7: 4287–4293.

Roghini G. C. and F. Enrichi 2020 Solar cells' evolution and perspective: A short review, in: Enrichi F. and G. C. Roghini (eds.), Solar Cells and Light Management: Materials, Strategies and Sustainability, Elsevier, Netherlands, pp. 1–32.

Romeo A., D. L. Bätzner, H. Zogg and A. N. Tiwari 2001 Potential of CdTe thin film solar cells for space applications, Proceedings of the 17th European Photovoltaic Conference and Exhibition, 22–26 October 2001, Munich, Germany, pp. 2183–2186.

Romeo A. and E. Artegiani 2021 CdTe-based thin film solar cells: Past, present and future, Energies, 14: 1684, pp. 1–24.

Rong Y., Y. Hu, A. Mei, H. Tan, M. I. Saidaminov, S. I. Seok, M. D. McGehee, E. H. Sargent and H. Han 2018 Challenges for commercializing perovskite solar cells, Science, 361(6408): eaat8235.

Seif J. P., A. Descoeudres, M. Filipič, F. Smole, M. Topič, Z. C. Holman, S. D. Wolf and C. Ballif 2014 Amorphous silicon oxide window layers for high-efficiency silicon heterojunction solar cells, Journal of Applied Physics, 115: 024502–1 to 024502–8.

Selvamanickam V. 2018 High efficiency, inexpensive thin film III-V photovoltaics using single-crystalline-like, flexible substrates: Final Report for DOE/EERE, DE-EE0006711, University of Houston, p. 15, www.osti.gov/servlets/purl/1508926 (Accessed on 23rd December 2021).

Shockley W. and H. J. Queisser 1961 Detailed balance limit of efficiency of p-n junction solar cells, Journal of Applied Physics, 32(3): 510–519.

Singh C. B., S. Bhattacharya, V. Singh, P. B. Bhargav, S. Sarkar, V. Bhavanasi and N. Ahmad 2013 Application of $Si_xN_y:H_z$ (SiN) as index matching layer in a-Si: H thin film solar cells, Journal of Renewable and Sustainable Energy, 5: 031605–1 to 031605–6.

Stoumpos C. C., D. H. Cao, D. J. Clark, J. Young, J. M. Rondinelli, J. I. Jang, J. T. Hupp and M. G. Kanatzidis 2016 Ruddlesden−Popper hybrid lead iodide perovskite 2D homologous semiconductors, Chemistry of Materials, 28: 2852–2867.

Taguchi M., A. Yano, S. Tohoda, K. Matsuyama, Y. Nakamura, T. Nishiwaki, K.i Fujita and E. Maruyama 2014 24.7% record efficiency HIT solar cell on thin silicon wafer, IEEE Journal of Photovoltaics, 4(1): 96–99.

Tang H., S. He and C. Peng 2017 A short progress report on high-efficiency perovskite solar cells, Nanoscale Research Letters, 12(410): 1–8.

Tumpa A. R., E. Sarker, S. Anjum, N. Sultana 2015 Analyze the effect of window layer (AlAs) for increasing the efficiency of GaAs based solar cell, American Journal of Engineering Research (AJER), 4(7): 304–315.

Zakery A., A. Shaker and M. Salem 2018 Chapter 1: Solar cells and arrays: Principles, analysis & design, in: Yahyaoui I. (ed.), Advances in Renewable Energies and Power Technologies, Vol. 1: Solar and Wind Energies, Elsevier, Amsterdam, Netherlands, p. 5.

2 Nanotechnology Fundamentals

In the previous chapter, we talked about solar cells only. This discussion painted one side of the picture. Since we are looking at the confluence of solar cell technology with nanotechnology, we must also pay attention to the other side of the picture represented by nanotechnology. Therefore, it is essential to become conversant with the elementary terms and learn the proven facts about nanotechnology, as we have done for solar cells. This knowledge will become the language to describe the interplay between nanotechnology and solar cells. Then, only we can appreciate the mutual cooperation between these two technologies and their engagement in teamwork.

2.1 NANOTECHNOLOGY

Nanotechnology is the science and engineering of matter in the dimensional range of 1 nanometer to 100 nanometers (Binns 2021). It strives to exploit the unique phenomena happening at that scale only and not realizable in bulk matter. This is accomplished by carefully investigating the properties of matter at the nanoscale and applying those properties to control, manipulate, and carry out processes for the fabrication of novel structures, devices, and systems exhibiting superior performance to their bulk-matter counterparts.

2.2 NANOMATERIALS

A *nanomaterial* is any chemical substance, either occuring naturally, engineered artificially in the laboratory, or incidentally produced during an industrial process, that has the characteristic feature of possessing at least one external dimension in the range 1–100nm (Roduner 2014).

Nanomaterials are classified as 0D, 1D, 2D, 3D nanomaterials. Table 2.1 briefly presents the characteristic features of the four classes of nanomaterials.

2.3 0D NANOMATERIALS

2.3.1 Nanoparticle

It is a nanoscale object whose all dimensions lie in the range 1–100nm; the lower limit is prescribed keeping in view the typical atomic bond lengths (0.1–0.2nm). The nanoparticle is distinguished from atoms or molecules having diameters ~ 0.1nm (diameter of gold atom ~ 0.144nm), and particulate matter ($PM_{2.5}$) in particle pollution from construction sites, unpaved roads, etc., with diameters ~ 10–2,500nm.

Nanoparticles show amazing properties, e.g., gold melts at 1,064°C, but a 1-nm gold nanoparticle will melt at room temperature, 20°C (Herr 2021). This melting point depression is ascribed to the enormous surface-to-bulk ratio of nanoparticles. The surface gold atoms behave differently

DOI: 10.1201/9781003215158-3

TABLE 2.1

Definitions and Examples of 0D, 1D, 2D, and 3D Nanomaterials

Sl. No.	Nanomaterial	Definition	Examples
1.	Zero-dimensional nanomaterials	All dimensions < 100nm; zero dimension > 100nm	Metal/dielectric nanoparticles, fullerenes (buckminsterfullerene C_{60}), dendrimers, quantum dots
2.	One-dimensional nanomaterials	Two dimensions < 100nm; one dimension > 100nm	Nanowires, nanofibers, carbon and inorganic nanotubes, nanorods, biopolymers (DNA molecules)
3.	Two-dimensional nanomaterials	One dimension < 100nm; two dimensions > 100nm	Graphene, hexagonal boron nitride, metal dichalcogenides; nanofilms, nanosheets, nanodisks, nanoplates, nanocoatings
4.	Three-dimensional nanomaterials	Zero dimension < 100nm; three dimensions > 100nm	Bulk powders, nanoparticle dispersions, nanowire bundles

from the embedded bulk atoms. Being exposed and unbound to their neighbors like the bulk atoms, the surface atoms differ in reactivity.

Nanoparticles are of various shapes, such as spherical, oval, prismatic, cubical, and others. A few nanoparticle shapes are displayed in Figure 2.1.

2.3.2 BUCKMINSTERFULLERENE (C_{60})

The buckminsterfullerene is an allotrope of carbon shaped like a hollow sphere. The C_{60} molecule has a cage-like structure, looking like a soccer ball (Figure 2.2). It is called a buckyball. It contains 60 sp^2 carbon atoms assembled together in 12 pentagons and 20 hexagons.

2.3.3 QUANTUM DOT

A quantum dot is a semiconductor nanoparticle having size in the range 1–10nm (Divsar 2020). In a quantum dot of size less than twice the exciton Bohr radius, the electrons are squeezed, resulting in quantum confinement. The exciton is a bound electron-hole pair, and the exciton Bohr radius is the distance between the electron and the hole in this pair. Quantum confinement leads to discrete energy levels for electrons, and the bandgap increases (Figure 2.3). The discretization of energy levels makes quantum dots similar to individual atoms. Hence, they are sometimes called artificial atoms.

The optoelectronic properties of quantum dots depend on their composition, size, and shape. Small-size quantum dots (2–3nm) emit high-frequency light, blue and green. Large-size quantum dots (5–6nm) emit low-frequency light, red or orange. Changes in the size of the quantum dot can be made to vary the emission frequency, and hence the color, of emitted light. Similarly, the absorption spectra of quantum dots can be tuned at discretion.

The quantum dots are synthesized by colloidal method from liquid solutions or by gas phase-based plasma processing. Self-assembled quantum dots grow epitaxially by spontaneous nucleation on lattice-mismatched substrates. The growth techniques include molecular beam epitaxy (MBE) and metal-organic chemical vapor deposition (MOCVD).

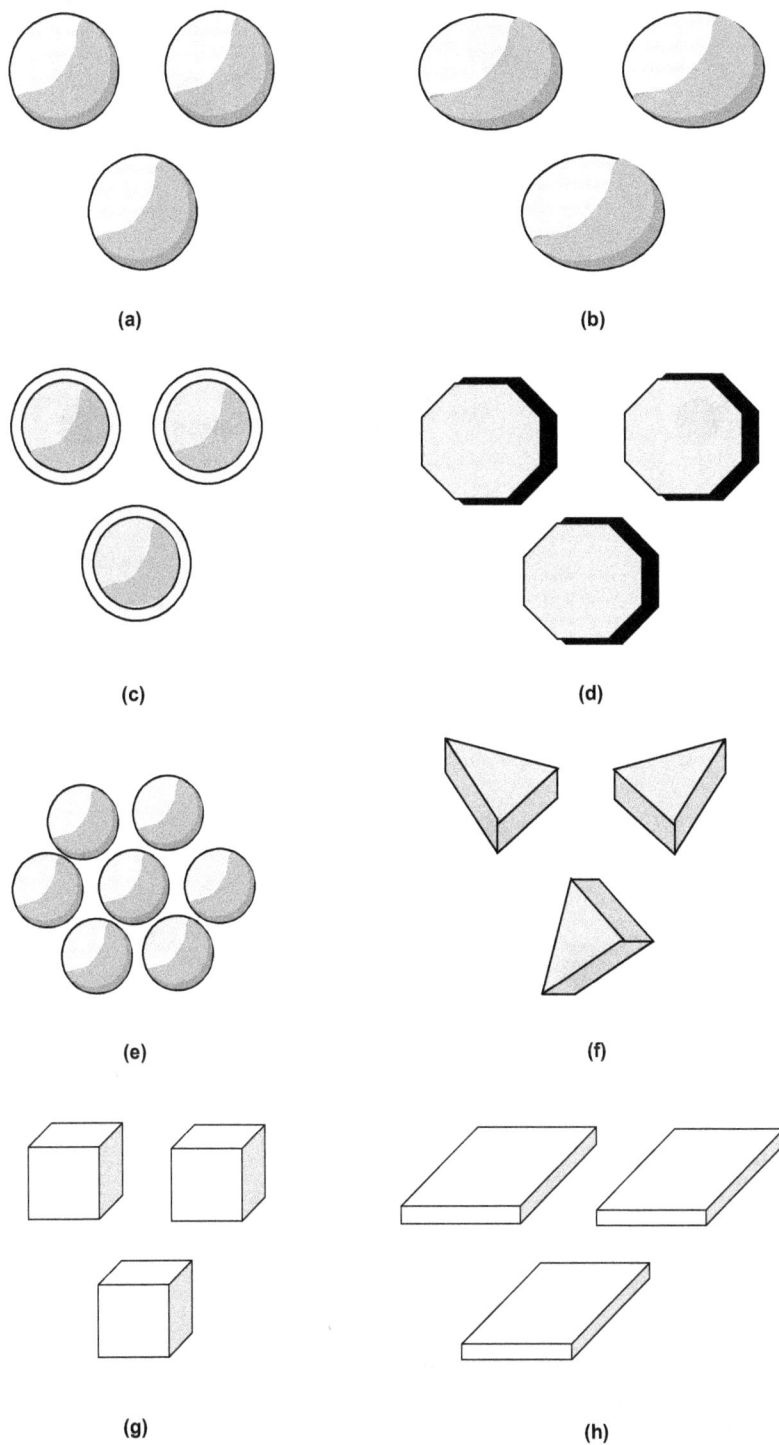

FIGURE 2.1 Nanoparticle shapes: (a) nanosphere, (b) nano oval, (c) core/shell nanoparticle, (d) nano-octahedron, (e) nanocluster, (f) nanoprism, (g) nanocube, and (h) nanoplate. The nanoparticles are made of metals or insulators.

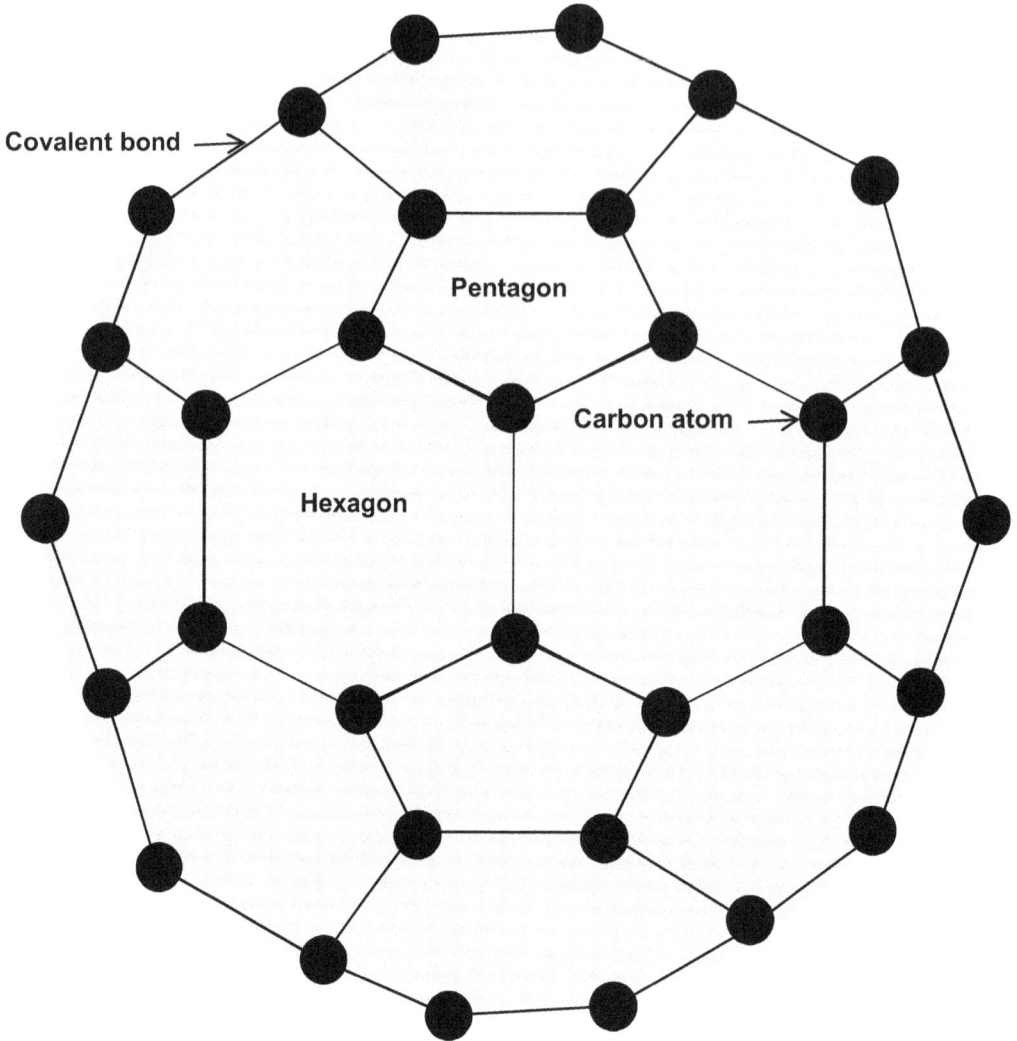

FIGURE 2.2 Buckminsterfullerene C_{60}, the smallest fullerene, showing the carbon atoms linked together in pentagonal and hexagonal rings through covalent bonds.

2.4 1D NANOMATERIALS

2.4.1 NANOWIRE

A *nanowire* is an extremely thin wire. It has a diameter less than 100nm but an unconstrained length so that it is a high–aspect ratio nanomaterial with the ratio of length to diameter exceeding 1,000 (Zhang et al 2018). The nanowire is made from various materials, including conducting (e.g., Au, Ag, Pt), semiconducting (e.g., Si, GaAs, GaN), and insulating materials (e.g., SiO_2, TiO_2). In a nanowire, the electrons are confined to move laterally but are free longitudinally. Due to quantum confinement, the electrons occupy discrete energy levels as opposed to a continuum of energy levels or energy bands in a wire of bulk material. Nanowires are fabricated by top-down approach, by lithographic and etching techniques, or by bottom-up approach, by synthesizing from constituent

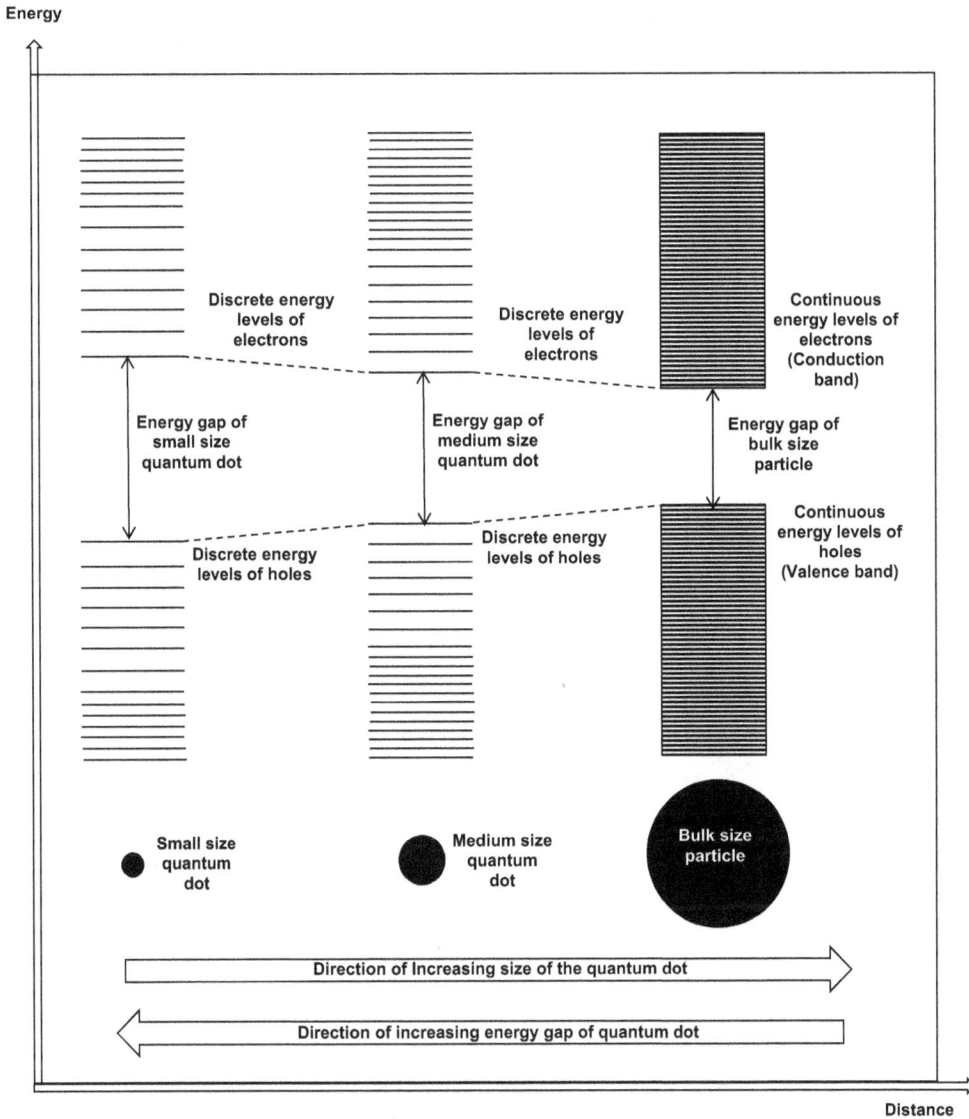

FIGURE 2.3 Influence of quantum dot size on the energy band structure showing discretized energy levels of a quantum dot and the decrease in energy gap along with the merging of discrete levels into continua, called energy bands, with increasing size of the quantum dot. Three particle sizes are taken for illustration: small, medium, and bulk. The smallest-size particle exhibits the largest energy gap. The energy levels are separated. The medium-size particle comes next with smaller energy gap and disassociated energy levels. The bulk-size particle shows the smallest energy gap with continuous energy bands.

atoms. The nanowire is one of the many nanoscale cylindrical shapes. Different cylindrical nanomaterial shapes are shown in Figure 2.4.

2.4.2 CARBON NANOTUBE

A carbon nanotube is a hollow cylindrical tube having a diameter of few nanometers that can be visualized as rolled-up sheets of graphene (Figure 2.5). A single-walled carbon nanotube

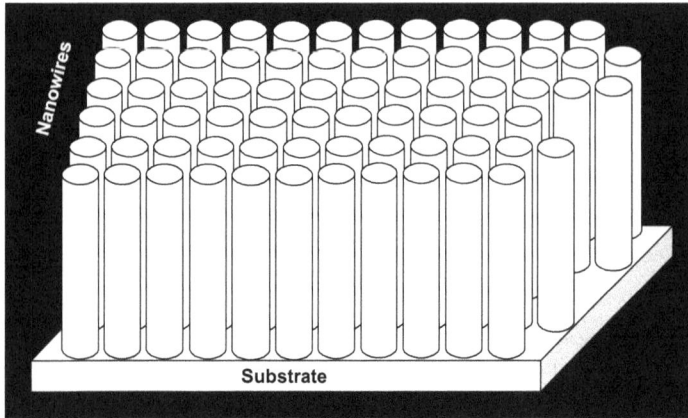

(a)

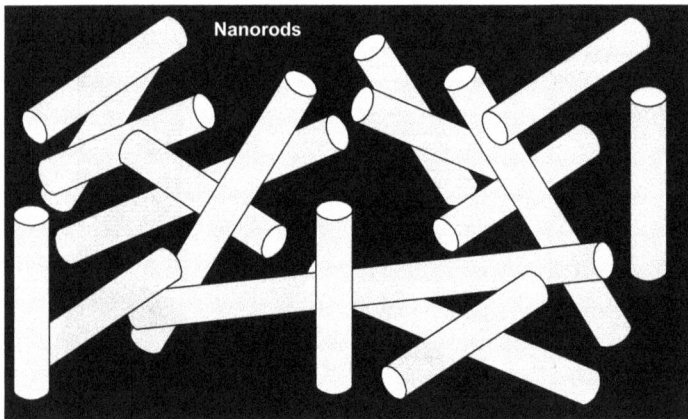

(b)

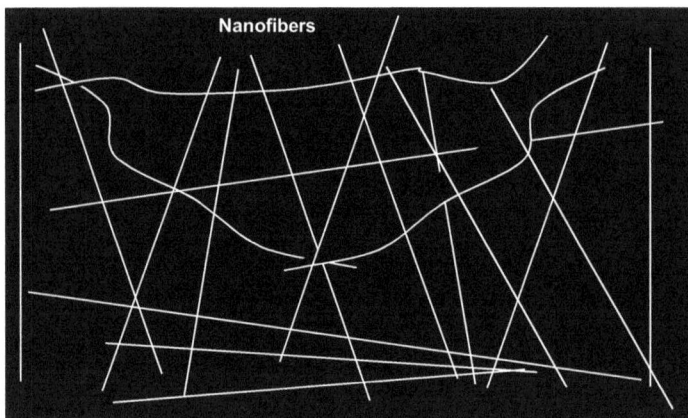

(c)

FIGURE 2.4 Cylindrical-shaped nanomaterials: (a) silicon nanowires, (b) zinc oxide nanorods, and (c) polymer nanofibers.

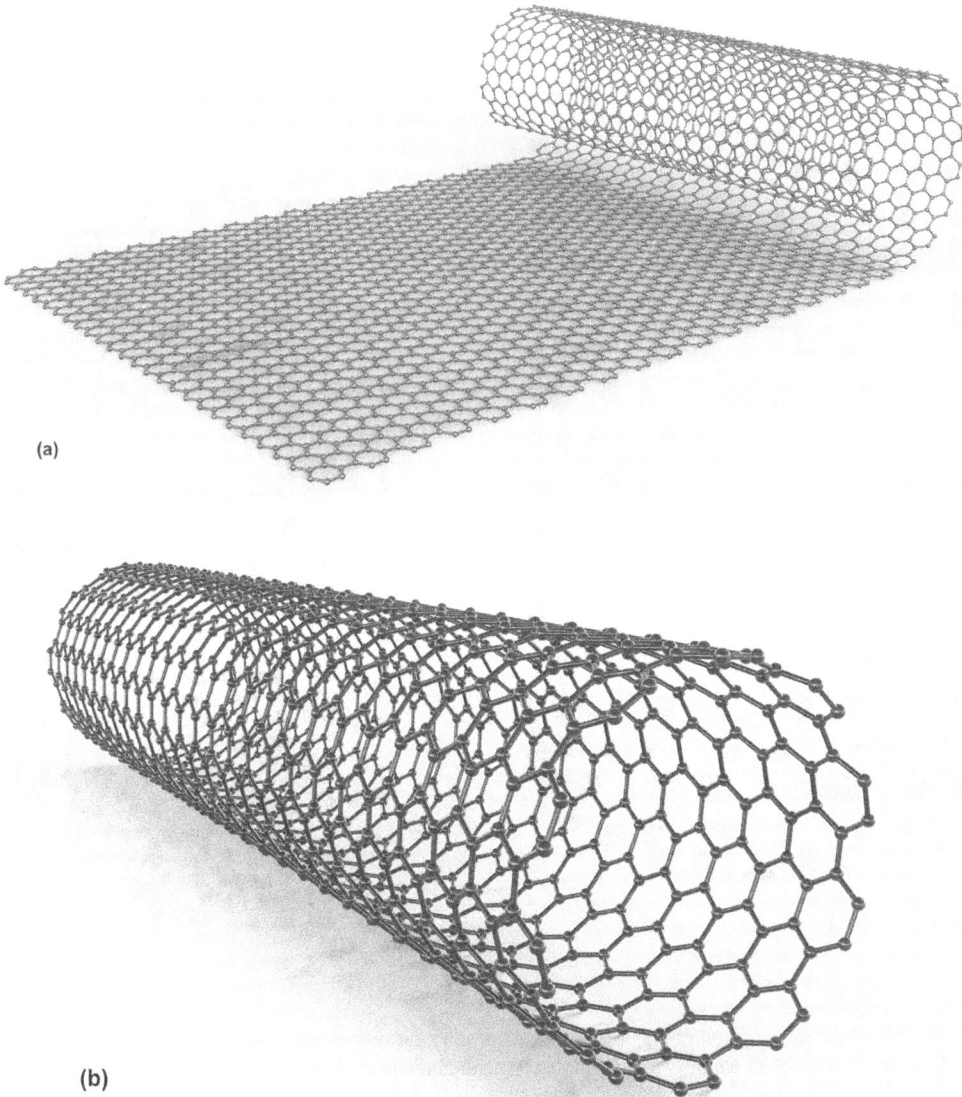

FIGURE 2.5 Imagining carbon nanotube as rolled-up graphene sheet: (a) Seamlessly rolling a graphene layer comprising six-membered carbon atom rings into a hollow tube whose walls are one atom thick. One or both ends of a nanotube may be capped with a hemispherical structure of fullerene molecule. (b) The generated carbon nanotube.

(SWCNT), shown in Figure 2.6(a), has a diameter ~ 0.4 to 2–3nm (Eatermadi et al 2014). A multiwalled carbon nanotube (MWCNT) consisting of assemblies of concentrically intertwined and nested single-walled nanotubes can have inner diameters 0.4 to few nm and outer diameters 20–30nm (Figure 2.6(b)). Lengthwise, a carbon nanotube may be several microns long, with aspect ratios ~ 1,000.

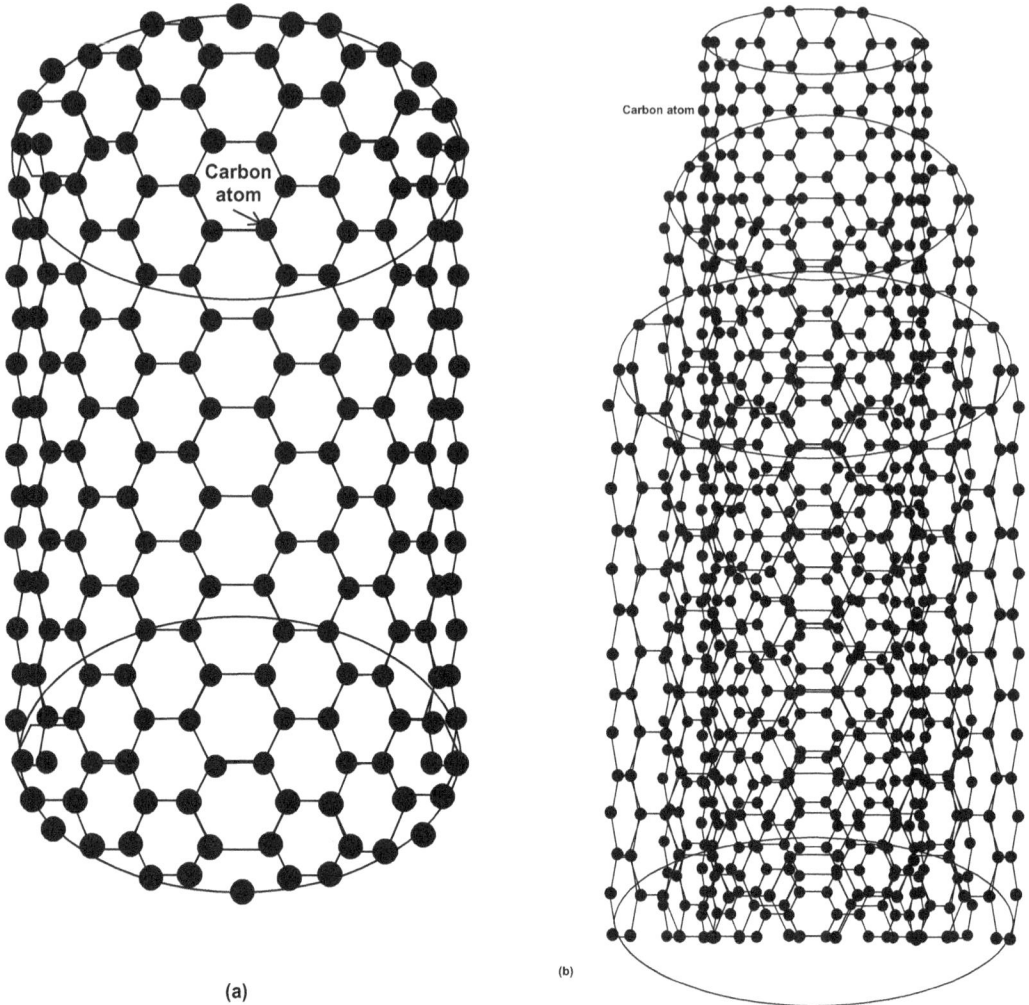

FIGURE 2.6 Carbon nanotubes: (a) single-walled carbon nanotube; (b) multiwalled carbon nanotube which is contemplated as several single-walled nanotubes arranged concentrically, one inside the other.

A SWCNT is described by a pair of indices (m, n), considering it as a graphene sheet rolled into a cylinder along a lattice vector (m, n) in the plane of graphene. The indices (m, n) are integers denoting the number of unit vectors along two directions in the graphene lattice. The electronic properties of the SWCNT are determined by the chiral vector (m, n) or chirality of the tube. For $m = 0$, the tube is referred to as a zigzag tube, for $n = m$ it is an armchair tube, and otherwise it is known as a chiral tube.

A carbon nanotube displays remarkable elasticity; Young's modulus = 1TPa for SWCNT. It can bend, twist, kink, and buckle. Its tensile strength (100GPa) is higher than that of steel. Its temperature-withstanding capability increases from 750°C in normal atmospheric pressure to 2,800°C in vacuum.

A CNT behaves as metallic or as a small bandgap semiconductor. Field-effect transistors have been fabricated using CNTs. From the transport characteristics, the field-effect mobility has been estimated as 7.9×10^4 cm^{2}V^{-1}s^{-1} and intrinsic mobility as $> 10^5$ cm^{2}V^{-1}s^{-1}; these values are well above those for all known semiconductors (Dürkop et al 2004).

CNTs are synthesized by arc discharge, laser ablation, and chemical vapor deposition.

2.5   2D NANOMATERIALS

2.5.1   Graphene

Graphene is essentially one atomic layer of graphite, the popular material used in pencils. It can be extracted from graphite using adhesive tape. It is an allotrope of carbon. The structural representation of graphene is depicted as a sheet made of a monolayer of carbon atoms bound together with sp^2 hybridization and arranged in a repeated pattern of hexagonal honeycomb lattice (Figure 2.7).

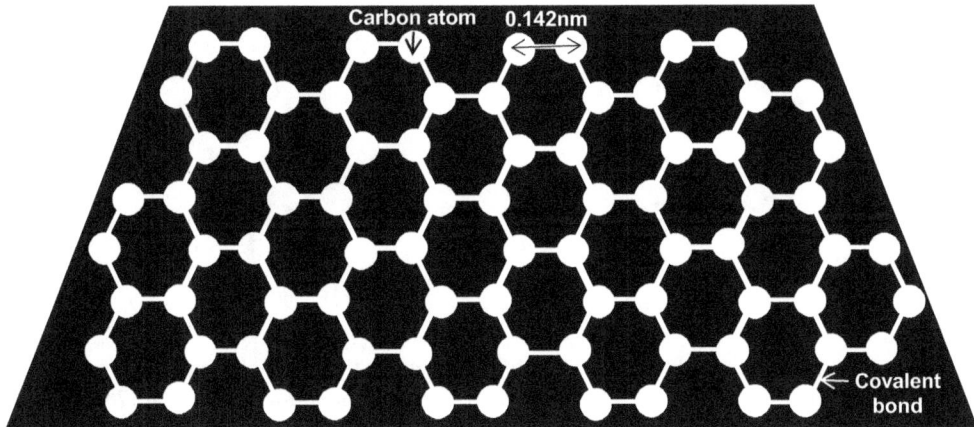

(a)

(b)

FIGURE 2.7   The 2D nanomaterial graphene: (a) a sheet made of a single layer of covalent bonded carbon atoms joined together in a repetitive pattern with spacing between atoms = 0.142nm, (b) 3D rendering of graphene molecular structure, and (c) the honeycomb structure made of beeswax secreted by worker bees and propolis, a plant resin made by bees to store honey. Note the similarity between the geometrical patterns of graphene made of hexagons, as shown in *a* and *b*, to the honeycomb structure in *c* made by bees.

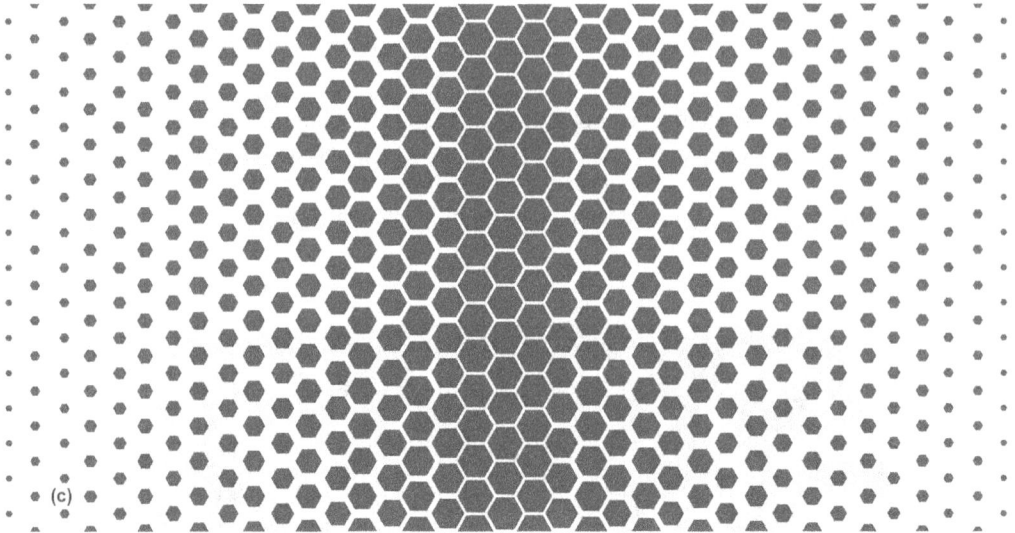

FIGURE 2.7 (Continued)

From this representation, it is clearly seen as a two-dimensional material because it extends in only length and breadth; the height, being the thickness of one carbon atom (diameter 0.17nm), is taken as zero. So it is known as a two-dimensional material.

The transparent graphene sheet is a good conductor of heat and electricity and strong absorber of light, accounting its black color. A zero-gap semiconductor, it has a high electron mobility $\sim 1.5 \times 10^4$ cm^{2}V^{-1}s^{-1}; the mobility is $> 2 \times 10^5$ cm^{2}V^{-1}s^{-1} in suspended graphene (Bolotin et al 2008). It shows ambipolar charge transport with nearly equal electron and hole mobilities. Its tensile strength of 130GPa is much larger than that of steel.

2.5.2 2D PEROVSKITES

Commercialization of 3D perovskite solar cells has been obstructed by poor stability of these devices (Elahi et al 2022). Naturally, the stability issue has been the focus of attention of researchers, and tremendous efforts have been made to develop stable perovskite solar cells. Research on perovskites has been pursued along different directions, namely, compositional engineering of perovskites, regulation of interfaces between solar cell layers, passivation of material defects, and encapsulation of the solar cell, as already discussed in Section 1.7.7.

2D perovskites have emerged as a leading nanomaterial to provide stability. These nanomaterials are derivatives of 3D perovskites formed by slicing the 3D frameworks into well-defined 2D slabs. They contain an organic layer of long-chain bulky hydrophobic alkylammonium cations, e.g., phenylethyl ammonium (PEA) or butylammonium (BA) cations separating inorganic layers of corner-shared $[BX_6]^{4-}$ octahedra, as shown in Figure 2.8. Coulombic forces between the layers maintain integrity of the structure. The cations prevent absorption of water vapor and invasion of the perovskite surface, imparting stability to these perovskites. However, these 2D perovskites have a wider bandgap than 3D perovskite, which limits harvesting of light. They also have a nonpreferred crystallographic orientation opposing vertical transport of photo-generated electrons and holes. Although 2D perovskites provide increased tolerance to humidity than 3D perovskites, they underperform with respect to 3D perovskites from the viewpoints of light absorption and charge dynamics. It is therefore desirable to follow a unified approach which seeks to blend the best properties of both materials, using 2D perovskite for preventing moisture attack and 3D perovskite as active

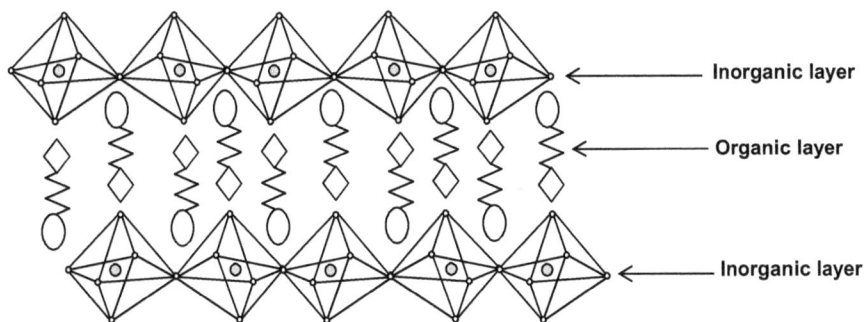

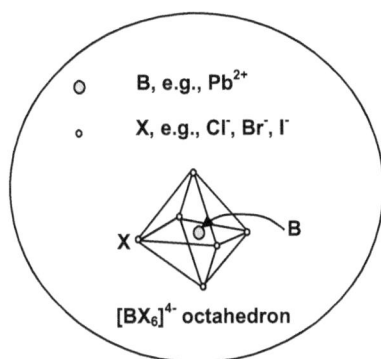

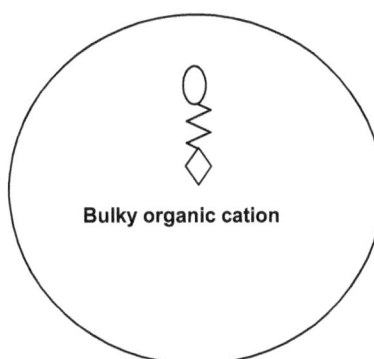

(b)   (c)

FIGURE 2.8 Hybrid inorganic-organic 2D halide perovskite: (a) Schematic structure consisting of inorganic layers separated by an organic layer. The inorganic layer is made of corner-shared $[BX_6]^{4-}$octahedrons, while the organic layer comprises bulky organic cation spacers. B is a metal ion, such as Pb^{2+}, while X represents a halide anion, Cl^-, Br^-, or I^-. (b) The $[BX_6]^{4-}$octahedron. (c) The bulky organic cation.

light-absorbing medium of solar cell. Such 2D/3D multidimensional solar cells show good stability together with high efficiency. Meanwhile, properties of 2D perovskites have also been improved, resulting in high-efficiency solar cells made with these nanomaterials only. 2D perovskites and their use in solar cell fabrication will be discussed in Chapter 8.

2.5.3 QUANTUM WELL

A *quantum well* is a potential well in which the potential energy of the electron is less than that outside. It comprises a flat thin layer (typical thickness 1–10nm) in which the motion of the electron perpendicular to the surface of the layer is confined but that in other two directions, i.e., along the planar surface of the layer is unrestricted. As a result, the freedom of the electron movement in three dimensions is reduced to two dimensions. The energy of the electron in the direction at right angles to the plane of the quantum well is quantized, i.e., it can vary in integral steps only.

The quantum confinement effects become operative when the thickness of the flat thin layer representing the quantum well is less than the de Broglie wavelength of electrons (Shik 1997). Practical realization of quantum well is achieved by packing a thin layer of narrow bandgap material, such as GaAs, less than de Broglie wavelength thick, between two layers of wide-bandgap material, like $Al_xGa_{1-x}As$ ($x = 0.15–0.35$) in the form $Al_xGa_{1-x}As/GaAs/Al_xGa_{1-x}As$. The central GaAs layer is the quantum well layer, while the border $Al_xGa_{1-x}As$ layers are called barrier layers. Another such structure is formed between InGaAs and GaAs as GaAs/InGaAs/GaAs. Here, the InGaAs quantum well layer lies between the GaAs barrier layers. Such structures made of two different semiconductors are known as heterostructures.

In the heterostructure $AlSb/GaAs_{1-x}Sb_x/AlSb$, shown in Figure 2.9(a), the $GaAs_{1-x}Sb_x$ layer is the quantum well layer. The AlSb layers are the barrier layers. Figure 2.9(b) shows the energy band

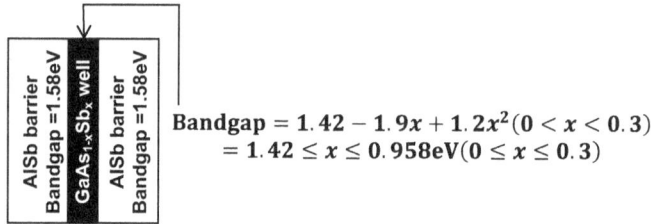

$$Bandgap = 1.42 - 1.9x + 1.2x^2 (0 < x < 0.3)$$
$$= 1.42 \le x \le 0.958eV (0 \le x \le 0.3)$$

(a)

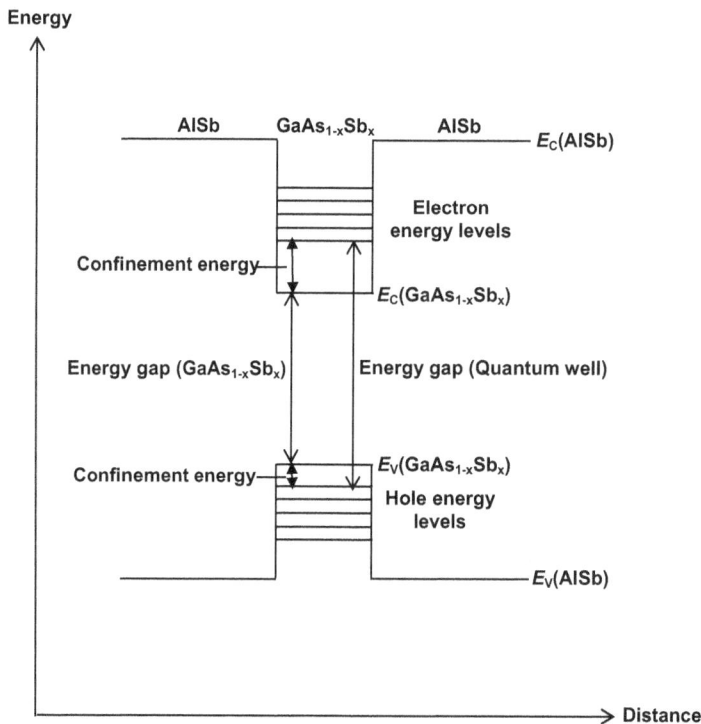

(b)

FIGURE 2.9 A quantum well and its energy band diagram: (a) The $GaAs_{1-x}Sb_x$ quantum well in the AlSb/$GaAs_{1-x}Sb_x$/AlSb heterojunctions between the AlSb wide-bandgap semiconductor and $GaAs_{1-x}Sb_x$ narrow-bandgap material. The thin $GaAs_{1-x}Sb_x$ quantum well region is enclosed within the AlSb barrier layers on both sides. (b) The energy band diagram of the AlSb/GaAsSb/AlSb structure showing the discrete energy levels in the quantum well region. The energy gap of the quantum well is larger than the energy gap of $GaAs_{1-x}Sb_x$ because of the addition of confinement energies on both conduction and valence band edges.

diagram of this heterostructure. Note that the energy levels do not display continuity. Furthermore, the energy gap of the quantum well is larger than the energy gap of $GaAs_{1-x}Sb_x$ in bulk form.

Heterostructrures are fabricated by molecular beam epitaxy or metal organic chemical vapor deposition capable of controlling thickness of layers up to a monolayer. The quantum well material system is of three types:

(i) Lattice-matched material system. The lattice constants of the well and barrier semiconductors match with that of the substrate, producing minimum dislocation.

(ii) Strain-compensated or strain-balanced material system. The effect of strain produced due to larger lattice constant of one semiconductor with respect to the substrate is counterbalanced by the effect of strain due to smaller lattice constant of the other semiconductor, i.e., the alternative layers of the well and barrier layers are under compressive and tensile strains so that the stress at successive interfaces is opposite in sign. Thus, it is possible to grow a large number of wells. The approach offers considerable design flexibility.

(iii) Strained material system. Here the quantum well and barrier layers have dissimilar lattice constants than the substrate, providing no strain balancing. Consequently, a strained structure results in which only a few wells can be formed.

2.6 SCOPE FOR NANOTECHNOLOGY APPLICATION IN SOLAR CELLS AND ORGANIZATIONAL STRUCTURE OF THE BOOK

Nanotechnology encompasses a wide diversity of fabrication processes, ranging in complexity from simple solution-based processes to complicated metal-organic vapor phase epitaxial growth. Nanomaterials surpass microparticles and macroparticles decisively when we talk about surface-area-to-volume ratio. The unique properties of matter at nanoscale, such as bandgap tuning in the 1–20nm range, offer flexibilities which have been used time and again in different types of solar cells.

The book is subdivided into five parts containing 15 chapters (Figure 2.10). The five parts are: Part I, "Preliminaries and Nanocrystalline Silicon Photovoltaics" (Chapters 1–3), Part II, "Nanotechnological Approaches to Sunlight Harvesting" (Chapters 4–6), Part III, "Electrochemical Photovoltaics Using Nanomaterials" (Chapter 7), Part IV, "Photovoltaics with 2D Perovskites and Carbon Nanomaterials" (Chapters 8–9), and Part V, "Quantum Well, Nanowire, and Quantum Dot Photovoltaics" (Chapters 10–15).

Each chapter of the book is designed with a specific goal. Meanings of specialized terms in solar cell technology presented in Chapter 1 serve as the vocabulary of the subject. Nanotechnology terms are recapitulated in Chapter 2, as was done for solar cells in Chapter 1.

After going through preliminaries of solar cells in Chapter 1 and essentials of nanotechnology in Chapter 2, each subsequent chapter focuses on a nanotechnology-enabled solar cell whose operation depends on a particular nanomaterial, nanoscale property or phenomenon, or groups of these entities.

2.6.1 Use of Nanocrystalline Silicon in Solar Cells

Chapter 3: An important beneficiary from nanotechnology is the nanocrystalline silicon-based solar cell. The nanocrystalline silicon layer is used as a replacement for amorphous silicon in heterojunction solar cells, not in the heterostructure, but as a carrier collector. Solar cells have also been made with this material as the main constituent.

2.6.2 Nanotexturing Solar Cell Surface

Chapter 4: Let us see how nanotexturing is helpful. To increase absorption of sunlight, reflection of light from the surface of the solar cell must be reduced. To decrease the reflection of light, the

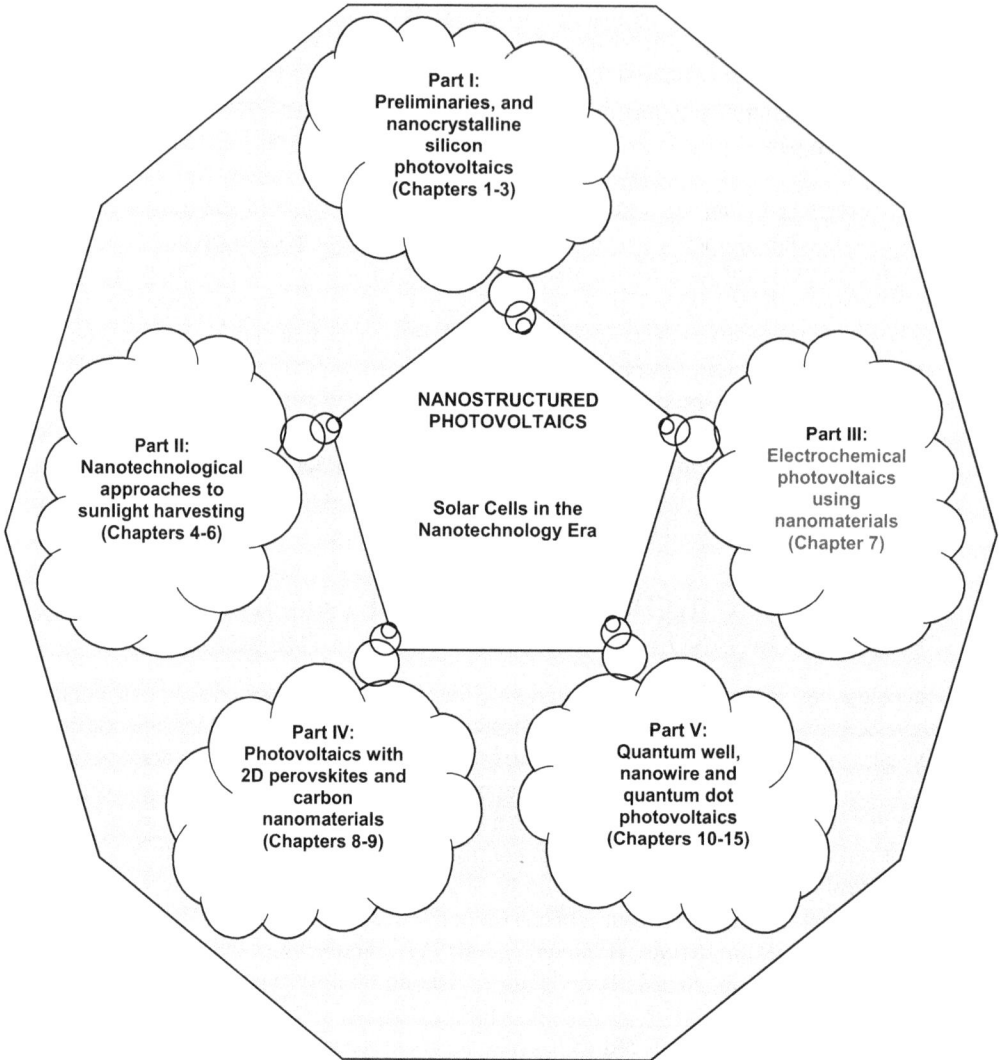

FIGURE 2.10 Organizational layout of the book showing its five parts and the chapter numbers in each part.

surface must be made antireflective. For making the surface antireflective, the refractive index must be gradually decreased from air to the semiconductor. The resulting surface is said to have a graded refractive index variation. The graded refractive index surface suppresses light reflection appreciably. The process used to produce graded refractive index variation is called nanotexturing of the surface.

Microtexturing has been extensively used to produce an antireflective surface in monocrystalline silicon solar cells made from thick silicon wafers. But in thin-film solar cells, where only a small thickness must fully absorb the sunlight, nanotexturing is mandatory.

2.6.3 Using Plasmonic Nanostructures for Maximizing Light Coupling in Solar Cells

Chapter 5: Resonant absorption of light takes place when the frequency v of the photon is such that its energy (hv) = quantum excitation energy of the absorbing material viz the energy necessary to excite the electrons in the atoms from a lower to a higher energy level. The phenomenon of resonant absorption can be applied in solar cells for redirection and confinement of light in the active medium of the solar cell, where it is beneficially utilized for cell operation.

Resonant absorption is obtained by employing metallic and dielectric resonators. Metallic resonators influence the absorption of light in three different ways. Firstly, they serve as scattering centers for light. Secondly, they act as concentrators of light. The concerned phenomenon is known as localized surface plasmon resonance (LSPR). Thirdly, they excite surface plasmon polaritons by surface plasmon resonance. The polaritons propagate along the metal/semiconductor or metal/dielectric interfaces.

When metallic nanoparticles are used as resonators, the effects of plasmon resonance become more pronounced and conspicuous. The LSPR induced by metallic nanoparticles increases the absorption of light in the semiconductor regions adjoining the nanoparticles. The size, distribution, and concentration of the metallic nanoparticles can be controlled to produce beneficial effects. Placement of nanoparticles in the vicinity of the P-N junction is more advantageous.

Briefly, localized surface plasmons can be excited in metal nanoparticles, surface plasmon polaritons can be excited at metal-semiconductor interfaces, and both modes of surface plasmon resonance can be applied to increase light trapping in solar cells. Metal nanoparticles can be adhered to the surface of a solar cell. They can also be embedded in the active layer.

2.6.4 FURTHER APPROACHES TO LIGHT INCOUPLING IN SOLAR CELLS

Chapter 6: It must be remembered that metallic nanoparticles reflect light, and their incorrect positioning in the solar cell may also be harmful because it may affect light in a way that increases optical losses. To avoid this disadvantage accompanying metallic nanoparticles, dielectric nanoparticles are sometimes used. They act as nanophotonic resonators. Like the commonly used metal nanoparticles, dielectric nanoparticles are employed for increasing absorption of light. They exploit Mie resonances to couple the leaky optical modes into the photoactive medium of the solar cell. Thus, dielectric nanoparticles work through Mie scattering.

Apart from positive nanotextures pointing outwards from the surface of the solar cell, negative nanotextures such as nanoholes are also used for intensifying absorption of light.

2.6.5 SENSITIZING METAL OXIDE SEMICONDUCTOR (TiO$_2$) NANOPARTICLES WITH DYE

Chapter 7: Dye-sensitized solar cells work with nanostructured metal oxide electrodes and metal-organic complex dyes, usually rhuthenium-based dyes. Titanium dioxide is a widely used semiconductor material for making the nanostructured electrode. TiO$_2$ nanoparticles are annealed to make mesoporous electrodes. The annealing conditions determine the surface morphology of these electrodes. A large surface-area-to-volume ratio of semiconductor nanoparticles makes charge transfer easy. Zinc oxide is another semiconductor which has received attention. A core/shell architecture with a TiO$_2$ core and ZnO shell provides a shorter path to carriers to reach contacts. Hence, it is effective in reducing current losses through carrier recombination.

2.6.6 PROMISES OF 2D PEROVSKITES NANOMATERIALS

Chapter 8: The 3D perovskite family of materials has shown remarkable progress in achieving high performance at low fabrication costs. These materials can be tuned to respond to different portions of the solar spectrum by changing the material composition. They are amenable to processing at low temperatures by easy methods. Despite offering bandgap flexibility and convenience in processing, they are unsuitable for commercial utilization because they are unable to withstand prolonged exposure to light, or intense heat, and also degrade in the presence of moisture or oxygen. Long operating lifetimes are of paramount importance for acceptance of a solar cell technology.

Perovskite layers separated by organic cation spacers, known as layered perovskites or 2D perovskites, have shown enhanced stability. The efficiencies of solar cells made from 2D perovskites are steadily rising. Solar cells made from these nanomaterials can compete with leading technologies. They have come to the vanguard of photovoltaics, assuring for large-scale terrestrial deployment in mainstream solar power generation (Omprakash et al 2021).

2.6.7 Applications of Carbon Nanostructures

Chapter 9: For solar cells, fullerenes, carbon nanotubes, and graphene are of interest.

Fullerenes are used in making heterojunction solar cells with conjugated polymers.

A plethora of applications and possibilities of deployment of CNTs in solar cells may be mentioned:

(i) A transparent electrode of a solar cell is made by depositing a conductive CNT film. Metal electrode-free perovskite solar cells have been made using laminated CNT networks.

(ii) An interpenetrating heterojunction is created by forming a dispersion of carbon nanotubes in a solution of electron-donating polymer. The CNT composite acts as a photoactive medium.

(iii) A P-N heterojunction diode is formed when carbon nanotube film is deposited on N-type crystalline silicon. A CNTs-Si hybrid solar cell is obtained in which CNTs harvest light.

(iv) Mixtures of CNTs offer an extensive range of bandgaps. Different bandgaps cater to different portions of the solar spectrum, so the availability of an all-inclusive range enables the proper matching of bandgaps with solar spectrum.

Unsurprisingly, CNTs are receiving an intensive research attention for integration in solar cells. Pervasive integration of carbon nanotubes in solar cells is on the cards.

Graphene absorbs only 2.3% light in the solar spectrum span from ultraviolet to infrared wavelengths. This makes graphene an ideal material. It can replace the costly indium tin oxide electrode, which is in widespread use today. Graphene has a high mechanical flexibility. It is very stable, both thermally and chemically. Abundance of carbon relative to scarcity of indium is a notable advantage in favor of graphene electrodes.

A Schottky junction can be formed between metallic graphene and the underlying semiconductor. Hence, graphene serves as the active medium for a Schottky barrier–type solar cell.

Graphene and graphene oxide are also used as electron and hole transport materials (Yin et al 2014).

2.6.8 Applications of Nanowires

Chapters 10 and 11: Semiconducting nanowires suffer low reflective losses than planar semiconductor films. Reflection of light is reduced and light trapping is increased. The nanowires increase the path length of light waves by factors of several tens.

Besides the optical advantage, let us also inquire about the use of nanowires from an electrical viewpoint. In a solar cell, the photocarriers are separated from each other; they are transported to the contacts and collected at the contacts to supply external current. Thus, all the three tasks of carrier separation, transport, and collection are equally crucial to realize the full capability of the solar cell. To ensure that these jobs are executed properly, two parameters are of utmost importance:

(i) The nature of the donor/acceptor interface. The junction should be free from any defects at the nanoscale because the defects act as recombination centers and many free carriers may be lost through the defect sites near the junction.

(ii) The distance travelled by the carriers to reach the contacts must be reasonably short. By *reasonably short distance*, what is meant is the distance determined by the diffusion length of minority carriers in the semiconductor. Beyond one diffusion length, very few carriers will be able to survive recombination. In this respect, core/shell nanostructures are very useful. Here the current flows radially and the distance traversed by the carriers can be very short. It is the thickness of the shell layer.

Nanowire solar cells are fabricated in axial and radial geometries. The radial geometry is a core/shell configuration and can be utilized for making solar cells using low-carrier-lifetime material. Hence, even a lower quality of semiconductor material is acceptable to make solar cells, which provides enormous economical gains. Further, the flexibility of bandgap tuning is available in the nanowire cells.

2.6.9 APPLICATIONS OF QUANTUM WELLS

Chapters 12 and 13: The lower-bandgap material introduced in the absorber region by the incorporation of multiple quantum wells extends the absorption to the infrared region to ameliorate the performance of the solar cell. Bandgap can be changed by using different materials for quantum well structures and different thicknesses of quantum well layer.

2.6.10 APPLICATIONS OF QUANTUM DOTS

Chapters 14 and 15: Photons in sunlight having greater energies than the bandgap of the material are able to produce electron-hole pairs in a solar cell. Hence, lower-bandgap materials produce more photocurrent. The large photocurrent leads to a high short-circuit current. But lower-bandgap materials are not suitable from open-circuit voltage consideration because the open-circuit voltage decreases with the bandgap. So a wider-bandgap semiconductor must be used to get a higher open-circuit voltage. To meet these conflicting requirements about bandgap, the optimum bandgap for solar cells is estimated at 1.2–1.4 eV. Owing to this basic limit, as discussed in Section 1.5.7, a large portion of the solar spectrum remains unutilized. By using quantum dots, it is possible to vary the bandgap by changing the size of the quantum dot. Thus, the bandgap range of the absorber medium in the solar cell can be broadened to tap photons from a wider portion of the spectrum of sunlight.

Quantum dots have many merits. They utilize hot photogenerated carriers. QDs form junctions with organic polymers. Such junctions display photovoltaic effects. Inorganic material quantum dots are dispersed in a polymer matrix to make a bulk heterojunction solar cell. The nanomaterials have high absorption coefficients. Small-size effects in nanomaterials can be utilized to tune the properties of quantum dots by varying their size. Since diffusion lengths in a conjugated polymer lie in the range of 5–20nm, the precise control of nanophase is imperative for working of this kind of device. The remarkable feature is that conjugated polymers can be processed in open atmosphere. Vacuum processing is not needed. They can be deposited by simple spin-coating procedures. So this solar cell has a low manufacturing cost.

QDs are used to sensitize nanocrystalline TiO_2 to work like dye-sensitized solar cells. Arrays of QDs can be used as photoelectrodes.

2.7 DISCUSSION AND CONCLUSIONS

This chapter provided a quick synopsis of the key terminology of nanotechnology for easy comprehension of the unification of solar cell technology with nanotechnology. Several examples of 0D, 1D, and 2D nanomaterials were cited. These nanomaterials are the ingredients that are rooted in the solar cell structure. With the inclusion of these nanomaterials, the nanoscale property associated with them begins to exert its influence on the behavior of the solar cell. However, a nanoscopic effect need not always come from the use of a new nanomaterial. Nanostructures can also be made in the body of the solar cell, e.g., nanotextures are formed on the surface of the solar cell to evoke a nanoresponse.

Incorporation of various forms of nanomaterials, e.g., nanoparticles, quantum dots, nanowires, nanotubes, nanorods, and quantum wells, providing large surface-to-volume ratios and facilitating new physical mechanisms, such as quantum confinement, bandgap tunability, etc., offers immense opportunities facilitating the relentless efforts to tap solar energy. Accordingly, the contents of the different chapters are chosen, and the layout plan of the book is sketched.

Questions and Answers

2.1   Define *nanotechnology* and *nanomaterials*. Answer: *Nanotechnology* is the technology dealing with properties and uses of materials in the dimensional range between 1nm and 100nm. A *nanomaterial* or *nanostructured material* is a material with one or more of its dimensions in the 1–100nm range.

2.2   Are the following nanomaterials: (a) oxygen molecule (diameter 292pm), (b) nitrogen molecule (diameter 300pm)? Answer: No.

2.3   Which type of nanomaterials are fullerenes: (a) 0D, (b) 1D, (c) 3D? Answer: 0D.

2.4   Carbon nanotubes are 3D nanomaterials. Right or wrong? Answer: Wrong, 1D.

2.5   Graphene is a nanomaterial of how many dimensions? Answer: 2D.

2.6   Nanowires are 0D nanomaterials. Correct or incorrect? Answer: Incorrect, 1D.

2.7   Which type of nanomaterials are quantum dots: (a) 0D, (b) 2D, (c) 3D? Answer: 0D.

2.8   Melting point of bulk gold is 1064°C. A nanoparticle of gold will melt at a lower or higher temperature than the melting point of bulk gold? Answer: Lower temperature.

2.9   How many carbon atoms comprise buckminsterfullerene molecule? What is the other name for buckminsterfullerene? Answer: 60, Buckyball.

2.10 What is an *exciton*? Define *exciton Bohr radius*. Answer: *Exciton* is the bound state of an electron-hole pair. *Exciton Bohr radius* is the electron-hole separation in an exciton (average distance between electron in the conduction band and hole left behind in the valence band) obtained by analogy to Bohr model for hydrogen atom consisting of an electron orbiting around a positively charged nucleus in a circular orbit in free space, with the difference that the exciton is in a dielectric medium.

2.11 Why are quantum dots often referred to as *artificial atoms*? Because electrons move in discrete energy levels (not in bands) in quantum dots in much the same way as they do in isolated atoms.

2.12 Does the color of light emitted by a quantum dot depend on its size? Which size of quantum dots emits high-frequency light? Small or big? Answer: Yes, small.

2.13 Into how many types of quantum dots can be classified based on composition and structure? Answer: Core type (single component with uniform composition), core-shell type (nanocrystals of one material embedded in a wider-bandgap material to improve efficiency and brightness), and alloyed type (multicomponent nanodots allowing to tune properties via compositional changes without changing crystallite size).

2.14 Lithographic technique of making nanowires falls under which class? Bottom-up or top-down method? Answer: Top-down.

2.15 In which direction is electron motion restricted in a nanowire? Lateral or longitudinal? Answer: Lateral.

2.16 Can a carbon nanotube be considered a rolled-up graphene sheet? Yes or no? Answer: Yes.

2.17 (a) Can a carbon nanotube be a micron long? (b) Can it have the diameter of a micron? Answer: (a) Yes, (b) no.

2.18 What are the possible values of the indices *m, n* in the chiral vector (*m, n*) of a SWCNT if it is a zigzag tube? Answer: *m*, 0, or 0, *n*, i.e., either *n* or *m* is zero.

2.19 Which will withstand a greater maximum stress before breaking when stretched? A carbon nanotube or a steel wire? Answer: Carbon nanotube.

2.20 Correct the statement, "Carbon nanotubes always exhibit metallic character." Answer: Exhibit metallic or semiconducting character.

2.21 What thickness of graphite sheet is graphene? Answer: One atomic layer.

2.22 Why is graphene said to be a two-dimensional nanomaterial? Answer: Because it has only two dimensions, length and breadth; the third dimension, or height, is diameter of carbon atom = 170pm, which may be assumed as zero.

2.23 What is the bandgap of graphene? Answer: Zero.

2.24 Which charge carriers transport electric current through graphene? Electrons or holes? Answer: Both electrons and holes. The transport mechanism is ambipolar.

2.25 Is graphene transparent? Answer: Yes.

2.26 Mention one advantage of 2D perovskites which encourages their use in solar cells. Answer: Stability in humid environment.

2.27 Point out two discouraging features of 2D perovskites for solar cells relative to 3D perovskites. Answer: Wider bandgap (hence inadequate light harvesting) and nonpreferred crystal orientation (therefore retarded vertical charge transfer).

2.28 In which direction electron motion is restricted in a quantum well? (a) Parallel to or (b) perpendicular to the plane of the well? Answer: (b) Perpendicular to the plane of the well.

2.29 In which direction is electron energy quantized in a quantum well? Parallel to or perpendicular to the plane of the well? Answer: Perpendicular to the plane of the well.

2.30 "Quantum confinement takes place when the flat thin layer representing a quantum well has a thickness larger than de Broglie wavelength." Right or wrong? Answer: Wrong, smaller than.

2.31 Write the quantum well and barrier layers in the following structures in the form barrier layer/quantum well layer/barrier layer: (a) AlGaAs and GaAs, (b) InGaAs and GaAs, and (c) GaAs$_{1-x}$Sb$_x$ and AlSb. Answer: (a) AlGaAs/GaAs/AlGaAs, (b) GaAs/InGaAs/GaAs, and (c) AlSb/GaAs$_{1-x}$Sb$_x$/AlSb.

2.32 Into how many types are material systems for quantum wells subdivided with regard to lattice matching and strain development? Name the systems. Answer: Three, lattice-matched, strain-balanced, and strained lattice systems.

2.33 Which material is replaced by nanocrystalline silicon in a silicon heterojunction solar cell? Monocrystalline silicon or amorphous silicon? Answer: Amorphous silicon.

2.34 A semiconductor surface will be antireflective to light if the refractive index changes rapidly from air to semiconductor. Yes or no? Answer: No, gradually.

2.35 Nanotexturing is compulsorily implemented in (a) thick monocrystalline silicon solar cells or (b) thin film solar cells. Answer: Thin film solar cells.

2.36 State the condition for resonant absorption of light by a material. Answer: Photon energy (dependent on its frequency) = energy difference between two consecutive energy levels in the atom of a given material.

2.37 What are the three ways in which metallic resonators affect the absorption of light by a material? Answer: Scattering, LSPR, surface plasmon polaritons.

2.38 Can dielectric nanoparticles be used for increasing light absorption in a solar cell? How do they help? Answer: Yes, by Mie scattering.

2.39 Name the semiconductor commonly used in a dye-sensitized solar cell. Is it a nanomaterial? Answer: TiO$_2$; yes, nanocrystalline.

2.40 Name another semiconductor material used in a dye-sensitized solar cell. Answer: ZnO.

2.41 Despite outperforming over traditional photovoltaic technologies, perovskite solar cells have lagged behind commercially. Why? Answer: Because of their undesirable operational instability in high-moisture environments and under strong irradiation.

2.42　Which perovskite nanomaterial helps in making stable solar cells? Answer: 2D perovskite.

2.43　Name two carbon-based nanomaterials used in solar cells. Answer: CNTs, graphene.

2.44　Which produces lower reflection losses: planar films or nanowires? Answer: Nanowires.

2.45　What are the two configurations in which nanowire solar cells are made? Answer: Radial and axial.

2.46　Radial nanowire solar cell is a core/shell structure. Yes or no? Answer: Yes.

2.47　Which semiconductor gives a high open-circuit voltage: (a) small-bandgap semiconductor, (b) large-bandgap semiconductor? Answer: large bandgap semiconductor.

2.48　How does the use of quantum wells help in improving solar cell performance? Answer: (i) For lattice-matched systems, the current and efficiency of the quantum well solar cells are greater than those for conventional cells having a bandgap = bandgap of the larger-bandgap barrier layer of quantum well solar cell by extending the absorption of light below this bandgap through the smaller-bandgap quantum well layer. (ii) The voltage of quantum well solar cells is greater than that of conventional cells having a bandgap = bandgap of the smaller-bandgap quantum well layer. (iii) Quantum well solar cells show good radiation tolerance and temperature dependence of efficiency characteristics (Barnham et al 2002).

2.49　What is the optimum bandgap of semiconductor for solar cells? Answer: 1.2–1.4eV.

2.50　Can the bandgap be changed by varying the size of the quantum dot? Answer: Yes.

REFERENCES

Barnham K. W. J., I. Ballard, J. P. Connolly, N. J. Ekins-Daukes, B. G. Kluftinger, J. Nelson and C. Rohr 2002 Quantum well solar cells, Physica E: Low-Dimensional Systems and Nanostructures, 14(1–2): 27–36.

Binns C. 2021 Introduction to Nanoscience and Nanotechnology, 2nd edition, John Wiley & Sons, Inc., NJ, pages 416.

Bolotin K. I., K. J. Sikes, Z. Jiang, M. Klima, G. Fudenberg, J. Hone, P. Kim and H. L. Stormer 2008 Ultrahigh electron mobility in suspended graphene, Solid State Communications, 146(9–10): 351–355.

Divsar F. (ed.) 2020 Quantum Dots: Fundamental and Applications, IntechOpen, London, pages 92.

Dürkop T., S. A. Getty, Enrique Cobas and M. S. Fuhrer 2004 Extraordinary mobility in semiconducting carbon nanotubes, Nano Letters, 4(1): 35–39.

Eatermadi A., H. Daraee, H. Karimkhanloo, M. Kouhi, N. Zarghami, A. Akbarzadeh, M. Abasi, Y. Hanifehpour and S. W. Joo 2014 Carbon nanotubes: Properties, synthesis, purification, and medical applications, Nanoscale Research Letters, 9(393): 13.

Elahi E., G. Dastgeer, A. S. Siddiqui, S. A. Patil, M. W. Iqbal and P. R. Sharma 2022 A review on two-dimensional (2D) perovskite material-based solar cells to enhance the power conversion efficiency, Dalton Transactions, 51: 797–816.

Herr D. J. C. 2021 What's so unusual about nanomaterial melting points?© Carolina Biological Supply Company, www.carolina.com/teacher-resources/Interactive/what%27s-so-unusual-about-nanomaterial-melting-points%3F/tr23010.tr (Accessed on 21st December 2021).

Omprakash P., P. Viswesh and P. Devadas Bhat 2021 A review of 2D perovskites and carbon-based nano-materials for applications in solar cells and photodetectors, ECS Journal of Solid State Science and Technology, 10(031009): 13.

Roduner E. 2014 Nanoscopic Materials: Size-Dependent Phenomena and Growth Principles, Royal Society of Chemistry, Cambridge, UK, pages 439.

Shik A. 1997 Quantum Wells: Physics and Electronics of Two-Dimensional Systems, World Scientific Publishing Co. Pte. Ltd., Singapore, pages 104.

Yin Z., J. Zhu, Q. He, X. Cao, C. Tan, H. Chen, Q. Yan and H. Zhang 2014 Graphene-based materials for solar cell applications, Advanced Energy Materials, 4(1): 1300574, 1–19.

Zhang A., G. Zheng and C. M. Lieber 2018 Nanowires: Building Blocks for Nanoscience and Nanotechnology, Springer International Publishing, AG Switzerland, pages 321.

3 Nanocrystalline Silicon-Based Solar Cells

Outstanding optoelectronic properties of nanocrystalline silicon have given impetus to its increasing utilization in heterojunction solar cells. Notwithstanding the advantages offered by this nanomaterial, the substitution of amorphous material with this nanomaterial must be implemented through a carefully planned scheme because the heterostructure between amorphous and crystalline phases of silicon is not changed. The nanocrystalline films are used for carrier collection. So the impact of material replacement on device performance is studied through experimentation, analysis, and evaluation. Besides the heterojunction devices, N-I-P solar cells made with nanocrystalline silicon constitute another significant application of this nanomaterial. Both heterojunction and N-I-P solar cells will be deliberated.

3.1 NANOCRYSTALLINE, POLYCRYSTALLINE, AND AMORPHOUS SILICON PHASES AS PHOTOVOLTAIC CELL MATERIALS

3.1.1 NANOCRYSTALLINE SILICON

Hydrogenated nanocrystalline silicon (nc:Si:H) is a biphasic or two-phase material (Chopra et al 2004; Adhikari et al 2019; Sharma et al 2021). The two phases are crystalline silicon and hydrogenated amorphous silicon (a-Si:H). Crystalline silicon islands of nanometer size embedded in the matrix of hydrogenated amorphous silicon constitute nanocrystalline silicon. So nanocrystalline silicon consists of silicon nanocrystallites dispersed in the hydrogenated amorphous silicon phase. The hydrogen is preferentially located either in the amorphous silicon or at the boundaries of the crystallites.

Nanocrystalline silicon being a mixture of crystalline and amorphous silicon, the properties of nanocrystalline silicon are restricted between the two extremes of crystalline and amorphous phases. In distinction to crystalline silicon having a bandgap = 1.1eV, absorption thickness required for solar cell application = 100μm, spectral range = 440–650nm, and capability of solar cell efficiency ~ 22.3–26.1%, the properties of a-Si:H are 1.75eV, 1–2 μm, 550–700nm, and 14.0%, respectively. In opposition to the bulk crystalline silicon having an electron mobility of 1,400 $cm^2V^{-1}s^{-1}$, hole mobility = 450 $cm^2V^{-1}s^{-1}$, carrier diffusion length ~ 100–300μm, and undoped conductivity < $10^{-5}\Omega^{-1}cm^{-1}$, the values for a-Si:H are 1 $cm^2V^{-1}s^{-1}$, 0.003 $cm^2V^{-1}s^{-1}$, 0.3μm, and $10^{-11}\Omega^{-1}cm^{-1}$ (Kang 2021; Street 2000, Si-Silicon: Electrical Properties).

3.1.2 NANOCRYSTALLINE VS. POLYSILICON

Nanocrystalline silicon differs from polycrystalline silicon (poly-Si) comprising grains of crystalline silicon separated by grain boundaries. Poly-Si is produced by chemical vapor deposition through pyrolysis of silane at 580–650°C:

$$SiH_4\left(gas\right) \rightarrow Si\left(solid\right) + 2H_2 \tag{3.1}$$

Its high deposition temperature is unfavorable for use of glass substrates. Compared to the high-temperature chemical vapor deposition or laser annealing-induced process of polysilicon, nanocrystalline silicon can be deposited by the low-temperature plasma-enhanced deposition process of a-Si:H, favoring its use in solar cell fabrication.

DOI: 10.1201/9781003215158-4

3.1.3 Amorphous Silicon

Amorphous silicon is noncrystalline silicon with disordered atomic structure forming continuous random network and no long-range order. Unhydrogenated a-Si has a high defect density with its numerous dangling bonds $\sim 10^{19}$ cm^{-3}, acting as recombination centers, degrading minority-carrier lifetime to the extent that Fermi level is pinned, thereby preventing its doping into N- or P-type. Passivation of dangling bonds with 10% atomic hydrogen produces hydrogenated amorphous silicon having lower defect density $\sim 10^{15}$ cm^{-3}, which can be doped for engineering applications. a-Si:H can be deposited at temperatures of 30–300°C by plasma-enhanced chemical vapor deposition, allowing the use of glass substrates. However, hydrogenated amorphous silicon undergoes changes in properties with time upon exposure to light. The phenomenon is known as the Staebler-Wronski effect. The thermodynamic metastability of a-Si:H, together with its low electron mobility, weaker infrared conversion ability owing to large bandgap, and lower solar cell efficiency are some of its drawbacks.

3.1.4 Advantages of Nanocrystalline Silicon over Amorphous and Polysilicon

Inclusion of crystalline regions in nc-Si:H relative to a-Si:H raises its doping efficiency, thereby increasing its electron mobility and conductivity above that of a-Si:H. The additional crystallinity also increases the transparency of nc-Si:H above a-Si:H. The absorption coefficient of nanocrystalline silicon layers is lower than that of a-Si:H throughout the visible light spectrum, causing less optical loss in these layers. The smaller bandgap of nc-Si:H (1.1eV) than a-Si:H (1.75eV) increases light absorption in red/infrared range. The smaller a-Si:H content and lower hydrogen concentration in nc-Si:H than a-Si:H makes it more stable. A lower deposition temperature required for nc-Si:H than poly-Si makes it more suitable for glass substrate processes.

3.2 PLASMA-ENHANCED CHEMICAL VAPOR DEPOSITION OF a-Si:H AND nc-Si:H FILMS

3.2.1 Effect of Hydrogen Dilution

a-Si:H films are usually prepared by a capacitively coupled plasma source consisting of two parallel plane electrodes and a mixture of silane with dopant gas diluted with hydrogen for hydrogenation of the film. Typical deposition pressure is 500–800mTorr, temperature is 150–350°C, and power density is 20–50mWcm^{-2} (Menéndez et al 2013).

The degree of hydrogen dilution of the plasma gas determines the morphology of the a-Si:H film. The hydrogen dilution ratio r is defined as:

$$r = \frac{\text{Hydrogen flow rate in the PECVD system}}{\text{Silane flow rate in the PECVD system}} = \frac{[\text{H}]}{[\text{SiH}_4]} \tag{3.2}$$

For low values of r, the film is mainly amorphous. On increasing r, crystalline regions appear in the film, and it contains a distribution of small crystallites suspended in a medium of amorphous material. Further increase of r results in a film made of almost-equal proportions of crystalline and amorphous phases. At still higher values of r, the crystalline portion becomes dominant and the film becomes principally crystalline.

Hydrogen incorporation in silicon films is systematically investigated by Kroll et al (1996) using a very high frequency process called VHF-PECVD. The plasma excitation frequency was 70MHz, leading to a high deposition rate. The hydrogen and silane gas flows were varied to get mixtures from 100% silane to 1.25% silane, for which r is 0/100 = 0 and r = (100–1.25)/1.25 = 98.75/1.25 = 79. The total gas flow was maintained at 50sccm. It was found that:

(i) In 100% SiH_4 to 9% SiH_4, i.e., r = 0 to r = 91/9 = 10.1, implying pure silane to silane diluted with 10.1 times hydrogen, the film is fully amorphous.

(ii) In 9% SiH_4 to 6% SiH_4, i.e., $r = 10.1$ to $r = 94/6 = 15.7$, meaning, silane diluted with 10.1–15.7 times hydrogen, the film structure is not clearly recognized, and a transition region between the amorphous and micro/nanocrystalline structures is observed.

(iii) In 6% to 1.25% SiH_4, i.e., $r = 94/6 = 15.7$ to $r = 98.75/1.25 = 79$, meaning, silane diluted with 15.7–79 times hydrogen, the film is micro/nanocrystalline.

This investigation clearly highlights the effects of dilution with hydrogen on the formation of amorphous and nanocrystalline phases. Examples of deposition parameters for (P-type) a-Si:H are (Zhao et al 2020) frequency 13.56MHz, temperature 180°C, pressure 675torr, power density 16mWcm^{-2}, H_2(sccm) = 24, SiH_4(sccm) = 8, B_2H_6(sccm) = 8 for which $r = 24/8 = 3$, which falls under case (i); the same for (P-type) nc-Si:H are 13.56MHz, 180°C, 1650.14mTorr, 90mWcm^{-2}, H_2(sccm) = 170, SiH_4(sccm) = 0.8, B_2H_6(sccm) = 10 for which $r = 170/0.8 = 212.5$, which according to case (iii) is trending towards increasing micro/nanocrystalline structure.

3.2.2 HIGH-PRESSURE DEPLETION (HPD) REGIME

It may be noted from the examples of a-Si:H and nc-Si:H deposition parameters that the pressure and power density for nc-Si:H are higher than that for a-Si:H. The reason is that lower absorption coefficient of nc-Si:H than a-Si:H makes it necessary to use comparatively thicker nc-Si:H films for solar cell applications, for which the power density has to be increased to raise the deposition rate by promoting the dissociation of silane. But upon increasing the power density, the voltage drop across the plasma sheath is also increased, whereby the impact of ion bombardment on the film creates more defects, producing inferior-quality film. For reduction of the ion energy, higher operating pressure is employed, in the range 1–10Torr. This deposition regime is therefore referred to as a high-pressure depletion regime (Kroely 2010).

3.3 SILICON HETEROJUNCTION (SHJ) SOLAR CELL

This solar cell was introduced in an elementary way in Section 1.7.4. Its discussion here will be continued as an extension of Section 1.7.4 with reference to Figure 1.10. The intent is to explain the genuine reason for a place of a layer of nanocrystalline silicon in SHJ solar cell where the layer of amorphous silicon layer was previously used.

The SHJ cell has the structure Ag/ITO/(P$^+$) a-Si:H/(I) a-Si:H/N-type c-Si/(I) a-Si:H/(N$^+$) a-Si:H/ITO/ Ag (De Wolf et al 2012; Mikolášek 2017). The top P$^+$ a-Si:H layer of the solar cell where the P-N junction is located is called the emitter, and the N-type crystalline silicon absorber is known as the base.

3.3.1 FABRICATION OF THE SOLAR CELL

Recall the description in Section 1.7.4 and note that the solar cell illustrated here is fabricated on an N-type crystalline silicon absorber, while in Section 1.7.4, it was made on P-type wafer. The absorber is coated on both sides with intrinsic a-Si:H passivation layers. Why are these intrinsic a-Si:H layers necessary? These layers passivate the dangling bonds protruding from the surfaces of crystalline silicon absorber and thus cause drastic reductions in their surface recombination velocities. On the front side, a P$^+$ a-Si:H layer is deposited over the intrinsic a-Si:H layer to form a P-N junction with N-type crystalline silicon absorber below the intrinsic a-Si:H layer. On the backside, the N$^+$ a-Si:H layer is deposited over the intrinsic a-Si:H layer to form N/N$^+$ junction for ohmic contact and for back surface field generation, to be enumerated later in this section. The a-Si:H layers are grown by PECVD.

ITO films are deposited by sputtering over the P$^+$ a-Si: layer on the front and N$^+$ a-Si:H layer on the back. Backside ITO-IR film is optimized to achieve low absorption in infrared wavelength segment. On the front side, Ag contacts are screen-printed, while on the backside, the Ag film is sputtered for back contact.

3.3.2 PROCESS SEQUENCE

The manufacturing process steps are (Figure 3.1) (i) chemical cleaning of silicon wafer, (ii) surface texturing in KOH, (iii) front-side (I) a-Si:H and (P$^+$) a-Si:H layers by PECVD at 200°C; the (P$^+$) a-Si:H layer is deposited by adding trimethyl boron to the process gases, (iv) rear-side (I) a-Si:H and (N$^+$)a-Si:H layers by PECVD at 200°C with phosphine (PH$_3$) added to the process gases during deposition of (N$^+$) a-Si:H layer, (v) front-side ITO physical vapor deposition (PVD), (vi) rear-side ITO-IR PVD, (vii) rear-side Ag sputtering, (viii) front-side Ag screen-printing and curing at 190°C on a belt furnace for ½ h., and (ix) testing.

3.3.3 BACK SURFACE FIELD (BSF)

A heavily doped N$^+$ region is introduced in a solar cell at the rear surface to reduce electron-hole recombination at this surface. How does it accomplish this task? The N$^+$/N high/low junction acts like a P-N junction. The electric field across this junction prevents minority carriers from flowing to the rear surface by electrostatic repulsion towards the absorber layer. This electric field is known as back surface field (BSF). Thus, the BSF reduces the surface recombination at the back surface by repelling away electrons that happen to wander towards or hang around the rear surface.

3.4 FRONT- AND REAR-EMITTER SILICON HETEROJUNCTION SOLAR CELL

SHJ solar cells use TCOs on both surfaces because the underlying a-Si:H layers have poor conductivity. The most commonly used TCO is indium tin oxide. This oxide is high priced, and there is always risk of shortage in supply because indium is a geologically scarce metal. So dependence of solar cell fabrication on ITO needs to be curtailed, especially for quality films.

3.4.1 FRONT-EMITTER SOLAR CELL

In the front-emitter SHJ solar cell (Figure 3.2(a)), the injecting P$^+$ layer is located on the front surface of the solar cell, while the N/N$^+$ junction is on the rear surface, producing the back surface field for preventing electron-hole recombination at this surface. Light enters the solar cell through the emitter/base junction: P$^+$a-Si:H/N-type crystalline silicon. Electron-hole pairs are produced in the N-type crystalline silicon absorber layer. The electrons move to the Ag film on the backside. The holes are not majority carriers in the N-type crystalline silicon. So they move straight upwards to the TCO layer. Inside the TCO, they move laterally to reach the Ag contact on the front surface. The Ag contact is fragmented into segments at a sufficient pitch to allow entry of light into the cell. So the hole may not be directly able to reach at a metallic spot and must travel some distance through the TCO. The finite resistance of the TCO film plays a vital role. Depending on the path length of holes through the TCO, transmission losses are unavoidably incurred. Such losses are avoided in the rear-emitter device as we shall see in Section 3.4.2.

3.4.2 REAR-EMITTER SOLAR CELL

The rear-emitter SHJ solar cell (Figure 3.2(b)) is the inverted form of the front-emitter geometry obtained by flipping this geometry upside down. The emitter is now located on the rear surface of the solar cell, whereas the N/N$^+$ junction is situated on the front side, providing front surface field, inhibiting carrier recombination adjoining this surface. Light has to travel across the full N-type crystalline silicon absorber layer to reach the emitter. *En route*, the electron-hole pairs are produced in the absorber layer by incident light. The holes move downward through the absorber

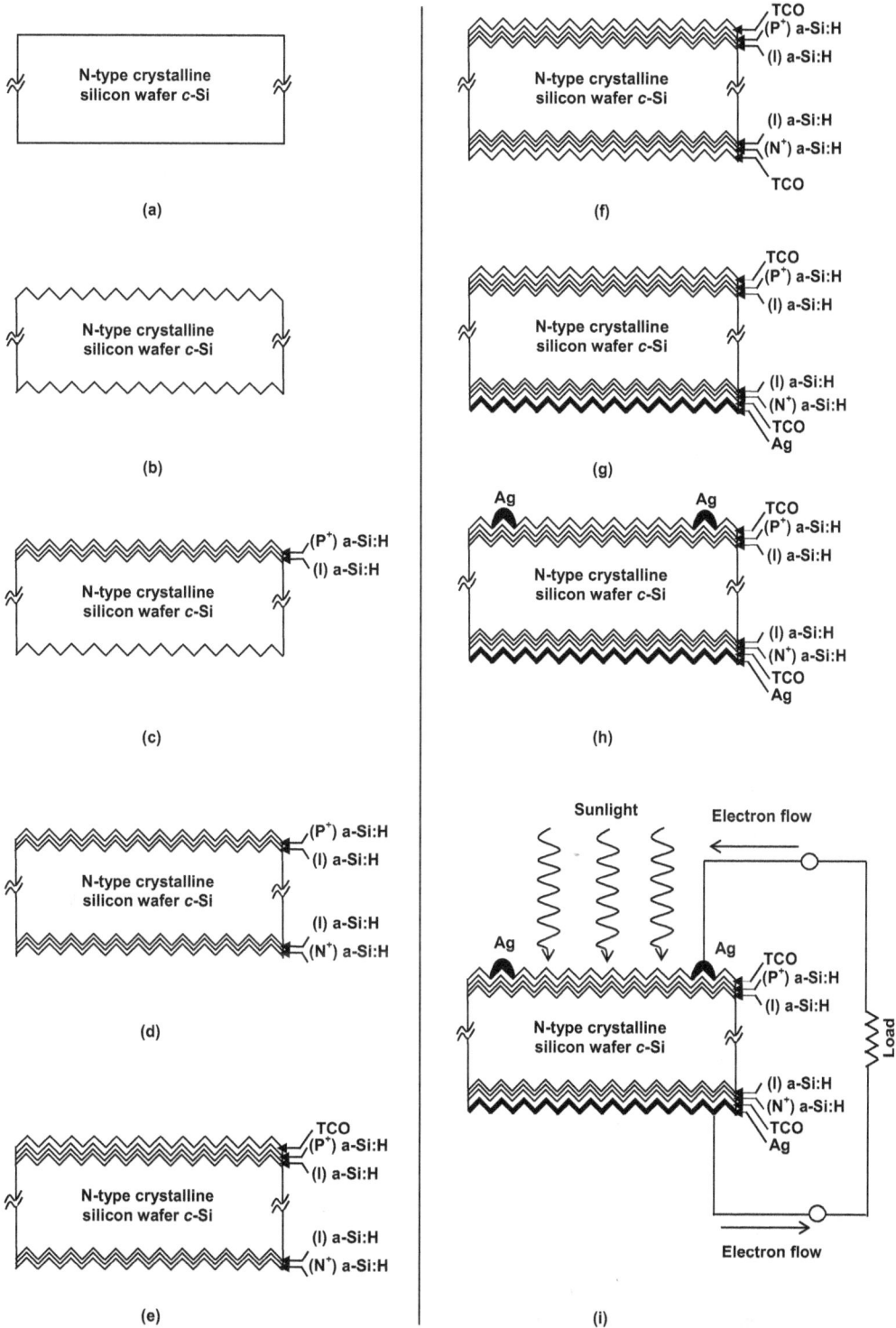

FIGURE 3.1 Process flow for fabrication of silicon heterojunction solar cell: (a) starting silicon wafer, (b) surface texturing, (c) front-side a-Si:H layers deposition, (d) rear-side a-Si:H layers deposition, (e) front-side TCO deposition, (f) rear-side TCO deposition, (g) rear-side Ag film deposition, (h) front-side Ag grid printing and curing, and (i) solar cell illumination with sunlight and measurement of current across Ag electrodes.

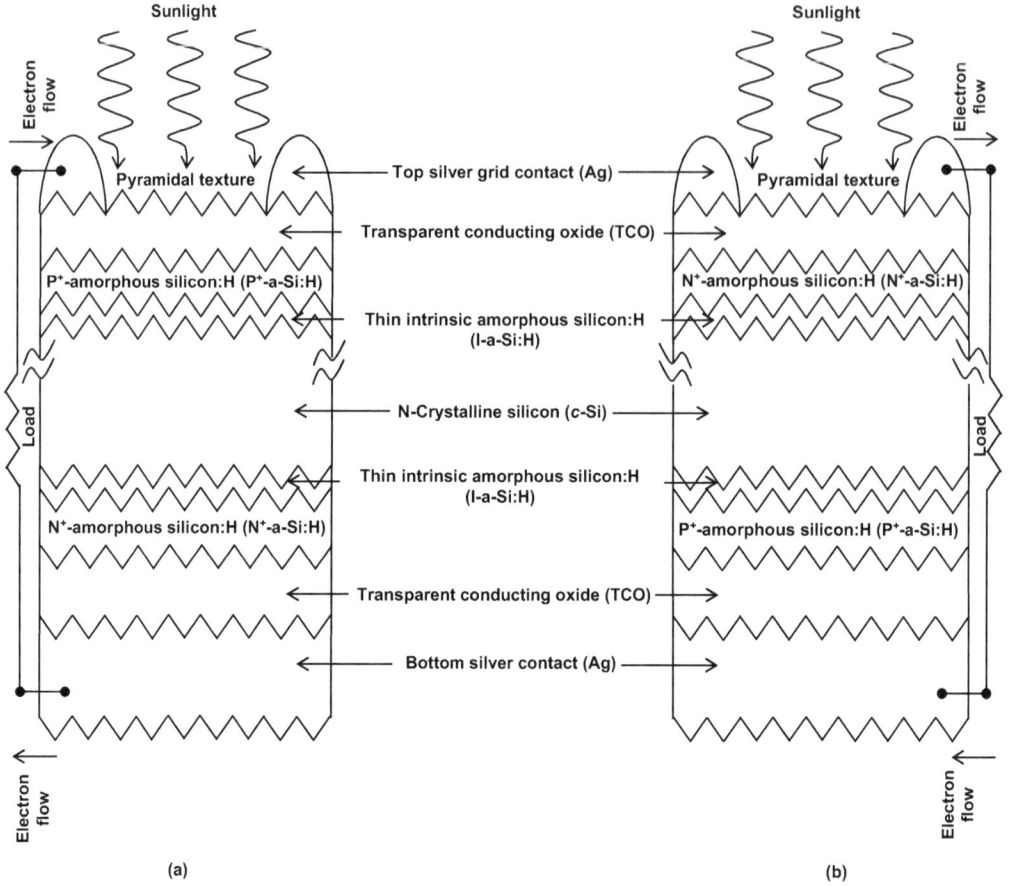

FIGURE 3.2 Two silicon heterojunction solar cell configurations: (a) front-emitter structure with P⁺ emitter located near the top surface of the cell and N/N⁺ junction near its bottom surface; (b) rear-emitter structure with the P⁺ emitter situated near the bottom surface of the cell and N/N⁺ junction near its top surface. Both the P⁺- and N⁺-doped amorphous silicon layers are separated from the crystalline silicon by thin intrinsic layers of amorphous silicon. Further, both the P⁺ and N⁺-doped amorphous silicon layers are coated with TCO layers. Silver is used for top as well as bottom contacts, but the top silver is in the form of a grid to allow sunlight to enter the solar cell. Current is delivered across the silver electrodes when the solar cell receives sunlight.

layer and are collected at the TCO and then the Ag metal film. The holes do not have to travel laterally through the TCO because the back surface is fully covered with a continuous metal film, unlike the fragmented contact metallization of forward emitter cell. The electrons are majority carriers in the N-type crystalline silicon absorber layer. So the electrons can travel straight through the absorber layer as well as laterally through this layer. In contrast to the front-emitter solar cell, the rear-emitter cell has an additional path for electron flow viz the lateral one. Recall that the front-emitter cell depends only on TCO for lateral hole flow. This means that the burden of carrying current on TCO is relieved in the rear-emitter cell, being shared with the absorber layer. Thus, the structural arrangement of layers in the rear-emitter cell is such that the electrons encounter a less-resistive path. Consequently, the path losses are reduced in comparison to the front-emitter geometry.

3.4.3 Advantages of Rear-Emitter Design

On the whole, the lower path losses and design flexibility offered by rear-emitter cell in terms of burden on the front-side TCO make it more convenient. Apart from relaxing the constraints on the optoelectronic properties of the front-side TCO and making the (P$^+$) a-Si:H/TCO contact easier, placement of the (P$^+$) a-Si:H emitter on the nonilluminated side of the solar cell also imposes less optical restrictions on the (P$^+$) a-Si:H emitter layer (Bivour et al 2014).

3.5 REPLACEMENT OF AMORPHOUS SILICON BY NANOCRYSTALLINE SILICON AS ELECTRON/HOLE COLLECTORS

3.5.1 Reasons for Replacement

Let us now inquire about the requirement for application of nanocrystalline silicon in the SHJ solar cell. The standard SHJ solar cell uses doped a-Si:H layers, (P$^+$) a-Si:H, and (N$^+$) a-Si:H layers over the intrinsic a-Si:H layers for carrier collection. Amorphous silicon suffers from the shortcoming that it cannot be heavily doped, because the higher the doping concentration, the greater is the number of defects introduced during doping, nullifying the effect of doping. This imposes a restriction on the conductivity of the carrier collection layers. So this layer needs to be substituted by a higher conductivity layer.

3.5.2 Effects of Replacement

The effects of replacement of (N$^+$) a-Si:H and (P$^+$) a-Si:H on the performance of solar cells were systematically investigated by fabricating three types of rear-emitter solar cells (Nogay et al 2016). The cross sections of these solar cells are drawn in Figure 3.3:

(i) Reference cell. This cell (Figure 3.3(a)) uses only amorphous silicon layers, no nanocrystalline silicon layer. It contains (N$^+$) a-Si:H layer at the front and (P$^+$) a-Si:H layer at the rear. The fill factor for this solar cell is 78.1, and efficiency is 20.7%.
(ii) Cell with (N$^+$) nc-Si:H layer on the front side. This cell (Figure 3.3(b)) contains (N$^+$) nc-Si:H layer at the front as electron collector. All the remaining layers are the same, as in the reference cell. The fill factor and efficiency for this solar cell are 78.4% and 20.9%, respectively.
(iii) Cell with (P$^+$) nc-Si:H layer on the rear side. In this cell (Figure 3.3(c)), the (P$^+$) nc-Si:H layer at the rear is used as the hole collector, keeping the rest of the layers the same, as in the reference cell. Fill factor of solar cell = 79.3%, and efficiency = 21.1%.

It is found that the fill factors and efficiencies of solar cells ii and iii are higher than the corresponding values of solar cell i, demonstrating the beneficial effect of substitution of amorphous silicon layers with nanocrystalline layers as electron or hole collectors. This improvement in performance of solar cells with nanocrystalline layers as electron/hole collection layers over amorphous silicon electron/hole collection is interpreted on the premise of lower contact resistivities of the nanocrystalline layers than amorphous silicon layers. Specific contact resistivities of (N$^+$) nc-Si:H layer and (P$^+$) nc-Si:H layer are respectively 0.83Ω-cm^2 and 1.01Ω-cm^2, while that for the best reference cell is 1.29Ω-cm^2, validating the statement regarding association of contact resistivity with the observed improvement. Thus, nanocrystalline silicon has carved a niche for itself in the SHJ solar cell due to its better conducting properties than amorphous silicon. But the advantage offered by this nanomaterial extends further when we look at its oxide.

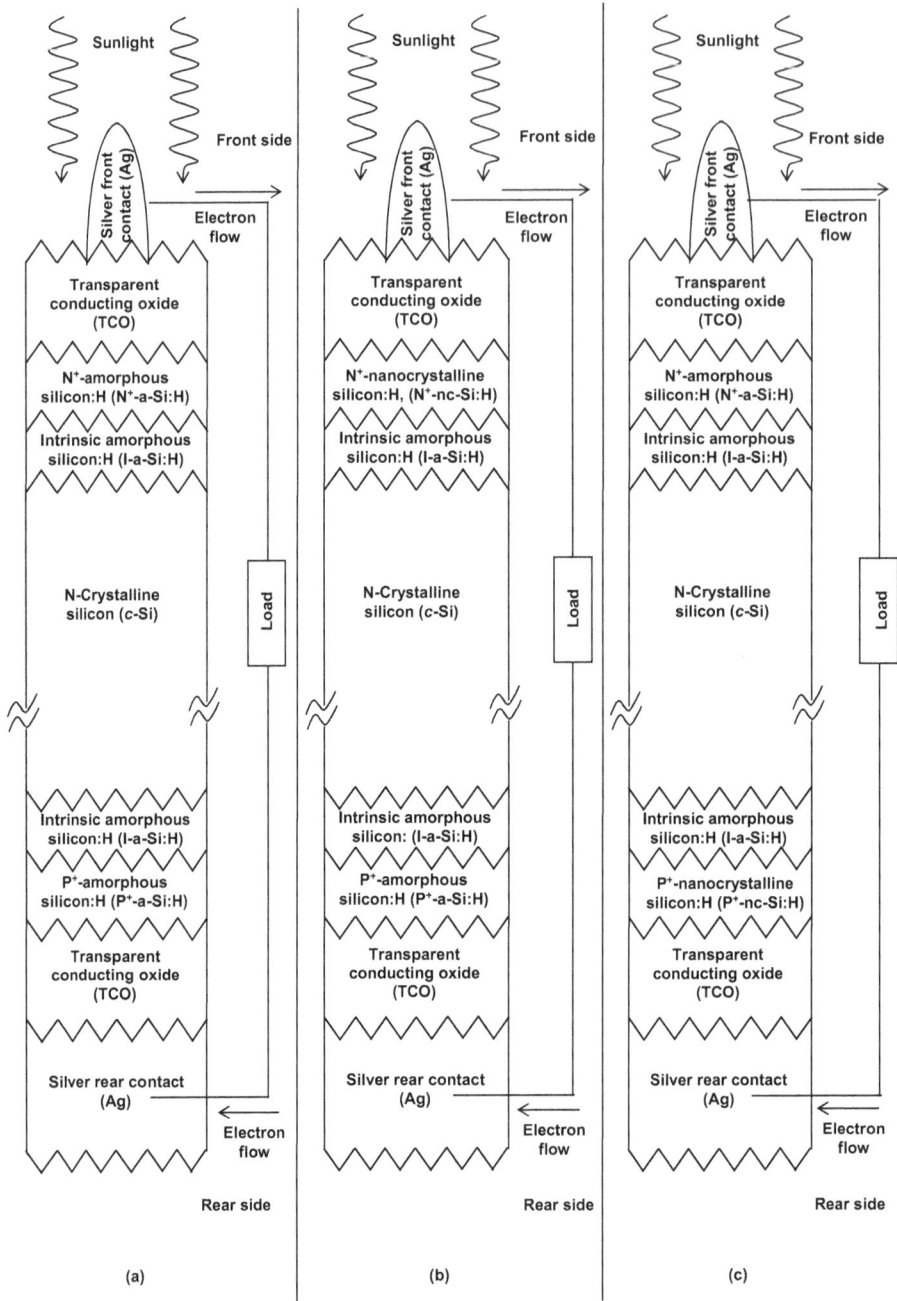

FIGURE 3.3 Three rear-emitter solar cells: (a) Reference cell containing only amorphous silicon layers, as in a standard heterojunction solar cell, (b) cell with (N⁺) nc-Si:H layer on the front side, and (c) cell with (P⁺) nc-Si:H layer on the front side. All three solar cells are made on N-crystalline silicon wafer textured on both surfaces. On both sides of the textured N-crystalline silicon wafer, I-a-Si:H films are deposited. In (a) the intrinsic I-a-Si:H, film is coated with N⁺-a-Si:H film on the front side, and P⁺-a-Si:H film on the rear side. In (b) the intrinsic I-a-Si:H, film is coated with N⁺-nc-Si:H film on the front side, and P⁺-a-Si:H film on the rear side. In (c) the intrinsic I-a-Si:H, film is coated with N⁺-a-Si:H film on the front side, and P⁺-nc-Si:H film on the rear side. In all cases, TCO is deposited on the front side and backside over the doped layers. A silver grid is made on the front side, as it will be exposed to sunlight, while a continuous Ag film is deposited on the rear side for back contact.

3.6 NANOCRYSTALLINE N-TYPE SILICON OXIDE FILMS AS FRONT CONTACTS IN REAR-EMITTER SOLAR CELLS

Returning again to the main optical causes undermining the efficiency of amorphous silicon SHJ solar cells, it is reiterated that there is loss of light through optical absorption predominantly in the front-side silicon stack, and additionally through the metal grid lines. This absorption is referred to as parasitic absorption.

3.6.1 EFFECT OF REFRACTIVE INDEX MATCHING OF TWO OPTICAL MEDIA UPON REFLECTION OF LIGHT AT THEIR INTERFACE

The refractive index of a-Si:H at 633nm is > 4 (Mazzarella et al 2018a), while that of nc-Si:H is 3.4 (Sharma et al 2021) and front-side TCO is 2 (Mazzarella et al 2018a). When light falls on the interface between two transparent media, it is partially reflected and partially transmitted. Reflectance, the ratio of reflected radiant flux to incident radiant flux, is determined by the difference in refractive indices n_1, n_2, of the two media. For the simplest case of normal incidence at the interface, the reflectance R is given by:

$$R = \left| \frac{n_1 - n_2}{n_1 + n_2} \right|^2 \tag{3.3}$$

When the refractive indices of the two media are equal, i.e., $n_1 = n_2 = n$, we have:

$$R = \left| \frac{n - n}{n + n} \right|^2 = 0 \tag{3.4}$$

Zero reflectance means that there is no reflection of light. For $n_1 >> n_2$, R approaches 1, which implies that light is almost completely reflected. Thus, the smaller the difference between the refractive indices of the two media, the smaller the reflectance.

3.6.2 COMPARING REFLECTANCES AT THE INTERFACES A-SI:H/TCO, NC-SI:H/TCO, AND NC-SIO$_2$:H/TCO

For reasons elaborated in the previous subsection about the refractive indices of two contacting layers, reflectance at the interface between nc-Si:H ($n_1 = 3.4$) and TCO ($n_2 = 2$), for which n_1-$n_2 = 1.4$, is less than for the interface between a-Si:H ($n_1 = 4$) and TCO ($n_2 = 2$), for which n_1-$n_2 = 2$.

A refractive index lower than that of nc-Si:H is achievable with oxygen-alloyed nc-Si:H or nc-SiO$_2$:H, providing better refractive index matching with the underlying TCO. This film has a refractive index ~ 2.8 (Sharma et al 2021) so that reflectance at the interface between nc-SiO$_2$:H and TCO (n_1-$n_2 = 0.8$) is lower than that at the interface of nc-Si:H with TCO (n_1-$n_2 = 1.4$). Hence, use of nc-SiO$_2$:H film is more beneficial than nc-Si:H to solar cell performance through reducing the loss of light. The nc-SiO$_2$:H is a nanomaterial consisting of crystallites of silicon of a few nanometers' size dispersed in the matrix of amorphous silicon dioxide. Films of refractive indices 2.1–2.7 also having high conductivities ~ 10^{-4}–10^{-1} S/cm are deposited (Mazzarella et al 2018a). Thus, these films have both optical and electrical merits, asserting their usefulness as front-side contacts in SHJ solar cells. The thicknesses of nc-SiO$_2$:H films are optimized accordingly as the surface is kept planar or textured. The planar surface is more reflecting. Hence, a less-transparent, thicker film of refractive index 2.7 can be used. For a textured wafer, the reflection is already low. So a more transparent thinner film of refractive index < 2.7 must be used.

3.6.3 DIFFICULTY IN DEPOSITION OF THIN nc-SiO₂:H FILM OVER (I)A-Si:H LAYER

A major technological hurdle faced here involves deposition of thin nc-SiO$_2$:H film over intrinsic amorphous silicon passivation layer without degrading the passivation property of the amorphous layer. Since crystalline film growth is invariably preceded by an amorphous incubation phase, deposition of a thin nc-SiO$_2$:H film requiring a time scale < 100 s. is done by accelerating the transition from amorphous to nanocrystalline phase. For speeding up this transition, a seed layer is used, and a large flux of atomic hydrogen is driven towards the substrate in a high-pressure, high-power PECVD system. The optimized solar cell (Figure 3.4(a)) with seed nc-Si:H film + 10nm-thick nc-SiO$_2$:H layer shows an open-circuit voltage of 730.7mV, short-circuit current density of 38.3mA cm^{-2}, fill factor of 80.6%, and efficiency of 22.6%. Parameters of the reference solar cell (Figure 3.4(b)) having 12nm-thick nc-Si:H film are 728.9mV, 37.8 mAcm^{-2}, 79.7%, and 21.9%. Thus, the advantage of nc-SiO$_2$:H over nc-Si:H is proven (Mazzarella et al 2018b).

3.7 NANOCRYSTALLINE SILICON THIN-FILM SOLAR CELL ON HONEYCOMB-TEXTURED SUBSTRATE

Textures with sharply rising or falling slopes and deep valleys are incongruous to nc-Si growth because they lead to defective porous regions (Sai et al 2014). To overcome this shortcoming, periodic textures known as honeycomb textures are produced by wet chemical etching on thermally oxidized silicon substrates after photolithographically defining the required pattern on the silicon oxide through a resist mask. If P is the period of the texture and H is the height of the peak, the aspect ratio is H/P. The period P is varied between 1.5 and 2.5µm. Aspect ratios lie between 0.1 and 0.25. Over the honeycomb-textured surface of silicon dioxide, silver film is deposited and covered with ZnO:Ga film. The silver and ZnO:Ga films together produce a reflective conducting coating serving as an efficient back reflector. A solar cell with N-I-P diode configuration (Figure 3.5) is fabricated by sequentially depositing the (n) nc-SiO$_x$:H, I-nc-Si:H, an I/P buffer, and (P) nc-SiO$_x$:H films using a PECVD system. Then, magnetron sputtering is employed to deposit an ITO film and an Ag finger-grid electrode. Reactive ion etching is performed to isolate the cell. Annealing is done at 175°C. For a solar cell of period 12µm and thickness 1.7µm, the open-circuit voltage is 542mV, short-circuit current density is 27.44mAcm^{-2}, fill factor is 0.738, and efficiency is 11.0%, using an antireflective film with moth-eye structure (Sai et al 2014). As opposed to a low deposition rate of 0.1nm s^{-1} for nc-Si:H films achieved with a laboratory scale RF-PECVD system operating at 13.56MHz, which is unacceptable as an industrial process, a large area VHF-PECVD system working at 60MHz is used to get a deposition rate of 1nm s^{-1} and a certified solar cell efficiency of 11.8 % (Sai et al 2015).

3.8 DISCUSSION AND CONCLUSIONS

Several useful conclusions can be drawn with regard to usage of nanocrystalline films in solar cells:

(i) Strategies are evolved for substitution of amorphous silicon with the nanocrystalline phase in heterojunction solar cells.

(ii) Rear-emitter SHJ solar cell geometry is less restrictive in design concerning TCO properties and with regard to path losses than front-emitter configuration (Bivour et al 2014).

(iii) The nc-Si:H film is better than the a-Si:H layer in optoelectronic properties because it has higher electrical conductivity and causes reduced parasitic absorption.

(iv) The nc-SiO$_2$:H film shows better refractive index matching with TCO than nc-Si:H layer and hence produces less reflection of light.

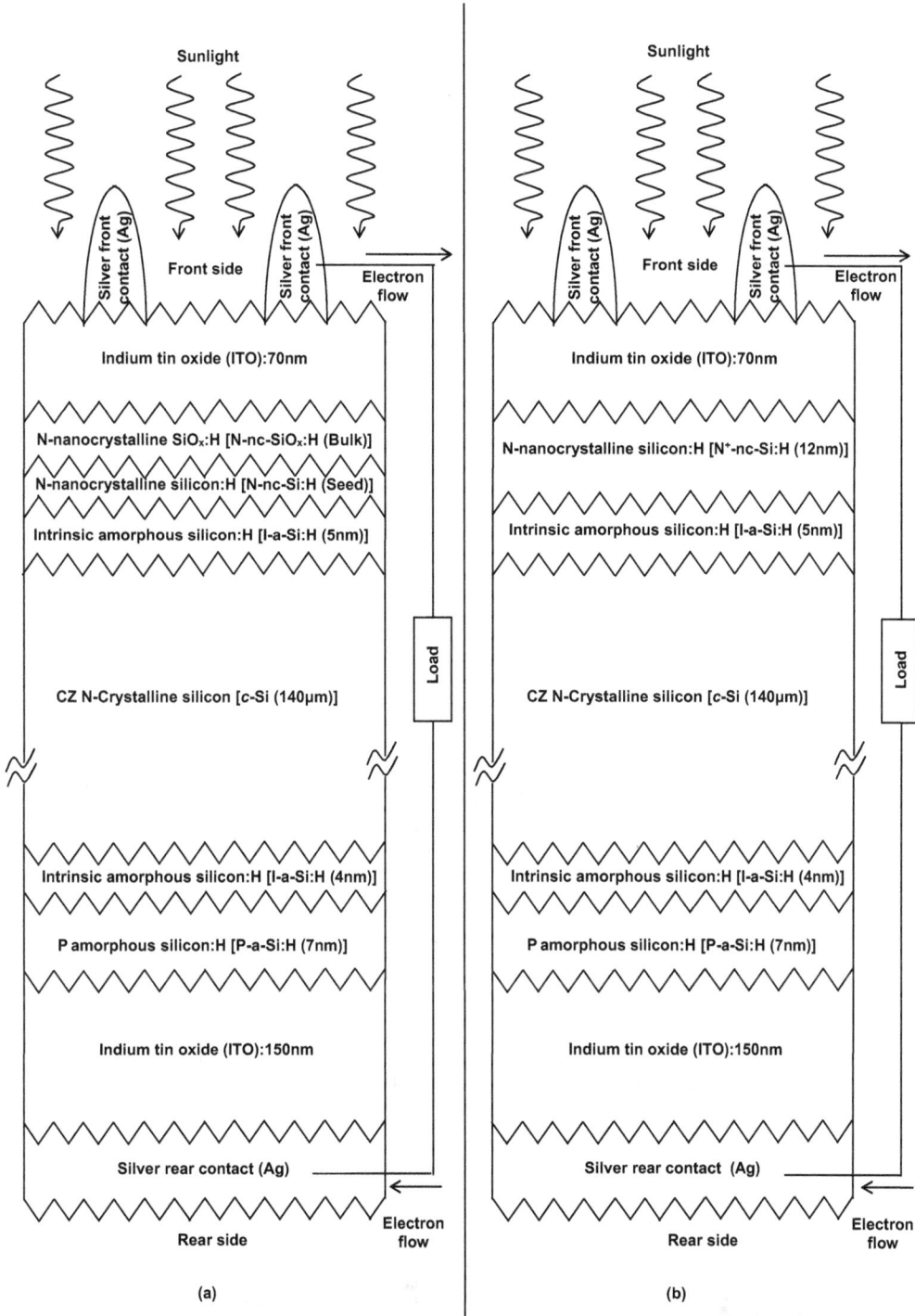

FIGURE 3.4 Overcoming the difficulty in deposition of thin nc-SiO$_2$:H film over (I) a-Si:H layer by comparing the performance of the (a) optimized rear-emitter solar cell with respect to (b) reference rear-emitter solar cell. In the optimized solar cell, deposition of nc-SiO$_2$:H film is preceded by deposition of a seed layer of nc-Si:H film, while the reference cell contains the nc-Si:H film only, no nc-SiO$_x$ layer. ITO is used as the TCO. Ag grid formed on the front side and continuous Ag film deposited on the rear side are used as metal layers. Sunlight is shown falling on the solar cell, and electronic current flow directions are indicated.

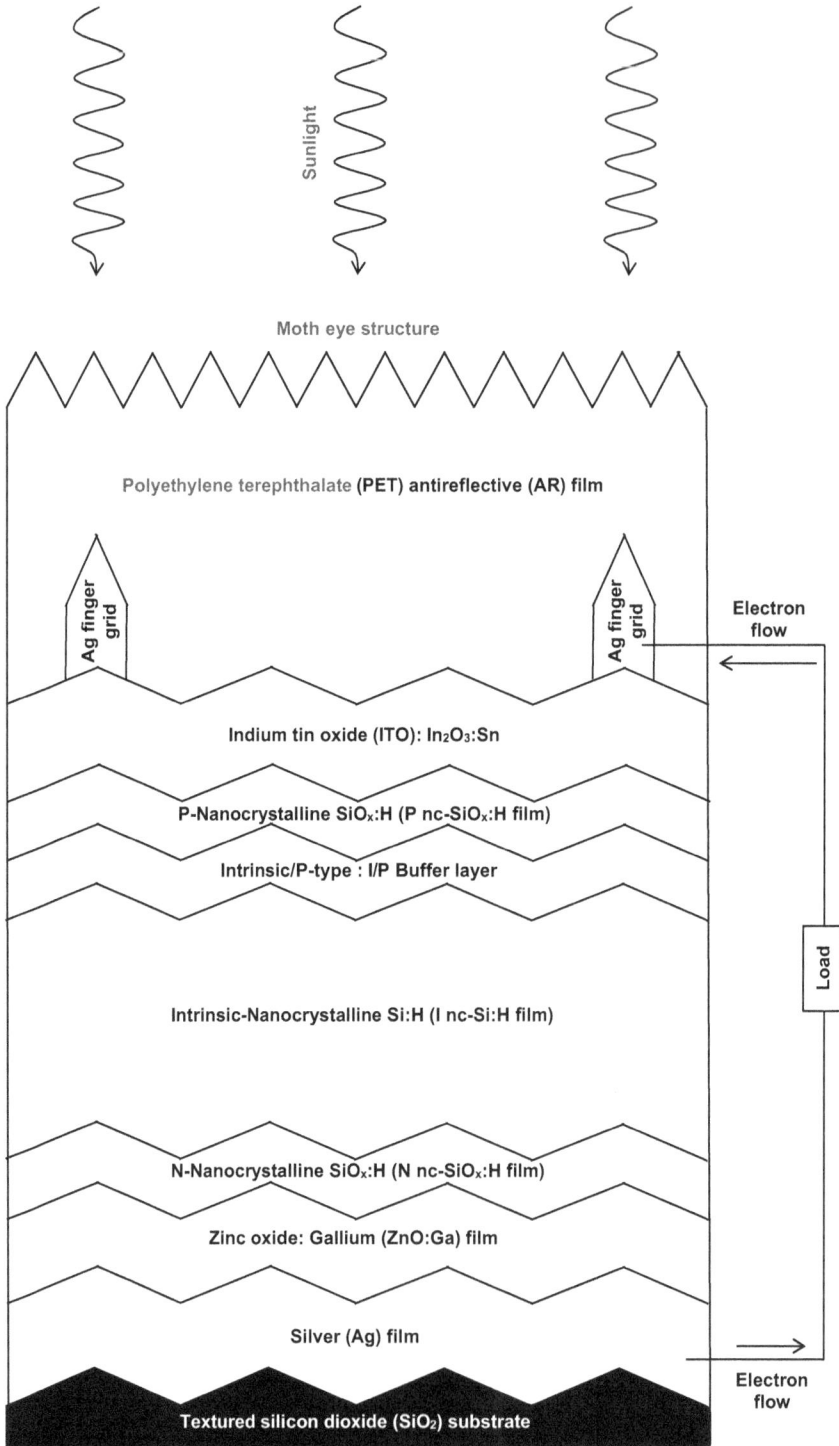

FIGURE 3.5 Nanocrystalline silicon thin-film solar cell with the N-I-P structure: honeycomb-textured SiO$_2$ substrate/Ag film/ZnO:Ga film/(N)nc-SiO$_x$:H film/(I) nc-Si:H film/(I/P) buffer layer/(P)nc-SiO$_x$:H/ITO/ Ag grid. An antireflective film with a moth-eye structure is applied over the front surface. It consists of a UV-curable polymer, a PET substrate, and an acrylic adhesive film. Sunlight illuminating the solar cell from the moth-eye structure side produces current flow between Ag electrodes.

(v) Substitution of a-Si:H films in solar cell structure with nc-Si:H or nc-SiO$_2$:H has led to desirable improvements in carrier collection and hence in characteristics and efficiencies.

(vi) Solar cells with N-I-P structure are fabricated entirely using nc-Si:H, i.e., without crystalline silicon or a-Si:H.

Questions and Answers

3.1   In a silicon heterojunction solar cell using nanocrystalline silicon, what are the heterostructure layers: (a) single-crystal and nanocrystalline silicon, (b) single-crystal and amorphous silicon? Answer: Single-crystal and amorphous silicon.

3.2   What are the two phases in hydrogenated nanocrystalline silicon? Answer: Crystalline silicon and a-Si:H.

3.3   Is it correct to say that nanocrystalline silicon is a mixture of crystalline and amorphous silicon? Answer: Yes, silicon nanocrystallites in a-Si.

3.4   Is silicon nanocrystal same as nanocrystalline silicon? Answer: No. A silicon nanocrystal is a piece of monocrystalline silicon having at least one dimension smaller than 100nm.

3.4   Bandgap of a-Si:H is (a) 1.1eV or (b) 1.75eV? Answer: 1.75eV.

3.5   Solar cell of which semiconductor requires smaller absorber layer thickness: (a) single crystal silicon or (b) a-Si:H? Answer: a-Si:H.

3.6   Electron mobility in a-Si:H is (a) 1400 cm^{2}V^{-1}s^{-1} or (b) 1cm^{2}V^{-1}cm^{-1}? Answer: 1cm^{2}V^{-1}cm^{-1}.

3.7. What is meant by pyrolysis of a material? Name a type of silicon which is deposited by a pyrolysis process. Answer: Pyrolysis is the thermal decomposition of a material at a high temperature in an inert atmosphere. Polysilicon is deposited by pyrolysis of silane releasing hydrogen gas.

3.8   Can unhydrogentated amorphous silicon be doped with impurities? Answer: No.

3.9   Discuss the advantages of nanocrystalline silicon over amorphous and polycrystalline silicon for use in solar cells. Answer: nc-Si has better doping efficiency, higher electron mobility, conductivity, and transparency and shows stronger IR absorption than a-Si. It is deposited at lower temperatures than polySi.

3.10 In a silicon PECVD process, a high value of hydrogen dilution ratio ~ 50 was used. Comment on the morphology of the silicon film obtained. Answer: Dominant crystalline portion.

3.11 A low value of hydrogen dilution ratio ~ 2 was used in a silicon PECVD process. Describe the morphology of the silicon film obtained. Answer: Fully amorphous.

3.12 Why is it necessary to raise the operating pressure for depositing nanocrystalline films at a high deposition rate? Answer: To lower defect creation by ion bombardments.

3.13 Identify the emitter and the absorber layers in a P$^+$-I-N$^+$ silicon heterojunction solar cell. Answer: Emitter layer, P$^+$ a-Si:H layer; absorber layer, N-type crystalline silicon (the intrinsic silicon layer).

3.14 What is the function of the intrinsic a-Si:H layers on both sides of the crystalline silicon absorber in a silicon heterojunction solar cell? Answer: Passivation.

3.15 What are the two functions of the a-Si:H layer on the backside of a silicon heterojunction solar cell? Answer: BSF and ohmic contact.

3.16 Which dopant gas is introduced for N$^+$ doping during PECVD? Answer: Phosphine.

3.17 How does the low/high (N/N$^+$) junction on the rear surface of a solar cell prevent carrier loss by recombination? Answer: By acting like a P-N junction with a built-in electric field, which repels and pushes back minority-carrier electrons towards the absorber layer, obviating their recombination.

3.18 Give two reasons for discarding ITO use in solar cells. Answer: High price, scarcity of indium.

3.19 Path losses suffered during carrier flow are lower in (a) front-emitter SHJ cell or (b) rear-emitter SHJ cell? Answer: Rear-emitter SHJ cell.

3.20 Nanocrystalline silicon is used to replace amorphous silicon in an SHJ solar cell for electron/hole collection because (a) it is more stable than amorphous silicon or (b) it can be heavily doped to increase its conductivity? Answer: It can be heavily doped to increase its conductivity.

3.21 Write nc-Si:H, a-Si:H and TCO in ascending order of their refractive indices. Answer: TCO (2) < nc-Si:H (3.4) < a-Si:H (> 4).

3.22 What is the reflectance at an interface between two optical media of equal refractive indices? Answer: Zero.

3.23 nc-Si:H has a refractive index of 3.4, while nc-SiO_2:H has a refractive index of 2.8. Which material will you prefer for interfacing with TCO (refractive index 2) to decrease reflectance? Why? Answer: nc-SiO_2:H. Because $2.8 - 2 = 0.8$ is less than $3.4 - 2 = 1.4$.

3.24 Which of the following statements is correct?
(a) SiO_2:H is a nanomaterial consisting of crystallites of silicon dioxide of a few nanometers' size dispersed in the matrix of amorphous silicon.
(b) SiO_2:H is a nanomaterial consisting of crystallites of silicon of a few nanometers' size dispersed in the matrix of amorphous silicon dioxide.
Answer: b.

3.25 Which solar cell performs better: (a) solar cell with nc-Si:H as front contact layer, (b) solar cell with nc-SiO_2:H film front contact layer? Answer: b.

3.26 Why are textured surfaces with sharply rising or falling slopes inappropriate for growth of nc:Si:H? Answer: Because of creation of defective porous regions.

3.27 The aspect ratio of a texture is the ratio (a) period/height or (b) height/period? Answer: Height/period.

3.28 Is it possible to make an N-I-P solar cell with all layers made of nc-Si:H? Answer: Yes.

REFERENCES

Adhikari D., M. M. Junda, C. R. Grice, S. X. Marsillac, R. W. Collins and N. J. Podraza 2019 *n-i-p* Nanocrystalline hydrogenated silicon solar cells with RF-magnetron sputtered absorbers, Materials, 12(1699): 13 pages.

Bivour M., H. Steinkemper, J. Jeurink, S. Schröer and M. Hermle 2014 Rear emitter silicon heterojunction solar cells: Fewer restrictions on the optoelectrical properties of front side TCOs, 4th International Conference on Silicon Photovoltaics, SiliconPV, March 25–27, 2014, s-Hertogenbosch, the Netherlands, Energy Procedia, 55: 229–234.

Chopra K. L., P. D. Paulson and V. Dutta 2004 Thin-film solar cells: An overview, Progress in Photovoltaics: Research and Applications, 12: 69–92.

De Wolf S., A. Descoeudres, H. Antoine, Z. C. Holman and C. Ballif 2012 High-efficiency silicon heterojunction solar cells: A Review, Green, 2(1): 7–24.

Kang H. 2021 Crystalline silicon vs. amorphous silicon: The significance of structural differences in photovoltaic applications, 2020 2nd International Conference on Resources and Environmental Research (ICRES 2020), 5th to 7th June 2020, Bangkok, Thailand, IOP Conf. Series: Earth and Environmental Science 726(012001): 5 pages.

Kroely L. 2010 Process and material challenges in the high rate deposition of microcrystalline silicon thin films and solar cells by matrix distributed electron cyclotron resonance plasma. Plasma Physics [physics.plasm-ph]. Ecole Polytechnique X, English. pastel-00550241, pp. 13–14.

Kroll U., J. Meier, A. Shah, S. Mikhailov and J. Weber 1996 Hydrogen in amorphous and microcrystalline silicon films prepared by hydrogen dilution, Journal of Applied Physics, 80(9): 4971–4975.

Mazzarella L., A. Morales-Vilches, M. Hendrichs, S. Kirner, L. Korte, R. Schlatmann and B. Stannowski 2018a Nanocrystalline n-type silicon oxide front contacts for silicon heterojunction solar cells: Photocurrent enhancement on planar and textured substrates, IEEE Journal of Photovoltaics, 8(1): 70–78.

Mazzarella L., A. B. Morales-Vilches, L. Korte, R. Schlatmann and B. Stannowski 2018b Ultra-thin nanocrystalline n-type silicon oxide front contact layers for rear-emitter silicon heterojunction solar cells, Solar Energy Materials and Solar Cells, 179: 386–391.

Menéndez A., P. Sánchez and D. Gómez 2013 Chapter 2: Deposition of thin films: PECVD process, in: Murri R. (ed.), Silicon Based Thin Film Solar Cells, Bentham Science Publishers Ltd., Sharjah, U.A.E., Oak Park, Bussum, The Netherlands, pp. 1–23.

Mikolášek M. 2017 Chapter 4: Silicon Heterojunction solar cells: The key role of heterointerfaces and their impact on the performance, in: Das N. (ed.), Nanostructured Solar Cells, IntechOpen, London, pp. 69–92.

Nogay G., J. P. Seif, Y. Riesen, A. Tomasi, Q. Jeangros, N. Wyrsch, F.-J. Haug, S. De Wolf and C. Ballif 2016 Nanocrystalline silicon carrier collectors for silicon heterojunction solar cells and impact on low-temperature device characteristics, IEEE Journal of Photovoltaics, 6(6): 1654–1662.

Sai H., K. Maejima, T. Matsui, T. Koida, M. Kondo, S. Nakao, Y. Takeuchi, H. Katayama and I. Yoshida 2015 High-efficiency microcrystalline silicon solar cells on honeycomb textured substrates grown with high-rate VHF plasma-enhanced chemical vapor deposition, Japanese Journal of Applied Physics, 54: 08KB05-1 to 08KB05-6.

Sai H., T. Matsui, K. Matsubara, M. Kondo and I. Yoshida 2014 11.0%-Efficient thin-film microcrystalline silicon solar cells with honeycomb textured substrates, IEEE Journal of Photovoltaics, 4(6): 1349–1353.

Sharma M., J. Panigrahi and V. K. Komarala 2021 Nanocrystalline silicon thin film growth and application for silicon heterojunction solar cells: A short review, Nanoscale Advances, 3: 3373–3383.

Si-Silicon: Electrical Properties, www.ioffe.rssi.ru/SVA/NSM/Semicond/Si/electric.html.

Street R. A. 2000 Technology and Applications of Amorphous Silicon, Springer-Verlag, Berlin Heidelberg, p. 4.

Zhao Y., L. Mazzarella, P. Procel, C. Han, G. Yang, A. Weeber, M. Zeeman and O. Isabella 2020 Doped hydrogenated nanocrystalline silicon oxide layers for high-efficiency c-Si heterojunction solar cells, Progress in Photovoltaics: Research and Applications, 28: 425–435.

Part II

Nanotechnological Approaches to Sunlight Harvesting

4 Nanotextured-Surface Solar Cells

A nanotextured surface is one covered with nanoscale structures, such as nanospheres, nanopyramids, nanocones, nanopillars, etc. These structures, resembling hillocks on the surface, are called photonic nanostructures. Nanotexturing is the process of producing such nanostructures. In a solar cell, nanotexturing is used as a strategy to improve energy conversion efficiency by reducing optical losses (Kim et al 2020).

4.1 OPTICAL LOSSES IN A SOLAR CELL AND LOSS-REDUCTION APPROACHES

4.1.1 OPTICAL LOSSES

By *optical losses in a solar cell*, what is meant is the portion of sunlight which is incident on the surface of the solar cell but is not used in the creation of electron-hole pairs. It is evident that light transmitted into the cell contributes to carrier generation, but light reflected from its surface is lost. Additionally, light falling on the portion of the top surface covered by contact metal finds it difficult to penetrate inside and represents an optical loss.

4.1.2 OPTICAL LOSS REDUCTION BY OPTICAL TRANSMITTANCE ENHANCEMENT

In order to gainfully utilize the light falling on a solar cell for the intended purpose of electricity production, transmittance of light into the cell must be maximized and its reflection from the cell surface must be minimized. So the top surface of the cell should be provided with antireflection capability (Figure 4.1). This method of reducing optical loss by increasing absorption of light is called optical transmittance enhancement.

4.1.3 OPTICAL LOSS REDUCTION BY OPTICAL PATH LENGTHENING

Light entering the cell must travel a sufficient distance in which it can interact with the absorber medium to produce free-charge carriers. Optical path length is the distance traversed by a photon during its journey through the solar cell before it escapes out of the cell. The longer the travel path of light through the absorber layer of the cell, the greater the chance of carrier generation by photon impact. However, carriers collected are essentially those which are produced at a distance from the contact metal less than the diffusion lengths of carriers; carriers produced beyond this distance may be lost by recombination. The method of optical loss reduction by extending the distance traversed by photons through the absorber layer is known as optical path lengthening.

Increase in the optical path is done by microtexturing or nanotexturing the surface of the solar cell, causing multiple reflections and total internal reflection whenever the angle of incidence becomes less than the critical angle. Texturing is accomplished by chemical etching. In this process, pyramid-shaped projections are formed on the surface of the solar cell. Geometrical optics is applied to find the effect of texturing on the length of the optical path.

DOI: 10.1201/9781003215158-6

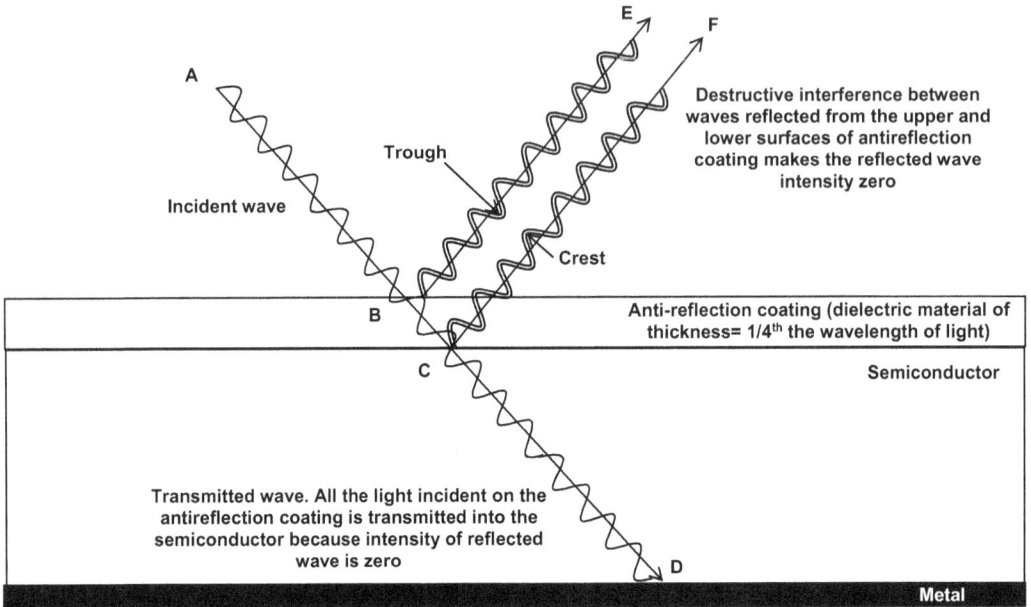

FIGURE 4.1 Diagram explaining the use of antireflection coating to minimize reflection losses in a solar cell. The coating is generally a dielectric material having thickness equal to a quarter wavelength of light to ensure that the path difference between the light beams reflected from the upper and lower surfaces of the antireflection coating is such that the two beams have their crests and troughs aligned opposite each other, i.e., out of phase, and cancel out the effect of each other so that there is no reflected wave.

The effect becomes evident by drawing the path of a ray of light in the solar cell, as shown in Figure 4.2.

The effect of texturing on the optical path length becomes evident when we realize that texturing makes the light fall nonperpendicularly on the surface of the solar cell (Sprafke and Wehrspohn 2015). Upon oblique incidence, light undergoes multiple reflections inside the absorber layer and is reflected back from the top surface into the absorber layer by total internal reflection. By total internal reflection, light is trapped inside the solar cell and suffers repeated reflections in moving through the cell several times. The optical path length increases from one device thickness for a solar cell with untextured surface to many tens of device thickness for a textured solar cell. Consequently, the ability of light to cause photogeneration of carriers is appreciably increased in a textured-surface solar cell than in one with an untextured surface.

4.2 OPTICAL TRANSMITTANCE ENHANCEMENT BY NANOTEXTURING

Let us now understand how nanotexturing of a solar cell surface helps in increasing transmittance.

4.2.1 REFLECTANCE AND TRANSMITTANCE EQUATIONS

The amplitude coefficients of transmission and reflection at the interface between two transparent homogeneous media of refractive indices n_1, n_2 are related through Fresnel's equations (Lvovsky

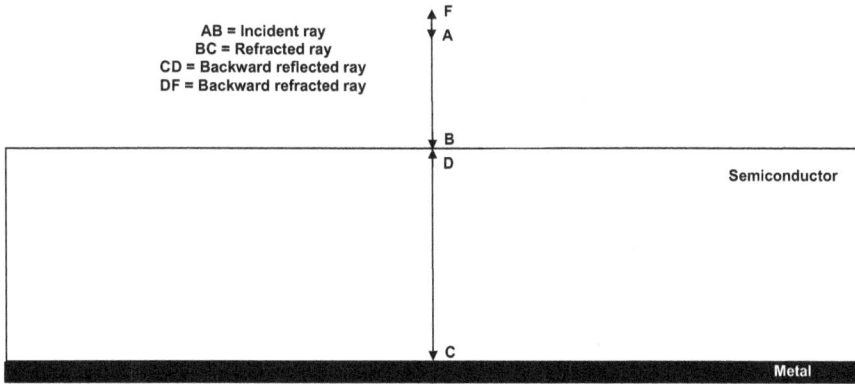

(a)

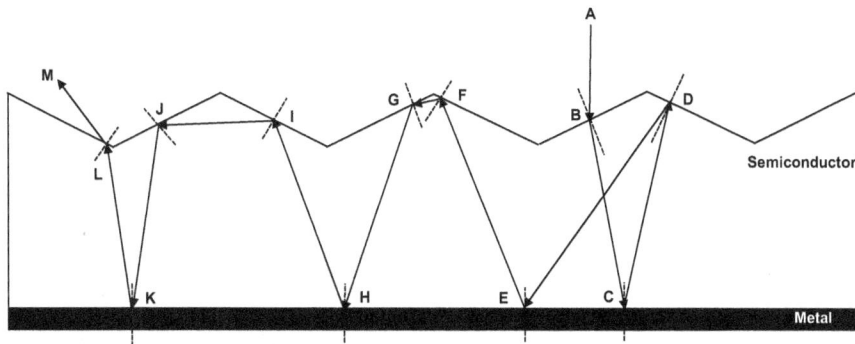

(b)

FIGURE 4.2 Tracing the path of light rays through a silicon solar cell: (a) without any surface texture on the front side, and (b) with front-side surface texturing. Paths followed by a single ray of light through the solar cell are compared in *a* and *b*. In *a*, a ray *AB* striking the surface of the solar cell is falling at normal incidence. So it moves straight along its path, producing the refracted ray *BC*. The ray *BC* is reflected back from the surface of the metal electrode on the backside of the cell along the path *CD*. At the point *D*, it is refracted along the straight-line path *DF* because it strikes the front surface of the cell perpendicularly. In *b*, the ray *AB* striking the textured surface of the solar cell is at oblique incidence. It is refracted at the point *B*, bending towards the normal. Then, it is repeatedly reflected internally inside the semiconductor layer of the solar cell between the front textured surface and the metal electrode on the backside of the cell. A typical path traced out by this ray is *ABCDEFGHIJKLM*.

2015). Recalling from Section 3.6.1, for normal incidence, the reflectance R (the ratio of reflected flux to incident flux) is given by

$$R = \left| \frac{n_1 - n_2}{n_1 + n_2} \right|^2 \tag{4.1}$$

while the transmittance T (the ratio of transmitted flux to incident flux) is

$$T = 1 - R = 1 - \left| \frac{n_1 - n_2}{n_1 + n_2} \right|^2 = \frac{\left(n_1 + n_2 \right)^2 - \left(n_1 - n_2 \right)^2}{\left(n_1 + n_2 \right)^2}$$

$$= \frac{n_1^2 + 2n_1 n_2 + n_2^2 - \left(n_1^2 - 2n_1 n_2 + n_2^2 \right)}{\left(n_1 + n_2 \right)^2} = \frac{4n_1 n_2}{\left(n_1 + n_2 \right)^2} \qquad (4.2)$$

Example 4.1

Compare and comment on the reflectance and transmittance of light for normal incidence at air-glass interface ($n_1 = 1$, $n_2 = 1.5$) with air-silicon interface ($n_1 = 1$, $n_2 = 3.45$).

Solution: Applying equation 4.1 for the air-glass interface:

$$R = \left| \frac{n_1 - n_2}{n_1 + n_2} \right|^2 \times 100\% = \left| \frac{1 - 1.5}{1 + 1.5} \right|^2 \times 100\% = \left| \frac{-0.5}{2.5} \right|^2 \times 100\% = \frac{0.25}{6.25} \times 100\% = 4\% \qquad (4.3)$$

From equation 4.2:

$$T = \frac{4n_1 n_2}{\left(n_1 + n_2 \right)^2} \times 100\% = \frac{4 \times 1 \times 1.5}{\left(1 + 1.5 \right)^2} \times 100\% = \frac{6}{6.25} \times 100\% = 0.96 \times 100\% = 96\% \qquad (4.4)$$

T can also be straightaway obtained from reflectance value in equation 4.3, as

$$T = 100 - R\% = 100 - 4\% = 96\% \qquad (4.5)$$

Similarly, equation 4.1 gives for the air-silicon interface

$$R = \left| \frac{n_1 - n_2}{n_1 + n_2} \right|^2 \times 100\% = \left| \frac{1 - 3.45}{1 + 3.45} \right|^2 \times 100\% = \left| \frac{-2.45}{4.45} \right|^2 \times 100\%$$

$$= \frac{6.0025}{19.8025} \times 100\% = 30.31\% \qquad (4.6)$$

and

$$T = 100 - R\% = 100 - 30.31\% = 69.69\% \qquad (4.7)$$

Comment: Silicon has higher reflectance and lower transmittance than glass because the difference between refractive indices (1–3.45 = –2.45) is larger for air-silicon than for air-glass (1–1.5 = –0.5). The larger the difference between refractive indices of two media, the higher the reflectance. A way to reduce reflectance is to decrease the difference in refractive indices of the two media.

4.2.2 EFFECTS OF SIZES OF STRUCTURES OF THE TEXTURED INTERFACE MORPHOLOGY ON ITS REFLECTANCE

Depending on the sizes of the geometrical features on the interface, there are two contrasting situations:

(i) Case I: Interface morphology contains structures of size larger than the wavelength of incident sunlight. Here, geometrical optics is valid and the angle of reflection equals the angle of incidence. The reflectance is large, and the transmittance is small (Figure 4.3(a)).

(ii) Case II: The structures in the interface morphology have sizes smaller than the wavelength of incident sunlight. In this circumstance, the interface behaves as a medium of graded refractive index. It has an effective refractive index $n_{Effective}$ which is lower than the nominal refractive index of the medium, i.e., the standard refractive index value for that medium. This lowering of refractive index results in a smaller reflectance and larger transmittance (Figure 4.3(b)). Hence,

$$R_{Effective} = \left| \frac{n_1 - n_{Effective}}{n_1 + n_{Effective}} \right|^2 < \left| \frac{n_1 - n_2}{n_1 + n_2} \right|^2 \tag{4.8}$$

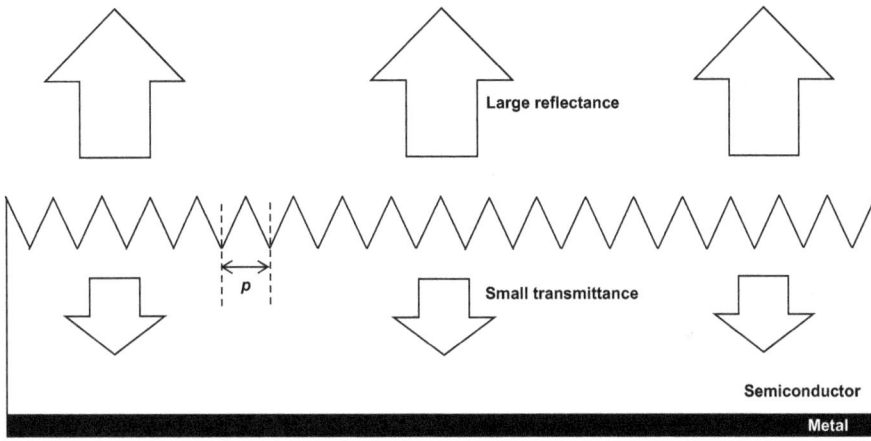

(a)

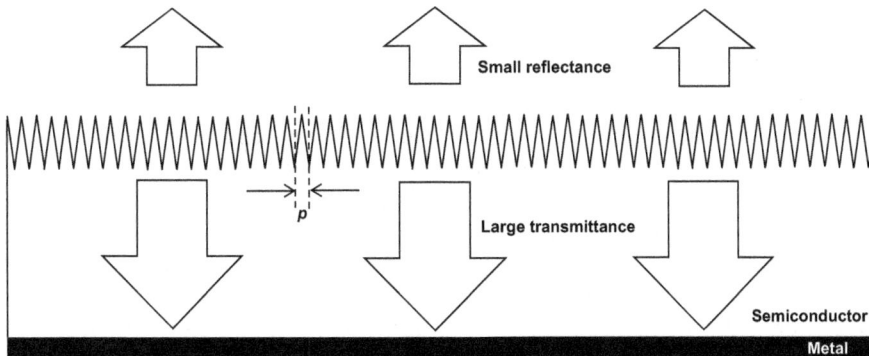

(b)

FIGURE 4.3 Effects of texture size on the optical behavior of the surface texturing: (a) microtextured surface and (b) nanotextured surface. In (a) the structure, size p (period of structure) > wavelength λ of light. The reflectance is large and transmittance is small. In (b) the structure, size p (period of structure) < wavelength λ of light. The reflectance is small and transmittance is large, i.e., the reverse of (a) happens.

because

$$n_{\text{Effective}} < n_2 \tag{4.9}$$

and therefore, as seen in the comparison between air-glass and air-silicon cases,

$$\left| n_1 - n_{\text{Effective}} \right| < \left| n_1 - n_2 \right| \tag{4.10}$$

$$\left| n_1 + n_{\text{Effective}} \right| < \left| n_1 + n_2 \right| \tag{4.11}$$

In both Cases I and II, for rays refracted into the semiconductor, the optical path is increased by multiple reflections, as already shown in Figure 4.2.

4.3   NANOTEXTURED SURFACE PROPERTIES, EXAMPLES IN NATURE, AND COMPARISON WITH MICROTEXTURING

4.3.1   SELF-CLEANING PROPERTY OF NANOTEXTURED SURFACES

A nanotextured surface is a self-cleaning surface because nanotexturing increases the hydrophobicity of a surface. Dust particle removal is ~ 41% for a smooth, hydrophilic surface. It increases to 98% for a hydrophobic nanotextured surface. The reason for this difference in behavior of the two types of surfaces is the decrease of adhesive force between the particle and the hydrophobic nanotextured surface with respect to the hydrophilic untextured surface. Further, on a nanotextured surface, the geometry of water-particle-air line tension acting on the particle changes. As a consequence, the force causing particle detachment from the surface upsurges. The larger force thus becoming available readily dislodges the particle (Heckenthaler et al 2019).

An example of a natural nanotextured surface is the lotus leaf. The nanotexturing keeps its surface free from dust, sand, and pathogens. In a dirty and dust-polluted environment, particularly in a desert, the solar cells are continuously exposed to lots of dust particles suspended in the atmosphere. As the dust particles do not stick well to a nanotextured surface, therefore a solar cell with nanotexturing does not allow dust to cover its surface. Due to its dust-repelling behavior, absorption of light does not worsen with prolonged falling of dust on a solar cell having this kind of surface.

4.3.2   MOTH-EYE NANOSTRUCTURED SURFACES

Nature has specially designed the surface of the eye of a nocturnal moth. On this surface, there are hexagonal arrays of needle-shaped pillars or bumps of lateral size 200nm (Figure 4.4), which build up a shallow refractive index gradient upon moving from the medium of incident light to the interior of the eye (Shen et al 2021). These subwavelength nanostructures are known as moth-eye nanostructures and are characterized by several useful properties for solar cell applications, notably, high mechanical strength and environmental tolerance, along with broadband and omnidirectional antireflection behavior, which is very effective for reduction of Fresnel reflections. Over and above, their hydrophobicity makes them dust resistant because they are self-cleaned by the lotus effect (Weiblen et al 2016). These useful features of subwavelength nanostructures are reinforced by several other qualities, e.g., requirement of minimal surface preparation, single material use, and good adhesion on surface, making them far superior to traditional antireflection coatings. Single material usage implies that there is no requirement of any additional material, such as a coating. Texturing modifies the surface of the material itself. Good adhesion is achieved because the nanostructures are built in the same material; they are not pasted to the surface. So there are no chances of their peeling off from the surface.

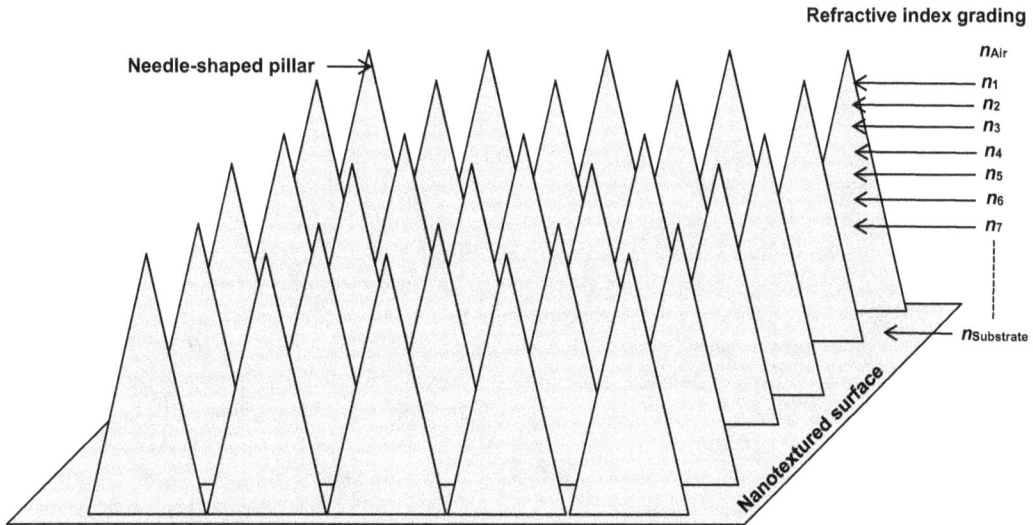

FIGURE 4.4 Moth-eye nanostructure consisting of array of vertical upright pointed nanopillars. As light travels downwards from the tip towards the bottom of the pillar, it encounters regions of slowly changing refractive indices n_{Air}, n_1, n_2, n_3, n_4, n_5, n_6, n_7, . . . , $n_{Substrate}$ accounting for the antireflection property of this nanostructure.

Similar to these natural antireflective nanostructures, biomimetic moth-eye nanostructures can be fabricated on the surfaces of solar cells by any one of the lithographic techniques, such as electron-beam lithography, nanoimprint lithography, or colloidal lithography. A roll-to-roll ultraviolet nanoimprint lithography (UV-NIL) technique uses a nickel mold to make a moth-eye nanostructure on a transparent polycarbonate substrate (Sun et al 2018).

4.3.3 Nanotexturing vs. Microtexturing

Nanotexturing is an essential requirement of thin-film solar cells. It fulfills their need because these cells have limited film thickness and so require intense antireflection property. Textures having submicron periodicity can provide the desired nanoscale graded refractive index interface with high antireflection property. Apart from providing antireflection property, it is accompanied by path length increase. So in a nanotextured surface, there are two mechanisms contributing to light trapping.

Prior to nanotexturing, microtexturing containing pyramids or inverted pyramid-shaped surface textures with sizes around 5–10 microns, as realized through alkali etching, has been extensively used in thick crystalline silicon solar cells. Undoubtedly, micron-size texturing is able to provide path length enhancement like the nanoscale texturing, although to a lesser extent. However, the antireflection property induced by graded refractive index is absent or much weaker in microtexturing than in nanotexturing. So nanotexturing provides both antireflection property and path length increase, while the effect of microtexturing is mainly the increase in path length. Absence of one mechanism in microtexturing makes this texturing less effective.

Keeping the features of the two types of texturing in view, microtexturing is a process used in solar cells with thick absorber layers, while thin-film solar cells employ nanotexturing because only a small thickness of absorber layer is available for photocarrier generation. Light has to be trapped within this small thickness.

4.4   NANOTEXTURED SILICON SOLAR CELL FABRICATION

4.4.1   INVERTED NANOPYRAMID CRYSTALLINE SILICON SOLAR CELL BY A MASKLESS TECHNIQUE (η = 7.12%)

Also called holographic lithography, *interference lithography* is a maskless technique in which two coherent laser beams from the same source intersect to form a standing wave pattern consisting of a periodic series of bright and dark interference fringes representing maxima and minima of intensity (Sivasubramaniam and Alkaisi 2014). This pattern of fringes is used to expose a photosensitive material to produce a regular array containing fine geometrical features. After processing, the repetitive pattern is visible in the photoresist. The technique avoids the use of photomasks and also eliminates the need of complicated exposure optics. This technique is useful for arrayed geometrical shapes but is not applicable to arbitrary geometries. Internal reflections are troublesome in this method, but these can be reduced by applying an antireflection coating (van Wolferen and Abelmann 2011).

Figure 4.5 shows the structure of the solar cell: Al front contact/N$^+$Si/P-Si/P$^+$Si/Al back contact. P-type crystalline silicon wafer of resistivity 0.1–5Ω-cm and thickness 350μm is taken (Figure 4.6). Silicon dioxide of thickness 100nm is thermally grown over the wafer. Then a three-layer stack is formed by successively depositing the layers. The layers in this stack are antireflection-coating AZ Barli II, evaporated silicon dioxide, and diluted AZMiR701 photoresist. Interference lithography is performed, in which the photoresist is exposed to interference fringes. Upon developing the photoresist, a series of nanosized holes is created in the photoresist at the intended places. Sequential etching of the layers is carried out until silicon is exposed at the locations of the holes. Silicon etching is done in KOH. After completion of silicon etching, the silicon dioxide masking layer is removed. A surface texture consisting of inverted pyramids of featured size 700nm at desired distances can be seen on the

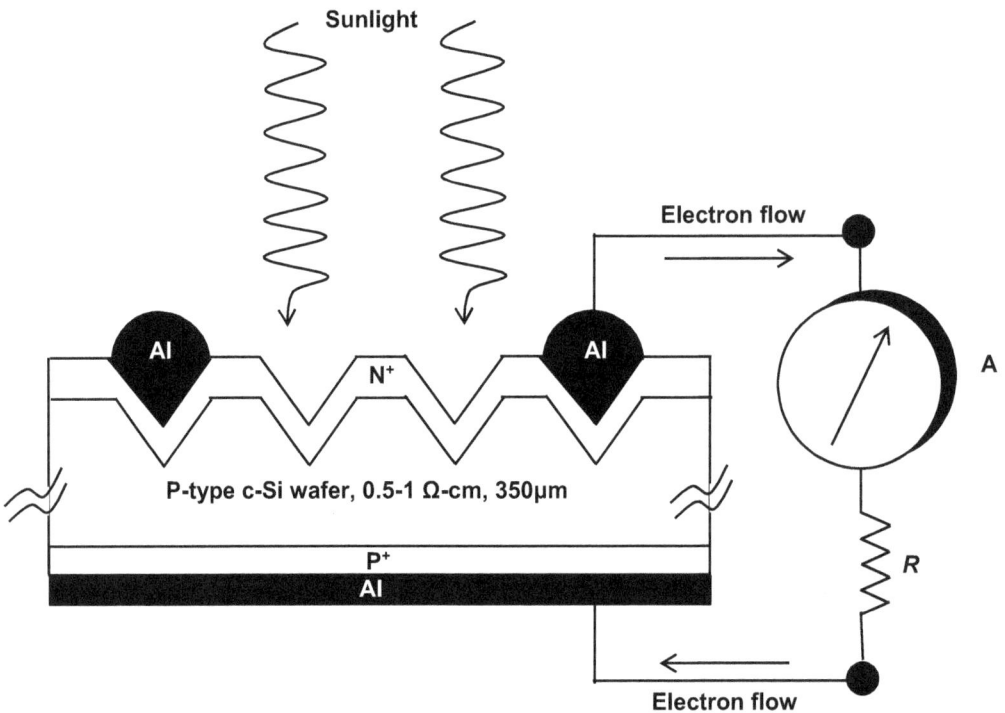

FIGURE 4.5   Inverted nanopyramid-textured solar cell fabricated by maskless interference lithography. The solar cell comprises N$^+$, P, and P$^+$ layers. Both the top and bottom contacts are made with aluminum. Sunlight falling on the front textured surface produces electric current, which is drawn across the aluminum electrodes.

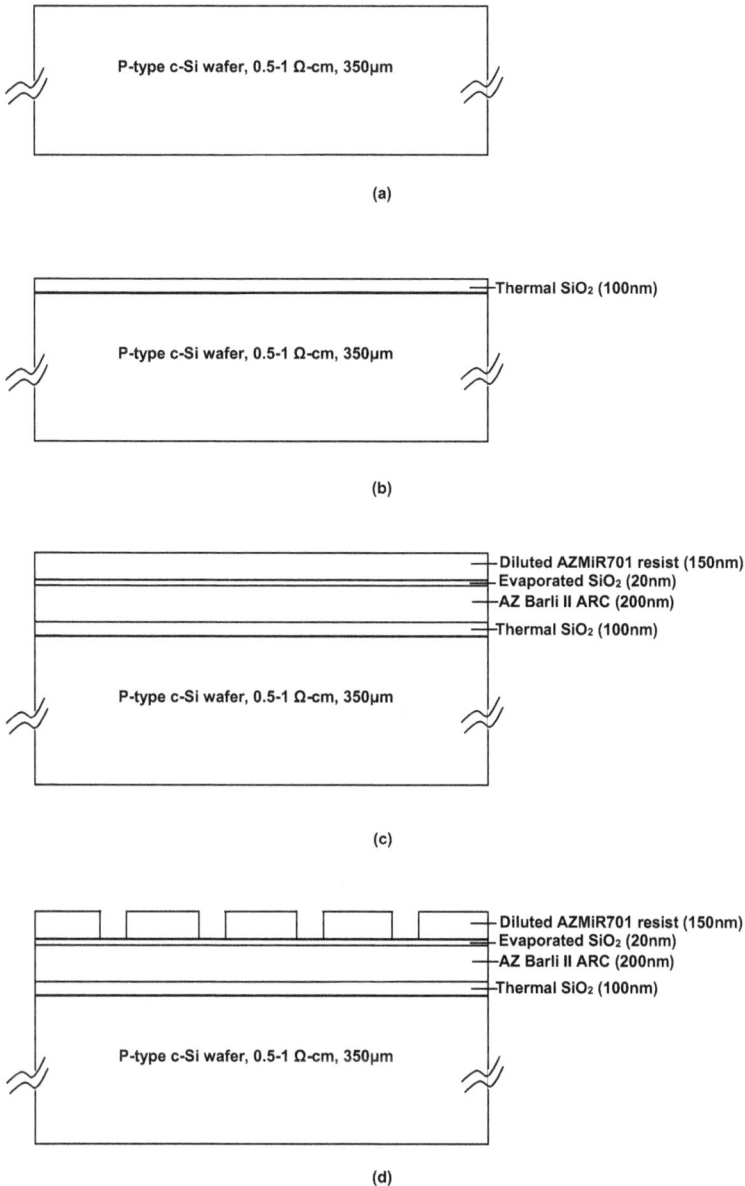

FIGURE 4.6 Maskless fabrication process of inverted nanopyramid-textured solar cell: (a) taking P-type crystalline silicon (c-Si) wafer of resistivity 0.5–1 Ω-cm, thickness 350μm, (b) thermal oxidation (thickness 100nm), (c) making trilayer stack consisting of AZ Barli II antireflection coating (thickness 200nm), thermally evaporated silicon dioxide (thickness 20nm) and diluted AZMiR701 resist (thickness150nm), (d) double exposure of resist with interference fringes using chosen underexpose dose to produce an array of holes in it; the technique used is laser interference lithography, a low-cost, large-area, maskless optical patterning technique in which two overlapping coherent laser beams produce high-resolution, periodic submicron features, (e) transference of the resist pattern to evaporated silicon dioxide interlayer by dry-etching in CHF₃/Ar plasma, (f) transference of the pattern to AZ Barli II ARC by etching in oxygen plasma, (g) transference of the pattern to thermally grown silicon dioxide by etching in CHF₃/Ar plasma to form a mask layer with array of holes, the patterned oxide mask, (h) wet-etching in KOH to form inverted nanopyramids, (i) removal of patterned oxide mask, (j) boron diffusion on backside, (k) phosphorus diffusion on front side using silicon pyrophosphate (SiP_2O_7) planar source, (l) aluminum (350nm) front contact grid, and (m) back Al metallization.

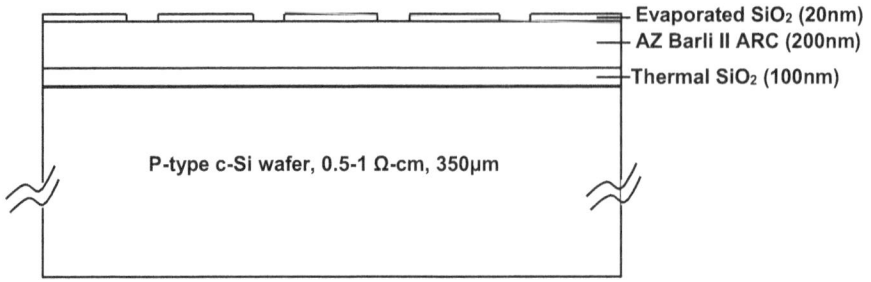

(e)

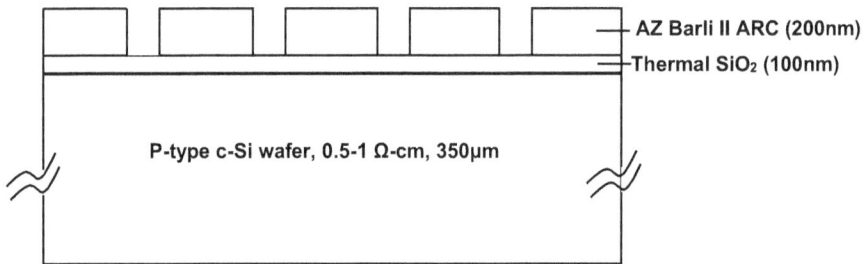

(f)

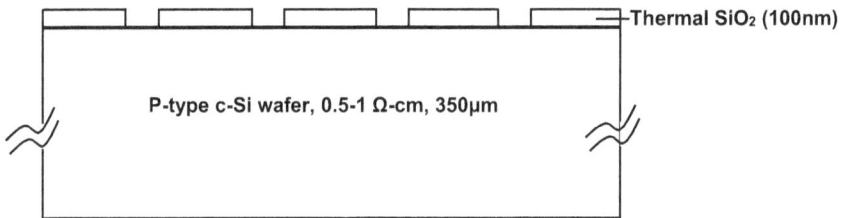

(g)

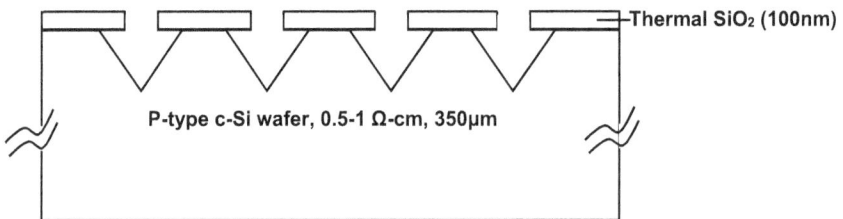

(h)

FIGURE 4.6   (Continued)

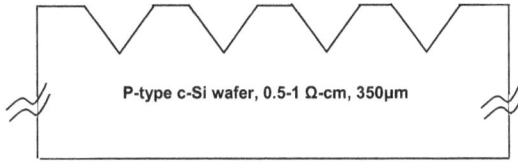

(i)

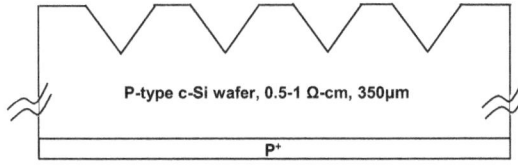

(j)

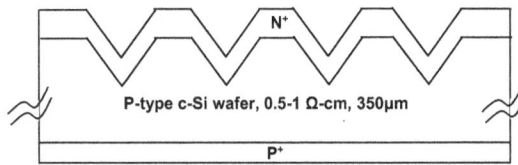

(k)

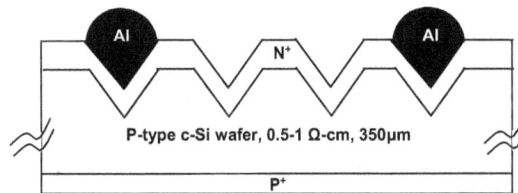

(l)

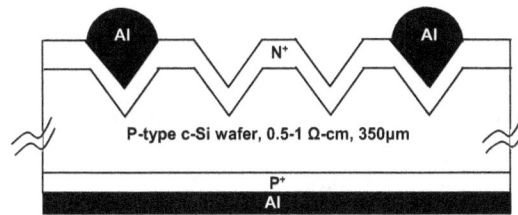

(m)

FIGURE 4.6 (Continued)

silicon wafer. The backside of the wafer is doped with boron to form a P/P$^+$ junction for producing the back surface field, while the front side is phosphorous-doped to form an N$^+$/P junction. An N$^+$/P/P$^+$ diode is formed. On the front side, aluminum grid contact is made. The grid contact pattern is defined using liftoff lithography. On the backside, a continuous aluminum film is deposited. DC sputtering technique is used for aluminum deposition. Dual ZnS/MgF$_2$ antireflection coating is applied.

For nanotextured cell with ARC, the short-circuit current density is 17.19mAcm^{-2}, open-circuit voltage is 0.54V, fill factor is 76.7%, and efficiency is 7.12%. For the bare nanotextured cell, the efficiency is 6.73%. For the planar (untextured) cell, the efficiency is 4.03% (Sivasubramaniam and Alkaisi 2014).

4.4.2 Ultrathin (Sub-10µm) Silicon Solar Cell with Silicon Nanocones and All Contacts on the Backside (η = 13.7%)

Monocrystalline silicon solar cells generally have thickness around 180 µm. This thickness is essential because light absorption coefficient for single-crystal silicon is low. But the volume of silicon used in solar cells adds to the cost. For cost reduction, the quantity of silicon used must be small. Thin or ultrathin solar cells must be made to decrease the quantity of silicon consumed. But thin crystalline silicon film will not be sufficient for absorbing light. So the solar cell will not be able to utilize light to its full capacity, resulting in low power conversion efficiency.

Therefore, if smaller quantity of silicon, as available in an ultrathin layer of 10µm, is to be used, it is imperative that the optical losses must be minimized. Proper nanotexturing is the answer to this hurdle.

An ultrathin all-back contact solar cell is fabricated with silicon nanocones on its front light-receiving surface to make a completely black surface (Jeong et al 2013). It is necessary to know why all the contacts are taken from the backside.

4.4.2.1 Carrier Recombination Problems Faced in a Front-Emitter Solar Cell

In the common, most widely used solar cell structure, the heavily doped emitter is located on the front side below the nanotextured surface. This emitter location is the cause of several problems and is therefore a bone of contention. Why? Because doping depth and concentration are dependent on surface morphology. So heavy-doping effects cannot be overlooked. Auger recombination becomes dominant in this regime. The higher the doping concentration, the shorter the minority-carrier lifetime, and so the greater the loss of free carriers by recombination. The second issue of concern is that the nanotexturing increases the surface area of this diffused region dramatically. The larger the surface area, the more the surface recombination, and therefore the more carriers are lost by recombination. Thus, both the Auger and surface recombination problems decrease the number of photogenerated carriers, resulting in lowering of efficiency of the solar cell. So how to solve these problems?

4.4.2.2 Solving the Recombination Problem

The remedy to this situation is relocation of the emitter junction away from the nanotextured front surface to the rear surface of the solar cell. Thus, the structure of the solar cell is revised to provide both the emitter and base contacts on the backside.

Figure 4.7 shows the solar cell structure which has all the Al contacts on the backside and has nanocones of height 400nm and diameter 450nm on its front surface to make it antireflective. The nanocone design criterion is to increase absorption of light with as low a surface area as possible to restrict surface recombination. As phosphorous diffusion is no longer being done in the nanostructured region, Auger recombination effects are also mitigated.

4.4.2.3 Fabrication of the Solar Cell

The solar cell is fabricated on an SOI wafer. The N-type device layer has a resistivity 1–5 Ω-cm and thickness 10µm (Figure 4.8(a)). A 300nm thick SiO$_2$ layer is grown by thermal oxidation (Figure 4.8(b)). Photolithography is done to open window for N$^+$ diffusion (Figure 4.8(c)). Phosphorous diffusion is done using liquid dopant source POCl$_3$ and the diffusion window is closed by re-oxidation (Figure 4.8(d)).

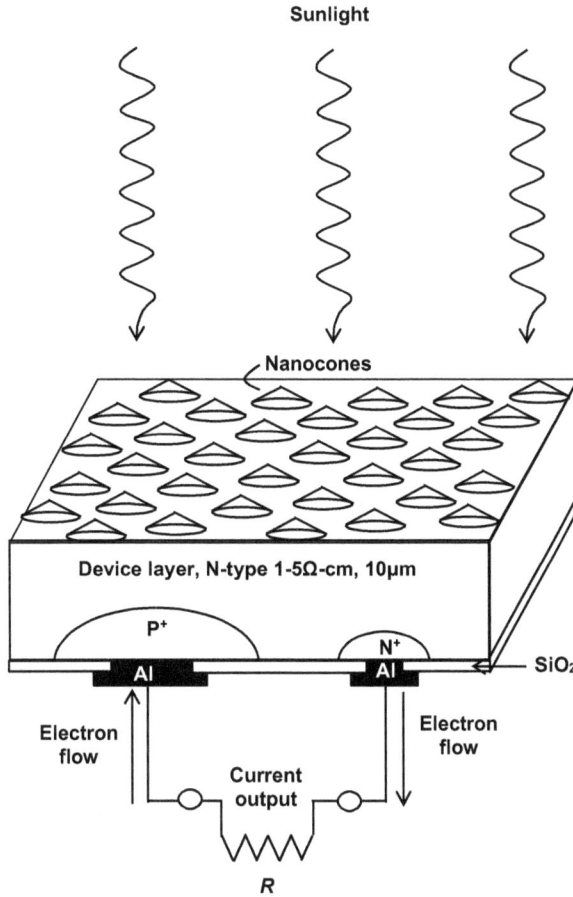

FIGURE 4.7 Ultrathin silicon nanocone solar cell with all-back contact. The diagram shows the P⁺- and N⁺-diffused regions formed in an N-type substrate. Aluminum is used as the contact metal for both regions. The nanocones can be seen on the front surface of the solar cell. The surface with nanocones is illuminated with sunlight. Current is extracted from the aluminum contacts on the back surface of the solar cell.

Again, photolithography is done to open window for P⁺ diffusion (Figure 4.8(e)). Boron diffusion is done using liquid dopant source BBr₃; as in case of phosphorous diffusion, the diffusion window is closed by reoxidation (Figure 4.8(f)). Before subsequent processing, the front side of the wafer is coated with a thick layer of photoresist for protection (Figure 4.8(g)).

Then the SOI wafer is flipped over and photolithography is done to the open region for etching on the backside (Figure 4.8(h)). Reactive ion etching is done to reach the buried oxide (Figure 4.8(i)). Photoresist is removed, and all exposed oxide portions are etched away by wet etchant (Figure 4.8(j)). Langmuir-Blodgett technique is applied to deposit an array of silica nanoparticles on the exposed backside of the wafer. 500nm-diameter silica nanoparticles are synthesized by hydrolyzing tetraethyl orthosilicate in (ethanol + ammonia) solution (Figure 4.8(k)). The diameter of silica nanoparticles is reduced to 200nm by reactive ion etching (Figure 4.8(l)). The layer of 200nm silica nanoparticles is used as a mask for reactive ion etching of silicon underneath. After silicon etching, the silica nanoparticles are removed by reactive ion etching (Figure 4.8(m)). Thermal oxide is grown on both sides of the wafer (Figure 4.8(n)). Oxide over silicon nanocones is etched to remove any surface defects, while on the opposite side of the wafer, windows are opened in the oxide for depositing contact metals for the emitter and base regions of the solar cell (Figure 4.8(o)). Electron-beam evaporation of aluminum is done (Figure 4.8(p)). Three-dimensional representation of the solar cell is drawn (Figure 4.8(q)).

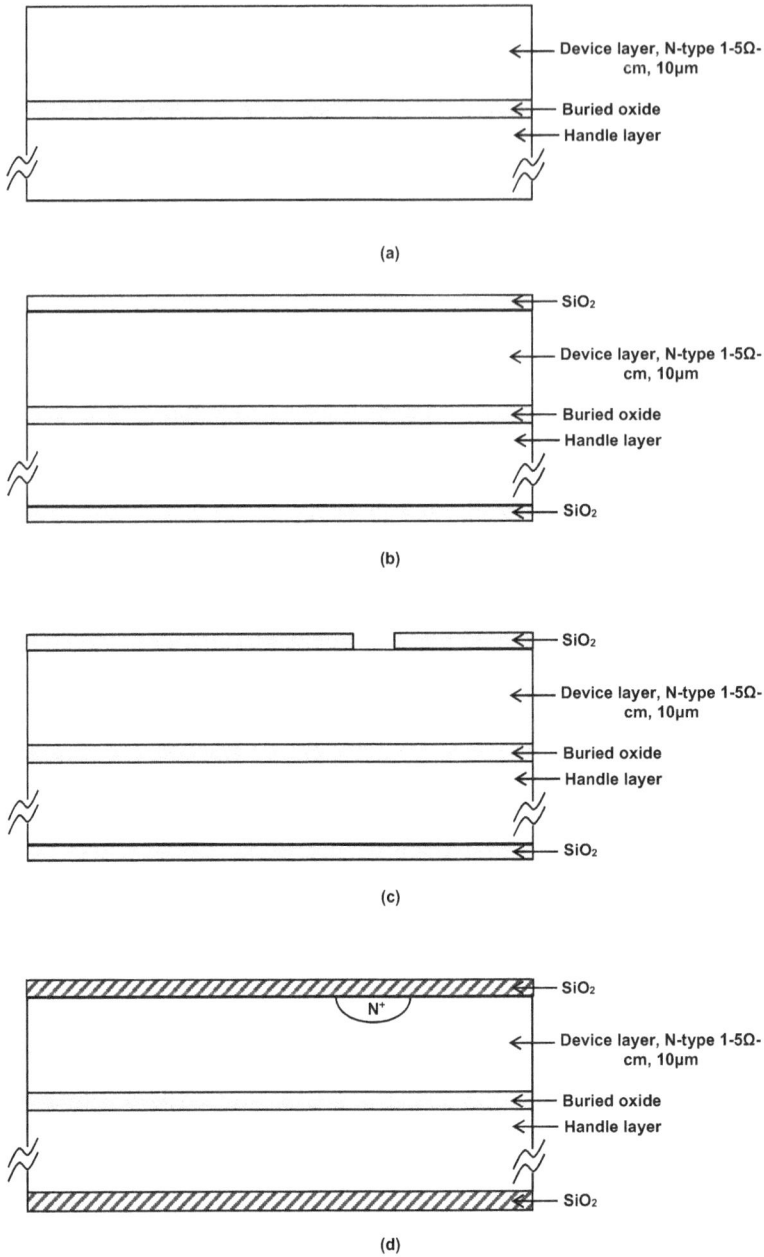

(a)

(b)

(c)

(d)

FIGURE 4.8 Ultrathin silicon nanocone solar cell with all-back contact: (a) taking an SOI wafer with N-type device layer of thickness 10μm and resistivity 1–5Ω-cm, (b) thermal oxidation (300nm), (c) photolithography for defining N+ region, (d) N+ doping using $POCl_3$, and oxidation, (e) photolithography for defining P+ region, (f) P+ doping with BBr_3, and oxidation, (g) protecting front side with thick photoresist (3μm), (h) flipping over the wafer and defining window pattern on the backside handle layer, (i) deep reactive ion etching to define area of the solar cell, (j) photoresist removal followed by wet-etching of the oxide on top surface as well as that remaining on the bottom surface, and the exposed buried oxide, (k) making a monolayer of SiO_2 nanoparticles by Langmuir-Blodgett method on silicon surface exposed after removing the buried oxide, (l) reactive ion etching of SiO_2 nanoparticles using CHF_3 and O_2 to reach 200nm diameter, (m) reactive ion etching of SiO_2 nanoparticles and underlying silicon in Cl_2 and HBr mixture to form nanocones, (n) thermal oxidation (80nm) of the whole surface to remove surface defects created in nanocone formation, (o) opening contact windows for N+/P+ contacts, (p) depositing and patterning aluminum (700nm), and (q) 3D view of a solar cell die.

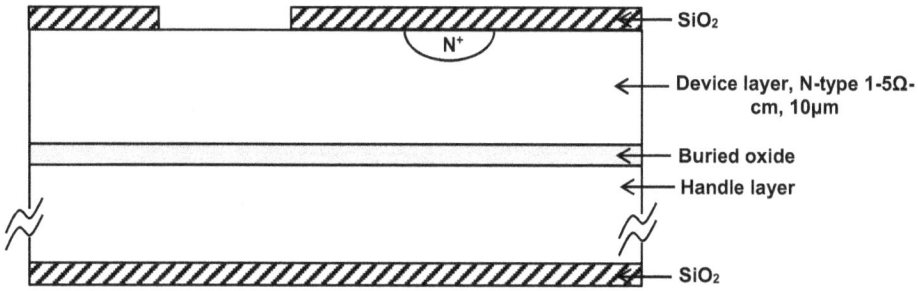

(e)

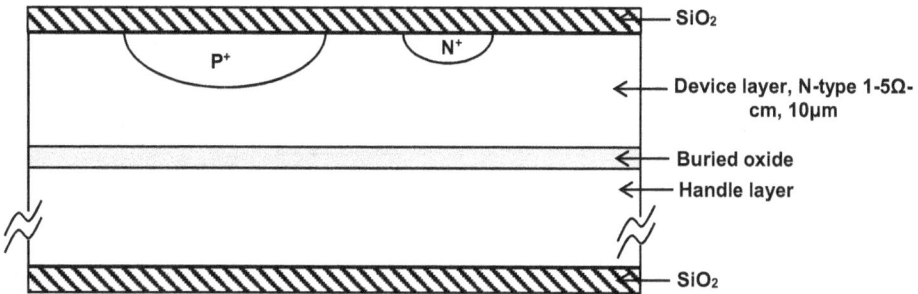

(f)

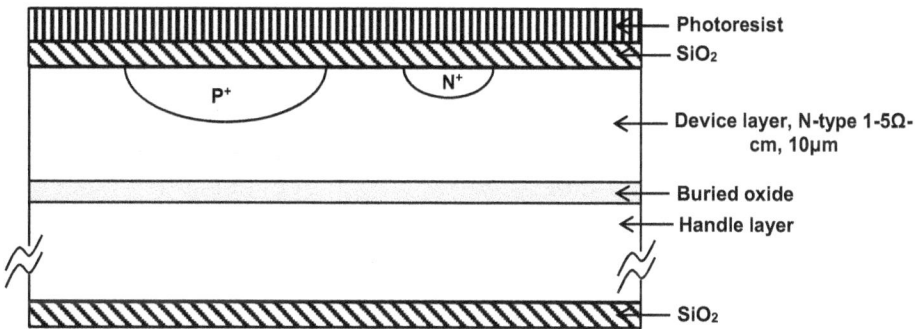

(g)

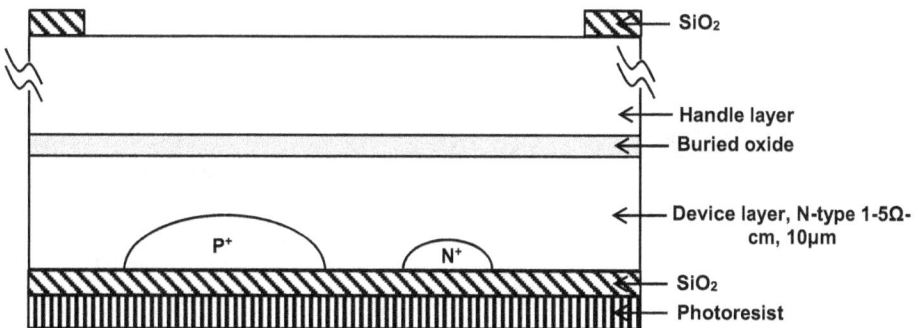

(h)

FIGURE 4.8   (Continued)

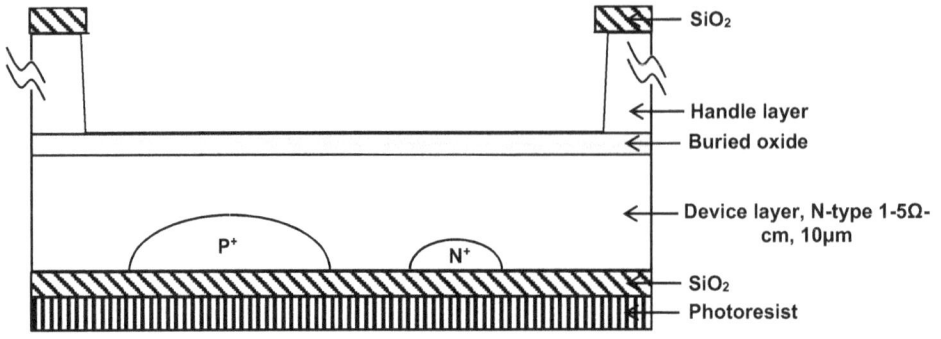

(i)

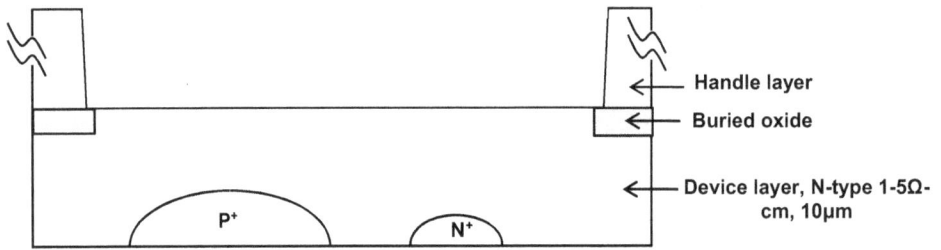

(j)

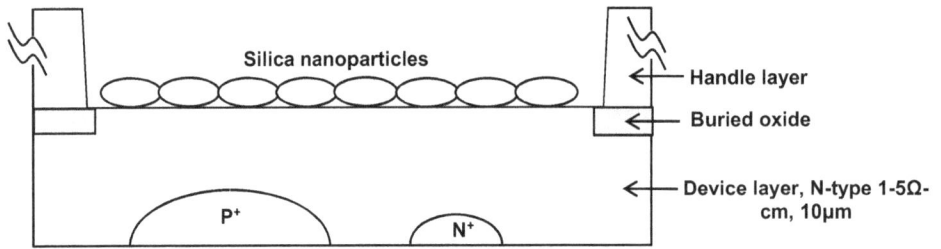

(k)

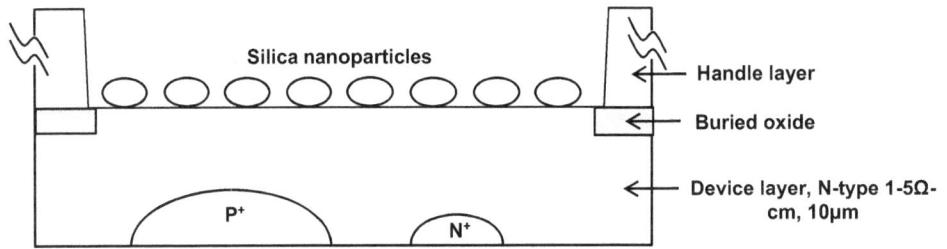

(l)

FIGURE 4.8   (Continued)

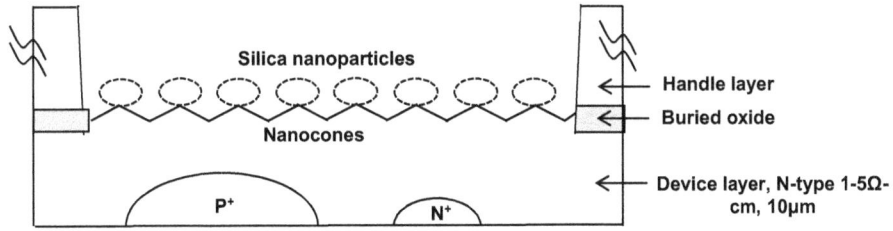

(m)

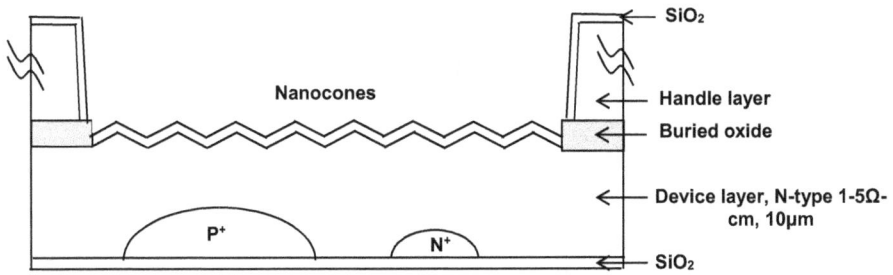

(n)

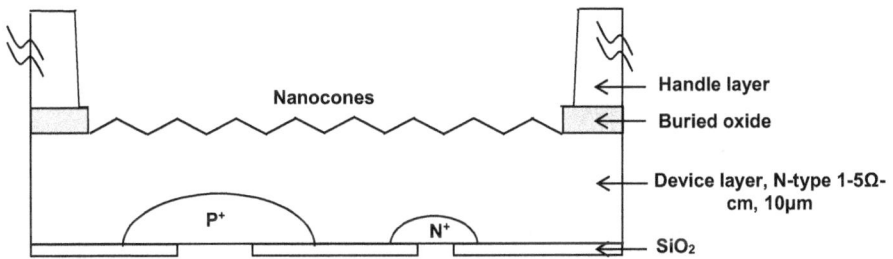

(o)

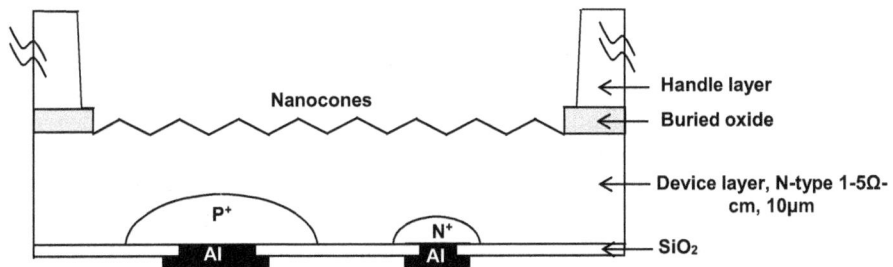

(p)

FIGURE 4.8 (Continued)

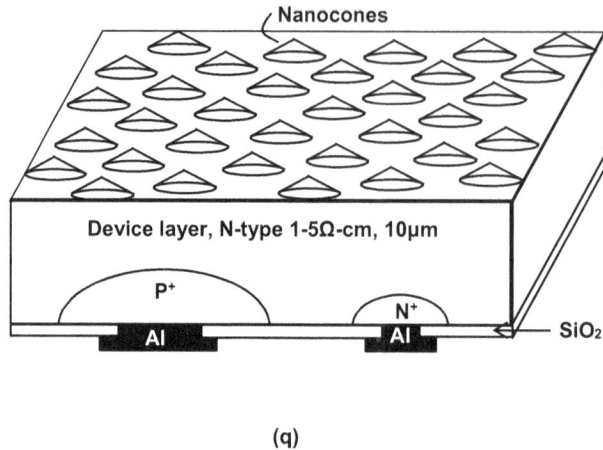

(q)

FIGURE 4.8 (Continued)

4.4.2.4   Planar Cell and Nanocone Cell Parameters

The short-circuit current density of the planar silicon solar cell having 10μm thickness and also an 80nm-thick silicon nitride reflection layer is 22.2mAcm^{-2}, its open-circuit voltage is 0.615V, fill factor is 80.2 %, and efficiency is 10.9%, whereas the same parameters for a silicon solar cell of identical thickness but with a nanocone structure are 29mAcm^{-2}, 0.623V, 76%, and 13.7%, respectively. The efficiency of nanostructured cell shows an increase = {(13.7–10.9) × 100}/10.9 = 25.69% with respect to that of the planar cell (Jeong et al 2013).

4.4.3   10-μm-Thick Periodic Nanostructured Crystalline Silicon Solar Cell (η = 15.7%)

Figure 4.9 shows the solar cell. The ultrathin solar cell has the structure (Branham et al 2015) (Ti/Pd/Ag) front contact/N-type/P-type/Al back contact.

A 2D photonic crystal of inverted nanopyramids on a 700nm pitch is formed on the upper surface of the solar cell. These nanopyramids have a 54.7° sloping angle. A PECVD silicon nitride ARC having a refractive index of 1.9 is also used. Hence, reflection from the upper surface of the solar cell is considerably inhibited. Aluminum metallic reflector on the lower surface of the solar cell provides additional suppression of optical loss. The structure takes advantage of physical optics effects to couple low-frequency photons into waveguide modes of the thin film. Probability of absorption of light is therefore increased. Overall, the solar cell combines together nanotexturing and ARC with a backside reflector to achieve performance at par with the widely used 180μm-thick crystalline silicon solar cells. The absorption is close to Lambertian over the optical spectrum from 500 to 1100nm. Lambertian distribution of light signifies an upper limit to enhancement of absorption. Completely diffused light is isotropic, whose flux through a surface shows a Lambertian distribution. An ideal, dull, and diffusely reflecting surface has a Lambertian reflectance.

This demonstration bears an eloquent testimony to the fact that by limiting optical losses, crystalline silicon solar cells competing in efficiency with thick wafer solar cells can be realized using an appreciable smaller quantity of silicon, thus cutting down the material cost effectively. We shall therefore describe its fabrication in detail.

The solar cell is fabricated on SOI wafer (Branham et al 2015). A nanopyramid pattern is formed on the surface of the wafer, as shown in Figure 4.10. An SOI wafer is taken with these specifications: active layer resistivity = 0.6–0.85Ω-cm, P-type doping, thickness = 10μm, and buried oxide thickness = 250nm. Silicon nitride film is formed by low-pressure chemical vapor deposition

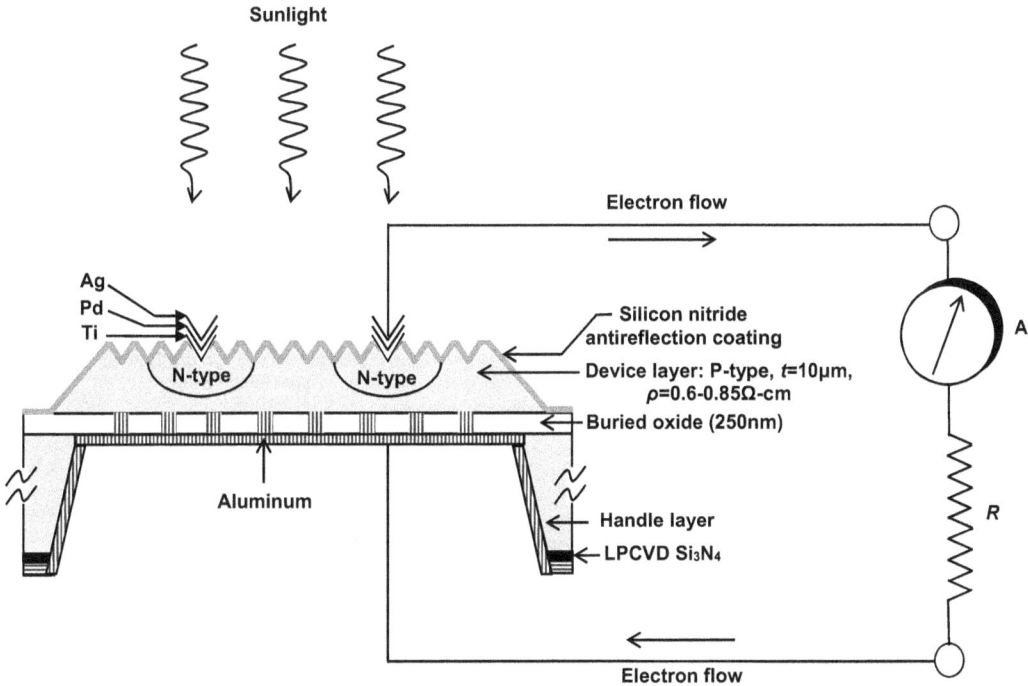

FIGURE 4.9 Ultrathin crystalline silicon solar cell (10μm thickness). The N- and P-diffused regions are shown. The front contact is made with Ti/Pd/Ag, while aluminum is used for the back contact. The front surface has an inverted nanopyramid texture with silicon nitride antireflection coating, while on the backside there is an Al metallic reflector. The solar cell is suspended over a cavity on silicon pillars. The textured surface of the solar cell is illuminated with sunlight. Current is measured across Ti/Pd/Ag and Al electrodes.

(LPCVD). A pattern of closely spaced holes is defined by photolithography. Dry-etching is done to create a silicon nitride hard mask. Anisotropic silicon etching is carried out through this hard mask. In this way, a nanopyramid pattern is produced on the surface of the wafer. After producing this pattern, the silicon nitride film is removed by chemical etching.

The diode area is defined by mesa etching (Figure 4.11). For definition of active area of the die, silicon nitride is redeposited by LPCVD and patterned. The nitride film surrounding the die area is removed followed by etching of silicon up to the buried oxide from these regions. The nitride is stripped from the upper surface of the wafer.

The window for ion implantation is opened. Ion implantation is performed, and antireflection coating is deposited. Vias are opened for the top contact, and the rear window is defined (Figure 4.12). These implementations are carried out as follows: Plasma-enhanced chemical vapor deposition (PECVD) technique is used to deposit silicon oxide on the front surface of the wafer. Photolithography is done to open windows, through which ion implantation is planned for N-type doping. Thin silicon dioxide is grown over the surface by dry thermal oxidation. Ion implantation is done. All the oxide is removed. Silicon nitride ARC is deposited. Vias are opened in the nitride ARC for contacting N-regions. Using double-side mask alignment photolithography system, rear window is defined and silicon nitride is etched to open the window.

The top contact is patterned, and the metal film is deposited (Figure 4.13). Liftoff photolithographic technique is applied for defining the metallization pattern on the front surface. This surface is coated with photoresist, and photolithography is done, during which the resist is removed from the via holes and retained elsewhere. Trilayer Ti/Pd/Ag metallization is done, and the resist is lifted off.

Etching is performed to release the membrane (Figure 4.14). ProTEK® B3 is a protective polymeric coating. It is applied by spin-coating on the front surface of the wafer, while the backside is

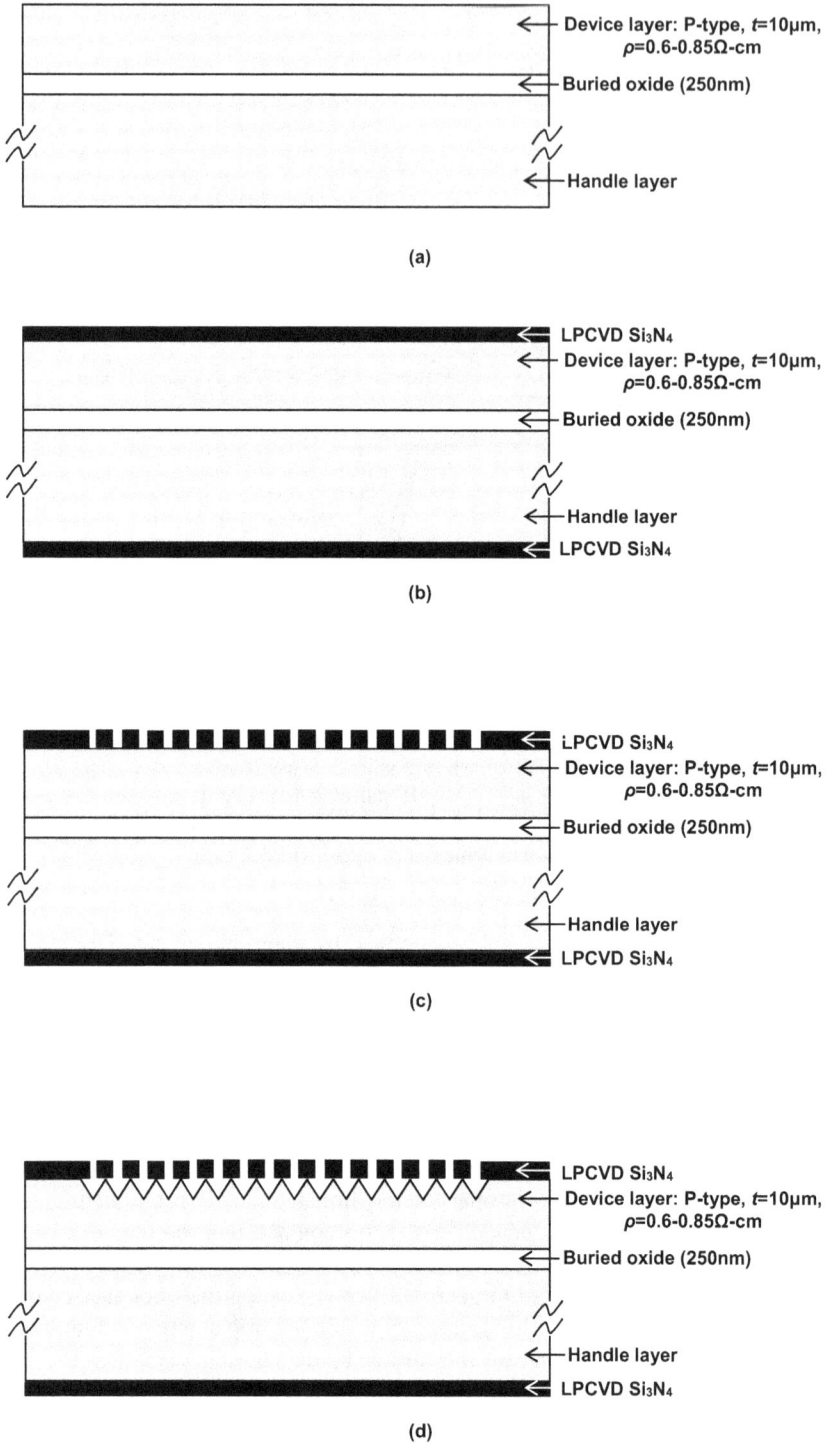

FIGURE 4.10 Fabrication of the nanopyramid pattern on the surface of silicon wafer: (a) starting silicon-on-insulator (SOI) wafer, (b) 50nm-thick silicon nitride film by LPCVD, (c) patterning 2D array of holes in silicon nitride by projection lithography and reactive ion etching of silicon nitride using SF_6/O_2, (d) anisotropic etching of silicon by potassium hydroxide through silicon nitride hard mask, and (e) dissolution of silicon nitride in hot phosphoric acid.

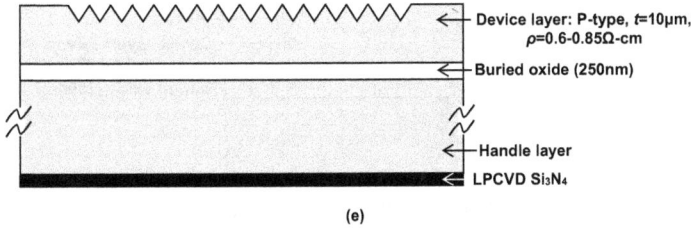

(e)

FIGURE 4.10 (Continued)

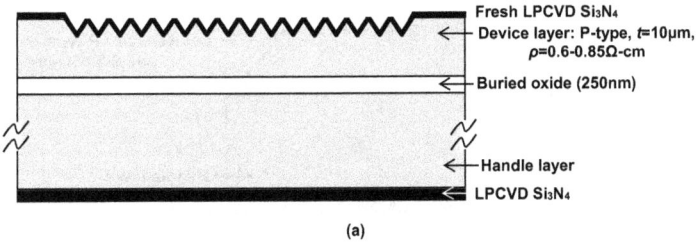

(a)

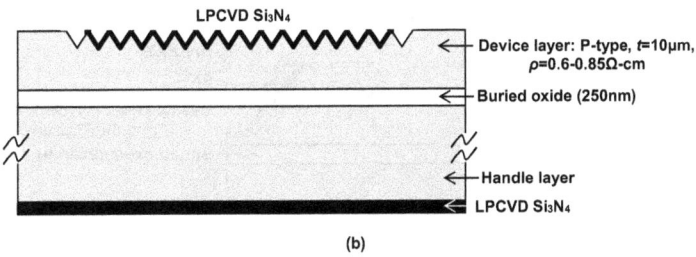

(b)

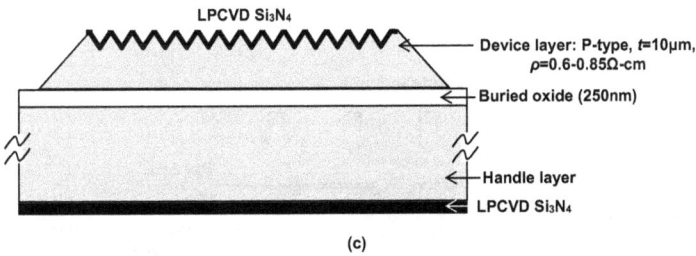

(c)

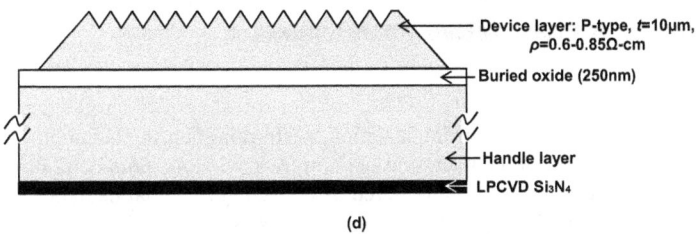

(d)

FIGURE 4.11 Definition of die area by mesa etching: (a) fresh LPCVD silicon nitride formation, (b) photolithography and reactive ion etching of silicon nitride by CF_4/O_2, (c) etching of silicon surrounding the die area, and (d) nitride stripping in hot H_3PO_4.

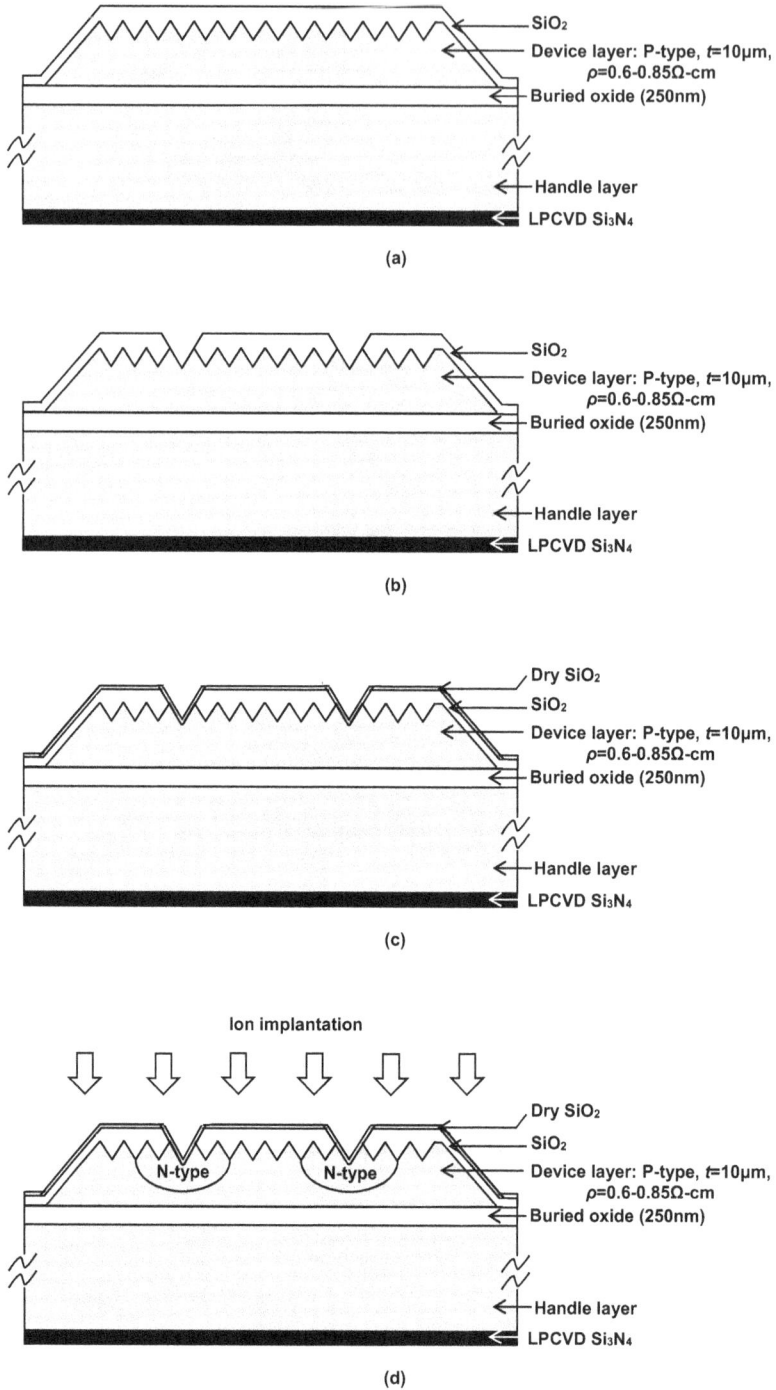

(a)

(b)

(c)

(d)

FIGURE 4.12 Opening windows for ion implantation, performing ion implantation, antireflection coating deposition, opening vias for top contact, and defining rear window: (a) silicon dioxide formation by PECVD, (b) patterning and etching oxide in buffered oxide etch, (c) thermal oxidation by dry method, (d) ion implantation in two steps: high-energy, low-dose phosphorous implantation and low-energy, high-dose arsenic implantation, (e) stripping implantation-damaged silicon dioxide and (f) formation of silicon nitride antireflection coating by PECVD using NH_3, SiH_4, and N_2, (g) photolithography for via opening and dry-etching of silicon nitride by magnetically enhanced RIE in CF_4/O_2, (h) definition of rear window by backside mask alignment, and dry-etching of silicon nitride in CF_4/O_2.

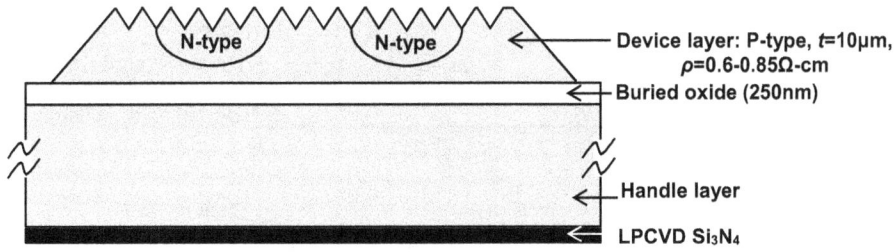

(e)

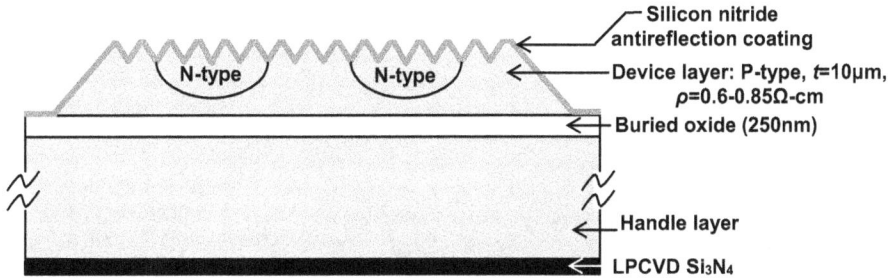

(f)

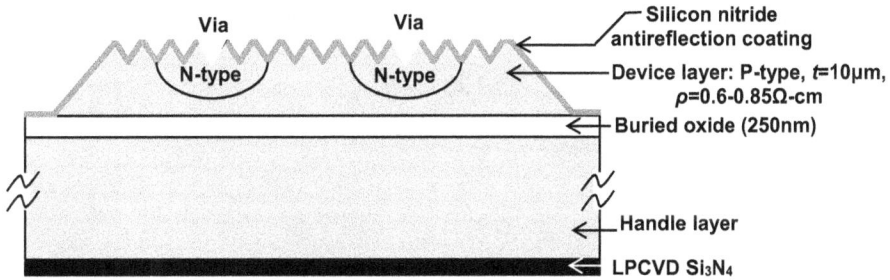

(g)

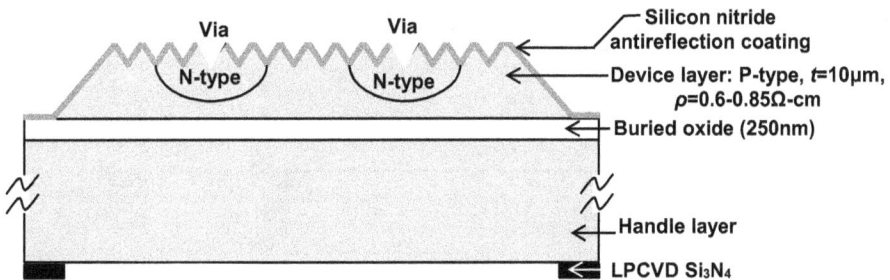

(h)

FIGURE 4.12   (Continued)

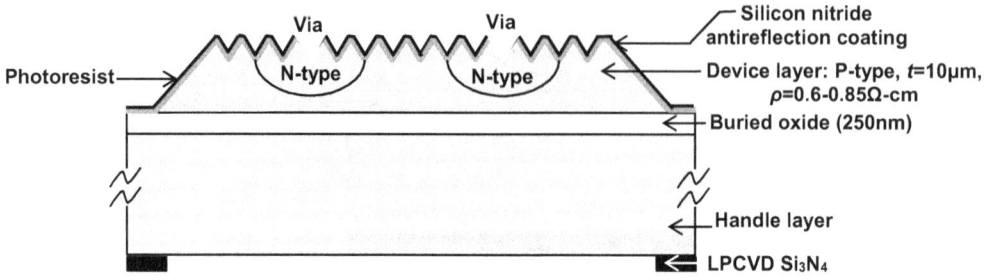

(a)

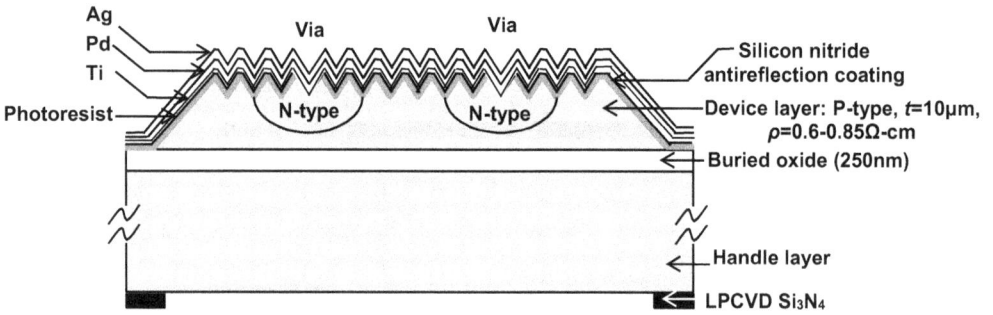

(b)

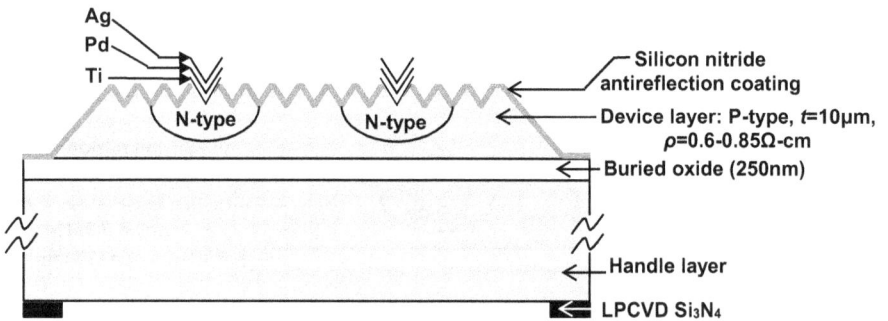

(c)

FIGURE 4.13 Patterning for top contact and metallization: (a) liftoff photolithography (photoresist on upper surface except at vias) and (b) electron-beam deposition of three-layer metal, Ti/Pd/Ag, and (c) lifting off the metal in acetone.

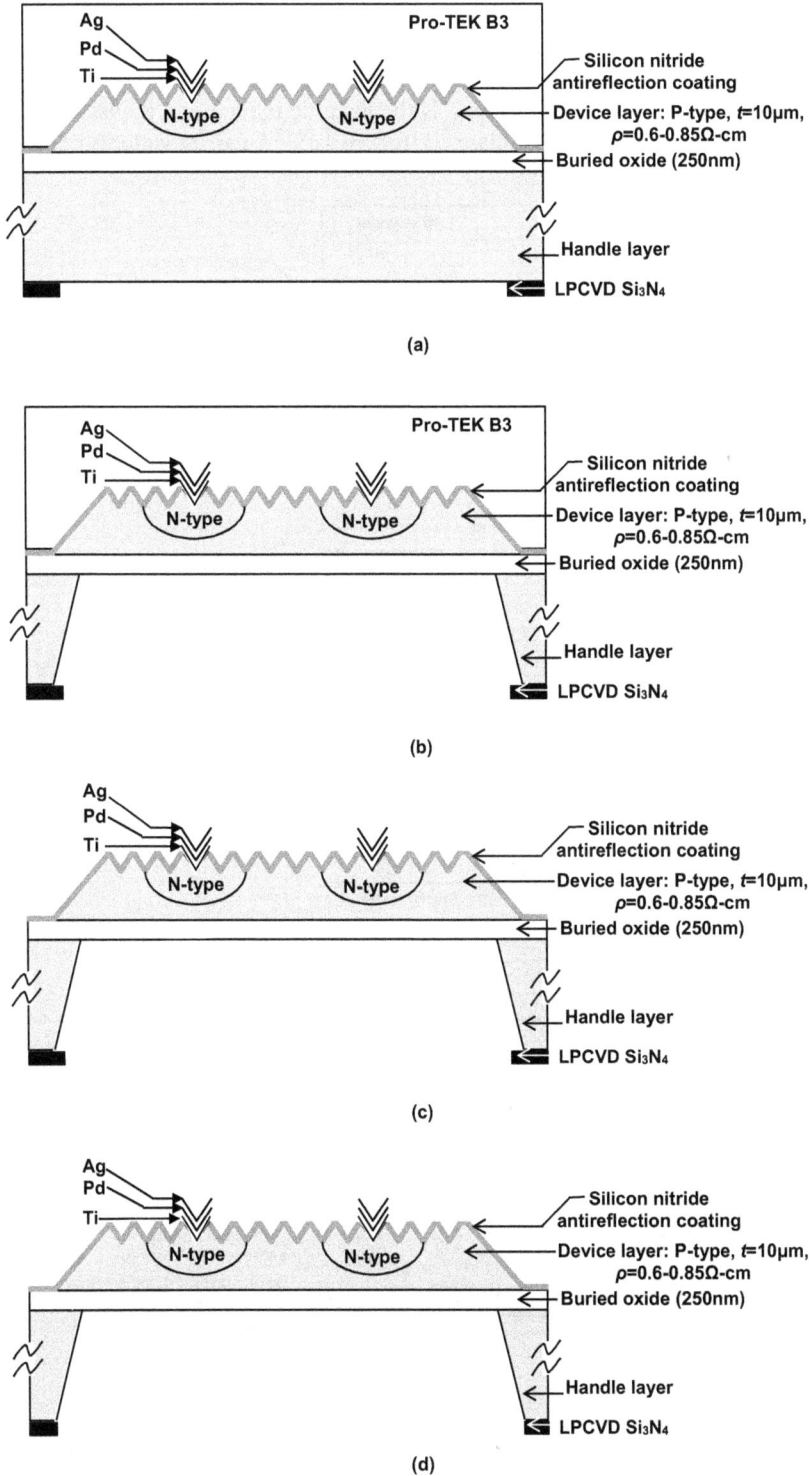

FIGURE 4.14 Etching and releasing the membrane: (a) protection of top side with Pro-TEK B3 and curing, (b) etching silicon in KOH, buried oxide acts as an etch stop, (c) removal of Pro-TEK B3 in acetone, and (d) scribing projected back side silicon nitride.

exposed to alkaline silicon etchant. Silicon is removed up to the buried oxide layer, where the etching stops. Then the protective coating is removed from the front side of the wafer. A small remnant projected portion of LPCVD silicon nitride leftover is scribed away.

The backside contact is still not made, nor is the metallic reflector on the backside. So the bottom contact must be formed to test the solar cell (Figure 4.15). A carrier wafer is fixed on the upper

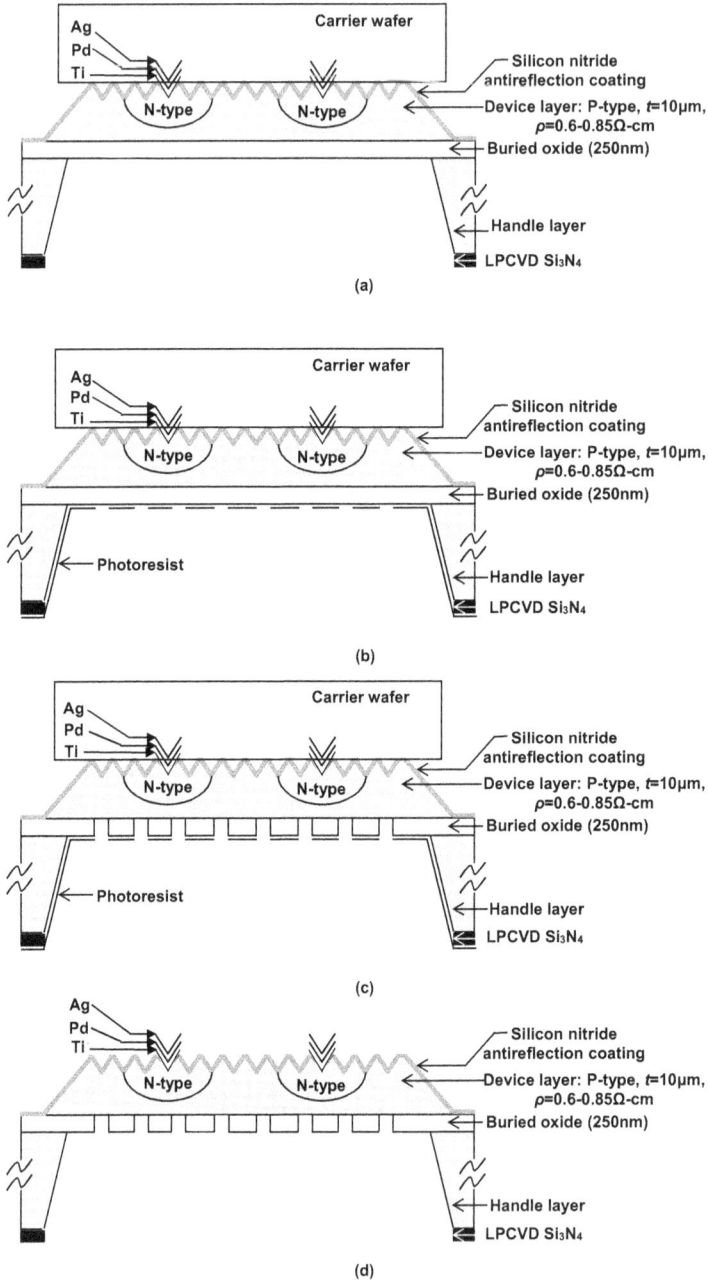

FIGURE 4.15 Formation of bottom contact to complete the fabrication of crystalline silicon solar cell (10μm thickness) with inverted nanopyramid texture on the surface and metallic reflector on the backside: (a) attaching carrier wafer on the upper surface, (b) dispensing photoresist into the wells of backside windows of dies, spinning and patterning holes with contact aligner, (c) etching silicon dioxide in buffered oxide etch, (d) removing photoresist and top carrier wafer, and (e) aluminum sputtering and annealing.

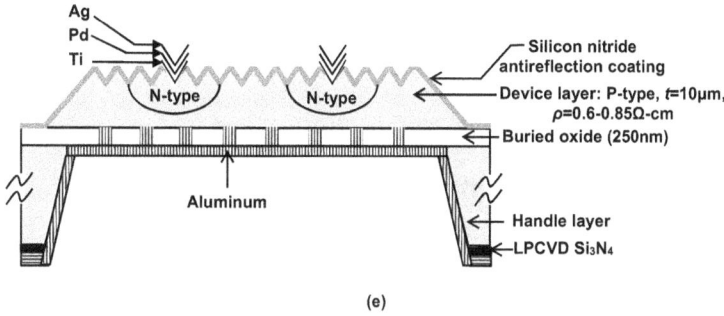

(e)

FIGURE 4.15 (Continued)

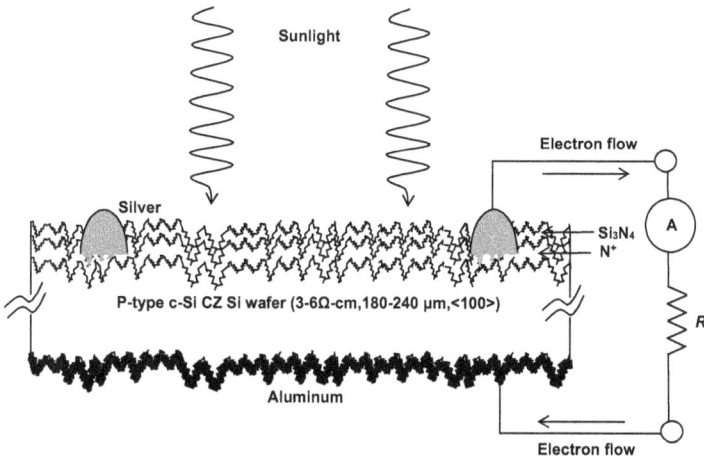

FIGURE 4.16 Structure of the micro/nanotexturized crystalline silicon solar cell showing the P-base and N^+ emitter with SiN_x passivation/antireflection layer. The solar cell has a silver pattern on the front surface and aluminum on the back surface. Sunlight falls on the antireflection layer on the front surface, while the electric current is taken out between the Ag and Al electrodes.

surface of the wafer. On the rear surface lying between the two slanting silicon supports, photoresist is coated and photolithography is performed to open holes for metal deposition. Aluminum is sputtered on the backside and annealed. The solar cell is ready for testing with the top contact through Ti/Pd/Ag and bottom aluminum contact. The thickness of the solar cell is the thickness of the active layer in the SOI wafer used for fabrication. It may be recalled as 10μm. An inverted nanopyramid texture is sculpted on the front side of the solar cell. It has a metallic reflector on the backside.

The photovoltaic parameters of the solar cell are short-circuit current density $33.9 mAcm^{-2}$, open-circuit voltage 0.589V, fill factor 78.5%, and efficiency 15.7% (Branham et al 2015).

4.4.4 TWO-SCALE (MICRO/NANO) SURFACE TEXTURED CRYSTALLINE SILICON SOLAR CELL ($\eta = 17.5\%$)

Micro-/nanotexture provides a better response towards the high-frequency blue segment of the spectrum than the conventional upright pyramid texture. Moreover, this surface is superhydrophobic in character, showing excellent self-cleaning property (Dimitrov and Du 2013).

Figure 4.16 shows the micro-/nanosurface textured solar cell. The solar cell fabrication follows the standard process for silicon solar cell with the inclusion of an additional nanotexturing step (Figure 4.17). First, the cleaned P-type wafer is subjected to microtexturing in an alkaline

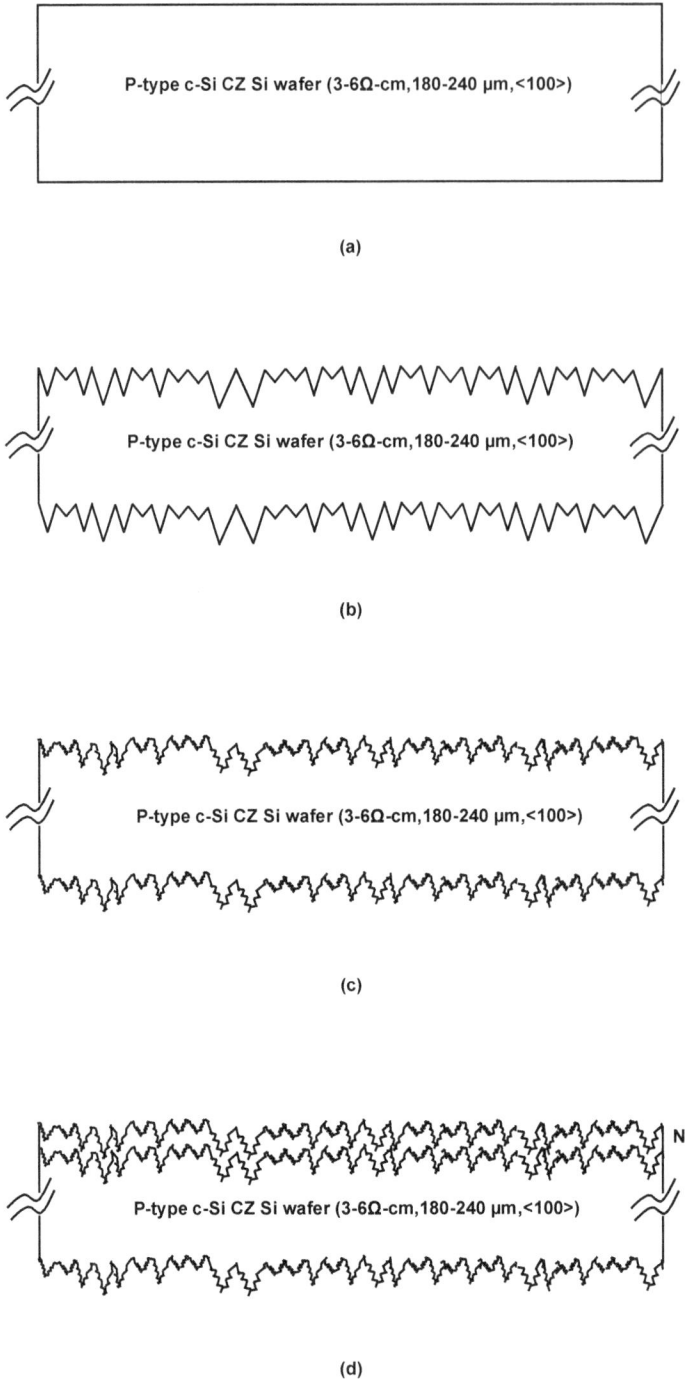

(a)

(b)

(c)

(d)

FIGURE 4.17   Fabrication of micro-/nanotexturized crystalline silicon solar cell: (a) c-Si CZ wafer (resistivity 3–6Ω-cm, thickness 180–240 μm, polarity P-type, orientation < 100 >), (b) random micron-sized pyramid texturization in KOH-IPA solution, (c) cleaning and texturization into nanopyramids through a two-step process in which the first step is electroless treatment in acidic ($Na_2S_2O_8$ + $AgNO_3$) solution in water and the second step is etching in (HF + H_2O_2) solution in water, (d) phosphorous diffusion in $POCl_3$, (e) SiN_x passivation/antireflection layer by PECVD after phosphosilicate glass removal, and (f) screen-printing and cofiring of silver pattern on the front surface, and aluminum on the back surface.

(e)

(f)

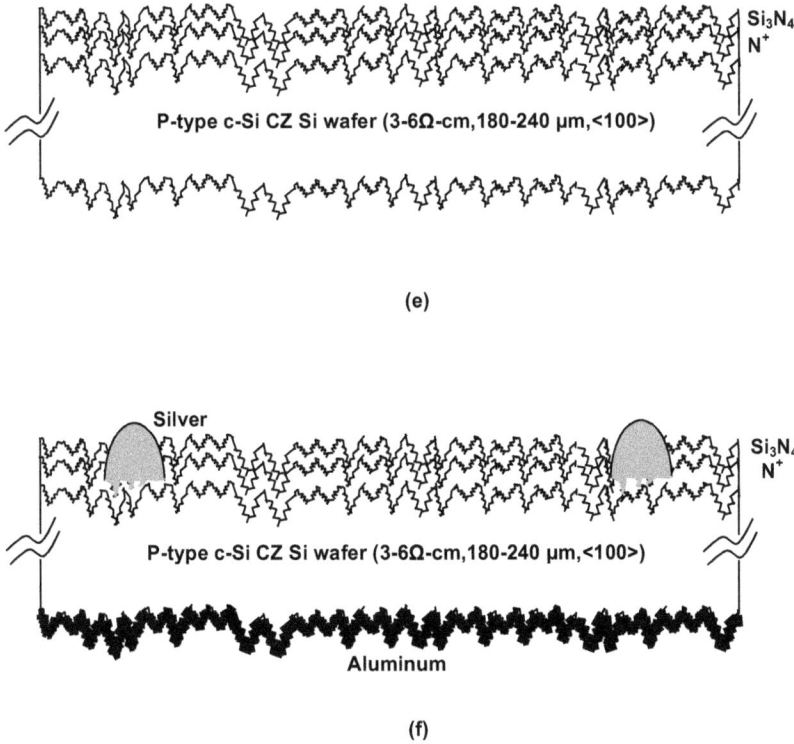

FIGURE 4.17 (Continued)

KOH-IPA solution. Then, it undergoes nanotexturing by first treating in aqueous $Na_2S_2O_8 + AgNO_3$ solution and then etching in aqueous $HF + H_2O_2$ solution. Phosphorous diffusion is done using liquid $POCl_3$ source to make a P-N junction diode. The front surface is passivated with PECVD silicon nitride, which also acts as an antireflection coating. Silver is screen-printed to make the front grid contact, while aluminum is screen-printed on the back surface. Co-firing is done on a belt furnace.

The short-circuit current density, open-circuit voltage, fill factor, and efficiency of the micro-/nanotextured solar cell are $35.5 mAcm^{-2}$, 623mV, 79.3%, and 17.5%, respectively, whereas the same parameters for a reference solar cell textured with conventional random upright pyramids are $33.9 mAcm^{-2}$, 615mV, 79.9%, and 16.7% (Dimitrov and Du 2013).

4.5 NANOTEXTURED SOLUTION-PROCESSED PEROVSKITE SOLAR CELL (η = 19.7%)

Light management using shallow nanostructures is done in a solution-processed perovskite solar cell (Tockhorn et al 2020). Different shallow nanostructures were investigated by these authors viz hexagonal negative sinusoidal or negative directed cosine lattice (Cos-), hexagonal positive sinusoidal, or positive directed cosine lattice (Cos+), with feature heights 220–380nm, inverted pyramidal (height 600nm), and pillar-shaped (height 350nm) structures.

The structure of the solar cell is NaF/glass substrate/UV-curable resist/ITO/MeO-2PACz HTL/mixed halide perovskite/C_{60} ETL/BCP hole-blocking layer/Cu electrode; ITO and Cu films act as contacts (Figure 4.18(a)). For comparison, the solar cell fabricated on a planar substrate is shown in Figure 4.18(b).

To make a perovskite solar cell by solution processing, a mold is produced in PDMS (Figure 4.19). The process starts with a master nanostructure. This nanostructure is made by interference

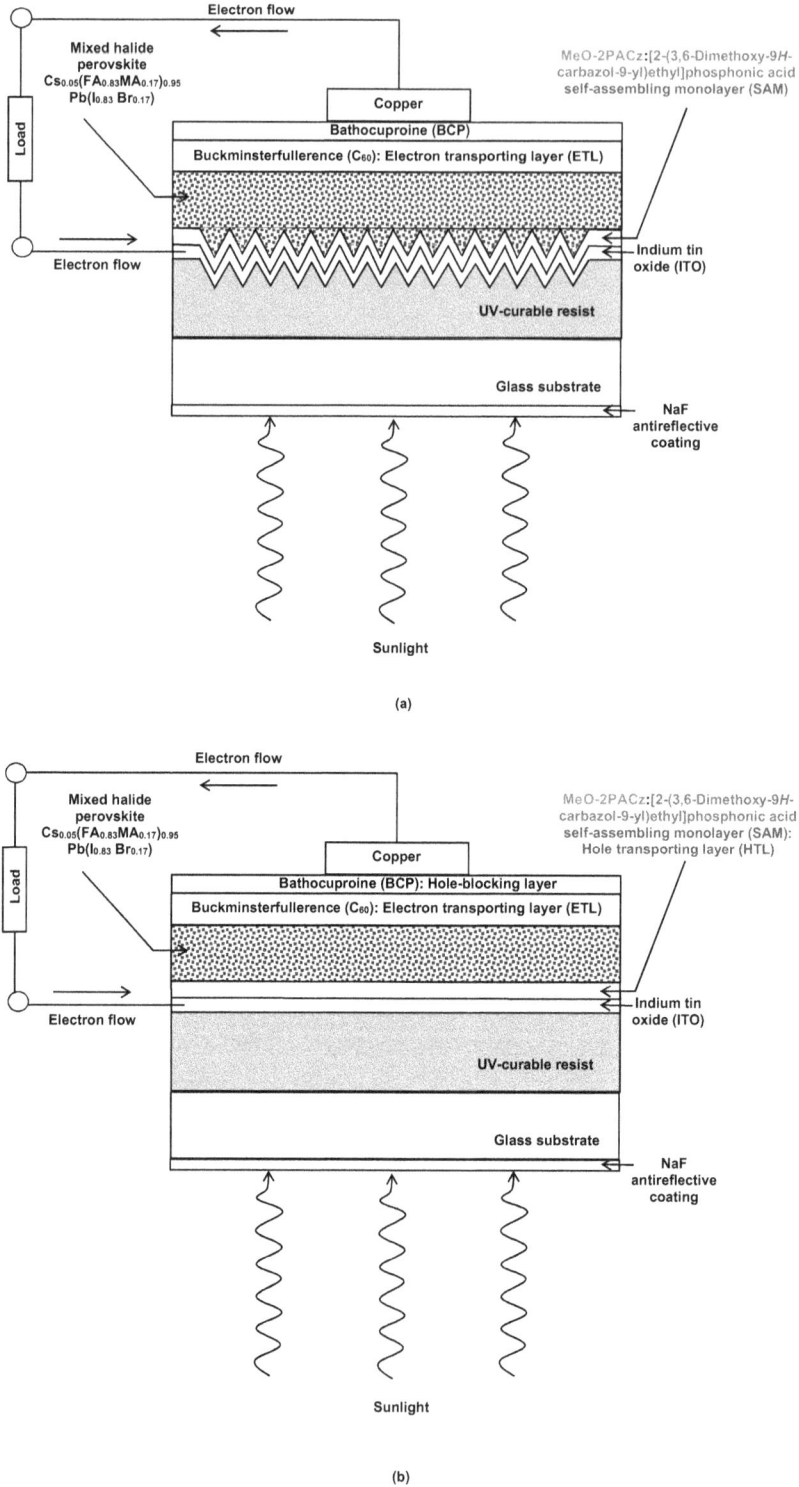

FIGURE 4.18 Two types of solar cells: (a) the cell fabricated on nanotextured substrate and (b) the cell made on planar structure. Both cells consist of an identical sequence of layers, which is bottom glass substrate, the resist layer, ITO, SAM layer, perovskite, C$_{60}$, BCP, and top copper electrode. The solar cells are illuminated from the glass substrate side. Electrical output is obtained through Cu and ITO electrodes.

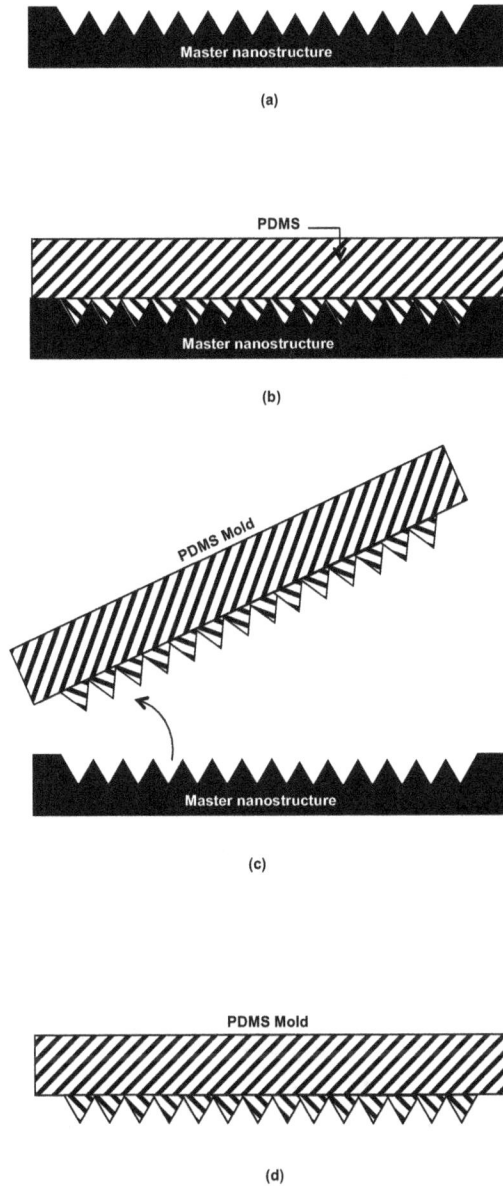

FIGURE 4.19 Production of PDMS mold: (a) the master nanostructure formed by interference lithography, (b) pouring the two-component PDMS into the master nanostructure to shape flowing PDMS according to projections in the master, followed by curing and hardening of PDMS, (c) peeling off the cured PDMS from the master nanostructure to create the PDMS mold having the inverse pattern of master, and (d) the PDMS mold showing the uneven surface generated.

lithography. It is used to make a mold with PDMS. For making this mold, PDMS is poured into the nanostructure, hardened, and peeled away. It is clear that the pattern transferred to PDMS will be opposite of that in the master nanostructure.

The inverse undulatory pattern to PDMS mold is transferred to the UV-curable resist film coated on a glass substrate. Nanolithographic imprinting is done for this transference (Figure 4.20). The process proceeds as: The glass substrate over which the solar cell is to be fabricated is coated with

(a)

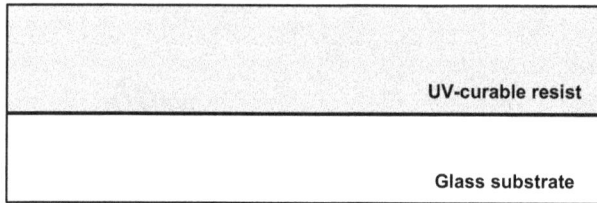

(b)

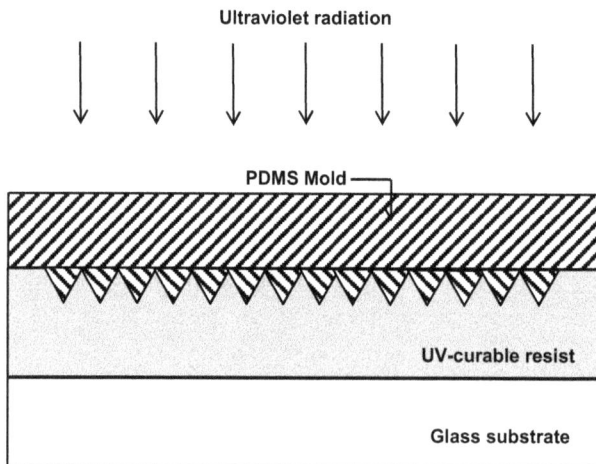

(c)

FIGURE 4.20   Transference of the inverse undulatory pattern to PDMS mold in the UV-curable resist film coated on the glass substrate by nanolithographic imprinting: (a) the glass substrate, (b) the glass substrate coated with resist layer, (c) pressing the PDMS mold with engraved pattern against the resist layer to shape the resist, and exposing the resist to ultraviolet rays, (d) lifting and removing the PDMS mold, followed by baking (annealing is done to improve transparency of the resist), and (e) the glass substrate with resist coating shaped by the undulations in the PDMS mold. Note that the inverse pattern here is the same as the pattern in the master nanostructure because the pattern in PDMS mold was inverse to the pattern in the master nanostructure.

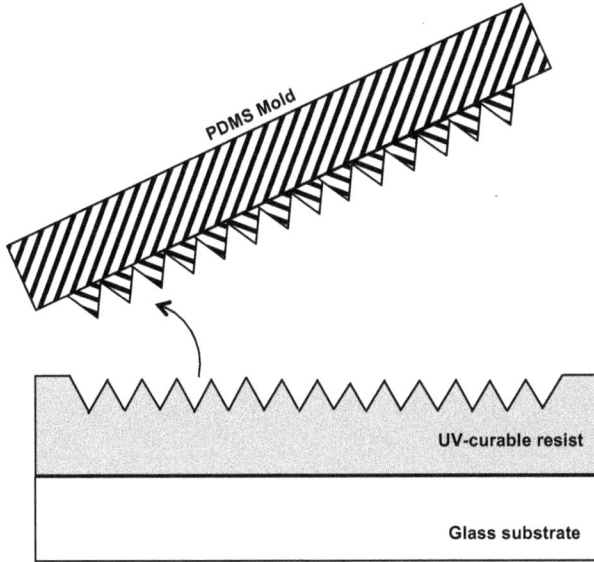

(d)

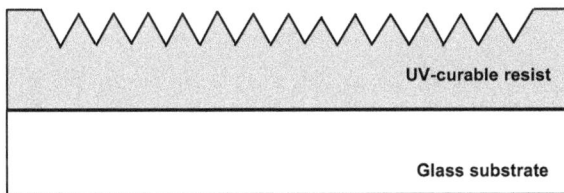

(e)

FIGURE 4.20 (Continued)

a UV-curable photoresist. The PDMS mold is slowly brought down and pressed against the photoresist when the PDMS mold sticks to the glass substrate with the UV-curable photoresist between them. The assembly of (PDMS mold + UV-curable photoresist + glass substrate) is exposed to ultraviolet radiation to cure the photoresist. The PDMS mold is slowly withdrawn, and the glass substrate with the photoresist is left behind. The photoresist has been shaped by the PDMS mold. It has acquired the inverse pattern to that in the PDMS mold. To increase the transparency of the photoresist, it is baked.

The engraved resist-coated glass substrate is used for fabrication of the solar cell (Figure 4.21). At this stage, we have a glass substrate coated with photoresist that has a nanostructure

(a)

(b)

(c)

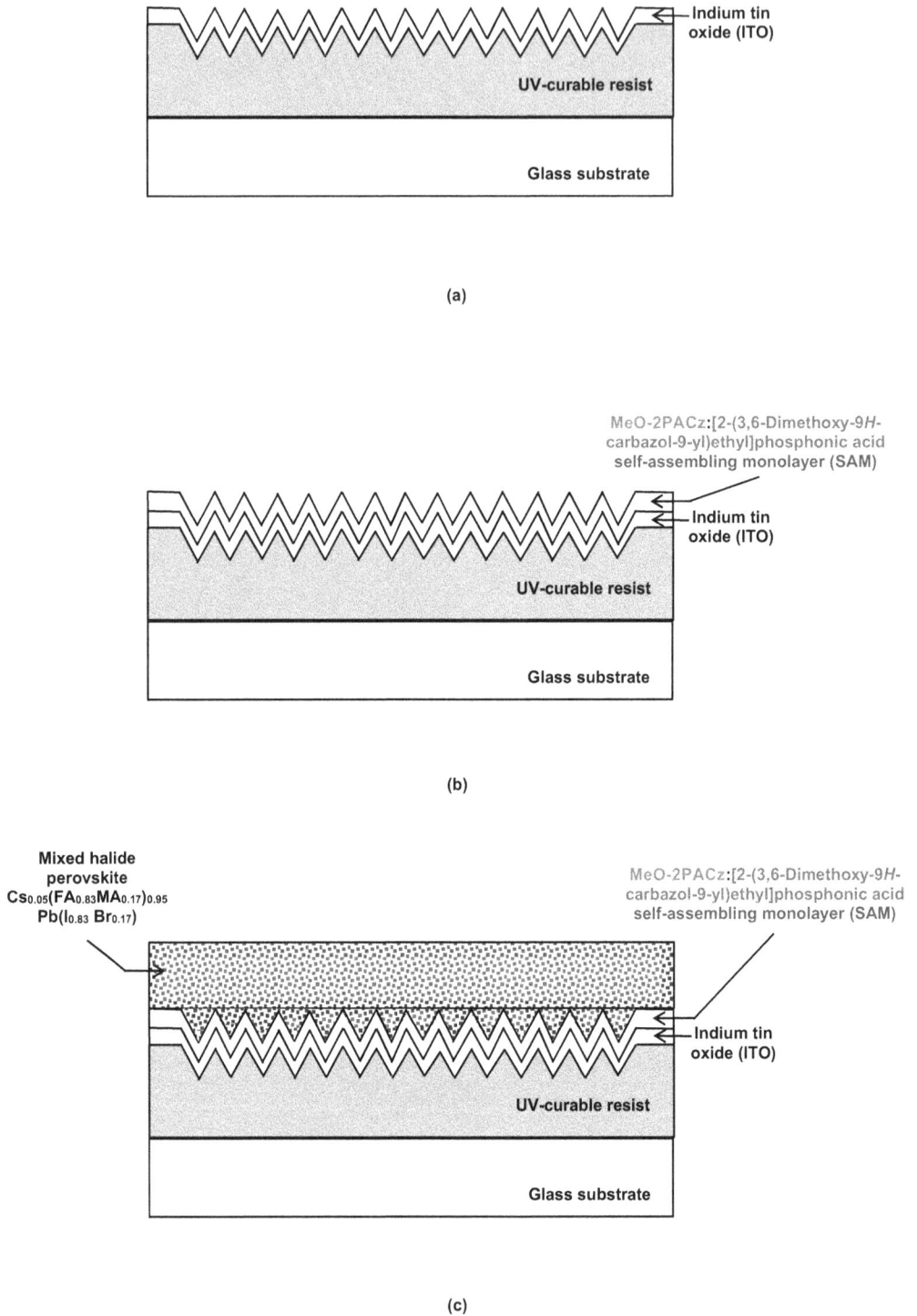

FIGURE 4.21 Solar cell fabrication on the engraved resist-coated glass substrate: (a) deposition of indium tin oxide film on the resist layer by RF sputtering, (b) MeO-2PACz self-assembled layer (SAM) formation on ITO, (c) mixed halide perovskite layer deposition on SAM, (d) evaporation of C_{60} film over perovskite, (e) evaporation of BCP film on C_{60} layer, (f) evaporation of copper over BCP film to form Cu electrode, and (g) deposition of antireflective coating of sodium fluoride on the glass substrate.

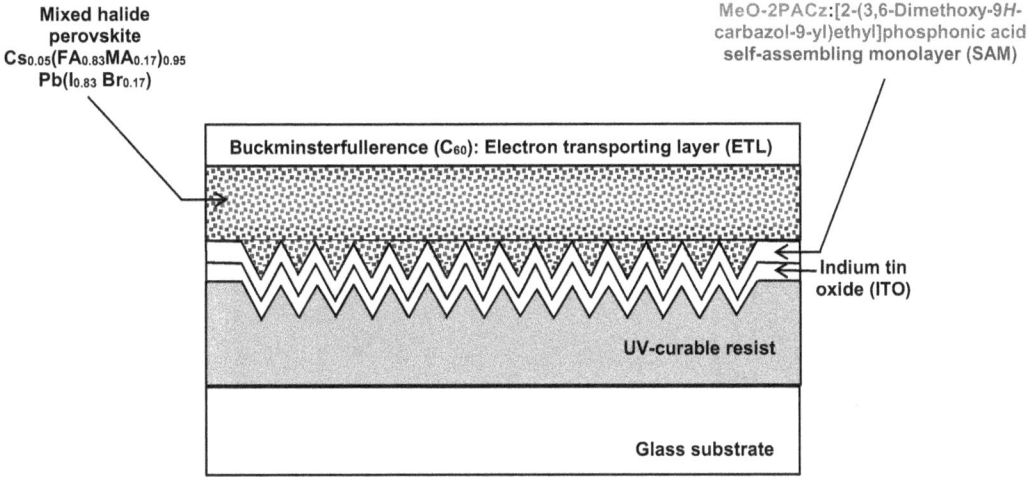

(d)

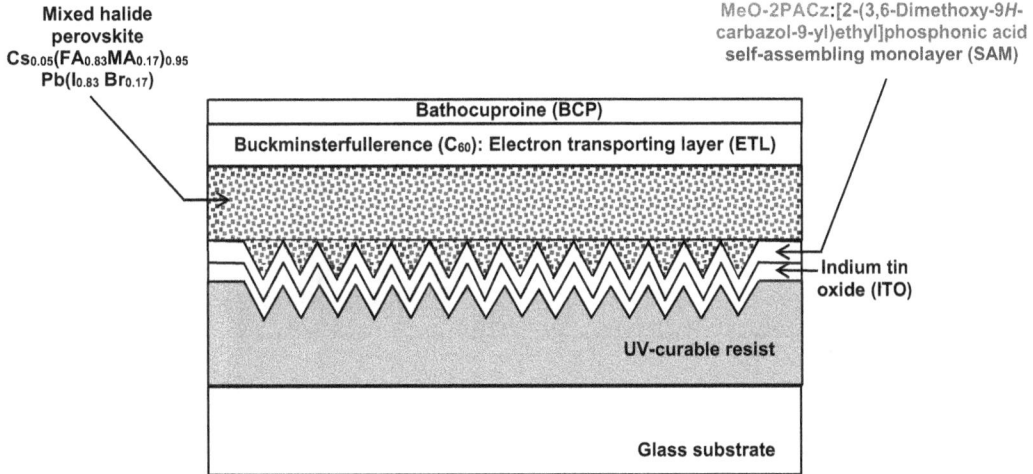

(e)

FIGURE 4.21 (Continued)

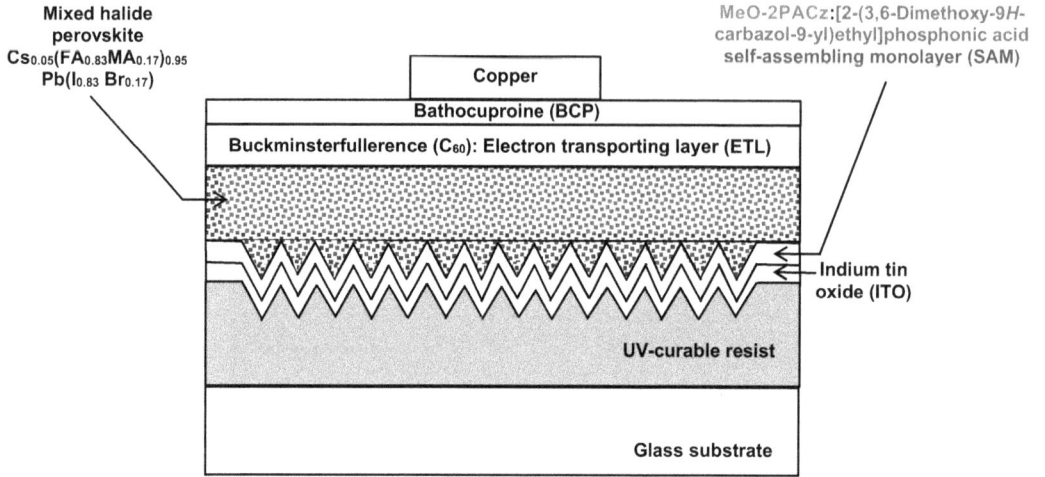

(f)

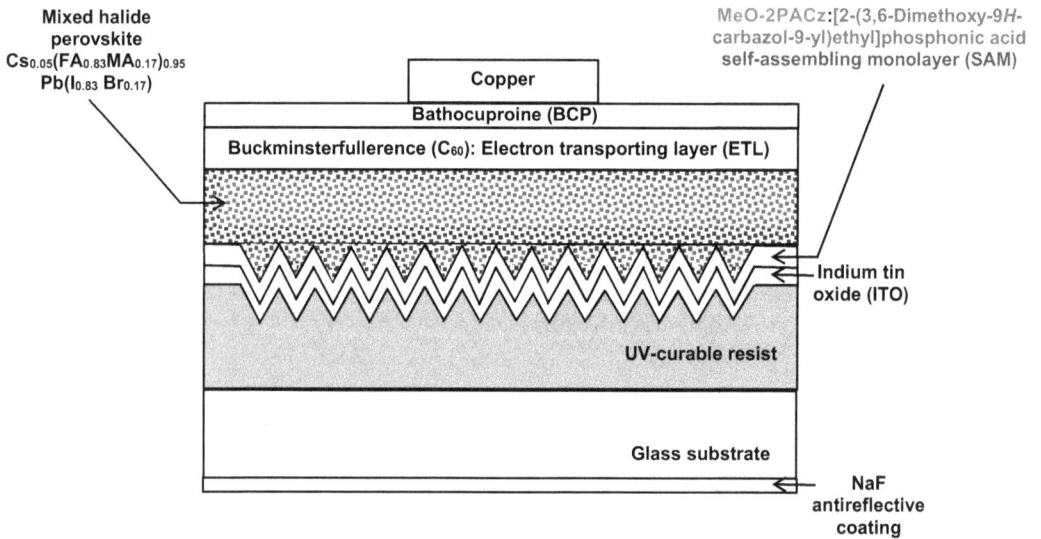

(g)

FIGURE 4.21   (Continued)

built in it. This is the substrate on which the perovskite solar cell will be built. Indium tin oxide is sputter-deposited on this substrate. Then a self-assembled monolayer of MeO-2PACz is conformally coated as a hole-transporting material. Over this SAM, a mixed halide perovskite is deposited by spinning. Then the C_{60} electron transporting layer is evaporated. Then two layers are evaporated. These are hole-blocking BCP layer and copper for contact. On the glass substrate, an NaF ARC is formed by evaporation. NaF has a lower refractive index (1.3255 at 0.6μm) than glass (1.5), providing good antireflection property than the glass/air interface.

The solar cell is ready for testing. One contact is made through copper film, and the other with ITO. Sunlight falls on the NaF layer on the glass substrate, and the output current is measured between the Cu and ITO electrodes.

The cell fabricated on a nanotextured substrate is compared with the one made on a planar substrate. The short-circuit current density, open-circuit voltage, fill factor, and power conversion efficiency of solar cell shown in Figure 4.18(a) are 23.1mAcm^{-2}, 1.13V, 76%, and 19.7%, respectively, while those of the solar cell of Figure 4.18(b) are 22.1mAcm^{-2}, 1.11V, 76%, and 18.7%, i.e., the efficiency of planar solar cell is less by 1% than that made on nanotextured substrate. The measurement condition is AM 1.5G illumination. The 19.7% efficiency is recorded for the light-trapping combination of Cos-structure and NaF ARC (Tockhorn et al 2020).

4.6 DISCUSSION AND CONCLUSIONS

The use of nanotexturing for light loss reduction is well appreciated when we realize that it employs a nanoscale property that does not involve use of any extra material. In terms of material requirement, it is more convenient to implement than the common technique of using antireflection coating, which requires deposition of the coating material on the surface of the solar cell. The advantage of nanotexturing over antireflection coatings is that it causes bending of the incident light beam so that trapping of light inside the cell is increased. ARC does not do so; it only decreases reflection (Maity et al 2014).

Table 4.1 gives examples of nanotextured solar cells.

TABLE 4.1
Effects of Nanotexturing on the Solar Cells

Sl. No.	Solar Cell	Efficiency (%)	Reference
1.	Inverted nanopyramid crystalline silicon solar cell	7.12 (nanotexturing + ARC), 6.73 (only nanotexturing, no ARC), and 4.03 (planar cell)	Sivasubramaniam and Alkaisi (2014)
2.	Sub-10μm silicon solar cell with nanocones and all contacts on the backside	13.7 (nanotextured) and 10.9 (planar cell)	Jeong et al (2013)
3.	10μm-thick periodic nanostructured crystalline silicon solar cell	15.7	Branham et al (2015)
4.	Two-scale (micro/nano) surface textured crystalline silicon solar cell	17.5	Dimitrov and Du (2013)
5.	Nanotextured solution-processed perovskite solar cell	19.7	Tockhorn et al (2020)

Questions and Answers

4.1   Light falling on the solar cell but not utilized in producing an electron-hole pair represents (a) an optical gain or (b) an optical loss? Answer: b.

4.2   To utilize sunlight for electricity generation, (a) transmittance of light into the cell must be minimized and its reflection from the cell surface must be maximized, (c) both transmittance of light into the cell and its reflection from the cell surface must be maximized, (c) transmittance of light into the solar cell must be maximized and its reflection from the surface of the solar cell must be minimized? Answer: c.

4.3   How does texturing enhance optical path length? Answer: By multiple reflections and total internal reflection from surfaces.

4.4   Texturing increases the optical path length because it makes light fall (a) normally on the solar cell or (b) obliquely on the solar cell? Answer: b.

4.5   What is the difference in the amount of reflection of a ray of light falling on a silicon wafer surface from air as compared to a light ray incident on a glass wafer from air? Answer: More light is reflected from silicon than glass.

4.6   What is the difference in the amount of reflection of a ray of light when (a) the interface morphology contains structures greater than the wavelength of light and (b) the interface morphology contains structures smaller than the wavelength of light? Answer: Reflection in b is less than that in a.

4.7   Why is a nanotextured surface a self-cleaning surface? Answer: Hydrophobicity.

4.8   Give an example of a nanotextured surface in natural environment. Answer: Lotus leaf.

4.9   What is a moth-eye nanotexture? Answer: Subwavelength-size needle-shaped pillars on the surface of the eye of a moth.

4.10 Is a moth-eye nanotexture self-cleaning structure? Yes.

4.11 Mention two lithographic techniques used for fabricating moth-eye nanotexture. Answer: Electron-beam lithography, nanoimprint lithography.

4.12 Explain how a graded refractive index helps in reducing reflection of light from the surface of a solar cell. Answer: Creation of a graded refractive index at the interface between air and semiconductor medium is equivalent to substitution of the original semiconductor medium without index gradation by a medium having an effective refractive index lower than that of the original medium. Reflectance is decided by this effective refractive index, not the original refractive index. Owing to the decrease in refractive index of the medium on which light is incident, the difference in refractive indices between the medium from which light is falling (air) and the medium on which it is falling (semiconductor) becomes smaller, leading to lower reflectance. See also Section 6.1.3.

4.13 Which type of texturing is essential for thin-film solar cells: (a) microtexturing, (b) nanotexturing? Answer: b.

4.14 (a) Name a lithographic technique which does not require masking. (b) Is this technique applicable to definition of arbitrarily shaped geometries? Answer: (a) Interference lithography. (b) No.

4.15 What is interference lithography? Answer: A lithographic technique in which the photoresist is exposed to a periodic interference pattern consisting of alternate bright and dark fringes produced from two coherent waves of light.

4.16 What is AZ Barli II? Answer: A bottom antireflective coating (BARC) for exposure of a highly reflective surface in lithography.

4.17 What is AZMiR701? Answer: A high softening point > 130°C, high-resolution photoresist used in etching of structures having submicron dimensions.

4.18 Justify: (a) Why a smaller thick solar cell is necessary for reducing cost; (b) why texturing becomes more critical as thickness of absorber layer is decreased. Answer: (a) To reduce material consumption, and (b) to use as much light as possible with the available thin absorber.

4.19 What is black silicon? Answer: Very low-reflectivity silicon.

4.20 (a) What are the two major problems faced in a front-emitter solar cell? (b) How are they solved? Answer: (a) (i) Heavy-doping induced defect creation and so increased recombination, and (ii) larger surface area of nanotextured surface resulting in high surface recombination. (b) By shifting the emitter junction to the backside of the solar cell.

4.21 What is the Langmuir-Blodgett technique? Answer: A method of forming a monomolecular layer of a water-loving (hydrophilic) and fat- or lipid-loving (lipophilic) substance on the surface of water and transferring it to a solid substrate.

4.22 What is a photonic crystal? Answer: An optical nanostructure with periodically varying refractive index.

4.23 What is ProTEK® B3? Answer: A polymeric coating temporarily applied by spinning on wafers for protection of device/circuit from damage during alkaline or acidic wet-etching.

4.24 What is a SAM? Answer: A *self-assembled monolayer* is a film that spontaneously forms on a surface as a higher-order structure by adsorption from a solution simply by immersion of the substrate into it.

4.25 What is the advantage of nanotexturing over ARC? Answer: Causes bending of light and hence increase in path length in addition to reducing reflection; ARC only lowers reflection.

4.26 What is the nanoscale property utilized by nanotexturing which does not happen in bulk matter? Answer: Graded refractive index creation.

REFERENCES

Branham M. S., W.-C. Hsu, S. Yerci, J. Loomis, S. V. Boriskina, B. R. Hoard, S. E. Han and G. Chen 2015 15.7% efficient 10-μm-thick crystalline silicon solar cells using periodic nanostructures, Advanced. Materials, 27(13): 2182–2188.

Dimitrov D. Z. and C.-H. Du 2013 Crystalline silicon solar cells with micro/nano texture, Applied Surface Science, 266: 1–4.

Heckenthaler T., S. Sadhujan, Y. Morgenstern, P. Natarajan, M. Bashouti and Y. Kaufman 2019 Self-cleaning mechanism: Why nanotexture and hydrophobicity matter, Langmuir, 35(48): 15526–15534.

Jeong S., M. D. McGehee and Yi Cui 2013 All-back-contact ultra-thin silicon nanocone solar cells with 13.7% power conversion efficiency, Nature Communications, 4(2950): 7 pages.

Kim M. S., J. H. Lee and M. K. Kwak 2020 Surface texturing methods for solar cell efficiency enhancement, International Journal of Precision Engineering and Manufacturing, 21: 1389–1398.

Lvovsky A. I. 2015 Fresnel equations, in: Hoffman C. and R. Driggers (eds.), Encyclopedia of Optical and Photonic Engineering, Vol. II, CRC Press, Boca Raton, 6 pages.

Maity S., A. Kundu, S. Das and P. Chakraborty 2014 Reduction of reflectance at c-silicon solar cell using nanotexturization, 2014 2nd International Conference on Devices, Circuits and Systems (ICDCS), 6–8 March 2014, Coimbatore, India, pp. 1–3.

Shen X., S. Wang, H. Zhou, K. Tuokedaerhan and Y. Chen 2021 Improving thin film solar cells performance via designing moth-eye-like nanostructure arrays, Results in Physics, 20(103713): 1–5.

Sivasubramaniam S., M. M. Alkaisi 2014 Inverted nanopyramid texturing for silicon solar cells using interference lithography, Microelectronic Engineering, 119: 146–150.

Sprafke A. N. and R. B. Wehrspohn 2015 Chapter 1: Current concepts for optical path enhancement in solar cells, in: Wehrspohn R. B., U. Rau and A. Gombert (eds.), Photon Management in Solar Cells, Wiley-VCH Verlag GmbH & Co. KGaA, Weinheim, Germany, pp. 1–20.

Sun J., X. Wang, J. Wu, C. Jiang, J. Shen, M. A. Cooper, X. Zheng, Y. Liu, Z. Yang and D. Wu 2018 Biomimetic moth-eye nanofabrication: Enhanced antireflection with superior self-cleaning characteristic, Scientific Reports 8(5438): 1–10.

Tockhorn P., J. Sutter, R. Colom, L. Kegelmann, A. Al-Ashouri, M. Roß, K. Jäger, T. Unold, S. Burger, S. Albrecht and C. Becker 2020 Improved quantum efficiency by advanced light management in nanotextured solution-processed perovskite solar cells, ACS Photonics, 7: 2589–2600.

van Wolferen H. and L. Abelmann 2011 Chapter 5: Laser interference lithography, in: Hennessy T. C. (ed.), Lithography: Principles, Processes and Materials, Nova Science Publishers, Inc., New York, USA, pp. 133–148.

Weiblen R. J., C. R. Menyuk, L. E. Busse, L. B. Shaw, J. S. Sanghera and I. D. Aggarwal 2016 Optimized moth-eye anti-reflective structures for As_2S_3 chalcogenide optical fibers, Optics Express, 24(9): 10172–10187.

5 Plasmonic-Enhanced Solar Cells

5.1 PLASMA, PLASMON, AND PLASMONICS

5.1.1 PLASMA

Plasma is an electrically conducting medium. It consists of negatively and positively charged particles known as ions formed by ionization of gas atoms or molecules by removal of electron(s) from their outermost shells. Plasma is often designated as the fourth state of matter. The first three states are the well-known states solid, liquid, and gas.

5.1.2 PLASMON

Plasmon is concerned with the collective oscillation of conduction electrons in a metal with respect to the static background of positive ion cores. It is a quantum of plasma oscillations in the same way as a photon is a quantum of optical oscillation known as light, or a phonon is a quantum of mechanical vibrations. A quantum is an elementary structural unit of a physical entity which cannot be broken down further.

Since plasmon is a particle-like object but is unlike a real particle, it is referred to as a quasiparticle.

5.1.3 PLASMONICS

Plasmonics is concerned with the investigation, generation, detection, manipulation, and application of electron density waves, called plasmons, by the resonant interaction between electromagnetic waves, particularly light waves, and free electrons in a metal at the interface between the metal and a dielectric. The interaction between electrons and electromagnetic waves occurs because electrons are charged particles and are susceptible to the influence of electric and magnetic fields of the electromagnetic waves. Resonant interaction is possible because the natural frequency of oscillation of electrons in a metal film of thickness < 100nm or metal nanoparticles of diameter < 100nm is in the same scale as the wavelength of visible light (350–700nm). Hence, this kind of interaction takes place in the nanometer scale of dimensions.

5.2 SURFACE PLASMONS, LOCALIZED SURFACE PLASMONS, AND SURFACE PLASMON POLARITONS

5.2.1 SURFACE PLASMONS

"Surface plasmons" is the name given to the collective oscillations of electrons or electron density waves. These waves propagate at the interface between a metal and a dielectric. A familiar example of such an interface is the interface between metal and air.

5.2.2 LOCALIZED SURFACE PLASMONS

Localized surface plasmon is the collective electron oscillation confined to the surface of a metal nanoparticle. It is a consequence of constraining the surface plasmon within the bounds of a nanoparticle having dimensions ≤ the wavelength of light used for excitation of surface plasmon.

DOI: 10.1201/9781003215158-7

5.2.3 SURFACE PLASMON POLARITONS

The motion of electronic charges in a surface plasmon produces electromagnetic fields so that the disturbance is a combination of electronic charge motion and electromagnetic field. Surface plasmon polariton is the total excitation, the electronic charge motion, plus the electromagnetic field bound to a planar metal-dielectric interface at the nanometer scale. Note that the electronic charge motion takes place inside the metal while the electromagnetic wave moves in the dielectric or air. The electronic charge oscillation constitutes the surface plasmon, while the quasiparticle resulting from the intense coupling of the charged particle, here the electronic charge, with the electromagnetic wave is the polariton.

5.2.4 LOCALIZED SURFACE PLASMON RESONANCE (LSPR) AND PROPAGATING SURFACE PLASMON RESONANCE (PSPR)

Surface plasmon resonance (SPR) is a generic term commonly used for resonance in light-matter interactions when the frequency of collective oscillation of electrons or plasmons matches with the frequency of oscillating electric field of incident light. SPR is of two types (Jatschka et al 2016).

5.2.4.1 Localized SPR

LSPR is a particular type of SPR related to metal nanoparticles. To understand how LSPR happens, we restate and apply basic knowledge to understand what happens when a beam of light strikes a metallic nanoparticle. We know that:

  (i) A metal contains a large population of free electrons undergoing incessant random motion.
 (ii) In the formulation of electromagnetism, light is an electromagnetic wave containing two types of fields, electric and magnetic fields. The two constituent fields of a light wave are oriented in directions mutually at right angles to each other. Their directions are also perpendicular to the direction of propagation of the light wave.

The sequential events taking place during LSPR are: The incident light passes through the metal nanoparticle. As the light wave has an electric field, the electric field moves through the metal nanoparticle. Under the influence of the electric field of light, the electrons in the metal nanoparticle suffer displacement relative to positive ion lattice. But the attractive Coulombic force from the atomic nuclei tries to restore the electrons to their equilibrium positions. Consequently, dipolar oscillations of electrons ensue. These oscillations constitute the surface plasmons. Because the nanoparticle is smaller in size than the wavelength of light, LSPR is restricted in space. The electron density oscillates around the surface of the nanoparticle, resonating with the frequency of light wave, leading to localized electromagnetic waves. Nonpropagating oscillations of free electrons of metal nanoparticles coupled to localized electromagnetic waves comprise LSPR. During resonance, the oscillations of the free electrons are dramatically increased, building a strong electromagnetic field around the metal nanoparticles.

Let us look into the situation more deeply. Figure 5.1 depicts the various events that take place in establishing LSPR on a metallic nanoparticle. A single nanoparticle is considered, and the effect of the changing electric field of light wave on this nanoparticle is shown for the case of oppositely directed electric fields.

Upon illumination with light, the electrons in the metallic nanoparticle come under the influence of the electric field of the light wave. As the electrons are negatively charged particles, they feel attractive and repulsive forces, depending on the direction of the electric field in the light wave. We know that electrons move in the opposite direction to the electric field. The directions of the electric field in the two half cycles of a light wave are shown in the diagram by drawing arrows. Looking at the directions of the electric field in the two half cycles, it is easy to see that the electron cloud of the metallic nanoparticle drifts downwards in the first half cycle and upwards in the second half cycle. In the first half cycle, the downward region of the nanoparticle becomes negatively charged by gain of electrons, and its opposite end is positively charged by loss of electrons. In the

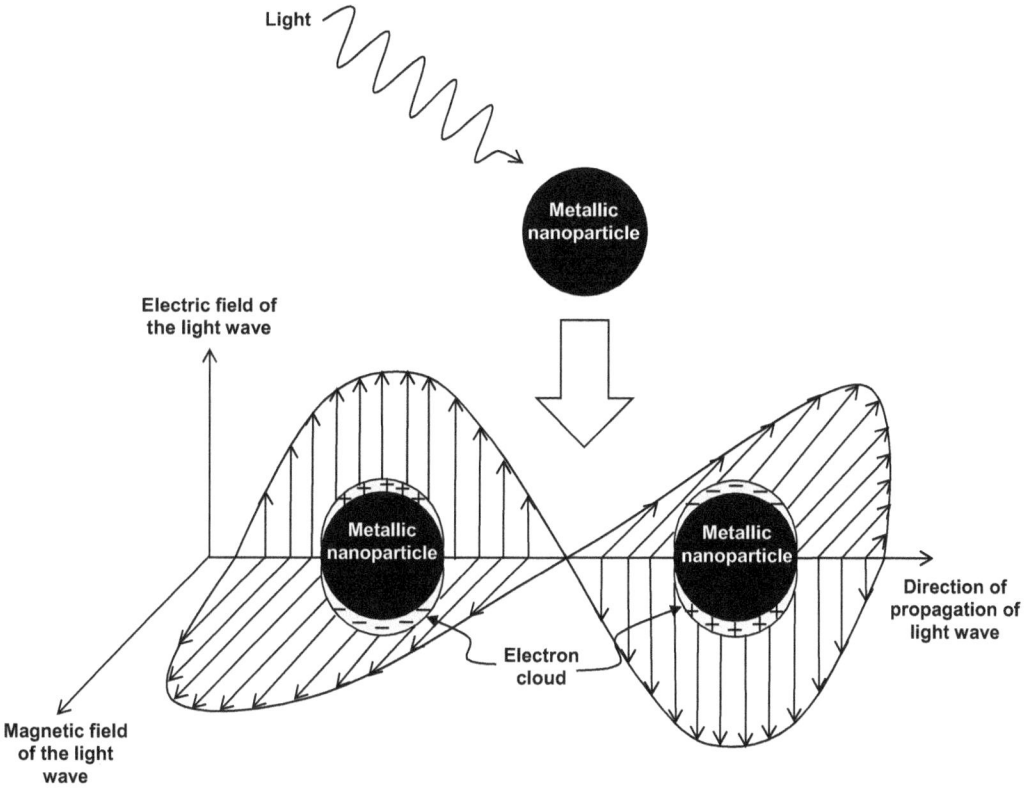

FIGURE 5.1 Localized surface plasmon resonance (LSPR) on a metallic nanoparticle showing the illumination of a nanoparticle with light, the relative directions of electric and magnetic fields of incident light and its direction of propagation, and the shifting of the electrons of the nanoparticle in opposite directions with the reversal of the direction of electric field of light. The consequent generation of positive charge in the electron-depleted region and negative charge in electron-augmented region of the electron cloud are indicated. Note that the electrons move in opposite direction to that of the electric field. When the electric field of light wave points in the upward direction, the electrons move downwards, and vice versa.

second half cycle, the downward region is positively charged and the upward region is negatively charged. Thus, the negative and positive charges are alternately produced on the opposite sides of the nanoparticle. This charge oscillation in the nanoparticle is restricted to the nanoparticle and does not move from one place to another. It is an immobile or localized surface plasmon.

5.2.4.2 Propagating SPR

Often referred to as PSPR, or simply SPR, this happens at a flat surface or thin film. When light illuminates a metal surface nearly parallel to it, the free electron density oscillates back and forth and moves along the surface due to the positive and negative potentials induced on the metal surface by the electric fields in the light wave along the direction of propagation. The disturbance consisting of electron density oscillations and the resulting electromagnetic fields, together called the surface plasmon polaritons, propagates along the smooth metal-dielectric interface. Note that the surface plasmon waves are longitudinal waves in which the displacements of the electrons relative to the background of positive ions are parallel to the direction of propagation of the wave.

Figure 5.2 illustrates the formation of the surface plasmon polariton wave when a beam of light falls on a metallic surface. The electrons near the surface of the metal, being negatively charged particles, experience the attractive and repulsive forces due to the electric field of light wave. The

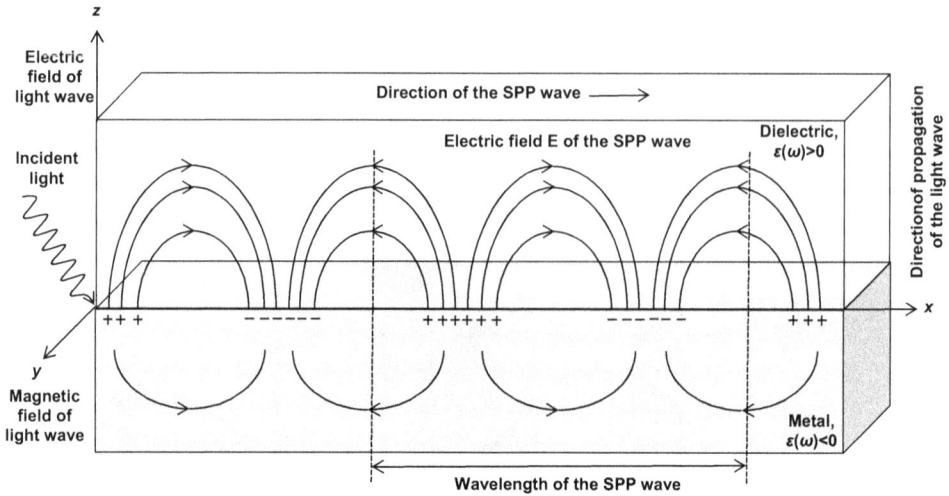

(a)

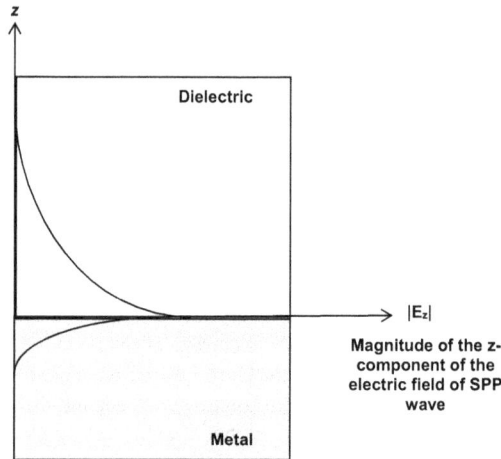

(b)

FIGURE 5.2 Propagating surface plasmon resonance (PSPR) at a metal-dielectric interface: (a) The incident light wave is shown. The electric and magnetic field directions and the direction of light propagation are indicated. The induction of negative and positive charges near the surface of the metal can be seen. The electric field lines associated with the induced charges are drawn. The direction of the surface polariton wave travelling along the metal-dielectric interface may be noted, as also the wavelength of this wave. (b) Fall of the electric field of surface plasmon polariton relative to the distance from the metal-dielectric interface, both inside the dielectric and inside the metal. Magnitude of the z-component of electric field of SPP wave is plotted against the distance z because this field changes direction, as clear from the opposite directions of successive electric fields lines in the diagram between one loop and the next. These directional changes arise from the oscillations in the electric field of light.

electric field of the light wave oscillates at a tremendously high frequency, ranging from 4×10^{14} to 8×10^{14} Hz. Electric charges are therefore induced by the oscillating electric field of light in the metal near the surface, as shown in Figure 5.2(a). The electrons are separated from the positive ion cores. The portion of metal from where the electrons move away due to the electric field of light becomes deficit of electrons, creating a net positive charge, and the portion where these electrons arrive becomes a surplus of electrons, resulting in a net negative charge. These positive and negative charges are shown in the diagram. They, in turn, produce their own electric fields. So a longitudinal wave motion is set up at the interface between the metal and the dielectric with the charges inside the metal responding to light wave and the electric fields due to the simultaneously travelling electromagnetic waves produced by oscillating charges. The combination of electron density waves with the electromagnetic waves from the positive and negative charges generated by electron density variations together constitute a wave motion known as the surface plasmon polariton wave.

The electric field remains confined near the metal-semiconductor interface up to a small distance from the interface. It falls off rapidly as one recedes away from the interface. The nature of decay of the electric field from the interface follows the exponential law, as seen in Figure 5.2(b).

The vital difference between PSPR and LSPR is in the spatial confinement. The surface plasmon polariton can propagate along the interface, while the localized surface plasmon is totally bounded and does not move at all. The localized surface plasmon is not accompanied by a polariton like the surface plasmon.

Due to differences in dimensions of interfaces, the electron oscillations in PSPR take place parallel to the light wave, whereas those in LSPR are perpendicular to the interface.

In LSPR, the evanescent field is highly localized at the nanoparticle and decays rapidly with distance, while in PSPR, the decay length of evanescent field is comparatively longer. Evanescent field is an electric/magnetic field whose energy is concentrated in the vicinity of the source. It does not move as an electromagnetic wave.

5.3   ABSORPTION AND SCATTERING OF LIGHT

5.3.1   ABSORPTION OF LIGHT

When a light wave strikes a particle, reflection, refraction/transmission, and absorption take place (Figure 5.3). When the frequency of light is same as the natural frequency of vibration of electrons

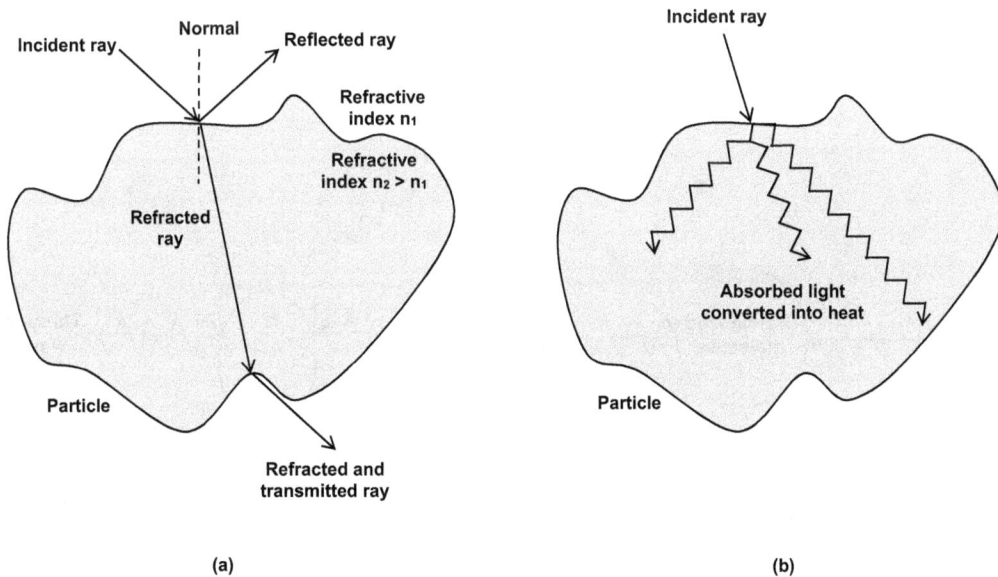

FIGURE 5.3 Interaction of light with a particle of matter: (a) reflection and refraction/transmission showing the incident, reflected, and refracted rays, and (b) absorption of light showing the transfer of the energy of light to the electrons, hence vibrational kinetic energy of neighboring atoms, and cessation of the photon existence accompanied by release of heat.

in the atoms of the particle, the electrons are set into vibrational motion. Through interaction of vibrating electrons with neighboring atoms, this vibrational energy is converted into thermal energy, and the light wave of the given frequency is absorbed by the particle.

5.3.2 SCATTERING OF LIGHT

Scattering is a phenomenon involving the interaction of light with a particle of matter. It is distinctly different from reflection or refraction. Scattering also differs from absorption, in which all the energy of light is taken up or soaked by the absorbing particle and the optical photon ceases to exist.

In elastic scattering, the light retains its energy but its direction is changed so that deflection of light takes place. Light is radiated in different directions. In inelastic scattering, a portion of the energy of incident light is absorbed by the particle and subsequently reradiated in different directions.

Rayleigh scattering is an elastic scattering which occurs for particles having size much smaller than the wavelength of light. This scattering is wavelength-dependent. The intensity of scattered light varies inversely as the fourth power of wavelength. The shorter the wavelength, the more the scattering. According to the theory of Rayleigh scattering (Figure 5.4), the interaction of a beam of light with its constituent electric and magnetic fields, with a particle of matter smaller than its wavelength, temporarily polarizes the molecules of the particle with a redistribution of electrons in such a way that one end of the molecule is weakly positively charged and its other end is weakly negatively charged. This separation of charges produces a dipole moment, and the oscillating electric and magnetic fields of light wave impinging on the molecule cause oscillations of the molecular dipole moment, accompanied by emission of light in all directions.

For particles of size similar to or slightly larger than the wavelength of light, the elastic scattering is named Mie scattering. This scattering gives white light when occurring from water droplets in clouds, fog, or mist because it is not strongly wavelength-dependent. Still, larger particles lead to optical scattering (Figure 5.5).

5.3.3 ABSORPTION AND SCATTERING CROSS SECTIONS OF A PARTICLE

Absorption cross section of a particle is the product of the probability of absorption of a photon by the particle with the average cross-sectional area of the particle. The absorption cross section

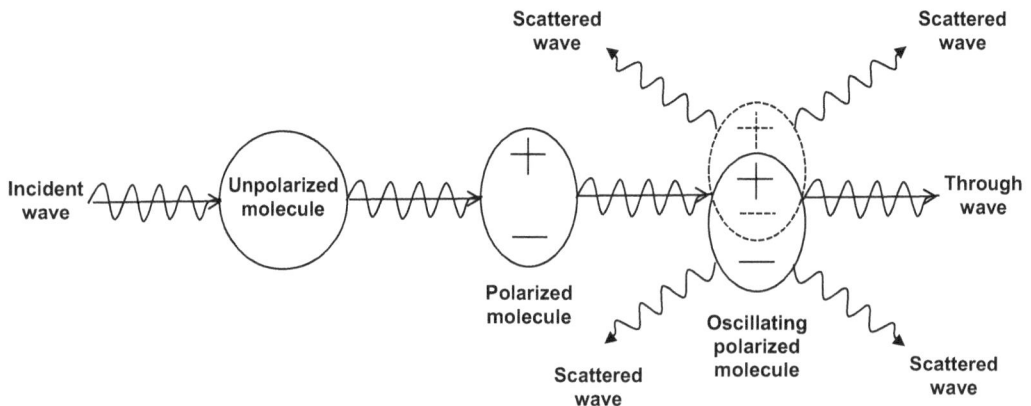

FIGURE 5.4 Mechanism of Rayleigh scattering in which an incident light wave polarizes the target molecule and the oscillations of the polarized molecule induced by incident light wave produce scattering of light in different directions, while the incident light wave passes through to the other side without loss of energy. Unpolarized molecule and oscillating polarized molecule along with incident, scattered, and through light waves are shown.

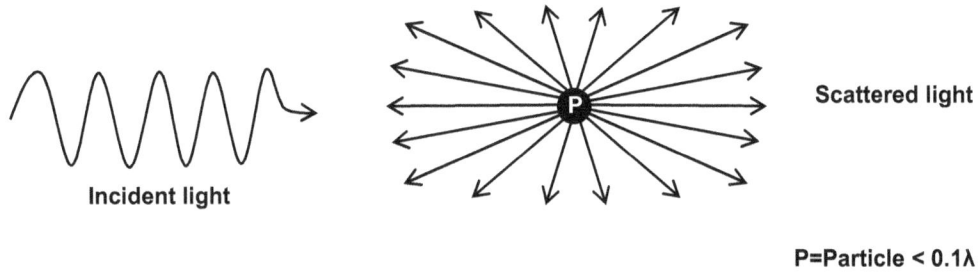

(a)

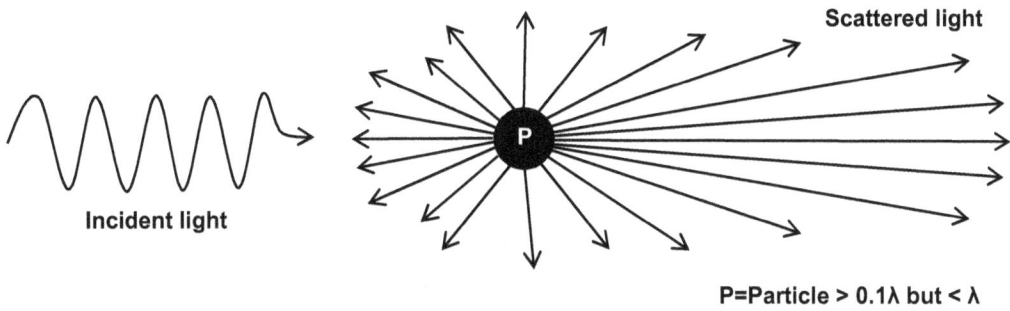

(b)

(c)

FIGURE 5.5   Different types of scattering: (a) Rayleigh scattering, for which particle size is < 0.1λ, (b) Mie scattering, for which 0.1λ < particle size is < λ, and (c) optic scattering, for which particle size is > λ. Typical sizes of particles for Rayleigh, Mie, and optic scattering are 1–50nm, 50–500nm, and ≥1,000nm, respectively. Distinguish between the distribution patterns of scattered light in the three cases. For comparison, wavelength of visible light extends from 400nm to 700nm.

is measured in units of area. The numerical value of this area is taken in such a way that if light strikes a circular area of this size perpendicular to its direction of propagation and is centered at the particle, it will be absorbed by the particle.

Similarly, scattering cross section of a particle = probability of scattering of light by the particle × cross-sectional area of the particle. It is specified in area units with the area chosen such that light incident normally on the given area will be scattered by the particle.

5.4 SURFACE PLASMON EFFECTS IN SOLAR CELLS

Both LSPR and PSPR effects are utilized for increasing light absorption in solar cells.

5.4.1 LSPR with Metal Nanoparticles

5.4.1.1 Device Structures Used

Two device structures are commonly used (Figure 5.6). In one structure, the metal nanoparticles are spread out and fixed on the top surface of the solar cell. In the other structure, they are embedded inside the photoactive medium of the solar cell.

5.4.1.2 Resonance Frequency Formula for LSPR

For a metal nanoparticle of size much smaller than the wavelength of light, i.e., a subwavelength nanoparticle, the interaction of light with the nanoparticle is accounted by dipole oscillation only. This is known as the dipole approximation. Further, quasistatic approximation is valid, implying that the electric field of incident light is constant. Let us consider a homogeneous, isotropic, spherical-shaped metal nanoparticle of radius a placed inside a static electric field $\mathbf{E}_0$ in a semiconductor medium. Since the dielectric function or frequency-dependent complex permittivity ε of the metal nanoparticle differs from the frequency-dependent complex permittivity $\varepsilon_{\text{Medium}}$ of the semiconductor medium, a charge will be induced on the surface of the metal nanoparticle and the original electric field $\mathbf{E}_0$ will be distorted by the spherical nanoparticle. The electric field $\mathbf{E}_1$ inside the spherical nanoparticle is derived from the scalar potential Φ_1 as:

$$\mathbf{E}_1 = -\nabla \Phi_1 \tag{5.1}$$

Similarly, the electric field $\mathbf{E}_2$ outside the spherical nanoparticle is obtained from the scalar potential Φ_2 as:

$$\mathbf{E}_2 = -\nabla \Phi_2 \tag{5.2}$$

The potential outside the spherical nanoparticle corresponds to the superimposition of the electric field $\mathbf{E}_0$ of incident light and that of a dipole located at the center of the spherical nanoparticle. The electric field $\mathbf{E}_0$ induces a dipole moment p inside the spherical nanoparticle, which is expressed as (Gangadharan et al 2017)

$$\mathbf{p} = 4\pi\varepsilon_0\varepsilon_{\text{Medium}}a^3 \frac{\varepsilon - \varepsilon_{\text{Medium}}}{\varepsilon + 2\varepsilon_{\text{Medium}}} \mathbf{E}_0 \tag{5.3}$$

where ε_0 denotes permittivity of free space. The equation relating the dipole moment p to the polarizability α of the nanoparticle is

$$\mathbf{p} = \alpha\varepsilon_0\varepsilon_{\text{Medium}}\mathbf{E}_0 \tag{5.4}$$

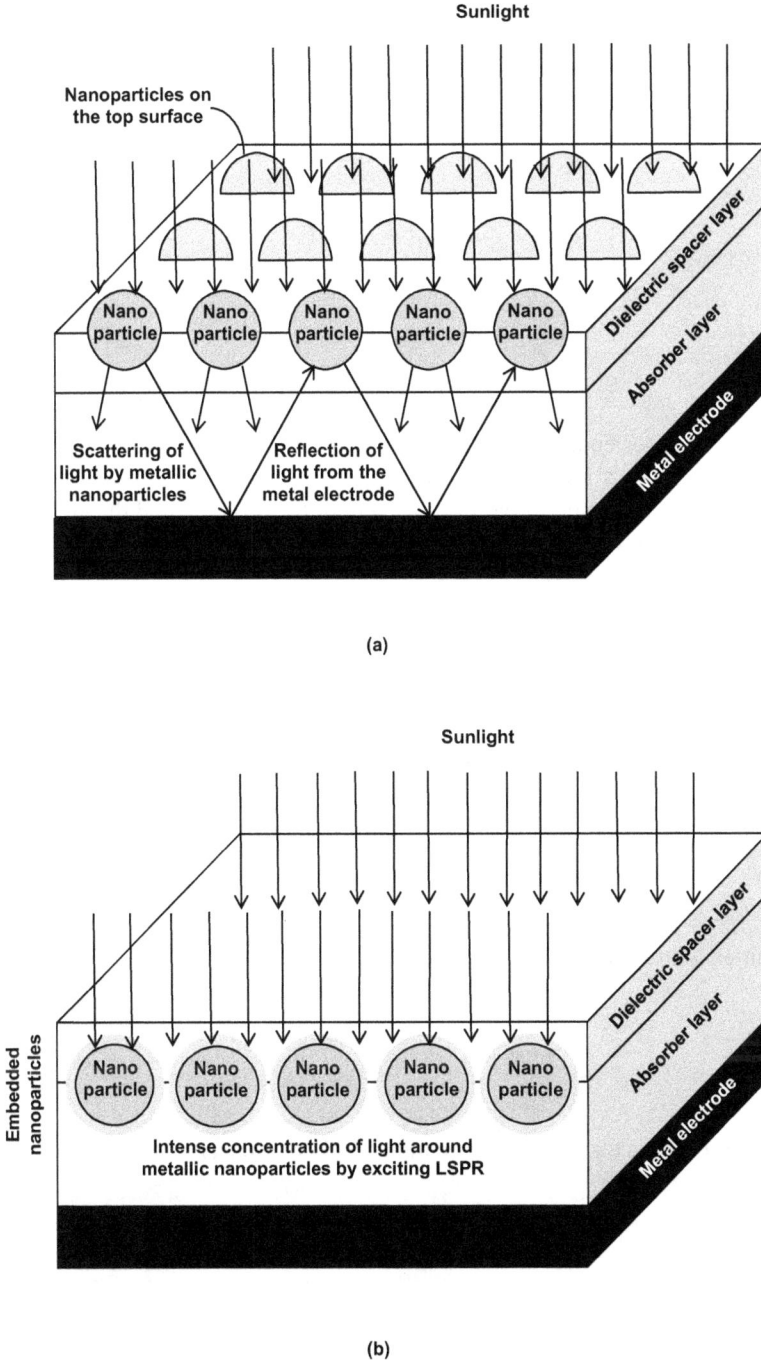

FIGURE 5.6 Enhancement of light trapping in a solar cell with the help of metallic nanoparticles: (a) nanoparticles glued to the top surface of the solar cell, and (b) nanoparticles embedded inside the cell (here on the surface of the absorber region). A dielectric spacer layer, an absorber layer, and the metal electrode of the solar cell are shown. In (a), the nanoparticles scatter light into the absorber layer. Any light reaching the metal electrode is reflected back to the metallic nanoparticle, from where it is rescattered into the absorber layer. In (b), LSPR is excited when the frequency of incoming light equals that of the surface plasmons. By this resonance effect, the intensity of light around the nanoparticles increases to provide efficient photocarrier generation.

Equating the right-hand sides of equations 5.3 and 5.4, we get

$$\alpha \varepsilon_0 \varepsilon_{\text{Medium}} \mathbf{E}_0 = 4\pi \varepsilon_0 \varepsilon_{\text{Medium}} a^3 \frac{\varepsilon - \varepsilon_{\text{Medium}}}{\varepsilon + 2\varepsilon_{\text{Medium}}} \mathbf{E}_0 \tag{5.5}$$

from which

$$\alpha = 4\pi a^3 \frac{\varepsilon - \varepsilon_{\text{Medium}}}{\varepsilon + 2\varepsilon_{\text{Medium}}} \tag{5.6}$$

When

$$\text{Re}\{\varepsilon(\omega)\} = -2\varepsilon_{\text{Medium}} \tag{5.7}$$

the polarizability α experiences a steep upswing, which is recognized as the resonance condition. According to Drude's model, the real component of the dielectric function $\text{Re}\{\varepsilon(\omega)\}$ of the metal nanoparticle is related to the frequency ω of the electric field through the plasma frequency ω_{Plasma} of the free electron gas in the metal nanoparticle and the relaxation time $\tau_{\text{Relaxation}}$ of electron as

$$\text{Re}\{\varepsilon(\omega)\} = 1 - \frac{\omega_{\text{Plasma}}^2 \tau_{\text{Relaxation}}^2}{1 + \omega^2 \tau_{\text{Relaxation}}^2} \tag{5.8}$$

Applying the resonance condition 5.7 to equation 5.8, we have

$$-2\varepsilon_{\text{Medium}} = 1 - \frac{\omega_{\text{Plasma}}^2 \tau_{\text{Relaxation}}^2}{1 + \omega^2 \tau_{\text{Relaxation}}^2} \tag{5.9}$$

which is rewritten with symbol ω_{LSPR} in place of ω as

$$-2\varepsilon_{\text{Medium}} = 1 - \frac{\omega_{\text{Plasma}}^2 \tau_{\text{Relaxation}}^2}{1 + \omega_{\text{LSPR}}^2 \tau_{\text{Relaxation}}^2} \tag{5.10}$$

to emphasize that LSPR is taking place. Equation 5.10 is written as

$$-2\varepsilon_{\text{Medium}} \left(1 + \omega_{\text{LSPR}}^2 \tau_{\text{Relaxation}}^2\right) = 1 + \omega_{\text{LSPR}}^2 \tau_{\text{Relaxation}}^2 - \omega_{\text{Plasma}}^2 \tau_{\text{Relaxation}}^2 \tag{5.11}$$

or

$$-2\varepsilon_{\text{Medium}} - 2\varepsilon_{\text{Medium}} \omega_{\text{LSPR}}^2 \tau_{\text{Relaxation}}^2 = 1 + \omega_{\text{LSPR}}^2 \tau_{\text{Relaxation}}^2 - \omega_{\text{Plasma}}^2 \tau_{\text{Relaxation}}^2 \tag{5.12}$$

or

$$-2\varepsilon_{\text{Medium}} \omega_{\text{LSPR}}^2 \tau_{\text{Relaxation}}^2 - \omega_{\text{LSPR}}^2 \tau_{\text{Relaxation}}^2 = 1 + 2\varepsilon_{\text{Medium}} - \omega_{\text{Plasma}}^2 \tau_{\text{Relaxation}}^2 \tag{5.13}$$

or

$$-\omega_{\text{LSPR}}^2 \left(2\varepsilon_{\text{Medium}} + 1\right) \tau_{\text{Relaxation}}^2 = 1 + 2\varepsilon_{\text{Medium}} - \omega_{\text{Plasma}}^2 \tau_{\text{Relaxation}}^2 \tag{5.14}$$

or

$$-\omega_{LSPR}^2 = \frac{1 + 2\varepsilon_{Medium}}{\left(2\varepsilon_{Medium} + 1\right)\tau_{Relaxation}^2} - \frac{\omega_{Plasma}^2 \tau_{Relaxation}^2}{\left(2\varepsilon_{Medium} + 1\right)\tau_{Relaxation}^2} \tag{5.15}$$

or

$$\omega_{LSPR}^2 = \frac{\omega_{Plasma}^2 \tau_{Relaxation}^2}{\left(2\varepsilon_{Medium} + 1\right)\tau_{Relaxation}^2} - \frac{1 + 2\varepsilon_{Medium}}{\left(2\varepsilon_{Medium} + 1\right)\tau_{Relaxation}^2} = \frac{\omega_{Plasma}^2}{\left(2\varepsilon_{Medium} + 1\right)} - \frac{1}{\tau_{Relaxation}^2}$$

$$= \omega_{Plasma}^2 \left\{ \frac{1}{1 + 2\varepsilon_{Medium}} - \frac{1}{\omega_{Plasma}^2 \tau_{Relaxation}^2} \right\} \tag{5.16}$$

The LSPR resonance frequency ω_{LSPR} is found to be

$$\omega_{LSPR} = \omega_{Plasma} \sqrt{\frac{1}{1 + 2\varepsilon_{Medium}} - \frac{1}{\omega_{Plasma}^2 \tau_{Relaxation}^2}} \tag{5.17}$$

5.4.1.3 Red Shifting of Resonance Frequency by Embedded Metal Nanoparticles

Equation 5.17 shows that the LSPR resonance frequency depends on the relative permittivity of the surrounding semiconductor medium. The plasma frequencies of noble metal nanoparticles lie in the deep UV range (1.1×10^{15} to 1.5×10^{15} Hz). When these nanoparticles are embedded in high relative permittivity semiconductors such as silicon (dielectric constant $\varepsilon_{Silicon} = 11.7$), gallium arsenide ($\varepsilon_{GaAs} = 12.9$), TiO$_2$ ($\varepsilon_{TiO2} = 85$), or PbS ($\varepsilon_{PbS} = 6.5$), it is clear from equation 5.17 and the dielectric constant values of these materials that the resonant frequencies will be relocated to red light range (4×10^{14} to 4.8×10^{14} Hz), which is a red shift of the resonance frequency. As a consequence of this red shift, the absorption of light will be stretched to include lower frequencies of the solar spectrum.

5.4.1.4 Intensification of Local Electric Field of Light at Resonance

At resonance, a strong electric field is created in the vicinity of the metal nanoparticle. The factor by which the electric field is increased by the nanoparticle is called the field enhancement factor η. It is defined as the ratio of electric field in the presence of the nanoparticle to the electric field in the absence of the nanoparticle, i.e., the electric field $\mathbf{E}_2$ outside the nanoparticle in the semiconductor medium to the electric field $\mathbf{E}_0$ in the semiconductor medium when the nanoparticle was not introduced in the medium. The field $\mathbf{E}_0$ is the electric field of the illuminating optical radiation. The factor η represents the degree of concentration of the electric field by the presence of the metal nanoparticle. It is given by (Tanabe 2022)

$$\eta = \frac{|\mathbf{E}_2|^2}{|\mathbf{E}_0|^2} = \left| 1 + \frac{\alpha}{2\pi r^3} \right|^2 \tag{5.18}$$

where α is the polarizability of the spherical nanoparticle and r is the distance of the point at which the field is observed from the center of the sphere.

The impact of this near field enhancement on the solar cell is that a higher electric field now becomes available for the generation of electron-hole pairs in the semiconductor. Since the carrier generation rate depends on the square of the electric field, the effect is manifested very prominently. It must be noted that the second term in equation 5.18 for field enhancement factor falls off quickly as the cube of the distance from the nanoparticle, which means that the effect of electric field intensification can be utilized only if the distance between the metal nanoparticle and the photoactive layer of the solar cell is a few nanometers. At larger distances, the effect is weakened.

5.4.1.5   Enhancement of Scattering of Light at Resonance

The scattering of light by the metal nanoparticle is increased when the resonance condition is fulfilled. The extent of this effect can be ascertained from the absorption and scattering coefficients of the metal nanoparticle. The absorption cross section $\sigma_{\text{Absorption}}$ of the spherical nanoparticle of radius a is given by (Chou and Chen 2014)

$$\sigma_{\text{Absorption}} = 4\pi k a^3 \text{Im}\left[\frac{\varepsilon - \varepsilon_{\text{Medium}}}{\varepsilon + 2\varepsilon_{\text{Medium}}}\right] \tag{5.19}$$

while its scattering cross section $\sigma_{\text{Scattering}}$ is

$$\sigma_{\text{Scattering}} = \frac{8\pi}{3} k^4 a^6 \left|\frac{\varepsilon - \varepsilon_{\text{Medium}}}{\varepsilon + 2\varepsilon_{\text{Medium}}}\right|^2 \tag{5.20}$$

where the symbol k denotes the wave vector related to the wavelength λ of light as

$$k = \frac{2\pi}{\lambda} \tag{5.21}$$

The absorption cross section is proportional to the third power of the radius of the nanoparticle (equation 5.19), whereas the scattering cross section varies as the sixth power of nanoparticle radius (equation 5.20) so that the latter reaches a large value quickly as the radius is increased. A useful insight is provided by the ratio of scattering and absorption cross sections. This ratio increases from 0.05 at 40nm nanoparticle diameter ($2r$) to 0.065 at 80nm diameter. Larger-size nanoparticles possessing high scattering cross section to absorption cross section ratios are better for solar cell application because a larger proportion of solar radiation will be scattered instead of being absorbed (Jain et al 2006; Gangadharan et al 2017).

5.4.2   PSPR at Metal-Semiconductor Interface

5.4.2.1   Necessity of Coupling Medium for Exciting Surface Plasmon Polaritons

Can the sunlight entering the solar cell and falling on the planar of the metal-semiconductor interface at the backside be used to trigger the production of surface plasmon polaritons? No. The hurdle faced in producing SPPs is the mismatching of momentum of sunlight with that of the electron oscillations or plasmons.

5.4.2.2   Approaches for Matching Momenta

Photons in free space cannot activate SPPs. Realizing this difficulty, how can matching of momenta be achieved? A medium is necessary to couple the incident solar radiation with the plasmons. One way to do this is to use a prism for coupling the sunlight with the plasmons. Another way is to build a periodic corrugated structure such as a grating or nanodome array at the absorption layer-metal electrode interface (Figure. 5.7). Such a structure is frequently adopted in solar cells for exciting SPPs due to its compatibility with thin-film deposition techniques.

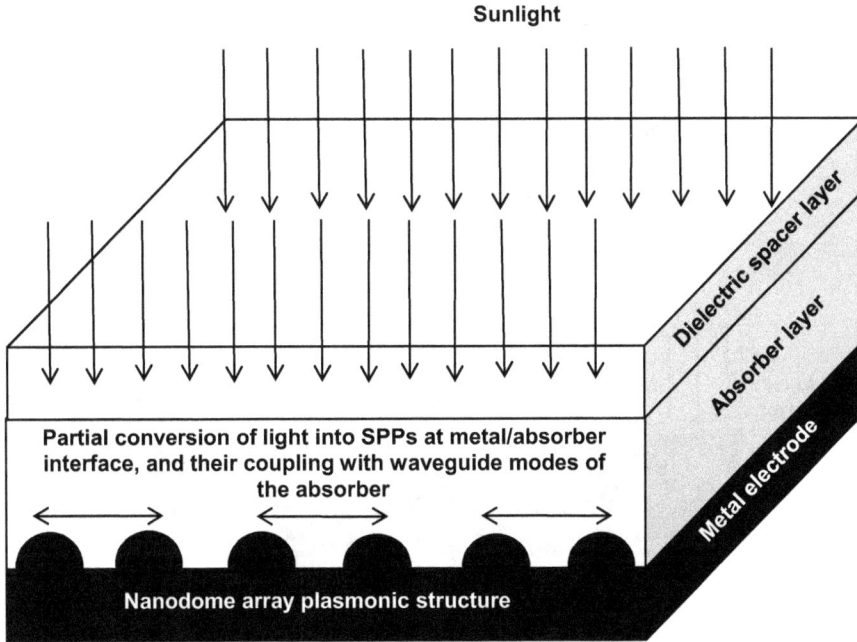

FIGURE 5.7 Effect of a plasmonic structure on the backside of the solar cell. The structure shown is a nanodome array formed in the metal electrode of the solar cell acting as the rear contact. Surface plasmon resonance occurs at the interface between the absorber layer and the metal electrode. A part of light is transformed to the surface plasmon polaritons, and the remaining part is coupled with the waveguide modes of the absorber. Consequently, the distance traversed by light is much greater than the physical thickness of the absorber layer. Thus, the metal electrode performs the task of light trapping in addition to its regular current collection work. The two effects (scattering and localization) are concomitantly present with the conversion of light into SPPs and its association with waveguide modes.

5.5 PLASMONIC-ENHANCED GaAs SOLAR CELL DECORATED WITH Ag NANOPARTICLES (η = 5.9%)

Improvement in performance of GaAs solar cell by sticking silver nanoparticles on the surface of the window layer is demonstrated (Nakayama et al 2008). As nanoparticles of controlled diameters must be deposited at predecided locations, a mask is necessary. This mask will have nanosized holes at known distances apart. It is made using aluminum sheet. A nanoporous film can be deposited on this aluminum sheet by anodic oxidation of aluminum in particular electrolytes (Figure 5.8).

Aluminum sheet is anodized in two steps: at 80V in (oxalic acid + malonic acid) solution and at 120V in malonic acid solution. Using saturated iodine solution in methanol, the aluminum sheet is etched away to reach the aluminum oxide layer underneath the pores. The aluminum oxide is removed in dilute phosphoric acid solution. Thus, a through-hole structure is obtained. This is the anodized aluminum oxide (AAO) template. It is used as the mask for deposition of Ag nanoparticles.

The GaAs solar cell is made on an N-GaAs substrate by depositing a stack of III-V layers by metal organic chemical vapor deposition (MOCVD) (Figure 5.9(a)). The solar cell has the structure Au bottom electrode/N-GaAs substrate/N-GaAs buffer layer/N-AlGaAs back surface field layer/ N-GaAs base layer/P-GaAs emitter layer/P-AlGaAs window layer/P-GaAs cap layer/Au front contact (Figure 5.9(b)). It is decorated with Ag nanoparticles (Figure 5.9(c)).

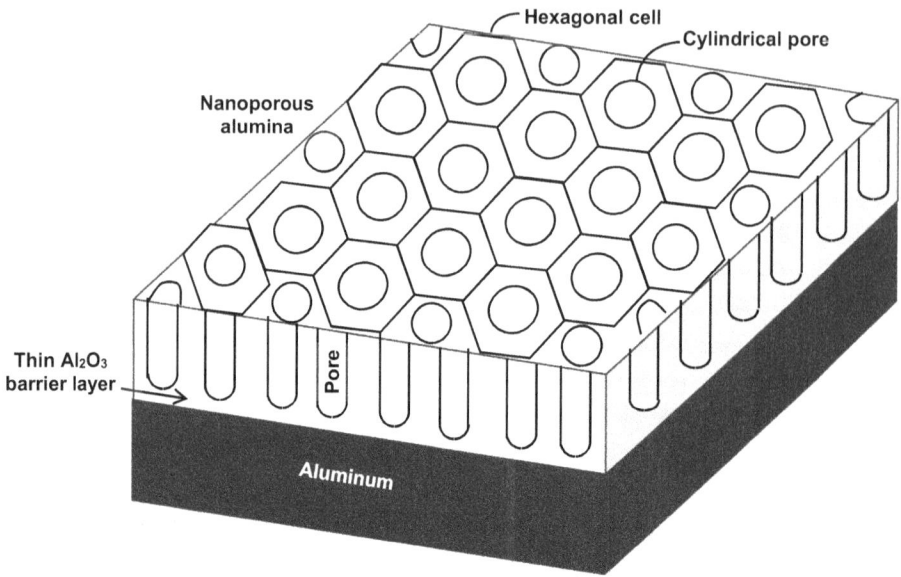

(a)

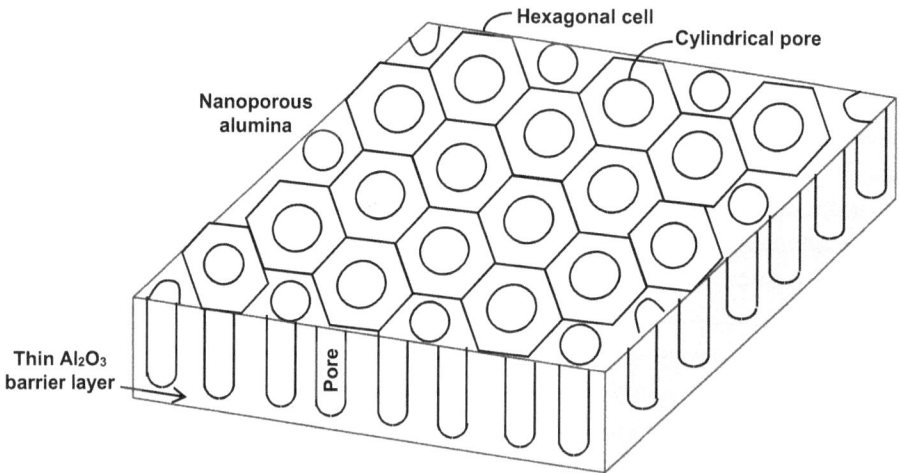

(b)

FIGURE 5.8 Fabrication of AAO template for use as a shadow mask for deposition of Ag nanoparticles by thermal evaporation: (a) anodic oxidation of aluminum sheet in (oxalic acid + malonic acid) mixture to form aluminum oxide film, (b) removal of aluminum sheet in saturated solution of iodine in methanol, and (c) removal of bottom aluminum oxide barrier layer in 5% dilute phosphoric acid solution to obtain a through-hole mask.

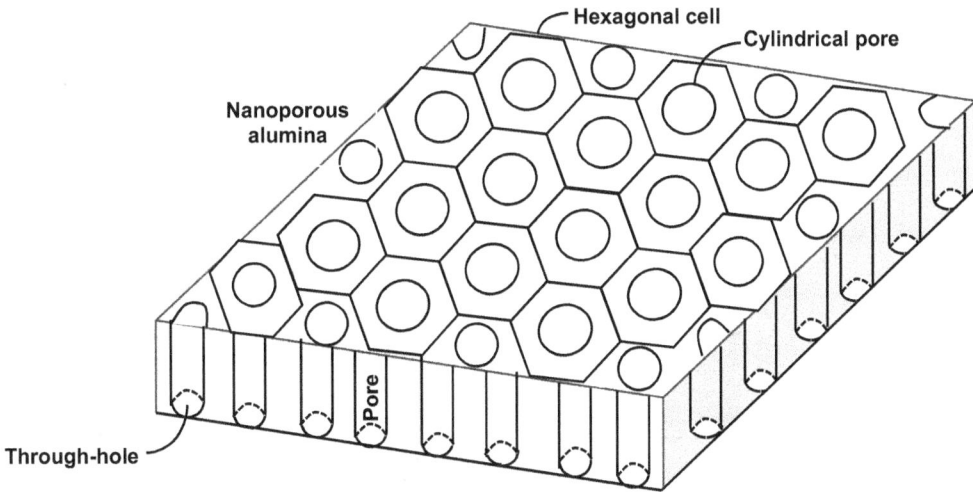

(c)

FIGURE 5.8 (Continued)

For deposition of silver nanoparticles, the top gold and P-GaAs cap layers are selectively etched to expose the underlying P-AlGaAs window layer of GaAs solar cell. The AAO template is placed as a mask over the window layer. Through this mask, silver is thermally evaporated over the window layer. Thus, an array of silver nanoparticles is formed on the window layer. The geometrical parameters diameter, distance, and height of the nanoparticles are ascertained from template structure and deposition conditions. The diameter of the nanoparticles obtained is determined by the diameter of the pores in the AAO template. The separation between the nanoparticles is the same as the interpore distance in the template. The height of the nanoparticles equals the thickness of the Ag film deposited. After the nanoparticle deposition has been completed, annealing is done at 200°C to fix the nanoparticles to the AlGaAs surface.

For solar cells decorated with densely formed high-aspect-ratio Ag nanoparticles of diameter 110nm having a height of 220nm and density of $3.3 \times 10^9 cm^{-2}$, the forward current density of the solar cell is 11.9mAcm^{-2}, open-circuit voltage is 0.76, fill factor is 0.65V, and the efficiency is 5.9%. The values of these parameters for the reference cell without any nanoparticles are 11.0 mAcm^{-2}, open-circuit voltage is 0.73, fill factor is 0.55V, and the efficiency is 4.7%, respectively. Thus, the plasmonic effect of Ag nanoparticles on solar cell is convincingly validated by this study.

The plasmonic nanoparticles have two effects:

(i) The optical path length is increased due to intense scattering by the surface plasmons in nanoparticles. This scattering effect becomes more pronounced due to the interactions among the surface plasmons in the nanoparticles. Consequently, more light is absorbed in the active layer of the solar cell.

(ii) The sheet resistance of the surface of solar cell is decreased by the lower sheet conductance of the array of nanoparticles (Nakayama et al 2008).

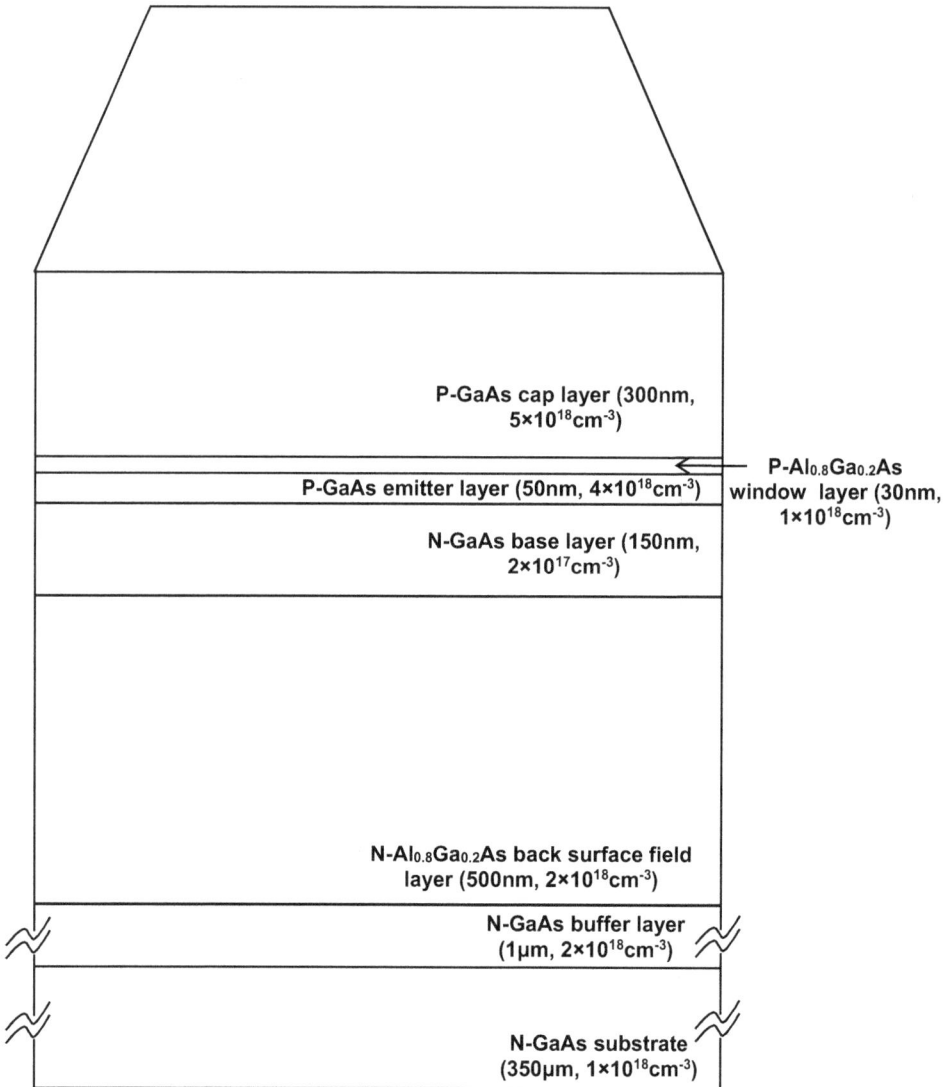

(a)

FIGURE 5.9 Fabrication of GaAs solar cell decorated with silver nanoparticles: (a) formation of stack of III-V semiconductor layers by metal organic chemical vapor deposition (MOCVD) using zinc for P-type doping and silicon for N-type doping, (b) deposition of gold films on both sides of the stack, (c) selective etching of Au layer and cap layer on the front side in citric acid: H_2O_2, placement of AAO template on III-V semiconductor layer stack, and thermal evaporation of silver with template as a shadow mask forming an array of Ag nanoparticles on the surface of the solar cell in which the diameters of the nanoparticles and the spacing between nanoparticles is same as in the mask; annealing is done at 200°C for ½ h. in (5%H_2 + 95%N_2) ambience.

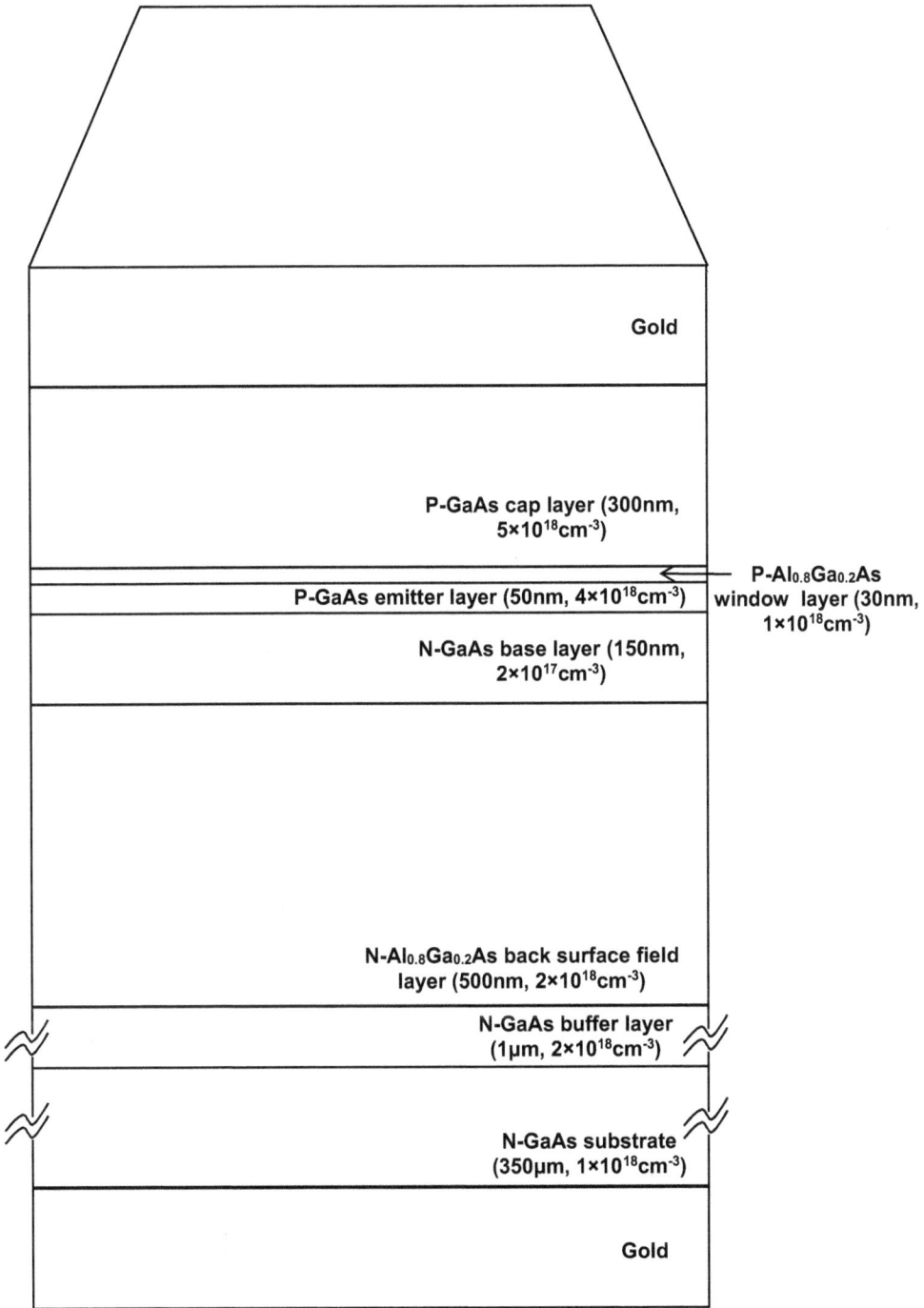

Gold

P-GaAs cap layer (300nm,
5×10^{18}cm^{-3})

P-GaAs emitter layer (50nm, 4×10^{18}cm^{-3})

P-Al$_{0.8}$Ga$_{0.2}$As
window layer (30nm,
1×10^{18}cm^{-3})

N-GaAs base layer (150nm,
2×10^{17}cm^{-3})

N-Al$_{0.8}$Ga$_{0.2}$As back surface field
layer (500nm, 2×10^{18}cm^{-3})

N-GaAs buffer layer
(1μm, 2×10^{18}cm^{-3})

N-GaAs substrate
(350μm, 1×10^{18}cm^{-3})

Gold

(b)

FIGURE 5.9   (Continued)

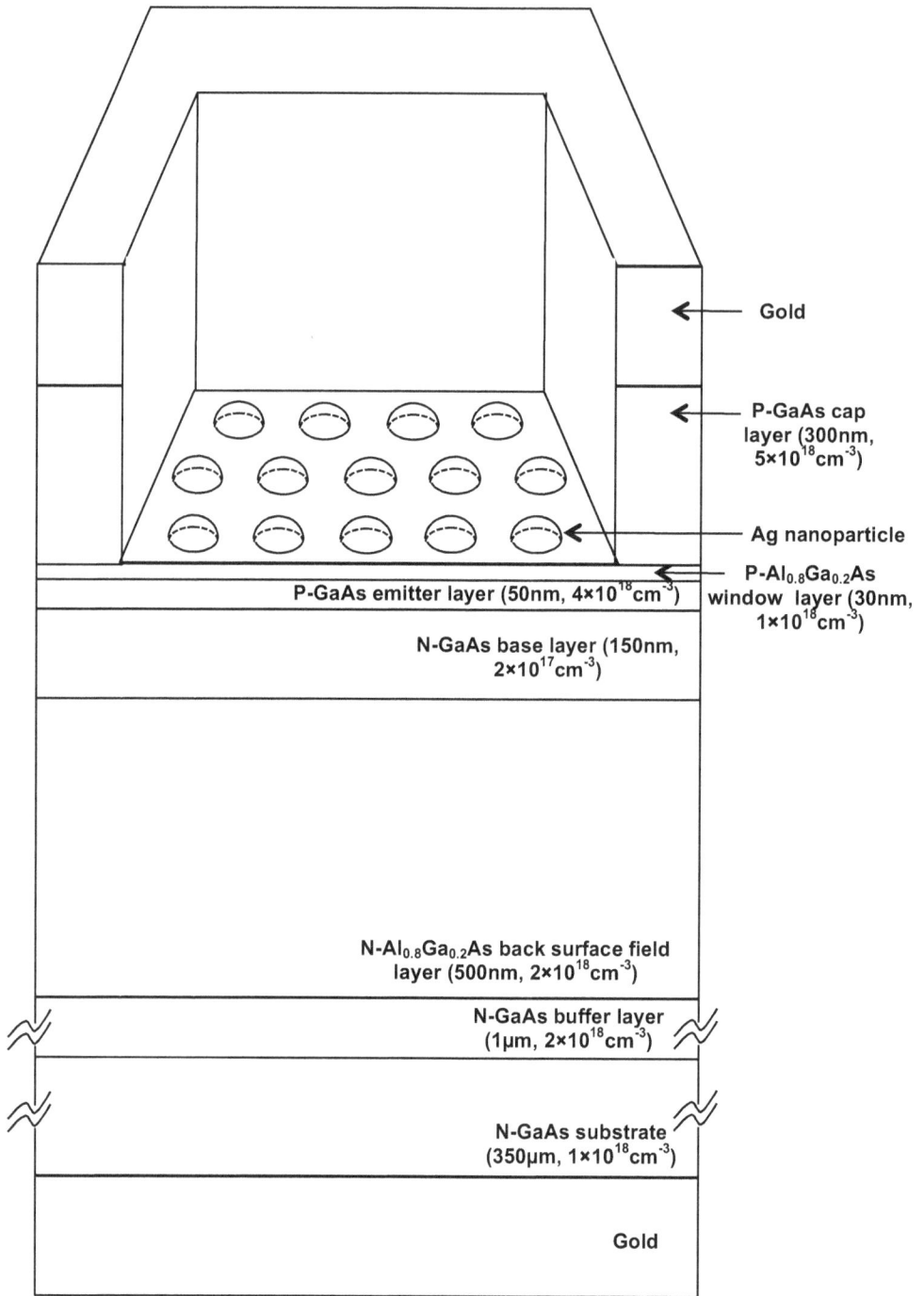

Gold

P-GaAs cap layer (300nm, $5\times10^{18}\text{cm}^{-3}$)

Ag nanoparticle

P-Al$_{0.8}$Ga$_{0.2}$As window layer (30nm, $1\times10^{18}\text{cm}^{-3}$)

P-GaAs emitter layer (50nm, $4\times10^{18}\text{cm}^{-3}$)

N-GaAs base layer (150nm, $2\times10^{17}\text{cm}^{-3}$)

N-Al$_{0.8}$Ga$_{0.2}$As back surface field layer (500nm, $2\times10^{18}\text{cm}^{-3}$)

N-GaAs buffer layer (1μm, $2\times10^{18}\text{cm}^{-3}$)

N-GaAs substrate (350μm, $1\times10^{18}\text{cm}^{-3}$)

Gold

(c)

FIGURE 5.9   (Continued)

5.6 PLASMONIC-ENHANCED ORGANIC SOLAR CELLS

5.6.1 LSPR EFFECT OF GOLD NANOSPHERES IN THE BUFFER LAYER ($\eta = 2.36\%$)

This solar cell has the structure (Qiao et al 2011) ITO-coated glass substrate/(PEDOT:PSS + Au NSs) buffer layer/(MEH-PP:PCBM) active layer/Al electrode; ITO and Al films act as contacts. The structure is shown in Figure 5.10.

Au nanospheres (NSs) of diameter 15nm are prepared by the sodium citrate reduction technique. PEDOT:PSS solution is doped with Au NSs. The Au NSs–doped solution is spin-coated on ITO-coated glass substrate to form a buffer layer. Stock solution prepared by dissolving MEH-PPV and PCBM in chlorobenzene is spun on the buffer layer. Aluminum electrode is made by thermal evaporation. The process is very appealing because of ease of fabrication.

For pristine PEDOT:PSS solar cells, the efficiency is 1.99%, whereas solar cells with 15nm-diameter Au nanospheres in the PEDOT:PSS layer show an efficiency of 2.36%, giving an improvement of $(2.36-1.99) \times 100/1.99 = 0.37 \times 100/1.99 = 18.59\%$ with respect to pristine solar cells.

Due to LSPR, the electromagnetic field around Au NSs is increased. Through scattering, absorption of light in the active layer is promoted. The stronger the electric field and the more the light harvested, the greater the probability of generation of excitons and their dissociation (Qiao et al 2011).

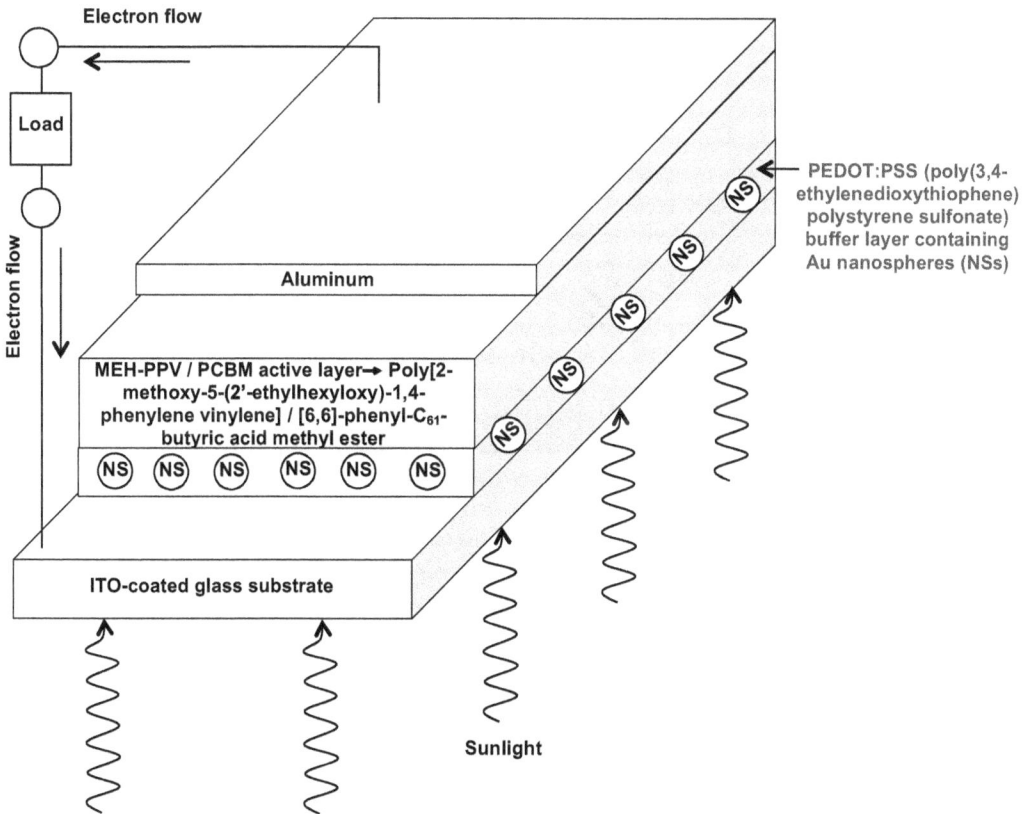

FIGURE 5.10 Organic solar cell with gold nanospheres in the PEDOT:PSS buffer layer. The device is made on an ITO glass substrate, upon which there is a PEDOT:PSS buffer layer doped with gold nanospheres. Over this buffer layer lies an active layer of MEH-PPV/PCBM. The active layer is covered with an aluminum electrode. When sunlight illuminates the glass substrate, the electrical output is drawn between the Al and ITO electrodes. LSPR in Au nanospheres increases the ability of solar cell to garner light.

5.6.2 COMBINED SURFACE PLASMON EFFECTS FROM AG NANODISKS IN HOLE TRANSPORT LAYER AND 1D-IMPRINTED AL GRATING OF A BULK HETEROJUNCTION SOLAR CELL (η = 3.59%)

This solar cell has the structure (Putnin et al 2019) ITO-coated glass substrate/(PEDOT:PSS + Ag NDs) layer/(P3HT:PCBM) active layer/Al grating; ITO and Al films act as contacts. The solar cell is shown in Figure 5.11.

PDMS is poured into the master template. After curing in vacuum oven and cooling to room temperature, the master template is detached from the PDMS mold. Silver nanodisks (Ag NDs) are synthesized. Solution of hole-transporting material PEDOT:PSS is mixed with silver nanodisks, spin-coated on ITO substrate and dried in a vacuum oven. P3HT and $PC_{61}BM$ are blended together in 1, 2-dichlorobenzene. The $(P3HT + PC_{61}BM)$ photoactive layer is coated by spinning over the (PEDOT:PSS + Ag NDs) layer. It is imprinted with the PDMS mold with the grating pattern in the vacuum oven. Thermal evaporation of aluminum is done. Aluminum is annealed in a vacuum oven.

The efficiency of control device without any grating or nanoparticles is 2.98%. When the grating is used, the efficiency rises to 3.16%. Without grating, introduction of silver nanodisks of diameter 70nm in HTL leads to a solar cell with efficiency of 3.37%, whereas for the device with (grating + silver nanodisks), the efficiency is 3.59%, showing a $\{(3.59-2.98) \times 100\}/2.98 = (0.61 \times 100)/2.98 = 20.47\%$ improvement with respect to the control device. The short-circuit current density of 3.59% efficiency solar cell is $7.41 mAcm^{-2}$, the open-circuit voltage is 0.63V, and the fill factor is 0.57. For 30nm-diameter silver nanodisks, the efficiency of (grating + silver nanodisks) device reduces to 3.42%.

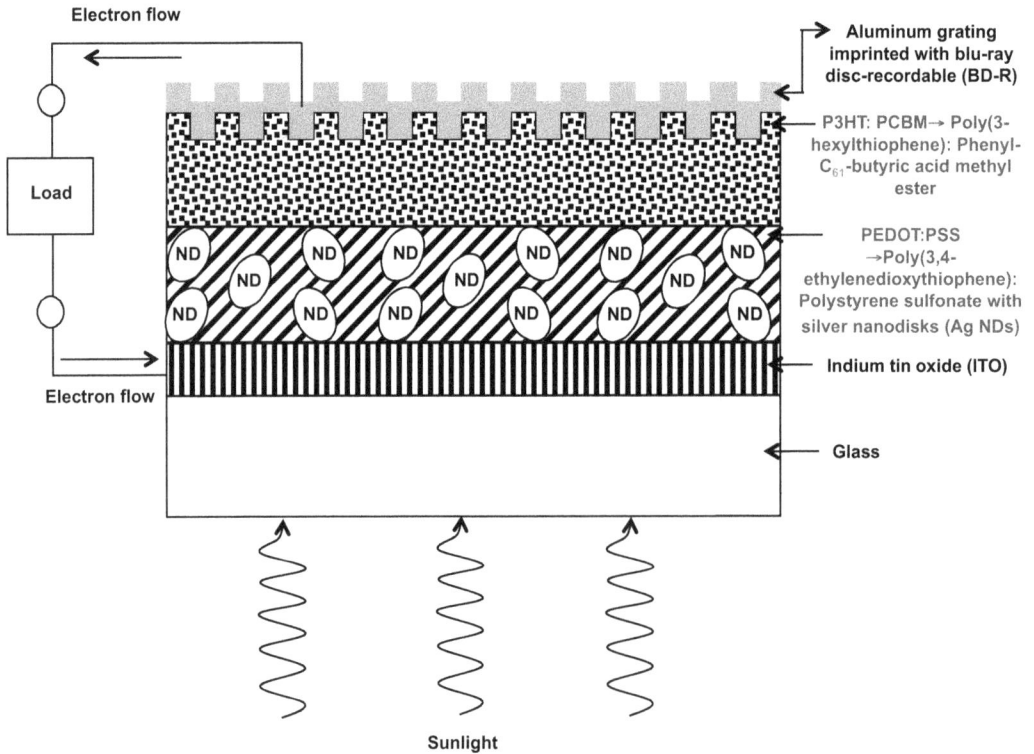

FIGURE 5.11 Aluminum grating-coupled organic thin-film solar cell comprising layers from bottom to top: indium tin oxide glass substrate, PEDOT:PSS hole transport layer (HTL) containing silver nanodisks, P3HT:PCBM photoactive medium, and aluminum electrode. When sunlight falls on glass, the current output is collected between the ITO and Al electrodes. Combined plasmonic action of Al grating and silver nanodisks enhances solar cell performance.

The plasmonic effects of aluminum grating and silver nanodisks act together cooperatively. They lead to broader enhancement of absorption of light. Exciton generation rate and efficiency of dissociation are increased. The contributions to plasmonic effects from Ag NDs and Al grating are revealed from the simulations of electric field distribution adjoining the surfaces of nanodisks and the grating (Putnin et al 2019).

5.6.3 Multiple Effects of Au Nanoparticles Embedded in the Buffer Layer of Inverted Bulk Heterojunction Solar Cell ($\eta = 7.86\%$)

This solar cell has the structure (Chi et al 2015) ITO-coated glass substrate/(Au NPs:ZnO) buffer layer/(PBDTTT-CF:PC$_{70}$BM)active layer/(MoO$_3$/Ag hole harvesting layer); ITO and Ag films act as contacts. The solar cell is shown in Figure 5.12.

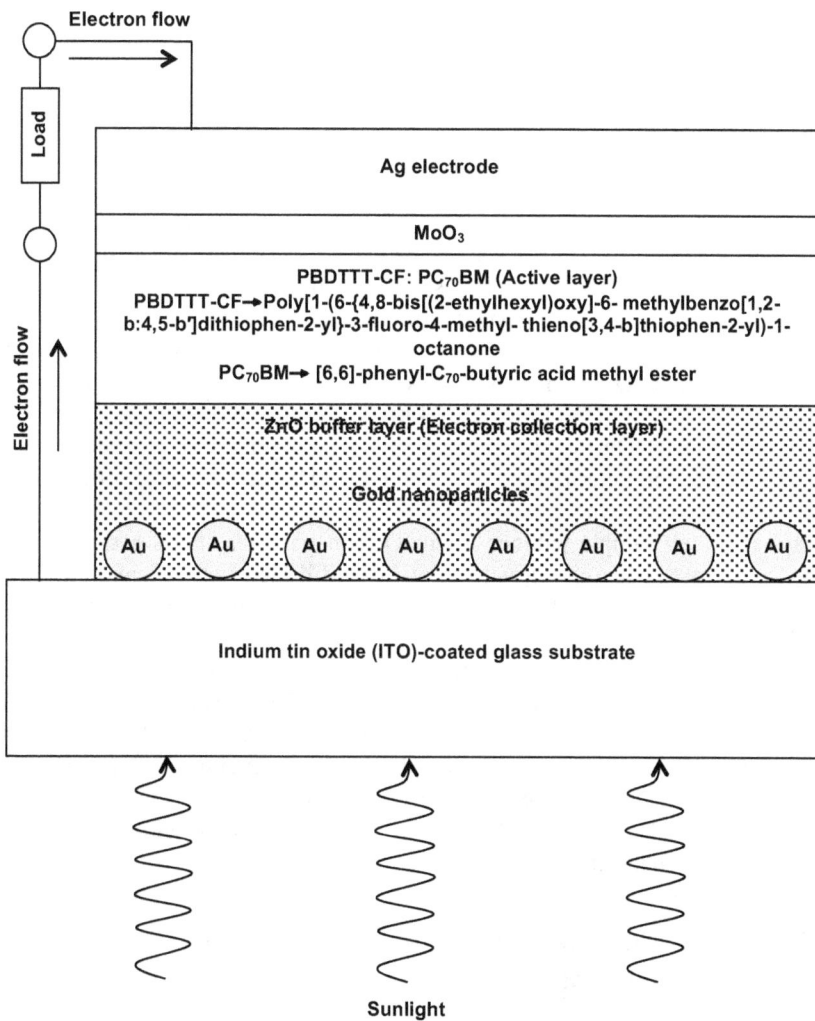

FIGURE 5.12 Inverted bulk heterojunction organic solar cell with gold nanoparticles embedded in the ZnO buffer layer for LSPR and scattering light. The diagram shows the sequence of layers from bottom to top: ITO-coated glass, the buffer layer made of AuNPs:ZnO, the PBDTTT-CF:PC$_{70}$BM active layer, and the MoO$_3$/Ag hole harvesting layer. Direction of incidence of sunlight is from the ITO glass side. The electrical output is derived between ITO and Ag electrodes.

Au nanoparticles are synthesized by reducing chloroauric acid with sodium citrate. Zinc acetate dihydrate and 2-ethanolamine are dissolved in methoxy-ethanol for preparing the ZnO precursor solution. After exposing ITO substrate to oxygen plasma, first the Au nanoparticles solution is spin-coated on the substrate, and then the ZnO solution. Annealing is done in air at 200°C for 10 min. PBDTTT-CF: $PC_{70}BM$ solution in o-dichlorobenzene (ODCB) is spun over the ZnO film. Finally, MoO_3 and Ag electrodes are deposited.

For solar cell fabricated without AuNPs, the short-circuit current density of the solar cell is 14.49mAcm^{-2}, the open-circuit voltage is 0.746V, the fill factor is 61.7%, and the efficiency is 6.67%.

When Au NPs are incorporated in the ZnO buffer layer of the solar cell, the efficiency increases with increasing diameter of nanoparticles up to 41nm, where it peaks at 7.86% with short-circuit current density = 15.81mAcm^{-2}, the open-circuit voltage = 0.751V, and fill factor = 66.2%.

LSPR and scattering effects of Au nanoparticles increase the absorption of light. Furthermore, the generation rate and efficiency of dissociation of excitons are augmented, facilitating the collection of carriers. The resistance of the solar cell is also decreased (Chi et al 2015).

5.7 PLASMONIC-ENHANCED PEROVSKITE SOLAR CELLS

5.7.1 REDUCED EXCITON BINDING ENERGY EFFECT IN PEROVSKITE SOLAR CELL WITH CORE-SHELL METAL NANOPARTICLES (η = 11.4%)

This solar cell has the structure (Zhang et al 2013) FTO-coated glass substrate/TiO_2 ETL/alumina layer with Au@SiO$_2$ core-shell NPs/perovskite absorber layer/spiro-OMeTAD HTL/Ag electrode; FTO and Ag films act as contacts. Figure 5.13 shows the solar cell.

A compact anatase TiO_2 layer is deposited on a fluorine-doped tin oxide (FTO) substrate. Au@SiO$_2$ core-shell nanoparticles (80nm-diameter Au NPs with 8nm-thick SiO$_2$ shells) are mixed with the colloidal solution of Al_2O_3. A (mesoporous Al_2O_3 + Au@SiO$_2$ NPs) layer is formed over the TiO_2 layer. The precursor solution of the perovskite is spin-coated on the alumina film and dried. This is the absorber layer. On the absorber layer, P-type organic hole conductor layer is spin-coated. The hole conductor material used is spiro-OMeTAD. The contact to spiro-OMeTAD layer is established with thermally evaporated silver film.

The control solar cell sample without Au@SiO$_2$ nanoparticles shows a short-circuit current density of 14.76 mAcm^{-2}, open-circuit voltage of 1.04V, fill factor of 0.67, and power conversion efficiency of 10.7%. These parameters for the solar cell with Au@SiO$_2$ nanoparticles are 16.91 mAcm^{-2}, 1.02V, 0.64, and 11.4%.

The improvement in solar cell performance is not attributed to increase in absorption of light, as commonly believed. Instead, a photoluminescence investigation shows that the effect of incorporation of core-shell nanoparticles in the perovskite absorber layer lowers the binding energy of excitons from 100meV to 35meV. As a result of this diminution in exciton binding energy, more free carriers are produced, yielding a larger photocurrent. This provides a new insight into the effect of inclusion of nanoparticles, which could serve as a facile method of tuning the exciton binding energy in a perovskite (Zhang et al 2013).

5.7.2 LSPR EFFECT OF GOLD NANORODS IN THE ELECTRON TRANSPORT LAYER OF INVERTED PEROVSKITE SOLAR CELL (η = 13.7%)

This solar cell has the structure (Cui et al 2016) FTO-coated glass substrate/NiO$_x$ HTL/Al_2O_3 film with Au@SiO$_2$ core-shell NRs/perovskite absorber layer/BCP ETL/Au electrode; FTO and Au films act as contacts. The structure of the solar cell is displayed in Figure 5.14.

Spray pyrolysis technique is applied for depositing P-type selective NiO$_x$ layer on fluorine-doped tin oxide (FTO)–coated glass substrate. Al_2O_3 nanoparticles are dispersed in isopropanol. This dispersion is thoroughly mixed with α-terpinol and ethanol solution of ethyl cellulose, forming

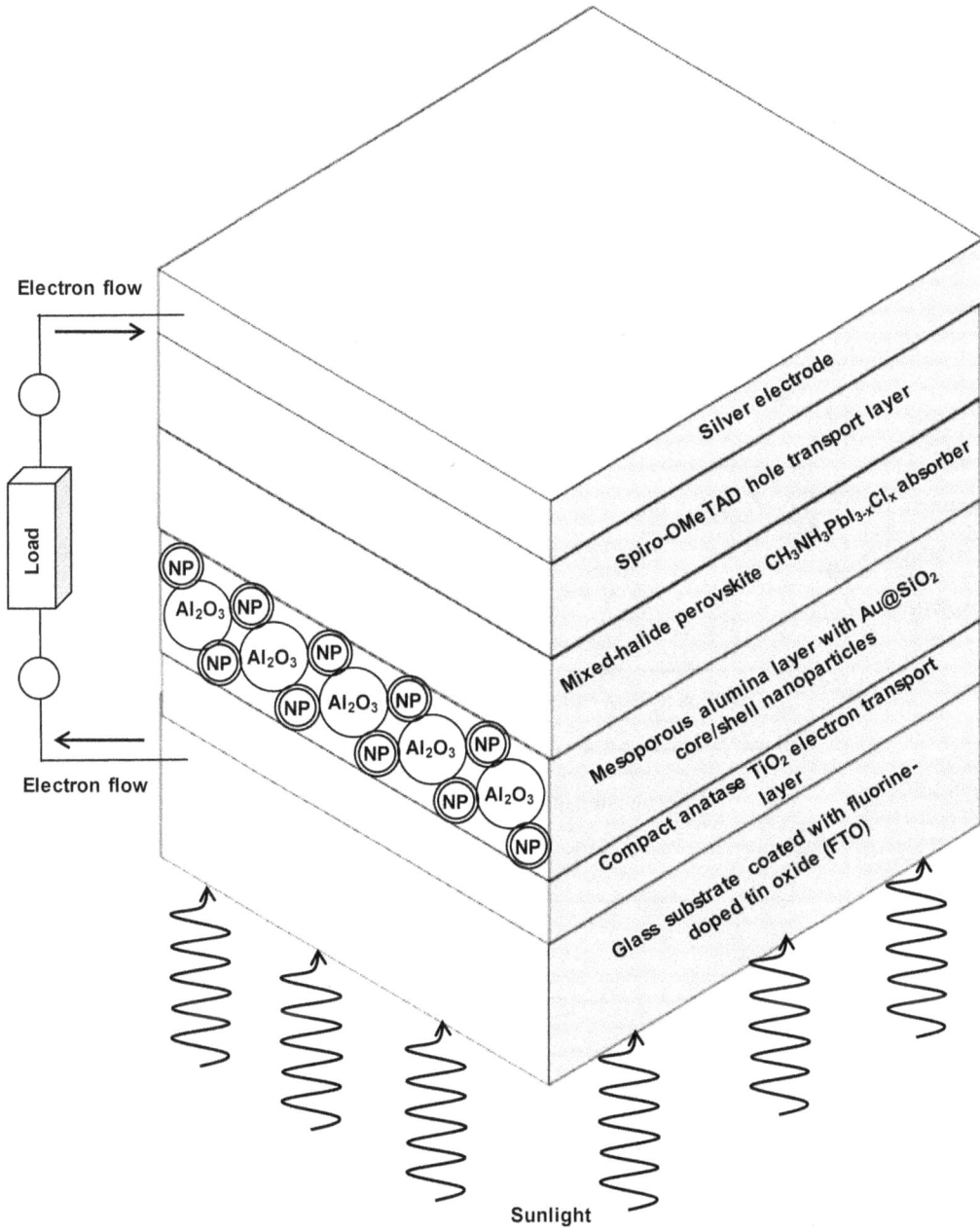

FIGURE 5.13 Schematic diagram of the perovskite-based solar cell with metal nanoparticles showing the layers from bottom upwards: the FTO-coated glass, the titania layer (ETM), the meso Al_2O_3 film with Au@SiO_2 nanoparticles, the perovskite layer, the spiro-OMeTAD layer (HTM), and the top silver electrode. Illumination of the FTO substrate with sunlight leads to electron-hole pair generation in the perovskite layer, and current is drawn across the FTO and silver electrodes. The Au@SiO_2 nanoparticles exert their effect by decreasing the binding energy of excitons, not by increase in absorption of light.

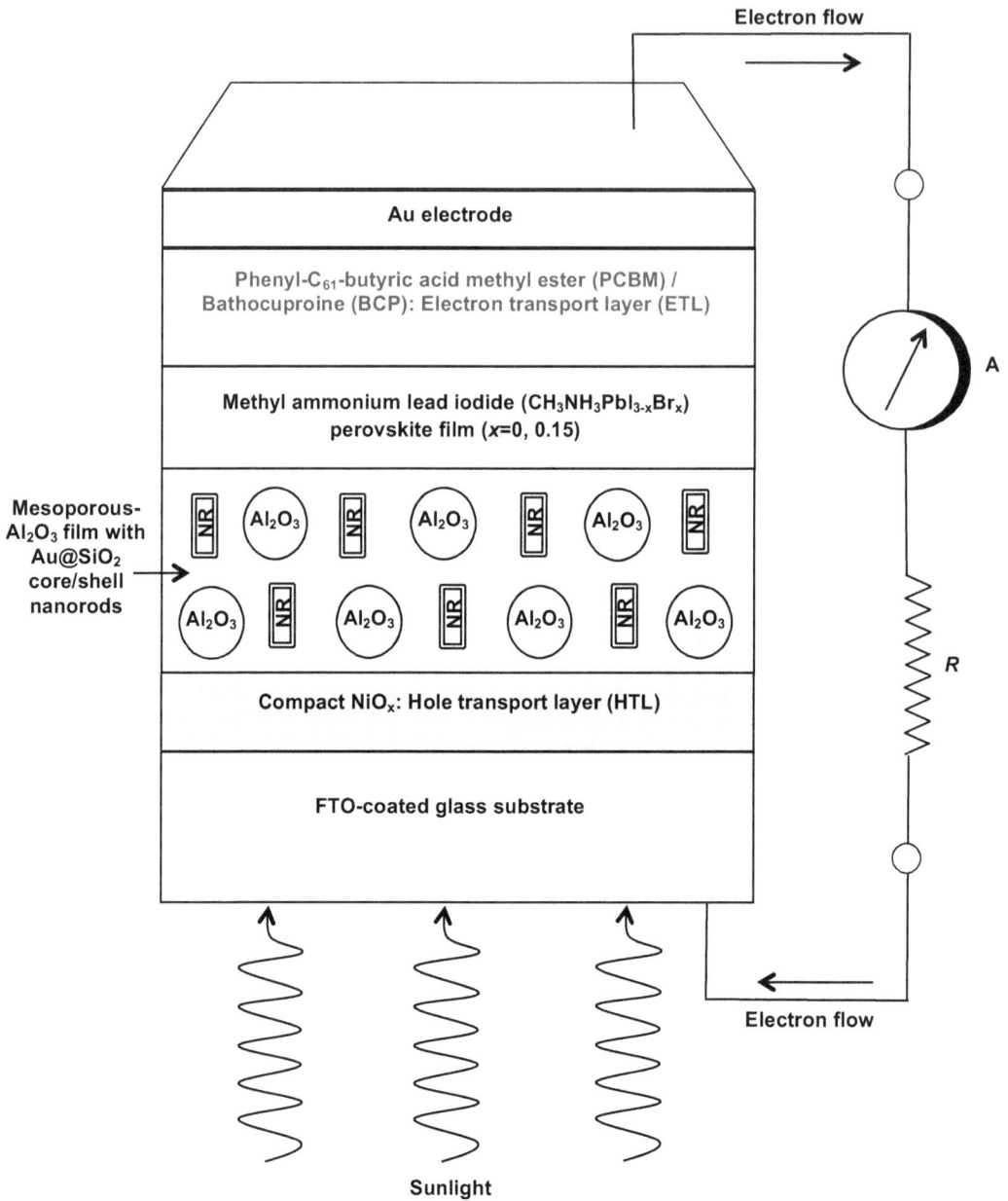

FIGURE 5.14 Diagrammatic representation of inverted perovskite solar cell (IPSC) consisting of the following layers from bottom upwards: glass substrate coated with fluorine-doped tin oxide (FTO), the electron transport layer made of NiO_x, transparent meso-Al_2O_3 film with Au@SiO_2 nanorods for LSPR effect, methyl ammonium lead iodide perovskite thin film acting as the absorber layer, PCBM/BCP electron transport layer, and the gold electrode. Sunlight falling on the FTO-coated glass substrate produces photogenerated carriers in the perovskite active layer. The current output is taken between FTO and Au electrodes.

a homogeneous paste. Au@SiO$_2$ NRs solution in ethanol is added. This Au@SiO$_2$ NRs/Al$_2$O$_3$ paste is spin-coated on NiO$_x$ layer and sintered at 550°C for ½ h. Over the Au@SiO$_2$ NRs/Al$_2$O$_3$ layer, PbI$_2$ in DMF solution is spin-coated. MAI/MABr in 2-propanol (MA = CH$_3$NH$_3^+$) solution is spin-coated on PbI$_2$ film. A compact CH$_3$NH$_3$PbI$_{3-x}$Br$_x$ perovskite film is formed after annealing at 100°C for 25 min. Spin-coating of PCBM and BCP film is done in a glove box. Thermal evaporation of gold is done in vacuum.

For $x = 0$, the perovskite is CH$_3$NH$_3$PbI$_{3-0}$Br$_0$ or CH$_3$NH$_3$PbI$_3$. The efficiency of solar cell fabricated using CH$_3$NH$_3$PbI$_3$ on Al$_2$O$_3$ only without AuNRs@SiO$_2$ is 11.3%. But the efficiency of solar cell made from CH$_3$NH$_3$PbI$_3$ on Al$_2$O$_3$ with AuNRs@SiO$_2$ is 12.2%.

For $x = 0.15$, the perovskite is CH$_3$NH$_3$PbI$_{3-0.15}$Br$_{0.15}$ or CH$_3$NH$_3$PbI$_{2.85}$Br$_{0.15}$. The efficiency of solar cell fabricated using CH$_3$NH$_3$PbI$_{2.85}$Br$_{0.15}$ on Al$_2$O$_3$ only without AuNRs@SiO$_2$ is 10.7%. But the efficiency of solar cell made from CH$_3$NH$_3$PbI$_{2.85}$Br$_{0.15}$ on Al$_2$O$_3$ with AuNRs@SiO$_2$ is 13.7%.

Excitation of LSPR in Au@SiO$_2$ nanorods increases the electric field locally. LSPR in the wavelength range 740–860nm is expected by resonant interaction (Cui et al 2016).

5.8 DISCUSSION AND CONCLUSIONS

Plasmonic nanoparticles are placed inside or outside the active layer of the solar cell. Plasmonic effects can be invoked as LSPR and surface plasmon polaritons. LSPR is associated with increase of electric field locally in the neighborhood of nanoparticles. Since the absorption of light is proportional to the intensity of field, the increase in near field increases the absorption of light in the areas proximate to the nanoparticles. Further, the nanoparticles act as scattering centers that scatter light in different directions inside the solar cell. As light propagates along a distribution of angles, the optical path length is enhanced so that adsorption of light in solar cell material is improved. The near-field effects combined with the far-field scattering enable use of thin absorber layers and therefore help in making thinner cells.

For evoking PSPR effects, Al gratings have been used on the backside of the solar cell. A glance of LSPR and PSPR plasmonic effects in solar cells is given by Table 5.1.

TABLE 5.1

Plasmonic and Related Effects in Solar Cells

Sl. No.	Solar Cell	Plasmonic/Other Effects	Efficiency (%)	Reference
1.	Plasmonic-enhanced GaAs solar cell decorated with Ag nanoparticles	Optical path length increased by silver nanoparticles fixed on the surface of the window layer of solar cell	5.9	Nakayama et al (2008)
2.	Organic solar cell with gold nanospheres in the buffer layer	LSPR	2.36	Qiao et al (2011)
3.	Bulk heterojunction organic solar cell with Ag nanodisks in HTL and 1D-imprinted Al grating	Joint SPR action of Ag nanodisks and Al grating	3.59	Putnin et al (2019)
4.	Inverted bulk heterojunction organic solar cell with Au NPs in the buffer layer	LSPR, light scattering, efficient dissociation of excitons, and reduction of cell resistance	7.86	Chi et al (2015)
5.	Perovskite solar cell with Au@SiO$_2$ core-shell metal nanoparticles	Decrease of exciton binding energy	11.4	Zhang et al (2013)
6.	Inverted perovskite solar cell with Au nanorods in ETL	LSPR	13.7	Cui et al (2016)

Questions and Answers

5.1   Is plasmon the same thing as plasma? Answer: No. Plasma is a state of matter, while plasmon is a quantum of plasma oscillations.

5.2   What is a surface plasmon? Answer: A collective electron oscillation (periodic charge density variation) propagating along a metal/dielectric interface in which electronic motions and electromagnetic waves coexist.

5.3   Are surface plasmons and surface plasmon polaritons the same thing? Answer: Yes.

5.4   What is a polariton? Answer: A hybrid particle comprising a photon strongly coupled to an electric dipole.

5.5   What is a localized surface plasmon? Answer: A plasmon localized on the surface of a metallic nanoparticle.

5.6   What is the difference between PSPR and LSPR? Answer: PSPR moves; LSPR remains fixed.

5.7   What is an evanescent field? Answer: An oscillating electric or magnetic field whose energy is not propagated as an electromagnetic wave but remains concentrated near the source.

5.8   What is an evanescent wave? Answer: A nonpropagating wave whose amplitude decays rapidly and which does not transport energy.

5.9   Rayleigh scattering is caused by particles having sizes much larger than the wavelength of light. Yes or no? Answer: No, much smaller.

5.10 Mie scattering is caused by particles having sizes similar to or larger than the wavelength of light. Yes or no? Answer: Yes.

5.11 Both Rayleigh and Mie scatterings are inelastic in nature. Yes or no? Answer: No, both are elastic in nature.

5.12 Distinguish between absorption and scattering of light. Answer: In absorption, all the energy of the photon is given to the absorber and the photon no longer exists. In scattering, light retains its energy (elastic scattering) or loses a portion of its energy (inelastic scattering), accompanied by its radiation in different directions.

5.13 Light of which wavelength is scattered more during Rayleigh scattering: 400nm or 500nm? Answer: 400nm. Rayleigh scattering is inversely proportional to the fourth power of wavelength of light, which means that shorter wavelengths (violet, blue) will scatter more than longer wavelengths (orange, red).

5.14 Does the wavelength of light change during elastic scattering? Answer: No. In elastic scattering, there is no change in energy E of the photon, and therefore its wavelength λ, because energy E is related to frequency ν by the relation $E = h\nu$, and frequency ν is linked to wavelength λ by the formula $\nu = c/\lambda$. The h is Planck's constant, and c is the velocity of light.

5.15 What are the units of absorption and scattering cross sections? Answer: Units of area (barn = 10^{-28}m^2).

5.16 Name three parameters on which the resonance frequency of LSPR depends. Answer: Plasma frequency ω_{Plasma} of the free electron gas, relaxation time $\tau_{Relaxation}$ of electron, and relative permittivity ε_{Medium} of surrounding semiconductor medium.

5.17 What is red shifting of LSPR resonance frequency by embedding metallic nanoparticles? Answer: Decrease in frequency due to the effect of dielectric constant of embedding medium.

5.18 What is the effect of increase in local electric field of light during LSPR? Answer: Increase in carrier photogeneration rate.

5.19 How is the scattering of light by a metal nanoparticle affected by LSPR? Answer: Increased.

5.20 How do the absorption and scattering cross sections of a spherical-shaped metallic nanoparticle vary with radius of the nanoparticle? Answer: Absorption cross section $\propto$ (Radius)^3 while scattering cross section μ(Radius)^6.

5.21 For enhanced scattering light, it is necessary to use metal nanoparticles with (a) small scattering-to-absorption cross-section ratio, (b) medium scattering-to-absorption cross-section ratio, (c) large scattering-to-absorption cross-section ratio? Answer: c.

5.22 Why can't free-space sunlight photons activate SPPs? Answer: Because of mismatching of momentum of sunlight photons with that of plasmons.

5.23 What is the advantage of sticking Ag nanoparticles on the surface of the window layer of a GaAs solar cell? Answer: They extend optical path length through strong scattering by surface plasmons to increase the short-circuit current and lower the sheet resistance of the surface as indicated by an improvement in fill factor.

5.24 What is gained by the incorporation of gold nanospheres in the buffer layer of an organic solar cell? Answer: Electric field around NSs is intensified, and scattering of light is increased. Both factors increase photocarrier generation.

5.25 Silver nanodisks are inserted in the hole transport layer and Al grating on the backside. What are the effects? Answer: PSPR effect of Al grating and LSPR effect of Ag nanodisks.

5.26 How does the introduction of an alumina layer with Au@SiO$_2$ core-shell NPs influence the working of a perovskite solar cell? Answer: By lowering the binding energy of excitons so that more excitons undergo dissociation, producing a larger number of free electrons and holes.

5.27 How does the incorporation of Au nanorods in the electron transport layer of an inverted perovskite solar cell help? Answer: By LSPR effect.

REFERENCES

Chi D., S. Lu, R. Xu, K. Liu, D. Cao, L. Wen, Y. Mi, Z. Wang, Y. Lei, S. Qu and Z. Wang 2015 Fully understanding the positive roles of plasmonic nanoparticles in ameliorating the efficiency of organic solar cells, Nanoscale, 7: 15251–15257.

Chou C.-H. and F.-C. Chen 2014 Plasmonic nanostructures for light trapping in organic photovoltaic devices, Nanoscale, 6: 8444–8458.

Cui J., C. Chen, J. Han, K. Cao, W. Zhang, Y. Shen and M. Wang 2016 Surface plasmon resonance effect in inverted perovskite solar cells, Advanced, Science, 3(1500312): 8 pages.

Gangadharan D. T., Z. Xu, Y. Liu, R. Izquierdo and D. Ma 2017 Recent advancements in plasmon-enhanced promising third-generation solar cells, Nanophotonics, 6(1): 153–175.

Jain P. K., K. S. Lee, I. H. El-Sayed and M. A. El-Sayed 2006 Calculated absorption and scattering properties of gold nanoparticles of different size, shape, and composition: Applications in biological imaging and biomedicine, Journal of Physical Chemistry B, 110: 7238–7248.

Jatschka J., A. Dathe, A. Csáki, W. Fritzsche and O. Stranik 2016 Propagating and localized surface plasmon resonance sensing—A critical comparison based on measurements and theory, Sensing and Bio-Sensing Research, 7: 62–70.

Nakayama K., K. Tanabe and H. A. Atwater 2008 Plasmonic nanoparticle enhanced light absorption in GaAs solar cells, Applied Physics Letters, 93: 121904–1 to 121904–3.

Putnin T., C. Lertvachirapaiboon, R. Ishikawa, K. Shinbo, K. Kato, S. Ekgasit, K. Ounnunkad and A. Baba 2019 Enhanced organic solar cell performance: Multiple surface plasmon resonance and incorporation of silver nanodisks into a grating-structure electrode, Opto-Electronic Advances, 2(7): 190010–1 to 190010–11.

Qiao L., D. Wang, L. Zuo, Y. Ye, J. Qian, H. Chen, S. He 2011 Localized surface plasmon resonance enhanced organic solar cell with gold nanospheres, Applied Energy, 88: 848–852.

Tanabe K. 2022 Chapter 2: Field enhancement around spherical metal nanoparticles and nanoshells, in: Plasmonics for Hydrogen Energy, Springer Nature, Cham, Switzerland AG, pp. 5–10.

Zhang W., M. Saliba, S. D. Stranks, Y. Sun, X. Shi, U. Wiesner and H. J. Snaith 2013 Enhancement of perovskite-based solar cells employing core-shell metal nanoparticles, Nano Letters, 13: 4505–4510.

6 Optically Improved Nanoengineered Solar Cells

6.1 INTROSPECTION ON LIGHT MANAGEMENT IN SOLAR CELLS

Looking back, the attention so far has been on some techniques for incoupling light into solar cell with the aim to get utmost light harvesting. These are briefly recollected in the following subsections.

6.1.1 ANTIREFLECTION COATING

It was seen that antireflection coatings having thickness equal to a quarter wavelength of incident sunlight are useful. But the solar spectrum is made of several wavelengths. So only those wavelengths, which are close to meeting the thickness criterion of coating, will be prevented from reflection, while upon receding away from this condition, the antireflective effect of the coating will become weaker. This is the reason that antireflective coatings are useful in a limited range of the optical spectrum. They cannot provide broadband antireflection commitment (Ahrlich et al 2016).

6.1.2 MICROPYRAMID-LIKE TEXTURING BY WET-ETCHING IN ALKALINE SOLUTIONS

Traditional solar cell manufacturing makes wide utilization of random pyramid texturing by anisotropic silicon etching. This texturing is limited to silicon wafers of < 100 > orientation. Its drawbacks include creation of surface defects and imperfections, which increase surface recombination. Residual etchant traces are difficult to remove. Additional pyramid smoothing treatments are necessary because sharp points may lead to short circuits. Contact formation by screen-printing is difficult due to gaps between the screen and the textured surface.

Random pyramid texturing is applicable to thick silicon wafers, but for thin silicon wafers, it loses significance because it only stretches and lengthens the path of light. It does not provide antireflection property arising from grading of refractive index. Thin wafers have much smaller path for light ways in which to produce carriers, and whatever is to be done is possible only with this smaller path, whereas in thicker wafers, the situation is not so critical.

6.1.3 NANOPYRAMID-LIKE TEXTURING BY LITHOGRAPHICAL TECHNIQUES

In Chapter 4, we talked about a light-trapping strategy in which nanotexturing of the front surface by forming pyramid-shaped or similar structures is combined with reflection from the rear contact acting as a back reflector. Light is unable to resolve the individual nanostructure of the array, and therefore a continuous grading of refractive index occurs, which inhibits reflection of light from the surface. In other words, the nanoscale features cushion the large difference between refractive indices of air and silicon by providing a tapered structural profile. Consequently, broadband antireflection and light trapping is observed (Chen and Hong 2016).

6.1.4 PLASMONIC EFFECTS OF METAL NANOPARTICLES OR THIN FILMS

In Chapter 5, plasmonic effects of metallic nanoparticles were presented as a method for increasing trapping of light. Absorption of light is intensified through LSPR and surface plasmon polariton excitation.

DOI: 10.1201/9781003215158-8

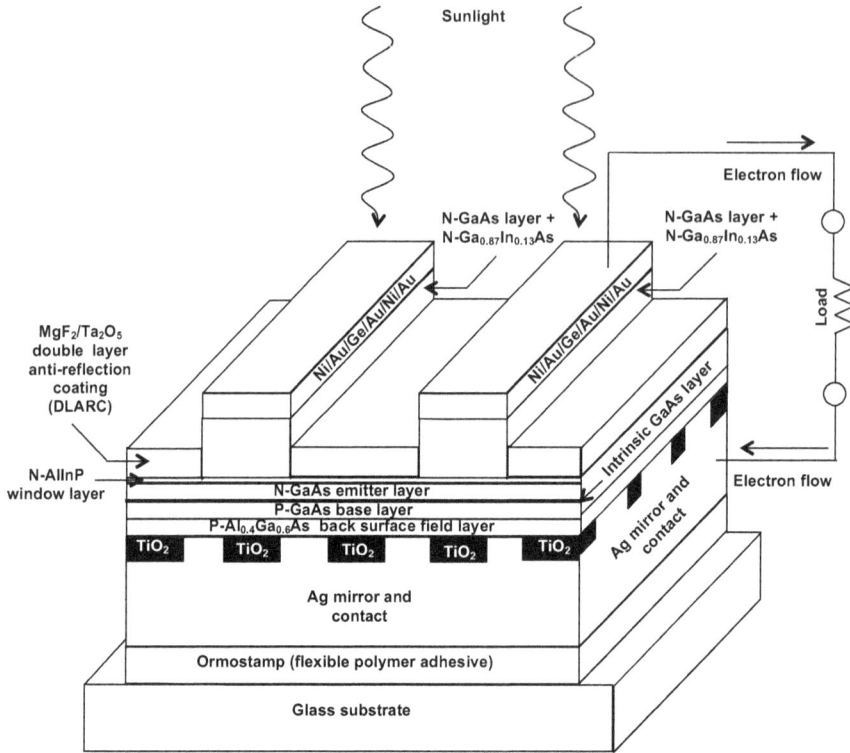

FIGURE 6.1 Ultrathin solar cell with GaAs absorber and nanostructured silver back mirror showing from bottom upwards: the glass substrate, the adhesive, the TiO_2/Ag mirror, the P-$Al_{0.4}Ga_{0.6}$As back surface field layer, the P-GaAs base layer, the N-GaAs emitter layer, the N-AlInP window layer, N-GaAs and N-$Ga_{0.87}In_{0.13}$As capping layers, the (Ni/Au/Ge/Au/Ni/Au) grid, and the MgF_2/Ta_2O_5 antireflection coating. Sunlight falls on the solar cell on the Ni/Au/Ge/Au/Ni/Au contact side, and the electrical output is delivered between the Ni/Au/Ge/Au/Ni/Au contact and silver mirror. The optical action of nanostructured back mirror plays a crucial role in this solar cell.

6.1.5 OTHER WAYS OF LIGHT TRAPPING

Now let us look forward and explore the opportunities that lie ahead. Let us inquire about novel ways of achieving higher light absorption in this chapter. Although nanotexturing has been delved into, carving nanostructures on the surface has not been considered. Let us first focus our attention towards the examination of the effects of nanostructuring of a surface on absorption of light.

6.2 ULTRATHIN GaAs ABSORBER (205NM) SOLAR CELL WITH TiO₂/AG NANOSTRUCTURED BACK MIRROR

This thin solar cell works on multiresonant absorption of light using a silver mirror (Chen et al 2019). Figure 6.1 presents the 3D diagram of the solar cell showing the various layers in the structure. The body of the solar cell has the structure N-AlInP window layer/N-GaAs emitter layer/ Intrinsic GaAs layer/P-GaAs base layer/P-$Al_{0.4}Ga_{0.6}$As back surface field layer/P$^+$-GaAs contact layer (from the front surface to back side). A multilayer metallization Ni/Au/Ge/Au/Ni/Au front contact is formed on the front surface of the solar cell, receiving sunlight. The front contact metallization layers are deposited over (N-GaAs + N-$Ga_{0.87}In_{0.13}$As) capping layers. MgF_2/Ta_2O_5 double-layer antireflection coating is applied. The back contact is made through a silver film on the backside of the solar cell. The film covers the spaces intervening TiO_2 nanopatterns on the P-$Al_{0.4}Ga_{0.6}$As back surface field of the solar cell. It is formed by conformally coating 200nm silver on the nanoimprinted TiO_2 portions to form a TiO_2/Ag mirror. Below the TiO_2/silver mirror is a glass substrate. The silver mirror is glued to the glass substrate.

In this solar cell, an antireflection coating deposited on the front surface restricts the reflection of incident sunlight. No nanotexturing is done on the front surface. Instead, the role of light management is entrusted to the nanostructured mirror on the rear surface.

6.2.1 Justification for Thinning of the Absorber Layer Together with Advanced Light Loss Reduction Technique

As already explained in Chapter 4, use of a thin absorber offers manifold advantages. Cost is a major factor of concern in solar cell fabrication. Although gallium arsenide is superior to silicon in many properties, production of large quantities of GaAs is exorbitantly expensive. So a lesser amount of GaAs must be consumed in solar cell fabrication. Making ultrathin cells is a way to material reduction. Moreover, if the absorber thickness is small, more carriers will be able to reach the contacts without recombination for a given minority carrier lifetime. Hence, lower-quality material can be tolerated, allowing for more cost savings.

Solar cells for space applications must be able to withstand the defects engendered by radiation. Ultrathin GaAs solar cells are shown to be intrinsically radiation-tolerant devices, eliminating the requirement of bulky cover glass. The efficiency of collection of carriers remains unimpaired in the presence of defects created from radiation bombardment in outer space. Choosing the proper light-trapping scheme helps to redeem any decrease in efficiency arising from creation of defects (Hirst et al 2017). From such arguments, it is clear that minimization of the absorber layer thickness must be accompanied with maximization of light trapping.

6.2.2 Multiresonant Absorption of Light

Multiresonant absorption is a light-trapping approach based on absorption of light through multiple resonant modes (Collin et al 2015). This technique is designed with the aspiration of reaching nearly ideal absorption of light in a wide range of optical spectrum, i.e., broadband absorption of light, which is an essential requirement for solar cell operation. The task of multiresonant absorption is accomplished by utilizing a large number of resonant modes produced in a periodical structure.

6.2.3 Location and Geometrical Parameters of the Nanostructured Mirror

For avoiding the optical losses through diffraction in free space, the periodic structure is positioned at the rear surface of the solar cell (Chen et al 2019). Then, diffraction losses in free space take place only after absorption. The period p of the periodic structure is taken as 700nm from the consideration of keeping the diffraction losses minimal. This period is decided because the diffraction losses increase when the period p is larger than the wavelength of light. Further, the larger the period p of the periodic pattern, the more the number of resonances. The value of period $p = 700$nm is chosen for provision of several resonances in the wavelength range 700–900nm. Regarding the material for the periodic structure, silver is selected as the material because it is the highest-reflectivity metal in this range of wavelengths. The width of the periodic structure is 420nm, and its height is 120nm. Thus, a nanostructured silver back mirror is designed.

6.2.4 Fabrication and Performance of the Solar Cell

On an N-type GaAs substrate, the different constituent layers in the solar cell structure are grown by MOCVD in the reverse order to the sequence of layers in the final fabricated solar cell. For the sake of clarity of presentation, the layers are drawn in Figure 6.2. As can be seen, the layers from bottom upwards are N-GaAs buffer layer/N-AlGaAs etch stop layer/(N-GaAs + N-$Ga_{0.87}In_{0.13}As$) capping layers/N-AlInP window layer/N-GaAs emitter layer/Intrinsic GaAs layer/P-GaAs base layer/P-$Al_{0.4}Ga_{0.6}As$ back surface field layer/P^+-GaAs contact layer. This series of layers contains two layers which will not be present in

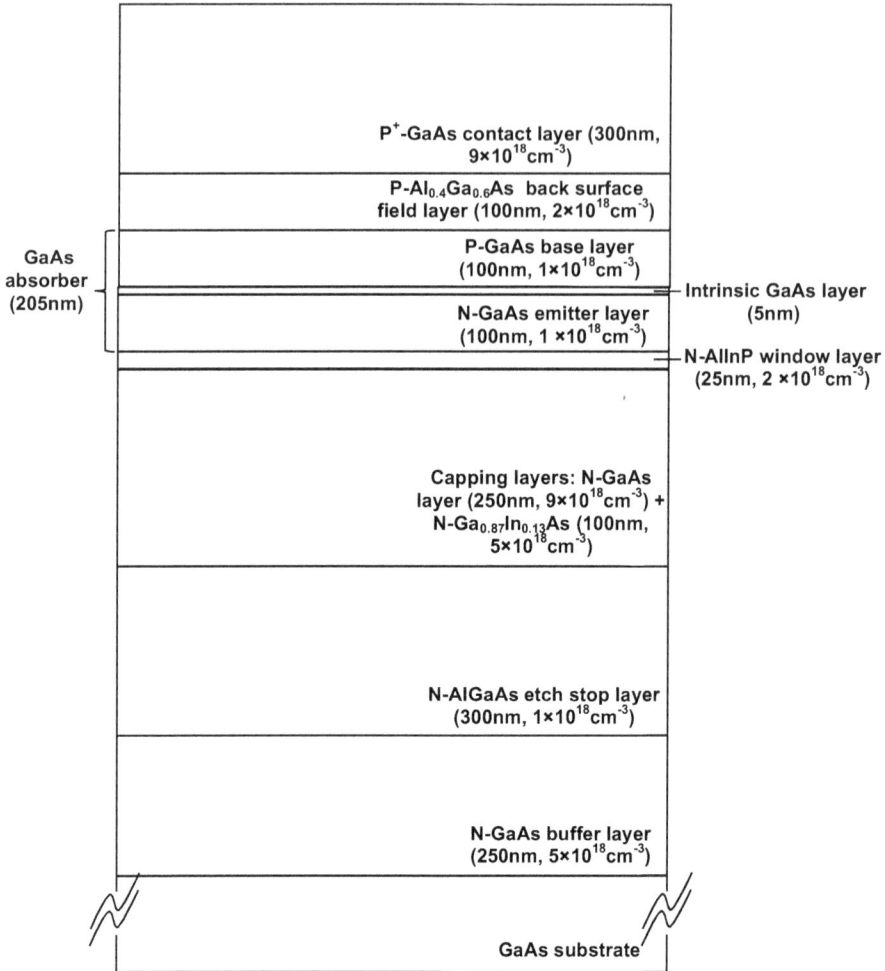

P⁺-GaAs contact layer (300nm, $9\times10^{18}cm^{-3}$)

P-Al$_{0.4}$Ga$_{0.6}$As back surface field layer (100nm, $2\times10^{18}cm^{-3}$)

P-GaAs base layer (100nm, $1\times10^{18}cm^{-3}$)

GaAs absorber (205nm)

N-GaAs emitter layer (100nm, $1\times10^{18}cm^{-3}$)

Intrinsic GaAs layer (5nm)

N-AlInP window layer (25nm, $2\times10^{18}cm^{-3}$)

Capping layers: N-GaAs layer (250nm, $9\times10^{18}cm^{-3}$) + N-Ga$_{0.87}$In$_{0.13}$As (100nm, $5\times10^{18}cm^{-3}$)

N-AlGaAs etch stop layer (300nm, $1\times10^{18}cm^{-3}$)

N-GaAs buffer layer (250nm, $5\times10^{18}cm^{-3}$)

GaAs substrate

FIGURE 6.2 The sequence of III-V semiconductor layers grown by metal organic vapor phase epitaxy for fabrication of the solar cell. Starting substrate is made of GaAs. Over this substrate, the successively grown layers are N-GaAs buffer layer, N-AlGaAs etch stop layer, (N-GaAs + N-Ga$_{0.87}$In$_{0.13}$As) capping layers, N-AlInP window layer, N-GaAs emitter layer, I-GaAs layer, P-GaAs base layer, P-Al$_{0.4}$Ga$_{0.6}$As back surface field layer, and P-GaAs contact layer. Thicknesses and doping concentrations of all the layers are mentioned in the diagram.

the final structure of the solar cell. These layers are the N-GaAs buffer layer and the N-AlGaAs etch stop layer. The buffer layer keeps the critical epitaxial layers away from the starting growth interface. The etch stop layer is introduced in the stack to stop the etching when the desired interface is reached. The material used in the etch stop layer has different etchant and etch characteristics than the material to be etched. Thus, the GaAs substrate can be safely separated from the overgrown stack of layers without damage to the stack. The etching will cease as soon as the etch stop layer is exposed.

The epitaxial layers grown over the N-GaAs substrate are shown in three-dimensional diagram drawn in Figure 6.3(a). In Figure 6.3(b), the P⁺-GaAs layer is deoxidized and Ti/Au is deposited. Localized P⁺GaAs/(Ti/Au) contact regions are defined, and the P⁺GaAs/(Ti/Au) portions are selectively etched to reach the underlying P-Al$_{0.4}$Ga$_{0.6}$As back surface field layer. Figure 6.3(c) shows the TiO$_2$ patterns formed on the P-Al$_{0.4}$Ga$_{0.6}$As with PDMS stamp using nanoimprint lithography with the TiO$_2$ over the P⁺ GaAs/(Ti/Au) contact regions removed. Photoresist is used for protecting the remaining TiO$_2$. Silver film is deposited by electron-beam evaporation to cover the surface geometry of TiO$_2$ and P⁺ GaAs/

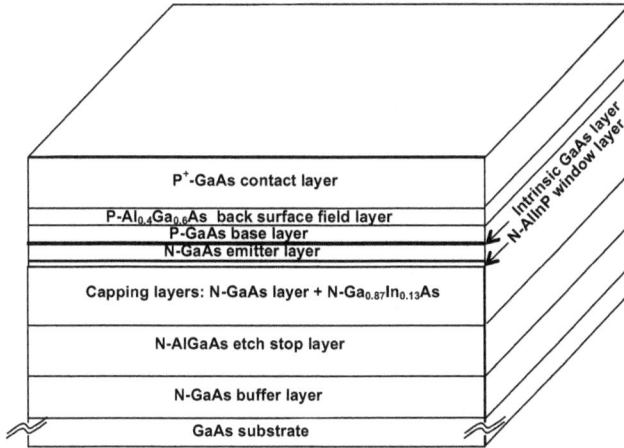

(a)

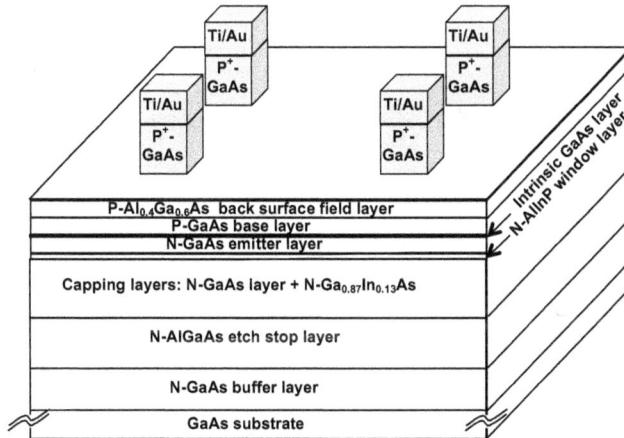

(b)

FIGURE 6.3   Fabrication of ultrathin GaAs solar cell: (a) the stack of group III-V semiconductors; (b) deoxidation of P+GaAs in dilute HCl, formation of localized Ti/Au contacts, and removal of P+ GaAs from remaining portions in citric acid: H_2O_2 etchant, etching stops after reaching the AlGaAs layer; (c) deposition of TiO_2 sol-gel film and its embossing by degassing-assisted patterning (DAP), a modified version of nanoimprint lithography; (d) removal of TiO_2 over Ti/Au contacts by etching in dilute HF and deposition of silver by electron-beam evaporation (200nm); (e) inversion of the stack of III-V layers and bonding of the silver-mirrored side with a glass host substrate using a hybrid organic/inorganic adhesive, followed by exposure to UV radiation; (f) removal of GaAs substrate and N-GaAs buffer layer by etching in $NH_4OH:H_2O_2:H_2O$ and removal of N-AlGaAs layer by etching in $HF:H_2O$; (g) deposition of front contacts (Ni/Au/Ge/Au/Ni/Au) with grid spacing; (h) removal of uncovered (N-GaAs and $N-Ga_{0.87}In_{0.13}As$) capping layers by etching in citric acid: H_2O_2; and (i) formation of double-layer antireflection coating of MgF_2/Ta_2O_5 by electron-beam deposition. (Step i is preceded by a step on delimiting solar cells by photolithography and mesa etching [removing AlInP window layer in dilute HCl and GaAs in $H_3PO_4:H_2O_2:H_2O$]; the delimiting step is not shown here.)

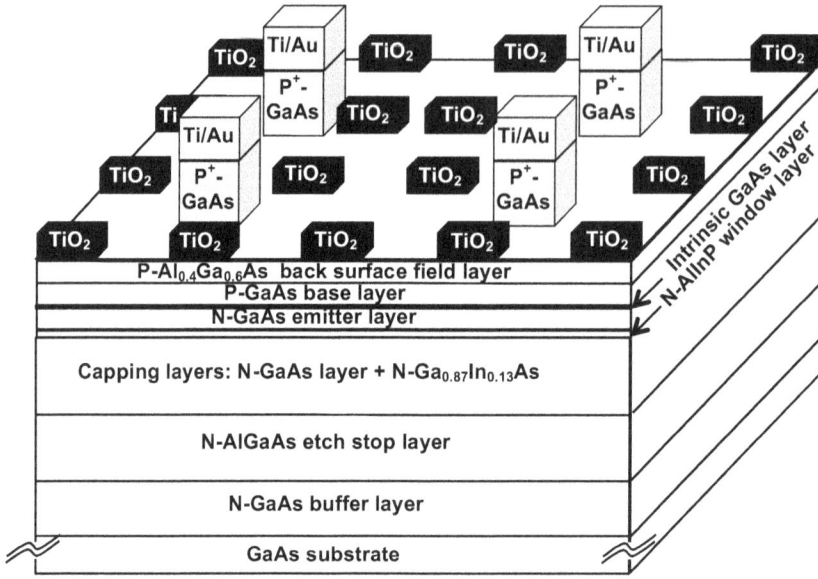

(c)

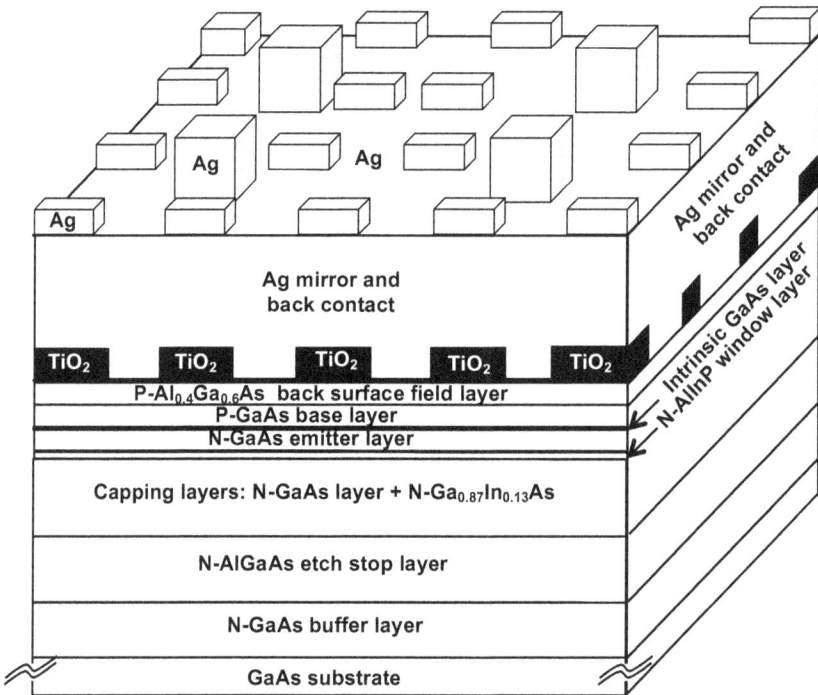

(d)

FIGURE 6.3 (Continued)

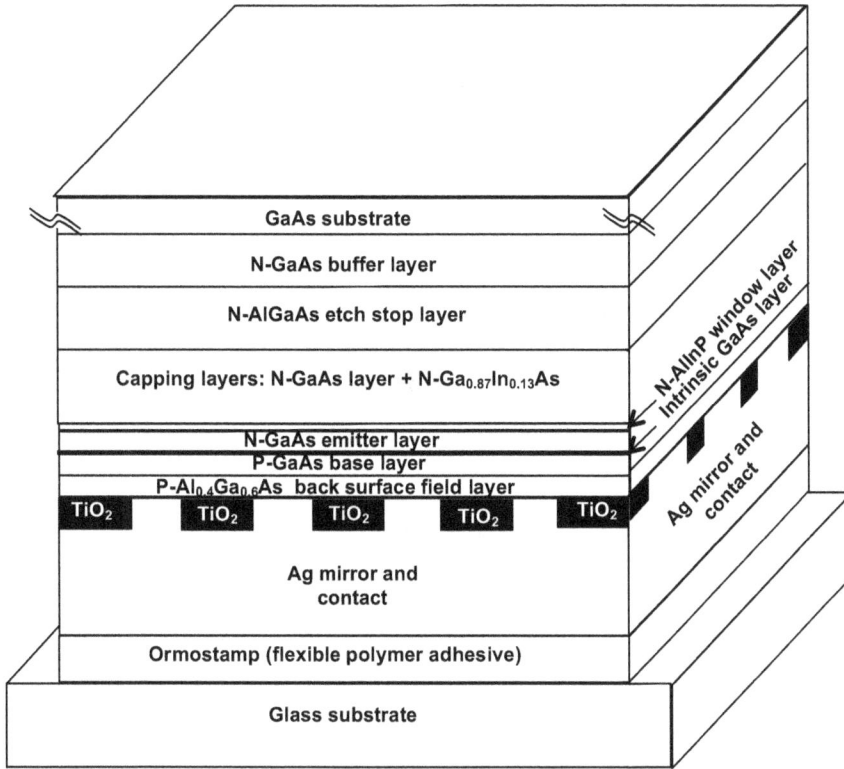

(e)

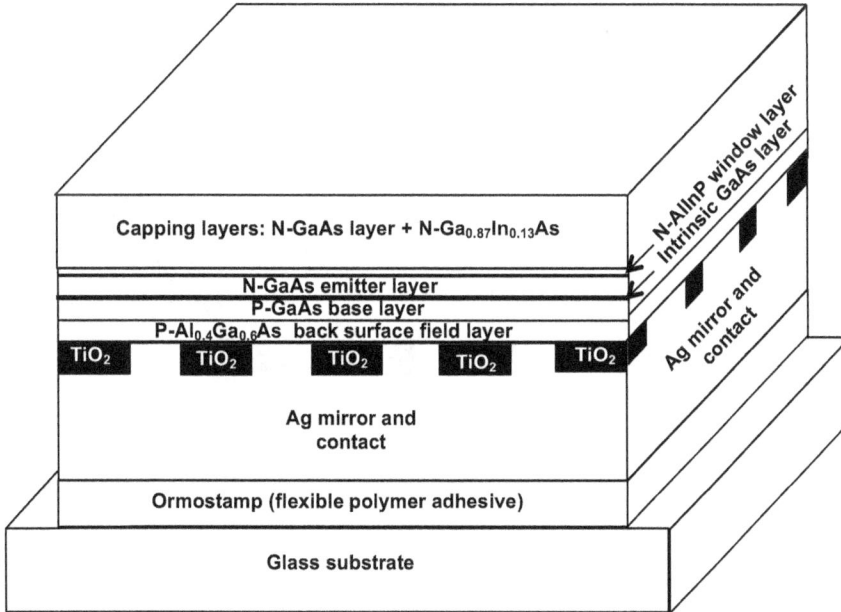

(f)

FIGURE 6.3 (Continued)

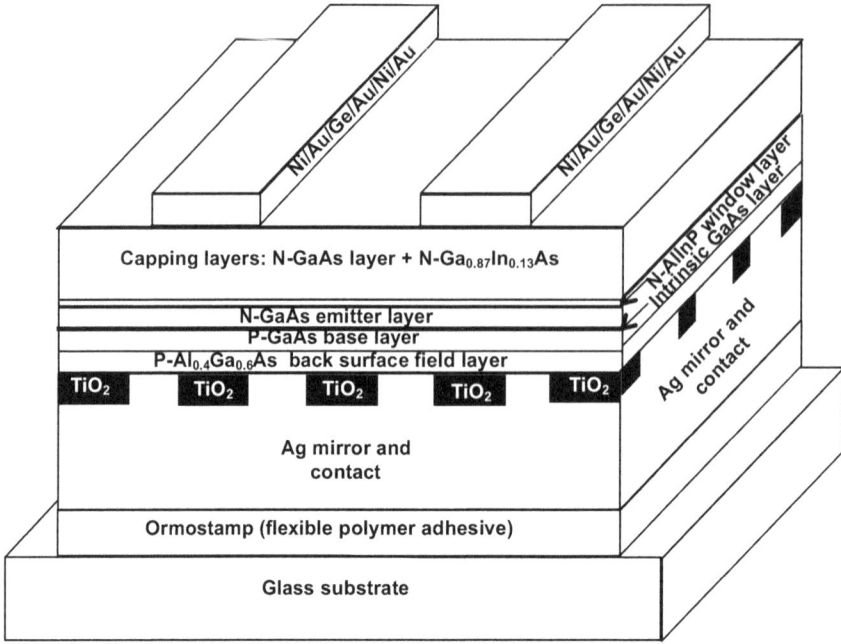

(g)

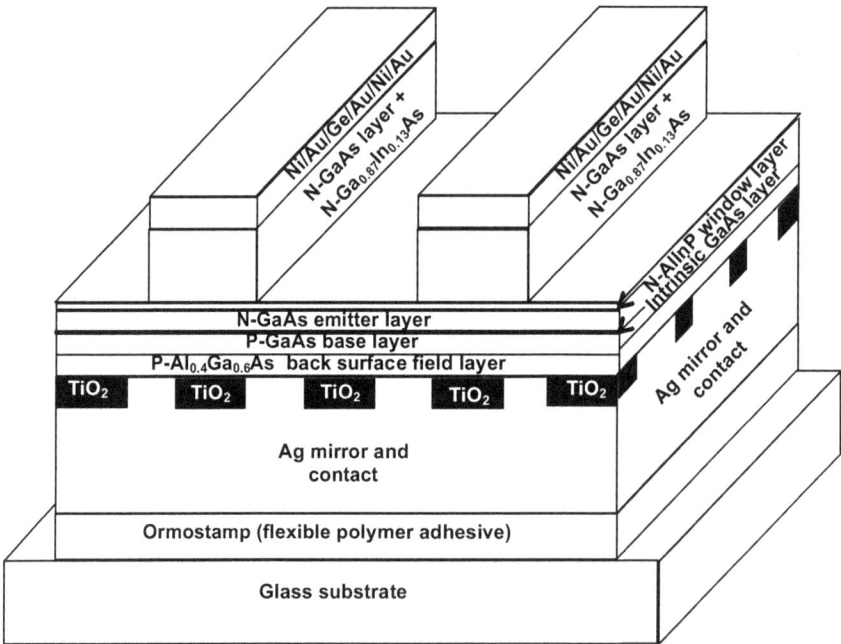

(h)

FIGURE 6.3 (Continued)

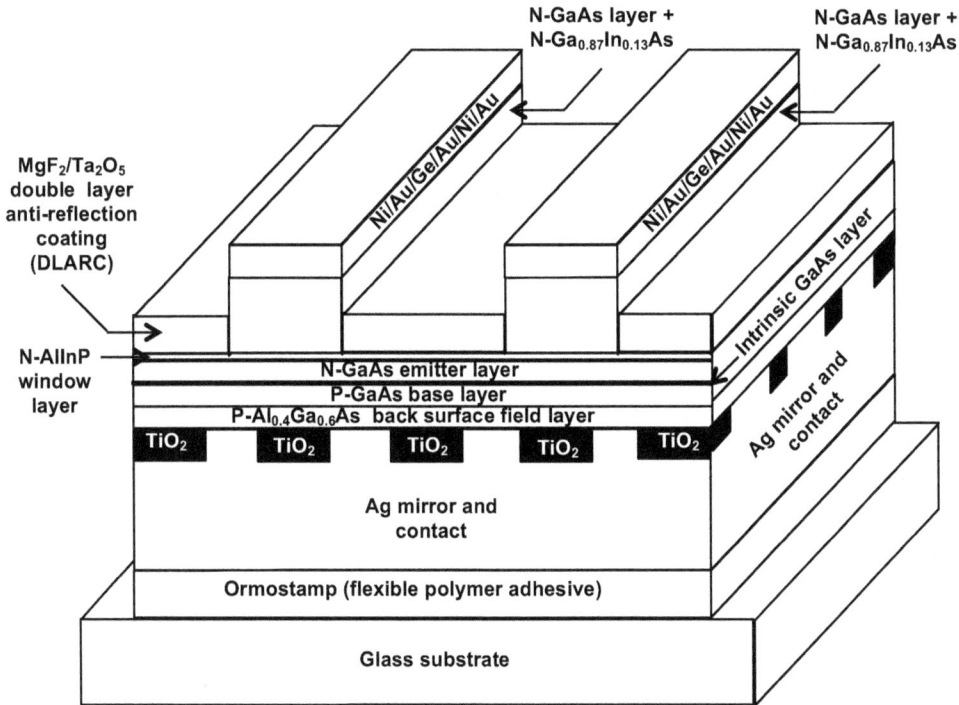

(i)

FIGURE 6.3 (Continued)

(Ti/Au) islands, as shown in Figure 6.3(d). In Figure 6.3(e), the structure of Figure 6.3(d) is inverted, placed upside down, and bonded to a glass substrate. In Figure 6.3(f), the N-GaAs substrate and the N-GaAs buffer layer are etched, and then the N-AlGaAs etch stop layer is removed. In Figure 6.3(g), the Ni/Au/Ge/Au/Ni/Au front contacts are deposited and patterned. In Figure 6.3(h), the (N-GaAs + N-Ga$_{0.87}$In$_{0.13}$As) capping layers are photolithographically defined and removed at all places except below the contact regions. In Figure 6.3(i), the MgF$_2$/Ta$_2$O$_5$ ARC is applied on the selected regions of the front surface of the solar cell. Finally, the solar cell is subjected to testing in sunlight.

The solar cell shows a short-circuit current density of 24.64mAcm^{-2}. It has an open-circuit voltage of 1.022V. Its fill factor is 0.792, and the efficiency is 19.9% under AM1.5 illumination (Chen et al 2019).

6.3 ULTRATHIN CIGSe ABSORBER (460 NM) SOLAR CELL WITH DIELECTRIC NANOPARTICLES

After discussion of multimode resonant absorption of light, let us focus our attention on the use of dielectric nanoparticles, bringing to the forefront the hitches of using metal nanoparticles for their plasmonic effects. Light trapping with silica nanoparticles on the rear surface and light incoupling with TiO$_2$ nanoparticles on the front surface of the solar cell are investigated by van Lare et al (2015).

6.3.1 STRUCTURE OF THE SOLAR CELL

The solar cell is displayed in Figure 6.4. Refer back to Section 1.7.6, Figure 1.12. At the bottom is a soda lime glass substrate. Sodium passivates the defects viz those at the surface and grain boundaries. The glass is coated with molybdenum, which is deposited by sputtering. Molybdenum provides

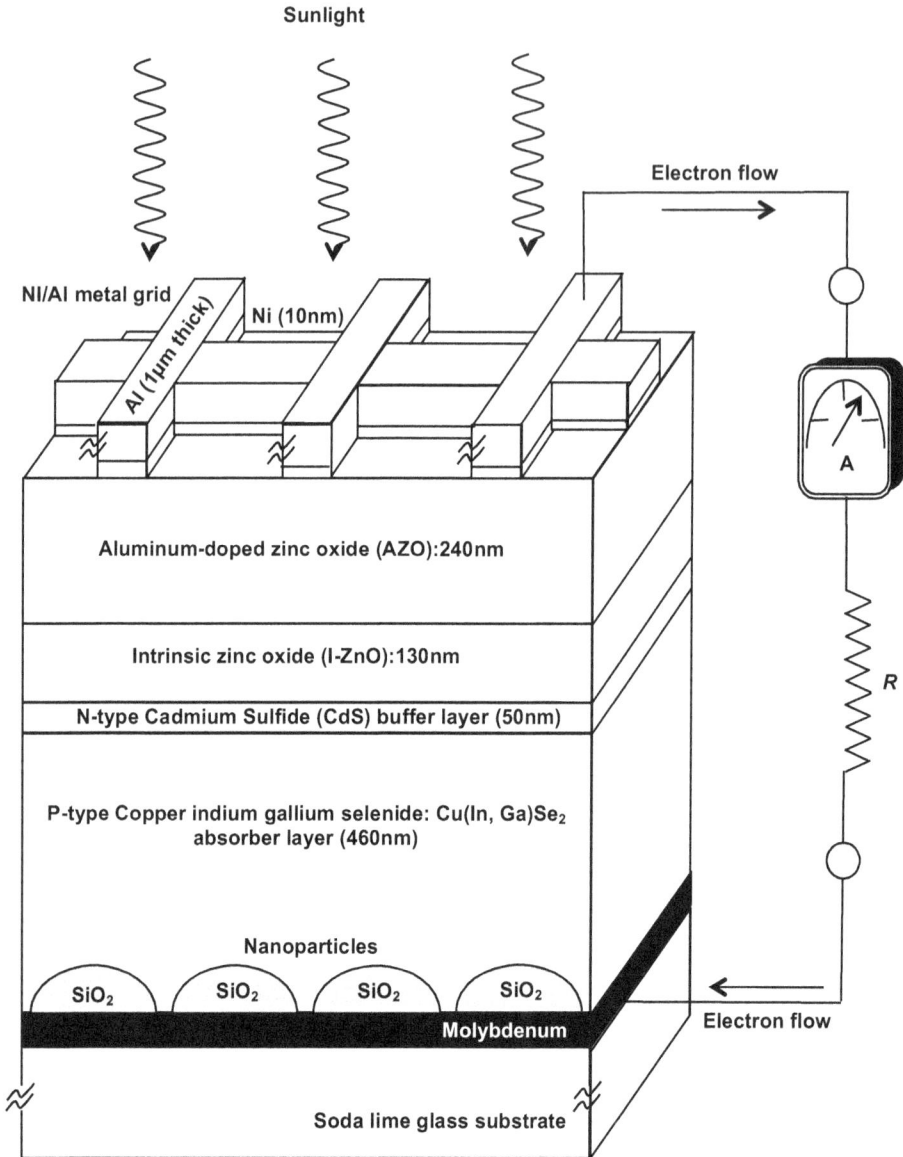

FIGURE 6.4 Cu(In, Ga)Se$_2$ solar cell of ultrasmall thickness with patterned silica dielectric nanoparticles showing from bottom to top: molybdenum-coated soda lime glass substrate, silica nanoparticles, P-type CIGSe film, N-type CdS layer, intrinsic and aluminum-doped ZnO layer, and Ni/Al contact metal grid. Sunlight falling on the top Ni/Al metal grid produces electric current which is collected between Ni/Al and Mo electrodes. The dielectric nanoparticles act as nanoscattering centers.

the back contact for the solar cell and also serves as a reflector for light back into the solar cell. Overlying the molybdenum film is the 460nm-thick P-type CIGSe absorber. At the interface between Mo and CIGSe absorber lie the silicon dioxide nanoparticles. An N-type CdS buffer layer deposited on P-type CIGSe absorber forms a P-N junction diode with the absorber. Transparent conducting oxide film covers the CdS layer. An Ni/Al rectangular grid makes the front contact. Looking simultaneously at Figures 1.12 and 6.4, it will be noticed that the two solar cells differ only with respect to silicon dioxide nanoparticles seen in Figure 6.4. Another noteworthy feature is that dielectric nanoparticles have been used for light trapping, whereas plasmonic effects of metal nanoparticles are commonly applied, as may be recalled from Chapter 5. So there must be convincing reasons for avoiding metal nanoparticles, and also for the use of dielectric nanoparticles. These reasons merit discussion and will now be spelled out.

6.3.2 DRAWBACKS OF PLASMONIC METAL NANOPARTICLES

Plasmonic metal nanoparticles are used through the exploitation of resonances by light and associated scattering phenomena. They can squeeze the optical energy into volumes smaller than the wavelength of light. In this way, energy is profoundly concentrated to improve optical processes (Khurgin 2017).

Notwithstanding their beneficial effects, it must not be forgotten that the metal nanoparticles themselves absorb light. So some fraction of incident light is lost through absorption by the metal nanoparticles. The loss via absorption by the metal is known as ohmic loss. Absorption by the metal leads to dissipation of plasmon energy through nonradiative relaxation processes. High ohmic loss is observed in all the metals, even the noble metals used in plasmonics.

Over and above, metal nanoparticles suffer from the disadvantage that their inclusion is incongruous with semiconductor device manufacturing, particularly noting that the fabrication process of some solar cells involves high-temperature operations in which the metal nanoparticles may diffuse into the structure. The metal deposition step is therefore planned at a process stage, after which there are no high-temperature cycles. So better options to metal nanoparticles are sought.

6.3.3 SCATTERING PROPERTIES OF DIELECTRIC NANOPARTICLES

Dielectric nanoparticles can cause a strong forward scattering of light. This scattering takes place at visible and near-infrared frequencies. It is caused by interactions between electric and magnetic dipolar resonances induced by the incoming light. Directional scattering occurs when the interactions between electric and magnetic modes comply with certain conditions. The interactions between these modes due to dielectric nanoparticles can also manipulate the angular distribution of scattering (Wang et al 2017).

In analogy to metallic nanostructures possessing large scattering cross sections at plasmon resonance, dielectric nanoparticles of subwavelength size act as efficient nanoscatterers. They exhibit geometrical Mie resonances whose scattering cross sections exceed the geometrical cross sections of the nanoparticles (Spinelli et al 2012). These scattering cross sections of resonantly excited Mie modes have comparable values to those for metallic nanoparticles. However, unlike metallic nanostructures, the parasitic absorption of light in dielectrics is much less (van Lare et al 2013). Lower absorption than metals is an obvious advantage in favor of dielectrics. Judging from these remarks, use of dielectric nanoparticles needs to be fostered for light trapping in solar cells by scattering of light and thereby increasing the length of light path.

6.3.4 FABRICATION OF THE SOLAR CELL WITH SILICA DIELECTRIC NANOPARTICLES AT THE REAR SURFACE

This cell is fabricated on a molybdenum-coated soda lime glass substrate (Figure 6.5(a)). Spin-coating is used to form a layer of PMMA on the substrate, followed by a sol-gel layer (Figure 6.5(b)). Substrate conformal imprint lithography (SCIL) is applied to form a pattern of holes at the positions planned for fixation of silica nanoparticles. By reactive ion etching, the PMMA in the holes is removed to reach the Mo surface (Figure 6.5(c)). Silicon dioxide film is deposited by electron-beam evaporation (Figure 6.5(d)). PMMA is removed in acetone at $50°C$ when the pattern of silicon dioxide nanoparticles on molybdenum is formed (Figure 6.5(e)). The CIGSe layer is deposited by three-stage coevaporation at a substrate temperature of $440°C$. CdS buffer layer is grown by chemical bath deposition using cadmium acetate, ammonia, and thiourea. Magnetron sputtering

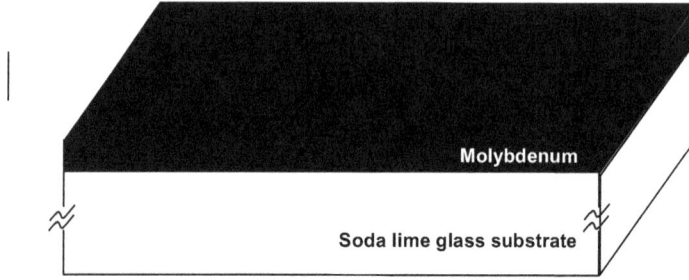

(a)

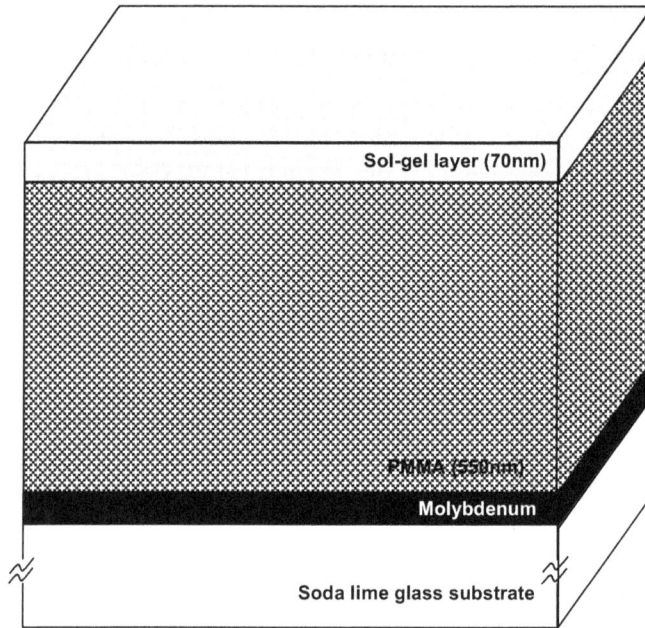

(b)

FIGURE 6.5 Fabrication of ultrathin Cu(In, Ga)Se$_2$ solar cell with patterned silica dielectric nanoparticles: (a) soda lime glass substrate coated with molybdenum, (b) deposition of a sol-gel layer (70nm) covering a PMMA (550nm) layer coated on molybdenum, (c) substrate conformal imprint lithography (SCIL) and reactive ion etching, (d) deposition of silica film by electron-beam evaporation, (e) creation of a pattern of silica nanoparticles by lifting off the mask, and (f) completion of the process by depositing the remaining layers: CIGSe film (460nm), CdS layer (50nm), intrinsic ZnO (I-ZnO) layer (130nm), aluminum-doped zinc oxide (AZO) layer (240nm), and the top contact metal grid consisting of Ni (10nm) and Al(1μm).

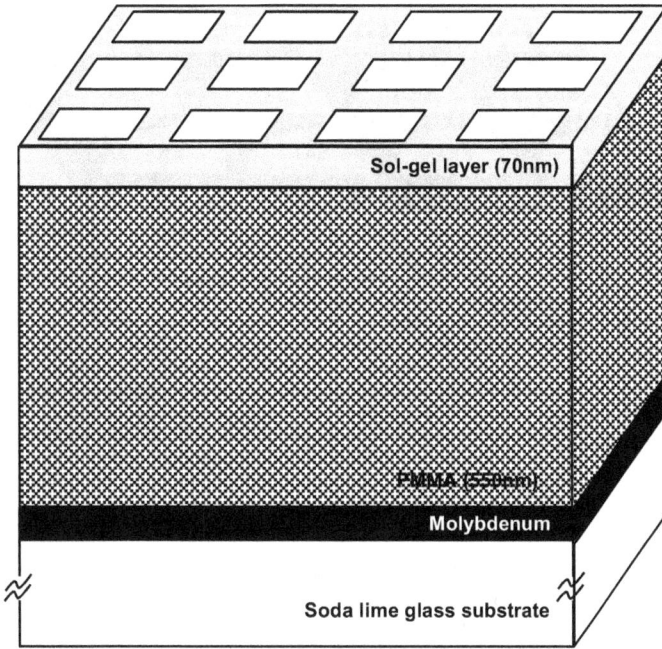

(c)

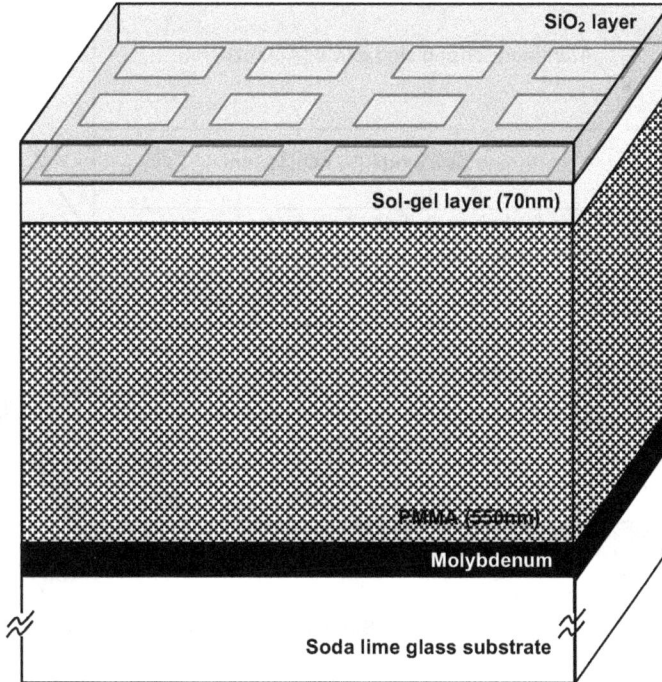

(d)

FIGURE 6.5 (Continued)

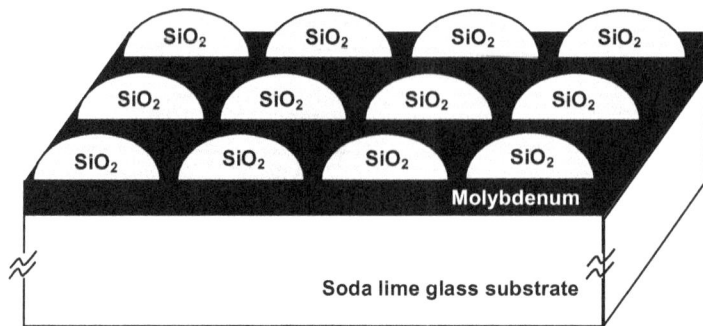

(e)

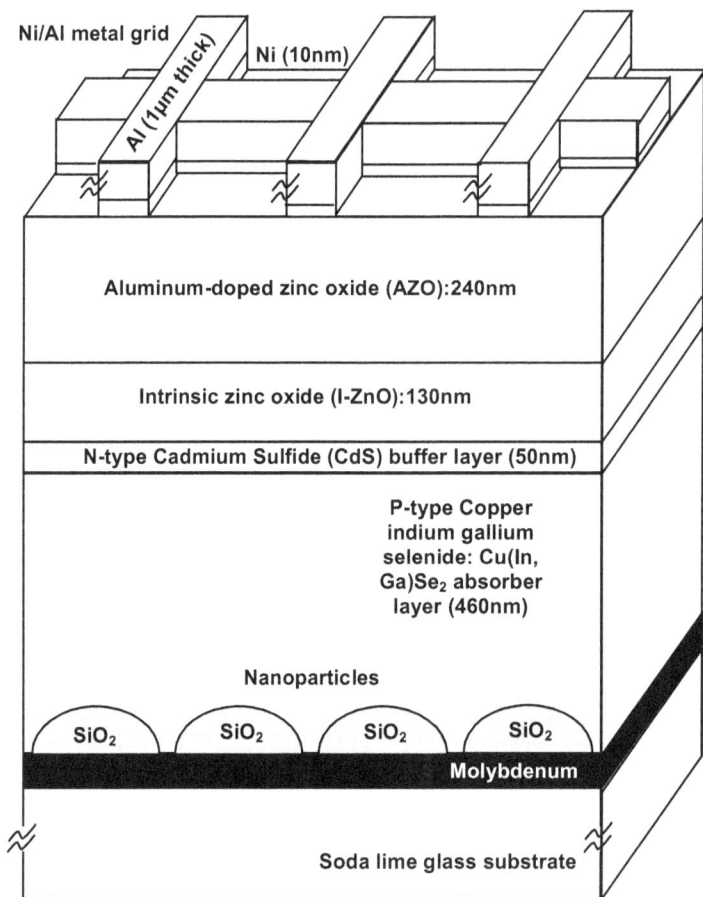

(f)

FIGURE 6.5   (Continued)

is used to deposit intrinsic and aluminum-doped zinc oxide layers. Ni/Al metal grid is formed by evaporation (Figure 6.5(f)).

6.3.5 Solar Cell with Silica Nanoparticles vs. Flat Solar Cell without Silica Nanoparticles

The solar cell with silica nanoparticles deposited at the interface between Mo and CIGSe exhibits a short-circuit current density of 30.6mAcm^{-2}, open-circuit voltage 0.592V, fill factor 68.2%, and efficiency 12.3%. Flat solar cell shows a short-circuit current density of 28.6mAcm^{-2}, open-circuit voltage 0.583V, fill factor 67.4%, and efficiency 11.1% (van Lare et al 2015).

6.3.6 Solar Cell with TiO$_2$ Nanoparticles on the Front Surface

By similar experimentation, efficiencies of solar cells without and with a periodic array of TiO$_2$ nanoparticles deposited on the front surface are determined. The efficiency of solar cell in the absence of TiO$_2$ nanoparticles is 10.1% and, with nanoparticles, increases to 10.9%, showing the incoupling of light into the solar cell by scattering by TiO$_2$ nanoparticles (van Lare et al 2015).

6.4 PERIODIC NANOHOLE ARRAY SOLAR CELL

6.4.1 Positive and Negative Textures

Regarding nanotexturing, the nanostructures considered so far, e.g., nanopyramids, nanocones, etc., are protruding outwards from the surface of the solar cell. Similar antireflection behavior is observed with other projecting nanostructures, such as nanowires, nanorods, etc. These outward-projecting nanostructures come under the class of positive textures (Zhang et al 2015). A negative texture will be one directed inward into the surface of the solar cell. Such a structure has not been treated so far. An example is a nanohole or nanocone. Here we shall talk about nanohole arrays as an example of negative texture class (Chen et al 2014).

6.4.2 Nanowires and Nanopores

A comparison of absorption of light in the solar spectrum has been made between silicon nanowires and nanoporous silicon (Xiong et al 2010). It is found that for silicon thickness < 10μm, keeping the thickness of silicon and ratio of filling silicon constant, the optical absorption by nanoporous silicon is larger than for silicon nanowires. When the filling ratio exceeds 0.25, the absorption of light by nanoporous silicon is even larger than for a silicon film of equal thickness.

Silicon nanoholes are much more rugged than freestanding silicon nanowires. From the point of view of mechanical strength, nanowires are easily prone to damage. They easily break and collect together. From the standpoint of efficiency of solar cell, the nanoholes can give the same ultimate efficiency as a silicon wafer of thickness 300μm while consuming 1/12th times less mass of silicon (Han and Chen 2010).

6.4.3 Fabrication of Nanohole Array Solar Cell

Figure 6.6 shows this solar cell, and Figure 6.7 illustrates the fabrication process flow of the solar cell. The silicon wafer has P-type polarity with orientation < 100 > (Peng et al 2010). Its resistivity is 8–12Ω-cm, and thickness is 180μm. For nanohole array fabrication, deep UV lithography is used. Negative photoresist is coated over the silicon wafer and thermally cured. The photoresist is exposed to UV radiation through a chromium mask containing the required periodic pattern of circular disks. The unexposed circular disk regions of the photoresist remain soft, while the exposed

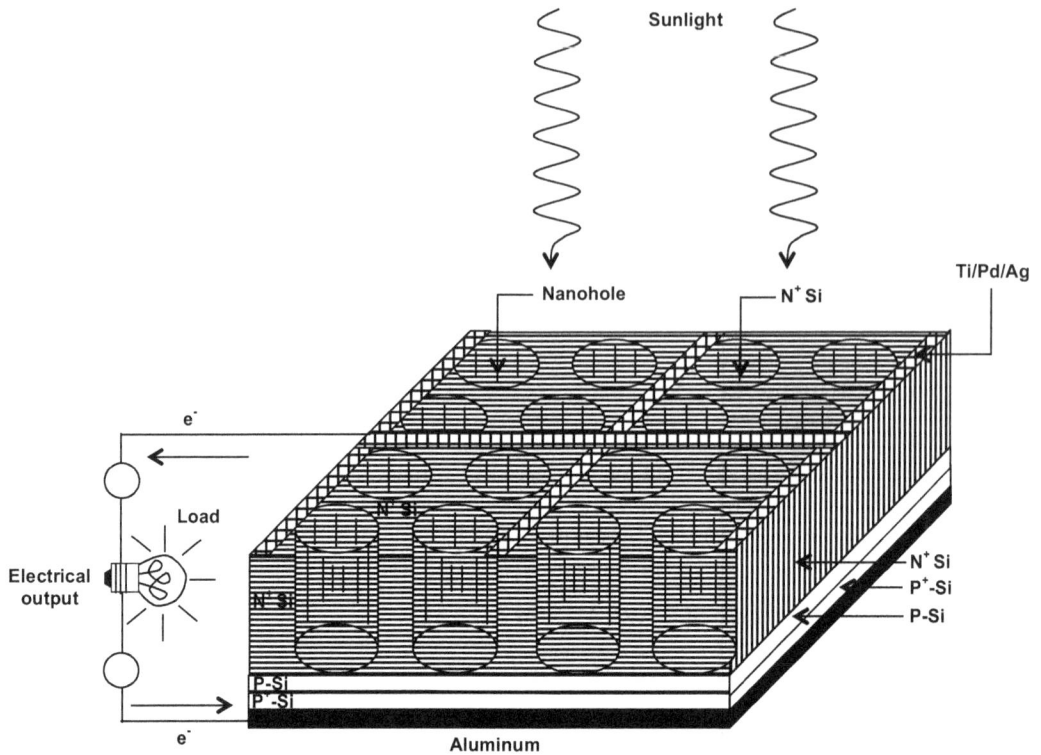

FIGURE 6.6 Periodic nanohole array solar cell showing the layers from bottom to top: aluminum backside contact, P+ back surface field layer, P substrate, N+ layer on inner surfaces of nanoholes, and front-side Ti/Pd/Ag grid. Sunlight falling on the nanoholes produces electric current flow between the Ti/Pd/Ag and Al electrodes. The nanoholes increase the absorption of light.

surrounding regions become hard. When the photoresist is developed, it is detached only from unexposed regions. So after developing, the unhardened photoresist in the circular disk regions is removed, while the remaining photoresist remains intact. In this way, the periodic pattern of circular disks in the mask is transferred to the photoresist. A thin silver film (10–50nm) is deposited on the photoresist by thermal evaporation. When the photoresist is finally stripped, an array of silver disks is left behind because the silver film is adhered to underlying silicon in the circular disk regions, whereas the remaining silver film deposited on hardened photoresist goes away with the photoresist. Silver-catalyzed electroless etching of silicon is done in (HF + H_2O_2) solution. H_2O_2 oxidizes silicon to silicon dioxide, and this silicon dioxide is etched by HF. Thus, silicon is etched in the spots in which the silver disks are placed, creating nanoholes at the positions of the silver disk regions. So the pattern of silver disks is transformed into a pattern of silicon nanoholes. Briefly, the pattern of circular disks on the mask becomes the pattern of silver disks on the wafer, and the pattern of silver disks on the wafer produces a pattern of nanoholes in silicon. The nanoholes have diameters of 500–600nm.

N+ diffusion in silicon nanoholes will make a radial N+P diode. But the silver disks must be removed before N+ diffusion. Concentrated nitric acid is used for silver etching. The wafer is thoroughly rinsed in DI water after etching and dried. Then N+ diffusion is performed using $POCl_3$ source at 910°C for 20 min. As a result, N+ regions are formed on the inner surfaces of the nanoholes as well as on their exposed external surfaces. After N+ diffusion, phosphosilicate glass is etched in dilute HF. For making the back contact, aluminum is thermally evaporated on the rear side of the wafer. But the rear side was doped N+ during phosphorous diffusion. This means that a

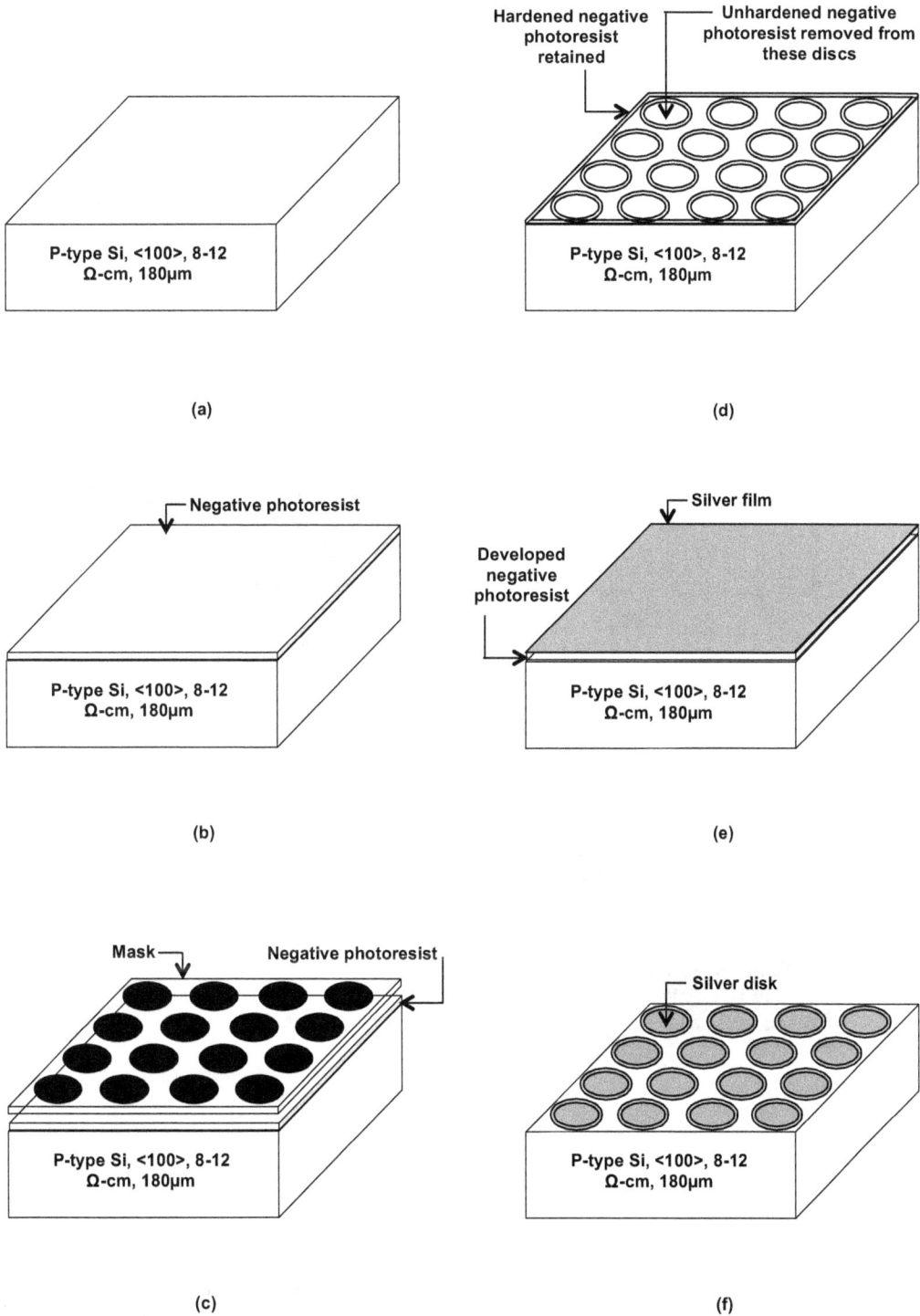

(a)

(b)

(c)

(d)

(e)

(f)

Labels in figure:
- Hardened negative photoresist retained
- Unhardened negative photoresist removed from these discs
- P-type Si, <100>, 8-12 Ω-cm, 180μm
- Negative photoresist
- Developed negative photoresist
- Silver film
- Mask
- Negative photoresist
- Silver disk

FIGURE 6.7 Fabrication of periodic nanohole array solar cell: (a) taking P-type silicon wafer, (b) spin-coating and curing negative photoresist, (c) exposing photoresist through mask and developing, (d) removing unhardened resist, (e) depositing silver film, (f) photoresist removal, (g) electroless Si etching, (h) removing remnant silver, (i) phosphorous diffusion (with N+ removed from top and side surfaces of silicon wafer but that in the nanoholes kept intact), (j) backside aluminum metallization, and (k) depositing and patterning front side Ti/Pd/Ag grid contact.

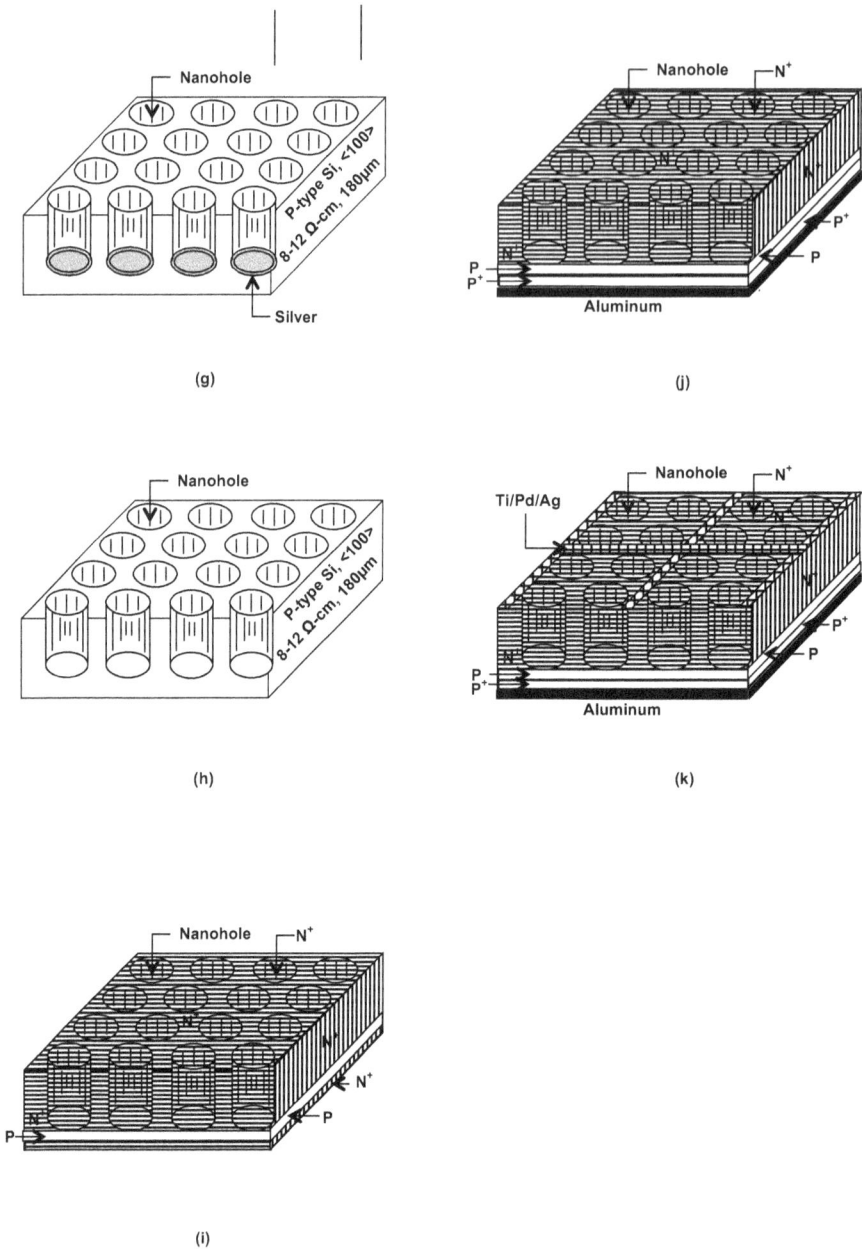

(g)

(h)

(j)

(k)

(i)

FIGURE 6.7   (Continued)

parasitic N^+P junction exists on the rear side. This parasitic junction is removed by sintering aluminum at a high temperature (980°C) when the rear surface becomes heavily P-doped, forming a P/P^+ low/high junction to provide the back surface field. For front ohmic contact, Ti/Pd/Ag grid is made on the front surface of the wafer and annealed at 380°C for ½ h. The solar cell is ready after the circumferential P-N junction is removed.

The solar cell shows an open-circuit voltage of 0.5666V, a short-circuit current density of 32.2 mAcm^{-2}, and efficiency of 9.51%. The testing condition is AM 1.5G illumination at 100 mWcm^{-2} (Peng et al 2010).

6.5   RANDOM NANOHOLE ARRAY SOLAR CELL

6.5.1   FABRICATION OF RANDOM NANOHOLE ARRAY

A facile process of random nanoholes fabrication employs low-melting-point indium nanoparticles as masks for silicon etching (Lee et al 2019). Figure 6.8 shows the sequence of process steps for making subwavelength-size nanoholes. Silicon wafer with silicon oxide layer is taken. The silicon wafer has orientation < 111 >. Indium is a low-melting-point metal (m.p. 156.6°C). Indium

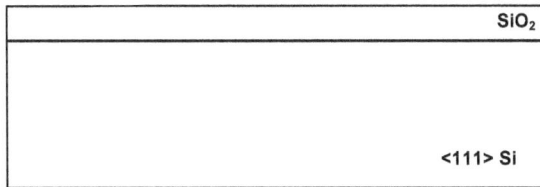

(a)

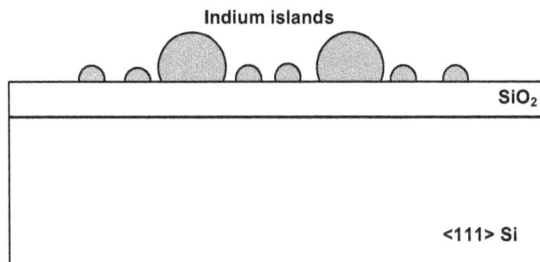

(b)

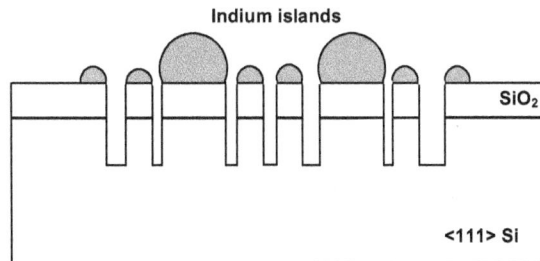

(c)

FIGURE 6.8   Fabrication of random nanoholes: (a) silicon wafer with silicon oxide layer, (b) indium island deposition, (c) reactive ion etching of silicon, (d) indium island etching, (e) isotropic etching of silicon, and (f) silicon oxide etching.

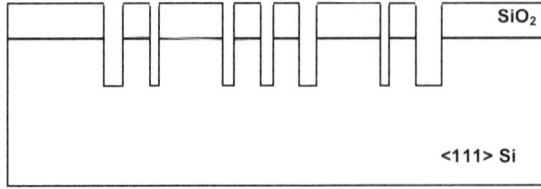

(d)

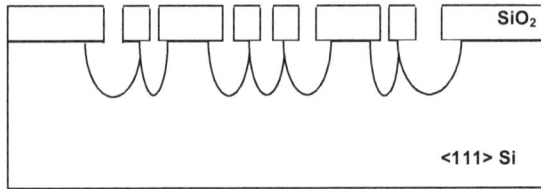

(e)

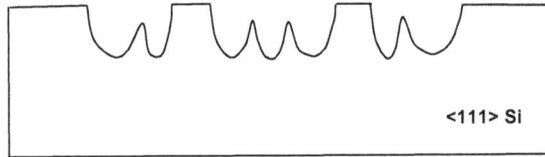

(f)

FIGURE 6.8 (Continued)

is deposited by thermal evaporation at a controlled thickness. By carefully adjusting the indium thickness, the indium deposition takes place in the form of islands. This island growth mode of indium is called the Volmer-Weber mode. Indium island growth takes place in a bimodal distribution of sizes containing larger- and smaller-size islands. Mostly, the islands are of larger size. Adjoining these larger-size islands, smaller-size islands are also formed.

The silicon oxide layer SiO_x is anisotropically removed by reactive ion etching using CF_4/O_2. A distribution of silicon nanopillars is formed in two modes, small size and large size. Indium etching is done in HCl to remove the indium islands. SiO_x etch masks are left behind. Now silicon is isotropically etched with an $HF:HNO_3:H_2O$ etchant. Here, HNO_3 oxidizes silicon to silicon dioxide, which is etched by HF so that silicon is removed. During this etching, the small-size silicon pillars are selectively etched. They combine together to produce large-size nanohole structures. The SiO_x etch masks are removed by etching. Random nanohole structures are thus obtained. The nanoholes are ellipsoidal-shaped.

6.5.2 Fabrication and Parameters of Solar Cell

PERC (passivated emitter and rear cell) devices are made on a 48μm-thick silicon foil produced by proton implant exfoliation (PIE) technique (Lee et al 2019). The open-circuit voltage, short-circuit current density, and fill factor for a solar cell with random nanohole arrays are 0.625V, 35.9mAcm^{-2}, 76%, and 17.1%, respectively. For planar solar cell, these parameters are 0.628V, 31.8mAcm^{-2}, 78.1%, and 15.6%.

6.6 SILICON NANOHOLE/ORGANIC SEMICONDUCTOR HETEROJUNCTION HYBRID SOLAR CELL

As repeatedly said, hybrid solar cells can lower the cost of fabrication and combine the advantages of silicon and organic semiconductor technologies. Fabrication of a silicon solar cell is complex, costly, and energy-intensive, but a hybrid solar cell process is much less demanding.

6.6.1 Fabrication of Hybrid Solar Cell

Figure 6.9 shows a hybrid solar cell (Hong et al 2014). The process for realization of this cell is shown in Figure 6.10; 25nm-thick silver film is deposited by electron-beam evaporation on an N-type < 100>

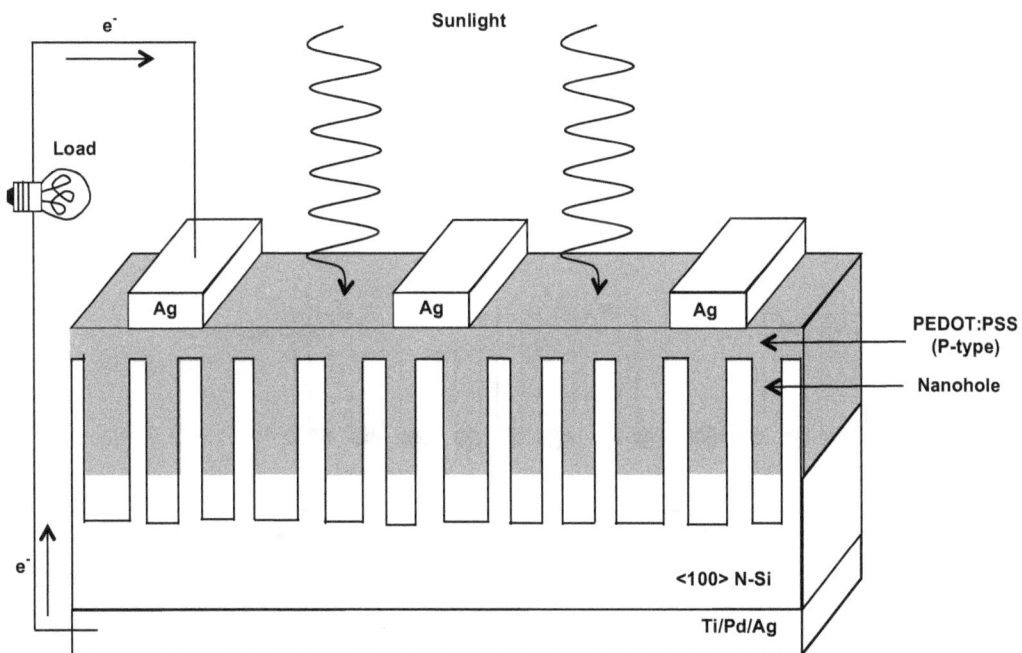

FIGURE 6.9 N-type Si/P-type PEDOT:PSS heterojunction solar cell showing from bottom upwards the Ti/Pd/Ag back contact film, the N-type silicon wafer, P-type PEDOT:PSS filled in the nanoholes, and top-side Ag grid contact. Upon illuminating the Ag grid contact surface with sunlight, electric current is produced by the solar cell. It is tapped across the Ag and Ti/Pd/Ag electrodes.

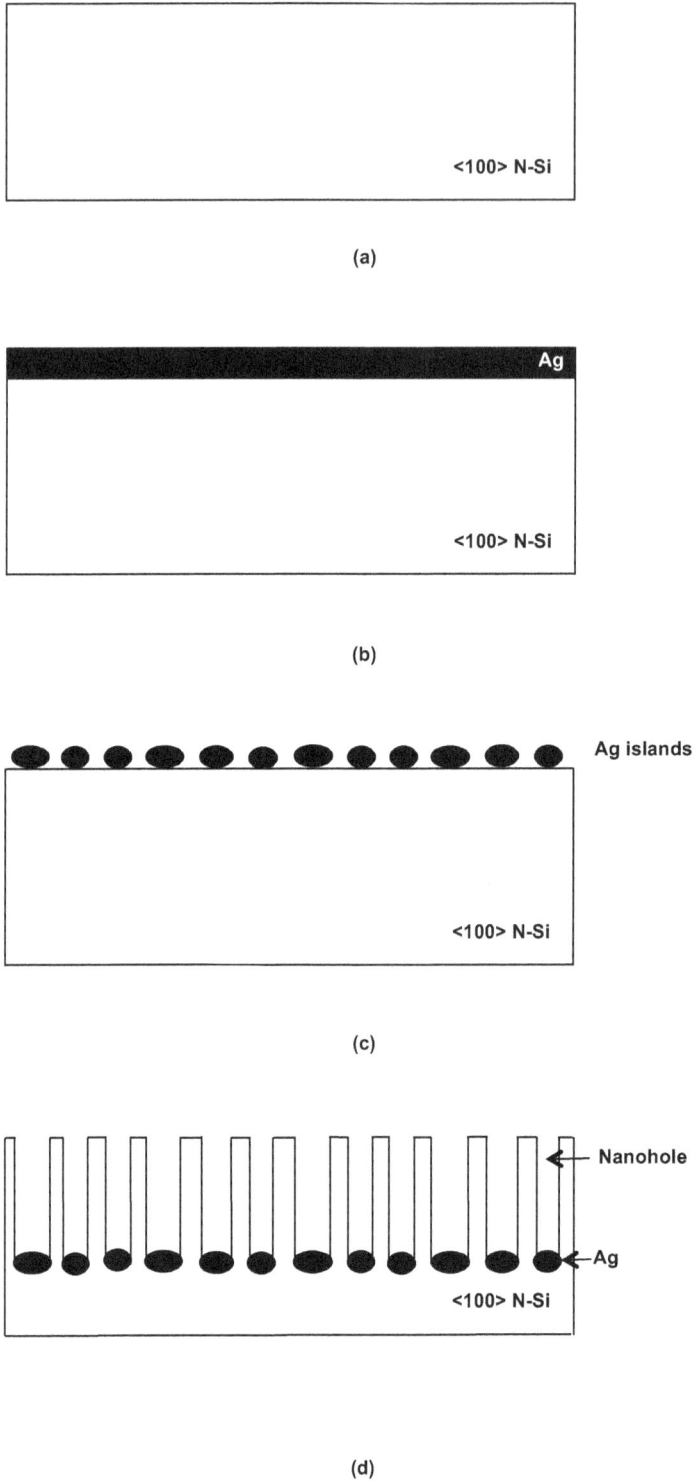

(a)

(b)

(c)

(d)

FIGURE 6.10 Fabrication of silicon nanohole/organic semiconductor solar cell: (a) silicon wafer, (b) deposition of silver film, (c) laser scanning for Ag island formation, (d) Ag nanoparticle-assisted silicon etching, (e) silver removal, (f) spin-coating PEDOT:PSS layer, (g) front-side Ag grid deposition, and (h) backside Ti/Pd/Ag film deposition.

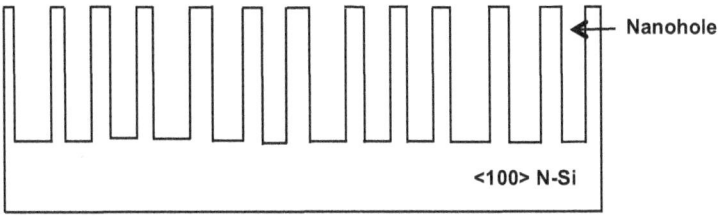

(e)

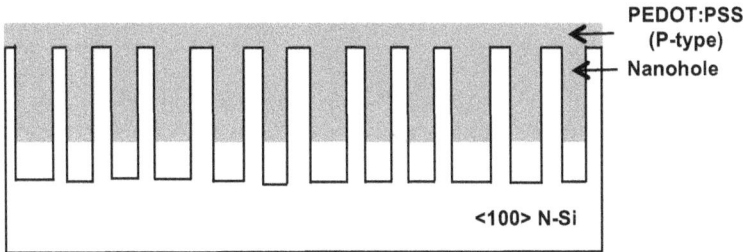

(f)

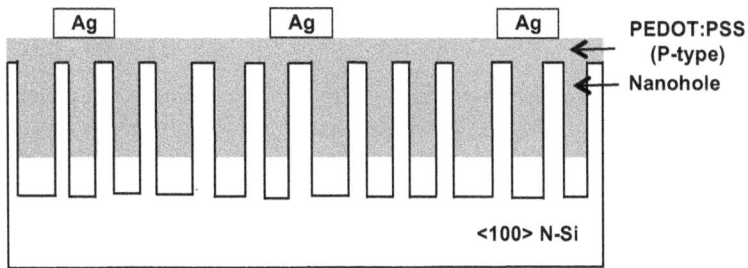

(g)

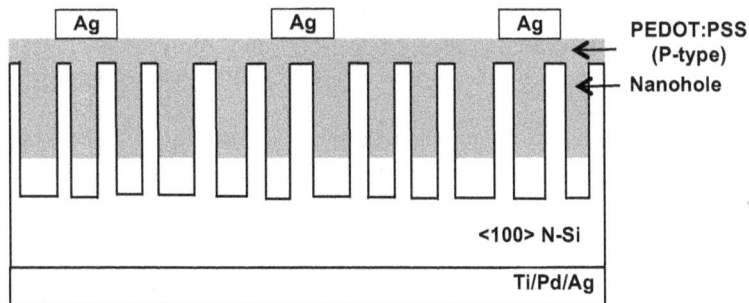

(h)

FIGURE 6.10 (Continued)

orientation, 625μm-thick silicon wafer. Placing the wafer on a movable platform, the silver film is annealed with a KrF excimer laser at 300mJcm^{-2} to form silver nanospheres by scanning in X- and Y-directions. These nanospheres have diameters in the range 200–400nm and lie at distances ~ 300–1,000nm. They act as catalysts for silicon etching. The etching is performed in a mixed etchant made with HF and H_2O_2. The Ag nanospheres are then removed by etching in HNO_3.

A P-N junction is formed by spin-coating poly (3,4-ethylenedioxythiophene) polystyrene sulfonate solution in dimethyl sulfoxide over the nanoholes. Ag grid is formed on the front PEDOT:PSS surface. Ti/Pd/Ag film is deposited on the rear surface of the silicon for back contact. The depth of the nanohole is optimized as 1μm because it shows the highest efficiency of 8.3% among the four nanohole structures of depths 0.5, 1, 2, and 4 μm.

6.6.2 Parameters of the Solar Cell

For an optimized hole depth of 1μm, this solar cell exhibits an open-circuit voltage of 0.55V, short-circuit current density of 25mAcm^{-2}, fill factor 60.4%, and efficiency 8.3% (Hong et al 2014).

6.7 DISCUSSION AND CONCLUSIONS

Fabrication of two solar cells involving construction of surface nanostructures was studied. In one structure, light is trapped by the principle of multiresonant absorption, and in the other structure, scattering of light by dielectric nanoparticles at the rear/front surface of the solar cell is exploited. It is observed that enhanced light trapping can be achieved without any front-side nanotexturing if a nanostructured silver-back mirror is used in a GaAs absorber solar cell. Light trapping is also improved when the Mo/CIGSe interface on the backside of a CIGSe solar cell is patterned using SiO_2 nanoparticles or TiO_2 nanoparticles are deposited on its front surface.

In contrast with positive surface textures of Chapter 4, negative textures such as nanoholes can be gainfully utilized. Silicon-organic semiconductor hybrid nanoarray solar cell is a more economical device than all-silicon solar cell. It is an illustrious representation of the combined strengths of inorganic and organic nanostructured photovoltaics. Table 6.1 presents various optical improvement methods that are applied to solar cells.

TABLE 6.1
Optically Improved Solar Cells Employing Different Light Management Techniques

Sl. No.	Solar Cell	Light Management Principle	Efficiency (%)	Reference
1.	Ultrathin GaAs absorber (205nm) solar cell with TiO_2/Ag nanostructured back mirror	Multiresonant absorption by silver mirror	19.9	Chen et al (2019)
2.	Ultrathin CIGSe absorber (460 nm) solar cell with dielectric nanoparticles	Light trapping by silica NPs on the rear surface and TiO_2 NPs on the front surface	(a) 12.3 with silica NPs and 11.1 without silica NPs on rear surface (b) 10.9 with TiO_2 NPs and 10.1 without TiO_2 NPs on front surface	van Lare et al (2015)
3.	Periodic Si nanohole array solar cell	Enhanced optical absorption by nanoholes	9.51	Peng et al (2010)
4.	Random Si nanohole array solar cell		17.1 with nanohole array and 15.6 for planar case	Lee at al (2019)
5.	Hybrid nanohole array solar cell (Si nanohole array heterojunction with PEDOT:PSS)		8.3	Hong et al (2014)

Questions and Answers

6.1 Why are antireflective coatings useful in a narrow wavelength range? Answer: Because they work by destructive interference, which requires that their thickness = 1/4th the wavelength of incident sunlight. The condition is met by a limited range of wavelengths = chosen wavelength ± a small wavelength range.

6.2 Why is random pyramid etching restricted to < 100 > crystalline silicon? Answer: Because it is silicon crystal orientation-dependent etching. KOH etches at 400 times faster rate in the < 100 > crystal directions as compared to < 111 > directions, resulting in a trapezoidal-shaped cavity. This cavity has a {100} plane at the bottom flanked with {111} planes on the sides. With respect to surface of the wafer, its sidewalls are inclined at an angle of 54.7°.

6.3 Does random pyramid etching produce any graded refractive index effect? Answer: No.

6.4 Why does nanotexturing provide a broadband antireflection? Answer: Because of the inability of light to resolve the nanostructure and consequent grading of refractive index.

6.5 Mention three advantages of using a thin absorber layer in a GaAs solar cell. Answer: (i) Cost reduction, (ii) a larger number of carriers able to reach the contacts escaping from recombination due to the short travel distance, (iii) increased radiation tolerance.

6.6 How are a large number of modes produced for multiresonant optical absorption? Answer: By using a periodic structure such as a nanostructured mirror.

6.7 Why is the location of periodic structure for multiresonant optical absorption chosen on the rear surface of a solar cell? Answer: To avoid light loss by diffraction in space.

6.8 What are the considerations in deciding the period p of the nanostructured mirror? Answer: A larger period provides a larger number of resonances but also leads to more diffraction losses. Therefore, a trade-off between the conflicting requirements is necessary.

6.9 In the GaAs solar cell with nanostructured mirror, (a) what is the function of the buffer layer, and (a) what is the role of the etch stop layer? Answer: (a) To keep the critical epitaxial layers away from the incipient growth interface and (b) to automatically stop etching when the material to be etched is removed so that the residual structure is left unharmed.

6.10 Mention two disadvantages of using plasmonic metallic nanoparticles. Answer: (i) Ohmic loss via absorption of light by the metal nanoparticles and (ii) incompatibility of metals with high-temperature processing.

6.11 In what respect are dielectric nanoparticles better than metal nanoparticle scatterers? Answer: They suffer from lower optical absorption loss than metal nanoparticles but are efficient in scattering light.

6.12 What is SCIL? Answer: Substrate conformal imprint lithography (SCIL) is a high-demand imprint technique using a soft stamp for large area patterning and rigid glass carrier for best resolution.

6.13 What is CIGSe? Answer: $Cu(In, Ga)Se_2$.

6.14 Differentiate between positive and negative textures. Answer: Positive texture is a texture projecting outwards from the surface of the solar cell, and negative texture is one directed inwards.

6.15 How is a nanohole texture better than a nanowire texture? Answer: Nanoholes are more robust, while nanowires are fragile and easily prone to breakage.

6.16 What is the mechanism of silicon etching using silver disks? Answer: Two simultaneous reactions take place: (i) Ag^+ ion is reduced to Ag by receiving electron from silicon: $4Ag^+ + 4e^- \rightarrow 4Ag$; (ii) Si is oxidized and etched by F^-: $Si + 6F^- \rightarrow [SiF_6]^{2-} + 4e^-$.

6.17 What is PERC solar cell technology? Answer: In this solar cell, a dielectric passivation layer is added on the rear of the solar cell to reduce electron recombination, reflect back any unabsorbed light for second absorption, and maintain temperature of the cell by reflecting heat-producing wavelengths out of the cell.

6.18 What is proton implant exfoliation (PIE) technique? Answer: Hydrogen is introduced into silicon wafer by ion implantation to required depth, and the wafer is thermally annealed. Formation of sub-subsurface hydrogen gas bubbles at high internal pressure induces cleavage of wafer.

REFERENCES

Ahrlich M., O. Sergeev, M. Juilfs, A. Neumüller, M. Vehse and C. Agert 2016 Improved light management in silicon heterojunction solar cells by application of a ZnO nanorod antireflective layer, 6th International Conference on Silicon Photovoltaics, SiliconPV, 7–9 March 2016, Chambéry, France, Energy Procedia, 92: 284–290.

Chen H.-L., A. Cattoni, R. De Lépinau, A. W. Walker, O. Hoehn, D. Lackner, G. Siefer, M. Faustini, N. Vandamme, J. Goffard, B. Behaghel, C. Dupuis, N. Bardou, F. Dimroth and S. Collin 2019 A 19.9%-efficient ultrathin solar cell based on a 205-nm-thick GaAs absorber and a silver nanostructured back mirror, Nature Energy, Nature Publishing Group, 4(9): 761–767.

Chen T.-G., P. Yu, S.-W. Chen, F.-Y. Chang, B.-Y. Huang, Y.-C. Cheng, J.-C. Hsiao, C.-K. Li and Y.-R. W 2014 Characteristics of large-scale nanohole arrays for thin-silicon photovoltaics, Progress in Photovoltaics: Research and Applications, 22: 452–461.

Chen W.-H. and F.C.N. Hong 2016 0.76% absolute efficiency increase for screen-printed multicrystalline silicon solar cells with nanostructures by reactive ion etching, Solar Energy Materials and Solar Cells, 157: 48–54.

Collin S., J. Goffard, A. Cattoni, C. Colin, C. Sauvan, P. Lalanne and J.-F. Guillemoles 2015 Multi-resonant light trapping: New paradigm, new limits, IEEE 42nd Photovoltaic Specialist Conference (PVSC), 14–19 June 2015, New Orleans, LA, IEEE, NY, pp. 1–3.

Han S. E. and G. Chen 2010 Optical absorption enhancement in silicon nanohole arrays for solar photovoltaics, Nano Letters, 10(3): 1012–1015.

Hirst L. C., M. K. Yakes, J. H. Warner, M. F. Bennett, K. J. Schmieder, S. Tomasulo, E. Cleveland, S. Maximenko, J. Moore, R. J. Walters, and P. P. Jenkins 2017 Ultra-thin GaAs solar cells: Radiation tolerance and space applications, IEEE 44th Photovoltaic Specialist Conference (PVSC), 25–30 June 2017, Washington, DC, IEEE, NJ, pp. 2091–2093.

Hong L., X. Wang, H. Zheng, L. He, H. Wang, H. Yu and Rusli 2014 High efficiency silicon nanohole/organic heterojunction hybrid solar cell, Applied Physics Letters, 104: 053104–1 to 053104–4.

Khurgin J. B. 2017 Replacing noble metals with alternative materials in plasmonics and metamaterials: How good an idea? Philosophical transactions Series A, Mathematical, Physical, and Engineering Sciences, 375(2090): 20160068.

Lee H.-S., J. M. Choi, B. Jung, J. Kim, J. Song, D. S. Jeong et al 2019 Random nanohole arrays and its application to crystalline Si thin foils produced by proton induced exfoliation for solar cells, Scientific Reports, 9(19736): 11 pages.

Peng K.-Q., X. Wang, L. Li, X.-L. Wu and S.-T. Lee 2010 High-performance silicon nanohole solar cells, Journal of the American Chemical Society, 132(20): 6872–6873.

Spinelli P., M. A. Verschuuren and A. Polman 2012 Broadband omnidirectional antireflection coating based on subwavelength surface Mie resonators, Nature Communications, 3(692): 1–5.

van Lare C., G. Yin, A. Polman and M. Schmid 2015 Light coupling and trapping in ultrathin Cu(In,Ga)Se$_2$ solar cells using dielectric scattering patterns, ACS Nano, 9(10): 9603–9613.

van Lare M., F. Lenzmann and A. Polman 2013 Dielectric back scattering patterns for light trapping in thin-film Si solar cells, Optics Express, 21(18): 20739, 9 pages.

Wang Z., N. An, F. Shen, H. Zhou, Y. Sun, Z. Jiang, Y. Han, Y. Li and Z. Guo 2017 Enhanced forward scattering of ellipsoidal dielectric nanoparticles, Nanoscale Res Lett, 12(58): 1–8.

Xiong Z., F. Zhao, J. Yang and X. Hu 2010 Comparison of optical absorption in Si nanowire and nanoporous Si structures for photovoltaic applications, Applied Physics Letters, 96: 181903–1 to 181903–3.

Zhang D., W. Ren, Z. Zhu, H. Zhang, B. Liu, W. Shi, X. Qin and C. Cheng 2015 Highly-ordered silicon inverted nanocone arrays with broadband light antireflectance, Nanoscale Research Letters, 10(9): 1–6.

Part III

Electrochemical Photovoltaics
Using Nanomaterials

7 Dye-Sensitized Solar Cells

7.1 CONSTRUCTION AND WORKING PRINCIPLE OF A DYE-SENSITIZED SOLAR CELL (DSSC)

7.1.1 THE NANOCONSTITUENT OF THE CELL

O'Regan and Grätzel (1991) reported the development of a solar cell using low- to medium-purity materials by inexpensive process. Based on a transparent TiO_2 film consisting of particles with size of a few nanometers coated with a charge-transfer dye, this solar cell yielded commercially convincing power conversion efficiency 7.1–7.9%.

7.1.2 CELL CONSTRUCTION

The DSSC consists of an anode made of a thin metal oxide film (TiO_2) deposited on a transparent conducting oxide-coated glass substrate (Figure 7.1). The TiO_2 film is sensitized with a dye. Platinum is deposited on a TCO-coated glass plate to form a cathode. After assembly of the components, an electrolyte, usually an iodide/triiodide solution, is filled in the cell.

7.1.3 CELL PRINCIPLE

The operating principle of a DSSC is fundamentally different from that of the conventional P-N junction diode (Grätzel 2003, 2004). In a P-N junction diode, the semiconductor performs the dual tasks of absorption of sunlight and transport of charge carriers. In a DSSC, two different materials are assigned these responsibilities:

(i) The dye or sensitizer absorbs light.
(ii) Charge separation is implemented by electron injection across the interface between the dye and the wide-bandgap semiconductor TiO_2.

7.1.4 MIMICKING THE NATURAL PHOTOSYNTHESIS PROCESS

Dye-sensitized solar cell is an environment-friendly cell. It works by photoelectrochemical principle. In many respects, its operation is similar to the biochemical process of photosynthesis by which plants manufacture food from water in the soil and carbon dioxide in the atmosphere. The plants synthesize their food using solar energy by absorption of blue and red parts of the sunlight through the green pigment in plant leaves, known as chlorophyll; the green and near-green parts are reflected so that the leaves look green. Carbon dioxide is reduced to glucose (sugar), while water is oxidized to oxygen, thereby transforming light energy into chemical energy:

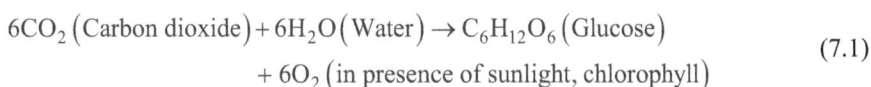

$$6CO_2\left(\text{Carbon dioxide}\right) + 6H_2O\left(\text{Water}\right) \rightarrow C_6H_{12}O_6\left(\text{Glucose}\right)$$
$$+ 6O_2\left(\text{in presence of sunlight, chlorophyll}\right) \tag{7.1}$$

DOI: 10.1201/9781003215158-10

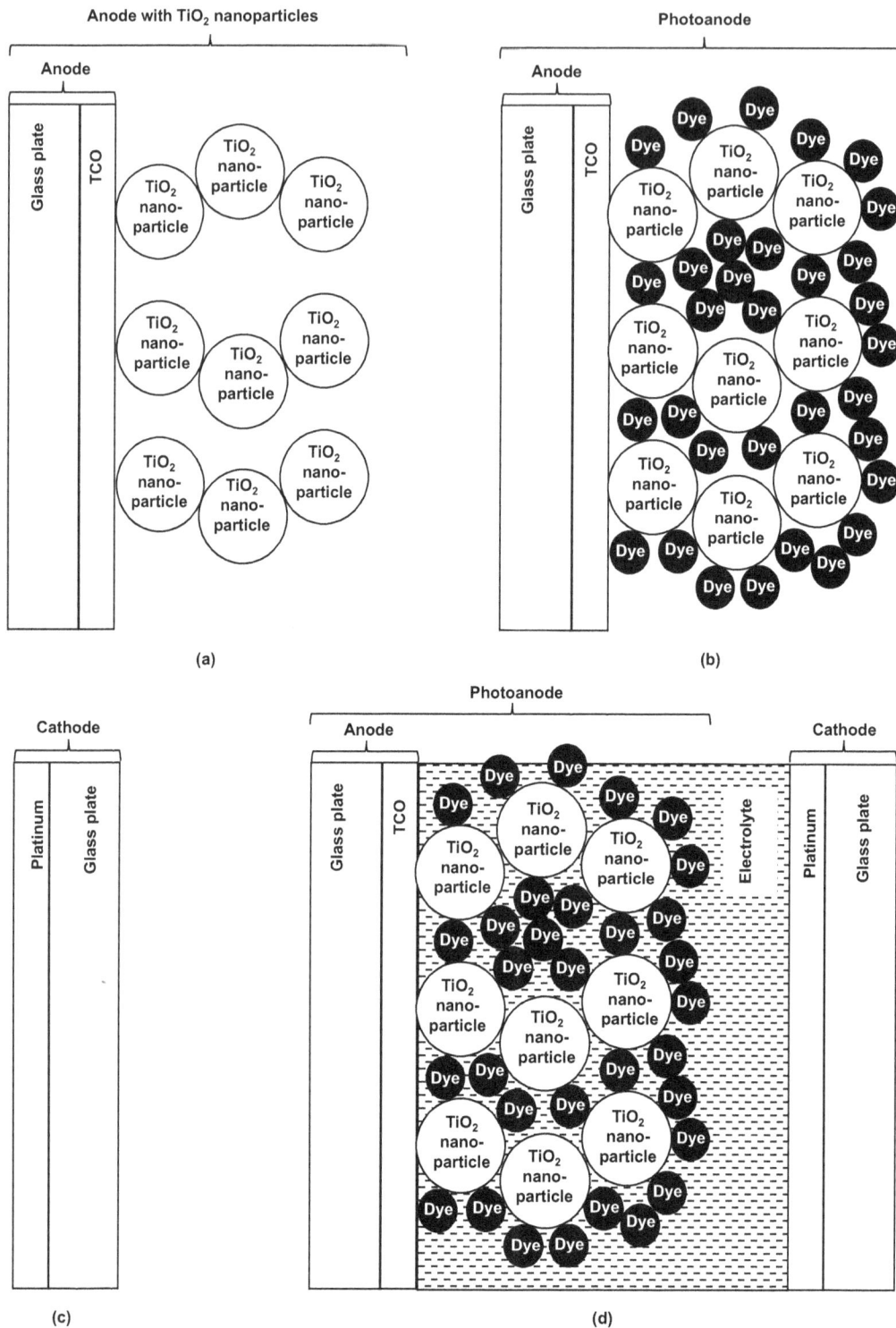

FIGURE 7.1 Components of a DSSC and their assembly: (a) TCO-coated glass plate covered with TiO$_2$ nanoparticles, (b) TCO-coated glass plate covered with TiO$_2$ nanoparticles and the TiO$_2$ nanoparticles soaked with dye molecules, (c) glass plate over which platinum film is deposited, and (d) parts shown in Figure b and Figure c assembled together, and intervening space filled with electrolyte.

In place of the pigment chlorophyll, the solar cell uses dye. Oxidation/reduction is done by an electrolyte in the cell.

The components of DSSC will now be explained in detail.

7.2 DSSC COMPONENTS

7.2.1 TRANSPARENT CONDUCTIVE SUBSTRATE

Desirable properties of substrate materials are high optical transparency and electrical conductivity together with good thermal stability up to 500°C and chemical inertness (Reddy et al 2014). The base material is soda lime glass. Common conductive coatings are thin films of fluorine-doped tin oxide (FTO, SnO_2: F), transmittance > 80%, sheet resistance $18\Omega cm^{-2}$, and indium-doped tin oxide (ITO, In_2O_3: Sn), transmittance ~ 75%, sheet resistance $8.5\Omega cm^{-2}$. Indium being a rare earth metal is scarce and costly. Among other substrate options, ITO-coated polymers such as polyethylene terephthalate (PET), polyethylene naphthalate (PEN), etc. form an important class due to mechanical flexibility but suffer from lack of thermal stability at high temperatures.

7.2.2 NANOSTRUCTURED SEMICONDUCTOR WORKING ELECTRODE (PHOTOANODE)

This electrode, called the photoanode, is made of wide-bandgap semiconductor metal oxide. N-type oxides are titanium dioxide, TiO_2 (E_G = 3–3.2eV); niobium pentoxide, Nb_2O_5 (E_G = 3.4eV); zinc oxide, ZnO (E_G = 3.37eV); and tin (IV) oxide, SnO_2 (E_G = 3.6eV). E_G is the energy gap. An example of P-type oxide is nickel (II) oxide, NiO (E_G = 3.7eV) (Sharma et al 2018).

TiO_2 has been the most widely used semiconductor because of its low cost, ready availability, nontoxicity, and biocompatibility (Andualem and Demiss 2018). Anatase allotrope of TiO_2 is preferred over the rutile allotrope because of the larger bandgap of anatase TiO_2 (3.2 eV) than rutile TiO_2 (3.0eV). Furthermore, electron transport through rutile TiO_2 is slow, owing to its high packing density. However, anatase TiO_2 is metastable and rutile TiO_2 is stable, with anatase TiO_2 transforming irreversibly to rutile TiO_2 at high temperature. Rutile TiO_2 is cheaper to produce and is also superior in light-scattering property.

7.2.3 DYE (PHOTOSENSITIZER)

This component absorbs solar energy. Essential properties of the dye are (Mehmood et al 2014):

(i) Surface absorption. It should adsorb strongly on the surface of the semiconductor with good adhesion to secure its firm grafting to the semiconductor.

(ii) Light absorption. It should absorb light intensely. It should possess a broad absorption spectrum covering the full visible spectrum (400–700nm) and extending into UV-visible and near-infrared regions. High extinction coefficient material is required. Extinction coefficient of a substance is a property which measures how strongly the substance absorbs light of a specific wavelength.

(iii) Stability in oxidized condition. It should be stable in oxidized state, permitting reduction by electrolyte.

(iv) Chemical stability. From chemical stability viewpoint, it should be sufficiently chemically stable because it has to execute 10^8 redox turnovers during 20 years of exposure to illumination spanning the solar cell life.

(v) LUMO and HOMO positions. The positions of the lowest unoccupied molecular orbital (LUMO) and highest occupied molecular orbital (HOMO) of the dye must be suitably located with respect to band edges of the semiconductor as well as the redox potential of the electrolyte.

When a photon is absorbed by the solar cell, the dye molecule is oxidized. During oxidation, an electron is promoted from the HOMO to the LUMO of the dye. From the LUMO of the dye, the electron is injected via the interfacial chemical bonds between the dye and the semiconductor into the conduction band of the semiconductor. For electron injection, the LUMO of the dye must be higher in energy than the conduction band edge E_C of the semiconductor.

Subsequently, the oxidized dye molecule is brought back to its original state, i.e., regenerated by gaining an electron from the electrolyte. In order that this can happen, the HOMO of the dye must be positioned at a lower energy relative to the redox potential of the electrolyte.

Locations of the HOMO, LUMO, and E_C are carefully studied for selection of the proper dye-semiconductor combination.

7.2.3.1 Naturally Occurring Dyes

They are easily available in abundance in leaves, flowers, and fruits of plants at low costs. Additionally, they are environment friendly and do not contain metals.

The leaves of plants contain three kinds of pigments:

  (i) the carotenoids (long-chain water-repellant molecules synthesized in the plastids of plant cells, giving yellow, red, and orange colors);
 (ii) the chlorophylls a and b produced in the chloroplasts in photosynthetic tissues of the leaves, helping plants to manufacture their food and giving them their color; and
(iii) the anthocyanins (water-soluble pigments with red, purple, blue, or black appearance, produced in the cytoplasm of the plant cells).

Red turnip, wild Sicilian, mangosteen, rhoeo spathacea, and shisonin are a few natural dye examples. The natural dyes have low efficiency and provide narrow optical absorption range.

7.2.3.2 Metal Complex Sensitizers

Ruthenium dyes consist of a family of complexes. This family has the chemical formula $\{(4,4'-CO_2H)_2-(bipy)\}_2RuX_2$ where bipy = 2,2'-bipyridyl, X = Cl, Br, I, CN, NCS (thiocyanato). For X = NCS, the dye is N3 (Ossila 2021).

N3 dye: Cis-bis(isothiocyanato)bis(2,2'-bipyridyl-4,4'-dicarboxylato)ruthenium(II), $C_{26}H_{16}N_6O_8RuS_2$, HOMO = −5.39eV, LUMO = −2.79eV. One of the earliest benchmarked ruthenium complex dyes and having two bipyridine and thiocyanato ligands, it absorbs light up to 800nm, owing to loosely bound thiocyanato groups (Sekar and Gehlot 2010).

N719 dye: Di-tetrabutylammonium cis-bis(isothiocyanato)bis(2,2'-bipyridyl-4,4'-dicarboxylato) ruthenium(II), $C_{58}H_{86}N_8O_8RuS_2$, HOMO = −6.01eV, LUMO = −3.64eV. It is the ammonium salt of N3 dye with tetrabutylammonium ion (TBA⁺) in place of H⁺ at two carboxyl groups.

N749 dye (black dye): Tris (N,N,N-tributyl-1-butanaminium)[[2,2″6',2″-terpyridine]-4, 4',4″-tricarboxylato(3-)-N1,N1',N1″]tris(thiocyanato-N)hydrogen ruthenate(4-), $C_{69}H_{117}N_9O_6RuS_3$, HOMO = −5.2eV, LUMO = −3.8eV. Developed for a wide range of spectral absorption, this dye provides absorption up to 860nm and behaves similarly to N3 and N719 dyes. However, it has a lower absorption coefficient than N3 and N719 dyes, requiring thicker TiO_2 electrodes. This degrades the short-circuit current and open-circuit voltage of the solar cell.

Z907 dye: *cis*-Bis(isothiocyanato)(2,2′-bipyridyl-4,4′-dicarboxylato)(4,4′-di-nonyl-2′-bipyridyl) ruthenium(II), $C_{42}H_{52}N_6O_4RuS_2$, HOMO = −5.4eV, LUMO = −3.8eV. This dye overcomes the short-comings of N3 and N719 dyes, which are prone to long-term degradation due to penetration of water molecules into the electrolyte, resulting in desorption of the dye from the surface of TiO_2. The hydro-phobic alkyl chains anchored to one of the bipyridine ligands in Z907 dye are able to isolate water molecules from the chemical bonds between the dye and TiO_2, thereby achieving stable performance.

7.2.3.3 Metal-Free Organic Dyes

Efforts are underway to develop ruthenium-free dyes because ruthenium is an expensive pre-cious metal. Some ruthenium-free photosensitive dyes are (Desilvestro and Hebting 2021) D102 ($C_{37}H_{30}N_2O_3S_2$), an indoline-based red organic dye; D205 ($C_{48}H_{47}N_3O_4S_3$), an indoline red-violet dye; and D358 ($C_{52}H_{53}N_3O_6S_3$), a red-violet dye.

The general design of ruthenium-free dyes consists of a donor-π-acceptor structure in which a π-conjugated spacer separates the substituents working as the donor and the acceptor. The perfor-mance of the dye depends on the proper tuning of the components and can be improved by match-ing the substituents in the donor-π-acceptor arrangement. Presently, the efficiency of these dyes is lower than that of ruthenium-based sensitizers.

7.2.4 Electrolyte

The electrolyte is a vital segment of the path of electron transfer.

7.2.4.1 Tasks Performed by the Electrolyte

The electrolyte performs two tasks:

(i) Its first task is regeneration of the dye because after injection of electron from the dye into the conduction band of the semiconductor, the dye molecule is deficit of one electron. The electrolyte removes this deficiency by supplying an electron to the dye.

(ii) The second task of the electrolyte is to act as a medium of charge transport. Positive charge is transferred through the electrolyte to the counter electrode.

7.2.4.2 Essential Properties of the Electrolyte

Critical dependence of operation of the DSSC on the performance of the electrolyte calls for its compliance with properties, such as:

(i) Fast electron diffusion, which is achieved with a high conductivity and low viscosity.
(ii) Good interfacial contact both with the working and the counter electrodes.
(iii) Noncorrosive nature, which is necessary because it should neither cause desorption of dye from the semiconductor nor react with the dye, hastening its degradation.
(iv) Efficient regeneration of the dye by the redox couple.
(v) Long-term thermal, chemical/electrochemical stability for reliable operation of DSSC.
(vi) Nonoverlapping absorption spectrum with the dye to prevent light absorption in the electrolyte.

7.2.4.3 Liquid Electrolytes

There are two subcategories of liquid electrolytes, organic solvent-based liquid electrolyte and room-temperature ionic liquid (RTIL). An ionic liquid is a nonmolecular compound. It consists of positive and negative ions only and has a melting point < 100°C.

Organic electrolytes: Three main components of organic electrolytes, namely, the redox couple, the solvent, and the additives, critically influence their working.

(i)   Redox couple. The redox couple is deemed as the most important component of the electrolyte. Among the redox couples studied, I_3^-/I^-, Br^-/Br_3^-, thiocyanate anion SCN^-/ thiocyanogen $(SCN)_2$, and substituted bipyridyl cobalt (III/II), the redox couple I_3^-/I^- has emerged as the most appropriate because of its useful properties. It displays excellent solubility. It accomplishes regeneration of the dye very fast. Its absorbance of visible light is low. Its redox potential is situated at the required level to facilitate charge movements. Further, the recombination process between the electron injected into the semiconductor and I_3^- ion is slow.

The chief disadvantage of I_3^-/I^- is its high volatility, causing the dye to desorb. It also lacks long-term-stability and corrodes glass/TiO_2/Pt.

It must be noted that an optimal concentration of I_3^-/I^- in the mixture is needed. When the concentration is low, the conductivity of the electrolyte decreases, and contrarily, at a high concentration, its optical absorption increases. So neither a low nor a high concentration can be allowed.

(ii) Solvent. The solvent dissolves the redox couple ions, such as I_3^-/I^-. Commonly used organic solvents are acrylonitrile (ACN), ethylenecarbonate (EC), propylene carbonate (PC), N-methylpyrrolidone (NMP), and 3-methoxypropionitrile (MePN). Mixtures of solvents, e.g., ACN/valeronitrile, have also been used. All these solvents have high relative permittivity. Donor number of the solvent, a quantitative measure of its Lewis basicity, is a staple parameter affecting the solar cell properties. Viscosity of the solvent must be low for easy diffusion of electrons.

(iii) Additives. The additive adsorbs on the interface between the semiconductor and the electrolyte. Its role is to hinder the surface recombination of back-transferred electrons with I_3^- ions. These are the electrons injected into the semiconductor but may be reverse transferred to the electrolyte and lost by surface recombination with I_3^- ions. Some of the additives used are 4-*tert*-butylpyridine (TBP), guanidinium thiocyanate (GuNCS), and N-methylbenzimidazole (NMBI).

Ionic Electrolytes: RTILs have been preferred against the more efficient covalent electrolytes, which, being highly volatile, suffer from poor stability over extended usage times. The RTILs are organic salts. They perform the functions of both iodine source and solvent. In these salts, the cations are alkyl imidazolium, alkyl pyridinium, and trialkylmethylsulfonium. The anions belong to the halide or pseudohalide family. The anions used with alkyl imidazolium are I^-, $(CFCOO)N^-$, BF^-, $N(CN)^-$, $B(CN)^-$, and PF-N. N′ bis-alkyl-substituted imidazolium iodides comprise a widely used ionic liquids group. 1-hexyl-3-methylimidazolium iodide, 1-methyl-3-(3,3,4,4,5,5,6,6,6,-nonafluorohexyl) imidazolium, and 1-butyl-3-methylimidazolium iodide are some IL examples. With shortening of the length of the alkyl chain, the viscosity increases, and so the conductivity falls. The low efficiency of ILs precludes their wide-ranging applications in solar cells.

7.2.4.4   Solid and Quasisolid Electrolytes

Problem of leakage of the electrolyte is a serious disadvantage when using liquid electrolytes. Electrolyte leakage decreases the life span of the solar cell. This limitation necessitates the search for solid electrolytes. A hole transport material (HTM) replaces the liquid electrolyte. Inorganic HTMs, such as good conductivity copper compounds, copper (I) iodide CuI and copper(I) thiocyanate CuSCN, are used. An organic HTM is: 2,2′,7,7′-tetrakis (N,N-di-p-methoxyphenylamine) 9,9′-spirobifluorene (OMeTAD). Solid electrolyte-based solar cells are lower in optical-to-electrical conversion efficiency than liquid-based cells. The primary reason of lower efficiency is the lack of intimacy of contact between the N-type semiconductor (TiO$_2$) and HTM. Intimacy is deficient because the solid electrolyte is unable to penetrate sufficiently into the pores of the mesoporous TiO$_2$ to establish a close contact like the liquid electrolyte. Moreover, the rate of charge recombination between the N-type semiconductor and HTM is high. Over and above, owing to the low intrinsic conductivity of the HTM, the electron transfer between the dye and the HTM is impeded.

Introduction of a redox couple into the solid electrolyte can be helpful in mitigating this issue. A polymer mixed with liquid electrolyte serves this purpose. An example is the polymer poly(ethylene glycol) (PEG) in the solvent (propyleneycarbonate + potassium iodide, iodine) (Mohmeyer et al 2004). Elevated temperature operation of polymeric electrolytes triggers a phase transformation from gel to solution state, affecting the performance.

7.2.5 COUNTER ELECTRODE (CE)

During operation of the solar cell, the electrolyte is oxidized. The oxidized electrolyte diffuses towards the counter electrode. The electrolyte must be regenerated for continued operation. For regeneration, it must receive electrons from the external circuit. A catalyst is necessary to accelerate the reduction of electrolyte. Platinum has served this purpose well. The platinum thin film is deposited on a transparent conducting oxide. Many techniques have been applied viz vapor deposition, sputtering, screen-printing, electro-deposition, and by thermal decomposition of a salt of hexachloroplatinic acid H$_2$PtCl$_6$ in isopropanol. The activity of platinum electrode degrades over time in the presence of I$_3^-$/I$^-$ redox couple. The redox couple affects the catalytic properties of platinum. Also, platinum is slowly removed by electrolyte action. Combined with the fact that platinum is an expensive and rare metal, alternative cost-effective materials have been searched, e.g., transparent counter electrode made of nanoporous FeSe binary alloy for bifacial DSSC (Liu et al 2015). Synthesized graphene nanosheets are dispersed in (terpineol + ethyl cellulose) mixture, screen-printed on FTO slides, and annealed to form a 3D network (Zhang et al 2011).

7.3 FORWARD AND BACKWARD ELECTRON TRANSFER PROCESSES IN DSSC

7.3.1 FORWARD ELECTRON TRANSFER PROCESSES

7.3.1.1 Receipt and Absorption of Sunlight by the Dye and Promotion of an Electron in the Dye from Its HOMO to the LUMO (Ground State to Excited State)

Sunlight falling on the glass plate of the solar cell crosses the transparent conducting oxide (Figure 7.2). Moving further, it passes through the transparent nanocrystalline TiO$_2$ and reaches the dye molecules. Photons in the sunlight have energies between 0.5eV and 3.5eV. Photons having energies $\geq$ the energy difference between the LUMO and HOMO of the dye are successful in uplifting the electron in the dye molecule from the HOMO to LUMO level. The dye molecule containing the electron in the LUMO level is said to be in the excited state and is represented with an asterisk mark. So the promotion of dye molecule to the excited state Dye^* under the influence of solar radiation is written as:

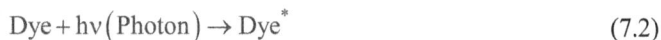

$$\mathrm{Dye} + h\nu\left(\mathrm{Photon}\right) \rightarrow \mathrm{Dye}^* \tag{7.2}$$

The higher-energy photons, which have more energy than required for this electron uplifting process, produce hot carriers. Photons with energies < the LUMO-HOMO energy difference do not produce any excitation.

7.3.1.2 Injection of an Electron from the LUMO of the Dye to the Conduction Band of the Semiconductor (TiO₂): Charge Separation

The electron sitting in the LUMO level of the dye finds itself at a higher energy than the energy of the edge E_C of the conduction band of TiO$_2$. Therefore, the electron is transferred from the dye molecule to the conduction band of TiO$_2$ through the chemical bonds at the dye/TiO$_2$ interface. This

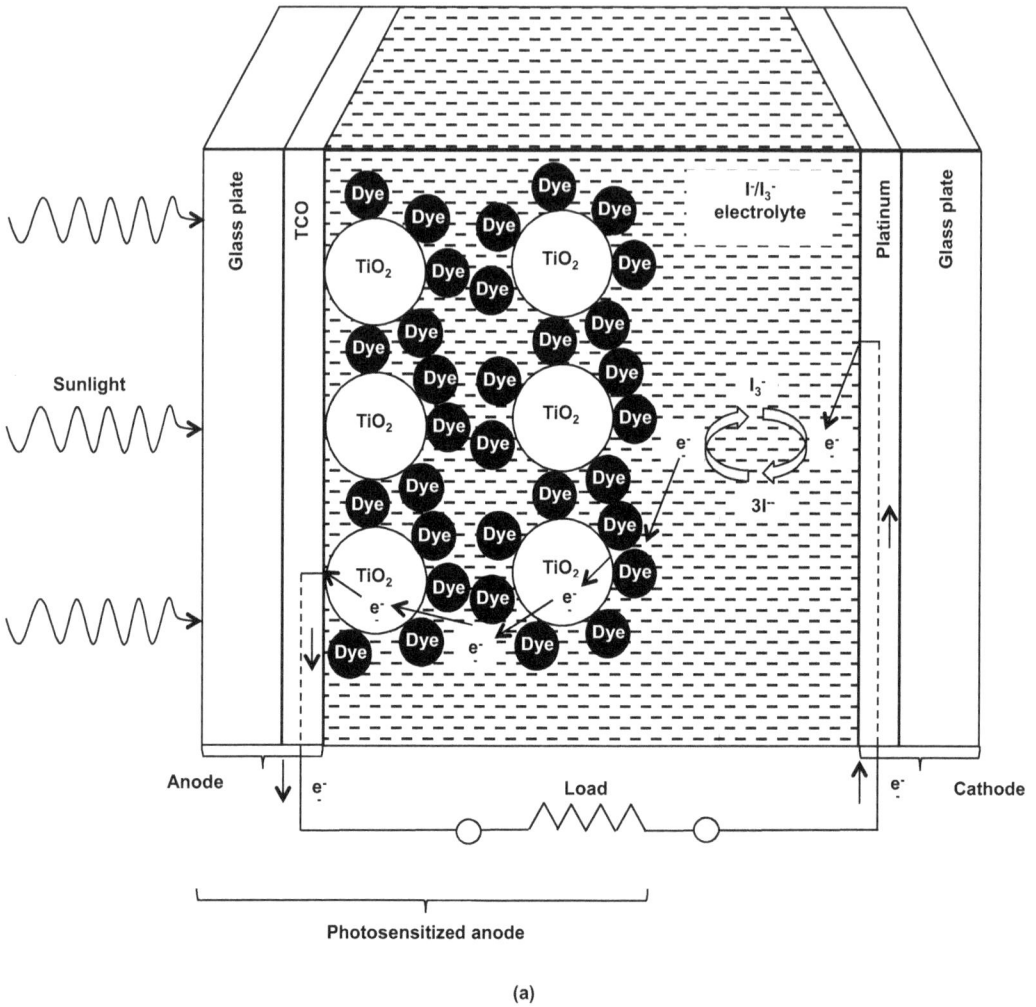

(a)

FIGURE 7.2 Dye-sensitized solar cell (DSSC), also called the Grätzel cell: (a) Cell construction: The anode is a transparent conducting oxide, such as indium tin oxide or fluorine-doped tin oxide, coated on a glass plate, and the cathode is a platinum film deposited on a glass plate. The anode is sensitized with a semiconducting TiO$_2$ nanoparticle film soaked in a solution of dye, usually a ruthenium complex, forming a photosensitized anode. An iodide/triiodide (I⁻/I$_3$⁻) electrolyte is filled in the cell to form a photoelectrochemical system. (b) Energy band diagram and working principle involving the steps of (i) (a) reception of energy by dye molecule ***Dye***, (b) excitation of the dye molecule ***Dye*** from the ground-state ***Dye*** to higher-energy state ***Dye**** above the bandgap energy of semiconductor TiO$_2$ by receiving energy from sunlight, (ii) oxidation of the excited dye molecule ***Dye**** to the state ***Dye***⁺ by losing electron *e⁻* and injection of lost electron *e⁻* into the conduction band of semiconductor TiO$_2$, from where the electron *e⁻* diffuses to the anode contact and flows into the external circuit, (iii) reduction of the oxidized dye molecule ***Dye***⁺ by the iodide *I⁻* in the electrolyte, with iodide converting to triiodide *I$_3$⁻*, and finally, (iv) regeneration of the iodide *I⁻* by electron extraction from cathode and its combination with *I$_3$⁻*.

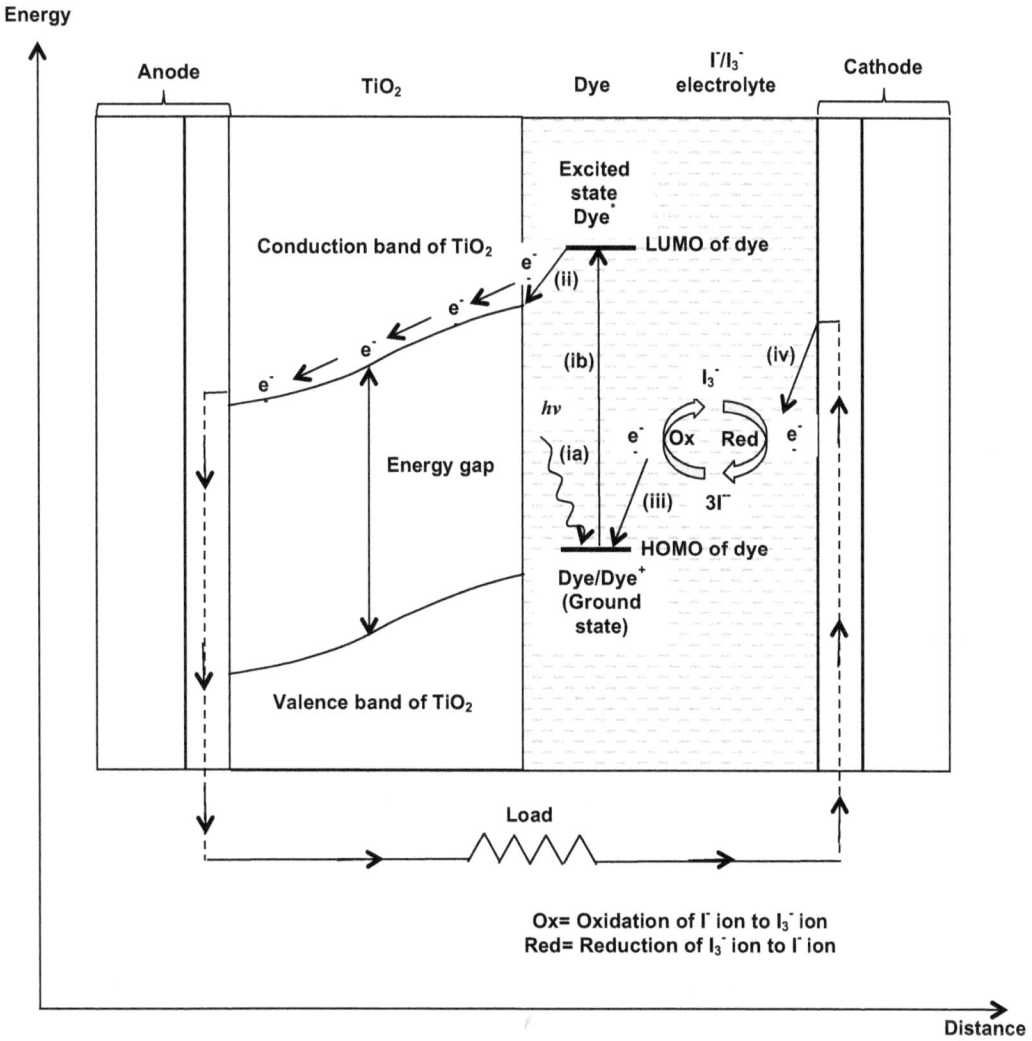

Energy

(b)

FIGURE 7.2 (Continued)

injection takes place within a time frame of picoseconds to femtoseconds of the electron reaching the LUMO level of the dye molecule. Upon losing the electron, the dye molecule becomes oxidized and positively charged. Thus, the excited dye molecule Dye^* is oxidized to Dye^+ by releasing an electron. The liberated electron is injected into the conduction band of TiO_2. From the conduction band of TiO_2, it reaches the anode:

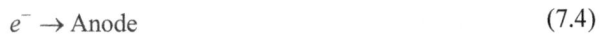

$$Dye^* \rightarrow Dye^+ + e^- \left(TiO_2 \right) \tag{7.3}$$

$$e^- \rightarrow Anode \tag{7.4}$$

This is the charge separation reaction with the positive charge on the dye and negative charge in titania.

7.3.1.3 Diffusion of the Electron through the TiO₂ Nanonetwork to Reach the TCO Layer

Diffusing across the TiO_2 nanoparticles, the electron comes to the TCO. The time scale for this process is in microseconds to milliseconds range. Unlike a P-N junction, where both electrons and holes move, there is only one carrier here, namely, the electron. Moreover, there is no built-in electric field to assist in electron transport. The solo mechanism of charge transfer is carrier diffusion. The diffusion coefficient of electron (D_n) depends on the quasi-Fermi level in the semiconductor and is therefore dependent on the intensity of incident light. It increases with the intensity of light. The lifetime of electron (τ_n) decreases in proportion. So the product of diffusion coefficient and carrier lifetime remains constant, irrespective of light intensity, and the diffusion length of electrons given by

$$L_n = \sqrt{D_n \tau_n} \tag{7.5}$$

is independent of light intensity. It is typically 5–20μm. The thickness of TiO_2 film should be less than the diffusion length to avoid loss of electron through recombination.

7.3.1.4 Flow of the Electron through the External Circuit Reaching the Counter Electrode

An electronic current flows through the circuit, and the electron arrives via the conducting wire at the counter electrode.

7.3.1.5 Reduction of I₃⁻ Ion in the Electrolyte to I⁻ Ion by the Arriving Electron at the Counter Electrode

In the I_3^-/I^- redox couple, the I_3^- ion diffuses to the counter electrode, where it meets with an electron from the counter electrode. The electron interacts with the I_3^- ion in the electrolyte, reducing it to I^- ion.

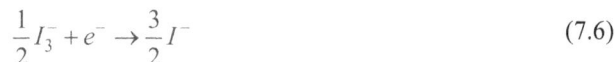

$$\frac{1}{2}I_3^- + e^- \rightarrow \frac{3}{2}I^- \tag{7.6}$$

This I^- ion reduces the oxidized dye molecule in the next step, itself becoming I_3^- ion again. It is once more brought back to its starting I^- state in step 7.3.1.7 below, ready for the next cycle of operations.

The difference between the Fermi level of the semiconductor and the redox potential of the electrolyte determines the potential difference across the DSSC. Accordingly, the open-circuit voltage of the solar cell is controlled by the difference between the bottom of the conduction band (E_C) of the semiconductor and the energy level of anion in the electrolyte, which is generally 0.5–0.7V.

7.3.1.6 Acceptance of an Electron by the Dye From the I⁻ Ion in the Electrolyte, Restoring It to Its Original State

The I^- ion donates an electron to the dye. By this donation, it oxidizes to I_3^- ion. This reaction takes place within nanoseconds.

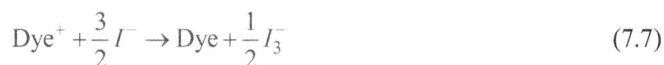

$$Dye^+ + \frac{3}{2}I^- \rightarrow Dye + \frac{1}{2}I_3^- \tag{7.7}$$

Thus, the oxidized dye molecule *Dye*⁺ is reduced to neutral dye molecule *Dye* by receiving two electrons donated by the iodide in the iodide/triiodide electrolyte:

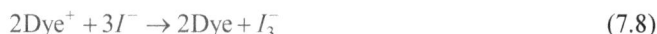

$$2Dye^+ + 3I^- \rightarrow 2Dye + I_3^- \tag{7.8}$$

because addition of two electrons to an I_3^- ion leads to the production of $3I^-$ ions, which is evident upon examination of the Lewis structures of the I_3^- and I^- ions.

7.3.1.7 Diffusion of I_3^- Ion Mediator towards the Counter Electrode and Its Reduction to I^- Ion by Receiving an Electron from the External Circuit Recovering Its Initial State

The iodide itself is regenerated through reduction of triiodide at the cathode, for which two electrons are pulled from the cathode.

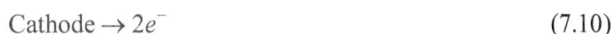

$$I_3^- + 2e^- \rightarrow 3I^- \tag{7.9}$$

$$\text{Cathode} \rightarrow 2e^- \tag{7.10}$$

Thus, the anode receives electrons and the cathode delivers electrons, leading to electronic current flow in the external circuit from anode to cathode.

Note that this is the same as Step 7.3.1.5 above. Thus, there is no net chemical reaction inside the DSSC.

7.3.2 BACKWARD ELECTRON TRANSFER PROCESSES: LOSS MECHANISMS

The backward electron transfer processes constitute the loss mechanisms, by which the efficiency of the solar cell is impaired. The electron produced during oxidation of dye molecule may be lost by recombination with the oxidized dye molecule itself. Another possible route of electron loss is by electron transfer restoring dye molecule in the excited state to ground state.

The effectiveness of the DSSC in the conversion of light into electricity depends on the relative speeds of the different participating reactions, i.e., the reaction kinetics. It is imperative that the recombination of electron with oxidized dye must be much slower than the reduction of oxidized dye by I^- ion in the electrolyte. Similarly, the relaxation rate of electron for decay of the dye molecule from excited to ground state must be significantly slower than the rate of electron injection into the semiconductor.

The main loss mechanism is the recombination with the electrolyte. The liquid electrolyte may be trapped within the pores of the nanocrystalline porous TiO_2. So the electron injected from the dye molecule into TiO_2 may be lost by recombination with the I_3^- ion in the electrolyte. Because the TCO/electrolyte interface is a vital recombination pathway, a thin layer of compact TiO_2 is deposited to separate the TCO and nanocrystalline porous TiO_2, thereby blocking this route. Furthermore, the nanocrystalline TiO_2 is subjected to a $TiCl_4$ treatment to augment the interfacial charge-transfer resistance of TCO/electrolyte interface.

7.4 EFFECT OF DOPING THE TiO_2 PHOTOANODE FILM WITH GOLD NANOPARTICLES ON DSSC PERFORMANCE

After becoming conversant with the main components and charge transfer processes in a DSSC, we now study specific examples from the literature involving investigations of the influence of nanoparticles of particular nature, e.g., metallic or upconversion nanoparticles, deliberately introduced in the photoanode. Although the TiO_2 film itself is nanocrystalline, the presence of more nanoparticles can produce extra effects. Let us see what happens when Au nanoparticles are added to TiO_2 film. This solar cell is fabricated by Wang et al. (2013).

Figure 7.3 shows the DSSC with Au nanoparticles–doped TiO_2 photoanode. For fabricating DSSCs with the TiO_2 film of the photoanode doped with the nanoparticles, the process is elaborated in forthcoming subsections.

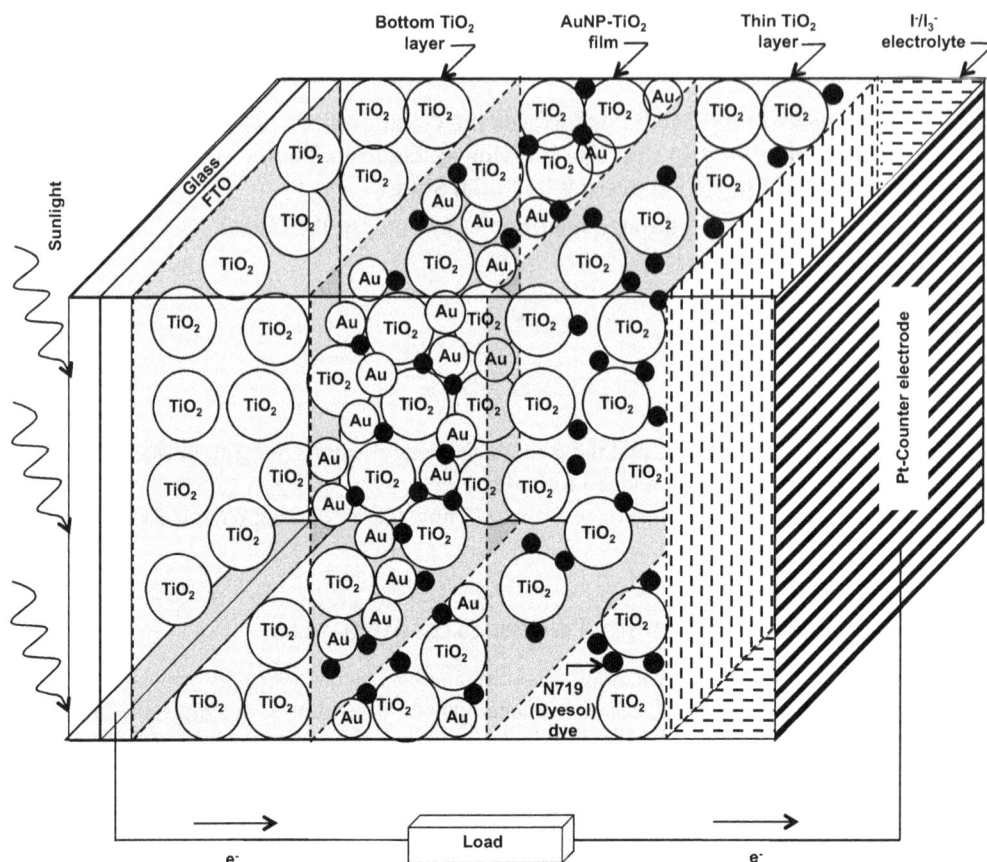

FIGURE 7.3 Au nanoparticles–doped TiO_2-photoanode DSSC consisting of a glass substrate coated with FTO, upon which three TiO_2 layers are successively deposited: a bottom TiO_2 layer, an Au nanoparticle–TiO_2 layer in the middle, and a thin covering TiO_2 layer to protect the gold nanoparticles from the corrosive action of the I^-/I_3^- electrolyte filled in the cell and, finally, the platinum counter electrode. The covering TiO_2 layer is sensitized with the N719 dye, which percolates to the Au NPs-TiO_2 layer. Sunlight falls on the solar cell from the side of FTO-coated glass substrate. The photovoltaic current is drawn by connecting wires to the FTO and Pt electrodes.

7.4.1 SYNTHESIS OF AU NANOPARTICLES

Au nanoparticles of three sizes, 5nm, 45nm, and 110nm, are synthesized using the reagents $HAuCl_4$, $3H_2O$ (hydrogen tetrachloroaurate [III] trihydrate), CTAB (cetyltrimethylammonium bromide) $Na_3C_6H_5O_7, 2H_2O$ (trisodium citrate), $NaBH_4$ (sodium borohydride), and $C_6H_8O_6$ (ascorbic acid).

7.4.2 TiO₂ FILM DEPOSITION

The TiO_2 film is deposited on FTO-coated glass substrate in three parts:

  (i) a bottom TiO_2 layer by doctor blading and heating at 550°C for ½ h.;
 (ii) an Au-TiO_2 film over the bottom layer with same thermal cycle and dipping in titanium organic sol; and
(iii) a final thin TiO_2 layer for increasing the loading with dye and shielding AuNPs from erosion due to iodine-based electrolyte followed by heating at 450°C for 50 min.

7.4.3 Sensitization of TiO₂ Film with Dye

This is done by keeping the TiO₂ photoanode with the three TiO₂ layers dipped in solution of N719 dye for 24 h. The solution concentration is 0.5mM. The solvent is a 1:1 mixture by volume of acetonitrile and *tert*-butanol.

7.4.4 Counter Electrode Fitting and Assembly of the Solar Cell

The counter electrode is made of platinum. The components of the cell are sealed with a 30µm-thick spacer of Surlyn. A drop of I^-/I_3^- electrolyte is introduced by vacuum back filling.

7.4.5 Characterization of Solar Cell

The best efficiency is achieved with 5nm spherical gold nanoparticles DSSC. For 5nm Au NPs-TiO₂, the open-circuit voltage is 0.797V, the short-circuit current density is 19.4 mAcm^{-2}, the fill factor is 0.698, and the efficiency is 10.8% under AM 1.5G 1 sun at 100mWcm^{-2} intensity. These parameters for the solar cell without Au NPs are 0.747V, 18.67mAcm^{-2}, 0.688, and 9.59%, respectively. The efficiency of solar cell using 45nm octahedral Au NPs is 10.52%, while that of solar cell with 110nm Au NPs is 10.2%, meaning, that efficiency decreases with increasing nanoparticle size (Wang et al 2013).

7.4.6 Dependence of Solar Cell Efficiency on Nanoparticle Dimensions and Shape

The photoconversion efficiency of DSSCs depends on both the size and shape of nanoparticles. The nanoparticles influence the performance of solar cells in three different ways. These are the photocharging effect, plasmonic effect, and scattering effect. All these three effects simultaneously act together in nanoparticles of various sizes, but the magnitude or degree of any effect is determined by the particle size. Photocharging effect is dominant in 5nm-size NPs, and the plasmonic effect dominates in 45nm NPs. In 110nm-size NPs, scattering effect is more influential. So the parameters of the solar cells fabricated using these nanoparticles are different. The photocharging effect increases the open-circuit voltage, while the other two effects raise the short-circuit current density of the solar cell. Overall, the three coexisting effects are able to increase the efficiency to the greatest extent in the small-size, 5nm-size Au NPs cell. Such differential behavior of nanoparticles arises from the fact that the small 5nm NPs interacting with high-energy, short-wavelength photons are more capable of photocharging. The medium-size 45nm NPs interacting with medium-energy and moderate-wavelength photons are more effective in plasmonic action, whereas the large-size 110nm NPs interact with low-energy, long-wavelength photons to produce more scattering (Wang et al 2013).

7.5 EFFECT OF INCLUSION OF BROADBAND NEAR-INFRARED UPCONVERSION NANOPARTICLES (UCNPs) IN THE TiO₂ PHOTOANODE OF DSSC ON ITS POWER CONVERSION EFFICIENCY

7.5.1 How Upconversion Nanoparticles Assist in Utilization of Low-Energy Photons?

The near-infrared part of the solar spectrum remains unutilized by DSSCs. The problem is circumvented by using upconversion nanoparticles, which are particles of nanoscale dimensions that increase the photon energies by absorption of two or more low-energy photons and emission of one higher-energy photon having energy equal to the sum of the energies of absorbed photons. Photon absorption takes place in the low-frequency, and hence smaller-energy, infrared portion of the spectrum, and photon emission occurs in the higher-frequency and larger-energy visible or ultraviolet portion. The solar cell is responsive to the higher-energy photons and can successfully convert their energy into electricity. The low-energy photons having energies in the infrared range are not squandered away, because higher-energy photons have been produced from them with the help of upconversion nanoparticles.

Transition metals doped with lanthanides (elements from lanthanum to lutetium, atomic numbers 57–71 in the periodic table) or actinides (elements with atomic numbers 89–103) are used for upconversion. Transition metals are those metals which have valence electrons in two shells instead of one, e.g., Yttrium, symbol Y identical to lanthanides and considered a rare earth element. Lanthanides include neodymium, Nd; Erbium, Er; and Ytterbium, Yb.

Solar cell with upconversion nanoparticles is investigated by Hao et al (2017). Figure 7.4 illustrates the schematic diagram of this solar cell.

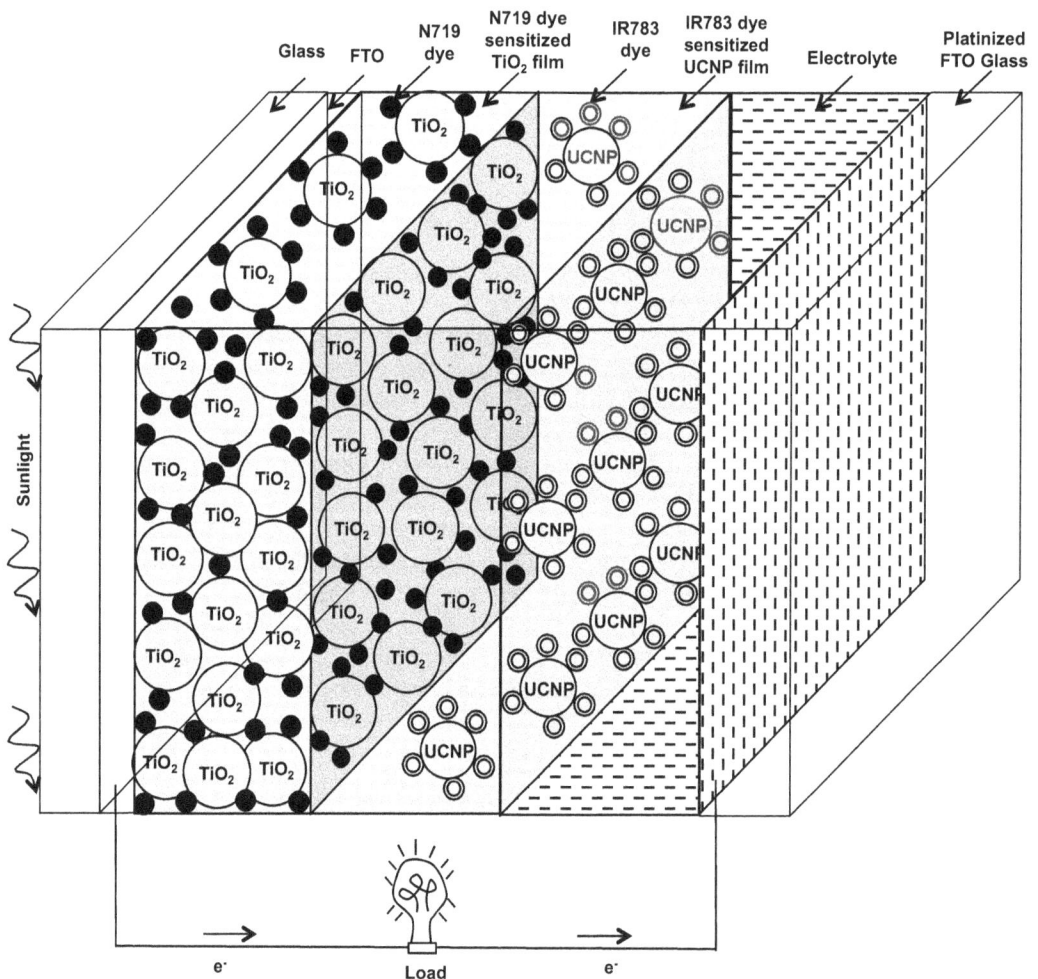

FIGURE 7.4 Dye-sensitized broadband NIR upconversion nanoparticles-based DSSC consisting of an FTO glass substrate and a TiO$_2$ photoanode sensitized with N719 dye and further coated with IR783 dye–sensitized UCNPs. The device has a platinized FTO glass counter electrode with electrolyte squeezed through a hole in the counter electrode, which is sealed before use. When sunlight is incident on the FTO glass substrate, current flows across the FTO glass and platinized FTO glass substrates.

7.5.2 Preparation of Upconversion Nanoparticles

Solvothermal method is applied for synthesizing $NaYF_4$ nanoparticles doped with 10 mol% Yb^{3+} and 2 mol% Er^{3+}, yielding $NaYF_4$:10% Yb^{3+}, 2% Er^{3+} core nanoparticles from which $NaYF_4$:10% Yb^{3+}, 2% Er^{3+}:$NaYF_4$:X% Nd^{3+} core/shell nanoparticles are obtained. The oleic acid ligands on the surfaces of these nanoparticles are replaced with sub-nm $NOBF_4$ (nitrosonium tetrafluoroborate) ligands. The nanoparticles are subsequently extracted into, purified, and redispersed in DMF (N,N-dimethylformamide).

7.5.3 Preparation of IR783 Dye-Sensitized Upconversion Nanoparticles

The IR783 is dissolved in DMF, and the nanoparticle dispersion in DMF is mixed with the IR783 dye solution to get IR783 dye-sensitized upconversion nanoparticles (DSUCNPs).

7.5.4 Reason for Sensitizing the Upconversion Nanoparticles with IR783 Dye

Upconversion nanoparticles without dye sensitization have low absorption at NIR frequencies. Also, the absorption occurs in a narrow-frequency range. Anchorage of IR783 dye molecules on the surfaces of the upconversion nanoparticles enables garnering the energy of NIR photons over a broader frequency range. The energy thus broadly collected by the IR383 dye is transferred to the upconversion nanoparticles. Consequently, the upconversion of energy to visible light is much larger with dye-sensitized nanoparticles than that obtained by the same nanoparticles without the attached IR783 dye molecules.

7.5.5 Making the N719 Dye-Sensitized TiO₂ Photoanode

The suspension of TiO_2 nanoparticles is mixed and sonicated with EC (ethyl cellulose) powder dissolved in ethanol, followed by drop-wise addition of terpineol, removal of ethanol on a rotary evaporator, screen-printing of TiO_2 film on FTO substrate, sintering at 450°C for 1 h. and 500°C for ¼ h. in nitrogen ambience, treatment with aqueous $TiCl_4$ solution, sintering at 500°C for ½ h., and finally immersion in solution of N719 dye for 24 h.

7.5.6 Deposition of IR783 DSUNPs on N719 Dye-Sensitized TiO₂ Photoanode

This is done by immersing the N719 dye-sensitized TiO_2 photoanode in 0.01M hexane solution of IR783 DSUNPs for time spans ranging from 5 to 30 min.

7.5.7 Making the Counter Electrode of Platinized FTO Glass

H_2PtCl_6 is deposited from 5M solution in isopropanol at 400°C for ½ h.

7.5.8 Sealing the IR783 DSUNPs@N719 Dye-Sensitized TiO₂ Photoanode with Counter Electrode

The photoanode and platinized FTO glass counter electrode are sealed together with Surlyn (25μm thickness) at 120°C. The electrolyte is introduced from a predrilled hole in the counter electrode. The hole is afterwards closed using Surlyn with thin glass.

7.5.9 DSUCNPs-Sensitized DSSC Testing

The efficiency of solar cell without IR783 dye sensitization of UCNPs is 7.573%, while that with IR783 dye sensitization of UCNPs is 8.568%, which represents an increment by (8.568–7.573) ×

$100/7.573 = 13.14\%$. The total increase of 13.1% is attributed to combination of contributions from broadband upconversion by DSUNPs and scattering by the DSUNPs. The first factor accounts for 7.1% of the total percentage increase, and the second factor is responsible for 6.0% of the total increase (Hao et al 2017).

7.6   DISCUSSION AND CONCLUSIONS

The dye-sensitized solar cell is a nanocrystalline system made up of photovoltaic and electro-chemical subsystems. It is a strong competitor to present-day solid-state solar cells, both from engineering viewpoint and cost considerations. Sensitizers with a broad spectral range of absorption combined with semiconductor films with crystallite size of a few nanometers enable harvesting of a significant part of sunlight. Additional nanoparticles can be incorporated to provide supplementary nanoeffects (Table 7.1).

TABLE 7.1

Effects of Inclusion of Additional Nanoparticles (Besides TiO_2 NPs) in Dye-Sensitized Solar Cells

Sl. No.	Type of Nanoparticle Inclusion	Effect on Efficiency of Solar Cell	Reasons	Reference
1.	Doping the nanocrystalline TiO_2 photoanode with Au NPs	Efficiency is 10.8% with 5nm spherical Au NPs, but 9.59% without Au NPs	Photocharging, plasmonics, and scattering	Wang et al (2013)
2.	Inclusion of broadband near-infrared upconversion NPs in TiO_2 photoanode	Efficiency is 8.568% with IR783 dye sensitization of UCNPs, and 7.573% without IR 783 dye sensitization of UCNPs	Broadband upconversion and scattering	Hao et al (2017)

Questions and Answers

7.1   How is a dye-sensitized solar cell different from a P-N junction solar cell? Answer: In a P-N junction solar cell, light is absorbed and carriers are transported in the same semiconductor material. In a dye-sensitized solar cell, light is absorbed by the dye while carriers are separated at the dye/TiO_2 interface.

7.2   How does a dye-sensitized solar cell qualify as a nanotechnology-based cell? Answer: Because it uses semiconductor TiO_2 nanoparticles.

7.3   In what respect is the operation of a dye-sensitized solar cell similar to photosynthesis in nature? Answer: Photosynthesis uses a green pigment known as chlorophyll. The dye-sensitized solar cell uses a dye.

7.4   Compare FTO-coated glass with ITO-coated glass. Answer: FTO-coated glass is cheaper, is formed by directly coating FTO on glass, has good conductivity and good transparency, is used up to 600°C maintaining conductivity, has higher IR reflectance, and is tolerable to physical abrasion. ITO-coated glass is expensive, is formed by coating ITO on a passivation layer on glass, has moderate conductivity and medium transparency, is used up to 350°C with decreasing conductivity, has lower IR reflectance, and is moderately tolerable to abrasion.

7.5   Give three examples of N-type semiconductor oxides used in dye-sensitized solar cells. Answer: TiO_2, ZnO, SnO_2.

7.6   Name a P-type semiconductor oxide used in dye-sensitized solar cells. Answer: NiO.

7.7   What are anatase and rutile? Answer: Two polymorphs of titanium oxide.

7.8 Which polymorph of titanium oxide is commonly used in dye-sensitized solar cells? Answer: Anatase.

7.9 The dye used in a dye-sensitized solar cell should have a high extinction coefficient. Yes or no? Answer: Yes.

7.10 What will happen if the LUMO of the dye is located at a lower energy than the conduction band edge E_C of TiO_2. Answer: Electron transference from dye to TiO_2 will become unlikely. Solar cell will not work.

7.11 What will be the result when HOMO of the dye lies at a higher energy than the redox potential of the electrolyte? Answer: Oxidized dye molecule will not recover its initial state by receiving an electron from the electrolyte. Solar cell will stop functioning.

7.12 What are carotenoids, chorophylls a and b, and anthocyanins? Answer: Pigments in leaves of plants.

7.13 Give an example of a natural dye. Answer: Red turnip.

7.14 Why are ruthenium complexes used as photosensitizers in DSSCs? Answer: They exhibit good photoelectrochemical properties and are highly stable in the oxidized state. Their photovoltaic stability shows no degradation signs in long-duration endurance tests.

7.15 What is "bipy" in rujhenium dyes? Answer: 2,2′-Bipyridyl or 2,2′-bipyridine, a colorless organic compound ($C_{10}H_8N_2$) used as a chelating ligand for forming complexes with metal ions.

7.16 How is the stability problem in N3 and N719 dyes overcome by Z907 dye? Answer: Desorption of Z907 dye from TiO_2 surface is avoided because this dye has hydrophobic alkyl chains. These chains isolate the chemical bonds between the dye and TiO_2, preventing penetration of water molecules.

7.17 Why should use of ruthenium in dyes be avoided? Answer: Because it is a costly metal in the platinum group of periodic table, being one of the rarest metals on Earth.

7.18 What are the two functions of the electrolyte in a DSSC? Answer: (i) Participation in dye regeneration and (ii) acting as a charge transport medium.

7.19 What will happen if the absorption spectrum of the electrolyte in a DSSC overlaps with that of the dye? Answer: Optical loss will take place because all the light will not be absorbed by the dye. Some light will be lost through absorption by the electrolyte.

7.20 What is an ionic liquid? Answer: It is an organic salt in liquid state under 100°C, usually consisting of an organic cation and a polyatomic inorganic anion.

7.21 Why can't a low concentration of redox couple of I_3^-/I^- in the electrolyte be used? Answer: Conductivity of the electrolyte will fall.

7.22 Why can't a high concentration of redox couple of I_3^-/I^- in the electrolyte be used? Answer: Absorption of light by the electrolyte will increase.

7.23 Give arguments disfavoring the use of HTM solid electrolytes in DSSCs. Answer: Low efficiency because of (i) poor contact between TiO_2 and HTM, (ii) high recombination rate, and (iii) low conductivity.

7.24 What do you mean by a dye molecule in the excited state? Answer: A dye molecule in which the electron is in the LUMO level.

7.25 What is the time scale in which the electron is transferred from the LUMO level of dye molecule to the conduction band of TiO_2? Answer: Pico- to femtoseconds.

7.26 What is the time scale in which the electron is transferred from TiO_2 to TCO? Answer: Micro- to milliseconds.

7.27 Why is the electron transference from TiO_2 to TCO slow? Answer: Because charge transfer is not aided by any built-in electric field. It occurs by diffusion.

7.28 What is the difference between an excited dye molecule and an oxidized dye molecule? Answer: The excited dye molecule is Dye*, one in which electron is promoted from

HOMO to LUMO level of the dye molecule. The electron is still in the dye molecule. Oxidized dye molecule is Dye$^+$, one which has lost an electron. Oxidation takes place when the electron leaves the dye molecule and moves to TiO$_2$.

7.29 In which step does charge separation take place in a DSSC? Answer: The step in which the Dye* becomes Dye$^+$ by liberating electron and the liberated electron is given to TiO$_2$. Positive charge is on the dye, and negative charge is on TiO$_2$. Hence separated.

7.30 What is photocharging effect in a DSSC in which semiconductor TiO$_2$ nanoparticles are doped with Au nanoparticles? Answer: The platinum electrode in a DSSC quickly discharges electrons, but in a DSSC with Au nanoparticles–doped TiO$_2$, the Au NPs store a portion of electrons captured from the photoexcited semiconductor nanoparticles. The density of these electrons varies with the size of the Au NPs. This electron storage is the photocharging effect and causes Fermi level shift. Charge equilibration between the stored electrons and photoexcited semiconductor nanoparticles shifts the quasi-Fermi level in the metal-semiconductor system to more negative levels.

7.31 What are the different effects of incorporation of Au nanoparticles in TiO$_2$ film of a DSSC? Answer: Photocharging, plasmonic effect, and scattering.

7.32 What is photon upconversion? Give an example. Answer: Sequential absorption of two or more low-frequency, and hence low-energy, photons accompanied by emission of a high-frequency or high-energy photon, e.g., absorption of infrared photons with emission of visible light or UV photons.

7.33 How is photon upconversion beneficial to solar cell performance? Answer: Low-energy photons cannot produce electron-hole pairs, but high-energy photons obtained by upconversion can cause photogeneration of carriers.

REFERENCES

Andualem A. and S. Demiss 2018 Review on dye-sensitized solar cells (DSSCs), Journal of Heterocyclics, 1(1): 29–34.

Desilvestro H. and Y. Hebting 2021 Ruthenium-based dyes for dye-sensitized solar cells, © 2021 Merck KGaA, Darmstadt, Germany, www.sigmaaldrich.com/IN/en/technical-documents/technical-article/materials-science-and-engineering/photovoltaics-and-solar-cells/dye-solar-cells (Accessed on 18th December 2021).

Grätzel M. 2003 Dye-sensitized solar cells, Journal of Photochemistry and Photobiology C: Photochemistry Reviews, 4: 145–153.

Grätzel M. 2004 Conversion of sunlight to electric power by nanocrystalline dye sensitized solar cells, Journal of Photochemistry and Photobiology A: Chemistry, 164: 3–14.

Hao S., Y. Shang, D. Li, H. Ågren, C. Yang and G. Chen 2017 Enhancing dye-sensitized solar cell efficiency through broadband near-infrared upconverting nanoparticles, Nanoscale, 9: 6711–6715.

Liu J., Q. Tang, B. He, L. Yu 2015 Cost-effective, transparent iron selenide nanoporous alloy counter electrode for bifacial dye-sensitized solar cell, Journal of Power Sources, 282: 79–86.

Mehmood U., S.-u. Rahman, K. Harrabi, I. A. Hussein and B. V. S. 2014 Reddy Recent advances in dye sensitized solar cells, Advances in Materials Science and Engineering 2014(Article ID 974782): 12 pages.

Mohmeyer N., P. Wang, H.-W. Schmidt, S. M. Zakeeruddin and M. Grätzel 2004 Quasi-solid-state dye sensitized solar cells with 1,3:2,4-di-O-benzylidene-D-sorbitol derivatives as low molecular weight organic gelators, Journal of Materials Chemistry, 14(12): 1905–1909.

O'Regan B. and M. Grätzel 1991 A low-cost, high-efficiency solar cell based on dye-sensitized colloidal TiO$_2$ films, Nature, 353: 737–740.

Ossila 2021, www.ossila.com/products/n3-dye, www.ossila.com/products/n719-dye, www.ossila.com/products/n749-black-dye, www.ossila.com/products/z907-dye (Accessed on 19th December 2021).

Reddy K. G., T. G. Deepak, G. S. Anjusree, S. Thomas, S. Vadukumpully, K. R. V. Subramanian, S. V. Nair and A. S. Nair 2014 On global energy scenario, dye-sensitized solar cells and the promise of nanotechnology, Physical Chemistry Chemical Physics, 16: 6838–6858.

Sekar N. and V. Y. Gehlot 2010 Metal complex dyes for dye-sensitized solar cells: Recent developments, Resonance, 15: 819–831.

Sharma K., V. Sharma and S. S. Sharma 2018 Dye-sensitized solar cells: Fundamentals and current status, Nanoscale Research Letters, 13(381): 46 pages.

Wang Q., T. Butburee, X. Wu, H. Chen, G. Liu and L. Wang 2013 Enhanced performance of dye-sensitized solar cells by doping Au nanoparticles into photoanodes: A size effect study, Journal of Materials Chemistry A, 1: 13524–13531.

Zhang D. W., X. D. Li, H. B. Li, S. Chen, Z. Sun, X. J. Yin, S. M. Huang 2011 Graphene-based counter electrode for dye-sensitized solar cells, Carbon, 49(15): 5382–5388.

Part IV

Photovoltaics with 2D Perovskites
and Carbon Nanomaterials

8 2D Perovskite and 2D/3D Multidimensional Perovskite Solar Cells

In Chapter 1, it was mentioned that widespread deployment of 3D perovskite solar cell technology is confronted with a challenging problem of short lifetime and stability issues. The obstacle is countered by the emergence of 2D perovskites. These nanomaterials have opened a new channel for development of reliable perovskite solar cells. In this chapter, we shall look at the 2D and 3D versions of the perovskites and give comparative analysis of solar cells made from them.

8.1 THE 3D PEROVSKITE

The reader may refer to Sections 1.7.7, 2.5.2. Any material having the crystalline structure of calcium titanate is designated as a perovskite. Recalling from the discussion there, the general formula of a metal halide perovskite is ABX_3 (A = a small organic or inorganic cation, B = Pb or Sn, X = Cl^-, Br^-, I^-) (Mao et al 2019). The small organic cation is formamidinium (FA^+) or methylammonium (MA^+). A small inorganic cation example is Cs^+.

The unit cell of a perovskite is represented as shown at the left in Figure 8.1. The A cations are placed at the edges of the cube. The B site is located at the center of the tetrahedron, while X sites are occupied by halogen ions. The tetrahedron formed is $[BX_6]^{4-}$. The corresponding crystal structure is illustrated at the right in Figure 8.1. The crystal consists of an inorganic backbone constituted by the corner-sharing $[BX_6]^{4-}$ octahedra. The void between the octahedra is filled by the A cation.

The reason for the evolution of this crystal structure is that the A cation is sufficiently small in size and can be accommodated in the space between the octahedra. The crystal structure is three-dimensional periodic repetition of the inorganic backbone, hence known as three-dimensional or bulk perovskite.

8.1.1 FAVORABLE PROPERTIES OF 3D PEROVSKITES FOR SOLAR CELL FABRICATION

Several desired properties expected of light-absorbing materials for solar cell fabrication are exhibited by 3D or bulk organic-inorganic hybrid perovskites. Notable properties are high optical absorption, low-temperature processability, long diffusion lengths of charge carriers, and ability of bandgap alteration. Certified power conversion efficiencies > 25% have been achieved with 3D perovskite solar cells.

8.1.2 SHORTCOMINGS OF 3D PEROVSKITES FOR USE IN SOLAR CELLS

Unfortunately, these advantages of 3D perovskites are accompanied by serious shortcomings. Toxicity of lead is a discouragement. Besides lead toxicity, solar cells made from 3D perovskites show operational instability in presence of oxygen, water vapor, and light. The perovskite degrades irreversibly. It undergoes decomposition by hydrolysis in the atmosphere, thereby returning to its precursors. This instability issue has become an impediment to the commercial viability of 3D perovskite solar cells.

DOI: 10.1201/9781003215158-12

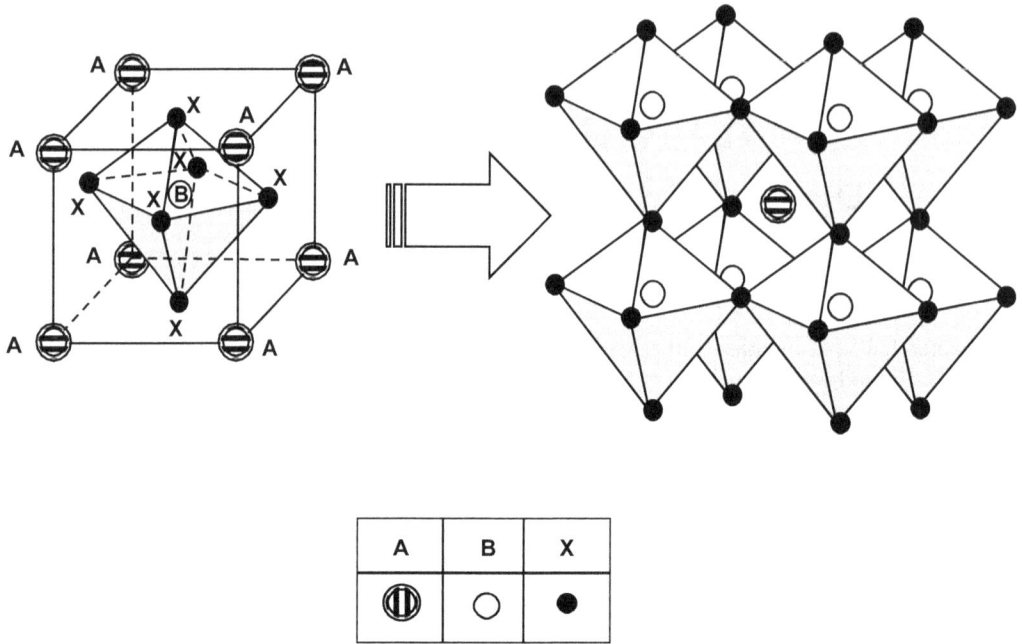

FIGURE 8.1 Two representations of a 3D perovskite with the formula ABX_3 (A = cation, B = metal cation, and C = halide anion). The "A cation" is a large monovalent alkali metal cation, e.g., cesium cation or small organic cations, e.g., methylammonium (MA) or formamidinium (FA) ions. The MA = $(CH_3NH_3)^+$, and FA = $HC(NH_2)_2^+$. The "B cation" is divalent, e.g., Pb^{2+}, Sn^{2+}. Accordingly, it is an alkali metal halide perovskite or organic metal halide perovskite. Left: The B cation–centered unit cell. The A cations are placed at the corners of the cube. One B cation and six X anions together form an octahedron $[BX_6]^{4-}$ in which the B cation is placed at the center and the X anions at the corners of the octahedron. Right: Extended network structure of a 3D perovskite. The linkages are made through the corner-shared $[BX_6]^{4-}$ octahedra.

8.2   THE 2D PEROVSKITE

8.2.1   What Happens When the A-Site Cation Is Large in Size?

Suppose the A-site cation is larger in size than the space between the octahedra in a 3D perovskite (Surrente et al 2021). Such a situation can arise in case of an organic molecule with aliphatic or aromatic chain. Then the crystal is formed in a different manner (Figure 8.2). It consists of thin sheets of $[BX_6]^{4-}$ octahedra separated by large organic molecules acting as spacers. These materials are given a new name. They are referred to as 2D perovskites or layered perovskites. They are classified as Ruddlesden-Popper (RP) and Dion-Jacobson (DJ) types accordingly as the large organic cation is monovalent or divalent. The RP 2D perovskites are represented by the general formula $(A')_2A_{n-1}BX_{3n+1}$, where A' is a large organic cation of valency 1, terminating with an amine group. The general formula of DJ 2D perovskitres is $(A')A_{n-1}BX_{3n+1}$, where A' is a large organic cation of valency 2, ending on both sides with an amine group. In these formulae, the variable n is an integer. It is the number of thin sheets of $[BX_6]^{4-}$ octahedra between any two large organic cation (A') layers. When n = 1, a monolayer 2D perovskite is obtained. Similarly, n = 2, 3, . . . are 2D perovskites of increasing number of sheets. But when n = ∞, the material becomes a 3D perovskite (Zhang et al 2020).

Here the difference between 2D perovskites and conventional 2D materials needs to be emphasized. In traditional 2D materials, the atoms are bound together by covalent bonding. In 2D perovskites, the chemical boding is ionic in nature. The ionic bonding is comparatively weaker than covalent bonding. An obvious result of this difference is observed in the growth pattern of the two types of materials. The 2D perovskites can be grown easily at low temperatures. Moreover, it is possible to grow them by solution-based and vapor transport techniques (Shi et al 2018).

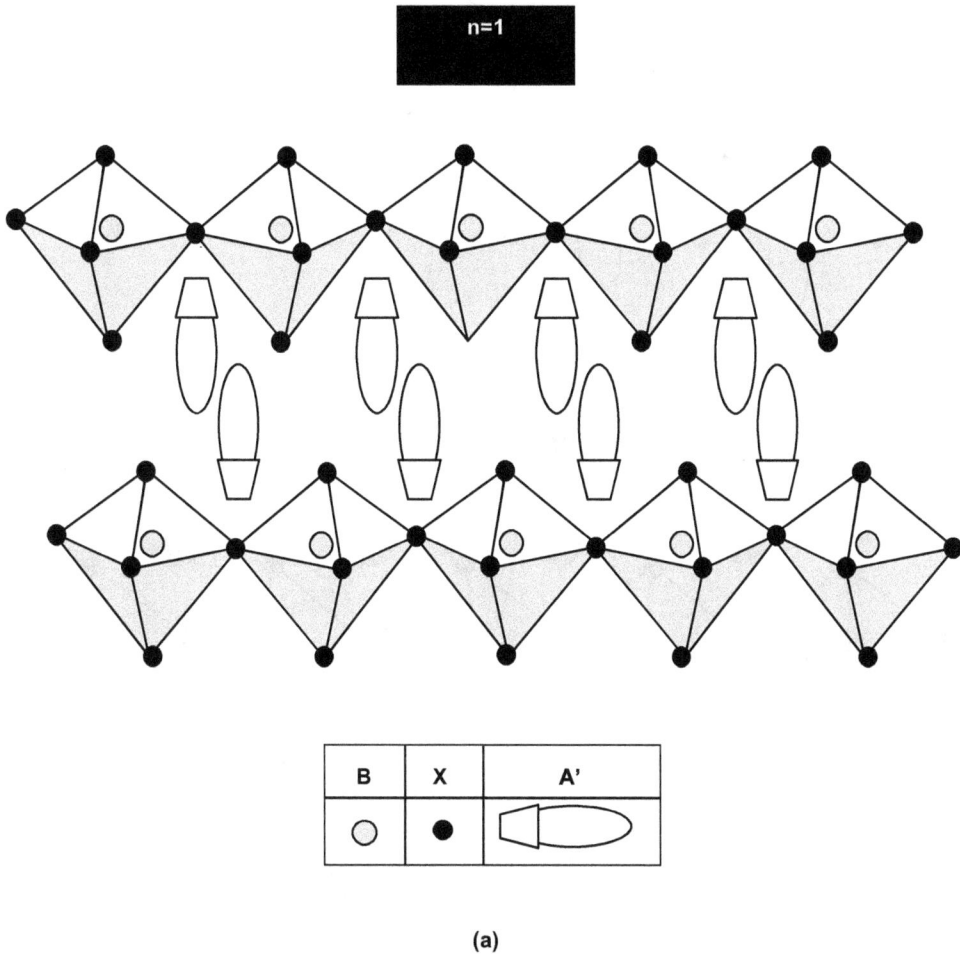

(a)

FIGURE 8.2 Evolution of 3D perovskite from 2D perovskite structure: (a) n = 1, pure 2D; (b) n = 2, quasi-2D; (c) n = 3, quasi-2D; and (d) n = ∞, 3D. A, B, X, and A' are symbols in 3D/2D perovskite generic formulae. Formula of 3D perovskite is ABX_3, where A = a cation of valency 1, e.g., MA$^+$ (methylammonium ion), FA$^+$ (formamidinium ion), Cs$^+$ (cesium ion), or Rb$^+$ (rubidium ion); B is a metal cation of valency 2, e.g. Pb^{2+}, Sn^{2+}, etc.; and X is a halide anion of valency 1, e.g., Cl$^-$, Br$^-$, I$^-$. Formula of 2D perovskite is $A'_m A_{n-1} B_n X_{3n+1}$, where A' = a bulky organic ammonium cation spacer separating the perovskite layers, e.g., the 2D (PEA)$_2$(MA)$_2$Pb$_3$I$_{10}$ perovskite is formed when the bulky phenylethylammonium (PEA) cation is incorporated into 3D MAPbI$_3$ (Niu et al 2020). 2D and quasi-2D perovskites are called layered perovskites. The bulky organic ammonium cation spacer is typically $C_m H_{2m+1} NH_3^+$. The symbol n denotes the number of octahedral [BX$_6$]$^{4-}$ sheets between two bulky organic cation spacers.

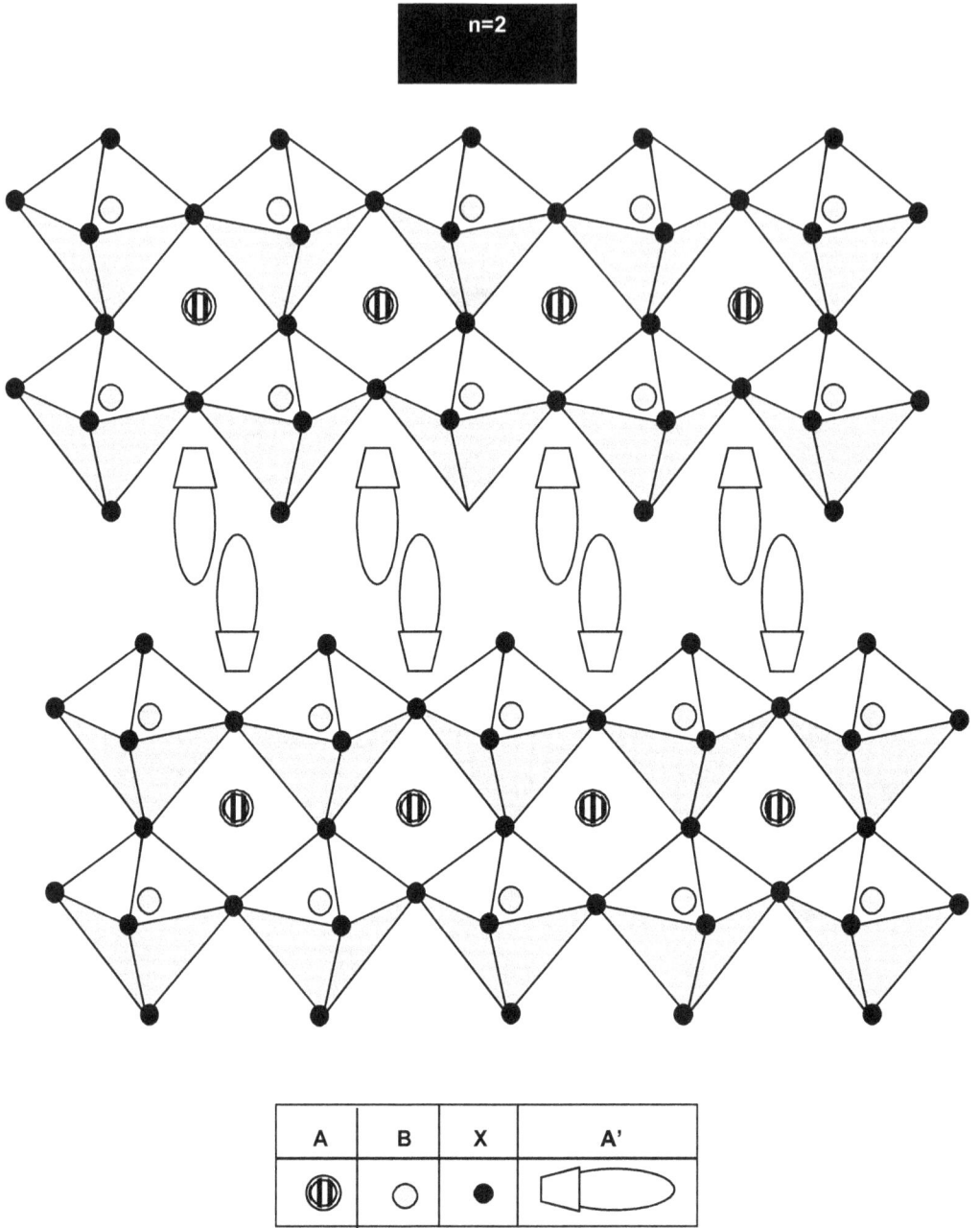

FIGURE 8.2 (Continued)

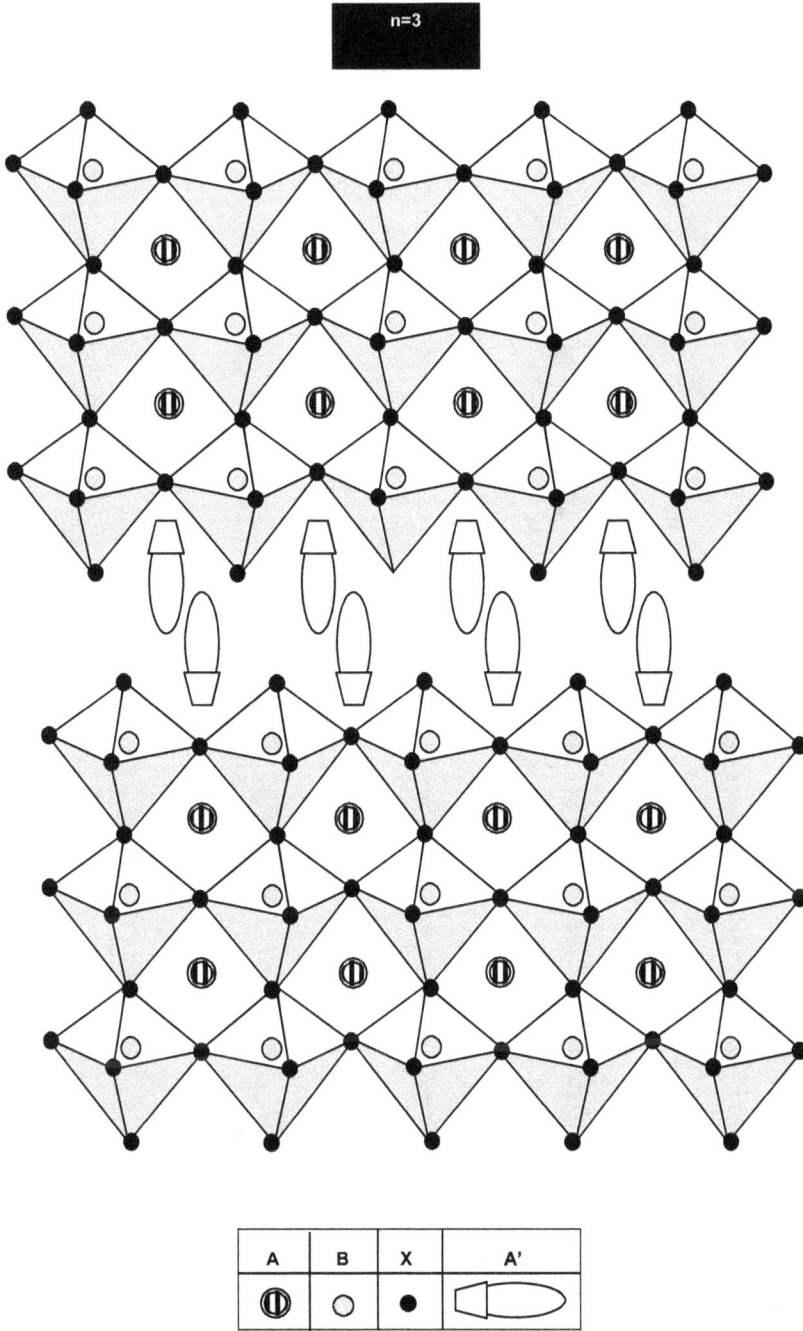

FIGURE 8.2 (Continued)

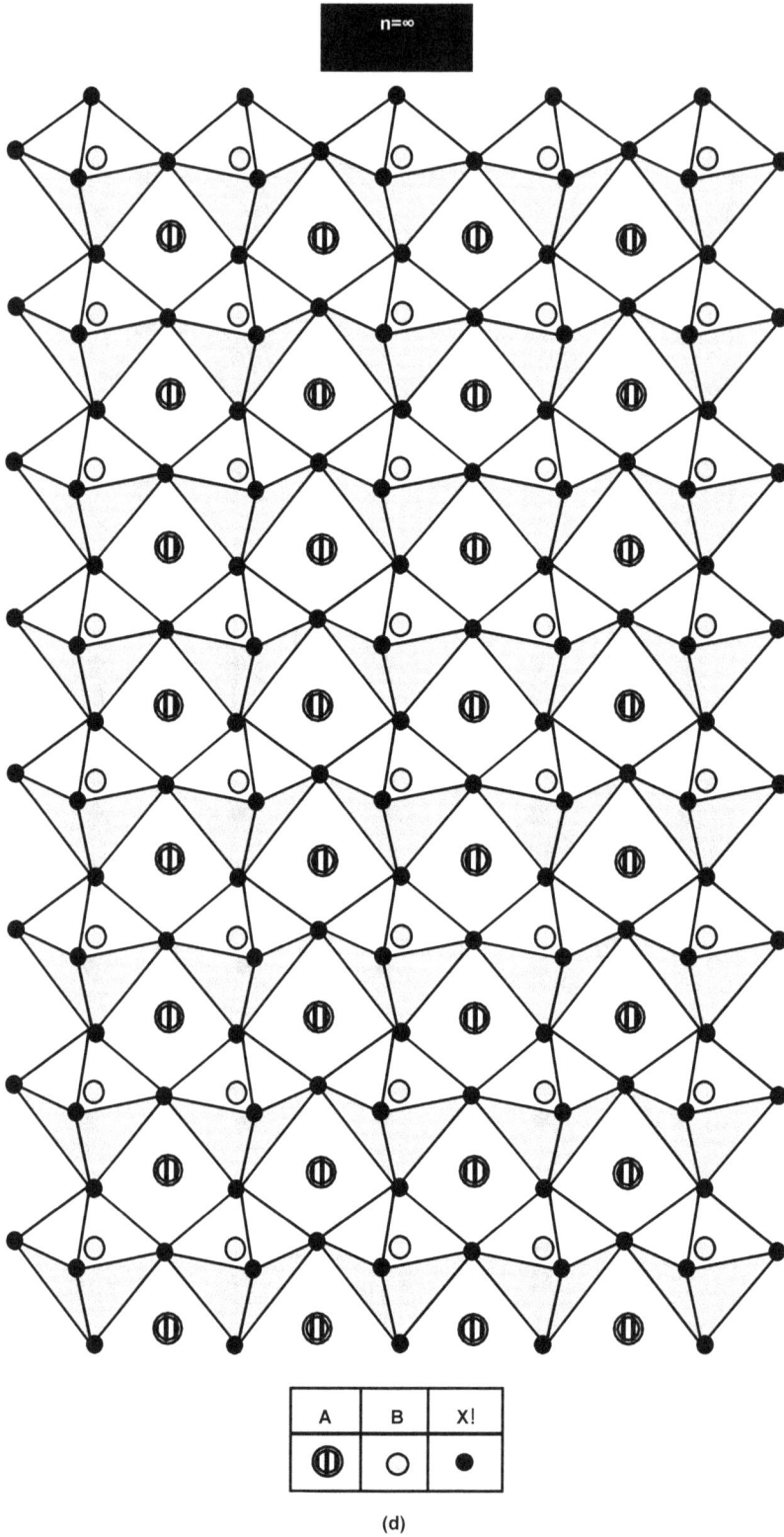

A	B	X!
◐	○	●

(d)

FIGURE 8.2 (Continued)

8.2.2 2D PEROVSKITES AS A PROMISING OPTION

We were talking about difficulties faced with 3D perovskites in solar cells. How to solve the instability problem? Necessity is the mother of invention. 2D perovskites come to the rescue. They can be looked upon as layers taken out from 3D perovskites and separated by large organic spacer cations. As we have seen, a 2D hybrid perovskite consists of inorganic layers of corner-shared $[BX_6]^{4-}$ octahedra, and between any two inorganic layers is placed a bulky organic cation A'. The bulky organic cation A' is the prime distinguishing feature between 3D and 2D perovskites. This cation is hydrophobic in nature. Its moisture-repellant property prevents water vapor from entering the perovskite structure and hydrolyzing it. Obviously, 2D perovskite is more resistant to moisture than the 3D perovskite. This is the property for which we were frantically rummaging around. This property provides a means of overcoming the innate instability problem jeopardizing the long-term reliability of 3D perovskite solar cell. An additional advantage is that the incorporation of bulky organic cation increases the energy of formation of 2D perovskite as compared to the 3D perovskite. In a 2D perovskite, the capping organic molecules and the $[BX_6]^{4-}$ units are held together by a strong van der Waals interaction (Carlo et al 2020). As a consequence, 2D perovskite is more stable to temperature variations than 3D perovskite. Moreover, migration of ions in 2D perovskite is improved relative to 3D perovskite. All these features contribute to longevity of 2D perovskite so that solar cells made with this material have a longer operating lifetime than those made with 3D perovskite variant (Kore et al 2021).

8.2.3 INFERIOR ASPECTS OF 2D PEROVSKITES TO 3D PEROVSKITES

2D perovskites have three shortcomings with respect to 3D perovskites:

(i) Insufficient optical absorption. The reason for insufficient optical absorption by 2D perovskite lies in its larger bandgap. The bandgap of 2D perovskite is larger than that of 3D perovskite, e.g., the bandgap of pure 2D perovskite is 3.1eV for n = 1; the bandgap of quasi-2D perovskite is 2.6 eV for n = 4, 2.7 eV for n = 3, and 2.9 eV for n = 2 (Xing et al 2018). On the other hand, the bandgap of 3D perovskite in best-performing solar cells has been 1.48–1.62 eV (Prasanna et al 2017). Owing to the larger bandgap of 2D perovskite, the absorption of light is reduced in the near-infrared wavelength range of the solar spectrum.

(ii) High exciton binding energy. 2D perovskites display high exciton binding energies. The exciton binding energy increases from 80meV for n = 4 quasi-2D perovskite with four inorganic layers between organic cations to ~ 190–400 meV for n = 1 pure 2D perovskite with a single inorganic layer separating the organic cations (Gélvez-Rueda et al 2020). The exciton binding energy of a 3D perovskite is approximately ten times lower. The implication of high exciton binding energy of 2D perovskite is that the exciton in this perovskite is more stable. A solar cell device works on charge separation. Free electrons and holes are required to produce a current, not excitons. This means that the stable excitons of 2D perovskite are unfavorable to the solar cell action.

(iii) Poor charge transport. The reason for poor charge transport in 2D perovskite lies in the effect of its growth direction on charge transport. Preferential direction of growth of 2D RP perovskite crystals is along the in-plane direction relative to the substrate. The natural consequence is that the charge transport in the out-of-plane direction between the conducting inorganic sheets of a 2D perovskite is hampered by the insulating interlayer bulky organic cations. Therefore, interlayer charge transport is inadequate (Shao et al

2021). Improvement in poor charge transport is possible by tuning the orientation of 2D perovskite crystal by application of various processing strategies and cation engineering techniques.

In view of disadvantages i–iii, the efficiency of a solar cell made from 2D perovskite is lower than that made from a 3D perovskite.

8.2.4   THE TWO ROUTES TO SUCCESS

Evidently, there are two paths to follow:

(i) Either fabricate solar cells entirely from 2D perovskite and try to improve efficiency to the maximum possible achievable by adopting appropriate measures. Here, 2D perovskite acts both as light absorber and protector from environmental effects.

(ii) Or fabricate solar cells, which gain from good properties of both 2D and 3D perovskites, i.e., embed 3D perovskite in 2D perovskite to get stability from 2D perovskite and efficiency from 3D perovskite. Here, 2D perovskite is not used as the primary light absorber. This role is left to 3D perovskite. 2D perovskite is the protector from atmospheric effects. Research strategies have been planned on both routes.

8.3   2D PEROVSKITE SOLAR CELLS

8.3.1   PbBr$_2$-INCORPORATED 2D PEROVSKITE SOLAR CELL (η = 12.19%)

Bromine is included in the 2D perovskite film of a solar cell with n-butylamine (BA) spacer (Han et al 2020). For bromine inclusion, lead (II) bromide PbBr$_2$ is used. Bromine is found to improve the quality of perovskite film by reduction of defect states. The optoelectronic properties and photovoltaic performance of 2D perovskite film are significantly enhanced. This leads to an increase in power conversion efficiency of the solar cell.

The solar cell is shown in Figure 8.3. The solar cell fabrication flow is as follows:

(i) ITO substrates are cleaned in detergent, acetone, ethanol, and DI water, dried in nitrogen, and subjected to UV-O$_3$ treatment.

(ii) Aqueous solution of PEDOT:PSS is spin-coated on ITO substrates and annealed at 150°C for ¼ h. in air.

(iii) PEDOT:PSS/ITO substrates are put in a glove box.

(iv) (a) The 2D perovskite is BA$_2$MA$_4$Pb$_5$I$_{16-10x}$Br$_{10x}$ (x = 0.1). The 0.8M precursor solution is prepared by mixing the ingredients n-butylammonium iodide (C$_4$H$_{12}$IN, BAI), methylamine iodide (CH$_3$NH$_3$I, MAI), lead (II) iodide (PbI$_2$), and lead (II) bromide PbBr$_2$ in a mixed solvent = [dimethyl sulfoxide (DMSO) + N,N-dimethylformamide (DMF) in 1:15 v/v ratio], taking the molar ratio BAI/MAI/PbI$_2$/PbBr$_2$ = 0.4/0.8/(1-x)/x.

(b) 2D perovskite precursor solution is spin-coated on PEDOT:PSS/ITO substrate and annealed at 100°C for 10 min., obtaining 2D perovskite/PEDOT:PSS/ITO substrate.

(v) 20mg/mL phenyl-C$_{61}$-butyric acid methyl ester (PC$_{61}$BM) solution in chlorobenzene is spin-coated on 2D perovskite film to get PC$_{61}$BM/2D perovskite/PEDOT:PSS/ITO substrate.

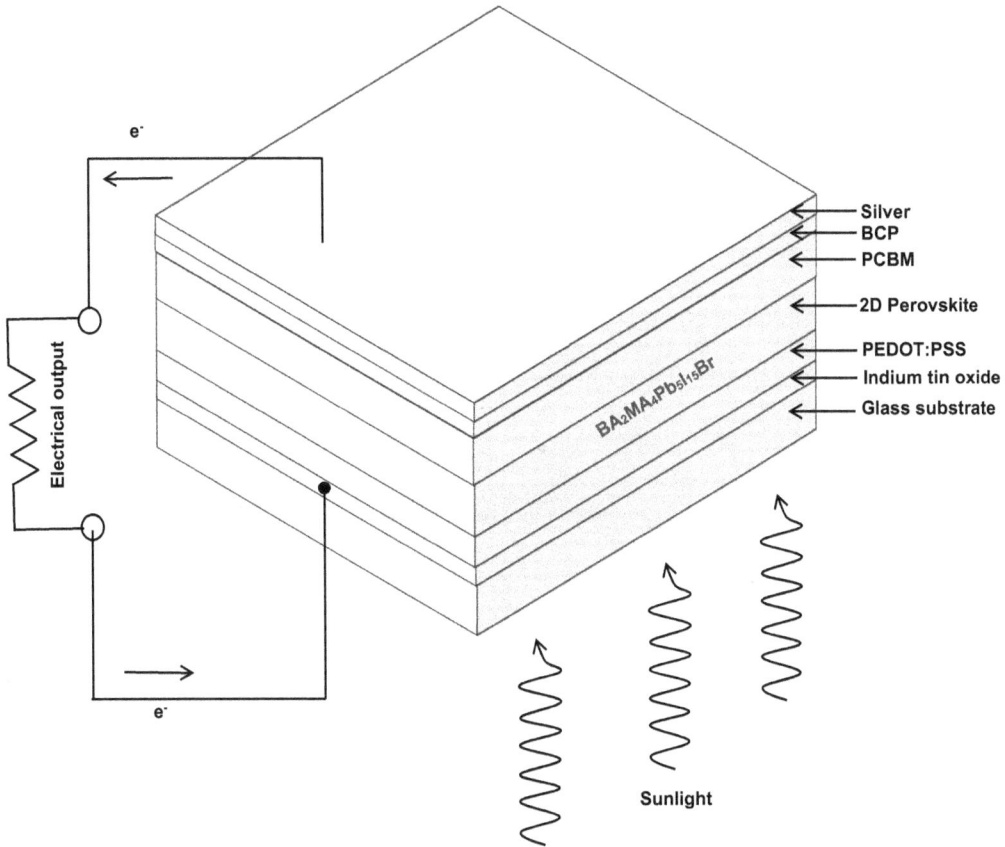

FIGURE 8.3 $PbBr_2$-incorporated 2D perovskite solar cell consisting of (from bottom upwards) glass substrate, indium tin oxide, PEDOT:PSS hole transport layer (HTL), $BA_2MA_4Pb_5I_{15}Br$ 2D perovskite, [6,6]-phenyl-C_{61}-butyric acid methyl ester (PCBM) electron transport layer (ETL), bathocuproine (BCP) hole-blocking layer, and silver film. The structure is ITO/PEDOT:PSS/$BA_2MA_4Pb_5I_{16-10x}Br_{10x}$/PCBM/ BCP/Ag. Light falls on the glass substrate, and electric current is drawn between the ITO and silver electrodes.

(vi)   0.5mg/mL bathocuproine (BCP) solution in isopropanol is spin-coated on $PC_{61}BM$ layer to get BCP/$PC_{61}BM$/2D perovskite/PEDOT:PSS/ITO substrate.

(vii) Silver electrode is deposited by thermal evaporation to get Ag/BCP/$PC_{61}BM$/2D perovskite/PEDOT:PSS/ITO substrate. Ag and ITO are electrode materials, BCP is a hole-blocking layer, $PC_{61}BM$ is an electron transport layer, 2D perovskite is light absorber and PEDOT:PSS is hole transport layer.

(viii) The open-circuit voltage of the solar cell fabricated with 2D perovskite $BA_2MA_4Pb_5I_{16-10x}Br_{10x}$ (x = 0.1), i.e., $BA_2MA_4Pb_5I_{15}Br$, is 1.02V, its short-circuit current density is 17.86mAcm^{-2}, fill factor is 66.91%, and efficiency is 12.19%. The same parameters for solar cell with 2D perovskite without bromine $BA_2MA_4Pb_5I_{16-10x}Br_{10x}$ (x = 0), i.e., $BA_2MA_4Pb_5I_{16}$, are 0.89V, 8.28 mAcm^{-2}, 40.79%, and 3.01% respectively (Han et al 2020).

8.3.2 2D GA$_2$MA$_4$Pb$_5$I$_{16}$ Perovskite Solar Cell Interface Engineered with GABr (η = 19.3%)

The 2D perovskite used is GA$_2$MA$_4$Pb$_5$I$_{16}$ (Huang et al 2021). Guanidinium bromide (GABr) is deposited over the perovskite. It has two effects:

(i) GABr optimizes the secondary crystallization of the perovskite. The GABr deposition induces rearrangement of the phase distribution of perovskite. The resulting perovskite has a smooth and shiny surface. It displays good optoelectronic properties.

(ii) When GA$^+$ enters the perovskite lattice, it stabilizes the structure of the lattice by forming strong hydrogen bonds with adjoining halide ions (Zhou et al 2019). This structural modification suppresses the formation of halide vacancy. Degradation of perovskite is thereby inhibited. Hence, the inclusion of GA$^+$ improves the environmental stability of perovskite as well as its photostability.

Figure 8.4 shows the solar cell. The process sequence of solar cell fabrication is as follows:

(i) ITO/glass substrate is cleaned in detergent, DI water, acetone, and ethanol, dried in nitrogen, and then for 20 min. in U V-ozone cleaner.

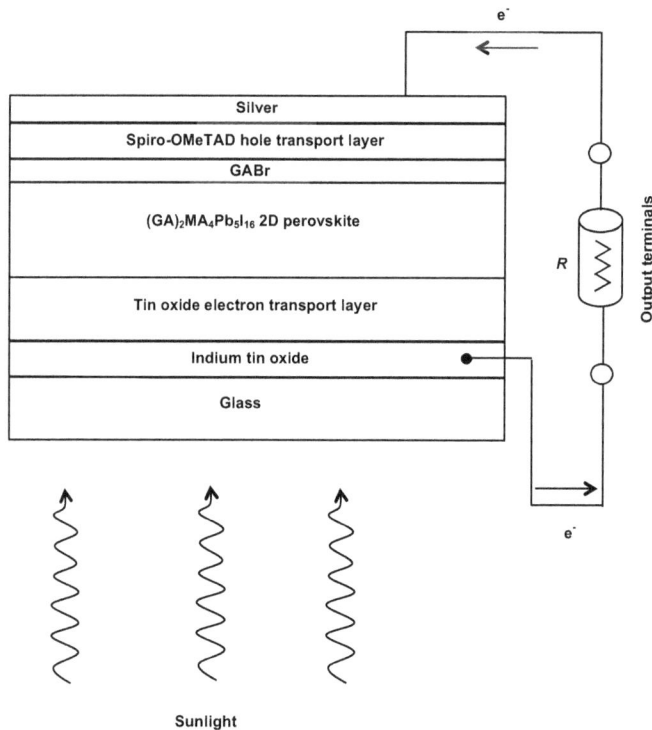

FIGURE 8.4 (GA)$_2$MA$_4$Pb$_5$I$_{16}$ 2D perovskite solar cell (interface engineered with GABr) consisting of glass substrate, ITO, SnO$_2$ electron transport layer (ETL), (GA)$_2$MA$_4$Pb$_5$I$_{16}$ 2D perovskite, GABr, spiro-OMeTAD hole transport layer (HTL), and silver layer. Light falls on the glass substrate. Electrical output is withdrawn across ITO and silver films.

(ii) SnO_2 nanoparticle film is spin-coated using its filtered aqueous colloidal dispersion and annealed on a hot plate at 180°C for ½ h. The SnO_2/ITO/glass is put in UV-ozone cleaner for 20 min.

(iii) (a) 2D perovskite precursor solution is prepared from guanidinium iodide [GAI, $C(NH_2)_3I$], methylammonium iodide (MAI, CH_3NH_2HI), and lead (II) iodide PbI_2. The ratio GAI/MAI/PbI_2 = 2/4/5 with concentration 1M Pb^{2+} in N, N-dimethylformamide (DMF)/dimethylsulfoxide (DMSO) = 4/1 (v/v). After heating (60°C, 1 h.), filtration is done.

(b) The precursor solution is deposited by spin-coating on SnO_2 film. Antisolvent ethyl acetate is pipetted near the end. Annealing is done at (100°C for 10 min. + 150°C for 10 min). We get $GA_2MA_4Pb_5I_{16}$ perovskite/SnO_2/ITO/glass.

(iv) GABr solution in isopropanol is spin-coated over the perovskite film. It is annealed at 100°C for 5 min. to get: GABr/$GA_2MA_4Pb_5I_{16}$ perovskite/SnO_2/ITO/glass.

(v) (a) Spiro-OMeTAD solution is made in chlorobenzene with 4-tert-butylpyridine (tBP) and lithium bis(trifluoromethanesulfonyl)imide (Li-TFSI) as additives in acetonitrile.

(b) Spiro-OMeTAD solution is deposited by spin-coating on perovskite. We get Spiro-OMeTAD/GaBr/$GA_2MA_4Pb_5I_{16}$ perovskite/SnO_2/ITO/glass.

(vi) Silver is thermally evaporated over Spiro-OMeTAD film to get Ag/Spiro-OMeTAD/GABr/$GA_2MA_4Pb_5I_{16}$ perovskite/SnO_2/ITO/glass.

(vii) The open-circuit voltage of the solar cell with 10mM GaBr is 1.17V, its short-circuit current density is 21.9mAcm^{-2}, fill factor is 0.75, and efficiency is 19.3%. The corresponding parameters for a pristine solar cell without GABr are, respectively, 1.14V, 20.2mAcm^{-2}, 0.68, and 15.6%.

Higher open-circuit voltage is achieved in solar cell using GABr than one without GABr because the GABr-tuned perovskite film has lower defect density. The explanation is provided by photoluminescence lifetime measurements. The photoluminescence lifetime of perovskite tuned with GABr is found to be longer than that of pristine perovskite. This shows that GABr inclusion reduces the defect density of perovskite and hence trap-assisted nonradiative recombination in it (Huang et al 2021).

8.3.3   FA-BASED 2D PEROVSKITE SOLAR CELL (η = 21.07%)

This solar cell achieves a high efficiency because of enhancement in optical absorption and facilitation of charge transport by virtue of a distinctive phase distribution in the 2D perovskite (Shao et al 2021):

(i) Better harvesting of sunlight. FA cation extends the range of light absorption. So a greater portion of optical spectrum is utilized.

(ii) Improved charge transport. The FA-based 2D perovskite has a unique phase distribution. In this distribution, a significant proportion of 3D perovskite-like phases is embedded. These phases penetrate through the full 2D perovskite film thickness. They furnish an electrical conduction network. The 2D-3D intermixing distribution provides a long diffusion length of 1.85μm.

To make the solar cell shown in Figure 8.5:

(i) ITO substrates are cleaned in detergent, DI water, acetone, and alcohol, dried, and exposed to oxygen plasma.

(ii)   PTAA in chlorobenzene solution is spin-coated on ITO substrate and annealed at 120°C for 1/3 h.

(iii)   (a) $(4FPEA)_2FA_4Pb_5I_{16}$ precursor solution is prepared by mixing 4FPEAI (0.4), FAI (0.8), PbI_2(1), MACl (0.3), and $PbCl_2$ (0.3) in DMF with 1M Pb^{2+}. The additives MACl and $PbCl_2$ prevent formation of nonperovskite δ phase.

(b) $(4FPEA)_2MA_4Pb_5I_{16}$ precursor solution is prepared by mixing 4FPEAI (0.4), MAI (0.8), PbI_2(1), MACl (0.1), and $PbCl_2$ (0.1) in DMF with 1M Pb^{2+}.

(c) $(4FPEA)_2(FA_{0.3}MA_{0.7})_4Pb_5I_{16}$ precursor solution is prepared by mixing 4FPEAI (0.4), FAI (0.24), MAI (0.56), PbI_2(1), MACl (0.1), and $PbCl_2$ (0.1) in DMF with 1M Pb^{2+}.

(d) Perovskite solution is spin-coated over the PTAA-coated ITO substrate preheated to 100°C. Annealing is done at 100°C for ¼ h. Perovskite solutions a, b, and c give three different solar cells.

(iv)   PC61BM solution in chlorobenzene is spin-coated over perovskite layer.

(v)   BCP solution in isopropyl alcohol is spin-coated on $PC_{61}BM$ layer.

(vi)   Silver is thermally evaporated to make the electrode.

(vii) The open-circuit voltage of the FA perovskite solar cell is 1.18V, its short-circuit current density is 22.45mAcm^{-2}, fill factor is 79.53%, and efficiency is 21.07% (reverse scan). The efficiency of MA perovskite solar cell is 17.05%, while that of FA/MA-mixed perovskite solar cell is 20.07%.

(viii) When FA perovskite solar cell is continuously heated at 85°C, 97% of its starting efficiency is retained after 1500 h. (Shao et al 2021).

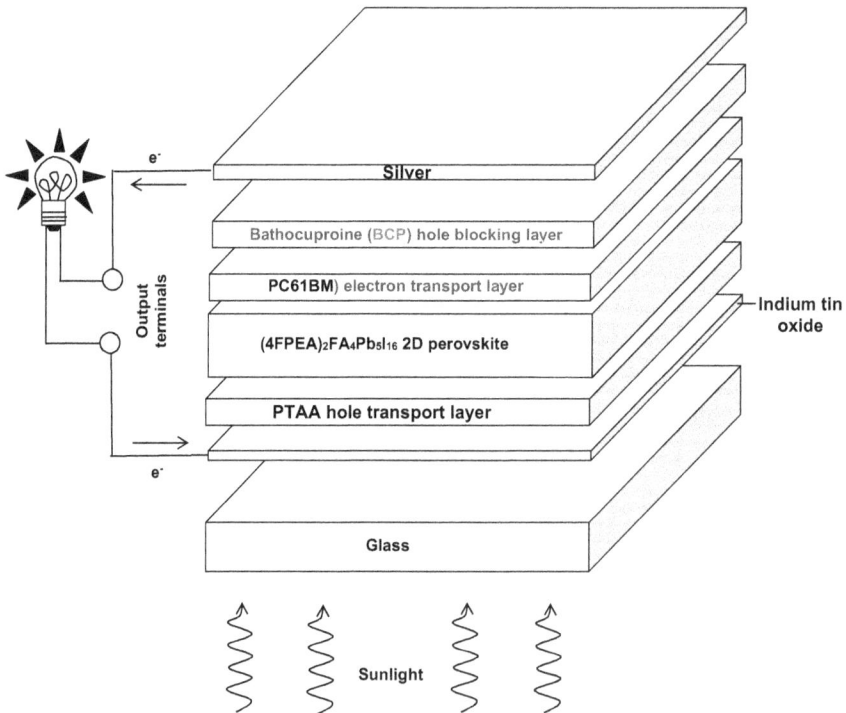

FIGURE 8.5 $(4FPEA)_2FA_4Pb_5I_{16}$ 2D perovskite solar cell consisting of glass, ITO, poly(triaryl)amine (PTAA) hole transport layer (HTL), $(4FPEA)_2FA_4Pb_5I_{16}$ 2D perovskite film, [6,6] phenyl-C_{61}-butyric acid methyl ester (PC61BM) electron transport layer (ETL), bathocuproine (BCP) hole-blocking layer, and silver layer. The structure is glass/ITO/PTAA/$(4FPEA)_2FA_4Pb_5I_{16}$/PC61BM/BCP/silver. Sunlight falling on glass produces electron-hole pairs in the perovskite film, which are collected at the ITO and silver electrodes.

8.4 2D/3D PEROVSKITE SOLAR CELLS

8.4.1 2D/3D (HOOC(CH$_2$)$_4$NH$_3$)$_2$PbI$_4$/CH$_3$NH$_3$PbI$_3$ Perovskite Interface Engineered Solar Cell (η = 14.6%)

This solar cell provides a combination of the good properties of 2D and 3D perovskites (Grancini et al 2017). As already remarked in Section 8.2.4, 2D perovskite is more stable, while 3D perovskite is better in charge transport and light absorption. So the two perovskites work together to bring stability of operation along with high-performance characteristics.

The 2D/3D perovskite growth takes place by formation of a bottom-up phase-segregated arrangement as the 2D/3D perovskite organizes into a gradual multidimensional junction in which the two phases are retained. The bandgap of 3D perovskite is enlarged at the interface by interaction with 2D perovskite. The 2D perovskite thin film does not oppose injection of electrons to TiO$_2$ but suppresses recombination of electrons. This happens because the conduction band of 2D perovskite is at a lower energy than that of 3D perovskite.

The solar cell (Figure 8.6) exhibits remarkable stability. The stability is observed as a constancy of efficiency over a long period exceeding 10^4h = 416.67 days when exposed to humidity and oxygen.

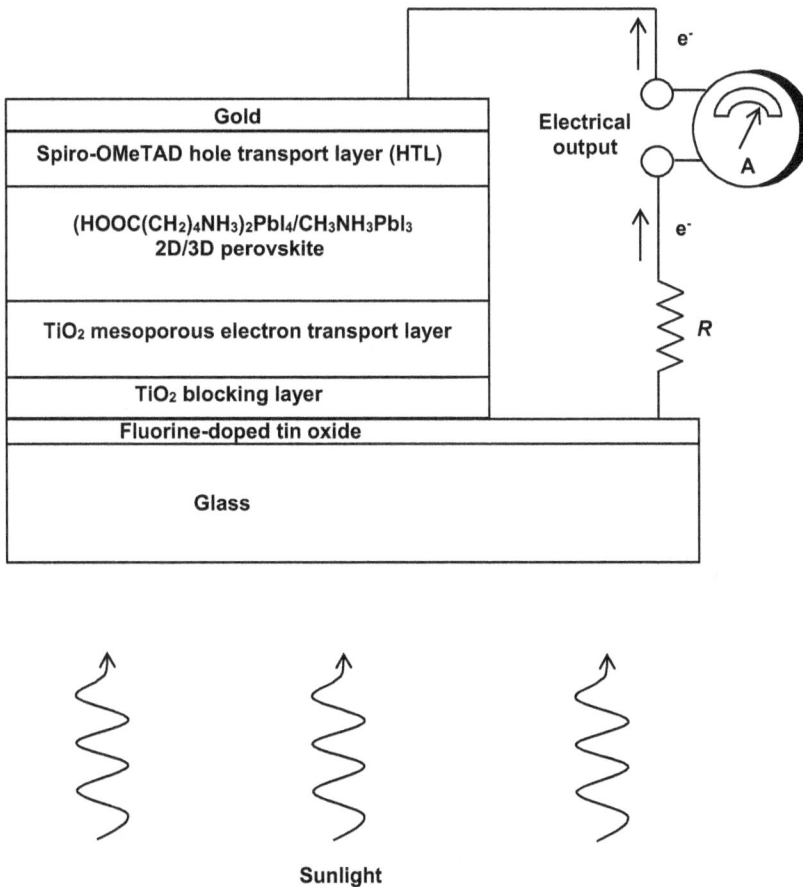

FIGURE 8.6 2D/3D (HOOC(CH$_2$)$_4$NH$_3$)$_2$PbI$_4$/CH$_3$NH$_3$PbI$_3$ perovskite solar cell consisting of glass substrate, fluorine-doped tin oxide, TiO$_2$ blocking layer, TiO$_2$ mesoporous electron transport layer (ETL), 2D/3D perovskite layer, spiro-OMeTAD hole transport layer (HTL), and gold film. Sunlight falling on glass produces electric current flow between FTO and gold electrodes.

The process steps for solar cell fabrication are:

(i)   Fluorine-doped tin oxide–coated glass substrates are washed in hellmanex solution (an alkaline liquid concentrate), distilled water, and isopropanol and treated with UV-ozone.

(ii) TiO_2 blocking layer is formed by spray pyrolysis. This is done using diisopropoxide bis(acetylacetonate) solution in isopropanol. The TiO_2 layer is sintered at 450°C.

(iii) Mesoporous TiO_2 layer is formed by spin-coating a solution of TiO_2 paste (Dyesol, 30NRD) in ethanol. The layer is annealed at 500°C for ½ h.

(iv) The 2D/3D perovskite precursor solution is made by mixing 1.1 M AVAI:PbI_2 (2:1) solution with MAI:PbI_2 solution [AVAI:PbI_2/(AVAI:PbI_2 + MAI:PbI_2) = 3% molar ratio], where MAI = CH_3NH_3I and AVAI = $HOOC(CH_2)_4NH_3I$.

(v)   The 2D/3D perovskite solution is spin-coated and sintered at 100°C for 1 h.

(vi) Spiro-OMeTAD is spin-coated from a chlorobenzene solution. In this solution, Li-TFSI {lithium bis(trifluoromethylsulfonyl)-imide}in acetonitrile, TBP(tert-butylpyridine), and Co(II)TFSI{tris(2-(1H-pyrazol-1-yl)pyridine)cobalt(II) di[bis(trifluoromethane)sulfonimide]}are dopants.

(vii) Gold is evaporated to make electrode.

(viii)The open-circuit voltage of the 2D/3D perovskite solar cell with 3% AVAI is 1.025V, its short-circuit current density is 18.84 mAcm^{-2}, fill factor is 75.5%, and efficiency is 14.6%. The same parameters for a reference 3D perovskite solar cell are 1.054V, 21.45mAcm^{-2}, 70.3%, and 15.95%. For making the 3D perovskite solar cell, PbI_2/MAI (1/1) solution in dimethyl sulfoxide (DMSO) is spin-coated in two successive steps of increasing RPM. Near the end, chlorobenzene is spin-coated on the perovskite by antisolvent method. The perovskite is sintered at 100°C for 1 h. (Grancini et al 2017).

8.4.2   2D PEROVSKITE-ENCAPSULATED 3D PEROVSKITE SOLAR CELL (η = 16.79%)

This solar cell uses the encapsulating layer of 2D perovskite $(C_nH_2n_{+1}NH_3)_2PbI_4$ (Kore and Gardner 2020; Kore et al 2021). The value of n = 18 gave the highest efficiency of solar cell. The 2D perovskite layer blends with the underlying 3D perovskite $(FA_{0.85}MA_{0.15})Pb(I_{0.85}Br_{0.15})_3$ used as the light-absorbing medium. This blending does not affect the light-absorbing and charge transport properties of 3D perovskite. Rather, it improves hydrophobicity of the thin film, making the solar cell resistant to humidity. Overall, the 2D perovskite plays a dual role: as an encapsulant and as a passivant. By passivation, the recombination losses are reduced.

To fabricate the solar cell shown in Figure 8.7:

(i)   Fluorine-doped tin oxide (FTO) substrates are patterned, cleaned in detergent, acetone, and ethanol, dried, and kept under UV-ozone lamp.

(ii) Compact TiO_2 film. This film is deposited by spray pyrolysis method. The spray solution consists of ingredients (0.2 M titanium (IV) isoporoxide + 2 M acetylacetone + isopropanol solvent). The substrates are preheated to 450°C. The spraying is done several times. The annealing is done for 20 min.

(iii) Mesoporous TiO_2 film. Dyesol, 30NRD TiO_2 paste suspension in ethanol is spin-coated over compact TiO_2 and heated at 80°C for 20 min. Sintering is done at 450°C for 1 h. followed by UV-ozone cleaning.

(iv) (a) 3D perovskite precursor solution preparation. This solution is made by dissolution of formamidinium iodide (FAI), lead iodide (PbI_2), methylammonium bromide(MABr), and lead bromide ($PbBr_2$) in N, N-dimethylformamide (DMF) (DMF), dimethyl sulfoxide (DMSO), with overnight stirring.

      (b) 2D perovskite crystals synthesis. First lead iodide (PbI_2) is dissolved in hydroiodic acid (HI). Then, n = 18 alkylammonium cation–based lead iodide perovskite

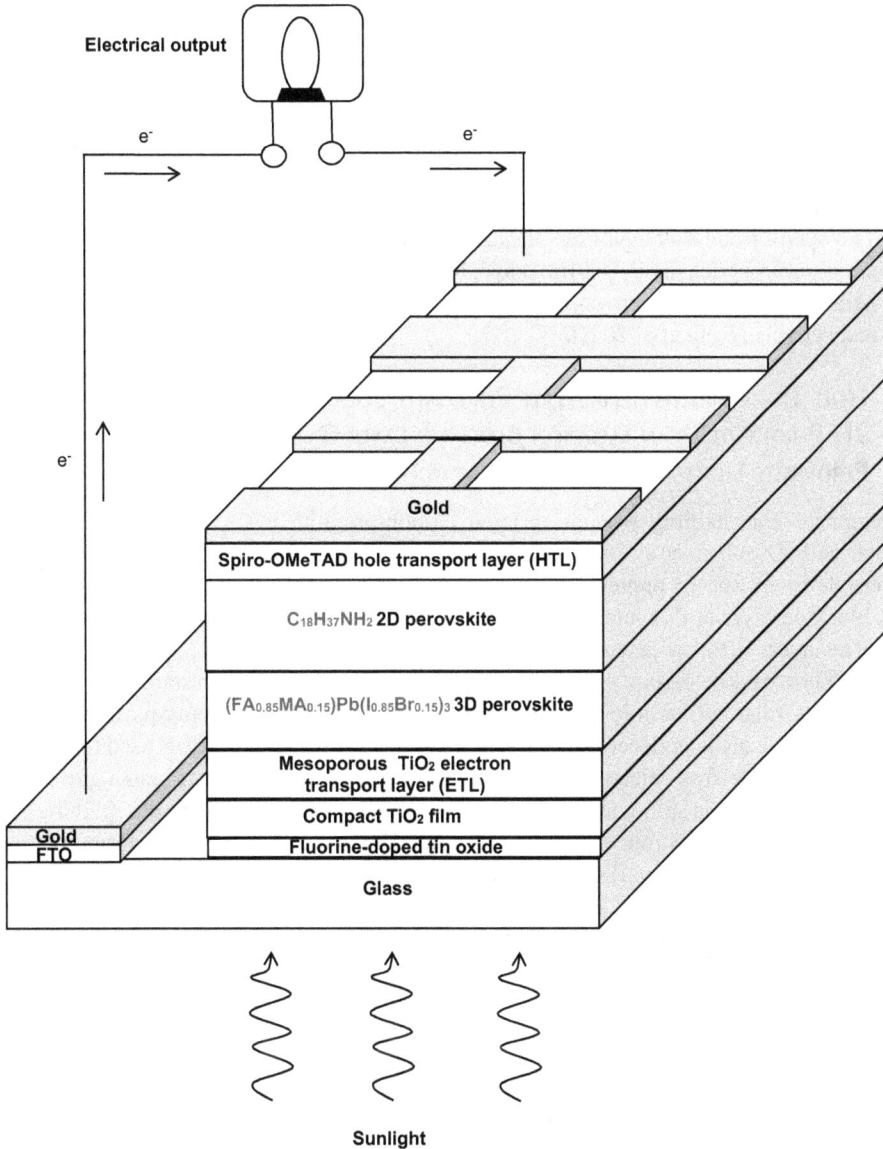

FIGURE 8.7 2D perovskite-encapsulated 3D perovskite solar cell consisting of glass, fluorine-doped tin oxide, compact TiO$_2$ film, mesoporous TiO$_2$ electron transport layer (ETL), (FA$_{0.85}$MA$_{0.15}$)Pb(I$_{0.85}$Br$_{0.15}$)$_3$ 3D perovskite, C$_{18}$H$_{37}$NH$_2$ 2D perovskite, spiro-OMeTAD hole transport layer (HTL), and gold film. Sunlight incident on glass produces electron current flow across FTO and gold electrodes.

(C$_{18}$H$_{37}$NH$_2$) is dissolved in (HI + ethanol) with stirring and heated to 60°C. Drops of lead iodide PbI$_2$ in HI are added with stirring. The mixture is refluxed for 45 min. and gradually cooled when needle-shaped crystals are formed. They are washed with diethyl ether and vacuum-dried.

(c) 3D perovskite film deposition. The precursor solution is spin-coated. Towards the end, chlorobenzene is dropped for rapid crystallization. Annealing is done at 100°C for 40 min.

(d) 2D perovskite deposition. 2D perovskite crystals are dissolved in tetrahydrofuran (THF) with stirring, spin-coated on 3D perovskite film, and annealed at 100°C for 15 min.

Perovskite film depositions are carried out inside a glove box.

(v) Hole transporting layer. Spiro-OMeTAD, lithium bis-(trifluoromethylsulfonyl)imide (Li-TFSI) solution in acetonitrile, 4-tert-butylpyridine (TBP), and Co(III)-complex (FK209) solution in acetonitrile are dissolved in chlorobenzene and spin-coated on the perovskite film.

(vi) Gold film is deposited by thermal evaporation.

(vii) The open-circuit voltage of the 3D@2D(n = 18) perovskite solar cell is 1.13V, its short-circuit current density is 20.01mAcm^{-2}, fill factor is 0.74, and efficiency is 16.79%. These parameters for a 3D perovskite solar cell are 1.09V, 21.98mAcm^{-2}, 0.76, and 18.19%, respectively (Kore et al 2021).

8.4.3   Hole Transport Material–Free Perovskite Solar Cell Using 2D Perovskite as an Electron Blocking Layer Over 3D Perovskite Light-Absorbing Layer (η = 18.5%)

For economical and stability reasons, printed carbon-graphite black electrode is attractive for these solar cells (Zouhair et al 2022). However, it causes severe performance degradation at the rear electrode interface. To prevent this degradation, 2D perovskite is used as a passivation and electron blocking layer at this interface. By this strategy, the recombination losses are appreciably reduced. The layers in this solar cell are arranged as FTO glass/compact TiO$_2$/mesoporous TiO$_2$/3D perovskite/2D perovskite/carbon electrode. The solar cell also shows high stability.

To make this solar cell, a hole-blocking TiO$_2$ layer is deposited by spray pyrolysis of titanium (IV) diisopropoxide bis(acetylacetonate) in IPA at 450°C. Diluted TiO$_2$ paste is used for spin-coating a mesoporous TiO$_2$ electron selective layer. Sintering is done at 500°C for ½ h. In a glove box, a 3D FAPbI$_3$ perovskite layer is spin-coated in two steps, dropping diethyl ether on the spinning substrate in the second step. The substrate is kept at 150°C for 13 min. 2D perovskite is formed by depositing octylammonium iodide (OAI) over the 3D perovskite and annealing at 100°C for 5 min. Then, carbon paste is blade-coated and 28μm-thick electrode is formed by curing at 120°C for 10 min.

The solar cell shows a short-circuit current density of 24.3mAcm^{-2}. Its open-circuit voltage is 1.03V, and fill factor is 73.9%. The efficiency achieved is 18.5% (Zouhair et al 2022)

8.4.4   Polycrystalline FAPbI$_3$ 3D Perovskite Solar Cell with 2D PEA$_2$PbI$_4$ Perovskite at Grain Boundaries (η = 19.77%)

2D phenylethylammonium lead iodide (PEA$_2$PbI$_4$) perovskite is incorporated into precursor solution to form phase-pure FAPbI$_3$ film (Lee et al 2018). Figure 8.8 shows the solar cell. The process steps for its fabrication are:

(i) ITO glass substrate is cleaned in detergent, DI water, acetone, and propanol and treated with UV-ozone.

(ii) 30 mM SnO$_2$ solution in ethanol is spin-coated on the substrate and baked at 150°C for ½ h.

(iii) Substrate is cooled, recoated with SnO$_2$ solution, and annealed at 180°C for 1 h.

(iv) (a) Phenylethylammonium iodide (PEAI) synthesis. Phenethylamine is dissolved in ethanol, placed in ice bath, and hydroiodic acid is added with stirring. After overnight stirring, the solvent is removed by a rotary evaporator. The solid obtained is washed with diethyl ether until it becomes white. After further purification by recrystallization in (methanol + diethyl ether) solvent, the white solid is filtered and dried.

(b) 2D perovskite–incorporated FAPbI$_3$ precursor solution. Perovskite precursor solution is made with HC(NH$_2$)$_2$I (FAI), phenylethylammonium iodide (PEAI), lead iodide (PbI$_2$), N-methyl-2-pyrrolidone (NMP), and N,N-Dimethylformamide (DMF).

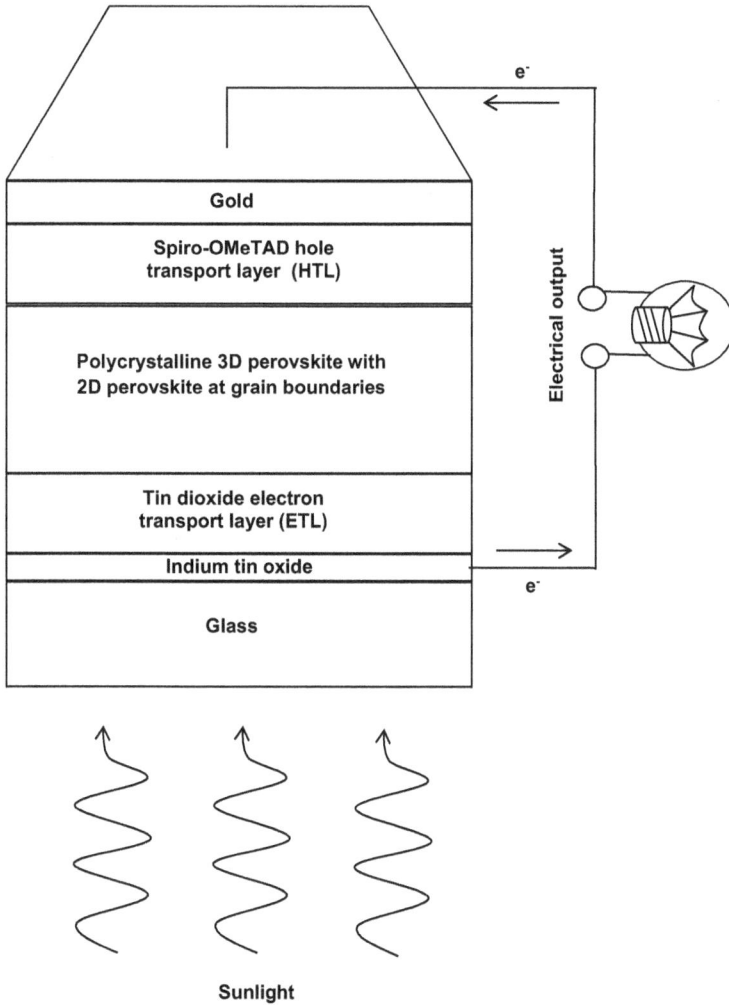

FIGURE 8.8 Polycrystalline FAPbI$_3$ 3D perovskite solar cell with 2D PEA$_2$PbI$_4$ perovskite at grain boundaries. At the bottom is an ITO-coated glass substrate. The substrate is covered with tin dioxide electron transport layer. Above tin oxide is 3D/2D perovskite and then spiro-OMeTAD hole transport layer. Finally, at the top, there is a gold contact film. Sunlight falls on the solar cell from the glass substrate side, while electrical output is taken between the ITO and gold electrodes.

(v) Perovskite layer is deposited by spin-coating the precursor solution and heat-treated at 150°C for 10 min. A small amount of diethyl ether containing stabilizer is dropped during spin-coating. At the grain boundaries of FAPbI$_3$, 2D perovskite is spontaneously formed, providing protection to it from water vapor.

(vi) Spiro-MeOTAD hole transporting layer is spin-coated.

(vii) Gold film is thermally evaporated.

(viii) The short-circuit current density of the solar cell is 24mAcm^{-2}, open-circuit voltage is 1.13V, and certified stabilized power conversion efficiency is 19.77%.

(ix) 2D perovskite at grain boundaries provides protection from humidity. It also subdues the migration of ions.

(x) The photoluminescence lifetime is found to increase by a factor of 10 due to inclusion of 2D perovskite in the precursor solution. The device shows enhanced stability during operation (Lee et al 2018).

8.5   DISCUSSION AND CONCLUSIONS

Perovskite materials enjoy an unrivalled reputation in the realm of photovoltaics because they have been at the forefront of the progress in two dimensional categories, 3D and 2D. First, the exploitation of 3D perovskites led by $CH_3NH_3PbI_3$ benefitted solar cells. 3D perovskite solar cells outperformed several prevalent photovoltaic technologies, such as multicrystalline silicon and CIGS (Ge et al 2020). The ensuing development was dominated by 2D perovskite solar cells and by concerted 2D/3D perovskite solar cells (Tables 8.1 and 8.2).

TABLE 8.1
2D Perovskite Solar Cells

Sl. No.	Type of Solar Cell	Specialty	Efficiency (%)	Reference
1.	$PbBr_2$-incorporated 2D perovskite solar cell	Defect reduction by $PbBr_2$ incorporation	12.19	Han et al (2020)
2.	2D $GA_2MA_4Pb_5I_{16}$ perovskite solar cell interface-engineered with GABr	Interface engineering with GABr improves optoelectronic properties, environmental, and photostability	19.3	Huang et al (2021)
3.	FA-based 2D perovskite solar cell	Provides FA-induced enhancement of light harvesting and better charge transport	21.07	Shao et al (2021)

TABLE 8.2
2D/3D Perovskite Solar Cells

Sl. No.	Type of Solar Cell	Specialty	Efficiency (%)	Reference
1.	2D/3D $(HOOC(CH_2)_4NH_3)_2PbI_4/$ $CH_3NH_3PbI_3$ perovskite interface-engineered solar cell	Interface engineering provides stability, better optical absorption and electrical transport	14.6	Grancini et al (2017)
2.	2D perovskite–encapsulated 3D perovskite solar cell	Encapsulation and passivation by 2D perovskite	16.79	Kore et al (2021)
3.	Hole transport material–free perovskite solar cell	2D Perovskite is used as a passivation and electron blocking layer	18.5	Zouhair et al (2022)
4.	Polycrystalline $FAPbI_3$ 3D perovskite solar cell with 2D PEA_2PbI_4 perovskite at grain boundaries	2D perovskite at grain boundaries protects from ambient moisture and quells ion migration	19.77	Lee et al (2018)

Questions and Answers

8.1 Give two examples of A-site small organic cations in a 3D perovskite. Answer: Methylammonium (MA) and formamidinium (FA) cations.

8.2 Cite two examples of A'-site large organic cations in a 2D perovskite. Answer: Phenylethyl ammonium (PEA) and butylammonium (BA) cations.

8.3 Mention two examples of B-site metals in 3D or 2D perovskites. Answer: Pb^{2+}, Sn^{2+}.

8.4 What kind of a perovskite is a material in which n = 3? Answer: Quasi-2D perovskite.

8.5 What type of perovskite is a material in which n = 1? Answer: Pure 2D perovskite.

8.6   Why is a 2D perovskite not a good choice as light absorber in a solar cell as compared to 3D perovskite? Answer: Because of its wider bandgap than 3D perovskite and limited charge transport.

8.7   What is the main drawback of 3D perovskite solar cells? Answer: Instability and short lifetime.

8.8   Which is more stable: (a) 3D perovskite or (b) 2D perovskite. Why? Answer: (b) 2D perovskite is more stable. The stability is provided to 2D perovskite by the large organic cation. This cation is hydrophobic. This water-repellant property does not allow ambient humidity to cause degradation of the perovskite film.

8.9   What is the general opinion about using 2D perovskites in solar cells? Answer: 2D perovskites can play two roles in solar cells. (i) They can be jointly used with 3D perovskites. Here, the 2D perovskite is a protective layer over the 3D perovskite light-absorbing layer, forming a 2D/3D perovskite solar cell. 2D perovskite is secondary light absorber, while 3D perovskite is the primary light absorber. (ii) They can be used as replacement materials for 3D perovskites. Here, the 2D perovskite is a stable light-absorbing medium. 2D perovskite is the primary light absorber. There is no secondary light absorber. This is a 2D perovskite solar cell.

8.10   How can multidimensional 2D–3D perovskites be used? Answer: To simultaneously provide good thermodynamic stability of the 2D perovskite material and high power conversion efficiency of the 3D perovskite material.

8.11   2D perovskites are considered as multiple quantum wells. What are the well layers in these perovskites? What are the barrier layers? Answer: The well layers are the inorganic slabs of $[BX_6]^{4-}$ octahedrons. The barrier layers are the large insulating organic cation spacer layers. Thus, we have inorganic well layers and organic barrier layers.

8.12   How are the carriers confined in the inorganic $[BX_6]^{4-}$ octahedrons of a 2D perovskite? By dimensional confinement in nanometer-thick layers and by dielectric confinement. The organic cation layer is the dielectric spacer.

8.13   Besides acting as insulating barriers confining charge carriers in two dimensions, what other function do the bulky organic cations perform in a 2D perovskite? Answer: They act as dielectric moderators determining the electrostatic forces exerted on the photo-generated electrons and holes.

8.14   How do the bulky organic cations influence the thermal stability and photostability of 2D perovskite? Answer: By passivating the defects and increasing the activation energy of ion migration.

8.15   Excitonic binding energy is a key parameter related to photovoltaic actions in solar cells. In the natural quantum well structures of 2D perovskites, spatial quantum confinement and dielectric confinement produce strongly bound excitonic states. For 2D perovskite $(C_6H_{13}NH_3)_2(CH_3NH_3)_{n-1}Pb_nI_{3n+1}$, the exciton binding energy increases from 100 to 361 meV with decreasing thickness (Gao et al 2021).

For comparison, the exciton binding energy in methylammonium lead iodide (MAPbI$_3$) 3D perovskite is ~ 10 meV (Chen et al 2018). This is 1/3rd the thermal energy (k_BT ~ 26 meV) at room temperature. From the given information, comment on the suitability of 3D and 2D perovskites for solar cell application.

Answer: The exciton binding energy is so small for 3D perovskite that excitons split up at room temperature into free electrons and holes which participate in electrical conduction, leading to the production of a photovoltaic current. Similar exciton decomposition is difficult in a 2D perovskite. As solar cell operation relies on the separation of charges, the high excitonic binding energy limits the application of 2D perovskites in solar cells, which require charge separation.

8.16   Suggest a possible solution to the lead toxicity problem of perovskites. Answer: Lead can be replaced by an environment-friendly metal such as tin (Sn). This technology is still in infancy stage.

8.17 How does incorporation of bromine in 2D perovskite impact the performance of the fabricated solar cell? Answer: It elevates the efficiency of the solar cell tremendously by decreasing the defect density of the perovskite material. The efficiency increases from 3.01% to 12.19%.

8.18 What improvements are observed in a 2D perovskite solar cell in which GABr is deposited over the perovskite? Answer: The solar cell has a smooth, shiny surface. It shows environmental stability and photostability.

8.19 When a cost-effective carbon-graphite black electrode is used in a hole transport material–free solar cell, how is performance impairment avoided at the interface of the 3D perovskite layer with the electrode? Answer: A 2D perovskite layer is interposed between the 3D perovskite layer and the carbon electrode for passivation purpose and for blocking electrons. This 2D perovskite layer inhibits the interfacial recombination losses and enhances stability.

8.20 How does UV-ozone treatment clean surfaces? Answer: It is a simple-to-use dry process. It is inexpensive to set up and easy to operate. The UV light breaks down many organic bonds, e.g., C-C, C = C, C-H, C-O, etc., in surface contaminant molecules. Thus, high-molecular-weight compounds are broken apart.

  The high-intensity UV lamp of the UV-ozone cleaner produces ultraviolet radiation at two wavelengths, 184.9nm and 253.7nm, each of which plays an assigned role. 185nm UV light dissociates molecular oxygen in the air into triplet atomic oxygen. The triplet atomic oxygen thus produced combines with molecular oxygen, forming ozone. 254nm UV light dissociates ozone to form molecular oxygen and singlet atomic oxygen. This singlet atomic oxygen has a strong oxidizing ability. When the singlet atomic oxygen falls on the surface to be cleaned, the decomposed organic contamination is gently removed from the surface as volatile byproduct molecules, such as CO_2, H_2O, and O_2.

8.21 How does oxygen plasma cleaning work? Answer: The oxygen plasma contains high-energy charged species in several forms, such as metastable excited oxygen, ionized ozone, and free electrons. These species react with organic residues on the surface to be cleaned. They decompose the contaminant molecules on the surface to form H_2O, CO, CO_2, and lower-molecular-weight hydrocarbons. The effluent gas stream carries away these molecules.

8.22 What is photoluminescence? What does a photoluminescence measurement reveal? Answer: Photoluminescence is the emission of light from a material induced by absorption of electromagnetic radiation. Photoluminescence measurements are done to assess the purity and crystal quality of a material and for quantification of disorder and impurities present.

8.23 What is light-activated interlayer contraction in 2D perovskites? How does it affect photovoltaic efficiency? Answer: Li et al (2022) reported that more than 1% contraction occurs in a 2D perovskite when continuously illuminated with light. This contraction is reversible. It is accompanied by increase in carrier mobility, and hence electrical conductivity, by a factor of 3. The contraction activates interlayer charge transport. The enhanced charge transport raises the solar cell efficiency up to 18.3%.

REFERENCES

Carlo A. D., A. Agresti, F. Brunetti and S. Pescetelli 2020 Two-dimensional materials in perovskite solar cells, Journal of Physics: Energy, 2(031003): 1–11.

Chen X , H. Lu, Y. Yang and M. C. Beard 2018 Excitonic effects in methylammonium lead halide perovskites, Journal of Physical Chemistry Letters, 9(10): 2595–2603.

Gao W., J. Ding, Z. Bai, Y. Qi, Y. Wang and Z. Lv 2021 Multiple excitons dynamics of lead halide perovskite, Nanophotonics, 10(16) 3945–3955.

Ge C., Y. Z. B Xue, L. Li, B. Tang and H. Hu 2020 October Recent progress in 2D/3D multidimensional metal halide perovskites solar cells, Frontiers in Materials, Energy Materials, 1–9, https://doi.org/10.3389/fmats.2020.601179.

Gélvez-Rueda M. C., M. B. Fridriksson, R. K. Dubey, W. F. Jager and W. van der Stam 2020 Overcoming the exciton binding energy in two-dimensional perovskite nanoplatelets by attachment of conjugated organic chromophores, Nature Communications, 11(1901): 1–9.

Grancini G., C. Roldán-Carmona, I. Zimmermann, E. Mosconi, X. Lee, D. Martineau, S. Narbey, F. Oswald, F. De Angelis, M. Graetzel and M. K. Nazeeruddin 2017 One-Year stable perovskite solar cells by 2D/3D interface engineering, Nature Communications, 8(15684): 1–8.

Han F., W. Yang, H. Li and L. Zhu 2020 Stable high-efficiency two-dimensional perovskite solar cells via bromine incorporation, Nanoscale Research Letters, 15(194): 1–8.

Huang Y., Y. Li, E. L. Lim, T. Kong, Y. Zhang, J. Song, A. Hagfeldt and D. Bi 2021 Stable layered 2D perovskite solar cells with an efficiency of over 19% via multifunctional interfacial engineering, Journal of American Chemical Society, 143: 3911–3917.

Kore B. and J. Gardner 2020 Moisture resistant 2D lead halide perovskites for high efficiency solar cells, Proceedings of International Conference on Perovskite Thin Film Photovoltaics and Perovskite Photonics and Optoelectronics (NIPHO20), 23rd–25th February, Sevilla, Spain, www.nanoge.org/proceedings/NIPHO20/5e21caed8abac34ab402df3b.

Kore B. P., W. Zhang, B. W. Hoogendoorn, M. Safdari and J. M. Gardner 2021 Moisture tolerant solar cells by encapsulating 3D perovskite with long-chain alkylammonium cation-based 2D perovskite, Communications Materials 2(100): 1–10.

Lee J.-W., Z. Dai, T.-H. Han, C. Choi, S.-Y. Chang, S.-J. Lee, N. De Marco, H. Zhao, P. Sun, Y. Huang and Y. Yang 2018 2D perovskite stabilized phase-pure formamidinium perovskite solar cells, Nature Communications, 9(3021): 1–10.

Li W., S. Sidhik, B. Traore, R. Asadpour, J. Hou, H. Zang et al 2022 Light-activated interlayer contraction in two-dimensional perovskites for high-efficiency solar cells, Nature Nanotechnololgy, 17: 45–52.

Mao L., C. C. Stoumpos and M. G. Kanatzidis 2019 Two-dimensional hybrid halide perovskites: Principles and promises, Journal of American Chemical Society, 141(3): 1171–1190.

Niu T., Q. Xue and H. L. Yip 2020 Advances in Dion-Jacobson phase two-dimensional metal halide perovskite solar cells, Nanophotonics, 10(8): 2069–2102.

Prasanna R., A. Gold-Parker, T. Leijtens, B. Conings, A. Babayigit, H. G. Boyen, M. F. Toney and M. D. McGehee 2017 Band gap tuning via lattice contraction and octahedral tilting in perovskite materials for photovoltaics, Journal of American Chemical Society, 139(32): 11117–11124.

Shao M., T. Bie, L. Yang, Y. Gao, X. Jin, F. He, N. Zheng, Y. Yu and X. Zhang 2021 Over 21% efficiency stable 2D perovskite solar cells, Advanced Materials, 2107211: 1–10.

Shi E., Y. Gao, B. P. Finkenauer, Akriti, A. H. Coffey and L. Dou 2018 Two-dimensional halide perovskite nanomaterials and heterostructures, Chemical Society Reviews, 47: 6046–6072.

Surrente A., M. Baranowski and P. Plochocka 2021 Perspective on the physics of two-dimensional perovskites in high magnetic field, Applied Physics Letters, 118: 170501–1 to 170501–10.

Xing J., Y. Zhao, M. Askerka, L. N. Quan, X. Gong, W. Zhao, J. Zhao et al 2018 Color-stable highly luminescent sky-blue perovskite light-emitting diodes, Nature Communications, 9(3541): 1–8.

Zhang F., H. Lu, J. Tong, J. J. Berry, M. C. Beard and K. Zhu 2020 Advances in two-dimensional organic—inorganic hybrid perovskites, Energy & Environmental Science, 13: 1154–1186.

Zhou Y., H. Xue, Y.-H. Jia, G. Brocks, S. Tao and N. Zhao 2019 Enhanced incorporation of guanidinium in formamidinium-based perovskites for efficient and stable photovoltaics: The role of Cs and B, Advanced Functional Materials, 29(1905739): 1–9.

Zouhair S., S.-M. Yoo, D. Bogachuk, J. P. Herterich, J. Lim, H. Kanda, B. Son, H. J. Yun, U. Würfel, A. Chahboun, M. K. Nazeeruddin, A. Hinsch, L. Wagner and H. Kim 2022 Employing 2D-Perovskite as an electron blocking layer in highly efficient (18.5%) perovskite solar cells with printable low temperature carbon electrode, Advanced Energy Materials: 2200837, https://doi.org/10.1002/aenm.202200837.

9 Carbonaceous Nanomaterials–Based Solar Cells

Three carbon nanomaterials, CNTs, C_{60}, and graphene, have pervaded solar cells as constituent layers. This ingress has brought in its wake advantages of making cell fabrication easier and less costly. The principal accomplishment of these advances is realization of equivalent or better performance than cells made by well-recognized processing.

9.1 USING CARBON NANOTUBES TO MAKE AN INEXPENSIVE COUNTER ELECTRODE FOR A DYE-SENSITIZED SOLAR CELL

9.1.1 REPLACING PLATINUM WITH CNTs-COATED NONCONDUCTIVE GLASS PLATE

Platinum-coated glass plate serves as the commonly used electrode in a dye-sensitized solar cell. Platinum is a costly metal mainly because of its rarity. The solar cells are required to be inexpensive. Therefore, large-scale use of platinum in solar cell fabrication is not recommended. The impracticability of bulk commercialization of platinum-coated electrodes compels us to look for materials, which are comparatively less-costly (Li et al 2021). Carbon nanotubes can be used in place of platinum because they possess a high electrical conductivity. So CNTs film-coated glass can be one electrode option. Carbon nanotubes, too, are expensive at present. But they are far less-expensive than platinum.

9.1.2 REPLACING PLATINUM WITH PT NPs/CNTs NANOHYBRID-COATED NONCONDUCTIVE GLASS PLATE

It will still be better if the beneficial effects of platinum can be provided without significant increase in cost. To this end, a Pt NPs/CNTs nanohybrid can be prepared. The Pt NPs are adsorbed over the CNTs. Due to their large surface area, CNTs can adsorb Pt NPs well. The nanohybrid of Pt NPs with CNTs is coated on glass to be used as an electrode. So Pt NPs/CNTs nanohybrid-coated glass is a second electrode option.

9.1.3 PERFORMANCE OF CNTs AND PT NPs/CNTs ELECTRODES IN A SOLAR CELL

The role and influence of carbon nanotubes in the counter electrode of a dye-sensitized solar cell is investigated to assess whether using them in place of a high-priced platinum thin film is a viable alternative. Dye-sensitized solar cells can be made separately with each electrode option. How these cells compare in performance with cells using standard Pt film-glass electrodes will tell about the capability of the electrodes.

9.1.4 PT NPs/CNTs NANOHYBRID

To make the Pt NPs/CNTs nanohybrid, a polymer-assisted dispersant is added to aqueous solution of CNTs. After ultrasonic agitation, hexachloroplatinic acid (H_2PtCl_6) solution in alcohol is added and the mixture is put in ultrasonic oscillator. Subsequently, it is placed in ice bath, sodium borohydride is added, and magnetic stirring is done to complete the reaction. Pt NPs/CNTs powder is obtained after rinsing with DI water and freeze-drying.

DOI: 10.1201/9781003215158-13

9.1.5 CNTs and Pt NPs/CNTs Dispersants

For preparation of CNTs dispersant and Pt NPs/CNTs dispersant, the CNTs and Pt NPs/CNTs are mixed with N-methylpyrrolidone and put in an ultrasonic oscillator. Then, an organic dispersant is added, followed by more ultrasonic agitation. Thus, dispersant liquids of CNTs and Pt NPs/CNTs are formed.

9.1.6 CNTs and Pt NPs/CNTs Electrodes

To make the CNTs and Pt NPs/CNTs electrodes, the respective dispersant liquids are drop-coated on glass substrates and dried at 110°C until a film is formed. This is done 10 times. Then the two electrodes are sintered at 350°C for 1 h. The two electrodes are shown in Figures. 9.1(a) and 9.2(a).

9.1.7 Dye-Sensitized Solar Cells with CNTs and Pt NPs/CNTs Electrodes

To fabricate dye-sensitized solar cells using these electrodes, the photoanodes are prepared by coating a titanium dioxide slurry containing smaller-size nanoparticles on the cleaned FTO-coated glass substrate at a controlled speed of the knife coater. Then, the deposited film is dried with a hot air gun. Coating is done for three times in this way. The TiO_2 film thus formed is sintered at 500°C for 2 h. This is the absorption layer of smaller-size TiO_2 nanoparticles. The scattering layer containing larger-size TiO_2 nanoparticles is identically deposited.

For sensitization of TiO_2 nanoparticles with dye, the sintered TiO_2 photoanode is immersed in N719 dye in (tert-butanol + acetonitrile) solution for 24 h.

For packaging the solar cells, the photoanode and the counter electrode are sealed together with heat seal adhesive. Electrolyte is injected through holes drilled in the electrode, which are finally sealed. The electrolyte is [lithium iodide (LiI) + dimethylimidazolium iodide (DMII) + iodine (I_2) + guanidine thiocyanate (GuSCN)]. Figures. 9.1(b) and 9.2(b) show the two solar cells.

9.1.8 Parameters of Solar Cells with CNTs and Pt NPs/CNTs Electrodes

The short-circuit current density, the open-circuit voltage, the fill factor, and the efficiency of the dye-sensitized solar cell (Figure 9.1(b)) using CNTs film-coated glass plate as counter electrode (Figure 9.1(a)) are respectively $J_{SC} = 10.40 \text{mAcm}^{-2}$, $V_{OC} = 0.62\text{V}$, FF = 0.63, and η = 4.03%. These parameters for a solar cell (Figure 9.2(b)) using Pt NPs/CNTs nanohybrid electrode (Figure 9.2(a)) are $J_{SC} = 14.63 \text{mAcm}^{-2}$, $V_{OC} = 0.68\text{V}$, FF = 0.63, and η = 6.28%. The same parameters for a solar cell made with Pt counter electrode are $J_{SC} = 10.04 \text{mAcm}^{-2}$, $V_{OC} = 0.64\text{V}$, FF = 0.68, and η = 4.37%. Looking at the values of efficiencies, the efficiencies of CNTs film electrode and Pt electrode are comparable. The CNTs film electrode is slightly inferior to the Pt electrode. The efficiency of Pt NPs/CNTs nanohybrid film electrode is better than that of the standard Pt electrode (Li et al 2021). Thus, use of one nanomaterial (CNTs) gives an acceptable electrode, and use of two nanomaterials (Pt NPs and CNTs) gives a superior electrode.

9.2 USING CARBON NANOTUBES TO IMPROVE THE PROPERTIES OF TiO₂-BASED ELECTRON TRANSPORT MATERIAL IN PEROVSKITE SOLAR CELLS

To investigate the influence of SWCNTs on the properties of electron transport layer, solar cells are fabricated using TiO_2 NPs-SWCNTs nanocomposites prepared with various concentrations of SWCNTs containing known ratios of semiconducting and metallic components in the SWCNTs mixture (Bati et al 2019). The nanocomposite composition for the solar cell showing the best performance is deemed as the optimal composition. Figure 9.3 shows the solar cell. Before proceeding further, we must recall the merits and shortcomings of TiO_2 and look into the need for replacement of TiO_2 with TiO_2 NPs-SWCNTs nanocomposite.

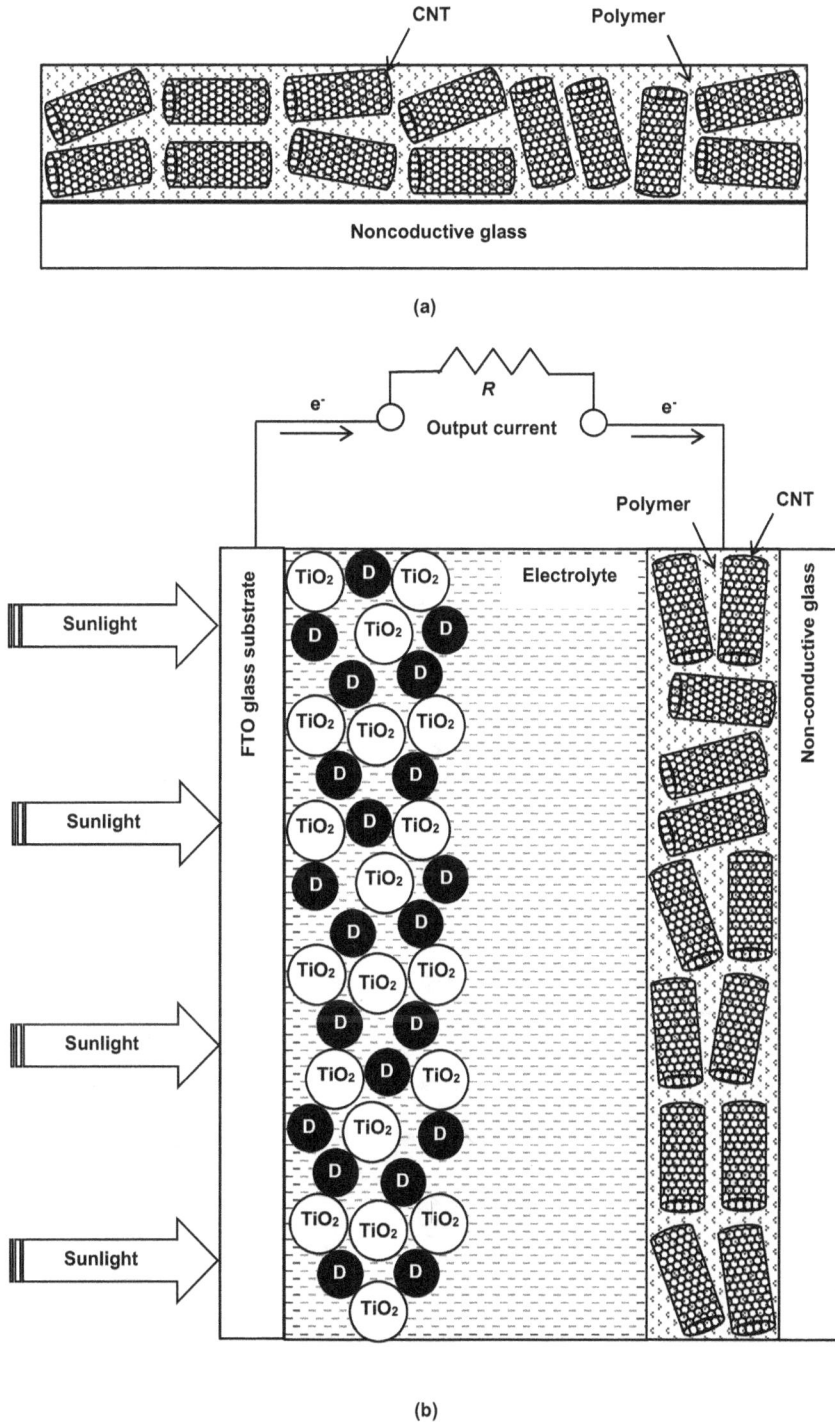

FIGURE 9.1 CNTs film-coated nonconductive glass electrode and a dye-sensitized solar cell using counter electrode made with CNTs film-coated glass: (a) the electrode showing the nonconductive glass substrate and CNTs dispersed in polymer, and (b) the dye-sensitized solar cell showing the TiO_2 nanoparticles covered with dye (D) molecules, the electrolyte, and the counter electrode made of CNTs nanohybrids film-coated nonconductive glass plate in place of a sputtered platinum film-coated glass electrode. Illumination of the solar cell from the side of FTO-coated glass electrode leads to current flow between the two electrodes.

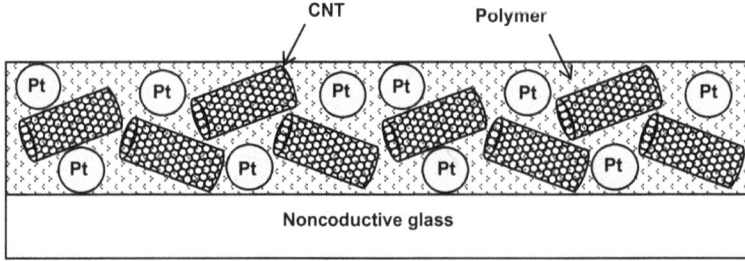

(a)

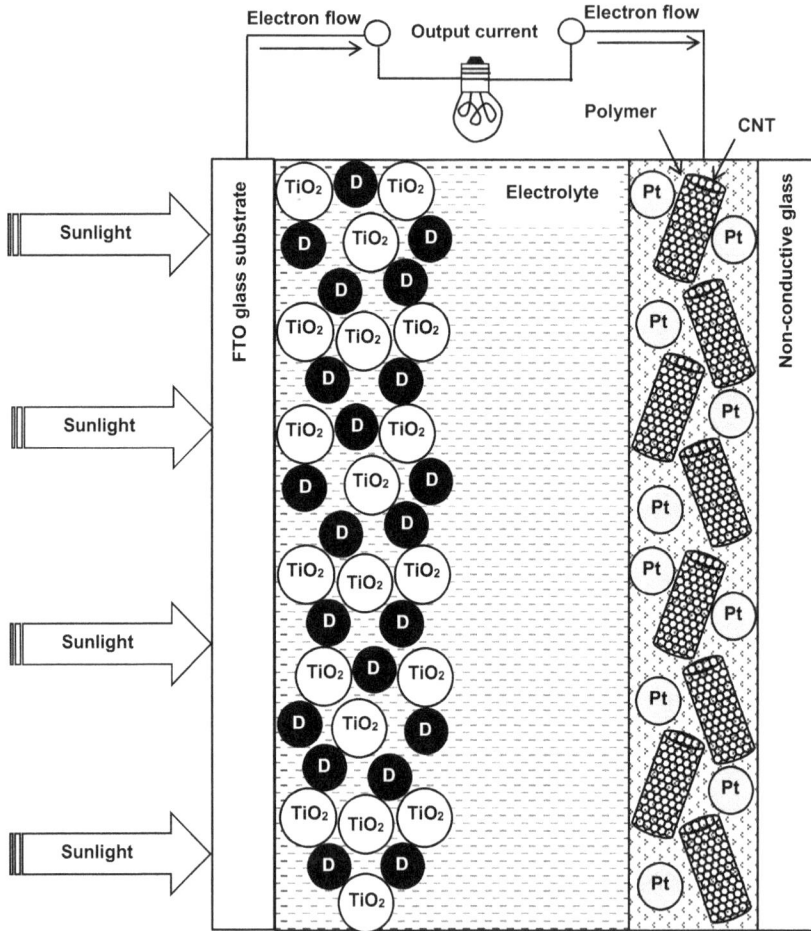

(b)

FIGURE 9.2 PtNPs/CNTs nanohybrids film-coated nonconductive glass electrode and a dye-sensitized solar cell using counter electrode made with Pt NPs/CNTs nanohybrids film-coated glass plate: (a) the electrode consisting of the nonconductive glass substrate and PtNPs dispersed in polymer, (b) the dye-sensitized solar cell in which the TiO_2 nanoparticles are covered with dye (D) molecules. Also, there is an electrolyte and a counter electrode made of PtNPs/CNTs nanohybrids film-coated nonconductive glass plate. The PtNPs/CNTs electrode substitutes the usual sputtered platinum film-coated glass electrode used in such a cell. When sunlight illuminates the FTO-coated glass substrate, an electric current flows between the two electrodes, and the directions of electron flow are shown.

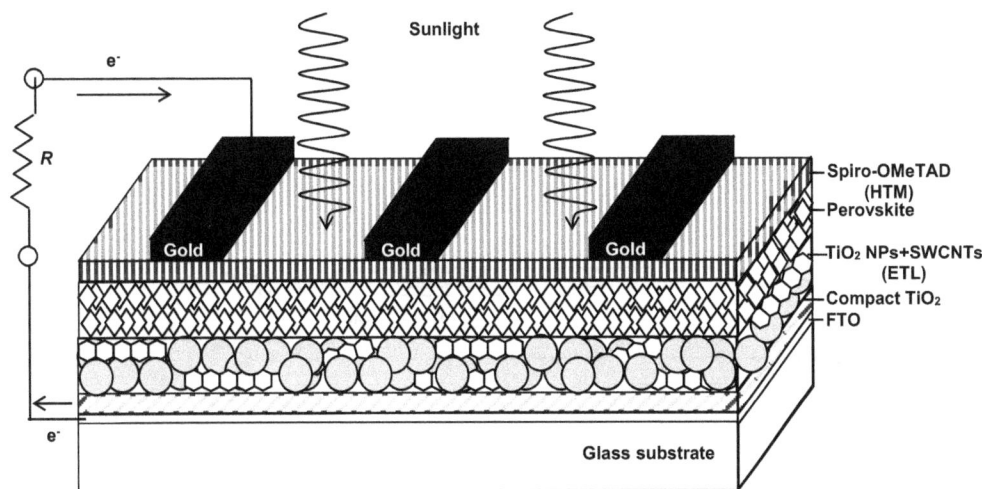

FIGURE 9.3 Perovskite solar cell made by integrating electrically sorted SWCNTs into TiO$_2$ electron transport layer (ETL) showing the fluorine-doped tin oxide (FTO)–coated glass substrate, TiO$_2$ compact layer, TiO$_2$ NPs-SWCNTs mesoporous layer, perovskite layer, spiro-OMeTAD layer, and the gold film forming the gold top electrode. Sunlight is incident on the glass substrate of the solar cell. Electrical output is extracted across FTO and gold electrodes.

9.2.1 Advantages and Limitations of TiO$_2$ as an Electron Transport Material

TiO$_2$ has been widely accepted as an ETL for solar cell because it is chemically stable and nontoxic. It combines a wide bandgap with a high degree of transparency. Notwithstanding, it offers a low mobility of charge carriers. The presence of innumerable grain boundaries together with heterogeneously spread network of TiO$_2$ nanoparticles randomizes the path of electron motion. The resulting rapid recombination processes degrade the performance of solar cells.

9.2.2 Choice of CNTs as TiO$_2$ Conductivity-Enhancement Nanomaterials

A nanomaterial that can be incorporated in TiO$_2$ must possess high electron mobility. It should also be optically transparent. CNTs fulfill these requirements and are also favored because of their high chemical and thermal stability and mechanical strength. Their large aspect ratios enable them to act as wiring channels. But the pristine CNTs consist of a mixture of semiconducting (s) and metallic (m) types. Properties of SWCNTs-incorporated TiO$_2$ nanomaterial for solar cell application depend on the ratio of semiconducting and metallic SWCNTs as well as the loading wt% of SWCNTs into TiO$_2$. Therefore, electrical sorting of two types of SWCNTs is necessary. A column gel chromatographic technique is used. SWCNTs are dispersed in sodium dodecyl sulfate (SDS). This dispersion is fed to a column containing beads of agarose gel. The beads trap the semiconducting SWCNTs. The metallic beads move through the column (Tanaka et al 2009). TiO$_2$ NPs-SWCNTs nanocomposite is prepared by addition of a precalculated volume of SWCNTs solution to the dispersion of TiO$_2$ NPs. The ratio of s-SWCNTs to m-SWCNTs is taken as 2:1 w/w.

9.2.3 Fabrication of the Solar Cell Using TiO$_2$ NPs-SWCNTs Nanocomposite

A glass substrate with a coating of fluorine-doped tin oxide is taken (Figure 9.4(a)). A compact TiO$_2$ film is deposited on the cleaned FTO-coated glass substrate (Figure 9.4(b)) by spin-coating a precursor solution containing titanium diisopropoxide bis(acetylacetonate) in 1-butanol. Annealing of the film is done at 150°C for ¼ h. The compact TiO$_2$ film is covered with a TiO$_2$ NPs-SWCNTs

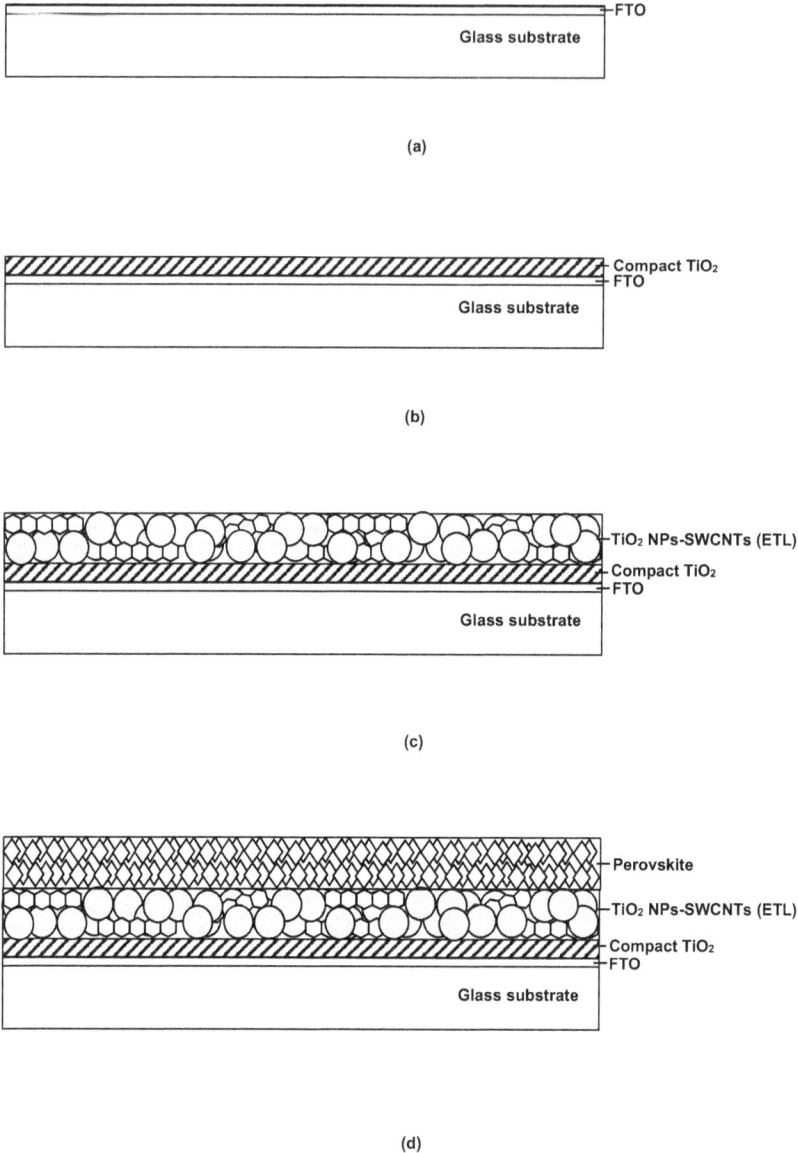

FIGURE 9.4 Fabrication of perovskite solar cell made by integrating electrically sorted SWCNTs into the TiO_2 electron transport layer (ETL): (a) fluorine-doped tin oxide (FTO)–coated glass substrate; (b) TiO_2 compact layer deposition by two-time spin-coating of precursor solution with 15 min. thermal annealing at 150°C each time; (c) spin-coating TiO_2 NPs-SWCNTs mesoporous layer containing SWCNTs in the ratio 2:1 at 0.4 weight % loading, followed by thermal annealing at 450°C for 1 h., immersion in 20mM $TiCl_4$ aqueous solution, and second 1 h. thermal annealing at 450°C; (d) deposition of perovskite solution by spinning at 1,000 RPM with ramping, dropping anhydrous chlorobenzene on the substrate, and again spinning at 5,000 RPM with a different ramping cycle followed by annealing at 100°C for 45 min.; the perovskite solution is prepared by mixing formamidinium iodide (FAI), lead iodide (PbI_2), methylamine bromide (MABr), and lead bromide ($PbBr_2$) in anhydrous dimethyl formamide: dimethyl sulfoxide (DMF: DMSO) and adding cesium iodide (CsI) in DMSO to the perovskite precursor, (e) spin-coating 2,2′,7,7′-tetrakis(N,N-di-pmethoxyphenylamine)-9,9′-spirobifluorene (spiro-OMeTAD) solution at 4,000 RPM and storing overnight in a desiccator to form a hole transport layer (HTL), and (f) thermal evaporation of gold forming the gold top electrode of thickness 50nm. Sunlight is incident on the glass substrate of the solar cell. Electrical output is extracted across FTO and gold electrodes.

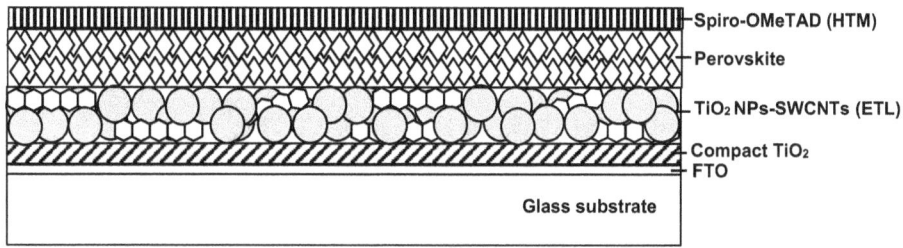

(e)

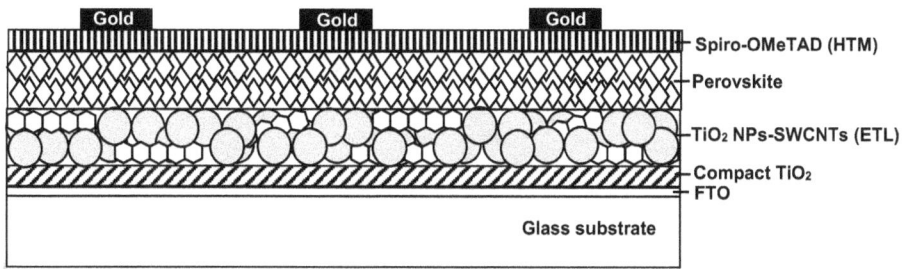

(f)

FIGURE 9.4 (Continued)

nanocomposite film (Figure 9.4(c)) formed by spin-coating. The nanocomposite film is annealed at 450°C for 1 h. The cooled film is dipped in TiCl$_4$ solution at 90°C for 20 min. Annealing is done again at 450°C for 1 h. The perovskite solution is prepared by mixing formamidinium iodide, lead iodide, methylamine bromide, and lead bromide in anhydrous dimethyl formamide: dimethyl sulfoxide, adding cesium iodide in dimethyl sulfoxide, stirring and filtering. Spin-coating is done (Figure 9.4(d)), first at 1,000 RPM, and then at 5,000 RPM, with few drops of anhydrous chlorobenzene dropped on the substrate before the end of the second spinning. The perovskite film is annealed at 100°C for 45 min. To make the solution of hole transport material (HTM), spiro-OMeTAD, 4-tert-butylpyridine, lithium bis(trifluoromethanesulfonyl)imide solution in acetonitrile, and tris(2-(1H-pyrazol-1-yl)pyridine)cobalt(II) di[hexafluorophosphate] solution in acetonitrile are taken and dissolved in chlorobenzene. The HTM solution is deposited over the perovskite film by spin-coating (Figure 9.4(e)). After overnight storage in desiccator, the top gold electrode is formed by thermal evaporation (Figure 9.4(f)).

9.2.4 Solar Cell with TiO$_2$ NPs-SWCNTs and Control Cell

The open-circuit voltage of the solar cell made with TiO$_2$ NPs-SWCNTs is 1.085 V. The short-circuit current density of the solar cell is 24.59 mAcm^{-2}. Its fill factor is 0.73, and efficiency is 19.35%. The control cell made without CNTs, i.e., with TiO$_2$ only shows an open-circuit voltage of 1.046 V. The short-circuit current density of the control cell is 22.5 mAcm^{-2}. Its fill factor is 0.72, and efficiency is 17.04%. A distinct improvement of characteristics of the solar cell using a TiO$_2$NPs-SWCNTs film is noticed over the control cell in which a bare TiO_2 film is employed. Hence, the significance of use of carbon nanotubes is proven (Bati et al 2019).

9.3   USING CNTs AND C_{60} TO MAKE A HIGH-STABILITY, COST-EFFECTIVE PEROVSKITE SOLAR CELL

9.3.1   Material Replacements in Traditional Structure

Materials in the traditional structure: FTO/compact TiO_2/Mesoporous TiO_2/$MAPbI_3$/spiro-MeO-TAD/Au of a perovskite solar cell are changed using nanomaterials as ITO/C_{60}/Perovskite/(CNTs + spiro-MeOTAD or other polymeric hole transport material) from considerations of stability upon exposure to atmosphere and illumination and fabrication cost (Ahn et al 2018). The structure of solar cell using nanomaterials is shown in Figure 9.5.

9.3.2   Fabrication of Perovskite Solar Cell with Replaced Materials

A vacuum thermal evaporator is used to deposit a C_{60} layer (thickness 35nm) on a cleaned ITO-coated glass substrate. For depositing the $MAPbI_3$ perovskite layer, a solution of lead iodide, methylammonium iodide, and dimethyl sulfoxide in dimethyl formamide is spin-coated on a

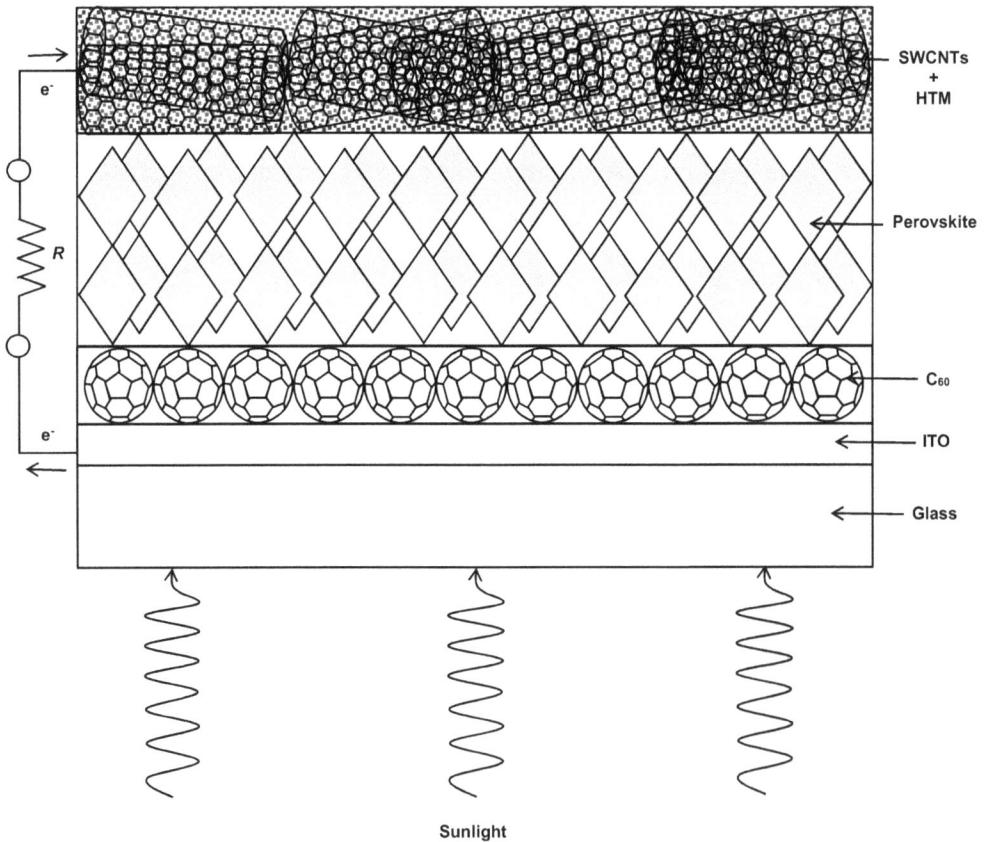

FIGURE 9.5   Perovskite solar cell fabricated by sandwiching the perovskite absorber layer with C_{60} on one side and (CNTs + hole transporting material) on the opposite side. The layers are, from bottom to top: glass, ITO, C_{60}, perovskite (CNTs + hole transporting material). Sunlight is incident on the glass substrate. Current output is obtained between ITO and CNTs electrodes.

PEDOT:PSS layer. Diethyl ether is dropped after few seconds of spinning. Upon heating, the green color of CH_3NH_3I-PbI_2-DMSO adduct film changes to dark brown. The CNT film made from CNTs synthesized by aerosol chemical vapor deposition (CVD) is mechanically transferred over the perovskite layer. Spiro-MeOTAD film is deposited by spin-coating the solution made by mixing spiro-MeOTAD, 4-*tert*-butyl pyridine, and lithium bis(trifluoromethanesulfonyl) solution in acetonitrile, in chlorobenzene, over the CNT film. A CNT film entangled with spiro-MeOTAD is created. Similarly, solar cells are made with (CNTs + PTAA), (CNTs + P3HT), and CNTs only.

9.3.3 SOLAR CELL PERFORMANCE VS. COST

Both the nanomaterials C_{60} and CNTs enhance stability with retention of 80% of incipient efficiency by the encapsulated solar cell after operating for 2,200 hours, while use of CNTs decreases the fabrication cost to < 13.5–3.2% of the conventional cell. Explicitly stated, for a solar cell made with (CNTs + spiro-MeOTAD), the efficiency is 17%; the open-circuit voltage, short-circuit current and fill factor are respectively 1.08V, 23.8mAcm^{-2} and 66.1%. The cost decreases to < 13.5%. For a solar cell made with (CNTs + PTAA), the efficiency is 15.30%, and the cost is < 6%. For a solar cell made with (CNTs + P3HT), the efficiency is 13.6%, and the cost is < 5.5%. For a solar cell made with CNTs only, the efficiency is 13.2%, and the cost is < 3.6%. The data provides a comparative assessment of the efficiency vs. cost of solar cells fabricated by such structural and compositional changes (Ahn et al 2018).

9.4 INTEGRATING CNTs IN A SILICON-BASED SOLAR CELL: Si-CNTs HYBRID SOLAR CELL

A layer of CNTs doped with dilute HNO_3 is combined with silicon substrate to fabricate a P-N heterojunction solar cell (Jia et al 2011). Figure 9.6 shows this solar cell.

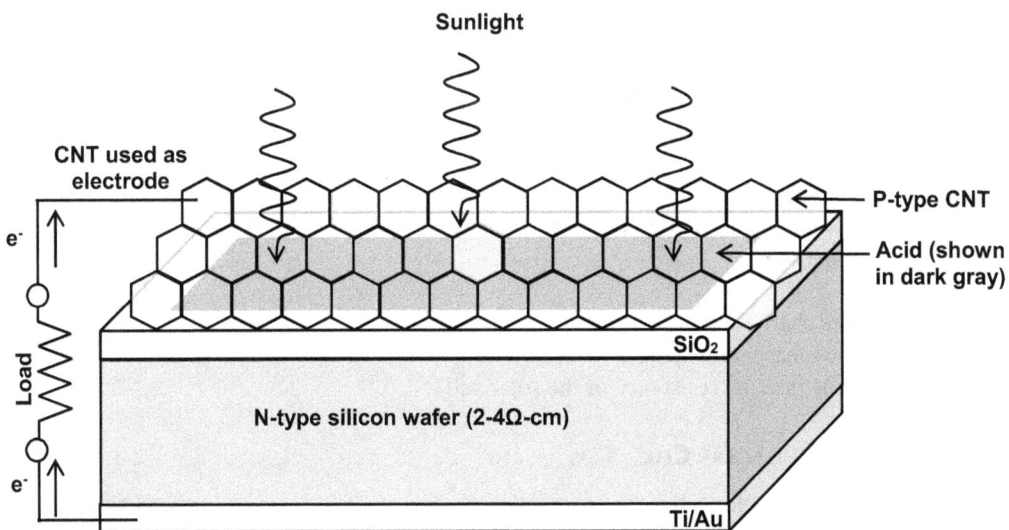

FIGURE 9.6 Silicon-carbon nanotube heterojunction solar cell showing Ti/Au film below and acid-doped CNT above the N-type silicon wafer. P-type CNT/N-type Si wafer form a heterojunction. Positive electrode of the solar cell is that contacting the top of CNT film on silicon dioxide. Negative electrode contact is made with Ti/Au film. Sunlight shines over the CNT film. Current is drawn between CNT and Ti/Au electrodes.

9.4.1 ADVANTAGES OF CNTs INTEGRATION

(i) CNTs act as the P-type layer of this solar cell, and silicon is the N-type layer. Thus, a P-N diode is formed without any high-temperature processing steps ~ 1,000°C or higher, which comprise an essential component of a silicon diode fabrication process. Avoidance of this high-temperature step simplifies process and saves fabrication costs. It allows harnessing the excellent photovoltaic properties of crystalline silicon by a simple process, which can be implemented at much lower temperatures than the established silicon solar cell processing.

(ii) The 2D network of carbon nanotubes laid over the silicon wafer constitutes a transparent film. It serves the dual functions of the front-surface electrode and the P-type layer of the diode. No extra contact metal grid needs to be provided for pulling together the charge carriers. The advantage gained becomes obvious when we recall that light falling over the contact metal grid portions in a silicon solar cell is lost. The same does not happen in this CNTs/Si solar cell, thereby preventing loss of sunlight.

(iii) Electronic and optical properties of CNTs can be varied to modify the characteristics of the solar cell, e.g., doping and thickness of the CNTs can be changed, or the CNTs can be chemically functionalized. So CNTs provide additional controlling elements of solar cell behavior.

(iv) The CNTs-Si solar cell is one example of using nanomaterial-silicon combinations for solar cell fabrication. A similar material combination such as Si-graphene can be used to construct solar cells. So a class of solar cells can emerge from such pairs of materials.

9.4.2 FABRICATION OF THE SOLAR CELL

The source solution (xylene + ferrocene + small quantity of sulfur) is injected in a quartz tube, vaporized at 200–220°C, and transported in an (argon + hydrogen) carrier gas stream to the reaction zone at 1,150–1,170°C. The CNTs produced are collected on a nickel foil and transferred to the surface of a prepatterned N-type silicon wafer.

This silicon wafer is already covered with a Ti/Au film on the backside, and SiO_2 film on the front side and photolithography has been done to open a window over which the CNT film will be spread, as shown in Figure 9.7, displaying the fabrication process of the solar cell step-by-step. After the CNT film is spread over the silicon surface, it is drenched in ethanol so that it closely touches the silicon surface, producing intimate contact. Ethanol is then dried.

Before doping the CNTs P-type with HNO_3, the native oxide on silicon surface is removed in HF, and HF is removed by rinsing in DI water. Then, few drops of dilute HNO_3 solution is poured over the CNT network and allowed to infiltrate through CNTs to reach the silicon surface. The acid performs three functions:

(i) It enhances charge separation.
(ii) It facilitates charge transport.
(iii) It lowers the internal resistance of the solar cell.

9.4.3 TESTING OF THE SOLAR CELL

Positive electrode of the solar cell is that contacting the top of CNT film and lying on silicon dioxide surface. Negative electrode contact is made with Ti/Au film. Sunlight shines over the CNT film. The short-circuit current density, fill factor, and efficiency of a pristine solar cell are 27.4 mAcm^{-2}, 47%, and 6.2%, respectively, under AM1.5G, 100 mWcm^{-2} illumination. These parameters increase to 36.3 mAcm^{-2}, 72%, and 13.8% after reduction of internal resistance of the solar cell by 0.5M/L acid infiltration and in wet condition. The large increase in solar cell efficiency by acid addition exceeding a factor of two clearly indicates that this step is extremely important (Jia et al 2011).

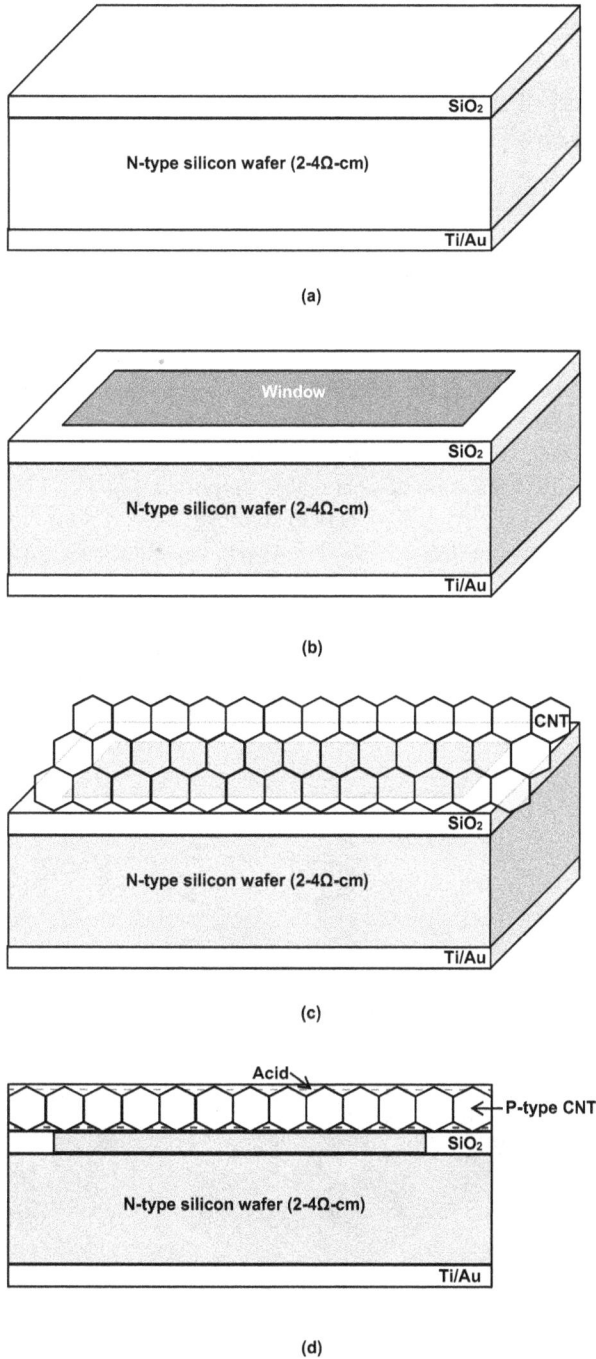

FIGURE 9.7 Fabrication of the silicon-carbon nanotube heterojunction solar cell with dilute HNO$_3$ doping: (a) N-type silicon wafer (resistivity 2–4Ω-cm) with 300nm-thick silicon dioxide on the front surface and Ti/Au metallization on the back surface, (b) etching silicon dioxide from a fixed segment to create a window for device area definition, (c) transference of a freestanding CNT film over the silicon on device area for device fabrication and spreading to the sides over silicon dioxide for external connection; good contact between silicon and CNT is established by dropping ethanol over CNTs and allowing its evaporation, and (d) native oxide removal from silicon surface by 2wt% HF treatment, DI water rinsing, delivering a droplet of dilute HNO$_3$ on CNT film, and its infiltration for covering the CNTs.

9.5   Si-CNTs HYBRID SOLAR CELL FABRICATION
BY SUPERACID SLIDING COATING

P-type SWCNT is deposited on N-type silicon by superacid sliding coating (Jung et al 2013). The solar cell is shown in Figure 9.8.

9.5.1   SUPERACID SLIDE CASTING METHOD FOR HIGH-QUALITY CNTs FILM PREPARATION

This is a method for depositing a high-quality SWCNT film in which a sliding technique is applied on superacid dispersed CNTs (Figure 9.9). Superacid is one having acidity > that of 100% sulfuric acid.

The film formed by this method has a high conductivity. Hence, charge carrier transportation in the film is improved. In addition, the film has a high transparency. Therefore, the absorption of light by solar cell increases. The roughness of the film surface is very low, providing a conformal contact with silicon surface. Also, the film is mechanically robust. So the junction is reliable. The process is easy. Hence, the fabrication of solar cell is simplified.

SWCNT ink is prepared by stirring SWCNTs in chlorosulfonic acid for three days in a nitrogen-filled glove box (Li et al 2013). A drop of SWCNT dispersion is put on a glass slide. Another glass slide is pressed against the glass slide carrying the drop. By pressing the glass slides together, the required thickness of SWCNT film is obtained. The two glass slides are slid against each other in opposite directions. By such shearing, SWCNT films are produced on each glass slide. These films are dried inside the glove box. They are floated on water and transferred to the silicon wafer.

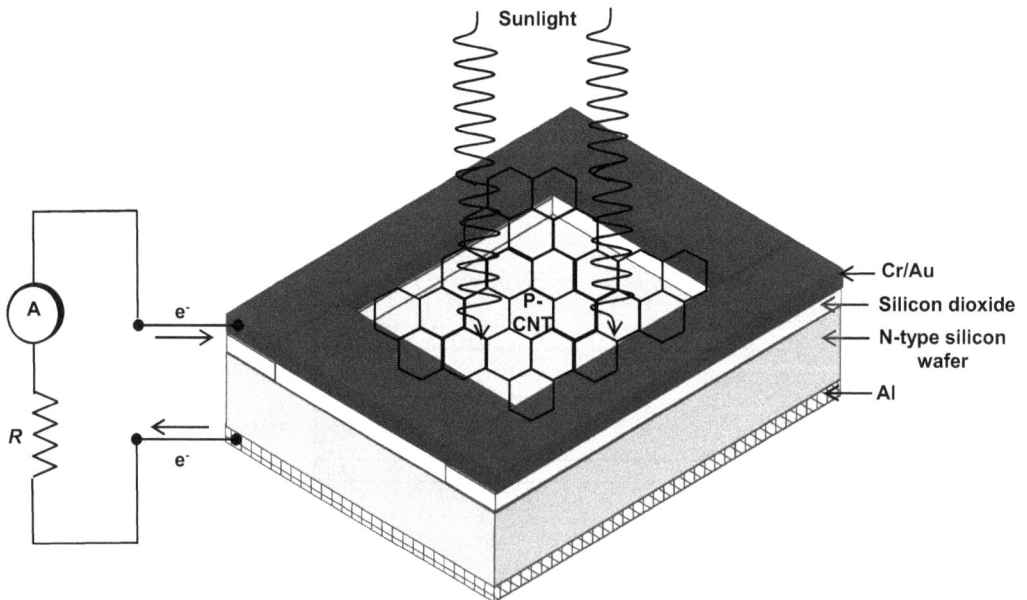

FIGURE 9.8   Carbon nanotube-silicon P-N junction solar cell showing N-type silicon wafer with silicon dioxide, and Cr/Au layers on the front surface, aluminum layer on the back surface, and a freestanding P-type SWCNT film over the front silicon surface. A P-N junction is formed between P-type SWCNTs and N-type silicon wafer. One electrode of the solar cell is Cr/Au film on the top of CNT film on silicon dioxide. The other electrode is Al film on the backside of the silicon wafer. Sunlight is incident over the CNT film. Electric current flows between Cr/Au and Al electrodes.

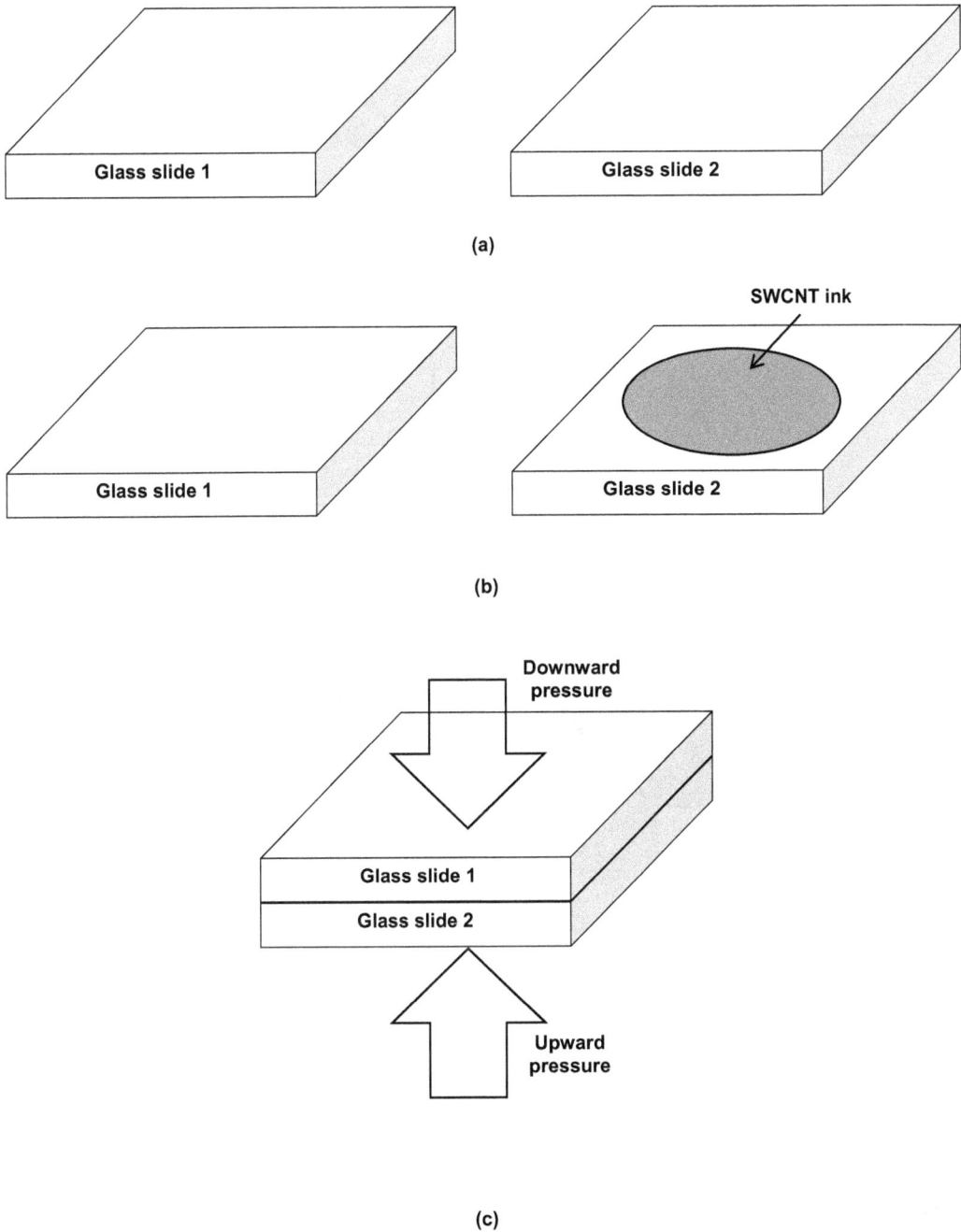

(a)

(b)

(c)

FIGURE 9.9 Superacid SWCNT sliding coating method: (a) two glass slides 1 and 2 are taken, (b) a drop of SWCNT ink is put on glass slide 2, (c) glass slide 1 is slid against glass slide 2, and the two slides are pressed against each other, (d) glass slide 2 is slid back over glass slide 1, (e) glass slide 1 is inverted while glass slide 2 is kept as it is, when SWCNT films are seen on the surfaces of both glass slides.

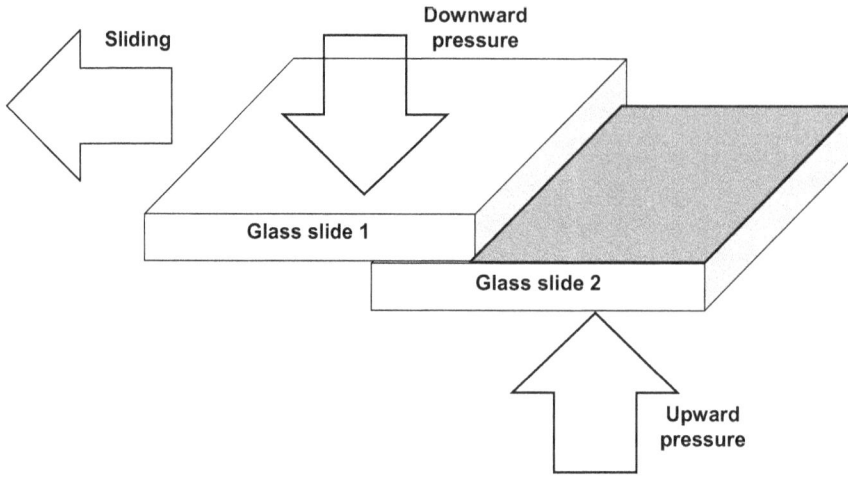

(d)

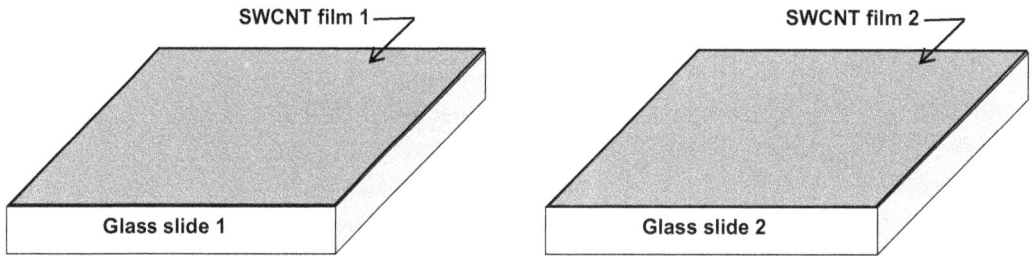

(e)

FIGURE 9.9   (Continued)

9.5.2   Process Sequence

N-type silicon wafer with doping concentration 10^{15}–10^{16}cm^{-3} is taken. The wafer has thermally grown silicon dioxide of thickness 500nm on its surface (Figure 9.10(a)). Cr/Au film is deposited over the thermal oxide layer and patterned photolithographically to open a window, where the SWCNT film will be transferred (Figure 9.10(b)). At the bottom surface of this window, there is an SiO$_2$ film which must be removed so that the SWCNT film to be brought here can make a contact with silicon. So, the oxide film in the window region is removed with HF treatment (Figure 9.10(c)).

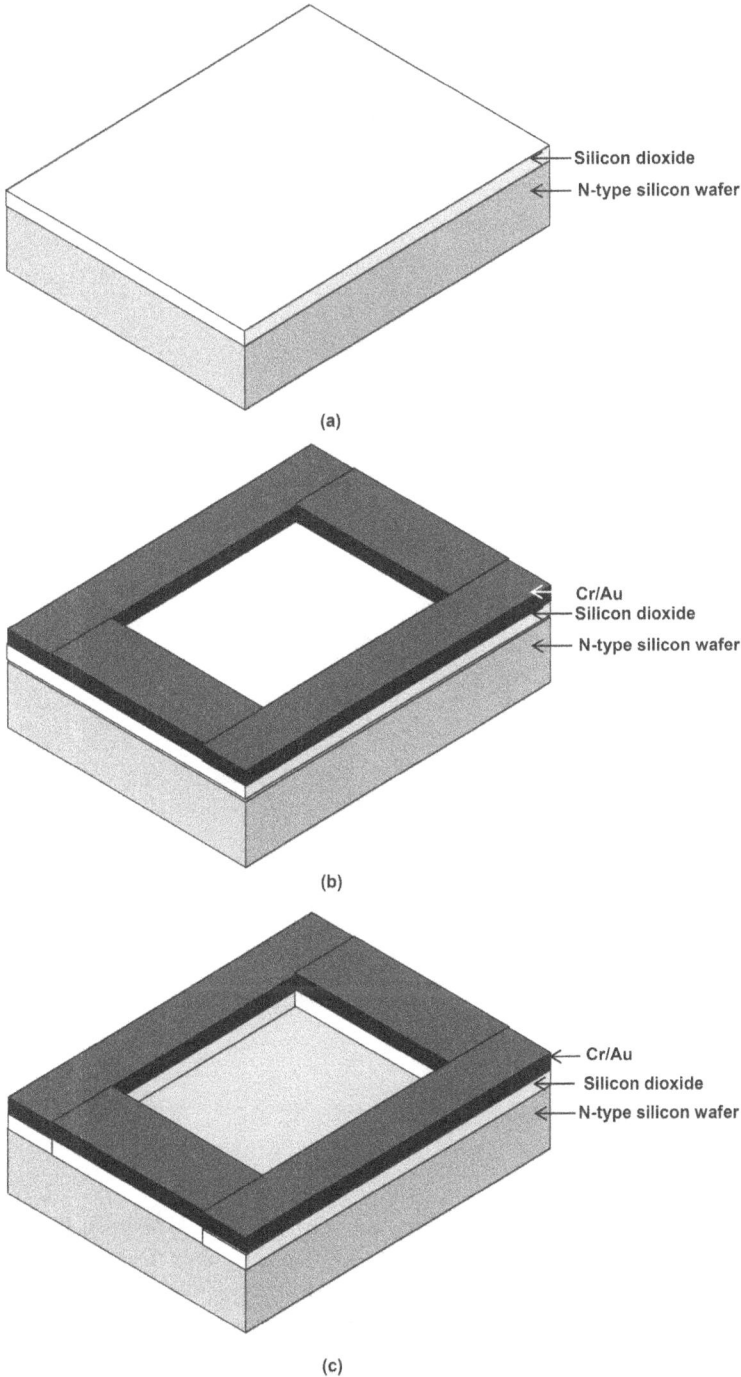

FIGURE 9.10 Carbon nanotube-silicon P-N junction solar cell: (a) N-type silicon wafer (dopant concentration 10^{15}–10^{16}cm^{-3}) with 500nm-thick silicon dioxide on the front surface, (b) Cr/Au deposition on the front surface and patterning using etch mask, (c) wet-etching of silicon dioxide to expose the silicon surface, (d) deposition of aluminum on the back surface for backside contact, and (e) transference of a freestanding P-type SWCNT film over the front silicon surface, treatment with HF, and subsequently HNO$_3$ for optimizing its photovoltaic properties, followed by treatment with AuCl$_3$ solution in nitromethane for enhanced P-doping of SWCNTs.

(d)

(e)

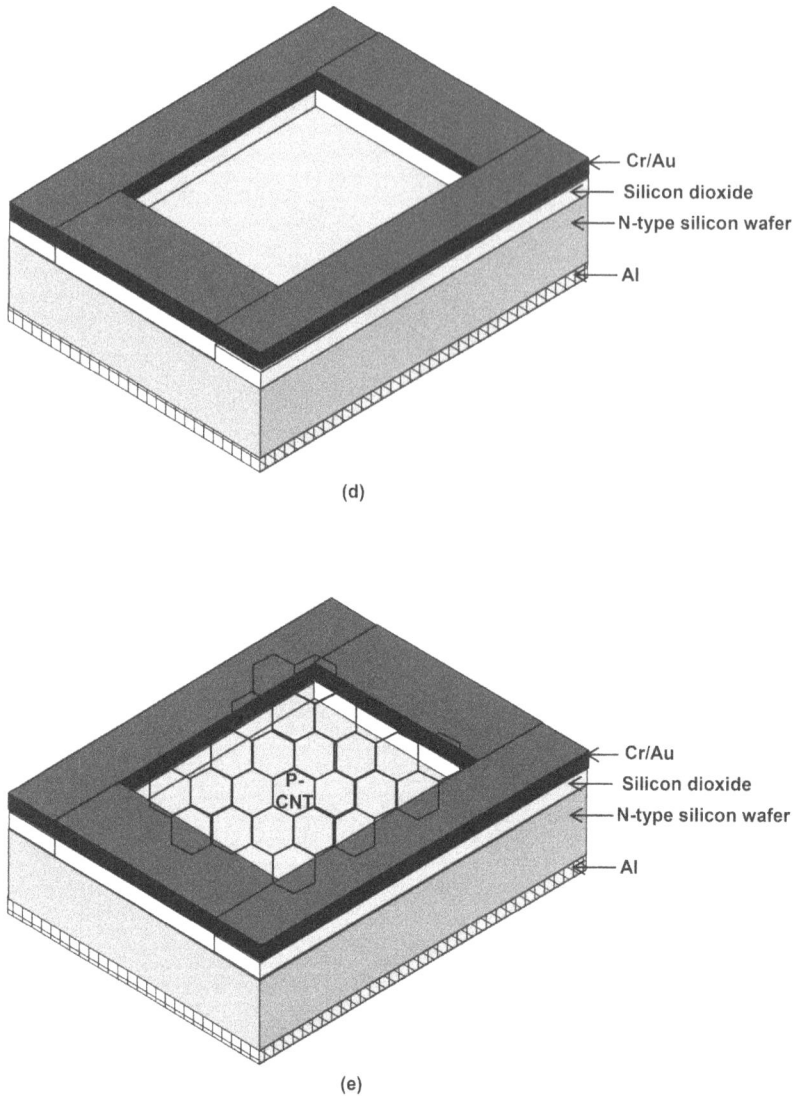

FIGURE 9.10   (Continued)

To form the back contact of the solar cell, aluminum film is deposited on the backside of the silicon wafer (Figure 9.10(d)). Following the procedure described in Section 9.5.1 for SWCNT film deposition, the SWCNT film is floated on water and moved to cover the window region created for it (Figure 9.10(e)). The solar cell fabrication is completed by three chemical treatments of the window region: first, HF treatment for any native oxide removal; secondly, HNO_3 treatment for P-doping; and thirdly, treatment with gold (III) chloride hydrate solution in nitromethane for improvement of P-doping.

A P-N junction is formed between P-type SWCNTs and N-type silicon wafer. One electrode of the solar cell is Cr/Au film on the top of CNT film on silicon dioxide. The other electrode is Al film on the backside of the silicon wafer. Sunlight is incident over the CNT film. The short-circuit current density, open-circuit voltage, fill factor, and efficiency of the SWCNTs/Si solar cell are 28.6mAcm^{-2}, 0.5301V, 74.1%, and 11.2%, respectively, under AM1.5G, 100mWcm^{-2} illumination (Jung et al 2013).

9.6 TiO$_2$-COATED CNTs-Si SOLAR CELL

So far, solar cells were considered with efficiency of 6.2% increasing to 13.8% after acid treatment, and with efficiency > 11% using superacid sliding coating of SWCNT. By spin-coating an additional TiO$_2$ colloidal antireflection layer over the CNT layer at 5,000–6,000 RPM for 1 minute, the reflection of light from the top surface of the solar cell decreases to < 10% (Shi et al 2012).

SWCNTs are synthesized by chemical vapor deposition using a source solution of ferrocene and sulfur dissolved in xylene. The solution injected by a syringe pump into the reactor is transported to the reaction zone by Ar/H$_2$ mixture. The CNTs grown at 1,160°C for ½ h. float downstream and are collected on substrates. TiO$_2$ colloid is synthesized by mixing Ti(OBu)$_4$ with ethanol, ethyl acetate to make a precursor solution, adding an (HNO$_3$ + DI water + ethanol) solution by droplets, and stirring for 12 h., followed by 12 h. aging. For front contact, Ag paste is applied to form a square window. As a backside contact, gallium-indium eutectic (E-GaIn) is used. Chemical doping of SWCNTs is done by exposure to vapors of HNO$_3$ and H$_2$O$_2$. At 100mW cm^{-2}, the short-circuit current density of this solar cell is 32mAcm^{-2}, the open-circuit voltage is 0.61V, the fill factor is 77%, and efficiency is 15.1% (Shi et al 2012).

9.7 USING GRAPHENE TO MAKE SEMITRANSPARENT PEROVSKITE SOLAR CELLS

9.7.1 SEMITRANSPARENT SOLAR CELLS AND SUITABILITY OF GRAPHENE FOR THESE CELLS

Semitransparent solar cells are mounted on windows and claddings of buildings and automobiles to supply power together with provision of aesthetically pleasing and decorative designs. Such solar cells have been made using electrodes of thin gold films or silver nanowires. While gold is costly, silver nanowires suffer from instability on perovskite due to formation of silver halide. Graphene electrodes offer manifold advantages owing to high transparency of graphene combined with its excellent conductivity and low cost. Stacked multilayer graphene is laminated on the surface of the spiro-OMeTAD layer.

9.7.2 PARTS OF SEMITRANSPARENT SOLAR CELL

This solar cell (Figure 9.11) consists of two parts (You et al 2015). Part I is made of the regular perovskite solar cell layers viz glass substrate, FTO coating, TiO$_2$ electron transport layer, CH$_3$NH$_3$PbI$_{3-x}$Cl$_x$ perovskite layer, and spiro-OMeTAD hole transport material. Part II has the structure PEDOT:PSS/graphene/PMMA/PDMS. Parts I and II are fabricated separately, then brought together and laminated to make the solar cell. PEDOT:PSS is used for enhancement of conductivity of graphene and to secure adhesion between parts I and II.

9.7.2.1 Making Part I

To make part I, FTO-coated glass substrate is spin-coated with a solution of titanium isopropoxide in ethanol acidified with HCl, followed by 1 h. annealing at 500°C. TiO$_2$ layer is formed. Methylammonium iodide and lead chloride are dissolved in N, N-dimethylformamide to make the precursor solution for the perovskite layer. This solution is spin-coated over the TiO$_2$ layer. Annealing is done in a glove box at 100°C for 45 min. Spiro-OMeTAD solution made in chlorobenzene with additives viz bis(trifluoro-methanesulfonyl)imide solution in acetonitrile and 4-tert-butylpyridine is deposited by spin-coating on the perovskite layer in a glove box and left overnight to dry.

9.7.2.2 Making Part II

To make part II, chemical vapor deposition process is applied for synthesizing graphene over a copper foil. 100nm-thick PMMA layer is deposited over graphene by spin-coating PMMA solution in

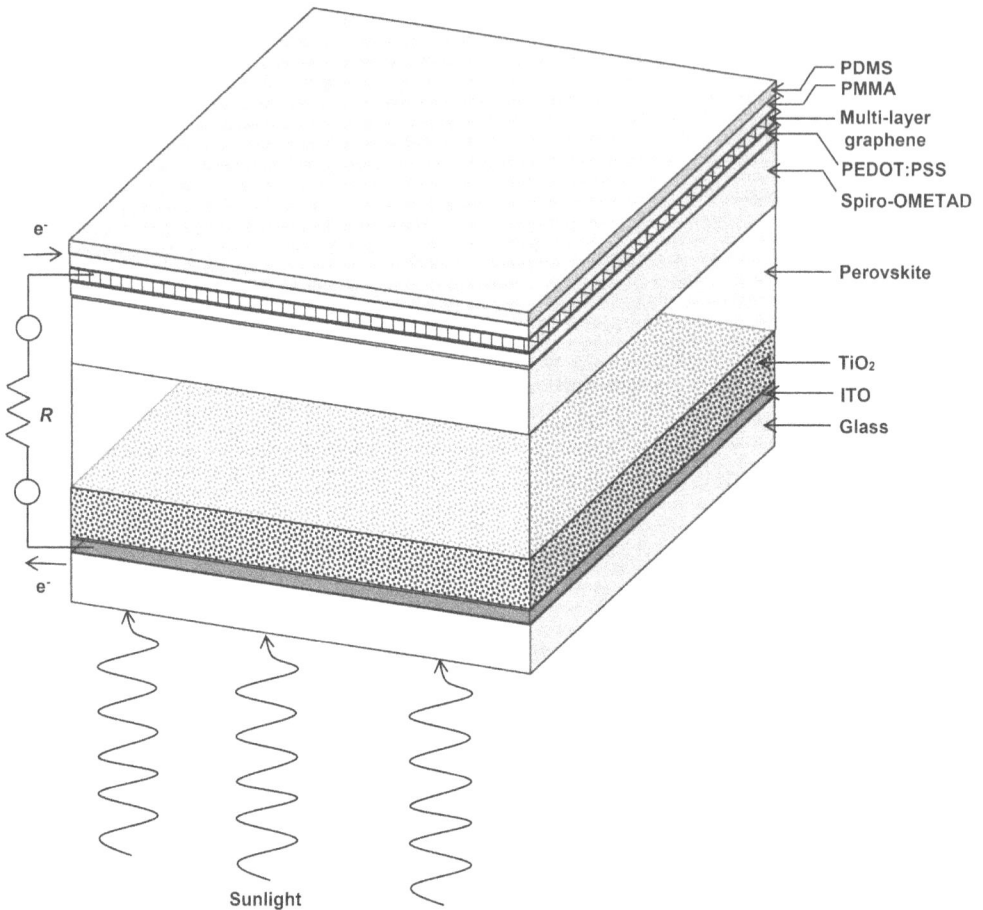

FIGURE 9.11 Perovskite solar cell with multistack graphene electrode showing the conventional layers: FTO-coated glass substrate/TiO$_2$/perovskite layer/spiro-OMeTAD layer, combined with the electrode layers PEDOT:PSS/multistack graphene/PMMA/PDMS. Electric current flows between the graphene and ITO electrodes when sunlight is incident on the FTO-coated glass substrate.

anisole with ½ h. curing at 90°C. 0.2mm-thick PDMS layer is laminated on PMMA. The Cu foil/ graphene/PMMA/PDMS structure is kept in aqueous iron chloride solution to etch copper. This etching takes several hours. We are left with a multilayer stack: graphene/PMMA/PDMS.

But graphene electrodes require multilayer graphene. So graphene/PMMA film is stacked with another graphene film on copper foil. The copper foil is etched away. The stacking process is repeated until the stack with desired number of layers has been obtained.

At the end of stacking process, PEDOT:PSS layer is deposited. The graphene surface is exposed to oxygen plasma. 1% Zonyl surfactant-doped PEDOT:PSS solution with D-sorbitol is spun over graphene and annealing is done at 120°C for 1 h. in a glove box. We get the structure PEDOT:PSS/ multilayer graphene/PMMA/PDMS.

9.7.2.3 Assembling Together Parts I and II

To assemble parts I and II together, the two parts are pressed together and heated at 65°C. By rolling a Teflon rod over the PDMS film, any air bubbles that have crept in at the interface between the two parts are removed.

9.7.2.4   Multilayer Graphene and Gold Electrode Solar Cells

When illuminated from the FTO side, the open-circuit voltage of the gold electrode solar cell is 0.985V, its short-circuit current density is 19.8mAcm^{-2}, fill factor is 73.58%, and average efficiency is 13.62%.

When illuminated from the FTO side, the open-circuit voltage of the two-layer graphene electrode solar cell is 0.960V, its short-circuit current density is 19.17mAcm^{-2}, fill factor is 67.22%, and average efficiency is 12.02%. For illumination from graphene side, these parameters are 0.945V, 17.75mAcm^{-2}, 71.72%, and 11.65% (You et al 2015).

9.8   GRAPHENE/N-TYPE Si SCHOTTKY DIODE SOLAR CELL

9.8.1   Doping Graphene with TFSA

Single-layer graphene/N-Si Schottky junction solar cell shows a steep increase in efficiency on chemical doping with bis(trifluoromethanesulfonyl)amide (CF$_3$SO$_2$)$_2$NH) (TFSA) (Miao et al 2012). TFSA has an acceptor nature and dopes graphene P-type. Concentration of holes in graphene increases, leading to a conductivity value equivalent to that of indium tin oxide (Tongay et al 2011). Transmittance of graphene in the visible and near-infrared regions increases. Additionally, TFSA makes the solar cell stable against environmental degradation owing to its hydrophobicity. Thus, TFSA doping impacts the conductivity and optical transparency of graphene film. It improves both these parameters and at the same time guarantees stability of operation.

The dramatic increase in solar cell efficiency with TFSA doping is ascribed to an increase in Schottky barrier height and therefore built-in potential of the diode along with a decrease in resistance of graphene layer, accounting for fall in resistive losses. The solar cell is shown in Figure 9.12.

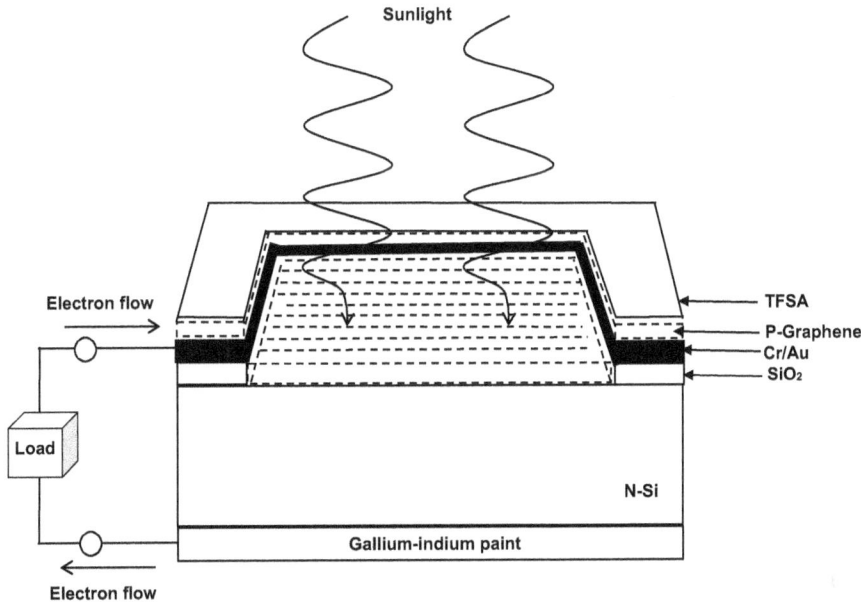

FIGURE 9.12   Graphene monolayer/N-type silicon Schottky junction solar cell showing the gallium-indium paint, N-Si wafer, SiO$_2$ layer, graphene, TFSA, and Cr/Au film. A Schottky junction is formed between single-layer graphene and N-type silicon wafer. One electrode of the solar cell is Cr/Au film on silicon dioxide with the graphene film overlapping the Au/Cr film. The other electrode is eutectic gallium-indium paint on the backside of the silicon wafer. Sunlight strikes the graphene layer over the upper surface of the cell. Current is drawn between gallium-indium paint and Cr/Au electrode.

9.8.2 Fabrication of Graphene/N-Si Solar Cell

The solar cell is fabricated on an N-type silicon wafer of orientation < 111 >, with doping concentration 8×10^{14} to $10^{15} cm^{-3}$ and 1-micron-thick pregrown silicon dioxide (Figure 9.13(a)). Over this silicon dioxide layer, Cr/Au films are deposited and patterned to open a window, where graphene will be transferred (Figure 9.13(b)). Buffered HF oxide etchant is used to remove silicon dioxide covering the window portion to expose silicon underneath (Figure 9.13(c)). A 2 h. gap is essential between oxide etching and laying down graphene in this window region. During this period, the

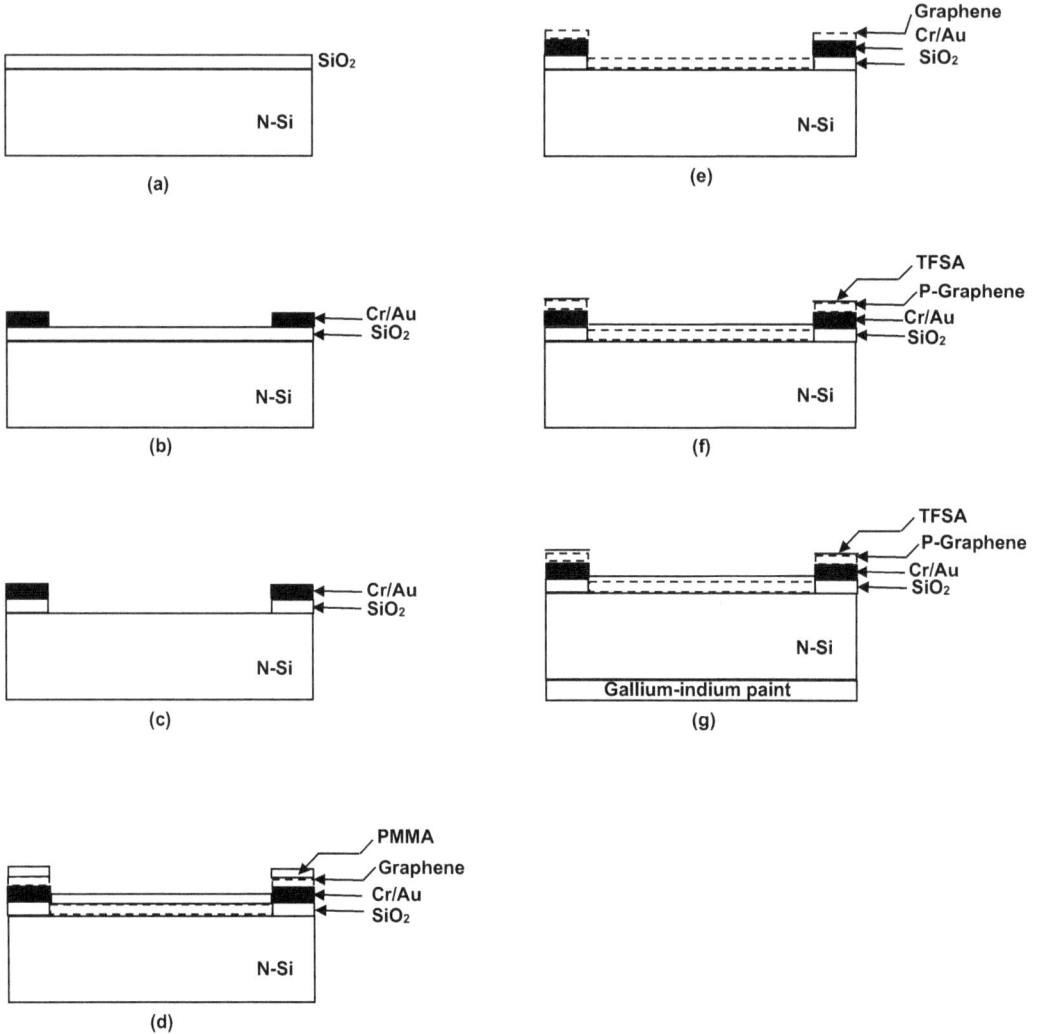

FIGURE 9.13 Fabrication of graphene monolayer/N-type silicon Schottky junction solar cell: (a) N-type silicon wafer (dopant concentration 8×10^{14} to $10^{15} cm^{-3}$) with 1μm-thick silicon dioxide on the front surface, (b) Au/Cr deposition on the front surface and patterning to form Au-Cr window, (c) exposure of silicon surface in the Cr/Au window by etching silicon dioxide in buffered HF, (d) transference of PMMA/graphene sheet onto silicon surface in the Cr/Au window, (e) dissolution of PMMA in acetone vapor followed by soaking in acetone solvent bath for 12 h., (f) spin-casting 20mM TFSA [bis(trifluoromethanesulfonyl) amide{$(CF_3SO_2)_2NH$}] in nitromethane over graphene for its doping, and (g) coating eutectic gallium-indium paint on the back surface of silicon wafer for back-side ohmic contact.

dangling bonds on the silicon surface are tied by passivation from atmospheric oxygen. Similar gap is required in CNT/Si devices.

Graphene is synthesized by chemical vapor deposition on copper foil to get graphene/Cu/graphene. PMMA is spin-cast over graphene, obtaining PMMA/graphene/Cu/graphene. The graphene on the reverse side of copper is removed by reactive ion etching. This gives PMMA/graphene/Cu. The copper below graphene film is removed by etching in $Fe(NO_3)_3$ etchant, leaving PMMA/graphene. The PMMA/graphene bilayer is moved over the window, which was made for fixing graphene, and graphene is shifted over the window region surrounded by Cr/Au metal boundary (Figure 9.13(d)).

PMMA supporting layer for graphene is removed by exposure to acetone vapor. Following this exposure, 12 h. soaking in acetone is necessary (Figure 9.13(e)). Graphene is doped by spin-casting a solution of bis(trifluoromethanesulfonyl)amide[$((CF_3SO_2)_2NH)$] (TFSA) in nitromethane over it (Figure 9.13(f)). Gallium-indium eutectic paint is applied on the undersurface of silicon wafer for backside ohmic contact of solar cell (Figure 9.13(g)).

9.8.3 Parameters of Solar Cell with and without Doping with TFSA

The short-circuit current density, open-circuit voltage, fill factor, and efficiency of the single layer graphene/Si solar cell with TFSA doping are $25.3 mAcm^{-2}$, $0.54V$, 0.63, and 8.6%, respectively, under AM1.5G illumination at an intensity of $100 mWcm^{-2}$. For the solar cell without TFSA doping, the same parameters are much lower, having the values $14.2 mAcm^{-2}$, $0.43V$, 0.32, and 1.9% (Miao et al 2012).

9.9 DISCUSSION AND CONCLUSIONS

CNTs, C_{60}, and graphene nanomaterials have been widely used in solar cell technology (Table 9.1). A few observations are:

TABLE 9.1
Effects of Carbonaceous Nanomaterials in Solar Cells

Sl. No.	Type of Solar Cell	Efficiency	Reference
1.	Dye-sensitized solar cell with CNTs counter electrode	4.03% (reference cell with Pt counter electrode: 4.37%)	Li et al (2021)
2.	Dye-sensitized solar cell with Pt NPs/CNTs nanohybrid counter electrode	6.28% (reference cell with Pt counter electrode: 4.37%)	Li et al (2021)
3.	Perovskite solar cell with TiO_2 NPs-SWCNTs nanocomposite ETL	19.35% (reference cell with TiO_2 only: 17.04)	Bati et al (2019)
4.	Perovskite solar cell using CNTs and C_{60} (CNTs mixed with HTL and C_{60} as ETL)	(CNTs + Spriro-OMeTAD 17%), CNTs only 13.2%	Ahn et al (2018)
5.	Si-CNTs hybrid solar cell	6.2% for pristine cell which increases to 13.8% after acid infiltration	Jia et al (2011)
6.	Si-CNTs hybrid cell by superacid sliding coating	11.2%	Jung et al (2013)
7.	TiO_2-coated CNTs-Si solar cell	15.1%	Shi et al (2012)
8.	Semitransparent perovskite solar cell using graphene	12.02%/11.65%	You et al (2015)
9.	Graphene/N-type Si Schottky diode solar cell	1.9% increasing to 8.6% after TFSA doping	Miao et al (2012)

(i) As revealed from a study on dye-sensitized solar cells, low-cost CNTs electrode can replace standard platinum electrodes with some sacrifice of performance, but Pt NPs/CNTs electrodes outperform the Pt electrodes.

(ii) SWCNTs inclusion in the TiO_2 NPs electron transport layer of a perovskite solar cell is beneficial for solar cell operation. A NPs-SWCNTs nanocomposite formed with an optimum ratio of semiconducting and metallic SWCNTs and incorporating a defined fraction of SWCNTs in the mixture brings about a significant improvement in performance of the solar cell with respect to a cell containing TiO_2 NPs only.

(iii) When CNTs electrode replaces gold electrode and C_{60} fullerene layer substitutes TiO_2 electron transport layer in a perovskite solar cell, the cell becomes cheaper and also exhibits better long-term stability.

(iv) CNTs/N-Si heterojunction solar cell is a panacea for many drawbacks of silicon technology, notably, the elimination of high-temperature processing. Deposition of an antireflection layer of TiO_2 colloid is an advantageous process step.

(v) Graphene electrodes help in making economical semitransparent solar cells.

(vi) Graphene/N-Si Schottky diode cells are easy to make at low cost, like the CNTs/N-Si devices.

Questions and Answers

9.1 What is the electron mobility in carbon nanotubes? Answer: Intrinsic mobility $> 10^5$ cm^2/V.s and field-effect mobility 79000 cm^2/V.s (Du1rkop et al 2004).

9.2 What is the electron mobility in graphene? Answer: 1.25×10^5 to 2.75×10^5 cm^2/V-s (Roch et al 2015).

9.3 What are electrical conductivities of CNTs and graphene? Compare with electrical conductivities of platinum, copper, and silver. Answer: CNTs, 10^6–10^7 Sm^{-1}; graphene, 10^8 Sm^{-1} (Wang and Weng 2018); platinum, 9.44×10^6 Sm^{-1}; copper, 5.98×10^7 Sm^{-1}; and silver, 6.3×10^7 Sm^{-1}.

9.4 Why is it necessary to make a counter electrode using CNTs? Answer: Because the generally used platinum counter electrode must be substituted from economical and platinum rarity considerations, and the high electrical conductivity CNTs offer a cheaper viable solution.

9.5 Which of the three DSSCs shows best efficiency: (a) CNTs film-coated glass plate, (b) Pt NPs/CNTs nanohybrid–coated glass electrode, and (c) Pt coated–glass electrode. Answer: (b). η = 4.03% for (a), η = 6.28% for (b), and η = 4.37% for (c).

9.6 Which of the two perovskite solar cells is more efficient: (a) solar cell made with bare TiO_2 film as ETL, and (b) solar cell using TiO_2 NPs-SWCNTs nanocomposite? Answer: (b) Because η = 19.5% for (b), whereas η = 17.04% for (a).

9.7 Why ETL made of TiO_2 NPs-SWCNTs nanocomposite is superior to bare TiO_2 ETL? Answer: Because bare TiO_2 NPs have a lower conductivity due to their heterogeneous network. The CNTs and TiO_2 NPs, having a large surface area, mix together thoroughly, closing any gaps in conduction pathways of TiO_2 to raise conductivity. The increased ETL conductivity results in better solar cell efficiency as the carriers are fast-transported and suffer less recombination.

9.8 Besides achieving high-power conversion efficiency with low cost, mention one prime metric that determines the translation of a solar cell technology from laboratory to market. Answer: High operational stability and long lifetime.

9.9 Do perovskite solar cells provide long-term stable operation? Answer: No, e.g., lifetime ~ 1 year is reported (Grancini et al 2017). Perovskite degrades with exposure to water, moisture, or UV and under thermal stresses. HTL shows instability on contact with water. Spiro-MeOTAD is highly hygroscopic and tends to crystallize. It is vulnerable to humidity and heat effects.

9.10 How can HTLs of perovskites be made more stable with the help of nanomaterials? Answer: Solar cells are made by replacing the HTL with (CNTs + HTL), e.g., (CNTs + spiro-MeOTAD), (CNTs + PTAA), (CNTs + P3HT), and with CNTs only achieving different efficiency values. Nanocomposites of these HTLs with CNTs have better stability.

9.11 What are the intents of using CNTs in hybrid solar cell? Answer:
(i) CNTs are used as an electrode material.
(ii) CNTs are used as one of the two semiconductors comprising the heterojunction, e.g., P-type CNTs and N-type silicon form a P-N heterojunction diode.

9.12 What are the advantages of using CNTs with silicon to make a hybrid solar cell? Answer:
(i) CNTs transparent electrode avoids the unavoidable loss of light that takes place under contact regions of a metallic grid electrode.
(ii) Using CNTs as one semiconductor material of the two materials in the P-CNTs/N-Si heterojunction allows making a junction with a low-temperature process, unlike conventional silicon processing techniques in which junctions are made by high-temperature impurity diffusion. By this process simplification, solar cells are made cheaper.

9.13 How is the close contact between CNTs and silicon surface established? Answer: By drenching the CNTs in ethanol so that they settle down on silicon surface and, subsequently, drying ethanol.

9.14 How are CNTs doped P-type? Answer: By chemical treatment, e.g., immersion in HNO_3 solution. The conductivity of CNTs film increases due to hole doping of CNT network accompanied by downward Fermi level shifting (Zhou et al 2005).

9.15 Why is HF treatment of silicon necessary before CNTs doping? Answer: For removal of native oxide on silicon surface. If this treatment is not given, a thin oxide layer will separate the CNTs from silicon surface, impairing the quality and reliability of CNTs-silicon junction.

9.16 Mention some advantages of superacid slide casting method for CNTs film preparation. Answer: It produces a high-conductivity, high-transparency, smooth, mechanically strong CNT film by a simple process.

9.17 What is a superacid? Give an example. Answer: Solution of a strong acid in a highly acidic, nonaqueous solvent. It has a strong property of adding proton to a molecule, e.g., fluorosulfuric acid (HsO_3F), which is 1,000 times stronger than H_2SO_4.

9.18 How does coating of the CNTs surface of CNTs-Si solar cell with TiO_2 colloidal layer improve cell performance? Answer: TiO_2 colloidal layer acts as an antireflection layer, preventing loss of light.

9.19 How is multilayer graphene made for graphene electrode fabrication? Answer: Graphene is synthesized on a copper foil. We get Cu-foil/graphene. Over graphene, PMMA is spin-coated, getting Cu-foil/graphene/PMMA. Cu-foil is etched away. Graphene/PMMA structure is left over. Stacking of graphene/PMMA is done with a new Cu-foil/graphene to get Cu-foil/graphene (two layers)/PMMA. Cu-foil is removed by etching. We get graphene (two layers)/PMMA, which is stacked with a new Cu-foil/graphene, obtaining C-foil/Graphene (three layers)/PMMA. Again, Cu-foil is etched away, and the process is repeated to get Graphene (*n* layers)/PMMA, where *n* is the required number of layers in the stack.

9.20 How does a solar cell with multilayer graphene electrode compare with a gold electrode solar cell? Answer: The solar cell with multilayer graphene electrode is slightly inferior to the gold electrode solar cell in efficiency, but it is better from cost standpoint.

9.21 State the advantages of doping graphene with TFSA. Answer: Firstly, TFSA doping increases conductivity of graphene through increase in hole concentration. Secondly, it makes optical transparency of graphene better. Thirdly, it makes graphene hydrophobic and therefore less vulnerable to atmospheric influence.

REFERENCES

Ahn N., I. Jeon, J. Yoon, E. I. Kauppinen, Y. Matsuo, S. Maruyama and M. Choi 2018 Carbon-sandwiched perovskite solar cell, Journal of Materials Chemistry A, 6: 1382–1389.

Bati A. S. R., L. P. Yu, S. A. Tawfik, M. J. S. Spencer, P. E. Shaw, M. Batmunkh and J. G. Shapter 2019 Electrically sorted single-walled carbon nanotubes-based electron transporting layers for perovskite solar cells, iScience, 14: 100–112.

Du1rkop T., S. A. Getty, E. Cobas and M. S. Fuhrer 2004 Extraordinary mobility in semiconducting carbon nanotubes, Nano Letters, 4(1): 35–39.

Grancini G., C. Roldán-Carmona, I. Zimmermann, E. Mosconi, X. Lee, D. Martineau, S. Narbey, F. Oswald, F. De Angelis, M. Graetzel and M. K. Nazeeruddin 2017 One-year stable perovskite solar cells by 2D/3D interface engineering, Nature Communications, 8: 15684.

Jia Y., A. Cao, X. Bai, Z. Li, L. Zhang, N. Guo, J. Wei, K. Wang, H. Zhu, D. Wu and P. M. Ajayan 2011 Achieving high efficiency silicon-carbon nanotube heterojunction solar cells by acid doping, Nano Letters, 11: 1901–1905.

Jung Y., X. Li, N. K. Rajan, A. D. Taylor and M. A. Reed 2013 Record high efficiency single-walled carbon nanotube/silicon p–n junction solar cells, Nano Letters, 13: 95–99.

Li X., Y. Jung, K. Sakimoto, T.-H. Goh, M. A. Reed and A. D. Taylor 2013 Improved efficiency of smooth and aligned single walled carbon nanotube/silicon hybrid solar cells, Energy & Environmental Science, 6: 879–887.

Li J.-W., Y.-S. Chen, Y.-F. Chen, J.-X. Chen, C.-F. J. Kuo, L.-Y. Chen and C.-W. Chiu 2021 Enhanced efficiency of dye-sensitized solar cells based on polymer-assisted dispersion of platinum nanoparticles/carbon nanotubes nanohybrid films as FTO-free counter electrodes, Polymers, 13(3103): 1–17.

Miao X., S. Tongay, M. K. Petterson, K. Berke, A. G. Rinzler, B. R. Appleton and A. F. Hebard 2012 High efficiency graphene solar cells by chemical doping, Nano Letters, 12(6): 2745–2750.

Roch A., M. Greifzu, E. R. Talens, L. Stepien, T. Roch, J. Hege, N. V. Nong, T. Schmiel, I. Dani, C. Leyens, O. Jost and A. Leson 2015 Ambient effects on the electrical conductivity of carbon nanotubes, Carbon, 95: 347–353.

Shi E., L. Zhang, Z. Li, P. Li, Y. Shang, Y. Jia, J. Wei, K. Wang, H. Zhu, D. Wu, S. Zhang and A. Cao 2012 TiO$_2$-coated carbon nanotube-silicon solar cells with efficiency of 15%, Scientific Reports, 2(884): 1–5.

Tanaka T., Y. Urabe, D. Nishide and H. Kataura 2009 Continuous separation of metallic and semiconducting carbon nanotubes using agarose gel, Applied Physics Express, 2: 125002–1 to 125002–3.

Tongay S., K. Berke, M. Lemaitre, Z. Nasrollahi, D. B. Tanner, A. F. Hebard and B. R. Appleton 2011 Stable hole doping of graphene for low electrical resistance and high optical transparency, Nanotechnology, 22(42): 425701.

Wang Y. and G. J. Weng 2018 Electrical conductivity of carbon nanotube and graphene-based nanocomposites, in: Meguid S. A. and G. J. Weng (eds.), Micromechanics and Nanomechanics of Composite Solids, Springer International Publishing AG, Cham, Switzerland, pp. 123–156.

You P., Z. Liu, Q. Tai, S. Liu and F. Yan 2015 Efficient semitransparent perovskite solar cells with graphene electrodes, Advanced Materials, 27: 3632–3638.

Zhou W., J. Vavro, N. M. Nemes, J. E. Fischer, F. Borondics, K. Kamaras and D. B. Tanner 2005 Charge transfer and Fermi level shift in p-doped single-walled carbon nanotubes, Physical Review B: Condensed Matter, 71: 205423–1 to 205423–7.

Part V

Quantum Well, Nanowire, and
Quantum Dot Photovoltaics

10 Quantum Well Solar Cells
Particle-in-a-Box Model and Bandgap Engineering

10.1 WHAT IS A QUANTUM WELL SOLAR CELL?

A quantum well solar cell is a solar cell in which the active region is made of a quantum well (QW) structure (Barnham et al 2002). The active region of a P-I-N or N-I-P solar cell is the intrinsic region. So in the quantum well solar cell, the intrinsic region consists of a QW region (Figure 10.1). It is this QW region which distinguishes a quantum well solar cell from the bulk solar cell. The QW region is usually a multiple quantum well (MQW) arrangement. The replacement offers the advantage of enhancing the efficiency of the solar cell, as was shown in the pioneering work of Barnham and Duggan (1990).

10.1.1 QW SOLAR CELL AS A WAY OF EXTENDING THE USEFUL RANGE OF SOLAR SPECTRUM UTILIZED FOR ENERGY CONVERSION

The QW solar cell is an attempt to obviate to the extent possible an arduous and challenging limitation of the conventional single-material solar cell. This limitation arises from the necessity of the energy of incident radiation to be greater than the bandgap of the material. The solar spectrum is very broad in extent. It stretches over a range of frequencies, and hence energies. So the single-bandgap cell fails to harness the radiations of various energies. It responds only to photons of energies exceeding the bandgap of the semiconductor from which the cell is fabricated. Photons of lower energies are wasted.

10.1.2 QW SOLAR CELL AS AN APPROACH TOWARDS REALIZING MULTIJUNCTION SOLAR CELLS WITH OPTIMAL BANDGAPS

Multijunction solar cells are made from semiconductor materials of different bandgaps to cater to the wavelength distribution of the solar spectrum. The QW solar cell concept provides a viable means of reaching this goal.

10.2 THE QW STRUCTURE

The QW structure (Figure 10.2(a)) is an example of thin-layered semiconductor heterostructures of high crystalline quality and low dislocation content having thicknesses in the range of few tens of atomic layers (Harrison 2009). It is recognized as one of the greatest inventions in the realm of compound semiconductors realized through accuracy and precision achieved by technological advancements in epitaxial crystal growth (Razeghi 2010). It offers many opportunities of observation of quantum mechanical phenomena and controlling these phenomena to build devices.

DOI: 10.1201/9781003215158-15

P-Al$_x$Ga$_{1-x}$As   Intrinsic-Al$_x$Ga$_{1-x}$As   N-Al$_x$Ga$_{1-x}$As

(a)

GaAs well

P-Al$_x$Ga$_{1-x}$As   Al$_x$Ga$_{1-x}$As barrier   N-Al$_x$Ga$_{1-x}$As

(b)

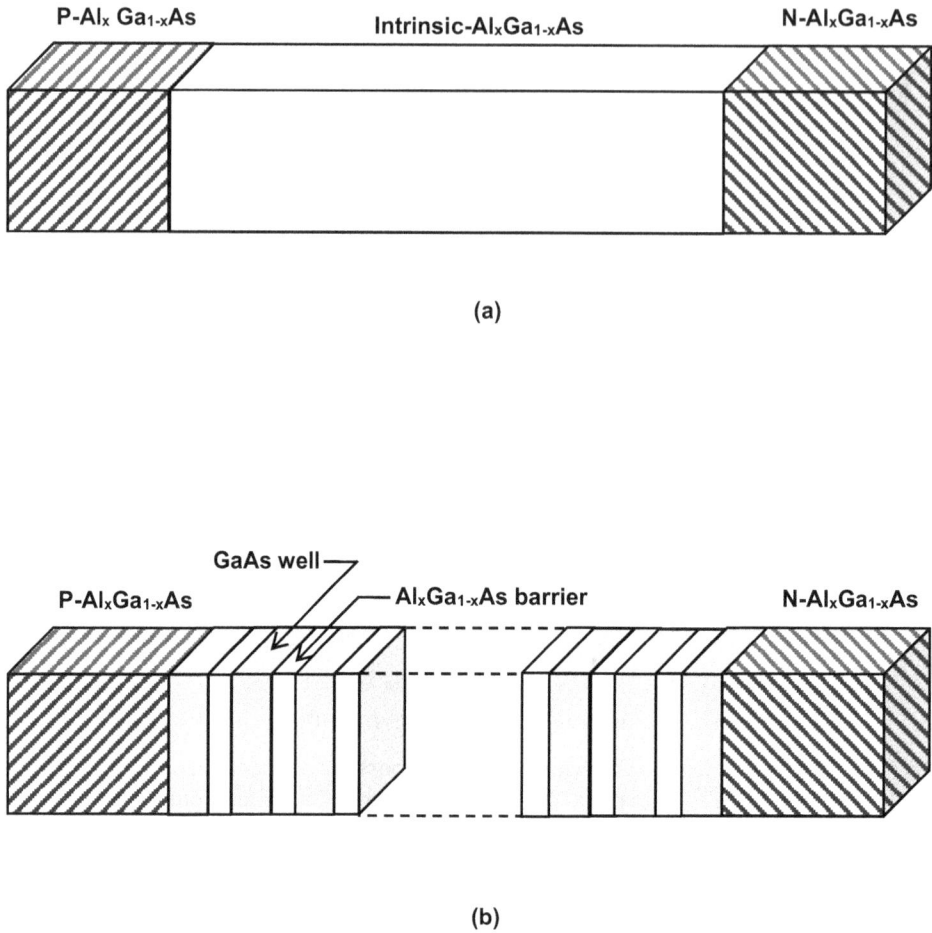

FIGURE 10.1 P-I-N junction diode: (a) without quantum wells and (b) the same device with intrinsic region replaced by multiple quantum wells. The diode in *a* contains three layers: a P-Al$_x$Ga$_{1-x}$As layer, an intrinsic Al$_x$Ga$_{1-x}$As layer, and an N-Al$_x$Ga$_{1-x}$As layer. The diode in *b* contains the same layers at the two ends, but the intrinsic region is subdivided into a series of stacks of GaAs quantum well layer/Al$_x$Ga$_{1-x}$As barrier layer, which together constitute the multiple-quantum well structure.

The MQW (Figure 10.2(b)) comprises a periodic stack of alternating layers of small-bandgap and large-bandgap materials of ultrasmall thickness. The layer of small-bandgap material serves as the quantum well, while the layer of large-bandgap material acts as the barrier layer. Several sandwiches of low-bandgap quantum well material with high-bandgap barriers are repeated to obtain an MQW. The thickness of the quantum well layer is comparable to or less than the de Broglie wavelength of electrons. The charge carriers, electrons and holes, are restricted in their motion along the direction perpendicular to the surface of the layer. However, they have complete freedom of movement in the remaining two directions as in a bulk material. The thickness of the quantum well layer is comparable to or less than the de Broglie wavelength of electrons. The de Broglie wavelength of electrons in semiconductors is typically a few tens of nanometers.

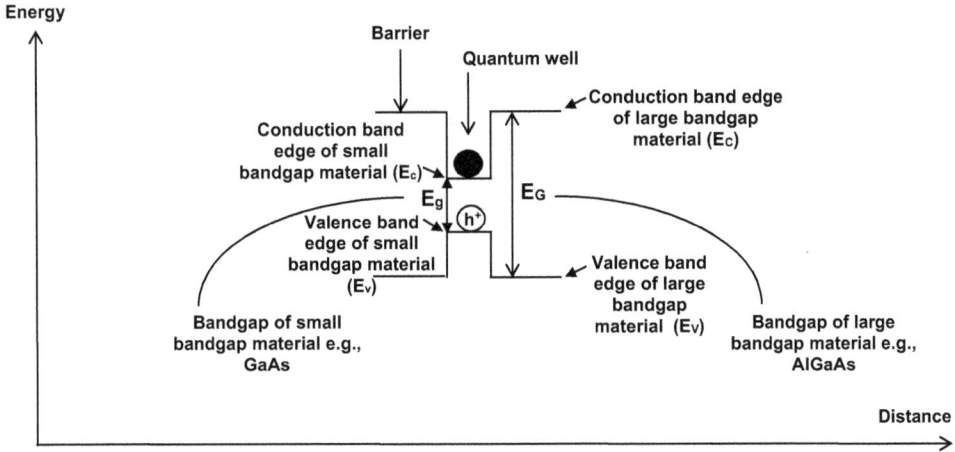

(a)

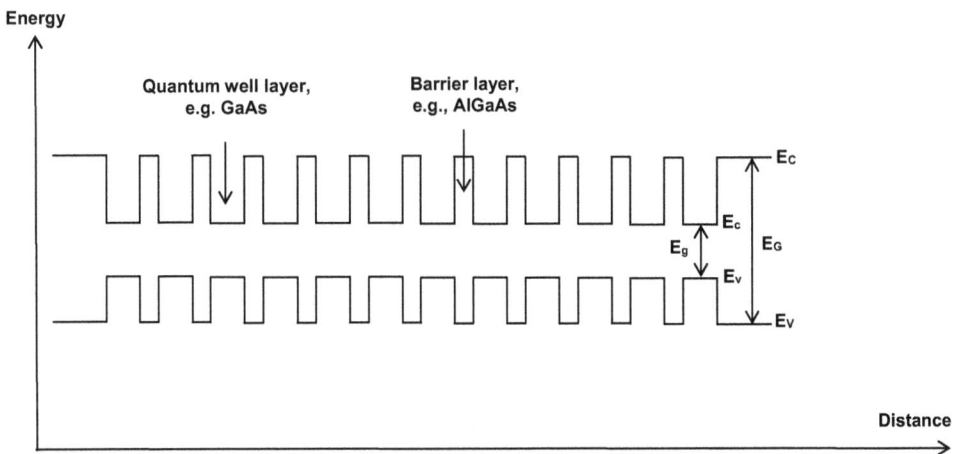

(b)

FIGURE 10.2 Energy band diagrams of (a) single–quantum well and (b) multiple–quantum well structures. The bandgaps E_g, E_G of the two semiconductors GaAs and AlGaAs are shown. The conduction and valence band edges are marked. In a, there is one quantum well bounded with barrier layers on the two sides. An electron and a hole trapped in the well are shown. In b, there are several quantum wells each enclosed between barrier layers on the two sides.

Example 10.1

(a) Even when no external electric field has been applied, the electrons incessantly execute random thermal motion in a crystal lattice. In GaAs, this velocity is $4.4 \times 10^7 \mathrm{cm\ s^{-1}}$. Determine the de Broglie wavelength $\lambda_{\text{de Broglie}}$ of an electron moving with this thermal velocity. (b) In an electric field, a drift velocity is superimposed on the randomly oriented thermal velocity of electrons. The drift velocity increases with electric field. At high electric fields, the drift velocity attains a peak or saturation value whose value is 2.1 × 10^{7}cm s^{-1} in GaAs. Ascertain $\lambda_{\text{de Broglie}}$ associated with the electron moving with saturation velocity. (c) What is the thermal de Broglie wavelength $\lambda_{\text{Thermal de Broglie}}$ of an electron? Find its value for GaAs at room temperature $T = 300$K. (d) Calculate $\lambda_{\text{de Broglie}}$ of an electron accelerated from rest by a potential difference of 5V in vacuum.

Solution: (a) The de Broglie wavelength of a particle of effective mass m^* moving with a velocity v is given by

$$\lambda_{\text{de Broglie}} = \frac{h}{m^*v} = \frac{6.62607 \times 10^{-34}}{0.067 \times 9.109 \times 10^{-31} \times 4.4 \times 10^7 \times 10^{-2}} = \frac{6.62607 \times 10^{-34+31-7+2}}{0.067 \times 9.109 \times 4.4} \quad (10.1)$$

$$= 2.4675 \times 10^{-8}\text{m} = 24.675 \times 10^{-9}\text{m} = 24.7\text{nm}$$

(b) Applying equation 10.1,

$$\lambda_{\text{de Broglie}} = \frac{6.62607 \times 10^{-34}}{0.067 \times 9.109 \times 10^{-31} \times 2.1 \times 10^7 \times 10^{-2}} = \frac{6.62607 \times 10^{-34+31-7+2}}{0.067 \times 9.109 \times 2.1} = \frac{6.62607 \times 10^{-8}}{1.2816} \quad (10.2)$$

$$= 5.17 \times 10^{-8}\text{m} = 51.7 \times 10^{-9}\text{m} = 51.7\text{nm}$$

(c) *Thermal de Broglie wavelength* is defined as the mean wavelength of electrons at a temperature T calculated by treating them as particles in an ideal gas. It is given by (Sacchetti 2010)

$$\lambda_{\text{Thermal de Broglie}} = \frac{h}{\sqrt{2\pi m^* k_B T}} \quad (10.3)$$

where h is Planck's constant, m^* is the effective mass of electron

$$m^* = 0.067 \times \text{Rest mass of electron} = 0.067 \times m_0 = 0.067 \times 9.109 \times 10^{-31} = 0.6103 \times 10^{-31}\text{kg} \quad (10.4)$$

and k_B is Boltzmann constant. Putting the values in equation 10.3,

$$\lambda_{\text{Thermal de Broglie}} = \frac{6.62607 \times 10^{-34}}{\sqrt{2 \times 3.14159 \times 0.6103 \times 10^{-31} \times 1.3806485 \times 10^{-23} \times 300}}$$

$$= \frac{6.62607 \times 10^{-34}}{\sqrt{2 \times 3.14159 \times 0.6103 \times 1.3806485 \times 300 \times 10^{-31-23}}} = \frac{6.62607 \times 10^{-34}}{\sqrt{1588.2807 \times 10^{-54}}} \quad (10.5)$$

$$= \frac{6.62607 \times 10^{-34}}{39.853 \times 10^{-27}} = 0.1663 \times 10^{-34+27} = 0.1663 \times 10^{-7} = 16.63 \times 10^{-9}\text{m}$$

$$= 16.63\text{nm}$$

(d) The equation for energy of the electron is written as

$$\text{Kinetic energy} = \frac{p^2}{2m} = \text{Potential energy} = eV \quad (10.6)$$

where p stands for the momentum of the electron, e is the electronic charge, and V is the accelerating potential. Equation 10.6 can be rewritten as

$$p^2 = 2meV \quad (10.7)$$

or

$$p = \sqrt{2meV}$$ (10.8)

Using equation 10.8, the de Broglie wavelength of electron is

$$\lambda = \frac{h}{p} = \frac{h}{\sqrt{2meV}}$$ (10.9)

Putting the given values in equation 10.9,

$$\lambda = \frac{h}{\sqrt{2meV}} = \frac{6.62607 \times 10^{-34}}{\sqrt{2 \times 9.109 \times 10^{-31} \times 1.602 \times 10^{-19} \times 5}} = \frac{6.62607 \times 10^{-34}}{\sqrt{2 \times 9.109 \times 1.602 \times 5 \times 10^{-31-19}}}$$

$$= \frac{6.62607 \times 10^{-34}}{\sqrt{2 \times 9.109 \times 1.602 \times 5 \times 10^{-50}}} = \frac{6.62607 \times 10^{-34}}{\sqrt{145.92618 \times 10^{-50}}} = \frac{6.62607 \times 10^{-34}}{12.08 \times 10^{-25}}$$

$$= 0.5485 \times 10^{-34+25} = 0.55 \times 10^{-9} \, \text{m} = 0.55 \, \text{nm}$$ (10.10)

10.3   PHYSICS OF QUANTUM WELLS

10.3.1   Particle-in-a-Box Model of the Quantum Well

Let us focus our attention on an electron in the conduction band of a large-bandgap semi-conductor. As it wanders and crosses the boundary of the semiconductor with another semi-conductor of lower bandgap, it sees lower energy. It cannot come out from this lower-energy region and move to a higher-energy, large-bandgap region as it does not have the required energy. It is said to be trapped in a potential well in analogy to someone fallen in an actual well (Zettili 2009). The electron cannot escape from the well until it receives sufficient energy to overcome the barrier. To analyze the behavior of the electron in the well, a useful model is to consider the electron to be enclosed in an infinite potential well (Figure 10.3). Obviously, the difference between the bandgaps of the semiconductors is too small to be treated as infi-nite. So the model is a gross oversimplification of the real situation. Nevertheless, it provides insights into and helps in making useful predictions about the behavior of the electron in the well (Messiah 2014).

Since the electron is confined in a region between $x = 0$ and $x = t_w$, the thickness of the quantum well, the interaction potential $V(x)$ is

$$V(x) = 0 \text{ for } 0 \leq x \leq t_w$$ (10.11)

$$V(x) = \infty \text{ otherwise}$$ (10.12)

The time-independent Schrodinger equation is

$$-\frac{\hbar^2}{2m} \frac{d^2 \psi(x)}{dx^2} + V(x) \psi(x) = E \psi(x)$$ (10.13)

where $\hbar$ is the reduced Planck's constant, m is the effective mass of the electron, $\psi(x)$ is its wave function, and E its energy. In the region outside the box, $V(x) = \infty$, making $V(x) \psi(x) = \infty$. But $V(x)$

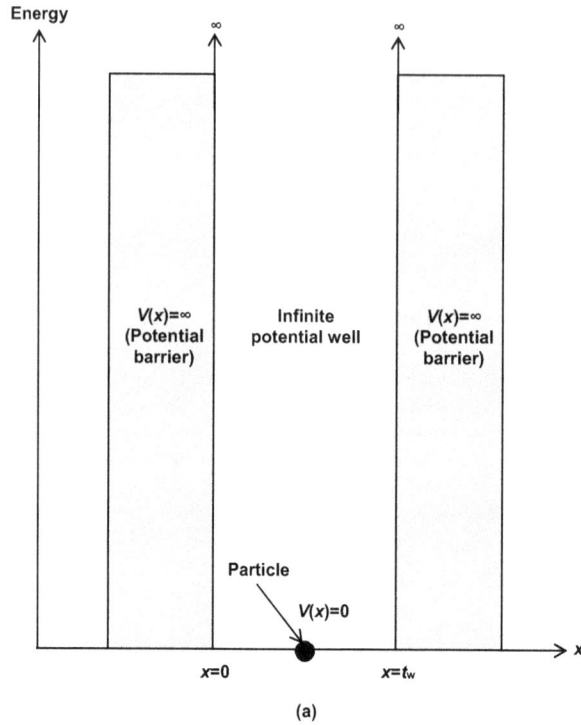

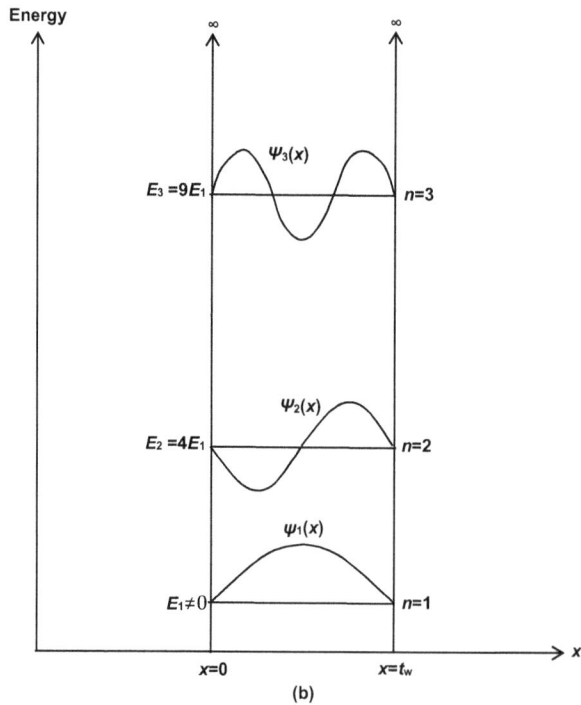

FIGURE 10.3 Infinite rectangular potential well: (a) potential-distance diagram showing that the potential $V(x)$ is zero inside the well and infinite everywhere outside the well, and the well extends from $x = 0$ to $x = t_w$; (b) the energy level and wave function plots showing the energy levels E_1, E_2, E_3, . . . occupied by electrons, and the variation of wave functions $\psi_1(x)$, $\psi_2(x)$, $\psi_3(x)$, . . . with position inside the well. Note that the wave functions become zero at the boundaries of the well so that there is no probability of electron being found outside this infinitely deep potential well.

$\psi(x)$ must have a finite value for finite energy. So outside the box, $\psi(x) = 0$, which implies that the electron cannot go outside the box. Inside the box, $V(x) = 0$ so that Schrodinger's equation 10.13 reduces to

$$-\frac{\hbar^2}{2m}\frac{d^2\psi(x)}{dx^2} + 0 \times \psi(x) = E\psi(x) \qquad (10.14)$$

or

$$-\frac{\hbar^2}{2m}\frac{d^2\psi(x)}{dx^2} = E\psi(x) \qquad (10.15)$$

This is a commonly encountered differential equation whose general solution is written using exponential functions as

$$\psi(x) = A\exp(ikx) + B\exp(-ikx) \qquad (10.16)$$

where we have introduced the variable k as

$$k = \pm\frac{\sqrt{2mE}}{\hbar} \qquad (10.17)$$

and symbols A, B denote arbitrary constants. Equation 10.16 can be recast in a form containing sine and cosine functions as

$$\psi(x) = C\sin(kx) + D\cos(kx) \qquad (10.18)$$

The constants C and D in equation 10.18 are determined from the boundary conditions

$$\psi(0) = 0 \text{ at } x = 0 \qquad (10.19)$$

$$\psi(t_w) = 0 \text{ at } x = t_w \qquad (10.20)$$

Application of the first boundary condition given by equation 10.19 at $x = 0$ to equation 10.18 gives

$$\psi(0) = C\sin(k0) + D\cos(k0) = C\sin(0) + D\cos(0) = C \times 0 + D \times 1 = D = 0 \qquad (10.21)$$

so that

$$D = 0 \qquad (10.22)$$

When the second boundary condition at $x = t_w$ given by equation 10.20 is applied on equation 10.18, we get

$$\psi(t_w) = C\sin(kt_w) + D\cos(kt_w) = C\sin(kt_w) + 0 \times \cos(kt_w) = C\sin(kt_w) = 0 \qquad (10.23)$$

yielding

$$C\sin(kt_w) = 0 \qquad (10.24)$$

which is possible if

$$kt_w = n\pi \tag{10.25}$$

where n is an integer acquiring the values 1, 2, 3, . . . Equation 10.25 is rewritten as

$$k = \frac{n\pi}{t_w} \tag{10.26}$$

Putting the value of $D = 0$ from equation 10.22 and k from equation 10.26 in equation 10.18, the wave function of the electron becomes

$$\psi(x) = C\sin(kx) + 0 = C\sin\left(\frac{n\pi x}{t_w}\right) \tag{10.27}$$

By application of the normalization condition on the wave function, the constant C is found to be

$$C = \sqrt{\frac{2}{t_w}} \tag{10.28}$$

Combining equations 10.27 and 10.28,

$$\psi(x) = \sqrt{\frac{2}{t_w}}\sin\left(\frac{n\pi x}{t_w}\right) \tag{10.29}$$

where n represents an integer, $n = 1, 2, 3, . . .$ as mentioned previously.

Again, putting the value of k from equation 10.26 in equation 10.17,

$$\frac{\sqrt{2mE}}{\hbar} = \frac{n\pi}{t_w} \tag{10.30}$$

Squaring both sides of equation 10.30,

$$\frac{2mE}{\hbar^2} = \frac{n^2\pi^2}{t_w^2} \tag{10.31}$$

or

$$E = \frac{n^2\pi^2\hbar^2}{2mt_w^2} \tag{10.32}$$

This equation has a profound significance because it proclaims that the energy E of the electron in the box is disallowed to vary over a continuous range of values. It is only permitted to take discrete values for different integral values of the quantum number n, meaning, that energy is quantized. Thus, the electron inside the box can move to a discrete set of energy levels E_1, E_2, . . . So it is more appropriate to write equation 10.32 in the form

$$E_n = \frac{n^2\pi^2\hbar^2}{2mt_w^2} \tag{10.33}$$

by attaching the subscript n with energy. Note that the minimum energy E_1 of the electron is not zero. Moreover, energy is measured from the bottom of the well upwards. We shall apply this equation due to its mathematical simplicity and elegance, always bearing in mind that the calculations are only rough conjectures because the potentials to be dealt with cannot be looked upon as infinite. Nonetheless, valuable information is obtained about the nature of changes, and the trend of variations can be predicted. Equation 10.29 for wave function can also be written with a subscript n as

$$\psi_n(x) = \sqrt{\frac{2}{t_w}} \sin\left(\frac{n\pi x}{t_w}\right) \tag{10.34}$$

10.3.2 Imagining Quantum Well as a Finite Potential Well

The finite barrier model provides a more realistic representation of the quantum wells in solar cells (Figure 10.4). This model is an extension of the infinite barrier model. The particle is still restricted by walls, but the walls are no longer impenetrable. Inconvenience is faced in applying this model because the finite well problem does not give closed-form analytical solutions. Transcendental equations result, and numerical solution is necessary to determine the energies. For the sake of comparison, the electron energy values calculated by the finite well model are a little lower than the corresponding values found from the infinite well model, e.g., the energy levels E_1, E_2 for a finite well of depth 1.0 eV are 0.068eV and 0.263eV against 0.094eV and 0.377eV for infinite well. Remember that a minimum of one bound state always exists for a finite potential quantum well. Regarding the nature of wave functions, they are still sine waves. They decline exponentially in the barrier regions.

A major difference is observed between the finite and infinite potential wells. In the finite potential well, the wave function extends into the classically prohibited region, unlike the infinite potential well, in which it becomes zero at the edge of the box. This extension of the wave function into the forbidden region and its not falling to zero at the boundary of the well have a far-reaching impact. It means that in a finite potential well, the probability of finding the particle outside the well in the forbidden regions is nonzero despite the particle possessing inadequate energy to surmount the barrier. The outcome of this possibility is the portrayal of quantum mechanical tunneling. Thus, the finite well model implies an infinite well model with lower electron energies together with quantum mechanical tunneling.

10.3.3 Energy States of a Quantum Well and Defining an Effective Bandgap of the Quantum Well

Background knowledge about the infinite and finite potential wells helps differentiate the bandgap and density of states in quantum well from the bulk semiconductor. Figure 10.5 picturizes this difference.

10.3.4 Difference between the Multiple–Quantum Well and Superlattice Structures

The superlattice (Figure 10.6) is a commonly used structure in quantum well solar cells. It is essentially a lattice of lattices. How does it differ from a multiple quantum well, and when do we call a structure superlattice, and when it is referred to as a multiple quantum well? When the wave functions of neighboring quantum wells in a multiple–quantum well structure interpenetrate into each other appreciably, the discrete states of the quantum wells blend together, forming minibands. Then the multiple–quantum well structure becomes a superlattice. When the wave functions of the individual quantum wells can be treated independently, i.e., the degree of interpenetration is low, we shall call the structure a multiple quantum well. So the superlattice can be looked upon as a special case of multiple–quantum well structure. Such a circumstance normally arises when the quantum wells are located in close proximity. Quantum wells placed far apart behave independently and exhibit distinct, discrete energy levels, not minibands.

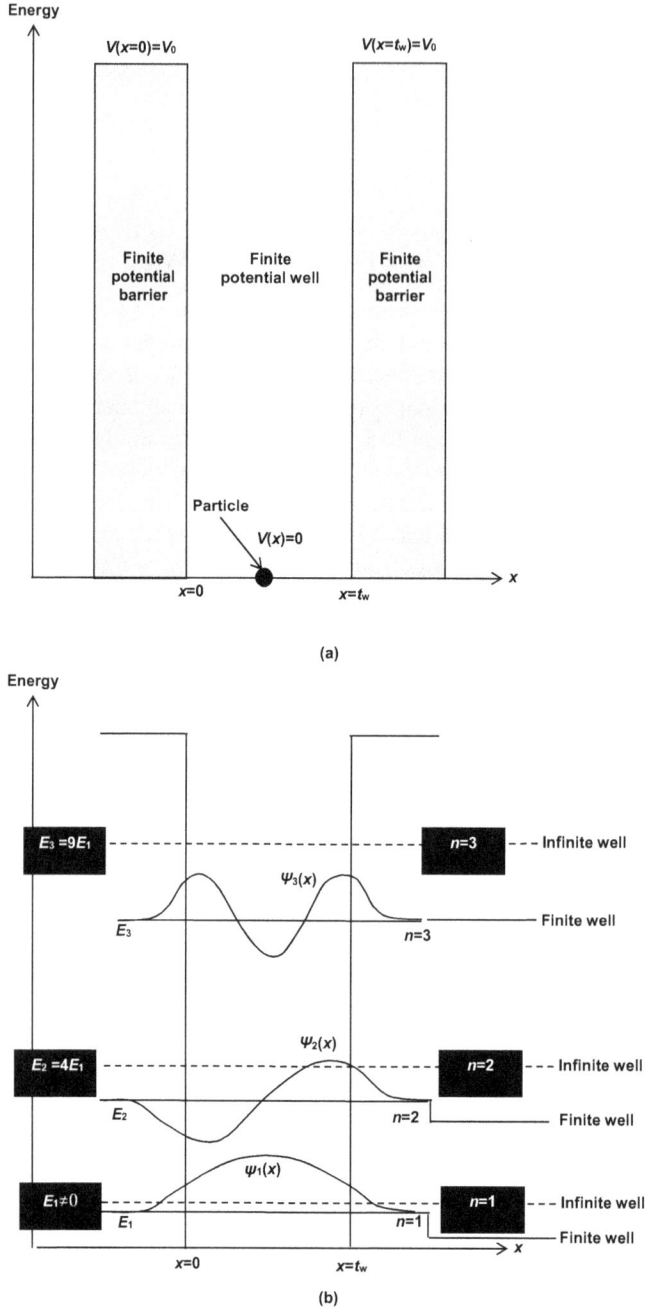

FIGURE 10.4 Finite rectangular potential well: (a) potential distance diagram showing that the potential $V(x)$ is zero inside the well but has a finite value V_0 everywhere outside the well, and the well extends from $x = 0$ to $x = t_w$; (b) energy level and wave function plots showing the energy levels E_1, E_2, E_3, ... occupied by electrons, and the wave functions $\psi_1(x)$, $\psi_2(x)$, $\psi_3(x)$, ... (full lines). For comparison, the energy levels for an infinite well are shown by dotted lines and labeled inside black boxes. Note that the wave functions for a finite well are sinusoidal inside the well but decay exponentially outside the well, falling asymptotically to zero. Unlike the infinite well, the wave functions do not become zero at the edges of the well but extend outside the well into the forbidden regions, indicating the possibility of presence of the electron in these regions. This possibility is the basis of tunneling phenomenon.

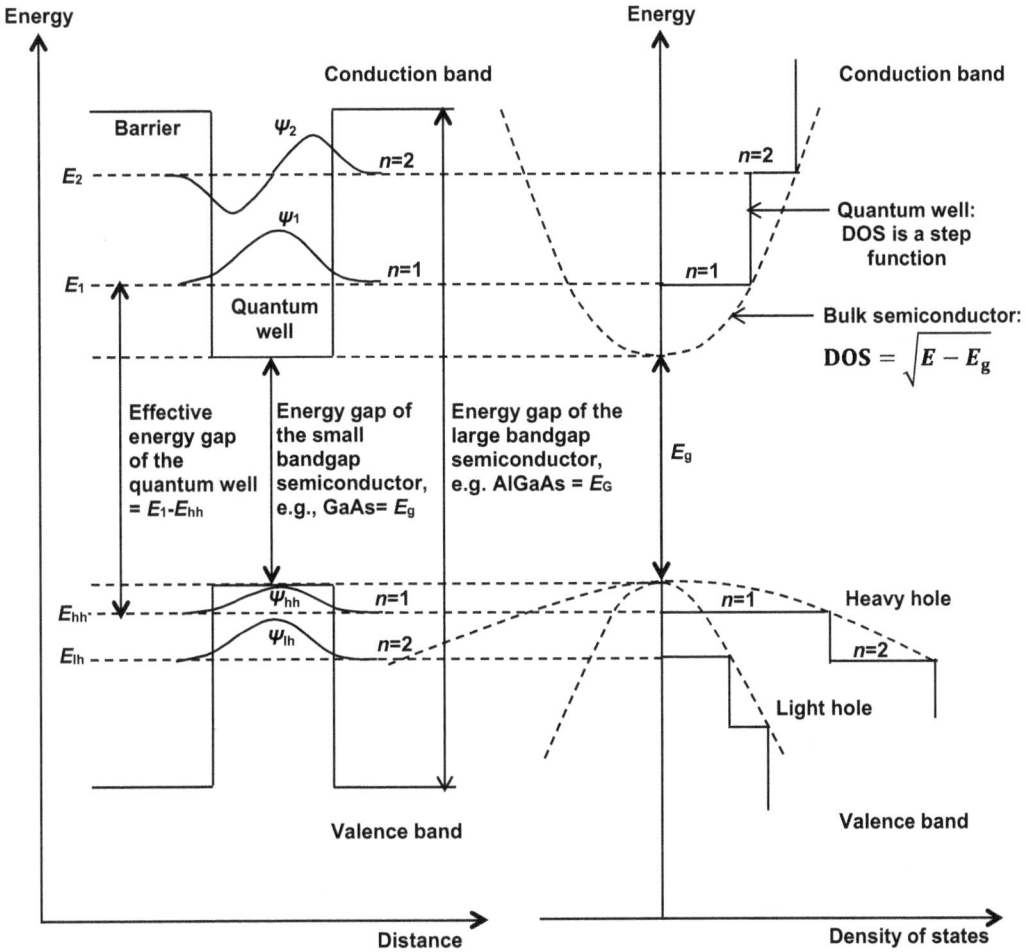

FIGURE 10.5 Energy states in the bulk semiconductor and quantum well structure as a function of distance: in the bulk semiconductor bandgap is the energy difference between conduction band edge and maximum energy of heavy hole/light hole levels, i.e., valence band edge but in the quantum well effective bandgap is the energy difference between first electron energy level in the conduction band and the heavy hole level in the valence band; this energy gap is larger than the gap in the bulk semiconductor. The wave functions of electrons and holes are also shown. On the right-hand side of the diagram are depicted the density of states in bulk semiconductor and in the quantum well. Unlike the bulk, the density of states in the quantum well is constant with energy. It has the appearance of a staircase built up of a series of steps in which each step is executed with the change of quantum number n. Further, the density of states increases sharply with energy as opposed to its variation with square root of energy in a bulk semiconductor. Hence, a larger number of energy states are available near the band edges in a quantum well than in bulk.

10.3.5 Charge Transport Mechanisms in the Quantum Well Solar Cell

Two basic processes continuously take place in a solar cell. One is carrier generation, and the other is carrier recombination. In carrier generation, an electron-hole pair is produced. Thermal energy at room temperature or sunlight falling on the solar cell produces these pairs. During carrier generation, an electron is promoted from the valence band to the conduction band, leaving behind a hole in

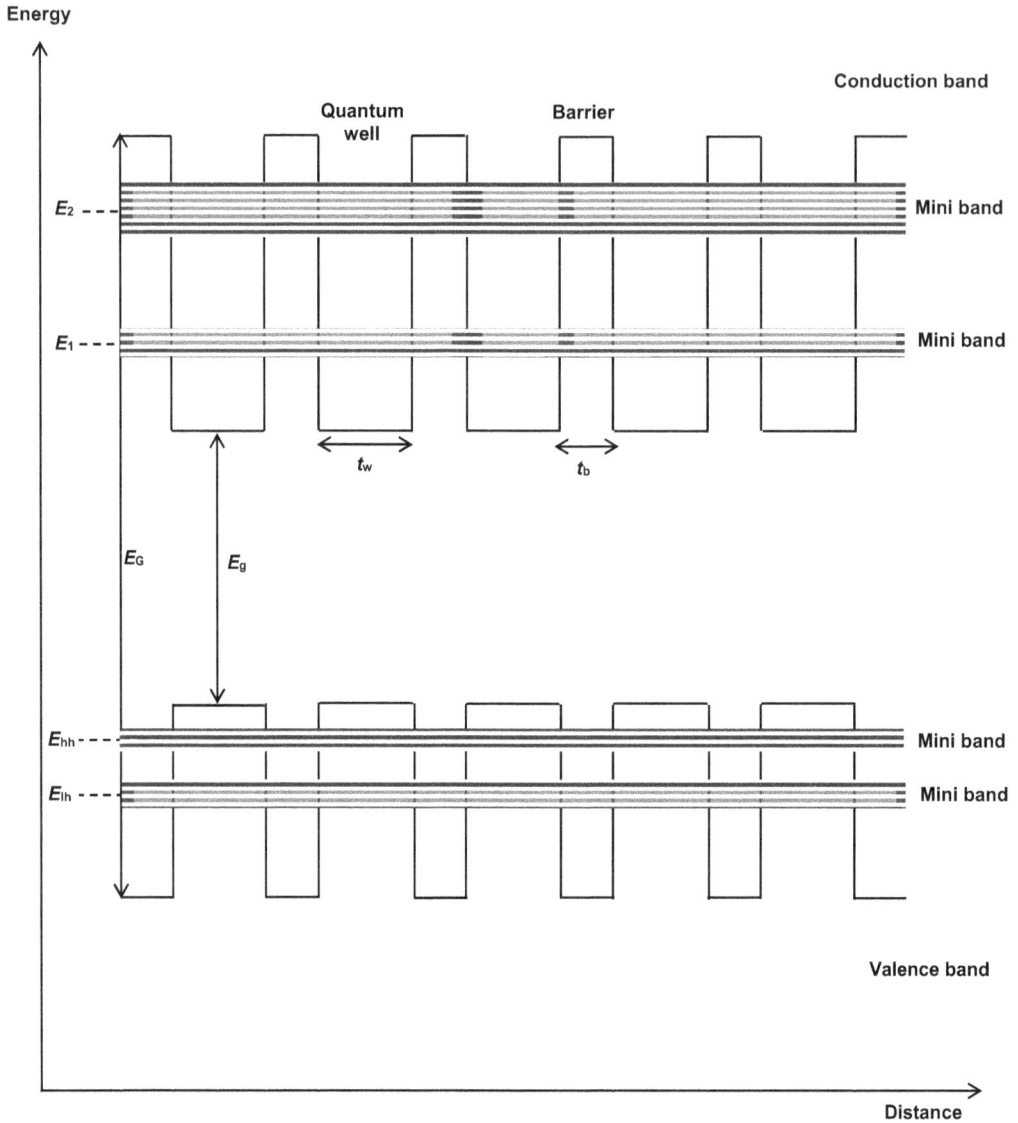

FIGURE 10.6 Energy state representation of a superlattice. It is a periodic structure in which the quantum wells are located in such close proximity to each other that they do not behave as independent wells. The wave functions of the individual quantum wells interpenetrate into each other, with the energy levels merging into minibands.

the valence band. This promotion occurs through the energy received thermally or optically. During carrier recombination, an electron-hole pair is annihilated. The electron falls from the conduction band and fills a hole in the valence band. The time elapsed before electron-hole recombination is the minority carrier lifetime in the given semiconductor. It is also called excess carrier lifetime. This lifetime is a crucial parameter affecting the working of the solar cell. It is a function of the concentration of impurities or defects in the material and the processing conditions it has undergone.

As we know, "drift mechanism" under the influence of the built-in potential is responsible for the carrier motion in a solar cell. Transport of charge through the multiple quantum wells occurs through the participation of two principal processes. These are thermionic emission and quantum mechanical tunneling. Let us digress briefly to learn about these processes. In thermionic emission, electrons are released from a material by heating effect. What is the source of heat here? It is the thermal energy of room temperature. Adequate energy must be imparted by heat to electrons to surmount the attractive force binding them to the semiconductor. The minimum energy necessary for liberation of electrons from the surface of the material is known as its work function.

In a quantum well solar cell, there are potential energy wells and energy barriers. Quantum mechanical tunneling is a process in which the probability of an electron crossing a barrier is nonzero under specific conditions even if it possesses lower energy than the height of the barrier. The wave function decays as we move past the barrier, but its finite value indicates the presence of the carrier on the opposite side of the barrier, the possibility decreasing with the barrier width and height.

Multiple–quantum well designs in solar cells use either thin wells and thick barriers or thick wells and thin barriers. In thin wells and thick barrier designs, carriers are transported by the built-in electric field in the cell mainly by thermionic emission (Figure 10.7), while in thick well and thin barrier designs, tunneling is the dominant mechanism of carrier transport by the built-in electric field of the cell (Figure 10.8). Figure 10.9 sketches a holistic scenario of carrier transport and generation-recombination processes incessantly occurring in a quantum well solar cell where carrier generation/recombination in quantum well and barrier regions as well as thermionic emission and tunneling are displayed.

10.3.6 Excitonic Model of Optical Absorption

The absorption spectrum of a quantum well contains a series of steps whose energies are expected to be provided by the particle-in-a box and finite potential well models. In practice, some peaks are observed in the absorption spectra, which are not accounted by these models. The origin of these peaks is interpreted by introducing the concept of excitons, which are bound states of electrons and holes mutually attracted towards each other by the Coulomb electrostatic force (Figure 10.10). A consequence of this excitonic model is that optical absorption can take place at a lower energy than the bandgap energy, and the extent of this lowering is equal to the binding energy E_B of the exciton.

Example 10.2

Optical transitions can take place at a lower photon energy $E_{Exciton}$ below the bandgap energy of the semiconductor under consideration. The energy $E_{Exciton}$ is expressed as the difference between the bandgap energy E_G and the binding energy E_B of the lowest energy $1S$ exciton and is given by

$$E_{Exciton} = E_G - E_B \tag{10.35}$$

where

$$E_B = \frac{\mu e^4}{8\varepsilon_0^2 \varepsilon_s^2 h^2} \tag{10.36}$$

In equation 10.36, μ is the reduced mass of the charge carriers expressed in terms of effective mass m_e^* of electrons and effective mass m_h^* of holes as

$$\mu = \frac{m_e^* m_h^*}{m_e^* + m_h^*} \tag{10.37}$$

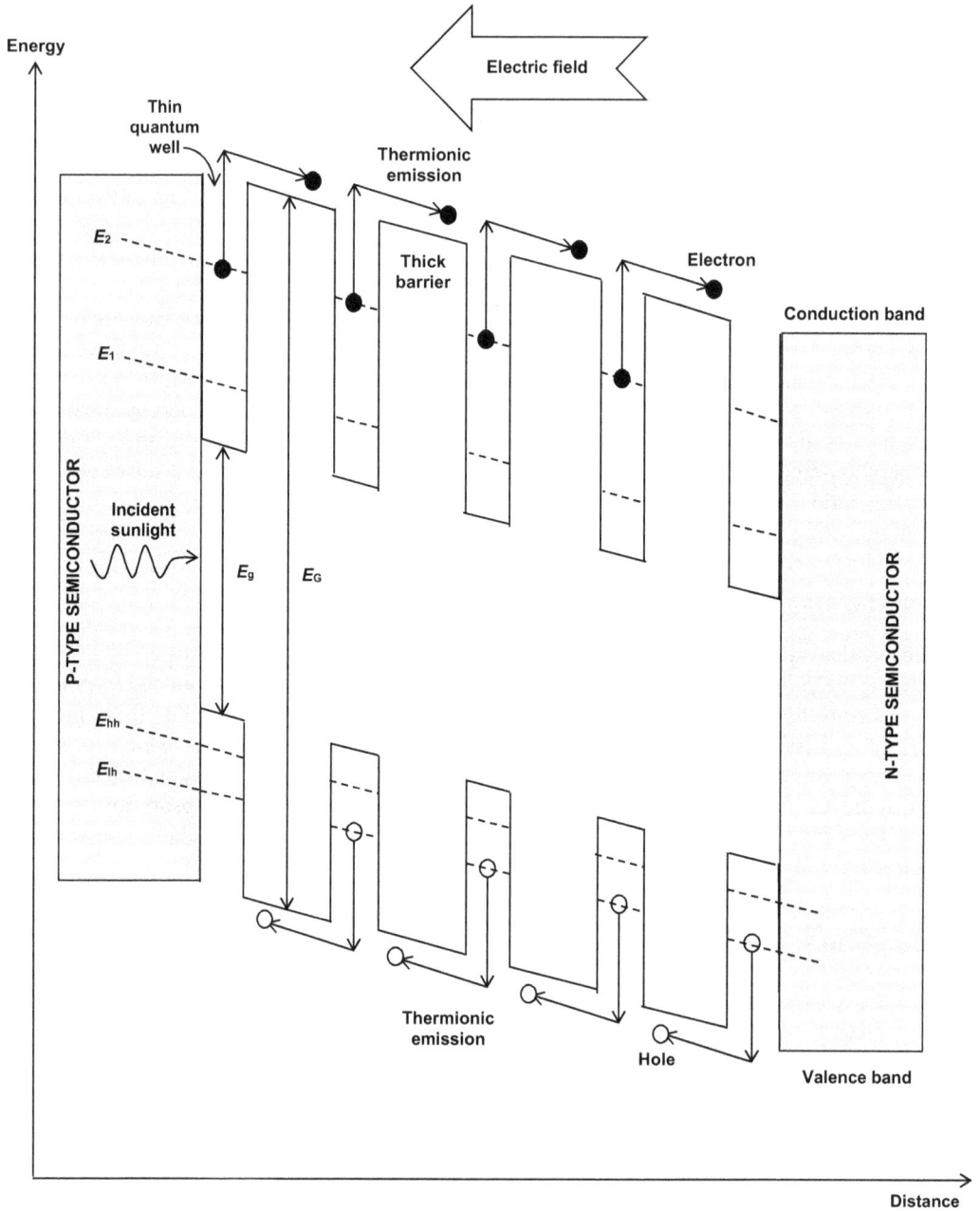

FIGURE 10.7 Carrier transfer mechanisms in an MQW structure having thin wells and thick barrier layers with low effective barrier height ~ a few k_BT. As the probability of tunneling through thick barriers is low, the carrier transport is dominated by thermal escape or thermionic emission over the barriers.

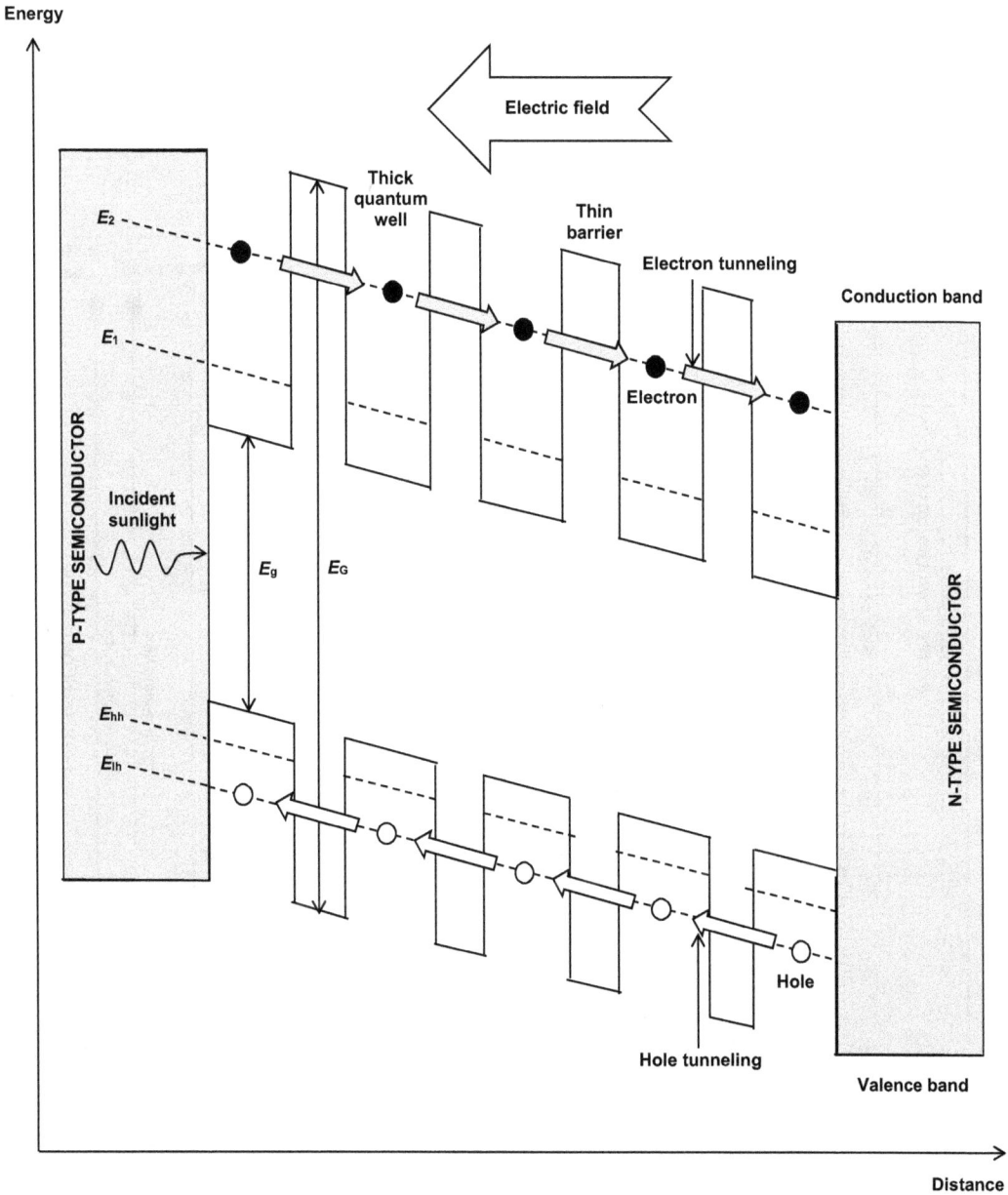

FIGURE 10.8 Carrier transfer mechanisms in a MQW structure with thick wells and thin barrier layers. As the probability of tunneling through thin barriers is high, carrier transport is dominated by tunneling through the barriers.

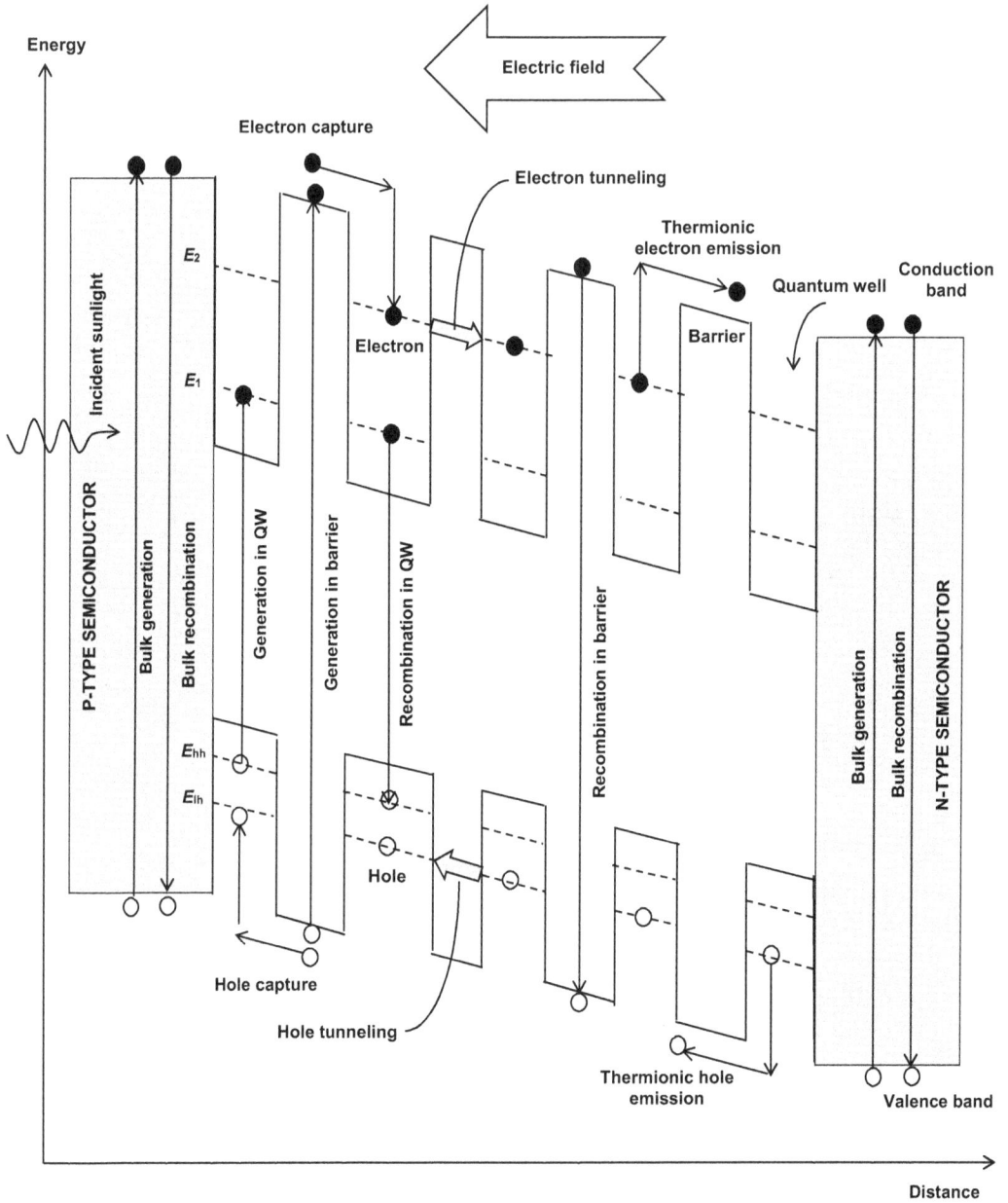

FIGURE 10.9 Overall picture of carrier transport and generation/recombination processes in an MQW solar cell. The carrier transfer mechanisms include thermal escape over the barriers and tunneling through the barriers. Generation and recombination of carriers in the quantum well and barrier regions are also shown.

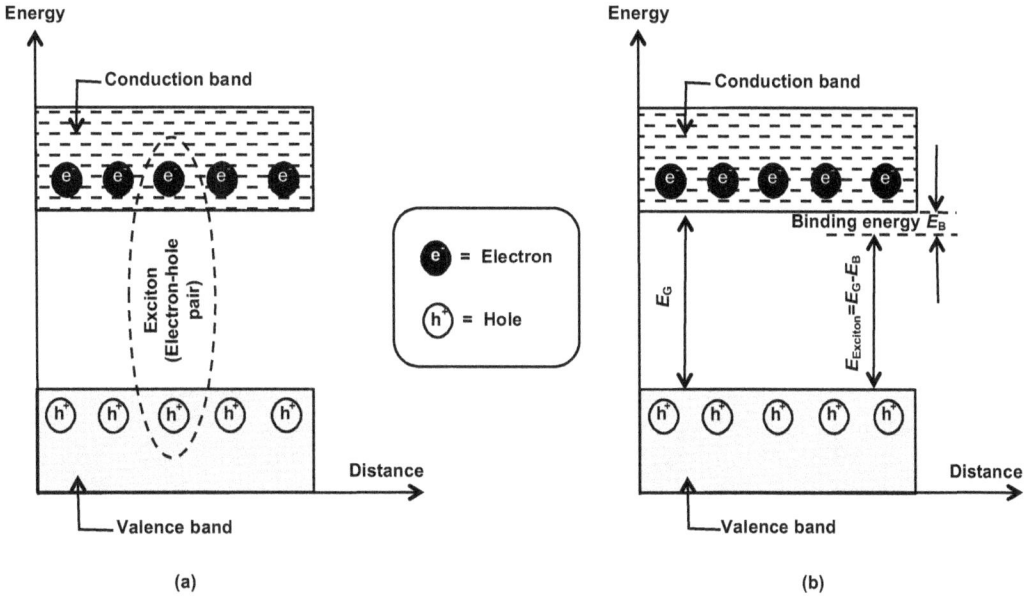

FIGURE 10.10 Exciton and related energies: (a) the exciton and (b) the exciton energy E_{Exciton}, the bandgap E_G, and the binding energy E_B of the exciton. Electrons in the conduction band are shown by filled black circles with e^- written inside and holes in the valence band by open circles with h^+ inside. An electron-hole pair comprising the exciton is enclosed within dashed borderline. The bandgap $E_G >$ exciton energy E_{Exciton} and E_B is the difference between these energies.

where e is the electronic charge, ε_0 is permittivity of free space, ε_S is the relative permittivity of the semiconductor, and h is Planck's constant. Find the binding energy for GaAs for both light and heavy hole cases, taking $m_e^* = 0.067 m_0$, $m_{\text{lh}}^* = 0.082 m_0$, and $m_{\text{hh}}^* = 0.45 m_0$. Dielectric constant of GaAs is 12.9. Also, find the corresponding optical transition energies.

Solution:

Putting $m_e^* = 0.067 m_0$ and for light holes, $m_{\text{lh}}^* = 0.082 m_0$ in equation 10.37,

$$\mu = \frac{m_e^* m_{\text{lh}}^*}{m_e^* + m_{\text{lh}}^*} = \frac{0.067 m_0 \times 0.082 m_0}{0.067 m_0 + 0.082 m_0} = \frac{0.067 \times 0.082 m_0^2}{(0.067 + 0.082) m_0} = \frac{0.067 \times 0.082 m_0}{0.067 + 0.082}$$

$$= \frac{0.005494 m_0}{0.149} = 0.03687 \times 9.109 \times 10^{-31} = 0.3358 \times 10^{-31} = 3.358 \times 10^{-32} \text{kg} \qquad (10.38)$$

From equations 10.36 and 10.38, the binding energy for the light hole case is

$$E_B = \frac{3.358 \times 10^{-32} \left(1.602 \times 10^{-19}\right)^4}{8 \left(8.854 \times 10^{-12}\right)^2 (12.9)^2 \left(6.62607 \times 10^{-34}\right)^2} \text{ J}$$

$$= \frac{3.358 \times 6.5864 \times 10^{-32-76}}{8 \times 78.3933 \times 166.41 \times 43.9048 \times 1.602 \times 10^{-19} \times 10^{-24-68}} \text{ eV} \qquad (10.39)$$

$$= \frac{22.11713 \times 10^{-108}}{7.34045 \times 10^{6} \times 10^{-111}} = 3.013 \times 10^{-108+111-6} \approx 3 \times 10^{-3} \text{ eV} = 3 \text{meV}$$

From equations 10.35 and 10.39, the energy for electron–light hole transition caused by photon absorption is

$$E_{Exciton} = 1.424 - 0.003 = 1.421 eV \tag{10.40}$$

representing the optical transition energy for light hole case.

Putting $m_e^* = 0.067m_0$ and, for heavy holes, $m_{hh}^* = 0.45m_0$ in equation 10.37,

$$\mu = \frac{m_e^* m_{hh}^*}{m_e^* + m_{hh}^*} = \frac{0.067m_0 \times 0.45m_0}{0.067m_0 + 0.45m_0} = \frac{0.067 \times 0.45m_0^2}{(0.067 + 0.45)m_0} = \frac{0.067 \times 0.45m_0}{0.067 + 0.45}$$

$$= \frac{0.03015m_0}{0.517} = 0.05832 \times 9.109 \times 10^{-31} = 0.5312 \times 10^{-31} = 5.312 \times 10^{-32} kg \tag{10.41}$$

From equations 10.36 and 10.41, the binding energy for heavy hole case is

$$E_B = \frac{5.312 \times 10^{-32} \left(1.602 \times 10^{-19}\right)^4}{8 \left(8.854 \times 10^{-12}\right)^2 (12.9)^2 \left(6.62607 \times 10^{-34}\right)^2} J$$

$$= \frac{5.312 \times 6.5864 \times 10^{-32-76}}{8 \times 78.3933 \times 166.41 \times 43.9048 \times 1.602 \times 10^{-19} \times 10^{-24-68}} eV \tag{10.42}$$

$$= \frac{34.98696 \times 10^{-108}}{7.34045 \times 10^6 \times 10^{-111}} = 4.7663 \times 10^{-108+111-6} = 4.77 \times 10^{-3} eV = 4.77 meV$$

From equations 10.35 and 10.42, the energy for electron-heavy hole transition caused by optical absorption is

$$E_{Exciton} = 1.424 - 0.00477 = 1.41923 eV \tag{10.43}$$

This is the required optical transition energy for heavy hole case.

Example 10.3

A quantum well is formed with a GaAs film of thickness 10nm coated on both sides with $Al_{0.4}Ga_{0.6}As$ barrier films. (a) Find the energy of an electron in the ground state in the quantum well if the effective mass of electron in GaAs is $0.067m_0$, where m_0 is the rest mass of electron. (b) Calculate the ground state energies for (i) light hole of effective mass $0.082m_0$ and (ii) heavy hole of effective mass $0.45m_0$. (c) Determine the energies of optical transition for (i) light and (ii) heavy holes. Apply the infinite potential quantum well model. Bandgap of GaAs is 1.424eV, and that of $Al_{0.4}Ga_{0.6}As$ is 1.92eV.

Solution:

(a) From equation 10.33, the energy E_n of the nth energy level in an infinite well is given by

$$E_n = \frac{h^2}{2m^*} \left(\frac{n\pi}{t_w}\right)^2 \tag{10.44}$$

where $\hbar$ stands for the reduced Planck's constant $= h/2\pi$, m^* denotes the effective mass of the particle (electron or hole) in the well, n is the quantum number $= 1, 2, 3, \ldots$ and t_w is the thickness of the well. Here, $\hbar = 1.05457 \times 10^{-34}$ J.s, $m^* = 0.067 m_0 = 0.067 \times 9.109 \times 10^{-31}$ kg, $t_w = 10$nm $= 10 \times 10^{-9}$m. Putting the values in equation 10.44, we obtain the energy for the ground state of the electron in the quantum well as

$$
\begin{aligned}
E_{1e} &= \frac{\left(1.05457 \times 10^{-34}\right)^2}{2 \times 0.067 \times 9.109 \times 10^{-31}} \left(\frac{1 \times 3.14159}{10 \times 10^{-9}}\right)^2 \frac{J^2 \times s^2}{kg \times m^2} = \frac{1.1121 \times 10^{-68}}{1.2206 \times 10^{-31}} \times \frac{9.86959}{10^{-16}} \frac{J^2}{kg \times m^2 \times s^2} \\
&= \frac{1.1121 \times 9.86959}{1.2206} \times 10^{-68+31+16} \frac{J^2}{J} = 8.9923 \times 10^{-21} J = \frac{8.9923 \times 10^{-21}}{1.602 \times 10^{-19}} eV \\
&= \frac{8.9923 \times 10^{-21+19}}{1.602} eV = 5.613 \times 10^{-2} eV = 0.05613 eV
\end{aligned}
\tag{10.45}
$$

Taking the energy of valence band edge as zero, the energy of the electron energy level in the ground state is

$$
\left|E_{1e}\right|_{VBE=0} = E_{G(GaAs)} + E_{1e} = 1.424 + 0.05613 = 1.48013 eV
\tag{10.46}
$$

where $E_{G(GaAs)}$ is the bandgap of GaAs and equation 10.45 has been applied.

(b) (i) From equation 10.44, the ground state energy of a light hole in the valence band is

$$
\begin{aligned}
E_{1lh} &= \frac{\left(1.05457 \times 10^{-34}\right)^2}{2 \times 0.082 \times 9.109 \times 10^{-31}} \left(\frac{1 \times 3.14159}{10 \times 10^{-9}}\right)^2 = \frac{1.1121 \times 9.86959}{1.4939} \times 10^{-21} J \\
&= 7.3472 \times 10^{-21} J = \frac{7.3472 \times 10^{-21}}{1.602 \times 10^{-19}} eV = 4.586 \times 10^{-2} eV = 0.04586 eV
\end{aligned}
\tag{10.47}
$$

With respect to the energy of valence band edge as zero, the energy of light hole energy level in the ground state is

$$
\left|E_{1lh}\right|_{VBE=0} = 0 - E_{1lh} = 0 - 0.04586 eV = -0.04586 eV
\tag{10.48}
$$

where equation 10.47 has been applied.

(ii) Applying equation 10.44, the ground state energy of a heavy hole in the valence band is

$$
\begin{aligned}
E_{1hh} &= \frac{\left(1.05457 \times 10^{-34}\right)^2}{2 \times 0.45 \times 9.109 \times 10^{-31}} \left(\frac{1 \times 3.14159}{10 \times 10^{-9}}\right)^2 = \frac{1.1121 \times 9.86959}{8.1981} \times 10^{-21} J \\
&= 1.3388 \times 10^{-21} J = \frac{1.3388 \times 10^{-21}}{1.602 \times 10^{-19}} eV = 0.8357 \times 10^{-2} eV = 0.008357 eV
\end{aligned}
\tag{10.49}
$$

Assuming the energy of valence band edge as zero, the energy of heavy hole energy level in the ground state is

$$
\left|E_{1hh}\right|_{VBE=0} = 0 - E_{1hh} = 0 - 0.008357 eV = -0.008357 eV
\tag{10.50}
$$

by using equation 10.49.

(c) (i) From equations 10.46 and 10.48, the energy level difference from light hole to conduction band is

$$E_{G-\text{Light hole-to-conduction}} = 1.48013\text{eV} - (-0.04586\text{eV}) = 1.48013\text{eV} + 0.04586\text{eV}$$
$$= 1.52599\text{eV} \tag{10.51}$$

(ii) From equations 10.46 and 10.50, the energy level difference from heavy hole to conduction band is

$$E_{G-\text{Heavy hole-to-conduction}} = 1.48013\text{eV} - (-0.008357\text{eV}) = 1.48013 + 0.008357\text{eV}$$
$$= 1.48849\text{eV} \tag{10.52}$$

Example 10.4

Consider a GaAs quantum well of thickness 5nm. Find all the parameters determined in parts a, b, and c of example 10.3 for this narrower quantum well.

Solution: (a) By equation 10.44

$$
\begin{aligned}
E_{1e} &= \frac{\left(1.05457 \times 10^{-34}\right)^2}{2 \times 0.067 \times 9.109 \times 10^{-31}} \left(\frac{1 \times 3.14159}{5 \times 10^{-9}}\right)^2 \frac{J^2 \times s^2}{kg \times m^2} \\
&= \frac{1.1121 \times 10^{-68}}{1.2206 \times 10^{-31}} \times \frac{9.86959}{25 \times 10^{-18}} \frac{J^2}{kg \times m^2 \times s^2} \\
&= \frac{1.1121 \times 9.86959}{30.515} \times 10^{-68+31+18} \frac{J^2}{J} = 0.35969 \times 10^{-19} J \\
&= \frac{0.35969 \times 10^{-19}}{1.602 \times 10^{-19}} \text{eV} = \frac{0.35969}{1.602} \text{eV} = 0.2245\text{eV}
\end{aligned}
\tag{10.53}
$$

and

$$|E_{1e}|_{VBE=0} = E_{G(\text{GaAs})} + E_{1e} = 1.424 + 0.2245 = 1.6485\text{eV} \tag{10.54}$$

by putting the value of E_{1e} from equation 10.53.

(b) (i) Applying equation 10.44,

$$
\begin{aligned}
E_{1lh} &= \frac{\left(1.05457 \times 10^{-34}\right)^2}{2 \times 0.082 \times 9.109 \times 10^{-31}} \left(\frac{1 \times 3.14159}{5 \times 10^{-9}}\right)^2 = \frac{1.1121 \times 10^{-68}}{1.4939 \times 10^{-31}} \times \frac{9.86959}{25 \times 10^{-18}} = \\
&= \frac{1.1121 \times 9.86959}{37.3475} \times 10^{-19} = 0.2939 \times 10^{-19} J = \frac{0.2939 \times 10^{-19}}{1.602 \times 10^{-19}} = 0.1835
\end{aligned}
\tag{10.55}
$$

and

$$|E_{1hh}|_{VBE=0} = 0 - E_{1lh} = 0 - 0.1835\text{eV} = -0.1835\text{eV} \tag{10.56}$$

where the value of E_{1lh} has been put from equation 10.55.

(ii) Equation 10.44 gives

$$E_{1hh} = \frac{\left(1.05457 \times 10^{-34}\right)^2}{2 \times 0.45 \times 9.109 \times 10^{-31}} \left(\frac{1 \times 3.14159}{2.5 \times 10^{-9}}\right)^2 = \frac{1.1121 \times 10^{-68}}{8.1981 \times 10^{-31}} \times \frac{9.86959}{25 \times 10^{-18}} = \tag{10.57}$$

$$= \frac{1.1121 \times 9.86959}{204.9525} \times 10^{-19} = 0.05355 \times 10^{-19} J = \frac{0.05355 \times 10^{-19}}{1.602 \times 10^{-19}} = 0.03343$$

and

$$|E_{1hh}|_{VBE=0} = 0 - E_{1hh} = 0 - 0.03343 eV = -0.03343 eV \tag{10.58}$$

by putting the value of E_{1hh} from equation 10.57.

(c) (i) From equations 10.54 and 10.56, the energy level difference from light hole to conduction band is

$$\begin{aligned} E_{G-\text{Light hole-to-conduction}} &= 1.6485 eV - \left(-0.1835 eV\right) \\ &= 1.6485 eV + 0.1835 eV = 1.832 eV \end{aligned} \tag{10.59}$$

(ii) From equations 10.54 and 10.58, the energy level difference from heavy hole to conduction band is

$$\begin{aligned} E_{G-\text{Heavy hole-to-conduction}} &= 1.6485 eV - \left(-0.03343 eV\right) \\ &= 1.6485 + 0.03343 eV = 1.68193 eV \end{aligned} \tag{10.60}$$

Note 1: The difference between energy levels obtained for the narrow well of Example 10.4 with respect to broader well of Example 10.3 is worthy of attention. All the energy levels of Example 10.4 are located at numerically bigger values of energy than that of example 10.3, and also, the energy transition differences are larger in Example 10.4 than Example 10.3.

Note 2: Since we are applying the approximate formula for infinite potential well, we must remember, as a point of caution, that any value of energy level greater than the bandgap 1.92 eV of the $Al_{0.4}Ga_{0.6}As$ barrier layer does not represent a bound state in the well and must be rejected.

Example 10.5

Show that in the infinite well approximation, the energy gap $E_{G, QW}$ of the quantum well of thickness t_w is the sum of energy gap $E_{G, \text{Well material}}$ of the well material, and a term representing the difference between the energies of the ground states of electrons and holes:

$$E_{G, QW} = E_{G, \text{Well material}} + \frac{\hbar^2 \pi^2}{2t_w^2} \left(\frac{1}{m_e^*} + \frac{1}{m_h^*}\right) \tag{10.61}$$

where m_e^* and m_h^* are the effective masses of electrons and holes in the well material.

Solution: Applying equation 10.33 for the ground state of the electron ($n = 1$), we write the energy of the ground state of electrons in the quantum well as

$$E_{1e} = \frac{\hbar^2}{2m_e^*}\left(\frac{\pi}{t_w}\right)^2 = \frac{\hbar^2 \pi^2}{2t_w^2}\left(\frac{1}{m_e^*}\right) \tag{10.62}$$

This energy is measured from the bottom of the conduction band. Taking the energy of the top of valence band (VBE) as the level of zero energy, the electron energy level in the ground state is placed at

$$\left| E_{1e} \right|_{\text{VBE}=0} = E_{\text{G, Well material}} + E_{1e} = E_{\text{G, Well material}} + \frac{\hbar^2 \pi^2}{2 t_w^2} \left(\frac{1}{m_e^*} \right) \tag{10.63}$$

where the expression for E_{1e} has been put from equation 10.62.

For the ground state energy level of holes in the valence band, an equation similar to equation 10.62 can be written by replacing the effective mass m_e^* of electrons with the effective mass m_h^* of holes

$$E_{1h} = \frac{\hbar^2}{2 m_h^*} \left(\frac{\pi}{t_w} \right)^2 = \frac{\hbar^2 \pi^2}{2 t_w^2} \left(\frac{1}{m_h^*} \right) \tag{10.64}$$

This energy is measured from the top of the valence band downwards. Referring to the top of the valence band as zero energy level, the energy level of holes in the ground state is

$$\left| E_{1h} \right|_{\text{VBE}=0} = 0 - E_{1h} = 0 - \frac{\hbar^2 \pi^2}{2 t_w^2} \left(\frac{1}{m_h^*} \right) = -\frac{\hbar^2 \pi^2}{2 t_w^2} \left(\frac{1}{m_h^*} \right) \tag{10.65}$$

using equation 10.64.

Equations 10.63 and 10.65 can be used to obtain the energy difference between the ground states of electrons and holes in the quantum well, which represents the energy gap of the quantum well. Thus,

$$E_{\text{G, QW}} = \left| E_{1e} \right|_{\text{VBE}=0} - \left| E_{1h} \right|_{\text{VBE}=0} = E_{\text{G, Well material}} + \frac{\hbar^2 \pi^2}{2 t_w^2} \left(\frac{1}{m_e^*} \right) - \left\{ -\frac{\hbar^2 \pi^2}{2 t_w^2} \left(\frac{1}{m_h^*} \right) \right\}$$

$$= E_{\text{G, Well material}} + \frac{\hbar^2 \pi^2}{2 t_w^2} \left(\frac{1}{m_e^*} \right) + \frac{\hbar^2 \pi^2}{2 t_w^2} \left(\frac{1}{m_h^*} \right) = E_{\text{G, Well material}} + \frac{\hbar^2 \pi^2}{2 t_w^2} \left(\frac{1}{m_e^*} + \frac{1}{m_h^*} \right) \tag{10.66}$$

Comment: $E_{\text{G,QW}}$ depends on $E_{\text{G, Well material}}$, and t_w^2.

Example 10.6

(a) A quantum well solar cell contains 80 quantum wells. The thermionic emission escape time is 10ns, the tunneling time is 180ps, and carrier recombination time is 0.1μs. Find the carrier escape probability. (b) If the barrier thickness is reduced to decrease the tunneling time to 2ps, what is the probability of carrier escape?

Solution: (a) The escape time τ_{Escape} of a carrier from the quantum well is related to the thermionic emission time $\tau_{\text{Thermionic}}$ and tunneling time $\tau_{\text{tunneling}}$ through the equation

$$\frac{1}{\tau_{\text{Escape}}} = \frac{1}{\tau_{\text{Thermionic}}} + \frac{1}{\tau_{\text{Tunneling}}} = \frac{1}{10 \times 10^{-9}} + \frac{1}{180 \times 10^{-12}} = 10^8 + 5.56 \times 10^9 \tag{10.67}$$

$$= 10^8 \left(1 + 55.6 \right) = 56.6 \times 10^8 \, \text{s}^{-1}$$

For a single quantum well, the probability of escape of excess carriers from the quantum well is given by

$$\left| P \right|_{\text{Single QW}} = \frac{\dfrac{1}{\tau_{\text{Escape}}}}{\dfrac{1}{\tau_{\text{Escape}}} + \dfrac{1}{\tau_{\text{Recombination}}}} \tag{10.68}$$

where $\tau_{\text{Recombination}}$ is the lifetime of recombination of carriers. Since $\tau_{\text{Recombination}} = 0.1 \, \mu s = 0.1 \times 10^{-6}$s, from equations 10.67 and 10.68 we get

$$|P|_{\text{Single QW}} = \frac{56.6 \times 10^8}{56.6 \times 10^8 + \dfrac{1}{0.1 \times 10^{-6}}} = \frac{56.6 \times 10^8}{56.6 \times 10^8 + 10^7}$$

$$= \frac{56.6 \times 10^8}{10^7 (566 + 1)} = \frac{56.6 \times 10^8}{56.7 \times 10^8} = 0.998236 \tag{10.69}$$

The carrier escape probability for an N quantum well solar cell is

$$|P|_{N \text{QWs}} = \left(|P|_{\text{Single QW}} \right)^N \tag{10.70}$$

Applying equations 10.69 and 10.70, the carrier escape probability for a solar cell containing 80 quantum wells is

$$|P|_{N=80 \text{ QWs}} = \left(|P|_{\text{Single QW}} \right)^{80} = (0.998236)^{80} = 0.8683 \tag{10.71}$$

(b) From equation 10.67, the escape time for a tunneling time of 2ps is obtained from

$$\frac{1}{\tau_{\text{Escape}}} = \frac{1}{\tau_{\text{Thermionic}}} + \frac{1}{\tau_{\text{Tunneling}}} = \frac{1}{10 \times 10^{-9}} + \frac{1}{2 \times 10^{-12}} = 10^8 + 5 \times 10^{11}$$

$$= 10^8 (1 + 5000) = 5001 \times 10^8 \, \text{s}^{-1} = 5.001 \times 10^{11} \, \text{s}^{-1} \tag{10.72}$$

Applying equation 10.68, the carrier escape probability from a single quantum well is

$$|P|_{\text{Single QW}} = \frac{5.001 \times 10^{11}}{5.001 \times 10^{11} + \dfrac{1}{0.1 \times 10^{-6}}} = \frac{5.001 \times 10^{11}}{5.001 \times 10^{11} + 10^7} = \frac{5.001 \times 10^{11}}{10^7 \left(5.001 \times 10^4 + 1 \right)}$$

$$= \frac{5.001 \times 10^{11}}{10^7 (50011)} = \frac{5.001 \times 10^{11}}{5.0011 \times 10^{11}} = 0.99998 \tag{10.73}$$

By equation 10.70, the carrier escape probability from an 80 quantum well solar cell is

$$|P|_{N=80 \text{ QWs}} = \left(|P|_{\text{Single QW}} \right)^{80} = (0.99998)^{80} = 0.9984 \tag{10.74}$$

10.4 BANDGAP ENGINEERING OF QUANTUM WELL ARCHITECTURES

It is the process of altering the bandgap of a material, namely, the energy segment in which no electron states are available. Alloyed semiconductors offer many possibilities in this context because their composition can be varied as desired. The size of a chunk of material also affects its bandgap, particularly at the nanoscale. Any in-built strain in the material, too, changes its bandgap. In this section, we shall solve many examples which will enlighten us on the different ways in which bandgap is controlled in quantum wells. These will be mainly based on the quantum size effect.

Example 10.7

In an InGaAs/GaAsP solar cell, the indium concentration is varied to produce quantum wells from two alloys: $In_{0.25}Ga_{0.75}As$ and $In_{0.45}Ga_{0.55}As$. What is the effect of increasing the indium content on the bandgap of the alloy if the bandgap of $In_xGa_{1-x}As$ is given by (Nahory et al 1978)

$$E_G(x) = 1.425 - 1.501x + 0.436x^2 \text{ eV} \tag{10.75}$$

Solution: Applying equation 10.75, the bandgap of $In_{0.25}Ga_{0.75}As$ ($In_xGa_{1-x}As$ with $x = 0.25$) is

$$\begin{aligned} E_G(x = 0.25) &= 1.425 - 1.501 \times 0.25 + 0.436(0.25)^2 \\ &= 1.425 - 0.37525 + 0.02725 = 1.077 \text{eV} \end{aligned} \tag{10.76}$$

Again, by equation 10.75, the bandgap of $In_{0.45}Ga_{0.55}As$ ($In_xGa_{1-x}As$ with $x = 0.45$) is

$$E_G(x = 0.45) = 1.425 - 1.501 \times 0.45 + 0.436(0.45)^2 = 1.425 - 0.67545 + 0.08829 = 0.83784 \text{eV} \tag{10.77}$$

Equations 10.76 and 10.77 show that the bandgap decreases from 1.077eV to 0.83784eV upon increasing the indium fraction x from 0.25 to 0.45.

Example 10.8

(a) The energy gap of $Ga_aIn_{1-a}As_bP_{1-b}$ at 300K is given in electron volts by the formula (Goldberg and Schmidt 1999)

$$\begin{aligned} |E_G|_{Ga_aIn_{1-a}As_bP_{1-b}} &= 1.35 + 0.668a - 1.068b + 0.758a^2 + 0.078b^2 \\ &\quad - 0.069ab - 0.332a^2b + 0.03ab^2 \end{aligned} \tag{10.78}$$

where the symbols a, b, $1-a$, and $1-b$ stand for the mole fractions of the respective atoms in the alloy. Find the energy gaps of $In_{0.32}Ga_{0.68}As_{0.60}P_{0.40}$ and $In_{0.49}Ga_{0.51}P$. Identify which of these two materials is suitable for making the quantum well and which one can be used as the barrier layer. (b) Calculate the bandgap of the alloys having the composition $In_{0.70}Ga_{0.30}As_{0.05}P_{0.95}$ and $In_{0.38}Ga_{0.62}As_{0.34}P_{0.66}$. How does the bandgap depend on composition of the alloy?

Solution: (a) If we write the formula of the alloy $In_{0.32}Ga_{0.68}As_{0.60}P_{0.40}$ in the form $Ga_{0.68}In_{0.32}As_{0.60}P_{0.40}$, we immediately see that a = 0.68, 1 − a = 0.32, b = 0.60, and 1 − b = 0.40. Putting the values of a and b in equation 10.78, we get

$$\begin{aligned} |E_G|_{In_{0.32}Ga_{0.68}As_{0.60}P_{0.40}} &= |E_G|_{Ga_{0.68}In_{0.32}As_{0.60}P_{0.40}} \\ &= 1.35 + 0.668 \times 0.68 - 1.068 \times 0.60 + 0.758(0.68)^2 + 0.078(0.60)^2 \\ &\quad - 0.069 \times 0.68 \times 0.60 - 0.332(0.68)^2 \times 0.60 + 0.03 \times 0.68(0.60)^2 \\ &= 1.35 + 0.45424 - 0.6408 + 0.3504992 + 0.02808 - 0.028152 \\ &\quad -0.09211008 + 0.007344 = 2.1901632 - 0.76106208 = 1.42910112 \approx 1.43 \text{eV} \end{aligned} \tag{10.79}$$

We write the alloy formula $In_{0.49}Ga_{0.51}P$ as $Ga_{0.51}In_{0.49}P$. We see that a = 0.51, 1 − a = 0.49, b = 0, and 1 − b = 1 − 0 = 1. Putting the values of a and b in equation 10.78, we get

$$\begin{aligned} |E_G|_{In_{0.49}Ga_{0.51}P} &= 1.35 + 0.668 \times 0.51 - 1.068 \times 0 + 0.758(0.51)^2 + 0.078(0)^2 \\ &\quad - 0.069 \times 0.51 \times 0 - 0.332(0.51)^2 \times 0 + 0.03 \times 0.51(0)^2 \\ &= 1.35 + 0.34068 - 0 + 0.1971558 + 0 - 0 - 0 + 0 = 1.8878358 \approx 1.89 \text{eV} \end{aligned} \tag{10.80}$$

From equations 10.79 and 10.80, it is evident that $In_{0.32}Ga_{0.68}As_{0.60}P_{0.40}$ is the narrow-bandgap material to be used for making the well, and $In_{0.49}Ga_{0.51}P$ is the wide-bandgap material to be used for the barrier layer.

(b) For $In_{0.70}Ga_{0.30}As_{0.05}P_{0.95}$ or $Ga_{0.30}In_{0.70}As_{0.05}P_{0.95}$, a = 0.3 and b = 0.05. Applying equation 10.78, we get

$$
\begin{aligned}
\left|E_G\right|_{In_{0.70}Ga_{0.30}As_{0.05}P_{0.95}} &= \left|E_G\right|_{Ga_{0.30}In_{0.70}As_{0.05}P_{0.95}} \\
&= 1.35 + 0.668 \times 0.30 - 1.068 \times 0.05 + 0.758(0.30)^2 + 0.078(0.05)^2 - 0.069 \times 0.30 \\
&\quad \times 0.05 - 0.332(0.30)^2 \times 0.05 + 0.03 \times 0.30(0.05)^2 \\
&= 1.35 + 0.2004 - 0.0534 + 0.06822 + 0.000195 - 0.001035 - 0.001494 + 0.0000225 \\
&= 1.6188375 - 0.055929 = 1.5629085 \approx 1.563eV
\end{aligned}
\tag{10.81}
$$

For $In_{0.38}Ga_{0.62}As_{0.34}P_{0.66}$ or $Ga_{0.62}In_{0.38}As_{0.34}P_{0.66}$, a = 0.62 and b = 0.34. Equation 10.78 gives

$$
\begin{aligned}
\left|E_G\right|_{In_{0.38}Ga_{0.62}As_{0.34}P_{0.66}} &= \left|E_G\right|_{Ga_{0.62}In_{0.38}As_{0.34}P_{0.66}} \\
&= 1.35 + 0.668 \times 0.62 - 1.068 \times 0.34 + 0.758(0.62)^2 + 0.078(0.34)^2 - 0.069 \times 0.62 \\
&\quad \times 0.34 - 0.332(0.62)^2 \times 0.34 + 0.03 \times 0.62(0.34)^2 \\
&= 1.35 + 0.41416 - 0.36312 + 0.2913752 + 0.0090168 - 0.0145452 - 0.043391072 \\
&\quad + 0.00215016 = 2.06670216 - 0.421056272 = 1.645645888 \approx 1.65eV
\end{aligned}
\tag{10.82}
$$

Comparing equations 10.79, 10.81, and 10.82 we note that the bandgap of InGaAsP is increased from 1.43 eV to 1.563 eV and then 1.65eV upon altering the composition of the alloy.

Example 10.9

(a) A quantum well is formed by interposing an 8nm-thick layer of $In_{0.5}Ga_{0.5}N$ low-bandgap material between layers of GaN wide-bandgap material. The dependence of energy gap of $In_xGa_{1-x}N$ on indium fraction x is given by Vegard's law (Vegard 1921; Nahory et al 1978), with the curvature correction applied through a bowing parameter:

$$
E_{G,InGaN} = xE_{G,InN} + (1-x)E_{G,GaN} - bx(1-x)
\tag{10.83}
$$

where $E_{G, InN}$ is the energy gap of indium nitride = 1.97 eV, $E_{G, GaN}$ is the energy gap of gallium nitride = 3.2 eV, and b is the bandgap energy bowing parameter. The bowing parameter b varies with composition (Berrah et al 2008); a mean value 2eV may be assumed for the calculations. The electron and hole effective masses for $In_{0.5}Ga_{0.5}N$ are $m_e^* = 0.12m_0$, $m_{lh}^* = 0.0583m_0$, and $m_{hh}^* = 0.6m_0$ (Anand et al 2009). Find the energy gap of the quantum well for both light and heavy hole cases.

(b) How does the contribution of the energy gap of the well material change when the composition of the $In_xGa_{1-x}N$ alloy is varied from $In_{0.2}Ga_{0.8}N$ to $In_{0.3}Ga_{0.7}N$?

Solution: (a) Applying equation 10.83, the energy gap of $In_{0.5}Ga_{0.5}N$ is

$$
\begin{aligned}
E_{G, In_{0.5}Ga_{0.5}N} &= 0.5 \times 1.97 + (1-0.5) \times 3.2 - 2 \times 0.5(1-0.5) \\
&= 0.985 + 1.6 - 0.5 = 2.085eV
\end{aligned}
\tag{10.84}
$$

From equation 10.61, the energy gap for light hole case is

$$
\begin{aligned}
\left| E_{G, QW} \right|_{lh} &= E_{G, \text{Well material}} + \frac{\hbar^2 \pi^2}{2 t_w^2} \left(\frac{1}{m_e^*} + \frac{1}{m_{lh}^*} \right) \\
&= 2.085 + \frac{\left(1.05457 \times 10^{-34}\right)^2 \times \left(3.14159\right)^2}{2 \times \left(8 \times 10^{-9}\right)^2 \times 9.109 \times 10^{-31}} \left(\frac{1}{0.12} + \frac{1}{0.0583} \right) \times \frac{1}{1.602 \times 10^{-19}} \\
&= 2.085 + \frac{10.976 \times 10^{-68}}{1165.95 \times 1.602 \times 10^{-31-18-19}} \left(8.333 + 17.15266\right) \\
&= 2.085 + 0.005876 \times 10^{-68+31+18+19} \left(8.333 + 17.15266\right) \\
&= 2.085 + 0.005876 \times 10^0 \times 25.4857 = 2.085 + 0.14975 = 2.235 \text{eV}
\end{aligned}
\tag{10.85}
$$

since $m_e^* = 0.12 m_0 = 0.12 \times 9.109 \times 10^{-31}$kg and $m_{lh}^* = 0.0583 m_0 = 0.0583 \times 9.109 \times 10^{-31}$kg. Also, by equation 10.61, the energy gap for heavy hole case is

$$
\begin{aligned}
\left| E_{G, QW} \right|_{hh} &= E_{G, \text{Well material}} + \frac{\hbar^2 \pi^2}{2 t_w^2} \left(\frac{1}{m_e^*} + \frac{1}{m_{hh}^*} \right) \\
&= 2.085 + \frac{\left(1.05457 \times 10^{-34}\right)^2 \times \left(3.14159\right)^2}{2 \times \left(8 \times 10^{-9}\right)^2 \times 9.109 \times 10^{-31}} \left(\frac{1}{0.12} + \frac{1}{0.6} \right) \\
&= 2.085 + 0.005876 \times 10^0 \left(8.333 + 1.67\right) = 2.085 + 0.005876 \times 10.003 \\
&= 2.085 + 0.058778 = 2.14378 \text{eV}
\end{aligned}
\tag{10.86}
$$

since $m_{hh}^* = 0.6 m_0 = 0.6 \times 9.109 \times 10^{-31}$ kg.

The energy gap of the quantum well is 2.2eV for the light hole case, and 2.14eV for heavy hole case.

(b) For $x = 0.2$, by equation 10.83

$$
\begin{aligned}
\left| E_{G, \text{In}_{0.5}\text{Ga}_{0.5}\text{N}} \right|_{x=0.2} &= 0.2 \times 1.97 + \left(1 - 0.2\right) \times 3.2 - 2 \times 0.2 \left(1 - 0.2\right) \\
&= 0.394 + 2.56 - 0.32 = 2.634 \text{eV}
\end{aligned}
\tag{10.87}
$$

For $x = 0.3$, application of equation 10.83 gives

$$
\begin{aligned}
\left| E_{G, \text{In}_{0.5}\text{Ga}_{0.5}\text{N}} \right|_{x=0.3} &= 0.3 \times 1.97 + \left(1 - 0.3\right) \times 3.2 - 2 \times 0.3 \left(1 - 0.3\right) \\
&= 0.591 + 2.24 - 0.42 = 2.411 \text{eV}
\end{aligned}
\tag{10.88}
$$

As seen from equations 10.87 and 10.88, the contribution of energy gap of the well material ($E_{G, \text{Well material}}$) towards the total energy gap of the quantum well ($E_{G, QW}$) decreases from 2.63eV to 2.41eV as x increases from 0.2 to 0.3.

10.5   INCLUSION OF STRAIN AND ELECTRIC FIELD EFFECTS FOR GENERALIZATION OF ENERGY GAP VARIATION EQUATION

The foregoing treatment of bandgap engineering is limited primarily to quantum size effect. But we know that the various material systems used in the fabrication of quantum well solar cells are susceptible to strain and electric field effects. The influence of different factors governing the variation of bandgap is expressed by the general equation for effective energy gap $E_{G, \text{Effective}}$, written as (Sayed and Bedair 2019)

$$E_G^{\text{Effective}} = E_G^{\text{Relaxed material}} + \Delta E_G^{\text{Quantum size effect}} + \Delta E_G^{\text{Strain effect}} + \Delta E_G^{\text{Quantum confined Stark effect}} \quad (10.89)$$

where the delta terms represent the incremental strains arising from the respective effects indicated. The quantum size effect, as have already explored, results in the discretization of energy into distinct levels when the thickness of the well and de Broglie wavelength of electrons are of the same order.

Strain modifies the bandgap in two ways:

(i) The difference ΔE between the conduction and valence band energies is a function of strain, the elastic stiffness coefficients, and the hydrostatic deformation potential. So strain alters the relative difference between the aforesaid energies.

(ii) The most significant effect of strain on the bandgap is manifested in the hole valence bands. Strain splits the degeneracy of heavy hole and light hole levels. The heavy hole level is transferred to a higher energy for material under compression, whereas the light hole level is shifted to a higher energy for a material under tension.

Quantum-confined Stark effect (QCSE) is concerned with the changes in optical absorption or emission spectra of quantum wells under an electric field when the discrete energy levels occupied by the electrons and the holes are affected (Miller et al 1984). The electron energy levels descend towards lower energies while the hole energy levels ascend to higher energies. This decent and ascent of electron and hole energy levels reduces the number of permitted frequencies for absorption or emission of light.

The magnitude of electric field is very high in InGaN/GaN structures (MV/cm) but comparatively lower in GaAs structures (kV/cm). So the absorption and emission processes are appreciably influenced by the QCSE in InGaN/GaN quantum well solar cells. However, it can be neglected in GaAs-based quantum well devices. Furthermore, in InGaN/GaN devices, the electric field increases with indium percentage, resulting in a red shift in the bandgap.

10.6 DISCUSSION AND CONCLUSIONS

The quantum well solar cell can be looked upon as an alternative architecture to tandem solar cell aimed at enhancing efficiency. The cell contains a multiquantum well in the intrinsic region of a P-I-N diode. The bandgap of a quantum well structure can be modulated to match with the solar spectrum. Photons having energies greater than the bandgap energy of the semiconductor used for the barrier layer create electron-hole pairs. In this respect, the quantum well cell behaves like a conventional cell. In addition, photons with energy less than the bandgap energy of the barrier layer also produce electron-hole pairs in the quantum well layer. In this aspect, it differs from the normal cell. If the material quality is good, the photocarriers escape from the quantum wells by thermal energy or tunneling processes and contribute to the photocurrent.

From the design perspective, the range of wavelengths in the solar spectrum to which the cell is sensitive is determined by the quantum well structure whose main parameters are composition of the quantum wells and the thicknesses of the well and barrier regions. The design of multiple–quantum well solar cell needs to be carefully optimized because it brings into play two opposing effects on solar cell performance. The performance of the cell is the resultant of these two counteracting effects (Abolghasemi and Kohandani 2018). The advantageous aspect of this structure is that the short-circuit current is increased. The disadvantageous aspect is that the open-circuit voltage is reduced. Therefore, the optimal design calls for reaping the benefits of increase in short-circuit current to the extent that the loss in open-circuit voltage is not substantially sacrificed, thus paving the way towards a higher overall efficiency.

Questions and Answers

10.1   Which region of a quantum well solar cell differs from a conventional cell? Answer: The intrinsic region of the P-I-N diode structure. Instead of being made from a single semiconductor, it comprises a quantum well structure, single or multiple.

10.2   What limitation of the conventional solar cell does the quantum well solar cell try to deal with? Answer: The bandgap of the intrinsic region of a solar cell C_1 made from a single semiconductor S_1 has one distinct value E_{G1}, and only photons having energies exceeding this energy E_{G1} can create electron-hole pairs in this solar cell. Incident photons with energies $< E_{G1}$ are not utilized. To utilize these photons, we need another solar cell C_2 made from a material with lower-bandgap E_{G2} semiconductor S_2. In this way, a combination of solar cells $C_1, C_2, C_3, \ldots$ made from semiconductors $S_1, S_2, S_3, \ldots$ with bandgaps $E_{G1}, E_{G2}, E_{G3}, \ldots$ are necessary.

In quantum well solar cells, the well layers are made from semiconductors of different bandgap than barrier layers. Additionally, varying thickness of the well layer allows the bandgaps to be changed without changing material. So a quantum well solar cell tries to perform the same function as a combination of solar cells made with different bandgap semiconductors.

Caution: Use of lower-bandgap material decreases the open-circuit voltage of the solar cell so that the sum total effect of quantum well is assessed from the combination of short-circuit current and open-circuit voltage changes, both of which must be taken into account. The next two exercises present the explanation.

10.3   What is the advantage gained by replacing the single semiconductor in the intrinsic region of a P-I-N diode solar cell by a quantum well structure? Answer: The short-circuit current of the solar cell increases. This happens because additional electron-hole pairs are produced by low-energy photons striking the smaller-bandgap material of the quantum well layer. These electrons add to the photocurrent, increasing its value beyond that in a single-semiconductor solar cell.

10.4   What is the disadvantage incurred by replacing the single semiconductor in the intrinsic region of a P-I-N diode solar cell by a quantum well structure? Answer: The open-circuit voltage of the solar cell decreases. This happens because the open-circuit voltage of the solar cell depends on the bandgap of the semiconductor. The quantum well layer of the quantum well solar cell has a smaller bandgap, while the barrier layer has a larger bandgap. So the resultant open-circuit voltage of the solar cell lies between the voltages obtained from the smaller-bandgap well layer and the larger-bandgap barrier layer of the quantum well.

10.5   Suppose the intrinsic region of the P-I-N diode solar cell made of semiconductor with bandgap $E_{Bandgap}$ is replaced by a quantum well made of two semiconductors, one larger-bandgap semiconductor (used in barrier layers) with energy gap $E_G > E_{Bandgap}$ and one smaller-bandgap semiconductor (used in the quantum well layer) with energy gap $E_g < E_{Bandgap}$. How does this solar cell behave with respect to the original cell? Answer: The open-circuit voltage is high for $E_G > E_{Bandgap}$ material of the quantum well, while short-circuit current is high for $E_g < E_{Bandgap}$ material. So the short circuit of the current of the quantum well solar cell is higher than that for a solar cell made with $E_{Bandgap}$ material, and the open-circuit voltage is lower than for the solar cell of $E_{Bandgap}$ material. Proper design, including material choices, barrier layer, well layer thicknesses, etc., leads to an optimized structure which increases short-circuit current by a factor greater than the decrease in open-circuit voltage. The electrons and holes escape from the wells, yielding a higher output current at a voltage between that of the barrier layer and well materials, i.e., a high output current is obtained at a lower output voltage than for the $E_{Bandgap}$ material.

10.6   What does the difference between solar cells with and without wells tell about wavelengths of light utilized by the two types of solar cells? Answer: The cell without quantum well cannot utilize lower-energy (lower frequency) or longer-wavelength light (infrared portion). The quantum well solar cell can utilize infrared wavelengths because the well material has a smaller bandgap than the bandgap of material in solar cell without well.

Comment: In exercises 10.5 and 10.6, the influence of quantum well on solar cell performance is explained so that the difference between a solar cell with quantum well and one without quantum well is clarified.

10.7   What happens if the single quantum well is replaced by multiple quantum wells such that the different wells have different layer thicknesses? Answer: As the bandgap of the well is determined by well layer thickness, we have a series of wells of different bandgaps. Therefore, a higher short-circuit current is obtained at a voltage intermediate between that of the lowest and the highest bandgaps in the structure. In other words, the benefits of quantum wells are extended over a broader portion of solar spectrum covering more frequencies/wavelengths of light than for single–quantum well case.

Comment: In exercise 10.7, the difference between single– and multiple–quantum well solar cells is explained.

10.8   Look at the energy band diagram of a single quantum well in Figure 10.2(a) and explain why an electron sitting on the conduction band edge of the small-bandgap material remains trapped inside the well. Answer: The conduction band edge of the smaller-bandgap material is at a lower energy than the conduction band edge of the larger-bandgap material. Since an electron moves from higher energy to lower energy in the band diagram, it cannot move to the larger-bandgap material and remains confined in the small-bandgap material.

Comment: Trapping of the electron in the well is explained.

10.9   Looking at the energy band diagram of a single quantum well in Figure 10.2(a), explain why a hole sitting on top of the valence band edge of the small-bandgap material remains trapped inside the well. Answer: The valence band edge of the smaller-bandgap material is at a higher energy than the valence band edge of the larger-bandgap material. Since a hole moves from lower energy to higher energy in the band diagram, it cannot move to the larger-bandgap material and remains confined in the small-bandgap material.

Comment: Trapping of the hole in the well is explained.

10.10   What is meant by de Broglie wavelength? Answer: The wavelength ascribed to a particle of matter = Planck's constant/Momentum of the particle (= mass × velocity).

10.11   How does the behavior of an electron trapped in a quantum well differ from that of an electron outside the well? Answer: The electron inside the well can acquire only discrete values of energy, while an electron outside the well can possess a continuous range of energies.

10.12   Mention one important outcome of the finite potential well model about electron behavior which is not apparent from the infinite potential well model. Answer: Quantum mechanical tunneling viz the existence of a finite probability of finding the electron outside the well if the barrier width and height are small.

10.13   How does a superlattice differ from a multiple quantum well? Answer: In a superlattice, the quantum wells are located close to each other so that the wave functions of various wells merge into each other, forming minibands. In a multiple quantum well, the different quantum wells are situated so much apart that the wave function of one well does not overlap with that of the adjoining well. Then each well is treated separately and energy levels of the wells are discretized without any miniband formation.

10.14 What event is observed in the band diagram when an electron-hole pair is produced by shining light on the solar cell? Answer: In the band diagram, an electron is seen rising from the valence band to the conduction band. The vacant site left behind in the valence band is the hole.

10.15 What event takes place in the band diagram when an electron recombines with a hole? Answer: An electron is seen falling from the conduction band to fill the hole in the valence band.

10.16 What is excess carrier lifetime? Answer: The time taken by an excess carrier in a semiconductor to recombine.

10.17 What is meant by thermionic emission? Answer: Liberation of electrons from the surface of a material by the effect of temperature.

10.18 What is tunneling? Answer: A quantum mechanical phenomenon in which a subatomic particle such as an electron is able to propagate and appear on the opposite side of a potential energy barrier which is larger than the kinetic energy of the particle. It is impossible for the particle to penetrate the barrier in classical mechanics parlance.

10.19 What is the electron transport mechanism under the built-in electric field in thin-well, thick-barrier multi–quantum well solar cells? Answer: Thermionic emission.

10.20 What is the electron transport mechanism under the built-in electric field in thick-well, thin-barrier multi–quantum well solar cells? Answer: Quantum mechanical tunneling.

10.21 What is an exciton? Answer: A combination of electron and hole which is free to move through a semiconductor as a single unit in which the electron and hole are bound together by electrostatic Coulomb force.

10.22 What does exciton binding energy mean? Answer: The energy of Coulomb attraction between electron and hole in an exciton.

10.23 Why is the excitonic model of optical absorption invoked? Answer: To explain that light can be absorbed in a material at an energy which is lower than the bandgap energy by an amount equal to the binding energy of the exciton.

10.24 When making a quantum well, it is desired to change the bandgap of the well material. Suggest three possible methods that can be applied to change the bandgap. Answer:
   (i)   By dimensional shrinkage, i.e., by diminishing the size of the material sufficiently to reach the nanoscale stage at which quantum confinement takes place; here, the thickness of quantum well layer is decreased.
   (ii) By altering the chemical composition of the material, e.g., by mixing different percentages of elements in an alloyed compound such as InGaN.
   (iii) By producing strain in the quantum well material, e.g., in InGaN well layer of InGaN/GaN quantum well solar cell.

10.25 How does the application of a strain on a material influence its bandgap? Answer: (i) By increasing the energy difference between conduction and valence band edges. (ii) By splitting the degeneracy of light and heavy hole levels of the valence band.

10.26 What does the quantum confined Stark effect do? Answer: It impacts the energy levels of electrons and holes; the former move downwards and latter are shifted upwards under the action of an electric field on the material.

10.27 Compare the relative magnitudes of electric fields in InGaN and GaAs solar cells. Answer: MV/cm in InGaN/GaN and kV/cm in GaAs-based quantum well solar cells.

10.28 How does the increase in indium content influence the performance of an InGaN/GaN quantum well solar cell? Answer: With increase in indium proportion, the electric field increases, resulting in a shift of the bandgap to cater to longer wavelengths, known as a red shift.

REFERENCES

Abolghasemi A. and R. Kohandani 2018 Numerical investigation of strain effects on properties of AlGaAs/ InGaAs multiple quantum well solar cells, Applied Optics, 57(24): 7045–7054.

Anand J., A. Buccheri, M. Gorley and I. Weaver 2009, November, 16 P2_12 InGaN quantum-well width w.r.t λ, Journal of Special Topics, Corpus ID: 56182252: 1–2, https://www.semanticscholar.org/paper/ of-Special-Topics-1-P-2-_-12-InGaN-quantum-well-w-.-Anand-Buccheri/ce070ebb3bfaed0cae6c0f5c8 ff53baab2d29b32#paper-header

Barnham K. W. J. and G. Duggan 1990 A new approach to high-efficiency multi-band-gap solar cells, Journal of Applied Physics, 67: 3490.

Barnham K. W. J., I. Ballard, J. P. Connolly, N. J. Ekins-Daukes, B.G Kluftinger, J. Nelson and C. Rohr 2002 Quantum well solar cells, Physica E: Low-Dimensional Systems and Nanostructures, 14(1–2): 27–36.

Berrah S., A. Boukortt and H. Abid 2008 The composition effect on the bowing parameter in the cubic InGaN, AlGaN and AlInN alloys, Semiconductor Physics, Quantum Electronics & Optoelectronics, 11(1): 59–62.

Goldberg Yu A. and N. M. Schmidt 1999 in: Levinshtein M, S. Rumyantsev and M. Shur (eds.), Handbook Series on Semiconductor Parameters, Vol. 2, World Scientific, London, pp. 153–179.

Harrison P. 2009 Quantum Wells, Wires and Dots: Theoretical and Computational Physics of Semiconductor Nanostructures, Wiley-Interscience, John Wiley & Sons Ltd., Chichesterm, England, pp. 2.1, 2.5.

Messiah A. 2014 Quantum Mechanics, Dover Publications, Inc., Mineola, New York, pp. 77–97.

Miller D. A. B., D. S. Chemla, T. C. Damen, A. C. Gossard, W. Wiegmann, T. H. Wood and C. A. Burrus 1984 Band-edge electroabsorption in quantum well structures: The quantum-confined Stark effect, Physical Review Letters, 53(22): 2173–2176.

Nahory R. E., M. A. Pollack, W. D. Johnston Jr. and R. L. Barns 1978 Band gap versus composition and demonstration of Vegard's law for $In_{1-x}Ga_xAs_yP_{1-y}$ lattice matched to InP, Applied Physics Letters, 33: 659, https://doi.org/10.1063/1.90455

Razeghi M. 2010 Technology of Quantum Devices, Springer Science+Business Media LLC, New York, pp. 1–40.

Sacchetti A. 2010 Electrical current in nanoelectronic devices, Physics Letters A, 374: 4057–4060.

Sayed I. and S. M. Bedair 2019 Quantum well solar cells: Principles, recent progress, and potential, IEEE Journal of Photovoltaics, 9: 402–423.

Vegard L. 1921 Die Konstitution der Mischkristalle und die Raumfüllung der Atome, Zeitschrift für Physik, 5(1): 17–26.

Zettili N. 2009 Quantum Mechanics: Concepts and Applications, John Wiley & Sons Ltd., Chichester, pp. 231, 234.

11 Quantum Well Solar Cells
Material Systems and Fabrication

Fabrication of quantum well solar cells involves considerations regarding the choice of equipment, material combinations, dimensional parameters (thicknesses of quantum well and barrier layers), and chemical composition of layers. The effects of dimensional parameters were examined in the previous chapter on the physics of quantum wells. In this chapter, the remaining topics are addressed.

11.1 TECHNIQUES FOR GROWTH OF QUANTUM WELL STRUCTURES

Two methods have found widespread use in the growth of quantum nanostructures. These are molecular beam epitaxy (MBE) and metal-organic chemical vapor deposition (MOCVD). Ideally, the lattice constants of materials constituting a heterostructure must be matched. In cases of lattice constant mismatch, materials can grow up to a critical thickness in a strained condition by adapting to the local lattice constant and preserving the epitaxial structure. Such strained materials have aroused technological interest because the bandgap is dependent on the strain. By proper choice of a pair of materials in which compressive stress is generated with respect to a substrate in one material and tensile stress in the other, a strain-balanced heterostructure is fabricated.

11.1.1 MOLECULAR BEAM EPITAXY

MBE is one of the cleanest techniques for layer-by-layer growth of films (Figure 11.1). It is an ultra-high vacuum technique carried out in vacuum $\sim 10^{-8}$ to 10^{-12} torr produced by turbo pumps. The films are grown at extremely slow rates $\sim 50\ nmmin^{-1}$. Elemental films as well as films of chemical compounds can be grown.

Different source materials are kept in elemental solid form in Knusden effusion cells or electron-beam evaporators on one side of the ultrahigh vacuum chamber. The materials are heated to respective sublimation temperatures, and the computer-controlled shutters of materials are selectively opened according to requirement, e.g., shutters in front of Ga and As sources are opened to grow film of binary alloy GaAs; additionally, the shutter-closing indium is opened for ternary alloy InGaAs growth. Furthermore, shutter for phosphorous is opened to get quaternary alloy InGaAsP. The molecular beams of materials interact directly with the substrate without any interaction among themselves or with any carrier gases *en route*, leading to epitaxial growth of ultrapure films. Hence, the name "molecular beam epitaxy" is assigned.

The source temperature determines the film growth rate, while the substrate temperature controls the film desorption rate. The temperature and flux ratios of the of individual source materials decide the composition and stoichiometry of the film of chemical compound. The substrate may be rotated at slow speed for better growth uniformity. Thickness precision to the extent of single atomic layer is achieved. Atomically flat heterointerfaces can be formed.

The films can be characterized *in situ* by showering electrons from a RHEED (reflection high-energy electron diffraction) gun upon them and observing their diffraction pattern on a RHEED fluorescent screen. Film quality is assessed, and thickness is measured.

DOI: 10.1201/9781003215158-16

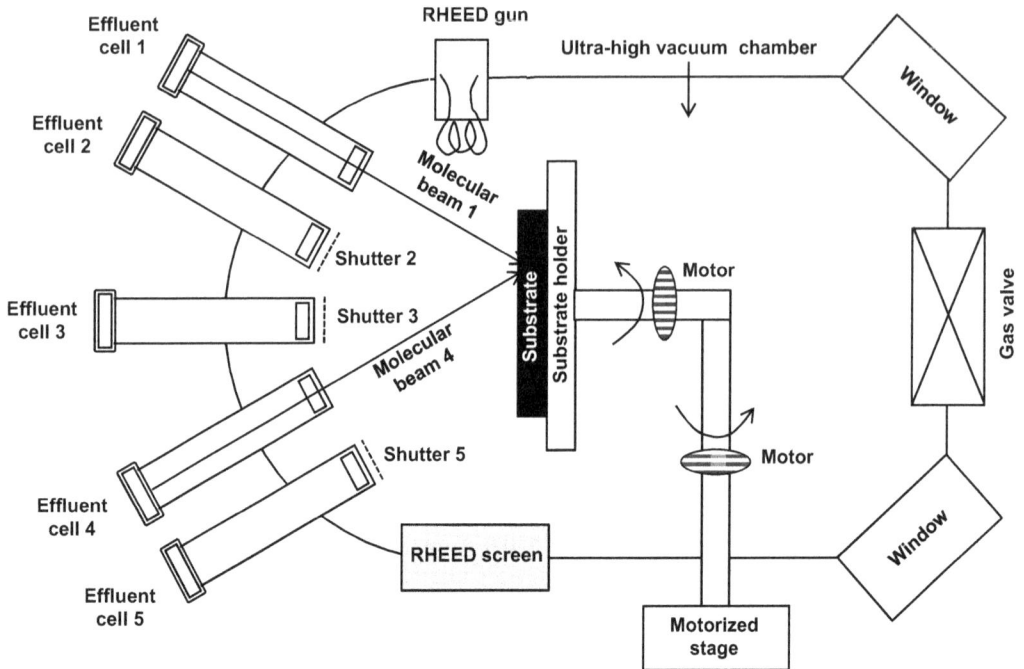

FIGURE 11.1 Molecular beam epitaxy system showing the effluent cells for producing molecular beams of different sources with a shutter attached to each cell to be opened and shut as required. RHEED gun and screen are connected for film characterization. The substrate is mounted on a rotating platform. The platform is energized by motors. A gas valve connects the working chamber with the vacuum system. Windows are provided for observation during experiment.

11.1.2 Metal-Organic Chemical Vapor Deposition

It is not a physical vapor deposition process. It involves a chemical reaction and is a variant of chemical vapor deposition. Unlike MBE, it is not performed in vacuum, uses carrier gases, and is carried out at pressures ~ 10–760 torr. MOCVD is done in a reactor made of stainless steel or quartz, either hot-wall or cold-wall, depending on whether the reactor walls are heated or only the substrate carrier is heated (Figure 11.2). The substrate is kept on a susceptor, usually made of graphite or some material which is chemically resistive to the corrosive gases used.

The method uses various precursors for depositing different materials, e.g., the precursor for aluminum is trimethyl or triethyl aluminum, precursor for gallium is trimethyl/triethyl gallium, the precursor for indium is trimethyl/triethyl indium, the precursor for antimony is trimethyl/triethyl antimony, and so on. These precursors are in liquid form, except trimethylindium, which is a solid. Precursors for phosphorous and arsenic are gases such as phosphine and arsine. In cases where precursors are in liquid form, they are transported to the reactor in vapor form by bubbling the carrier gases through them. In the reactor, chemical reactions take place between the reactants by maintaining the desired elevated temperature, and the reaction products are deposited on the substrate to form the film. The flow rate of the carrier gas and the temperature of the bubbler determine the quantity of vapors entering the reactor. The MOCVD system has an efficient exhaust and scrubber

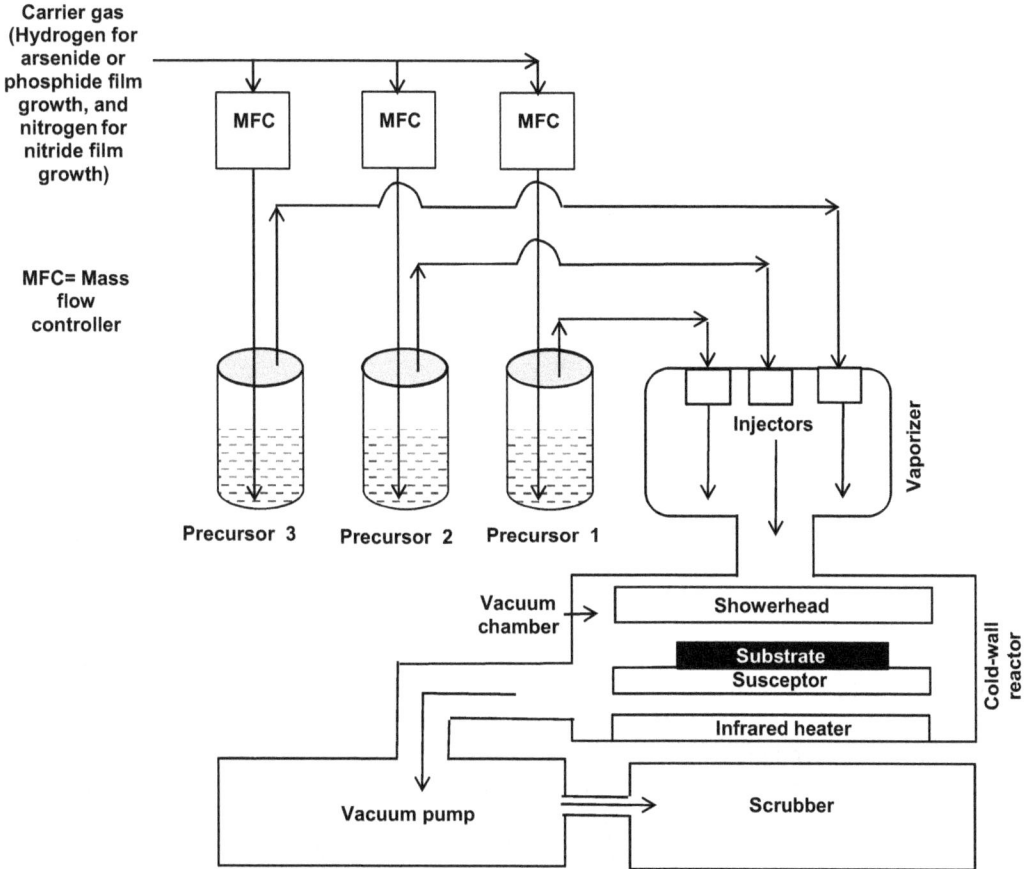

FIGURE 11.2 Setup of MOCVD system in which carrier gas is bubbled through selected precursor liquid(s) for transporting to the evaporator. The metal-organic vapors of reactant gases are fed to a showerhead in a vacuum chamber. The substrate is placed below the showerhead on a susceptor, which is heated by an infrared source. After deposition of the film on the substrate, the products of the reaction are sucked by the vacuum pump to the exhaust system and conveyed to a scrubber for cleaning.

assembly because the outgoing toxic gaseous products must be removed and inactivated for safe disposal.

11.1.3 DIFFERENCE BETWEEN MBE AND MOCVD

MBE is a physical deposition process, while MOCVD is a chemical deposition process. MBE works in ultrahigh vacuum ~ 10^{-8} torr or better, while MOCVD works under a high pressure of reactant gases ~ 10 torr or higher. MBE is more precise than MOCVD in controlling the thickness and composition of the deposited film. MBE can provide thickness control up to fractional monolayers, whereas MOCVD is able to do so with nanometer resolution. MBE is more expensive than MOCVD. MBE is useful for optimization of process parameters in the laboratory environment. MOCVD is used in large-scale commercial production. The explosive and toxic precursor gases used in MOCVD require strict enforcement of precautions, along with proper disposal, which are difficult in small laboratories.

11.2 MATERIALS SYSTEMS AND STRUCTURES FOR QUANTUM WELL SOLAR CELLS

The material systems are categorized under three principal heads: lattice-matched, stain-balanced, and strained systems.

11.2.1 LATTICE-MATCHED QUANTUM WELL SOLAR CELLS

In these solar cells, the lattice constants of the quantum well material, the barrier layer material, and the substrate are nearly equal. Owing to the matching of lattice constants, a layer of one semiconductor of arbitrary thickness can be grown on the other without any change in crystalline structure.

An important subclass of lattice-matched solar cells is the GaAs/AlGaAs quantum well solar cells (Figure 11.3). Lattice parameter of AlGaAs is 0.56533nm, and that of GaAs is 0.565325nm. Regarded as ideal from the standpoint of matching of lattice constants of the epitaxial layer with the substrate, the AlGaAs/GaAs material system generally does not demand special technological skills for controlling the mechanical stresses during film growth (Fardi 2012; Ladugin et al 2019).

Conversion efficiency calculations on the GaAs/AlGaAs quantum well solar cells were performed by Rimada et al (2005) for various widths of the wells and bandgaps of the barrier layer. The calculations revealed that the efficiencies of quantum well solar cells were superior to those of homogeneous P-I-N diode cells.

GaAs/AlGaAs quantum well solar cells were fabricated by embedding GaAs quantum wells of 3nm thickness between indirect bandgap $Al_{0.78}Ga_{0.22}As$ barrier layers (Noda et al 2014). The cells showed significant increase in photocurrent. This indicates that the recombination in the quantum wells was suppressed and the extraction of electrons was improved through the use of indirect gap material. The X-valley of AlGaAs was accredited for this improvement.

11.2.2 STRAIN-BALANCED QUANTUM WELL SOLAR CELLS

Strain balancing is exploited for pseudomorphic growth of materials with mismatched lattices while restricting defect densities to acceptably low levels. A pseudomorph, meaning "false form," is a compound obtained by a substitution process preserving the appearance and dimensions. However, the original material is replaced by another material. A strain-balanced semiconductor is a combination of layers of compressive and tensile stresses, such as compressive stress in the well layer and tensile stress in the barrier layer. Consequently, the heterostructure is locally strained. However, no force is exerted on the substrate.

Strain-balanced InGaAs/GaAsP multi–quantum wells are grown on GaAs substrate in a P-I-N diode structure to fabricate strain-balanced quantum well solar cells (Ekins-Daukes et al 1999; Zhang et al 2015). These solar cells exhibit comparable power conversion efficiency to good GaAs cells. Figure 11.4 illustrates the idea of strain balancing.

A strain-balanced and lattice-matched material system is the $In_xGa_{1-x}As_{1-y}P_y/In_{0.49}Ga_{0.51}P$ material system in which $In_xGa_{1-x}As_{1-y}P_y$ wells and $In_{0.49}Ga_{0.51}P$ barrier layers are grown in the intrinsic region of $In_{0.49}Ga_{0.51}P$ solar cells on GaAs substrates (Figure 11.5). Adopting a multipronged strategy, a subbandgap peak value of external quantum efficiency ~ 80% is achieved at 700nm (Sayed et al 2016, 2017). The procedures implemented include:

(i) Thickness of the low-doped InGaAsP wells is taken to be larger than that of the InGaP barrier layers. As a result, a large proportion of the space-charge region permeates

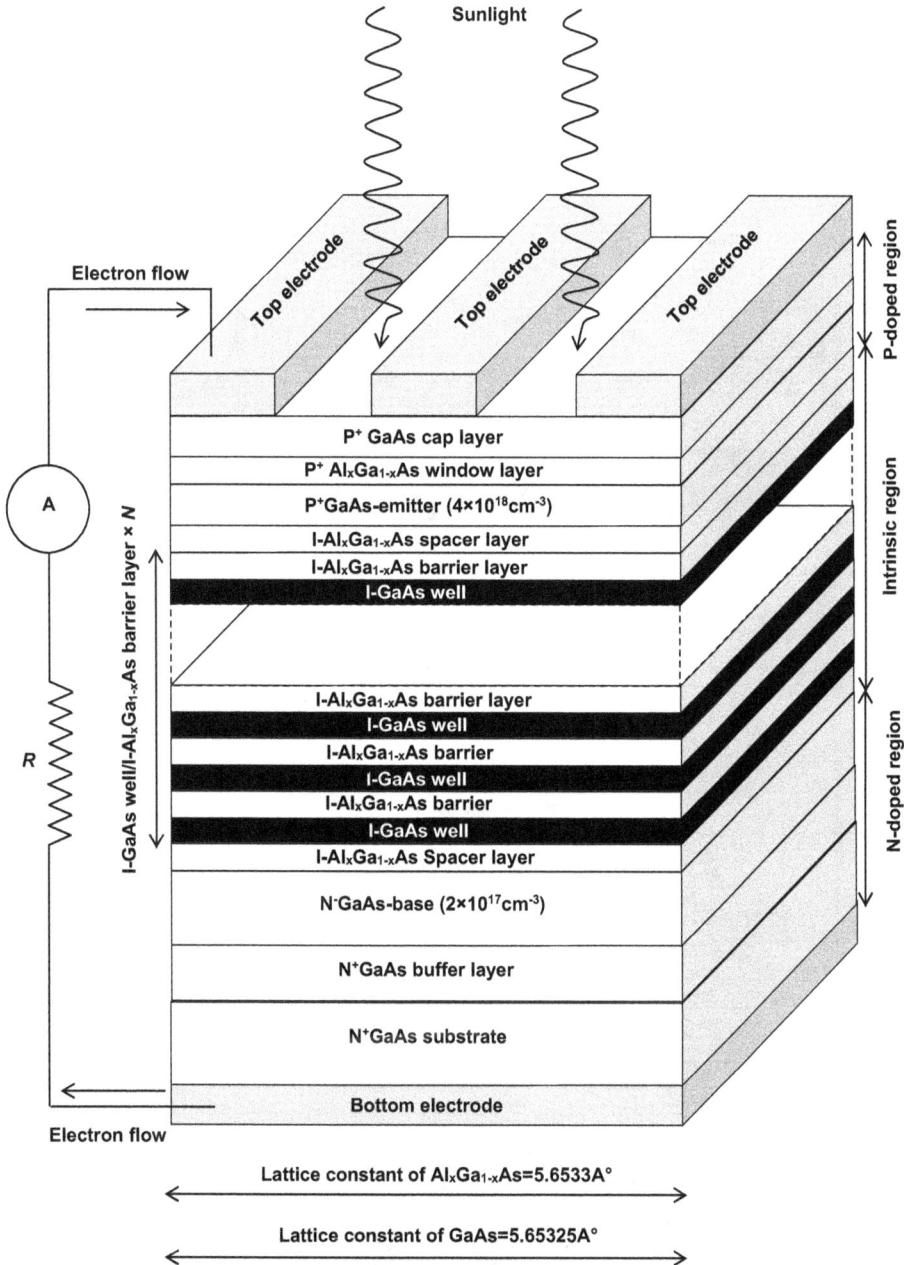

FIGURE 11.3 GaAs quantum well/AlGaAs barrier layer solar cell. From the bottom upwards, the solar cell consists of a metal electrode, an N⁺ GaAs substrate, N⁺ GaAs buffer layer, N-GaAs base, and I-Al$_x$Ga$_{1-x}$As spacer layer. After the spacer layer are placed, a succession of I-GaAs quantum well/I-Al$_x$Ga$_{1-x}$As barrier layer × N stacks. Beyond the quantum well/barrier layer stacks, there is an I-Al$_x$Ga$_{1-x}$As spacer layer. Then comes the P⁺ GaAs emitter. Following the emitter are the P⁺ Al$_x$Ga$_{1-x}$As window layer and P⁺ GaAs cap layer. At the top of these layers, the metal electrode is deposited. The N-doped part of the diode is made of N⁺ GaAs substrate, N⁺ GaAs buffer layer, and N⁺ GaAs base. The P-doped part is built of P⁺ GaAs emitter, P⁺Al$_x$Ga$_{1-x}$As window layer, and P⁺ GaAs cap layer. All the remaining semiconductor layers are undoped or intrinsic. Sunlight enters the cell through the transparent metal-top electrode patterned in the form of a grid. (Quantum wells/barriers are shown on enlarged scale for clarity.)

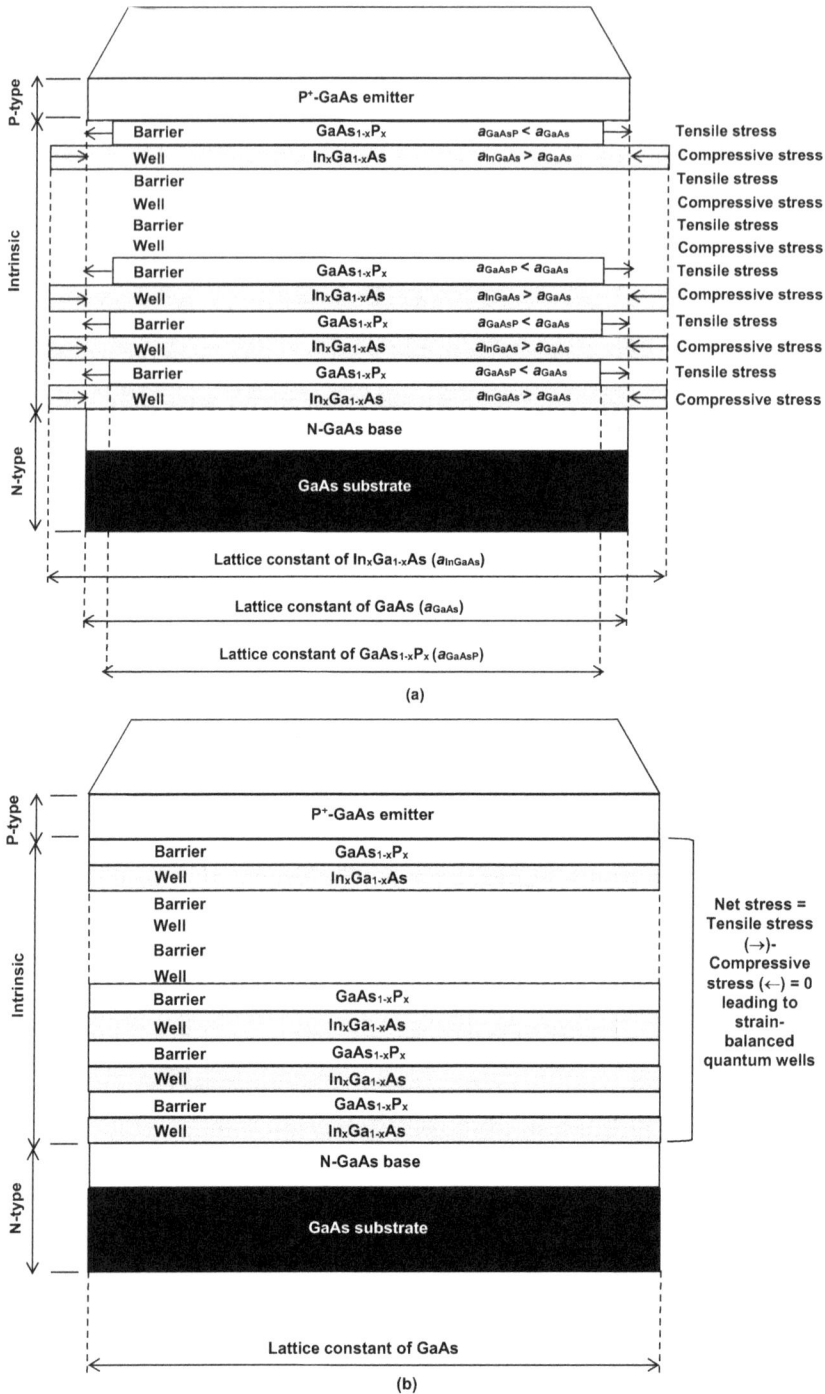

FIGURE 11.4 Strain balancing of $In_xGa_{1-x}As$ and $GaAs_{1-x}P_x$ layers with respect to GaAs substrate for fabricating a strain-balanced quantum well solar cell: (a) layers before strain balancing and (b) layers after strain balancing. Lattice constant of $In_xGa_{1-x}As$ is > lattice constant of GaAs, while the lattice constant of $GaAs_{1-x}P_x$ is < lattice constant of GaAs. So $In_xGa_{1-x}As$ layers experience compressive stress, while $GaAs_{1-x}P_x$ layers undergo tensile stress. The former layers undergo contraction, while the latter layers expand to equalize dimensionally with GaAs substrate. The outcome of oppositely directed compressive and tensile stresses is zero, yielding a strain-balanced device. (Quantum wells/barriers are shown on enlarged scale for clarity).

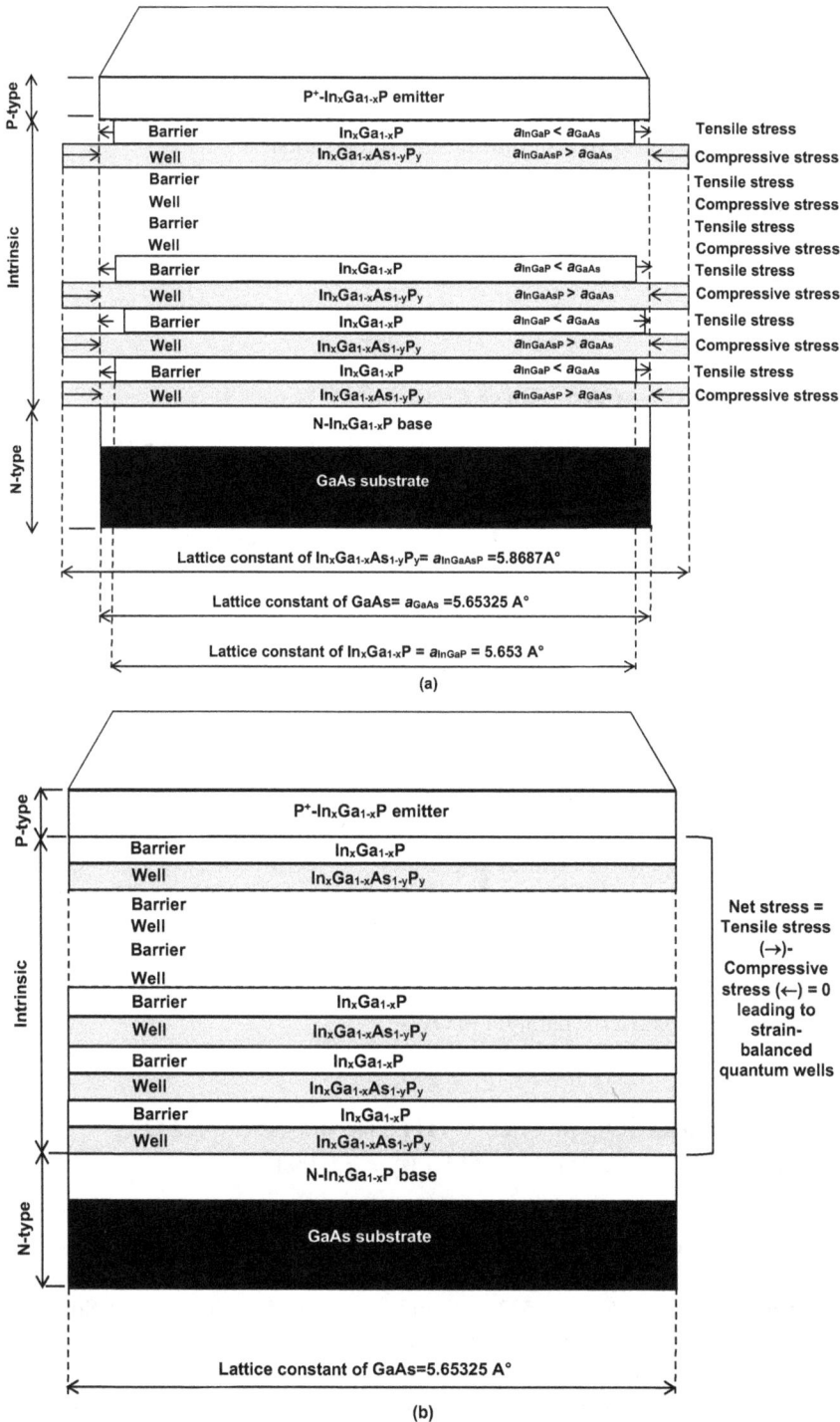

FIGURE 11.5 Strain balancing of $In_xGa_{1-x}As_{1-y}P_y$ and $In_xGa_{1-x}P$ layers with respect to GaAs substrate for fabricating a strain-balanced quantum well solar cell: (a) layers before strain balancing and (b) layers after strain balancing. Lattice constant of $In_xGa_{1-x}As_{1-y}P_y$ is > lattice constant of GaAs, while the lattice constant of $In_xGa_{1-x}P$ is < lattice constant of GaAs. So $In_xGa_{1-x}As_{1-y}P_y$ layers experience compressive stress, while $In_xGa_{1-x}P$ layers undergo tensile stress. The former layers are contracted, while the latter layers expand to equalize dimensionally with GaAs substrate. Balancing of compressive and tensile stresses results in null stress, giving a strain-balanced solar cell. (Quantum wells/barriers are shown on enlarged scale for clarity.)

the thicker InGaAsP wells, permitting the tunneling to take place through the thinner InGaP barrier layers. By this design, absorption of light is increased and collection of carriers is facilitated.

(ii)   The accumulated stress is low because of strain balancing, allowing the growth of 100 period QW solar cell with insignificant stress relaxation.

(iii) Etching away the GaAs substrate, followed by deposition of a planar back surface reflector, ensures better absorption of light.

(iv)   Reduction of carbon background doping concentration enables the insertion of 100 depleted QWs because the depletion region extends to a larger distance in a low-doped region.

11.2.3 STRAINED QUANTUM WELL SOLAR CELLS

InGaN/GaN quantum well solar cells fall under the strained solar cell group (Figure 11.6). Lattice mismatch between GaN and InN hinders the growth of InGaN films of high indium content and sufficient thickness (Choi et al 2013). Intolerable dislocation density in InGaN film resulting from this mismatch degrades the efficiency of the solar cell. Under 1.2 suns AM1.5G illumination, 30QW-period solar cells displayed external quantum efficiency of 70.9% at 700nm, which decreased to 39% at 450nm (Farrell et al 2011). Solar cells fabricated with 15–30 period $In_xGa_{1-x}N/$ GaN MQWs, $x = 0.19$ exhibited a conversion efficiency of 2% under 1 sun AM1.5G equivalent illumination (Valdueza-Felip et al 2014). Application of an external stress on InGaN/GaN quantum well solar cells increases the conversion efficiency by partial compensation of lattice mismatch–induced internal stress, thereby flattening the energy band diagram, which in turn increases the spatial overlapping of wave functions of electrons and holes. The increased spatial overlap enhances optical absorption. The maximum conversion efficiency rises by 11% for an applied strain = 0.134% (Jiang et al 2017).

$In_{0.08}Ga_{0.92}As/$GaAs superlattice containing strained, defect-free $In_{0.08}Ga_{0.92}As$ QWs with GaAs barriers is incorporated in a P-on-N $In_{0.49}Ga_{0.51}P/$GaAs heterojunction device to achieve > 26% 1-sun efficiency under standard AM1.5 illumination (Welser et al 2019). The structure (Figure 11.7) subdues radiative recombination and promotes carrier collection, giving better results than strain-balanced cells limited to 25% efficiency. Thin barrier layers of thickness below 4nm improve performance by augmenting carrier separation in QWs.

11.3   INVERTED GaAs SOLAR CELL WITH STRAIN-BALANCED GaInAs/GaAsP QUANTUM WELLS (η = 27.2%)

The bandgap of GaAs (1.41eV) is not optimal for the solar spectrum. The desired optimum bandgap is 1.34. So it is necessary to stretch the bandgap towards lower values (Steiner et al 2021). This can be achieved by incorporating GaInAs (bandgap 1.262eV) wells in the GaAs solar cell. But the lattice constants of GaAs (0.565325nm) and GaInAs (0.58687nm) differ. So a compressive strain is built up in the GaInAs layer. The stress increases with the growth of each successive well. The result is that dislocations are formed, and the structure contains a high dislocation density. Compensation of the compressive stress with tensile stress helps in minimization of defect generation. GaAsP (bandgap 1.7eV, lattice constant 0.56nm) (Strömberg et al 2020) barrier layers are used to provide the required tensile stress that cancels out the compressive effect of GaInAs well layers. By adopting this strain balancing strategy, a zero total stress condition is attained, avoiding the creation of defects. Therefore, a solar cell made by inclusion of GaInAs/GaAsP quantum wells is able to utilize photons having energies less than the GaAs bandgap and perform better than the GaAs solar cell.

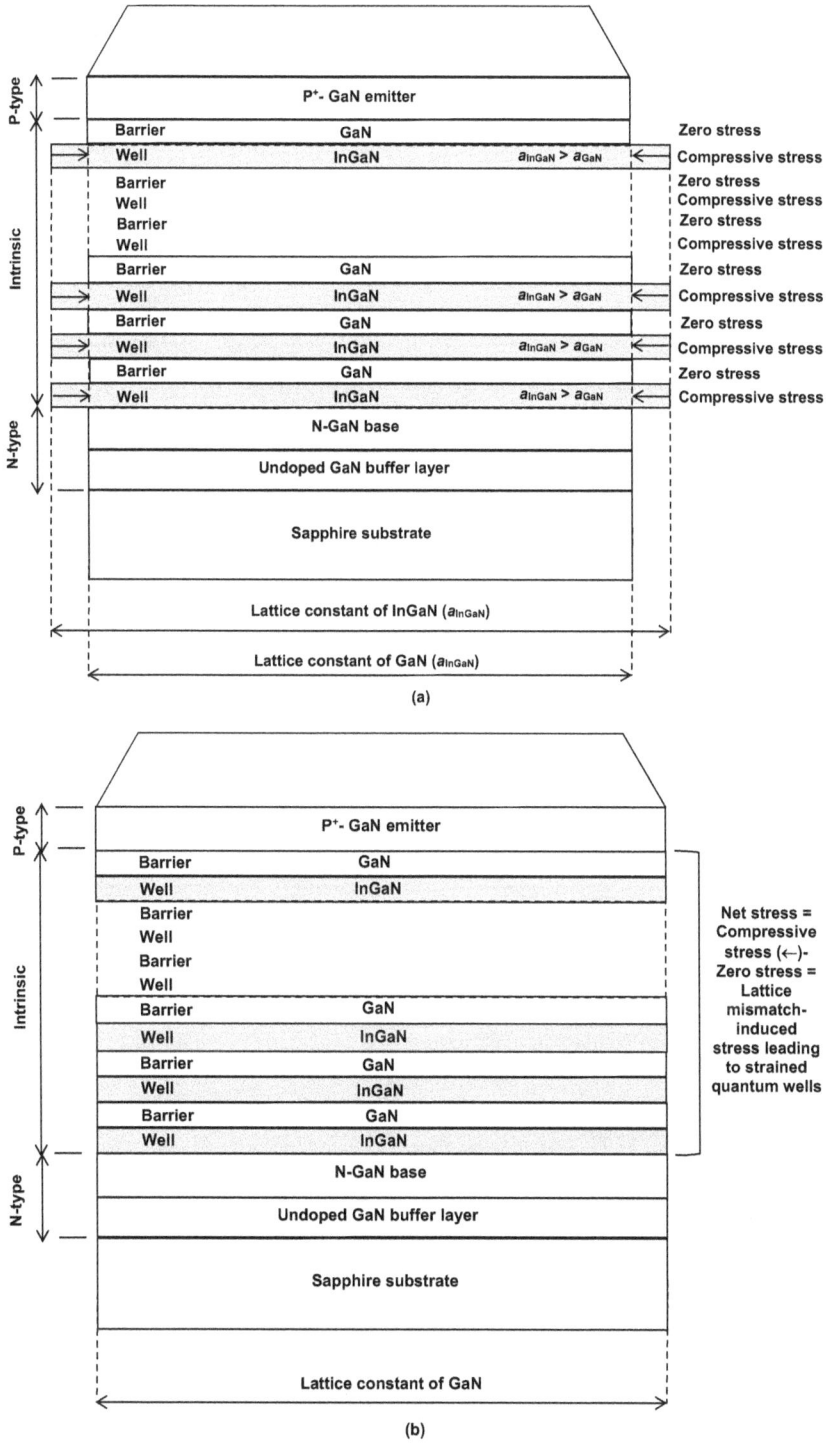

FIGURE 11.6 Strained solar cell with InGaN quantum well and GaN barrier layer showing that (a) initially InGaN film extends over the underlying GaN film due to its larger lattice parameter than GaN film, (b) finally InGaN film shrinks to fit over the GaN film with the InGaN/GaN stack equilibrating according to the lattice parameter of GaN, and (c) strained quantum well solar cell. The strain remains unbalanced because there is only compressive stress on InGaN film and there is no layer providing compensating tensile stress, resulting in the formation of a strained solar cell. (Quantum wells/barriers are shown on enlarged scale for clarity.)

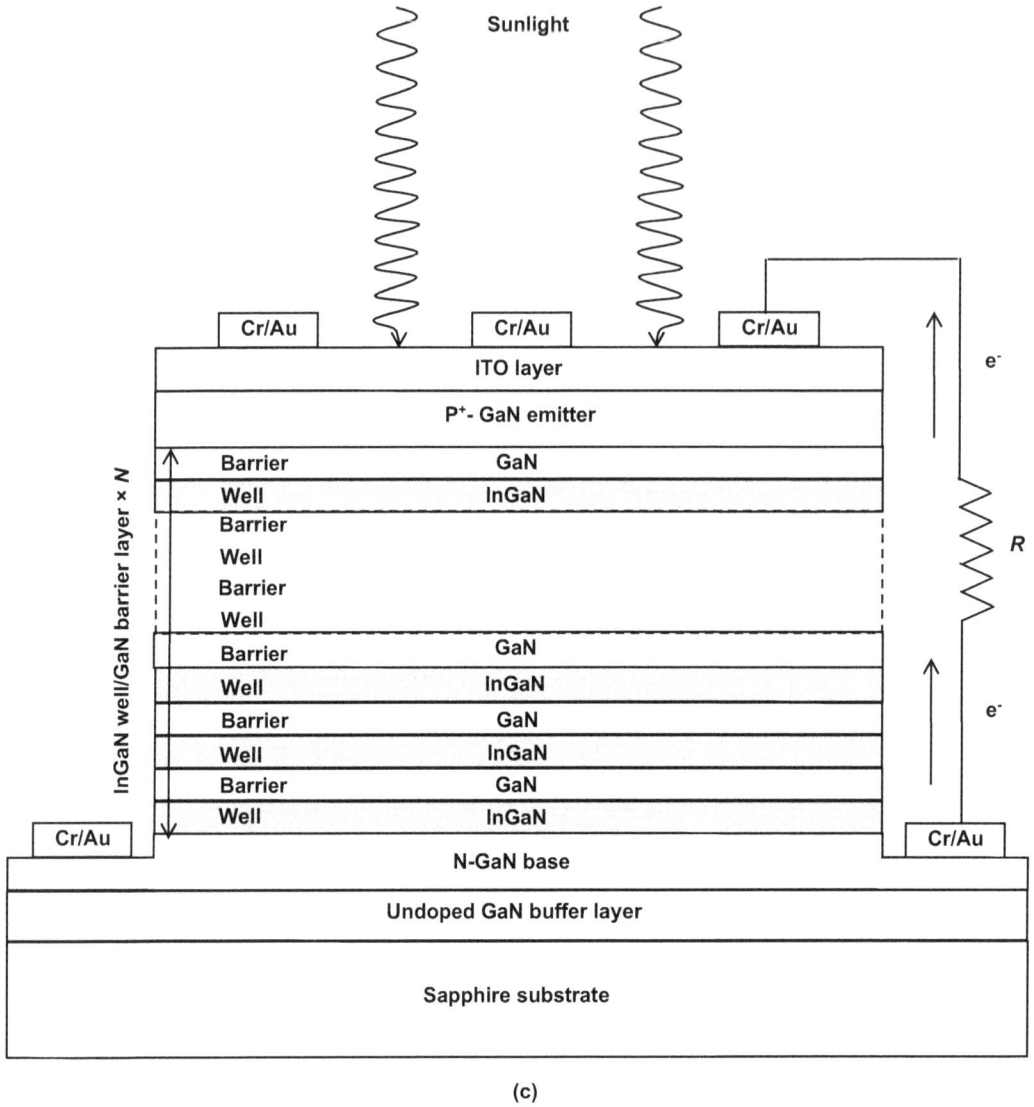

(c)

FIGURE 11.6   (Continued)

The quantum well solar cell is grown in inverted form over 350µm-thick GaAs substrate. Towards the end of the process, the GaAs substrate is removed by etching in $NH_4OH:H_2O$. The GaInP etch stop layer is removed after substrate etching in concentrated HCl.

The technique used for growth of the stack of III-V films in the solar cell is atmospheric pressure metal-organic vapor phase epitaxy (MOVPE). After the GaInP etch stop layer, an AlInP window layer is grown for surface passivation (Figure 11.8); GaInP etch stop layer is not shown in the figure. Then an N-type GaAs emitter layer of thickness 1µm is grown. 50–80 GaInAs/GaAsP quantum wells are formed. The next layer is an undoped GaAs buffer layer,

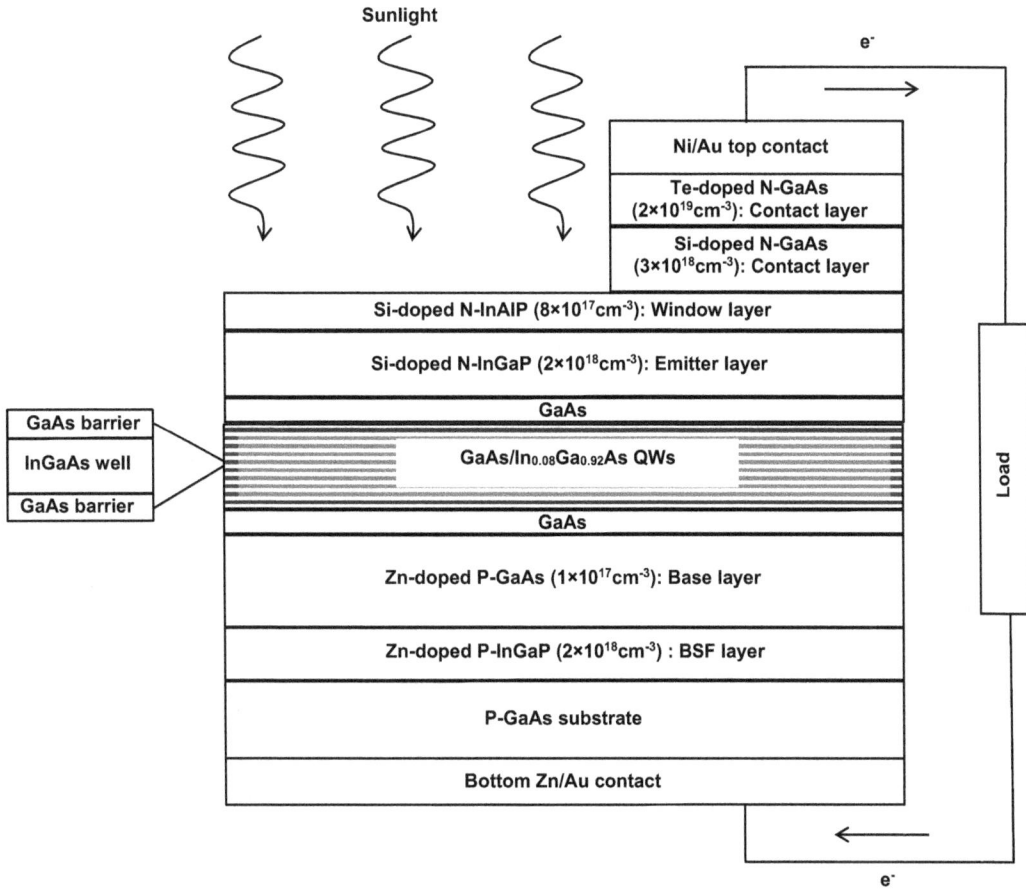

FIGURE 11.7 Heterojunction emitter GaAs solar cell with a thin strained quantum well superlatttice structure. The solar cell is made on a P-GaAs substrate. The bottom contact is a Zn/Au film. The solar cell has a P-InGaP BSF layer, a P-GaAs base layer, InGaAs quantum wells with GaAs barriers, an N-InGaP emitter layer, an N-InAlP window layer, N-GaAs contact layers, and Ni/Au top contact. Sunlight falls on the top contact side. Electron flow directions are shown.

following which a P-type base layer is grown, and finally an AlGaAs BSF layer. The BSF layer is electroplated with gold. Ni/Au electroplating is employed for front-grid metallization. MgF_2/ZnS antireflection coating is thermally evaporated over the front surface of the solar cell.

The open-circuit voltage of the quantum well solar cell is 1.04V. It is 60mV less than that for the GaAs solar cell without quantum wells. This voltage reduction arises from the lowering of the band edge from 1.41eV for GaAs to 1.35eV for the quantum well. At 1 sun, the short-circuit current density of the solar cell is 31.5 $mAcm^{-2}$. It is higher than the 29.5$mAcm^{-2}$ value for GaAs solar cell without quantum wells. The fill factor of the quantum well solar cell is 83%. The efficiency of this solar cell under AM1.5G is 27.18% at 1000Wm^{-2} (Steiner et al 2021).

Sunlight

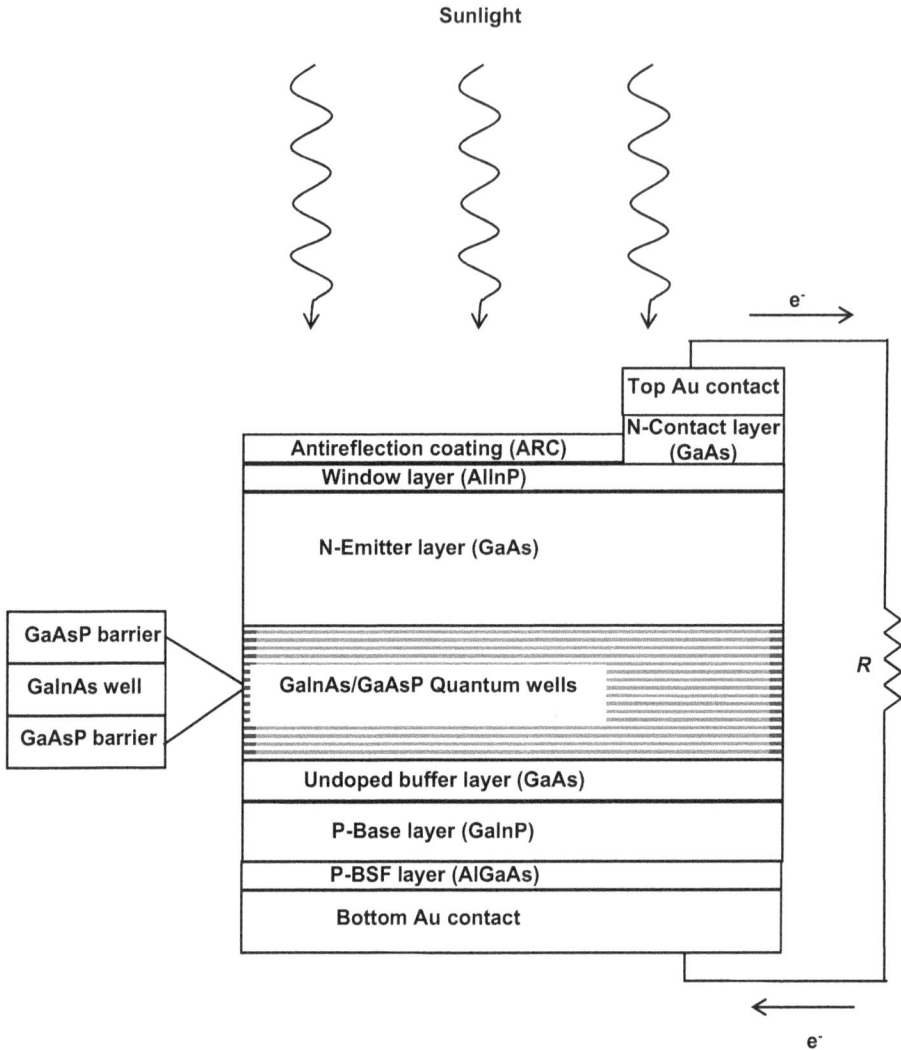

FIGURE 11.8 Single-junction GaAs solar cell with GaInAs/GaAsP quantum wells. Diagram shows the constituent layers from top to bottom: gold front contact, GaAs contact layer-ARC, AlInP window layer, N-type GaAs emitter layer, GaInAs/GaAsP QWs, undoped GaAs buffer layer, P-type GaInP base layer, P-type AlGaAs BSF layer, and gold back contact. When the cell is illuminated with sunlight, electron current flows through the load in the direction indicated on the diagram.

11.4 GaInP/GaAs DUAL-JUNCTION SOLAR CELL WITH STRAIN-BALANCED GaInAs/GaAsP QUANTUM WELLS IN THE BOTTOM CELL (η = 32.9%)

Single-junction solar cells are incapable of capturing the solar energy from photons having energies lower than the bandgap of the semiconductor used for the quantum well layer. Therefore, multiple-junction solar cells are developed. They are made from different semiconductors to tap energy from a broader chunk of the solar spectrum. In these solar cells, the junctions are stacked with larger-bandgap semiconductor receiving the incident sunlight, and the bandgap decreasing with successive junctions. By this arrangement, the low-energy photons which fail to produce electron-hole pairs from the large-bandgap semiconductor pass through it and are transmitted to the next junction made from smaller-bandgap semiconductor. Here, they are able to cause photogeneration. Still, further lower-energy photons move on to the next semiconductor of smaller bandgap than its predecessor. They are successful this time. Thus, this solar cell provides additional chances to photons which fail in the first junction.

The film growth direction is from top vertically downwards (Steiner et al 2021). The films are grown on a GaAs substrate with an etch stop layer. As mentioned above for the single-junction GaAs quantum well solar cell, the substrate and etch stop layer are removed postprocessing.

The top solar cell (Figure 11.9) has the structure, from top downwards, Au grid contact-MgF$_2$/ZnS ARC/AlInP window layer/N-type GaInP emitter/P-type GaInP base/P-type AlGaInP confinement layer. The confinement layer acts as a barrier to electron flow. It suppresses the leakage of electrons to the underlying layer.

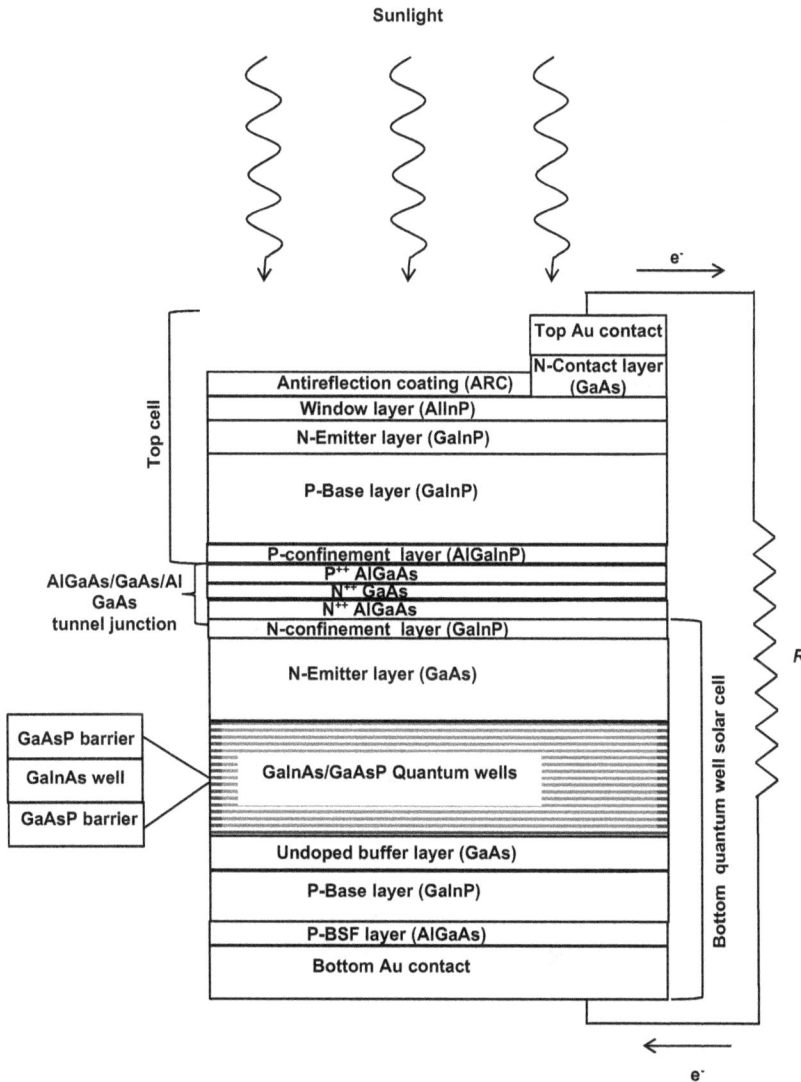

FIGURE 11.9 Two-junction GaInP/GaAs solar cell with GaInAs/GaAsP quantum wells in the bottom cell. Diagram shows the layers, from top downwards: (i) top cell consisting of the layers gold front contact, GaAs contact layer-ARC, AlInP window layer, N-type GaInP emitter layer, P-type GaInP base layer, P-type AlGaInP confinement layer, P^{++} AlGaAs-N^{++}GaAs-N^{++}AlGaAs tunnel junction; (ii) bottom cell comprising the layers N-type GaAs emitter layer, GaInAs/GaAsP QWs, undoped GaAs buffer layer, P-type GaInP base layer, P-type AlGaAs BSF layer, and gold back contact. Sunlight is incident on the top cell, and current is delivered across the gold contacts in the direction shown in the diagram.

The structure of the bottom solar cell is N-type GaInP confinement layer/N-type GaAs emitter/ GaInAs-GaAsP QWs (50–80)/undoped GaAs buffer layer/P-type GaInP base/P-type AlGaAs BSF layer/Au back contact.

The top and the bottom solar cells are connected through a $P^{++}AlGaAs/N^{++}GaAs/N^{++}AlGaAs$ tunnel junction. Electrons pass through this junction by quantum-mechanical tunneling to establish an ohmic contact between the two cells.

The efficiency of the GaInP/GaAs quantum well tandem solar cell is 32.9% (Steiner et al 2021).

11.5   TRIPLE-JUNCTION SOLAR CELL WITH GaInAs/GaAsP QUANTUM WELLS IN THE MIDDLE CELL (η = 39.5%)

This solar cell is a modified version of a monolithic GaInP/GaAs/GaInAs solar cell (Wanlass et al 2005a, b). The original cell has an efficiency of 37.9%. This cell is an inverted series-connected solar cell made from materials of three different bandgaps: GaInP (1.87eV), GaAs (1.42eV), and GaInAs (1eV). The layers are grown on a GaAs substrate. Being made from semiconductors of different lattice parameters, the multijunction solar cell is prone to dislocation defects arising from lattice mismatching. Although the top cell is a lattice-matched cell, the middle and bottom cells are lattice-mismatched cells. A technique known as metamorphic epitaxy helps in keeping dislocations away from the active regions of the growth material by relieving the strain around lattice-mismatched interfaces.

In this original cell design, the bandgap of 1.42eV of GaAs is nonideal, leaving a scope for improvement in efficiency. A bandgap smaller than GaAs is more appropriate for both terrestrial and space spectra. A multiple–quantum well structure is included in the middle cell to achieve a small reduction in bandgap below the 1.42eV value for GaAs. As we know, the quantum well structure consists of a smaller-bandgap well layer sandwiched between larger-bandgap barrier layers. The absorption edge of the quantum well structure is dictated by the smaller-bandgap well layer. This smaller bandgap is incremented by quantum size and strain effects. So all these factors need to be adjusted to produce an effective bandgap which is lower than that of GaAs. When correctly designed, the quantum wells can give the needed small lowering of bandgap without creating unwanted dislocations and other defects. They are therefore favored above the metamorphic material. For these reasons, the quantum well design is chosen for implementing this bandgap engineering of material.

The revised triple-junction solar cell (Figure 11.10) comprises three solar cells joined by two tunnel junctions as under (France et al 2022):

(i) A GaInP homojunction top cell having the structure MgF_2-ZnS ARC/Au front contact/ Se-doped N-type GaInNAs contact layer/Se-doped N-type AlInP window layer/ Se-doped N-type GaInP emitter layer/Zn-doped P-type GaInP base layer/Zn-doped P-type AlGaInP BSF layer.

(ii) An AlGaAs-GaAs tunnel junction.

(iii) GaInAs/GaAsP strain-balanced quantum well middle cell with the structure N-type Se-doped GaInP window layer/N-type Si-doped GaAs emitter/thin GaAs layer/GaInAs-GaAsP QWs/thin GaAs layer/P-type Zn-doped GaAs base layer/P-type Zn-doped GaInP BSF layer. The cell contains 184 GaInAs quantum wells. Thickness of the GaInAs well layer is 8.5nm. The GaAsP barrier layers are 2.5 nm thick.

(iv) An AlGaAs-GaAs tunnel junction.

(v) A transparent GaInP compositionally graded buffer (CGB) in which the dislocations are glided during metamorphic epitaxy to restrict dislocations in inactive regions so that nonradiative recombination in active regions is decreased. In the CGB, the lattice constant is gradually varied over several microns of thickness of the layers.

(vi) A metamorphic GaInAs bottom cell having the structure N-type Si-doped GaInAs emitter layer/P-type Zn-doped GaInAs base layer/P-type Zn-doped GaInP BSF layer/P-type Zn-doped GaInAs contact layer/bottom Au contact.

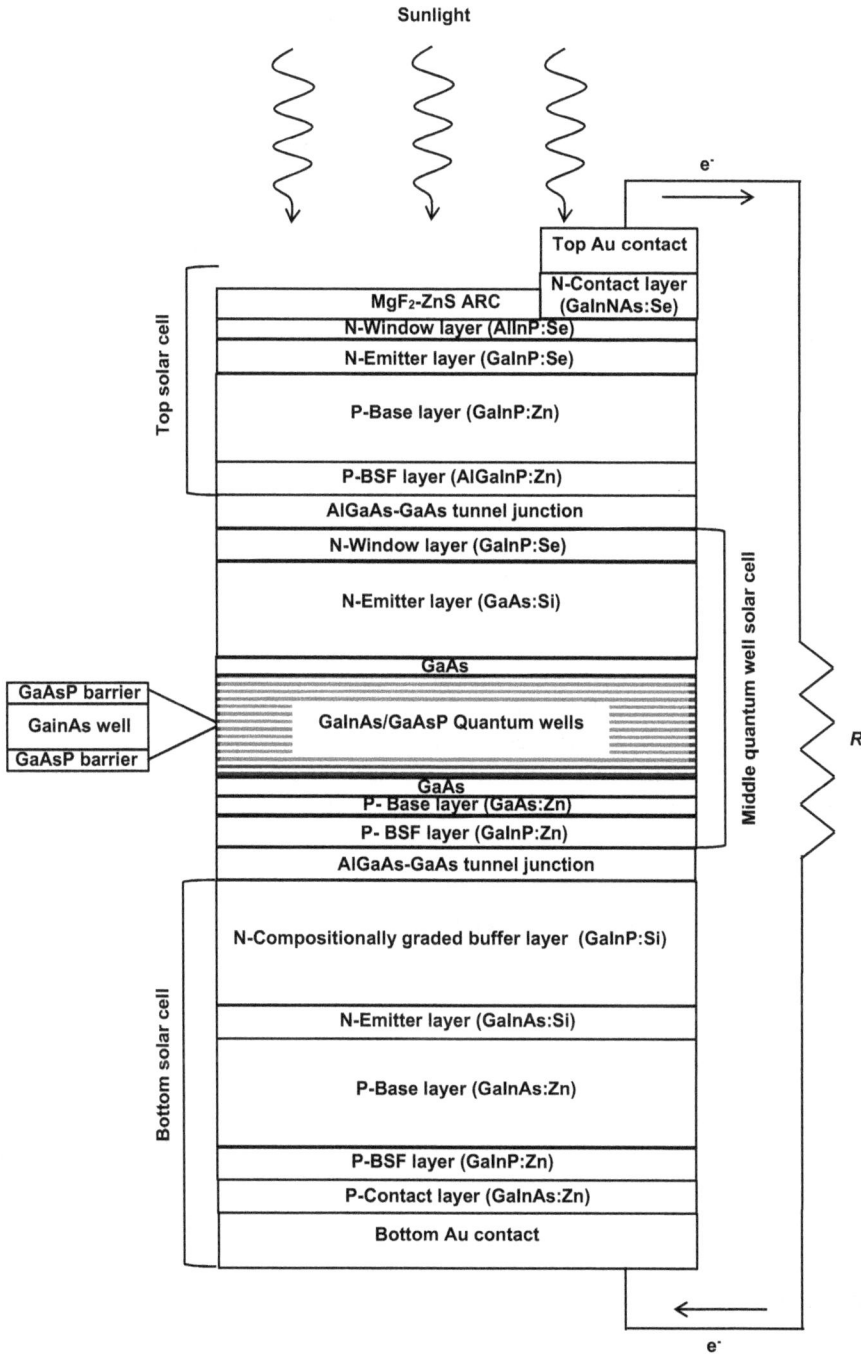

FIGURE 11.10 Using quantum wells in the middle cell of a three-junction solar cell. Diagram shows the various layers in the combined cell consisting of (i) top cell (MgF$_2$-ZnS ARC, Au front contact, N-type GaInNAs contact layer, N-type AlInP window layer, N-type GaInP emitter layer, P-type GaInP base layer, and P-type AlGaInP BSF layer), (ii) middle cell (N-type GaInP window layer, N-type GaAs emitter, thin GaAs layer, GaInAs-GaAsP QWs, thin GaAs layer, P-type GaAs base layer, P-type GaInP BSF layer), and (iii) bottom cell (N-type GaInAs emitter layer, P-type GaInAs base layer, P-type GaInP BSF layer, P-type GaInAs contact layer, bottom Au contact). The three solar cells are connected through AlGaAs-GaAs tunnel junctions. A compositionally graded buffer (CGB) is used between the middle and bottom cells. Sunlight falls on the top cell, and the output current is obtained across the Au contacts.

The triple-junction solar cell is fabricated by inverted process sequence. The III-V layers are epitaxially grown by MOVPE at temperatures between 620°C and 700°C using standard organometallic precursors and dopant sources for III-V layers (triethyl gallium, TEG, for gallium; trimethyl aluminum, TMA, for aluminum; trimethyl indium, TMI, for indium; dimethylhydrazine for nitrogen; phosphine, PH_3, for phosphorous; Arsine, AsH_3, for arsenic; hydrogen selenide, H_2Se, for selenium; dimethyl zinc, DMZ, for zinc). The tunnel junctions and contact layers are formed at lower temperatures between 570°C and 620°C. The back contact Au film is electroplated; the wafer is inverted and bonded with epoxy to a silicon handle for etching the GaAs substrate. Ni/Au front contact is electroplated and defined by photolithography. The MgF_2/ZnS ARC is deposited by thermal evaporation.

The efficiency of the triple-junction solar cell is 39.5% under AM1.5G spectrum. It is 34.2% under AM0 space spectrum (France et al 2022).

11.6   DISCUSSION AND CONCLUSIONS

Concerning the equipment used for fabrication of quantum well solar cells, MBE and MOCVD facilities are widely employed, depending on the requirement of accurate or economical processing. From the materials viewpoint, lattice matching, strain balancing, or deliberate introduction of strain occurs according as the lattices of chosen constituent layers are equivalent or intended to counterbalance strain or produce strain in the resultant stack (Table 11.1). For bandgap control, it is emphasized that the effective bandgap is mainly governed by quantum confinement effects, but the effects of strain and electric field must also be kept in mind.

Briefly, efficiencies of GaAs/AlGaAs quantum well solar cells exceed those of P-I-N diode cells without quantum wells. InGaAs/GaAsP quantum well solar cells show efficiencies at par with GaAs solar cells. InGaAsP/InGaP quantum well solar cells show a subpeak bandgap external quantum efficiency ~ 80% at 700nm. 300-period InGaN/GaN quantum well solar cells exhibit external quantum efficiency of 70.9% at 700nm. A glance at Table 11.2 gives information on state-of-the-art quantum well solar cells.

TABLE 11.1

Examples of Three Types of Quantum Well Solar Cells

Sl. No.	Type of Solar Cell	Quantum Well and Barrier Layers	Base and Emitter Layers
1	Lattice-matched quantum well solar cell	GaAs quantum well and AlGaAs barrier	N⁻GaAs base and P⁺GaAs emitter
2	Strain-balanced quantum well solar cell	(a) InGaAs well and GaAsP barrier	N⁻GaAs base and P⁺GaAs emitter
		(b) InGaAsP well and InGaP barrier	N⁻InGaP base and P⁺InGaP emitter
3	Strained quantum well solar cell	(a) InGaAs well and GaAs barrier	N⁻GaAs base and P⁺GaAs emitter
		(b) InGaN well and GaN barrier	N⁻GaN base and P⁺GaN emitter

---

TABLE 11.2

Efficiencies of Different Quantum Well Solar Cells

Sl. No.	Solar Cell	Quantum Well Layers/Barrier Layers	Efficiency (%)	Reference
1	Heterojunction emitter GaAs solar cell	Strained $In_{0.08}Ga_{0.92}As$ QWs with GaAs barriers	26.3	Welser et al (2019)
2	Inverted GaAs Solar Cell	Strain-balanced GaInAs quantum wells with GaAsP barriers	27.2	Steiner et al (2021)
3	GaInP/GaAs dual-junction solar cell	Strain-balanced GaInAs quantum wells with GaAsP barriers in the bottom cell	32.9	Steiner et al (2021)
4	Triple-junction solar cell	Strain-balanced GaInAs quantum wells with GaAsP barriers in the middle cell	39.5	France et al (2022)

Questions and Answers

11.1   Which technique requires an ultrahigh vacuum: (a) MBE or (b) MOCVD? Answer: *a*.

11.2   Which of the two techniques is cleaner: (a) MOCVD or (b) MBE? Answer: *b*.

11.3   Which technique is slow in film deposition: (a) MOCVD or (b) MBE? Answer: *b*.

11.4   Which of the two techniques, MOCVD and MBE, is physical and which one is chemical in action? Answer MOCVD is chemical, and MBE is physical.

11.5   Can films of chemical compounds be deposited by MBE? Answer: Yes.

11.6   Do molecular beams of materials in MBE interact among themselves prior to depositing on the substrate? Answer: No, they interact only with the substrate.

11.7   Why is molecular beam epitaxy called by this name? Answer: Because the molecular beams of materials to be deposited do not undergo any interaction among themselves or with any carrier gases before deposition on the substrate.

11.8   In MBE, what parameter does the source of molecular beam control? Answer: Film growth rate.

11.9   During MBE, how is the composition and stoichiometry of chemical compounds controlled? Answer: By varying the temperatures and flux ratios of different source materials used for the concerned compound.

11.10   What level of precision in thickness of deposited film does MBE provide? Answer: Precision up to a single layer of atoms.

11.11   What level of flatness of deposited film is achieved in MBE? Answer: Atomically flat film.

11.12   What is a RHEED system? Answer: RHEED is a real-time monitoring tool for deposition of single atomic layers during MBE process in which an electron gun sends high-energy electron beam at a shallow grazing angle on the studied surface so that the electrons do not penetrate deep enough in the specimen. The electrons experience diffraction in the top few atomic layers of the specimen. After diffractions, constructive interference takes place between electrons in accordance with the structure of the crystalline surface, the atomic spacing, and the wavelength of electrons. The resultant electron beam falls on photoluminescent detector screen to produce a pattern containing information about the surface of the film.

11.13   What is a RHEED screen? A phosphorescent screen displaying diffraction patterns about the surface features of the film sample under test. These patterns are analyzed to reveal surface configuration and kinetics.

11.14   In what typical pressure range are MOCVD operations carried out? Answer: 10–760 torr.

11.15   What are the materials from which an MOCVD reactor is made? Answer: Stainless steel or quartz.

11.16   What is the material used for making the susceptor on which the substrate is placed during MOCVD? Answer: Graphite.

11.17   What is meant by a hot-wall CVD reactor? Answer: A reactor of which the walls are heated during film deposition.

11.18   What is meant by a cold-wall CVD reactor? Answer: A reactor of which the walls remain cold during film deposition and the substrate is kept on a susceptor, which is heated during film deposition.

11.19   What is meant by precursor of a material? Answer: A chemical compound containing the element to be deposited in MOCVD and which undergoes reaction during the process to deposit the required material film on the substrate.

11.20   Mention one precursor used in MOCVD in solid form? Answer: Trimethylindium, the precursor for indium.

11.21   Give two examples of precursors used in MOCVD in liquid form. Answer: Trimethyl or triethyl aluminum, precursor for aluminum; trimethyl/triethyl gallium, precursor for gallium.

11.22   Give two examples of precursors used in MOCVD in gaseous form. Answer: Phosphine, precursor for phosphorous; arsine, precursor for arsenic.

11.23   What principal difference do you find between the deposition mechanisms of MOCVD and MBE films? Answer: During MOCVD, the precursors (if in liquid form) are carried to the reactor by bubbling carrier gases through them. So the reactor contains the vapors of reactant gases. If a hot-wall reactor is used, the temperature of reactor walls is maintained at the required value. In a cold-wall reactor, the temperature of the substrate carrier is adjusted at the desired value. At the high temperature available, a chemical reaction takes place between the reactant gases. The products of the chemical reaction are deposited in the form of films on the substrate. In MBE, the source materials are vaporized to form separate molecular beams in an ultrahigh vacuum chamber. They do not interact mutually before reaching the substrate. No chemical reaction takes place between them. The materials are directly deposited on the substrate, which is usually heated to a predecided temperature.

11.24   Why must the gases coming out from an MOCVD reactor be delivered to an exhaust and scrubber system? Answer: Because of the hazardous nature of gases used in MOCVD reactions, they must be carefully removed and converted into harmless products before releasing them into the atmosphere to avoid human exposure to dangerous gases.

11.25   Which is less expensive: (a) MOCVD or (b) MBE? Answer: *a.*

11.26   Which is suitable for solar cell manufacturing: (a) MBE or (b) MOCVD? Answer: (b) MOCVD.

11.27   Which is good for process optimization in research laboratories: (a) MOCVD or (b) MBE? Answer: (a) MBE.

11.28   Which system of materials used in quantum well solar cell fabrication allows film growth to arbitrary thickness values without disturbing the crystalline form? Answer: Lattice-matched material system.

11.29   Give an example of a pair of lattice-matched materials used in solar cells. Answer: GaAs/AlGaAs.

11.30   Are GaAs and AlGaAs direct-bandgap materials? Answer: GaAs is a direct-bandgap semiconductor with bandgap 1.422eV, while the bandgap of $Al_xGa_{1-x}As$ depends on aluminum content x, increasing from 1.422eV for $x = 0$ to 2.16eV for $x = 1$. It is a direct-bandgap semiconductor for $x < 0.45$ but becomes indirect bandgap when $x > 0.45$.

11.31   What is a strain-balanced structure? Answer: A structure made up of two materials such that the compressive strain in one material balances the tensile strain in the other material.

11.32   Is any strain produced on the substrate by a strain-balanced structure? Answer: No, local strains at the interface of the two materials counteract each other.

11.33   Give one example of a pair of strain-balanced materials. Answer: InGaAs/GaAsP.

11.34   In which of the two materials of the InGaAs/GaAsP pair is compressive strain produced, and which one experiences tensile strain? Answer: Compressive strain develops in InGaAs, and tensile strain in GaAsP.

11.35   In a quantum well solar cell, the doping concentration of the well region is lower than that of the barrier layer. Towards which side will the depletion layer stretch to a larger thickness? Answer: The low-doped well region.

11.36   Give two examples of material pairs used to fabricate strained quantum well solar cells. Answer: InGaAs/GaAs, InGaN/GaN.

11.37   Which material property does strain influence in a strained-material system for use in solar cell? Answer: Bandgap of a material is dependent on strain.

11.38   What limitation does the strain production impose on growing layers? Answer: Materials can be grown only up to a certain limited thickness without disturbing the crystalline structure. This thickness is the critical thickness of material growth.

11.39   Have strained quantum well solar cells been fabricated? Answer: Yes, using InGaAs/ GaAs or InGaN/GaN pairs of layers.

11.40   How is a compressive strain produced in InGaN layer in InGaN/GaN quantum well solar cell? Answer: Lattice constant of InGaN is larger than that of GaN. So when the InGaN film is deposited over the GaN film, the deposited film adjusts itself in two stages. In the initial stage, the InGaN film sits on top of the GaN film. Afterwards, it shrinks to fit over the lattice of the GaN film. The InGaN film tries to equalize its lattice parameter with that of the underlying GaN film. The shrinkage of InGaN film is associated with the production of a compressive strain in this film. No counterbalancing tensile strain exists in the GaN film. So the compressive strain persists in the InGaN film. Persistence of strain leads to the formation of a strained quantum well solar cell.

11.41   Does GaAs have the optimum bandgap for making a solar cell? Answer: No. GaAs has a bandgap of 1.424eV, whereas the optimal required value of bandgap is 1.34eV.

11.42   How does the incorporation of a quantum well in the GaAs solar cell structure help in optimizing the bandgap for making solar cells? Answer: The quantum well contains a smaller-bandgap GaInAs well layer (1.262eV) crammed between larger-bandgap GaAsP barrier layers (1.7eV). The smaller bandgap of the well layer provides the benefit of lowering the bandgap. Nonetheless, it must be emphasized that the layer thicknesses must be properly chosen because as the thickness of quantum well layer decreases, quantum confinement effect comes into play, increasing the bandgap. Furthermore, during solar cell fabrication, stress buildup in quantum wells should be avoided because bandgap increases with strain. Unless these factors are duly accounted for, the required reduction in bandgap will not be obtained. So the quantum wells will be useful only when correct well layer thickness is chosen and satisfactory strain balancing has been achieved.

11.43   Which type of strain is produced by growing GaInAs on GaAs substrate? Answer: Compressive strain.

11.44   Which type of strain is created on growing GaAsP on GaAs substrate? Answer: Tensile strain.

11.45   How are the junctions stacked in a multiple-junction solar cell? Answer: The junction of the solar cell made with large-bandgap semiconductor should receive the incoming light. Light coming out from this solar cell should fall on the junction of solar cell with smaller bandgap. In the same way, junctions of successive solar cells are made in descending order of bandgaps.

11.46   Argue to explain why materials are arranged in descending order of bandgaps in a multijunction solar cell. Answer: When considering the application of a semiconductor material in a solar cell, there are three groups of photons in sunlight:

(i)   Photons with energies below the bandgap of the semiconductor. These photons do not produce any electron-hole pairs. They are transmitted unused through the semiconductor and lost.

(ii)   Photons having energies above the bandgap. These photons are absorbed and used in the creation of electron-hole pairs. The excess energy of photons above the bandgap produces hot carriers. This energy is lost by thermalization processes and not used.

(iii)   Photons with energies much larger than the bandgap. These photons are also utilized like group ii photons. But they have a large energy in excess of bandgap. This large excess energy is wasted by thermalization processes. So in case of these photons, energy wastage is greater than for group ii photons.

The multijunction solar cell has a twofold objective: (i) to maximize the part of solar spectrum utilized, i.e., to produce electricity from a more diverse distribution of photon energies, and (ii) to minimize thermalization losses (Philipps and Bett 2014).

If materials are stacked in ascending order of bandgaps with small bandgap semi-conductor first, the first smaller solar cell will absorb a large fraction of the incident photons to produce electricity. The photons absorbed by the first cell consists of (i) photons having energies slightly larger than its energy bandgap as well as (ii) photons with energies much larger than the bandgap. The latter photons, i.e., group ii photons, will produce hot carriers. The energy of these hot carriers will be dissipated in the lattice. Thus, a lot of energy will be lost. The remaining low-energy photons with energies less than the bandgap will pass through to the higher-bandgap semiconductor. Here they will not be able to produce any electron-hole carriers and so become useless.

Let us reverse the stacking order and place larger-bandgap semiconductor first and smaller-bandgap semiconductors behind. By this stacking arrangement, a systematic photon sorting effect is created. The photons are sorted into different groups according to their energies. The higher-energy photons are absorbed by the large-bandgap semiconductor and utilized more fully than if they were absorbed by the small-bandgap semiconductor. Here, "utilized more fully" means that these photons do not have much excess energy above the large-bandgap semiconductor, and so energy portion spent towards hot carriers is insignificant as compared to the situation in which light had fallen on the small-bandgap semiconductor first.

The lower-energy photons pass through the large-bandgap semiconductor and fall on the ensuing semiconductor of smaller bandgap. Here they are absorbed and exploited more effectively than if they were absorbed by the subsequent still-smaller-bandgap semiconductor. "Exploited more effectively" means that a comparatively smaller portion of photon energy is used in hot carrier production.

Thus, although the amount of light absorbed is the same in both the stacking orders, the photon utilization is far better when the semiconductors are placed in decreasing order of bandgaps.

11.47   What is a tunnel junction? Explain why we need a tunnel junction to connect two solar cells of a dual-junction solar cell. Answer: A tunnel junction is an optically transparent, electrically conductive junction between two layers of materials. To explain its action, suppose two P-N diodes representing two solar cells are directly connected in a series without a tunnel junction separating them. The resultant structure is P-N-P-N. In this structure, there are three diodes: P-N, N-P, and P-N. The third diode is formed naturally where the two corner diodes are joined. The third central diode is of opposite polarity to the two corner diodes. The effect of this third central diode is therefore opposite to that of the corner diodes. This effect comes from the depletion region across it. To counteract the effect of the central diode, the effect of its depletion region must be annulled. This is done by replacing the structure of two diodes by the structure P-N-N^{++}-P^{++}-P-N. It can be seen that two heavily doped regions, N^{++} adjoining the N-layer and P^{++} adjacent to the P-layer, have been introduced. As we know, the depletion region thickness is a function of doping concentrations in the two contacting layers. The higher the doping concentrations, the thinner the depletion region. Since very high doping concentration layers N^{++} and P^{++} have been formed, the depletion region across the N^{++}/P^{++} junction will be very thin. Carriers can flow across this thin depletion region by quantum mechanical tunneling. So the effect of the depletion region of the central diode is mitigated, and carriers can flow very easily across the junction. A low-resistance interconnection has been formed. It is known as a tunnel junction.

11.48   Why is the tunnel junction a good choice for connecting two diodes without incurring any drop of voltage? Answer: The current-voltage characteristic of a tunnel junction consists of three regions: a tunneling region, a negative differential resistance region, and a normal diode region. In the tunneling region of operation, the energy states of the electrons on one side of the depletion region are aligned with the vacant energy states on the opposite side. The current flowing across the junction is high, and the slope of the current-voltage characteristic is steep, leading to a low resistance, and hence a small voltage drop.

REFERENCES

Choi S.-B., J.-P. Shim, D.-M. Kim, H.-I. Jeong, Y.-D. Jho, Y.-H. Song and D.-S. Lee 2013 Effect of indium composition on carrier escape in InGaN/GaN multiple quantum well solar cells, Applied Physics Letters, 103: 033901–1 to 033901–4.

Ekins-Daukes N. J., K. W. J. Barnham, J. P. Connolly, J. S. Roberts, J. C. Clark and G. Hill 1999 Strain-balanced GaAsP/InGaAs quantum well solar cells, Applied Physics Letters, 75: 4195.

Fardi H. 2012 Design and simulation of multiquantum-well AlGaAs/GaAs single junction solar cell with back surface reflector, International Scholarly Research Notices, Article ID 859519: 5 pages.

Farrell R. M., C. J. Neufeld, S. C. Cruz, J. R. Lang, M. Iza, S. Keller, S. Nakamura, S. P. DenBaars, U. K. Mishra and J. S. Speck 2011 High quantum efficiency InGaN/GaN multiple quantum well solar cells with spectral response extending out to 520 nm, Applied Physics Letters, 98: 201107–1 to 201107–3.

France R. M., J. F. Geisz, T. Song, W. Olavarria, M. Young, A. Kibbler and M. A. Steiner 2022 Triple-junction solar cells with 39.5% terrestrial and 34.2% space efficiency enabled by thick quantum well superlattices, Joule, 6: 1121–1135.

Jiang C., L. Jing, X. Huang, M. Liu, C. Du, T. Liu, X. Pu, W. Hu and Z. L. Wang 2017 Enhanced solar cell conversion efficiency of InGaN/GaN multiple quantum wells by piezo-phototronic effect, ACS Nano, 11(9): 9405–9412.

Ladugin M. A., I. V. Yarotskaya, T. A. Bagaev, K. Y. Telegin, A. Y. Andreev, I. I. Zasavitskii, A. A. Padalitsa and A. A. Marmalyuk 2019 Advanced AlGaAs/GaAs heterostructures grown by MOVPE, Crystals 9(305): 1–11.

Noda T., L. M. Otto, M. Elborg, M. Jo, T. Mano, T. Kawazu, L. Han and H. Sakaki 2014 GaAs/AlGaAs quantum wells with indirect-gap AlGaAs barriers for solar cell applications, Applied Physics Letters 104(12): 122102.

Philipps S. P. and A. W. Bett 2014 III-V Multi-junction solar cells and concentrating photovoltaic (CPV) systems, Advanced Optical Technologies, 3(5–6): 469–478.

Rimada J. C., L. Hernandez, J. P. Connolly, K. W. J. Barnham 2005 Quantum and conversion efficiency calculation of AlGaAs/GaAs multiple quantum well solar cells, Physica Status Solidi, 242(9): 1842–1845.

Sayed I. E. H., C. Z. Carlin, B. G. Hagar, P. C. Colter and S. M. Bedair 2016 Strain-balanced InGaAsP/GaInP multiple quantum well solar cells with a tunable bandgap (1.65–1.82 eV), IEEE Journal of Photovoltaics, 6(4): 997–1003.

Sayed I. E. H., N. Jain, M. A. Steiner, J. F. Geisz and S. M. Bedair 2017 100-period InGaAsP/InGaP superlattice solar cell with sub-bandgap quantum efficiency approaching 80%, Applied Physics Letters, 111: 082107.

Steiner M. A., R. M. France, J. Buencuerpo, J. F. Geisz, M. P. Nielsen, A. Pusch, W. J. Olavarria, M. Young and N. J. Ekins-Daukes 2021, High efficiency inverted GaAs and GaInP/GaAs solar cells with strain-balanced GaInAs/GaAsP quantum wells, Advanced Energy Materials, 11(4): 2002874.

Strömberg A., G. Omanakuttan, Y. Liu, T. Mu, Z. Xu, S. Lourdudoss and Y.-T. Sun 2020 Heteroepitaxy of GaAsP and GaP on GaAs and Si by low pressure hydride vapor phase epitaxy, Journal of Crystal Growth, 540(125623): 9 pages.

Valdueza-Felip S., A. Mukhtarova, L. Grenet, C. Bougerol, C. Durand, J. Eymery and E. Monroy 2014 Improved conversion efficiency of as-grown InGaN/GaN quantum-well solar cells for hybrid integration, Applied Physics Express, 7: 032301–1 to 032301–4.

Wanlass M. W., S. P. Ahrenkiel, D. S. Albin, J. J. Carapella, A. Duda, K. Emery, J. F. Geisz, K. Jones, Sarah Kurtz, T. Moriarty and M. J. Romero 2005a GaInP/GaAs/GaInAs monolithic tandem cells for high-performance solar concentrators, International Conference on Solar Concentrators for the Generation of Electricity or Hydrogen, 1–5 May 2005, Scottsdale, Arizona, 4 pages.

Wanlass M. W., J. F. Geisz, S. Kurtz, R. J. Wehrer, B. Wernsman, S. P. Ahrenkiel, R. K. Ahrenkiel, D. S. Albin, J. J. Carapella, A. Duda and T. Moriarty 2005b Lattice-mismatched approaches for high-performance, III-V photovoltaic energy converters, Conference Record of the Thirty-first IEEE Photovoltaic Specialists Conference, 3–7 January, 2005, Lake Buena Vista, FL, IEEE, NY, pp. 530–535.

Welser R. E., S. J. Polly, M. Kacharia, A. Fedorenko, A. K. Sood and S. M. Hubbard 2019 Design and demonstration of high-efficiency quantum well solar cells employing thin strained superlattices, Scientific Reports, 9(13955): 10 pages.

Zhang B., W. Xie and Y. Xiang 2015 Development and prospect of nanoarchitectured solar cells, International Journal of Photoenergy Volume 2015(Article ID 382389): 11 pages.

12 Nanowire Solar Cells
Configurations

As already mentioned in Section 2.4.1, the nanowire is a one-dimensional nanostructure (Zhang et al 2016). It is a high-aspect-ratio structure having a diameter in the range 1–100nm and length of a-few-microns-to-millimeter range. A nanowire solar cell is one in which nanowire(s) is (are) used as the optical absorber medium.

12.1 REASONS FOR INTEREST IN NANOWIRE SOLAR CELLS

Nanowire geometry is beneficial in many respects (Garnett et al 2011):

(i) It provides cost-effectiveness relative to thin-film solar cell by reducing the amount and quality of semiconductor material used in solar cell fabrication. The reduction in amount of material is achieved because nanowires consume less quantity of semiconductor than bulk material. Lower quality of material is tolerated because these solar cells are more tolerant to defects in material than their thin-film counterparts, particularly so in the radial configuration discussed in Section 12.2.2, which follows.

(ii) It allows fabrication of low-cost, lightweight, flexible solar cells.

(iii) It provides better light absorption because of the antireflective property of nanowires and also gives good light trapping by scattering. The connection of the origin of antireflective property of nanowires to refractive index must be remembered. This property is endowed to the nanowires by their ability to provide a graded refractive index. Time and again, the role of refractive index in reflection of light at an interface between two optical media has been emphasized. It has been reiterated that the extent of reflection of light at an interface between two media of different refractive indices is determined by the difference in the refractive indices of the media. The larger this difference, the more the quantity of light reflected. For a solar cell without an antireflecting coating, 10–50% of incident sunlight in the wavelength range 400–2,000nm is lost by reflection so that an average of 30% of incident light remains unutilized. When a solar cell uses an antireflection coating, the reflection of incident light of a particular wavelength is avoided via destructive interference between incident and reflected light. Since one antireflective coating provides only one refractive index, the strategy works only for a single wavelength. To cover more wavelengths, multiple antireflection coatings are necessary. So a large number of antireflection coatings will be able to provide good antireflection behavior over a broad wavelength range by building a gradual variation in refractive index. An array of nanowires can provide the requisite graded refractive index in a wider range of wavelengths because they have dimensions comparable to or smaller than the wavelength of light. This graded refractive index is a weighted average between the refractive indices of the material and air.

(iv) It allows tuning of bandgap by varying the nanowire diameter.

(v) Nanowires can be grown by heteroepitaxy, the process of growth of a material on a substrate of a different material. This becomes possible because nanowires provide a facile relaxation of strain owing to their extremely small areas of contact with the

DOI: 10.1201/9781003215158-17

bulk material. The heteroepitaxial process is rewarding because it can be used to grow nanowires on low-cost nonepitaxial substrates, allowing fabrication of inexpensive solar cells.

12.2 BROAD CLASSIFICATION OF NANOWIRE SOLAR CELLS

12.2.1 Two Types of Solar Cells According to the Number of Nanowires

The two kinds of cells are:

- (i) Single-nanowire solar cell
- (ii) Nanowire array solar cells

Single-nanowire cells are useful for operating nanoelectronic devices with power consumption in nanowatt range (Dutta et al 2018). Large-area nanowire array solar cells offer a viable means for commercial manufacturing of economical solar cells to fulfill high power demands. Feasibility of self-powered logic circuits is demonstrated by integrating nanowire photovoltaics (Tian et al 2007).

12.2.2 Two Types of Solar Cells According to Direction of Charge Separation

Depending on the axis about which charges are separated in a nanowire solar cell, they are subdivided into two classes:

- (i) Radial junction solar cell. This solar cell is fabricated as a coaxial structure. Again, looking at a vertically upright nanowire, the P$^+$-region is the core of the nanowire. A shell of I-region surrounds this P$^+$-core, and a shell of N$^+$-region surrounds the I-region shell. So the charge separation occurs along the radius of the nanowire. In Figures 12.1 and 12.2, single-nanowire and nanowire array solar cells of radial design are shown.
- (ii) Axial junction solar cells. This solar cell is fabricated as a P-I-N structure along the length of the nanowire. Considering a vertically standing nanowire on a substrate, the N$^+$-region is at the top of the nanowire, the I-region is in its middle, and the P$^+$-region is at its bottom. So the charge separation takes place along the length of the nanowire. Figures 12.3 and 12.4 show the single-nanowire and nanowire array solar cells of axial geometry.

Thus, the nanowire solar cells are termed axial or radial accordingly as charge separation takes place along the length or radius of the nanowire.

12.2.3 Radial vs. Axial Junction Solar Cell

One pertinent fact immediately striking the mind is that the carriers have to travel a shorter distance from the depletion region around the junction to the contacts in a radial junction solar cell than in an axial junction cell (Raj et al 2019). So a smaller diffusion length of material is adequate for collection of a large number of carriers. Very few carriers are lost by recombination in this

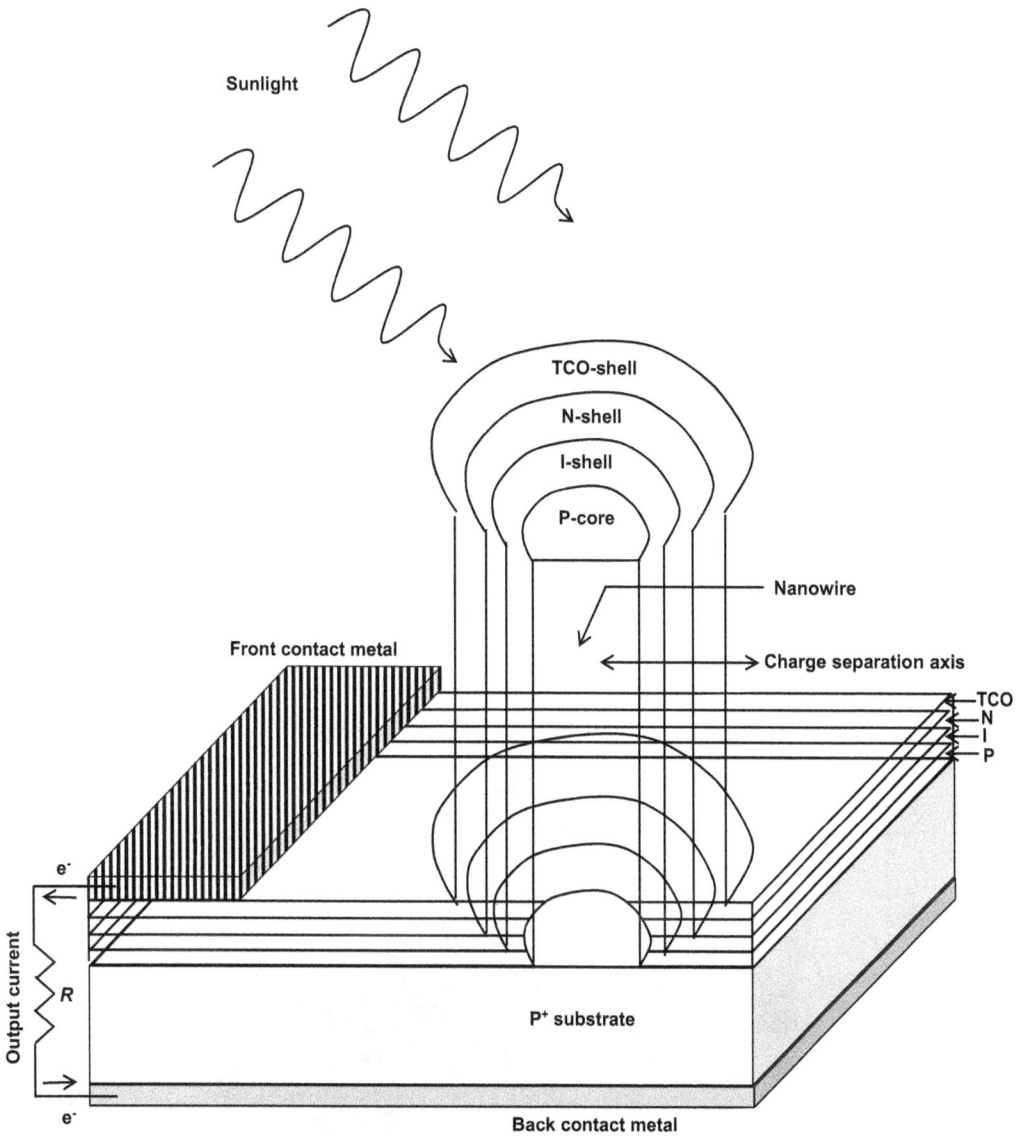

(a)

FIGURE 12.1 Radial single-nanowire solar cell: (a) 3D and (b) 1D views showing the back contact metal, P⁺ substrate, P-core, I-shell, N-shell, transparent conducting oxide (TCO)-shell, and front contact metal. Charge separation takes place along the charge separation axis. The direction of this axis points along the radius of the nanowire. Sunlight striking the nanowire produces current flow between the two metal electrodes.

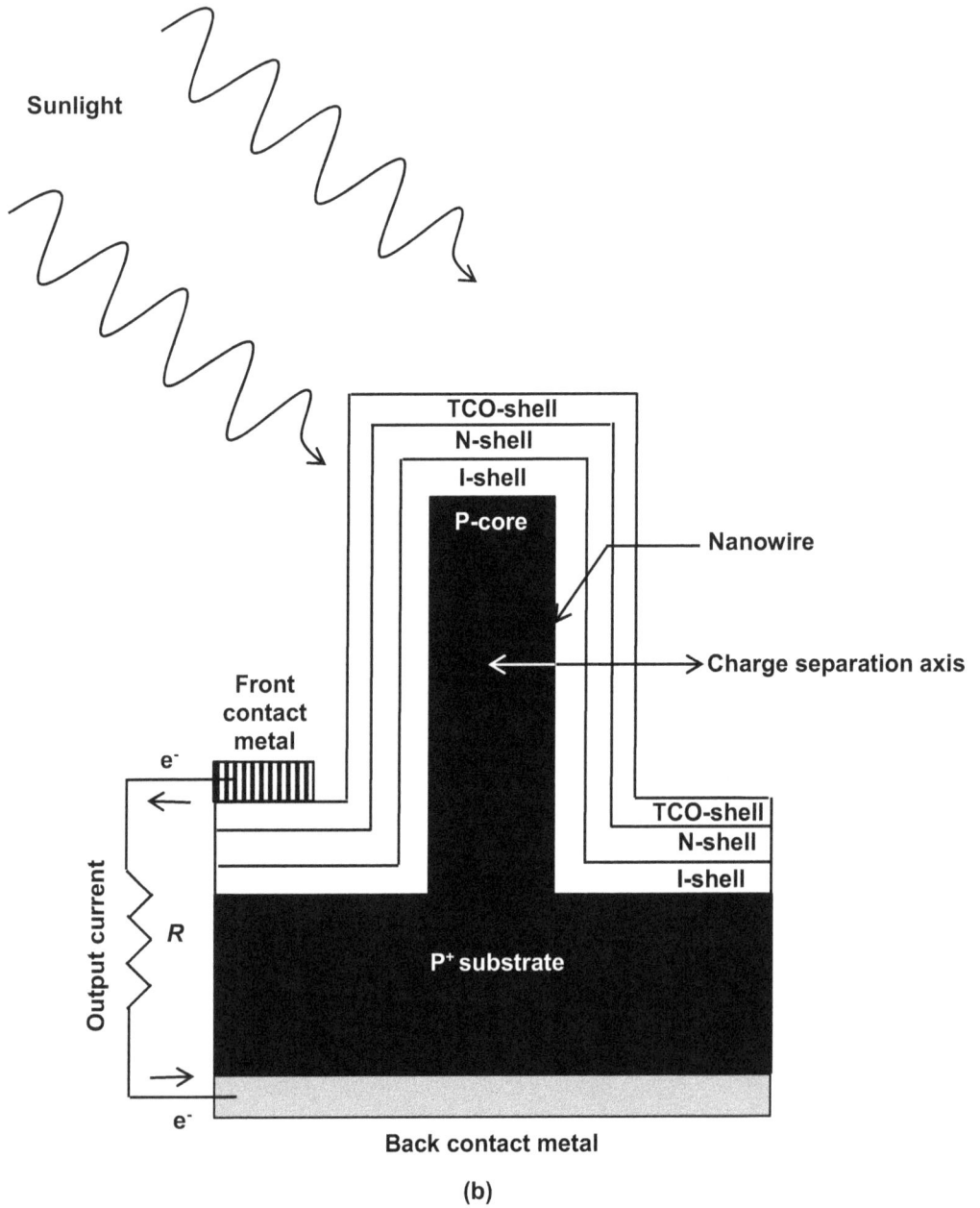

FIGURE 12.1 (Continued)

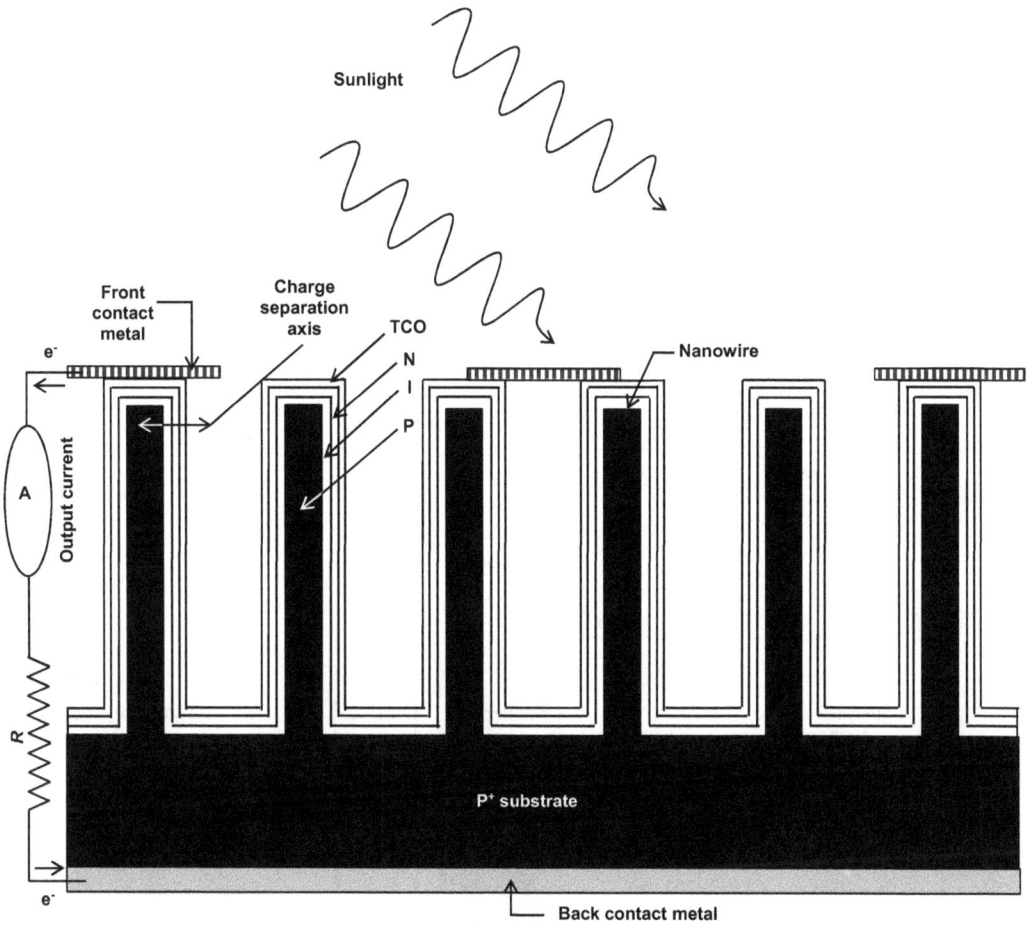

FIGURE 12.2 Radial nanowire array solar cell showing the back contact metal, P⁺ substrate, P-core, I-shell, N-shell and TCO layers, and front contact metal. Charge separation takes place along the radius of the nanowire. Sunlight reception surface and electrodes for drawing current are indicated.

smaller diffusion length. But a smaller diffusion length means a low carrier lifetime, which in turn implies a material with greater defect density or a material of low quality. Such a low-quality material is cheaper in cost. So the radial junction provides cost savings in raw material. Carrier lifetime preservation requires careful semiconductor processing. So processing requirements are also relaxed.

Thus, a radial junction nanowire solar cell provides excellent light management with increased absorption, provides the advantage of nanowire synthesis on low-cost nonepitaxial substrates, offers interfacial strain relaxation due to small diameter, and gives the benefit of charge separation with greater leniency to material defects. The axial nanowire solar cell provides all these advantages except for the last one of defect forbearance.

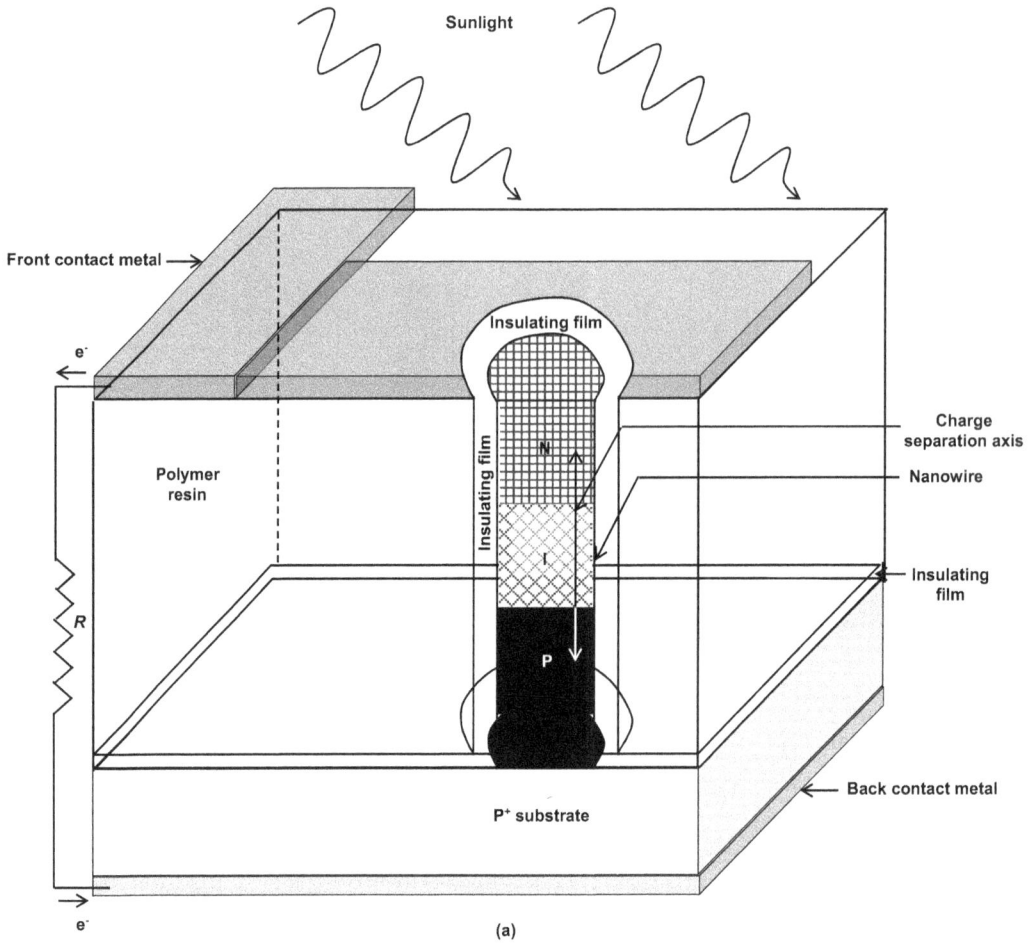

FIGURE 12.3 Axial single-nanowire solar cell: (a) 3D and (b) 1D views showing the back contact metal; P⁺ substrate; P-, I-, and N-doped layers along the length of the nanowire; insulating film covering the nanowire exposed at the tip; and front contact metal. Charge separation takes place along the charge separation axis, whose direction points along the length of the nanowire. When sunlight falls on the nanowire, an electric current flows between the two metal electrodes.

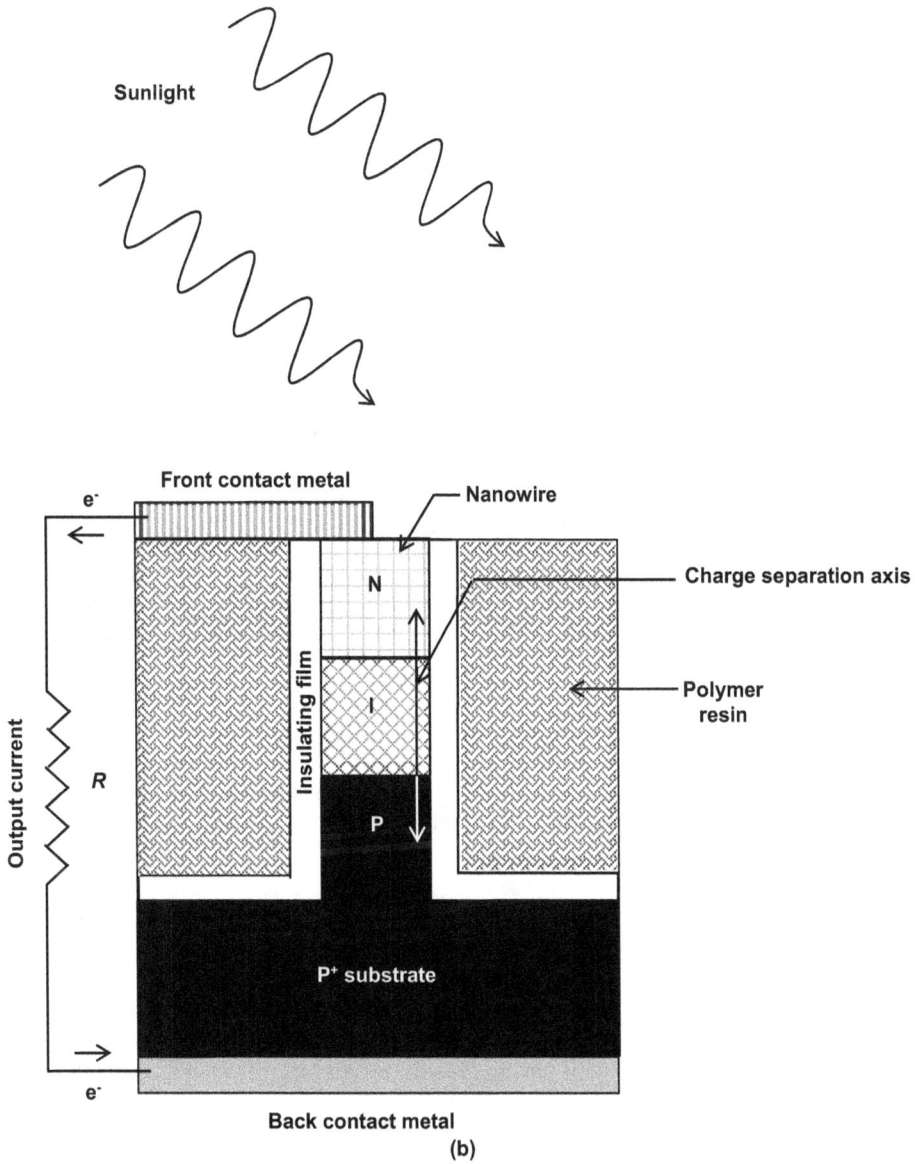

FIGURE 12.3   (Continued)

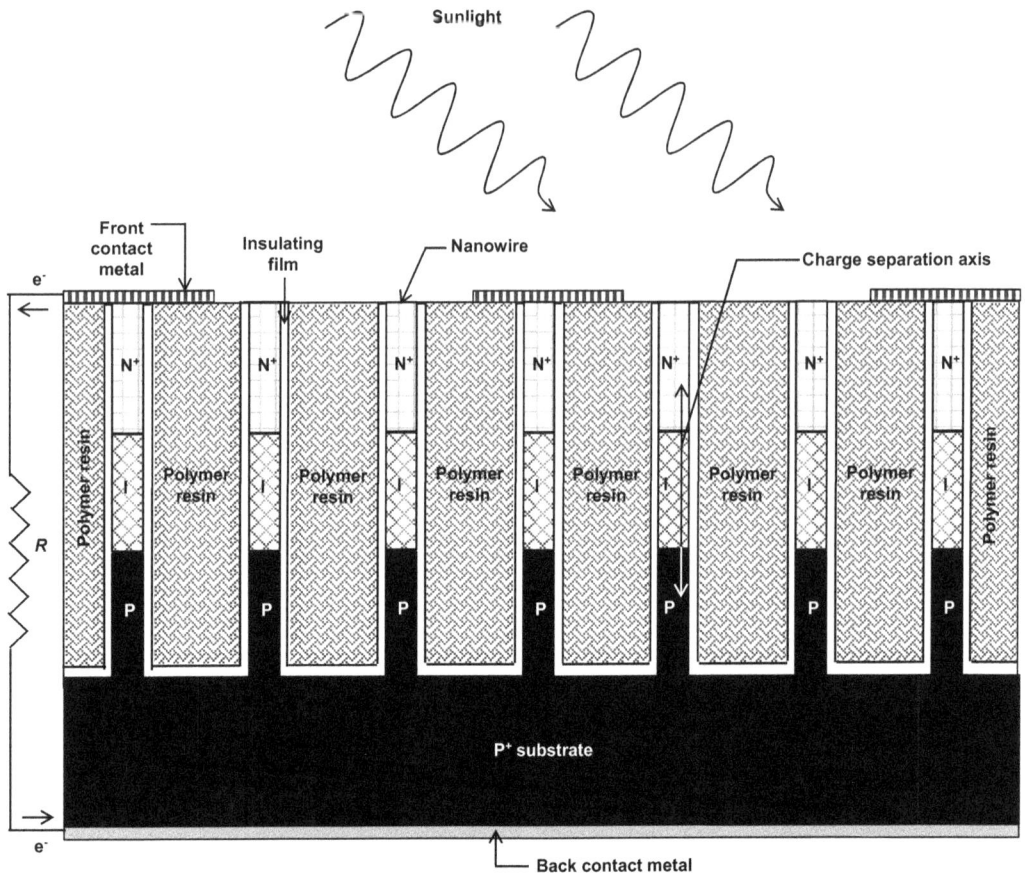

FIGURE 12.4 Axial nanowire array solar cell showing the back contact metal; P⁺ substrate; P-, I-, and N-doped layers along the length of the nanowire; insulating film covering the nanowire exposed at the tip; and front contact metal. Charge separation takes place along the charge separation axis, whose direction points along the length of the nanowire. Sunlight striking the nanowire produces current flow between the two metal electrodes.

12.3 NANOWIRE SOLAR CELL PROPERTIES AND OPERATION THROUGH EXAMPLES

The vital differences between radial and axial geometrical designs of nanowire solar cells and the important properties of these solar cells can be grasped by working through illustrative examples.

Example 12.1

Efficiency dependence on nanowire height for radial and axial junction solar cells: efficiency of radial junction cell falls after reaching maximum, whereas that of axial junction solar cell remains constant after attaining maximum.

Argue to show that the efficiency of a radial junction nanowire solar cell will attain a maximum value as the height of the nanowire is increased and then will start falling, whereas the efficiency of an axial junction nanowire solar cell will increase with the height of the nanowire to reach a peak saturation value without falling with further increase in nanowire height.

Solution: Let us consider a radial solar cell with nanowire of height h (Figure 12.5(a)). Let $J_{Light}(0)$ be the light-generated current density from the photons impacting at the top surface of the nanowire. As the light travels downwards into the nanowire, its intensity diminishes. The intensity $I(x)$ of light of wavelength λ at a depth x below the top surface is

$$I(x) = I(0)\exp\{-\alpha(\lambda)x\} \tag{12.1}$$

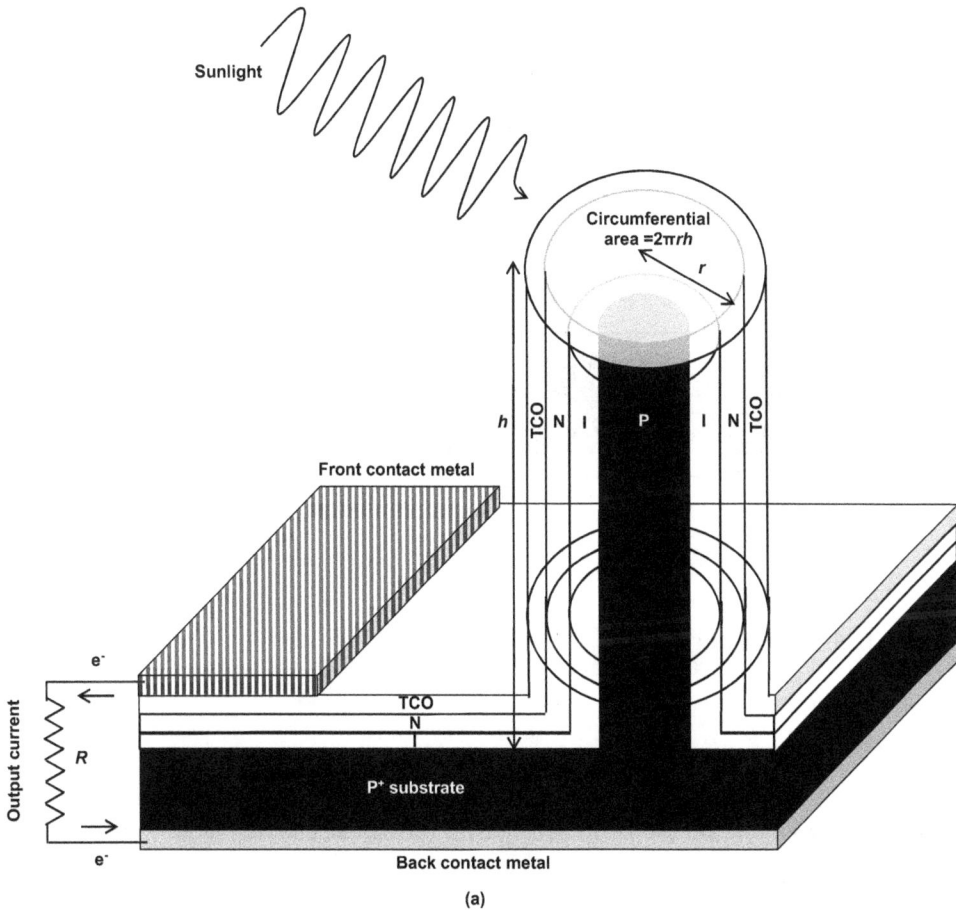

(a)

FIGURE 12.5 Nanowire solar cells: (a) radial and (b) axial showing for the radial cell: the back contact metal, the P⁺ substrate, the P-core, I-shell, N-shell, the TCO, and the front contact metal; and for the axial cell: the P⁺ substrate, the P-I-N diode along the nanowire length, the insulating film surrounding the nanowire, the polymer resin, and the front contact metal. Sunlight falling on these solar cells as well as the output current terminals are shown. The radius r of the radial junction nanowire cell and radius R of the axial junction nanowire cell are marked. The heights h of the nanowires are also marked. The junction areas in the two cases are $2\pi rh$ and πR^2, respectively.

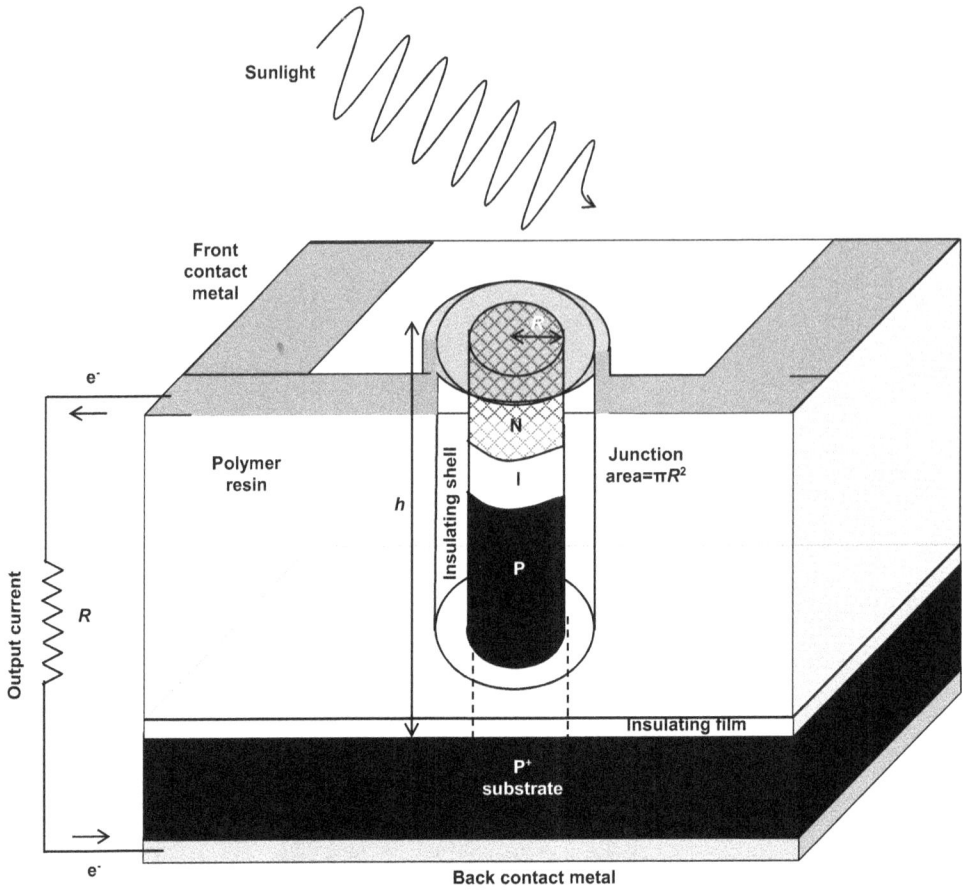

(b)

FIGURE 12.5 (Continued)

where $\alpha(\lambda)$ is the absorption coefficient of the light of a given wavelength. The light-generated current density also falls as the distance below the nanowire surface increases, and its variation follows a relation similar to equation 12.1 for attenuation of intensity of light with depth

$$J_{\text{Light}}(x) = J_{\text{Light}}(0)\exp\{-\alpha(\lambda)x\} \tag{12.2}$$

This equation shows that the nanowire portion near the surface contributes significantly to this current, while the portion distant from the surface has not much effect. The absorption coefficient has a fixed value for a particular material at a specified wavelength. During the initial stage of increasing nanowire height, the current density $J_{\text{Light}}(x)$ increases because more volume is accessible to light for producing carriers. Therefore, increasing the height h of the nanowire will raise the current density $J_{\text{Light}}(x)$. But after a certain height has been reached, the current will not increase because light will not be able to travel up to the longer distance. So the extra height will not be beneficial by providing more current, i.e., the excess portion of the nanowire will not be used. Hence, the efficiency of the solar cell should become constant at a certain value of the nanowire height. This will not happen in a radial solar cell because there is a competitive mechanism depending on the nanowire height, and this competitive mechanism is related to the dark saturation current.

The dark saturation current density $J_0(x)$ prevents the efficiency from becoming constant. It depends on the cross-sectional area of the P-N junction, and for the radial junction, the area is a function of height. The current flows radially across this junction. If the junction is located at a radius r of the cylindrical-shaped nanowire, the cross-sectional area of the junction is

$$\left|A_{\text{Junction}}\right|_{\text{Radial cell}} = 2\pi rh \tag{12.3}$$

As the cross-sectional area increases with nanowire height h, the dark saturation current also increases. So for larger values of h after the light-generated current has ceased to increase, the increase in dark current will upset the constancy of efficiency. The efficiency will decrease from its maximum value beyond a fixed height.

The axial nanowire solar cell (Figure 12.5(b)) differs from the radial cell in the respect that the current flows along the axis of the cylindrical wire. The cross-sectional area of the junction for the axial cell is

$$\left|A_{\text{Junction}}\right|_{\text{Axial cell}} = \pi R^2 \tag{12.4}$$

where R is the radius of the axial junction. As expected, this equation does not contain the nanowire height, showing that the dark current density of an axial junction solar cell will not increase with the nanowire height. Thus, the mechanism responsible for degradation of efficiency of solar cell being absent in the axial junction case, the efficiency remains constant as the nanowire becomes higher.

Example 12.2

Relation connecting excess carrier parameters (effective and bulk lifetime, and surface recombination velocity) and diameter of nanowire for an axial junction nanowire solar cell.

For an axial junction nanowire solar cell, derive an equation connecting the effective lifetime $\tau_{\text{Effective}}$ of the minority carriers in the nanowire, the bulk lifetime τ_{Bulk} of the minority carriers, the surface recombination velocity S of the carriers, and the diameter d of the nanowire.

Solution: (a) Volume of the nanowire of diameter d and height h is

$$V_{\text{NW}} = \pi \left(\frac{d}{2}\right)^2 h \tag{12.5}$$

If the recombination rate per unit volume of the nanowire (bulk + surface) is R_{NW}, the recombination rate for the nanowire is

$$R = R_{\text{NW}} \times V_{\text{NW}} = R_{\text{NW}}\pi \left(\frac{d}{2}\right)^2 h \tag{12.6}$$

by substituting for V_{NW} from equation 12.5. Recombination occurs in the bulk of the nanowire as well as on its surface. So the recombination rate R of the carriers in the nanowire is the sum of the components arising from its bulk and surface, R_1 due to bulk, and R_2 from the surface:

$$R = R_1 + R_2 \tag{12.7}$$

Let the recombination rate per unit volume of the bulk of nanowire be R_{Bulk}, then

$$R_1 = R_{\text{Bulk}}V_{\text{Bulk}} = R_{\text{Bulk}}V_{\text{NW}} = R_{\text{Bulk}}\pi \left(\frac{d}{2}\right)^2 h \tag{12.8}$$

where equation 12.5 has been used.

Suppose the recombination rate per unit volume of the surface of the nanowire is $R_{Surface}$, then

$$R_2 = R_{Surface} V_{Surface} = R_{Surface} (\pi d) h \qquad (12.9)$$

Combining together equations 12.6 to 12.9, the recombination dynamics of the P-N junction nanowire comprising components from the bulk and surface components can be written as

$$R_{NW} \pi \left(\frac{d}{2}\right)^2 h = R_{Bulk} \pi \left(\frac{d}{2}\right)^2 h + R_{Surface} (\pi d) h \qquad (12.10)$$

Dividing both sides of equation 12.10 by $\pi(d/2)^2 h$, we get

$$R_{NW} = R_{Bulk} + R_{Surface} \frac{\pi d h}{\pi \dfrac{d^2}{4} h} = R_{Bulk} + R_{Surface} \frac{4}{d} \qquad (12.11)$$

If $\tau_{Effective}$ is the effective lifetime in the nanowire, taking into account the bulk and surface contributions, τ_{Bulk} is the lifetime in the bulk of the nanowire, S is the surface recombination velocity, and Δn is the excess electron concentration:

$$R_{NW} = \frac{\Delta n}{\tau_{Effective}} \qquad (12.12)$$

$$R_{Bulk} = \frac{\Delta n}{\tau_{Bulk}} \qquad (12.13)$$

and

$$R_{Surface} = S \Delta n \qquad (12.14)$$

Putting the values of the different recombination rates from equations 12.12 to 12.14 into equation 12.11, we have

$$\frac{\Delta n}{\tau_{Effective}} = \frac{\Delta n}{\tau_{Bulk}} + S \Delta n \frac{4}{d} \qquad (12.15)$$

Dividing equation 12.15 throughout by Δn,

$$\frac{1}{\tau_{Effective}} = \frac{1}{\tau_{Bulk}} + S \frac{4}{d} = \frac{1}{\tau_{Bulk}} + \frac{4S}{d} \qquad (12.16)$$

which is the required equation interrelating $\tau_{Effective}$, τ_{Bulk}, S, and d.

Example 12.3

Checking whether incident photons will cause photogeneration of carriers at a given distance below the top of a silicon vertical nanowire solar cell and examining the likelihood of collection of produced carriers by contacts for axial and radial junction devices.

(a) In an axial P-I-N single vertical nanowire solar cell made from silicon, the thickness of intrinsic region is 4μm, while the end P- and N-regions are each 0.5μm thick. The diameter of the nanowire is 200nm. Electron lifetime in bulk silicon is 10μs, and the surface recombination velocity is 8,700cms^{-1}, which reduces to 500cms^{-1} after passivation with silicon dioxide. Absorption coefficient of light in silicon is 1×10^4 cm^{-1} at 500nm and 2×10^3 cm^{-1} at 700nm. Check both before surface passivation and after passivation, whether the incoming photons will be able to excite electron-hole pairs at a midpoint distance of 2.5μm below the top of the nanowire. Will all the electrons generated at the center of the intrinsic region be successful in reaching the contacts? Electron diffusion coefficient in silicon is 36 cm²s⁻¹.

(b) A radial P-I-N single vertical nanowire solar cell is designed with nanowire of diameter 200 nm, height 5μm. The core *P*-region has a diameter of 50nm. The annular intrinsic region has internal diameter of 50nm and external diameter of 150nm. Likewise, the internal and external diameters of the annular *N*-region are 150nm and 200nm, respectively. How will this solar cell respond to the circumstances described in example *a*?

Solution: (a) The ability of light to reach a given depth below the top end of the nanowire depends on its absorption coefficient. The dimensional diagram of silicon axial nanowire solar cell is given in Figure 12.6(a). The absorption depth *l* of light in silicon is the reciprocal of its absorption coefficient α in silicon. At λ = 500nm, the absorption depth

$$l = \frac{1}{\alpha} = \frac{1}{1 \times 10^4 \, cm^{-1}} = 1 \times 10^{-4} cm = 1 \times 10^{-4} \times 10^{-2} m = 1 \times 10^{-4-2} m = 1 \times 10^{-6} m$$

$$= \frac{1 \times 10^{-6}}{10^{-6}} \mu m = 1 \mu m \tag{12.17}$$

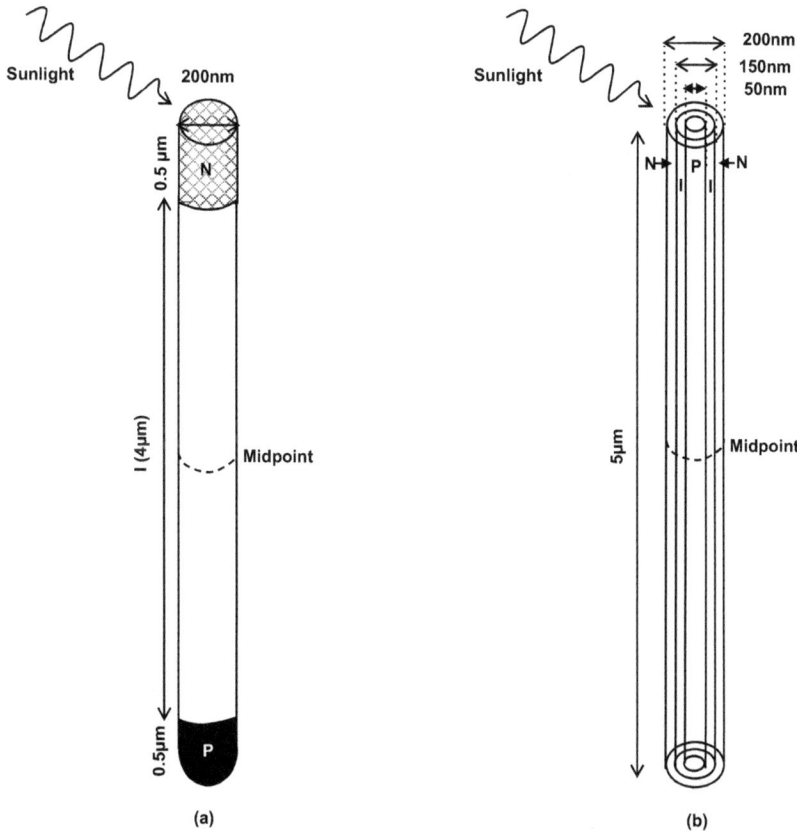

(a)   (b)

FIGURE 12.6 Nanowire dimensions for silicon solar cells: (a) axial junction and (b) radial junction. The length of the intrinsic region of axial solar cell extends over 4μm. At its opposite ends are N- and P-doped regions. These N- and P-doped regions extend over lengths of 0.5μm each. The length of the nanowire of radial solar cell is 5μm. It has a P-core of diameter 50nm. Annular rings of diameters 150nm and 200nm are drawn to indicate the intrinsic and N-regions, respectively. The sunlight is falling on the cells.

Penetration of light into the nanowire is governed by the optical thickness L of the material. Since the midpoint distance $L = 2.5\mu m > l$, the 500nm light is unable to reach up to this point, and so it is unable to generate carriers in the cell.

At $\lambda = 700$nm,

$$l = \frac{1}{\alpha} = \frac{1}{2 \times 10^3 \text{cm}^{-1}} = 0.5 \times 10^{-3} \text{cm} = 5 \times 10^{-4} \times 10^{-2} \text{m} = 5 \times 10^{-4-2} \text{m} = 5 \times 10^{-6} \text{m}$$

(12.18)

$$= \frac{5 \times 10^{-6}}{10^{-6}} \mu m = 5\mu m$$

Since

$$L < l$$

(12.19)

700nm light reaches up to the midpoint $L = 2.5\mu m$ and can excite electron-hole pairs.

The ability of carriers to reach the electrodes without recombining is determined by the effective lifetime. For electrons, the effective lifetime τ_n in the nanowire of diameter d is obtained from the bulk lifetime τ_{Bulk} and surface recombination velocity S from the formula given in equation 12.16:

$$\frac{1}{\tau_n} = \frac{1}{\tau_{\text{Bulk}}} + \frac{4S}{d}$$

(12.20)

For unpassivated nanowire, $\tau_{\text{Bulk}} = 10$ μs and $S = 8700$ cms^{-1} = 8700×10^{-2}ms^{-1}, $d = 200$nm = 200×10^{-9}m, so equation 12.20 gives

$$\frac{1}{\tau_n} = \frac{1}{10 \times 10^{-6}} + \frac{4 \times 8700 \times 10^{-2}}{200 \times 10^{-9}} = \frac{1}{10^{-5}} + \frac{4 \times 8700 \times 10^{-2+9}}{200} = 10^5 + 174 \times 10^7$$

$$= 1.7401 \times 10^9 \text{s}^{-1}$$

(12.21)

$$\therefore \tau_n = \frac{1}{1.7401 \times 10^9 \text{s}^{-1}} = 5.747 \times 10^{-10} \text{s}$$

(12.22)

Since the diffusion coefficient of electrons $D_n = 36$ cm^{2}s^{-1} = 36×10^{-4}m^{2}s^{-1}, their diffusion length is

$$L_n = \sqrt{D_n \tau_n} = \sqrt{36 \times 10^{-4} \times 5.747 \times 10^{-10}} = 14.38 \times 10^{-7} = 1.438 \times 10^{-6} \text{m}$$

$$= \frac{1.438 \times 10^{-6}}{10^{-6}} \mu m = 1.438 \mu m$$

(12.23)

The center of the cell is at $L = 2.5\mu m$. Since

$$L_n < L$$

(12.24)

and diffusion length is the average distance traversed by a carrier between its generation and recombination, a large number of electrons are unable to reach the contact.

After covering the surface of the nanowire with an oxide layer, we assume that the bulk lifetime of electrons remains the same, but the surface recombination velocity falls to 500cms^{-1} = 500×10^{-2}ms^{-1}. Hence, from equation 12.20

$$\frac{1}{\tau_n} = \frac{1}{10 \times 10^{-6}} + \frac{4 \times 500 \times 10^{-2}}{200 \times 10^{-9}} = \frac{1}{10^{-5}} + \frac{4 \times 500 \times 10^{-2+9}}{200} = 10^5 + 10 \times 10^7$$

$$= 10^5 + 10^8 = 1.001 \times 10^8 \text{s}^{-1}$$

(12.25)

$$\therefore \tau_n = \frac{1}{1.001 \times 10^8 \text{s}^{-1}} = 9.99 \times 10^{-9} \text{s}$$

(12.26)

But

$$L_n = \sqrt{D_n \tau_n} = \sqrt{36 \times 10^{-4} \times 9.99 \times 10^{-9}} = 5.997 \times 10^{-6} m = \frac{5.997 \times 10^{-6}}{10^{-6}} \mu m$$

$$= 5.997 \mu m \qquad (12.27)$$

Since

$$L_n > L \qquad (12.28)$$

many electrons are collected at the contact.

(b) The response of the radial nanowire solar cell (Figure 12.6(b)) is similar to the axial solar cell insofar as the excitation of electron-hole pairs at a depth of 2.5µm below the top surface of the nanowire is concerned; only 700nm light excites photocarrier generation. However, it differs from the axial solar cell regarding the collection of generated charge carriers. Here, the charge carriers produced at the center of the core P-region have to travel a smaller distance of 100nm. Radially, no distance exceeds 200nm. In the unpassivated nanowire solar cell, the diffusion length was 1.438µm, while in the passivated device, it was 5.997µm. So in both cases, the electrons have to cover a much smaller distance than the diffusion lengths, which means that a large number of electrons will reach the contacts.

Example 12.4

Similar problem to example 12.4 for a GaAs nanowire solar cell.

(a) An axial P-I-N single vertical nanowire solar cell is made from GaAs. The end N- and P-regions are each 250nm thick, while the intrinsic region has a thickness of 1,000nm. The diameter of the nanowire is 150nm. The electron lifetime in GaAs is 32ns, and the surface recombination velocity is $10^5 cms^{-1}$, which decreases to $10^3 cms^{-1}$ after passivating with AlGaAs. Light of wavelength 400nm strikes the cell, producing electron-hole pairs. Will it be able to produce electron-hole pairs at the central point of the cell located at 750nm depth from the top surface of the nanowire? Will light of wavelength 880nm be able to produce electron-hole pairs at this point? Will all the electrons produced be collected at the contacts? Comment on the working of the solar cell under the different situations mentioned. Absorption coefficient of light in GaAs is 7 × $10^5 cm^{-1}$ at 400nm and $1 cm^{-1}$ at 880nm. Diffusion coefficient of electrons in GaAs is $200 cm^2 s^{-1}$.

(b) A single nanowire of diameter 150nm, height 1,500nm is used to fabricate a radial P-I-N single vertical nanowire solar cell. In this nanowire, the core P-region has a diameter of 50nm. The intrinsic region is an annular ring having internal diameter of 50nm and external diameter of 100nm. The annular N-region has internal and external diameters of 100nm and 150nm, respectively. Discuss the response of this solar cell under situations similar to example a.

Solution: (a) Look at Figure 12.7(a).

At λ = 400nm, the absorption depth

$$l = \frac{1}{\alpha} = \frac{1}{7 \times 10^5 cm^{-1}} = 0.143 \times 10^{-5} cm = 1.43 \times 10^{-6} \times 10^{-2} m = 1.43 \times 10^{-6-2} m$$

$$= 1.43 \times 10^{-8} m = \frac{1.43 \times 10^{-8}}{10^{-9}} nm = 14.3 nm \qquad (12.29)$$

Since the given distance L = 750nm > l, the 400nm light does not reach this distance and fails to cause carrier generation.

At λ = 880nm,

$$l = \frac{1}{\alpha} = \frac{1}{1 cm^{-1}} = 1 cm = 1 \times 10^{-2} m = \frac{1 \times 10^{-2}}{10^{-9}} nm = 10^7 nm \qquad (12.30)$$

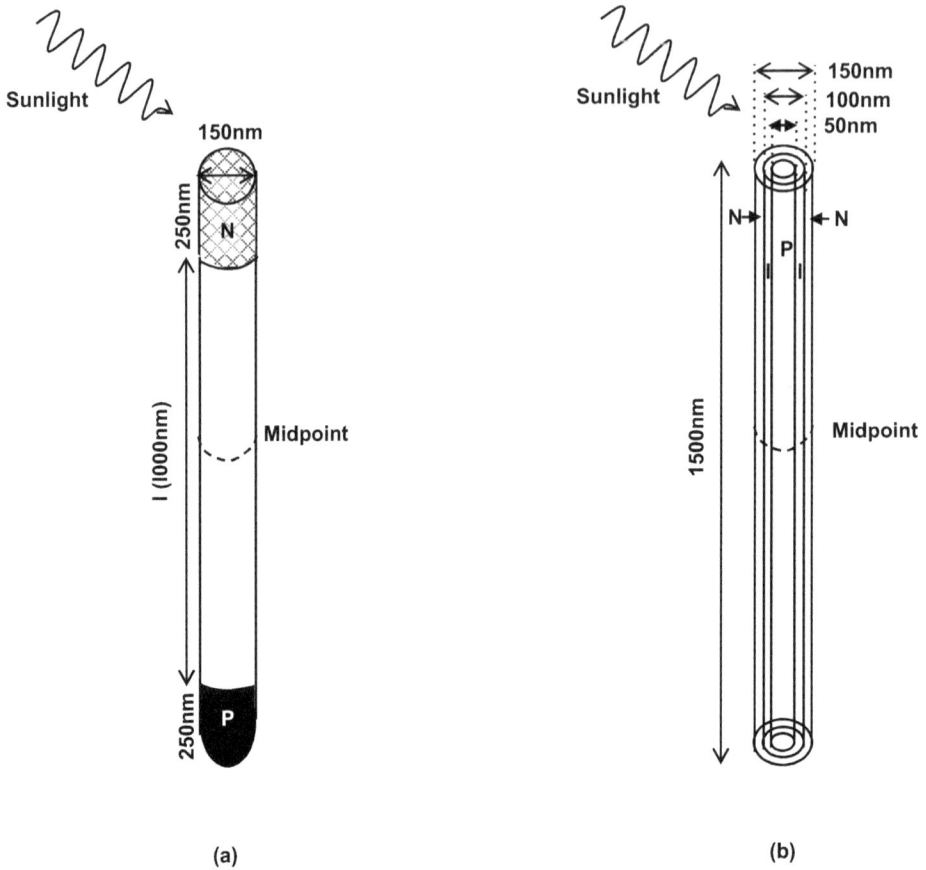

FIGURE 12.7 Dimensions of nanowires for GaAs solar cells: (a) axial junction and (b) radial junction. The axial solar cell has 250nm-long P- and N-regions at the opposite ends of a 1,000nm-long intrinsic region, while the radial solar cell has a P-core of length 1,000nm and diameter 50nm surrounded successively by annular rings of I-region of diameter 100nm and N-region of diameter 150nm. The sunlight is shown falling on the cells.

Since

$$L < l \tag{12.31}$$

880nm light reaches up to the point $L = 750$nm. It therefore succeeds in exciting electron-hole pairs.

When the GaAs nanowire is used as such without passivation, $\tau_{Bulk} = 32$ ns and $S = 10^5$ cms^{-1} $= 10^5 \times 10^{-2}$ ms^{-1}, $d = 150$nm $= 150 \times 10^{-9}$m. From equation 12.20, we have

$$\frac{1}{\tau_n} = \frac{1}{32 \times 10^{-9}} + \frac{4 \times 5 \times 10^5 \times 10^{-2}}{150 \times 10^{-9}} = 0.03125 \times 10^9 + \frac{4 \times 5 \times 10^{5-2+9}}{150}$$

$$= 3.125 \times 10^7 + 0.133 \times 10^{12} = 3.125 \times 10^7 + 1.33 \times 10^{11}$$

$$= 1.3303125 \times 10^{11} \text{s}^{-1} \tag{12.32}$$

$$\therefore \tau_n = \frac{1}{1.3303125 \times 10^{11} \text{s}^{-1}} = 0.7517 \times 10^{-11} \text{s} \tag{12.33}$$

Since $D_n = 200$ cm²s⁻¹ $= 200 \times 10^{-4}$ m²s⁻¹

$$L_n = \sqrt{D_n\tau_n} = \sqrt{200\times10^{-4}\times7.517\times10^{-12}} = 38.7737\times10^{-8}\,\text{m} = \frac{38.7737\times10^{-8}}{10^{-9}}\,\text{nm}$$

$$= 387.737\,\text{nm} \tag{12.34}$$

The center of the cell is at $L = 750$nm. Since

$$L_n < L \tag{12.35}$$

it is evident that many of the electrons will recombine before arriving at the contact.

When the surface of the GaAs nanowire is protected with a film of AlGaAs, it is assumed that the bulk lifetime remains unaffected. However, the surface recombination velocity decreases to 10^3 cms⁻¹ $= 10^3 \times 10^{-2}$ms⁻¹. Then, from equation 12.20 we get

$$\frac{1}{\tau_n} = \frac{1}{32\times10^{-9}} + \frac{4\times10^3\times10^{-2}}{150\times10^{-9}} = 0.03125\times10^9 + \frac{4\times10^{3-2+9}}{150}$$

$$= 0.03125\times10^9 + 0.0267\times10^{10} = 3.125\times10^7 + 2.67\times10^8$$

$$= 2.9825\times10^8\,\text{s}^{-1} \tag{12.36}$$

$$\therefore \tau_n = \frac{1}{2.9825\times10^8\,\text{s}^{-1}} = 0.3353\times10^{-8}\,\text{s} \tag{12.37}$$

From equation 12.34,

$$L_n = \sqrt{D_n\tau_n} = \sqrt{200\times10^{-4}\times0.3353\times10^{-8}} = 8.189\times10^{-6}\,\text{m} = \frac{8.189\times10^{-6}}{10^{-9}}\,\text{nm}$$

$$= 8189\,\text{nm} \tag{12.38}$$

The center of the cell being at $L = 750$nm,

$$L_n \gg L \tag{12.39}$$

most of the electrons are able to reach the contact without recombining.

(b) The radial nanowire solar cell is shown in Figure 12.7(b). 400nm light is unable to reach the given point. However, 880nm light is able to reach this point and produce carriers, as in a. The diffusion length for the unpassivated nanowire solar cell is 387.737nm, while that for the passivated nanowire solar cell is 8,189nm. In both cases, the diffusion lengths are much larger than the distances to be covered to reach the electrodes. As the distance from the center to the boundary is only 75nm and the maximum radial distance is only 150nm, the electrons generated can travel to the contacts without recombination. So a large number of electrons will be available at the contacts of the radial solar cells, unpassivated or passivated, unlike the axial solar cells, in which this happens only in the passivated devices.

Example 12.5

Variation of intensity of light with vertical distance of penetration into nanowires of different materials.

Absorption coefficients of light in GaAs, crystalline Si, amorphous silicon, InP, and CdS at 600nm are 5.5×10^4cm⁻¹, 4.5×10^3cm⁻¹, 3×10^4cm⁻¹, 7×10^4cm⁻¹, and 40cm⁻¹, respectively. Solar cells are made with vertical nanowires of height 1,000nm from these materials. Find by what ratios the intensity of incoming light incident vertically downwards on the solar cell decreases at the middle point of these nanowires.

Solution: The intensity $I(x)$ at a distance x is

$$I(x) = I(0)\exp(-\alpha x) \tag{12.40}$$

where $I(0)$ is the intensity at $x = 0$ and α is the absorption coefficient of the material. Equation 12.40 can be written as

$$\frac{I(x)}{I(0)} = \exp(-\alpha x) \tag{12.41}$$

The unit

$$\mathrm{cm}^{-1} = \frac{1}{\mathrm{cm}} = \frac{1}{10^{-2}\,\mathrm{m}} = 10^2\,\mathrm{m}^{-1} \tag{12.42}$$

For GaAs, equation 12.41 gives

$$\left.\left|\frac{I(x=500\mathrm{nm})}{I(0)}\right|\right|_{\mathrm{GaAs}} = \exp\left(-5.5\times10^4 \times 10^2 \times 500\times10^{-9}\right)$$

$$= \exp\left(-5.5\times5\times10^{4+2+2-9}\right)$$

$$= \exp\left(-5.5\times5\times10^{-1}\right) = \exp\left(-27.5\times10^{-1}\right) = \exp(-2.75) = 0.0639 \tag{12.43}$$

For crystalline Si, using equation 12.41,

$$\left.\left|\frac{I(x=500\mathrm{nm})}{I(0)}\right|\right|_{\mathrm{c-Si}} = \exp\left(-4.5\times10^3 \times 10^2 \times 500\times10^{-9}\right)$$

$$= \exp\left(-4.5\times5\times10^{3+2+2-9}\right)$$

$$= \exp\left(-4.5\times5\times10^{-2}\right) = \exp\left(-22.5\times10^{-2}\right) = \exp(-0.225) = 0.7985 \tag{12.44}$$

For amorphous Si, by equation 12.41,

$$\left.\left|\frac{I(x=500\mathrm{nm})}{I(0)}\right|\right|_{\mathrm{a-Si}} = \exp\left(-3\times10^4 \times 10^2 \times 500\times10^{-9}\right)$$

$$= \exp\left(-3\times5\times10^{4+2+2-9}\right)$$

$$= \exp\left(-3\times5\times10^{-1}\right) = \exp\left(-15\times10^{-1}\right) = \exp(-1.5) = 0.223 \tag{12.45}$$

For InP, by applying equation 12.41,

$$\left.\left|\frac{I(x=500\mathrm{nm})}{I(0)}\right|\right|_{\mathrm{InP}} = \exp\left(-7\times10^4 \times 10^2 \times 500\times10^{-9}\right)$$

$$= \exp\left(-7\times5\times10^{4+2+2-9}\right)$$

$$= \exp\left(-7\times5\times10^{-1}\right) = \exp\left(-35\times10^{-1}\right) = \exp(-3.5) = 0.0302 \tag{12.46}$$

For CdS, application of equation 12.41 leads to

$$\left. \frac{\left| I\left(x = 500\text{nm}\right)\right|}{I\left(0\right)} \right|_{\text{CdS}} = \exp\left(-40 \times 10^2 \times 500 \times 10^{-9}\right)$$

$$= \exp\left(-40 \times 5 \times 10^{2+2-9}\right)$$

$$= \exp\left(-40 \times 5 \times 10^{-5}\right) = \exp\left(-200 \times 10^{-5}\right) = \exp\left(-0.002\right) = 0.998 \qquad (12.47)$$

Example 12.6

Effective bandgap of a nanowire of a given diameter.
What is the effective bandgap in an InP nanowire of diameter 80nm as predicted by the particle-in-a-box model? Bandgap of bulk InP is 1.344eV. Effective mass of electrons in InP is $m_e^* = 0.08m_0$, and that of holes is $m_{hh}^* = 0.6m_0$, $m_{lh}^* = 0.089m_0$.

Solution: Applying equation 10.61 for the particle-in-a box model to the nanowire, the bandgap of a nanowire of diameter d is given by

$$E_{G, \text{NW}} = E_{G, \text{Bulk}} + \frac{\hbar^2 \pi^2}{2\mu d^2} \qquad (12.48)$$

where μ is the electron-hole reduced mass. For electron-heavy hole case, the reduced mass is

$$\left. \mu \right|_{\text{Electron-Heavy hole}} = \frac{m_e^* m_{hh}^*}{m_e^* + m_{hh}^*} = \frac{0.08m_0 \times 0.6m_0}{0.08m_0 + 0.6m_0} = \frac{0.048m_0^2}{0.68m_0} = 0.070588m_0 \qquad (12.49)$$

$$= 0.070588 \times 9.109 \times 10^{-31}\text{kg} = 0.642986 \times 10^{-31}\text{kg} \approx 6.43 \times 10^{-32}\text{kg}$$

Applying equation 12.48, we get the effective bandgap of the nanowire material as

$$\left. \left| E_{G, \text{NW}} \right| \right|_{\text{Electron-Heavy hole}} = 1.344\text{eV} + \frac{\left(1.05457 \times 10^{-34}\right)^2 (3.14159)^2}{2 \times 6.43 \times 10^{-32} \times \left(80 \times 10^{-9}\right)^2}\text{J}$$

$$= 1.344 + \frac{1.112 \times 10^{-68} \times 9.8695}{2 \times 6.43 \times 10^{-32} \times 6.4 \times 10^3 \times 10^{-18}} \times \frac{1}{1.602 \times 10^{-19}}\text{eV}$$

$$= 1.344 + \frac{10.97488 \times 10^{-68}}{2 \times 6.43 \times 6.4 \times 10^{-32-18-19+3}}\text{eV} \qquad (12.50)$$

$$= 1.344 + \frac{10.97488 \times 10^{-68}}{82.304 \times 10^{-66}}\text{eV} = 1.344 + 0.1333456 \times 10^{-68+66}\text{eV}$$

$$= 1.34533\text{eV} \approx 1.345\text{eV}$$

For electron-light hole case, the reduced mass is

$$\left. \mu \right|_{\text{Electron-Light hole}} = \frac{m_e^* m_{lh}^*}{m_e^* + m_{lh}^*} = \frac{0.08m_0 \times 0.089m_0}{0.08m_0 + 0.089m_0} = \frac{0.00712m_0^2}{0.169m_0} = 0.04213m_0 \qquad (12.51)$$

$$= 0.04213 \times 9.109 \times 10^{-31}\text{kg} = 0.38376 \times 10^{-31}\text{kg} \approx 3.84 \times 10^{-32}\text{kg}$$

Using equation 12.48, the effective bandgap of the nanowire material is

$$\left|E_{G,\,NW}\right|_{Electron-Light\,hole} = 1.344eV + \frac{\left(1.05457\times10^{-34}\right)^2\left(3.14159\right)^2}{2\times3.84\times10^{-32}\times\left(80\times10^{-9}\right)^2}\,J$$

$$= 1.344 + \frac{1.112\times10^{-68}\times9.8695}{2\times3.84\times10^{-32}\times6.4\times10^3\times10^{-18}}\times\frac{1}{1.602\times10^{-19}}\,eV$$

$$= 1.344 + \frac{10.97488\times10^{-68}}{2\times3.84\times6.4\times10^{-32-18-19+3}}\,eV$$

$$= 1.344 + \frac{10.97488\times10^{-68}}{49.152\times10^{-66}}\,eV = 1.344 + 0.2232845\times10^{-68+66}\,eV$$

$$= 1.34623eV \approx 1.346eV$$

$$(12.52)$$

Example 12.7

Finding open-circuit voltage of a nanowire solar cell from its experimentally measured short-circuit current density and correcting for fractional coverage of solar cell area by nanowires.
The short-circuit current density of an InP nanowire solar cell is measured as 20mAcm^{-2}. Find the open-circuit voltage of the cell. Only 10% of the solar cell area is covered by the nanowires. Correct the determined open-circuit voltage by accounting for the fact that the carrier flow is restricted to nanowires only and that the carriers do not flow through the full area of the cell. How much is the incremental V_{OC} due to this correction?
Solution: The open-circuit voltage of the solar cell is given by

$$V_{OC} = \frac{k_B T}{e}\ln\left(\frac{J_{SC}}{J_0}\right)$$

$$(12.53)$$

where k_B is the Boltzmann constant, e is the elementary charge, J_{SC} is the short-circuit current density, and J_0 is the saturation current density. The saturation current density is given by the Shockley equation,

$$J_0 = A\exp\left(-\frac{E_G}{k_B T}\right)$$

$$(12.54)$$

where A is a material-independent constant:

$$A = 2.95\times10^5\,Acm^{-2}$$

$$(12.55)$$

E_G denotes the bandgap of the semiconductor, and T represents the temperature in Kelvin scale. For InP, $E_G = 1.344eV$. Hence, the value of saturation current density at $T = 300K$ is

$$J_0 = A\exp\left(-\frac{E_G}{k_B T}\right) = 2.95\times10^5\exp\left(-\frac{1.344\times1.602\times10^{-19}}{1.38065\times10^{-23}\times300}\right)$$

$$= 2.95\times10^5\exp\left(-\frac{1.344\times1.602\times10^{-19+23}}{1.38065\times300}\right)$$

$$(12.56)$$

$$= 2.95\times10^5\exp\left(-5.198\times10^{-3}\times10^4\right) = 2.95\times10^5\exp\left(-51.98\right)$$

$$= 2.95\times10^5\times2.663\times10^{-23} = 7.856\times10^{-18}\,Acm^{-2}$$

Putting the values in equation 12.53, we get the open-circuit voltage:

$$V_{OC} = \frac{k_B T}{e} \ln\left(\frac{J_{SC}}{J_0}\right)$$

$$= \frac{1.38065 \times 10^{-23} \times 300}{1.602 \times 10^{-19}} \ln\left(\frac{20 \times 10^{-3}}{7.856 \times 10^{-18}}\right)$$

$$= \frac{1.38065 \times 300 \times 10^{-23+19}}{1.602} \ln\left(\frac{20 \times 10^{-3+18}}{7.856}\right) \tag{12.57}$$

$$= 258.5487$$

$$\times 10^{-4} \ln\left(2.5458 \times 10^{15}\right) = 258.5487 \times 10^{-4} \times 35.47 = 0.91707 V$$

This V_{OC} value needs to be corrected because in a nanowire solar cell, the current only flows through the portions in which the nanowires are grown. No current flows through the regions without nanowires. The portion containing nanowires is much smaller than the total area of the cell. Here it is given as only 10% of the cell area.

The short-circuit current being given as a measured value is the outcome of the actual area through which the current flows. But the saturation current density J_0 has been calculated by taking the full area of the cell. Therefore, this current density must be reduced to equalize the areas in J_{SC} and J_0. The current flowing through the nanowire portion will be 10% of the total current. Hence, the saturation current must be multiplied by 10/100 = 0.1. The corrected open-circuit voltage is

$$|V_{OC}|_{Corrected} = \frac{1.38065 \times 10^{-23} \times 300}{1.602 \times 10^{-19}} \ln\left(\frac{20 \times 10^{-3}}{0.1 \times 7.856 \times 10^{-18}}\right)$$

$$= \frac{1.38065 \times 300 \times 10^{-23+19}}{1.602} \ln\left(\frac{20 \times 10^{-3+18}}{0.7856}\right) \tag{12.58}$$

$$= 258.5487$$

$$\times 10^{-4} \ln\left(25.458 \times 10^{15}\right) = 258.5487 \times 10^{-4} \times 37.7758 = 0.976688 V$$

The difference between the two voltages, the corrected and uncorrected values of V_{OC}, is found from equations 12.57 and 12.58 as

$$|V_{OC}|_{Corrected} - V_{OC} = 0.976688 - 0.91707 = 0.059618 V = 0.059618 \times 10^3 \, mV \tag{12.59}$$

$$= 59.618 \, mV$$

12.4   DISCUSSION AND CONCLUSIONS

Nanowire architecture offers a convenient way of controlling the optical and electrical properties of solar cells than the planar junction devices (Tsakalakos 2010; Yang et al 2015). Planar technology faces criticism owing to the high material and production costs incurred. Large quantities of high-quality material are tremendously high-priced. Nanowires, too, have pitfalls. The large surface area of nanowires results in a high surface recombination. The surface recombination enforces limitations on nanowire designs. Apart from this, growing nanowires cores/shells with low defect density is a challenging task.

Questions and Answers

12.1   Argue in favor of using nanowires in solar cells.

Answer from economical considerations: Nanowires enable fabrication of low-cost solar cells at comparable performance with standard technologies because they consume smaller quantity of semiconductor material and can be fabricated with inferior quality of semiconductor. They can be grown by heteroepitaxial process on cheaper substrates.

Scientific answer: Nanowires offer bandgap tuning, enable optical loss reduction by virtue of their antireflection property, and can be used for making mechanically flexible solar cells.

12.2   How are nanowires superior to antireflection coatings? Answer: The antireflective property of nanowires originates from a nanoscale effect by which they exhibit broadband antireflection behavior covering a greater part of the range of wavelengths in the solar spectrum. The antireflective property of antireflection coatings arises from destructive interference effect, in which the path length traversed by light, as determined by coating thickness, plays the central role. As one thickness of ARC can produce destructive interference for one wavelength or at best a narrow band around that wavelength, the antireflectivity behavior of coatings is observed over a short range of solar spectrum. Light of remaining wavelengths suffers reflection and is therefore not utilized by the solar cell for electricity production.

12.3   What is the use of making single-nanowire solar cells? Answer:
   (i)   They are useful for driving nanoelectronic circuits working at extremely low power levels in nanowatts region.
   (ii) They are used for research investigations to study the properties, capabilities, and limitations of a single nanowire as a solar cell because the multinanowire cells are upscaled versions of this basic unit. Such studies can be applied to make better nanowire array solar cells.

12.4   Of what use are nanowire array solar cells? Answer: To meet the commercial power requirements for domestic and industrial applications.

12.5   What are the directions of light propagation and charge separation in a vertical axial nanowire solar cell? Answer: Light propagation, vertically downwards; charge separation, along the length of the wire. Hence, light propagates parallel to charge motion.

12.6   Does light propagate parallel to the direction of charge separation in a radial vertical nanowire solar cell? Answer: No. Light moves vertically downwards, while charge separation takes place along the radius of the nanowire. The two directions are mutually orthogonal to each other.

12.7   Both material and fabrication costs are reduced by choosing the radial nanowire solar cell configuration over the axial geometry. Why? Answer: Material costs decrease because semiconductor absorber with smaller carrier lifetime can be used. The reason that smaller lifetime semiconductor is adequate for making this type of solar cell is that the smaller lifetime means a shorter diffusion length. The photogenerated charge carriers have to travel a short distance through the absorber material in the radial cell to reach the contacts, and chances of their being lost by recombination on the way are very remote. Fabrication costs diminish because carrier lifetime must be maintained by adhering to careful cleaning and thermal processing protocols. These labor- and capital-intensive procedures need not be strictly followed during fabrication of radial nanowire solar cell.

12.8   What is the advantage derived from the fact that nanowires allow facile strain relaxation? Answer: The area of contact of nanowire with the growth substrate being very small, any strain produced at the nanowire/substrate interface is easily relaxed. Easy strain relaxation allows to grow nanowires on a less-expensive, nonepitaxial substrate. The growth process is known as heteroepitaxy.

12.9   Two radial junction nanowire solar cells are fabricated with equal-diameter nanowires. The nanowire length in cell A is 1μm, while in cell B, it is 2μm. Can we say that more photocurrent will be produced in cell B upon shining sunlight? Answer: Not necessarily. It depends on the length up to which sunlight travelling through the nanowire has sufficient energy to create electron-hole pairs. After the light has weakened, the extra length will not contribute to photocurrent but will only add to dark current.

12.10   How do cells A and B of exercise 12.9 behave if they are axial junction nanowire solar cells? Answer: There is no dependence of photocurrent on length after the peak current is reached nor will dark current increase with length.

12.11   What is meant by surface recombination velocity? Answer: A measure of the carrier recombination at the surface of a semiconductor, expressed in velocity units (ms^{-1}), that relates the recombination rate at the surface to the minority carrier density at the surface.

12.12   What is the surface recombination velocity for a surface on which there is no recombination? Answer: Zero.

12.13   What is meant by absorption coefficient of light in a material? What is its unit? How is it related to absorption depth? Answer: Absorption coefficient α is a measure of the decrease in intensity of light as it travels through a medium given by the natural logarithm ratio of the intensity of light at a depth x below the surface to the intensity of incident light at the surface ($x = 0$) divided by the depth x, i.e., $\alpha = -\ln\{I(x)/I(0)\}/x$. The unit of absorption coefficient is $1/\text{Length} = 1/\text{cm} = \text{cm}^{-1}$. Absorption depth $= 1/$ Absorption coefficient.

12.14   What do the terms "extinction coefficient" and "attenuation coefficient" of light in a material mean? Answer: They are synonyms of *absorption coefficient.*

12.15   Nanowires are made from silicon and gallium arsenide. Absorption coefficient of silicon is less than that of gallium arsenide. In which of these two semiconductor nanowires will the intensity of light at a given depth below the surface be lower? Answer: GaAs nanowire, because light incident on the top surface of this nanowire will be weakened faster in this material than silicon.

12.16   What is meant by effective bandgap of a nanowire? Answer: Bandgap of the bulk material from which the nanowire is made + contribution to bandgap by quantum confinement effect.

12.17   Which has a larger bandgap: (a) larger-diameter nanowire or (b) smaller-diameter nanowire? Answer: *b.*

12.18   How does dependence of bandgap of nanowire on its diameter help the solar cell technology? Answer: It is used as a method to tailor bandgap of the nanowire to adapt to the wavelengths in sunlight that are to be used.

REFERENCES

Dutta M., L. Thirugnanam and N. Fukata 2018 Si Nanowire solar cells: Principles, device types, future aspects, and challenges, in: Ikhmayies S. (ed.), Advances in Silicon Solar Cells, Springer International Publishing AG, Cham, Switzerland, pp. 299–329.

Garnett E. C., M. L. Brongersma, Y. Cui and M. D. McGehee 2011 Nanowire solar cells, Annual Review of Materials Research, 41: 269–295.

Raj V., H. H. Tan and C. Jagadish 2019 Axial vs radial junction nanowire solar cell, Asian Journal of Physics, 28(7–9): 719–746.

Tian B., X. Zheng, T. J. Kempa, Y. Fang, N. Yu, G. Yu, J. Huang and C. M. Lieber 2007 Coaxial silicon nanowires as solar cells and nanoelectronic power sources, Nature Letters, 449: 885–890.

Tsakalakos L. 2010 Chapter 6: Nanowire and nanotube-based solar cells, in: Tsakalakos L. (ed.), Nanotechnology for Photovoltaics, CRC Press, Boca Raton, USA, pp. 211–252.

Yang P., S. Brittman and C. Liu 2015 Chapter 6: Nanowires for photovoltaics and artificial photosynthesis, in: Lu W. and J. Xiang (eds.), Semiconductor Nanowires: From Next-Generation Electronics to Sustainable Energy, The Royal Society of Chemistry, Cambridge, pp. 277–311.

Zhang A., G. Zheng and C. M. Lieber 2016 Nanowires: Building Blocks for Nanoscience and Nanotechnology, Springer International Publishing AG, Switzerland, p. 228.

13 Nanowire Solar Cells
Fabrication

13.1 SINGLE-NANOWIRE SOLAR CELLS

A single nanowire must be able to work as a solar cell. A nanowire made of good-quality material is necessary for realization of high-performance solar cell. A nanowire of high-quality III-V material can be made on a low-cost silicon substrate. But the lattices of silicon and III-V materials do not match well, and so also their coefficients of thermal expansion. Nonetheless, the strain at the nanowire-silicon interface is relieved within the first few monolayers, owing to the small contact area between the substrate and the nanowire having a large surface-area-to-volume ratio. In other words, small footprints of nanowires make them capable of accommodating more strain than planar structures, reducing the unavoidable restrictions faced in heteroepitaxial growth.

There are two principal directions in which the nanowire points relative to the substrate, vertical and horizontal. In a vertically standing nanowire, the absorption of light is determined by waveguide modes, whereas in a horizontally lying nanowire, it depends on leaky modes or Mie resonances supported by it (Cao et al 2009). An efficiency of 40% has been achieved in a vertical nanowire solar cell, and 10.2% in a horizontal nanowire solar cell (Zhang and Liu 2019).

13.1.1 SINGLE GaAs NANOWIRE SOLAR CELL IN VERTICAL CONFIGURATION ($\eta = 40\%$)

This device consisting of a single standing nanowire shows the colossal light concentration property of the GaAs nanowire arising from the resonant increase in its absorption cross section and gives an idea of the high efficiency achievable with this material in this configuration (Krogstrup et al 2013). Figure 13.1 shows the solar cell.

13.1.1.1 Fabrication Plan Outline

A ~ 2.5μm-long GaAs core-shell nanowire of diameter ~ 425nm is anchored in vertical configuration to a P+ Si substrate, serving as one contact. It is embedded in SU-8 photoresist for firm fixation. A P-I-N radial junction is formed on nanowire core by making shells of P+, intrinsic, and N+ regions. The N+ region is contacted through a transparent top contact.

The fabrication starts with substrate preparation. Growth of nanowires, their doping to form a diode, providing mechanical support to the nanowire for standing vertically, and making ohmic contacts with nanowire are the vital steps. The process of growth and doping of nanowires is shown in Figure 13.2, while Figure 13.3 illustrates the fabrication of a single vertical nanowire solar cell.

13.1.1.2 Preparation of Oxidized P+ Silicon (100) Substrate with Apertures of 50–70nm Size

P+ (100) silicon substrate is taken. The substrate has a thermally grown silicon dioxide layer of thickness 30nm on it. Electron-beam lithography is performed to define holes of size 50–70nm in the silicon dioxide layer. The oxide is etched in buffered HF to make the holes.

13.1.1.3 Ga-Assisted VLS Growth of P-type GaAs Nanowire Core

On the prepared Si substrate, GaAs nanowire cores are grown in a molecular beam epitaxy system. The gallium and arsenic fluxes are simultaneously opened with V/III ratio of 4, Ga rate is 0.3nms^{-1},

DOI: 10.1201/9781003215158-18

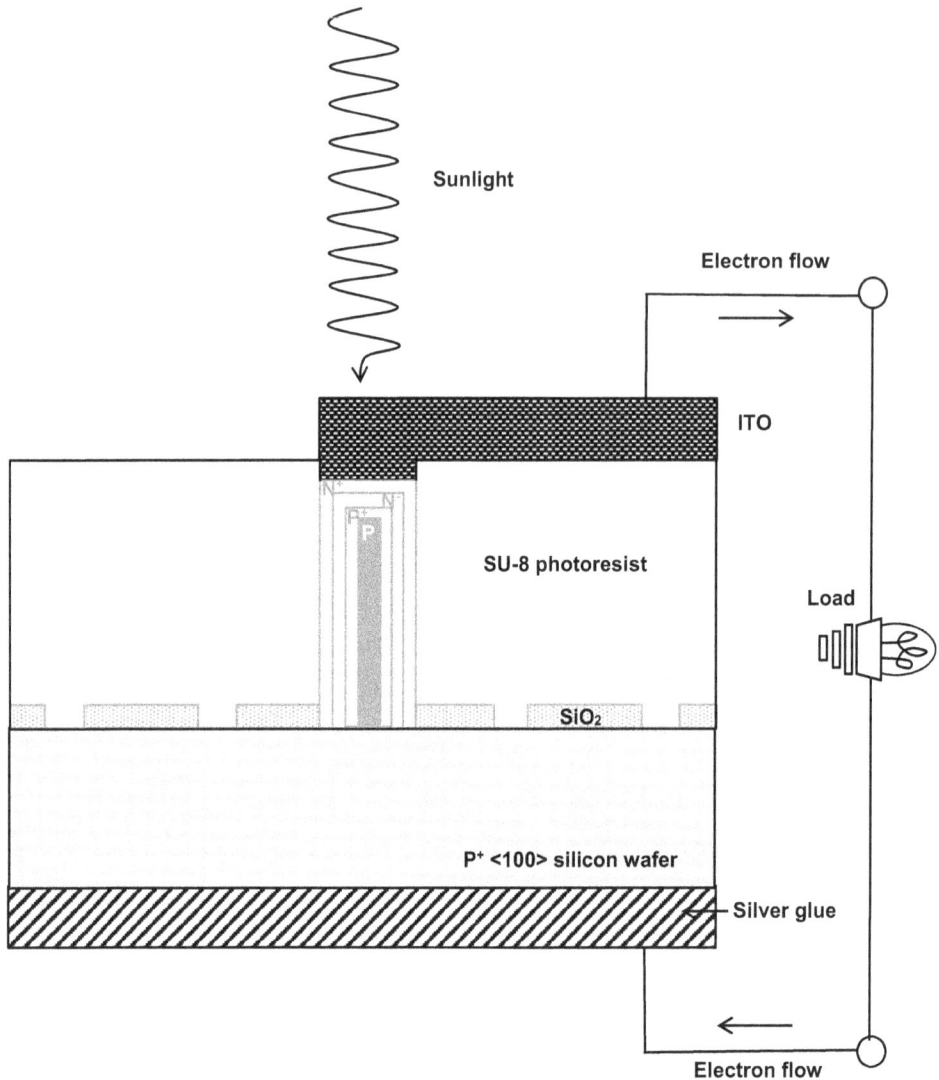

FIGURE 13.1 A single vertically oriented GaAs nanowire solar cell showing the single GaAs P-core/ P$^+$shell/N$^-$shell/N$^+$shell nanowire on silicon in the patterned SiO$_2$/Si substrate, the SU-8 photoresist, ITO top electrode, and silver glue applied on the underside of SiO$_2$/Si substrate for bottom connection. Sunlight is incident on the top ITO electrode side. Electric output is obtained between the ITO and Ag electrodes.

the growth temperature is 630°C, and the growth time is 1 h. The growth takes place by self-catalyzed method without any external metal catalyst. Nanocraters are formed by interaction between reactive gallium and holes in SiO$_2$ acting as a prelude to nucleation of nanowires. During growth of the nanowire core, a flux of beryllium is added for P-doping. Consequently, the nanowire core is P-type, and the flux of P-type impurity beryllium is equivalent to a planar doping concentration $= 3.5 \times 10^{17}$ cm^{-3}.

13.1.1.4   Growth of P$^+$-type GaAs Nanowire Shell

After the P-type nanowire core has been grown, the P$^+$-type shell is formed by radial growth at a lower temperature (460°C), keeping the V/III ratio at 50. The flux of P-type impurity beryllium used in shell growth is equivalent to a planar doping concentration $= 7 \times 10^{18}$ cm^{-3}. Annealing is done at 630°C for 10 min. for diffusion of Be atoms.

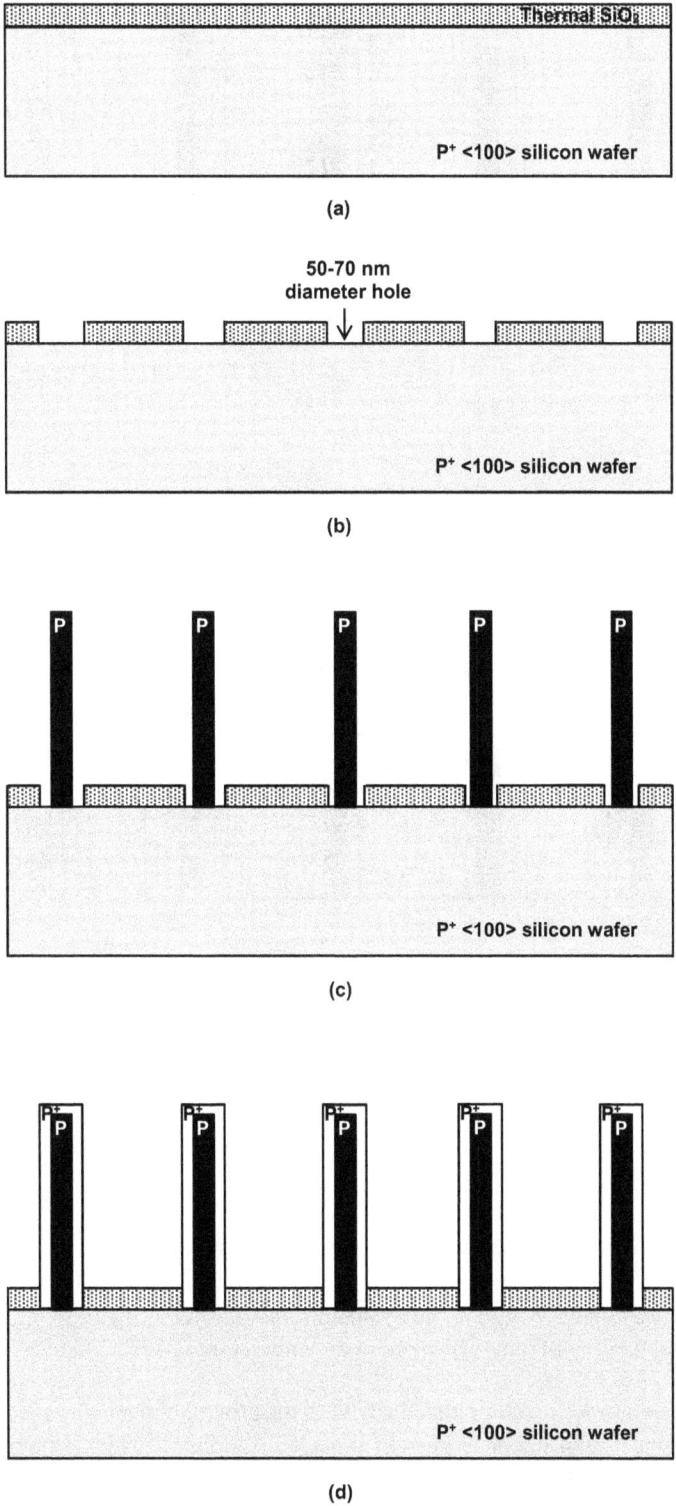

FIGURE 13.2 Growth and doping of nanowires for fabrication of a single vertically oriented GaAs nanowire solar cell: (a) SiO$_2$/Si substrate, (b) etching nanometer size holes in the oxide, (c) growth of P-type GaAs nanowire core, (d) growth of P$^+$GaAs nanowire shell, (e) growth of N$^-$GaAs nanowire shell, and (f) growth of N$^+$GaAs nanowire shell.

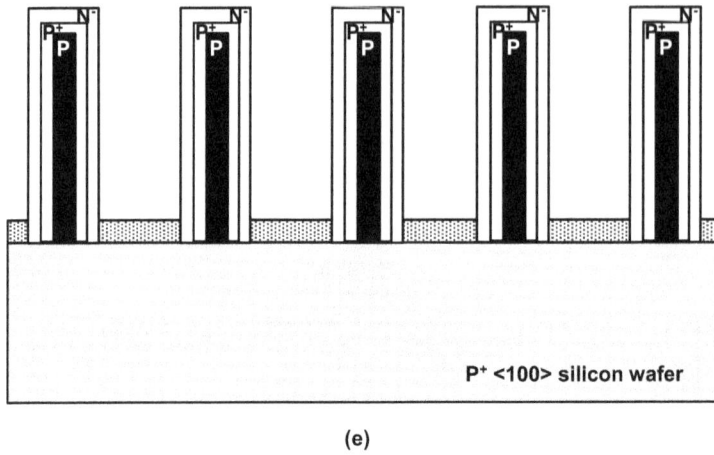

(e)

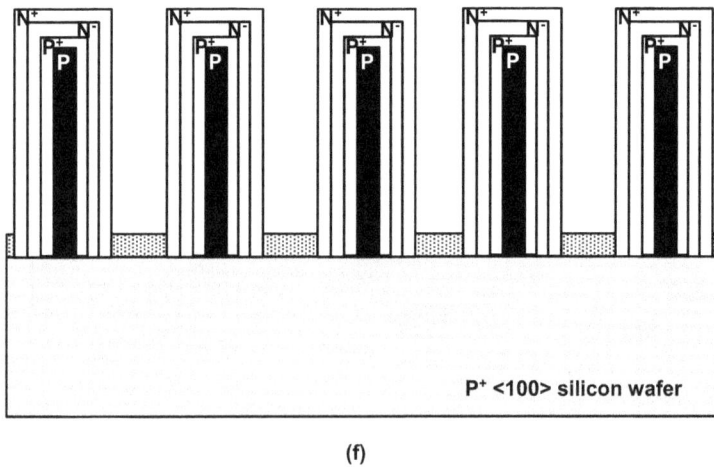

(f)

FIGURE 13.2   (Continued)

13.1.1.5   Growth of an Undoped and N-type GaAs Nanowire Shell

The P-type shell is covered with an undoped and an N-doped shell. Addition of silicon flux makes the shell N-type. The flux of N-type impurity silicon used in shell growth is equivalent to a planar doping concentration $= 7 \times 10^{18} \text{cm}^{-3}$. Shell growth temperature is 460°C. The N-doped layer has a thickness of 50nm.

Thus, dopants are introduced simultaneously with the growth of nanowire core and shell, beryllium as P-type impurity, and silicon as N-type impurity. A radial P-I-N junction is formed between the core and the shell with the core P-type and shell N-type.

13.1.1.6   Making Electrical Contacts with the Nanowire

SU-8 photoresist is coated on the P$^+$ Si substrate by spinning at 4,000RPM for 45 s. The photoresist is cured with UV and baked on a hot plate. For making the top contact with the nanowire, its tip is

(a)

(b)

(c)

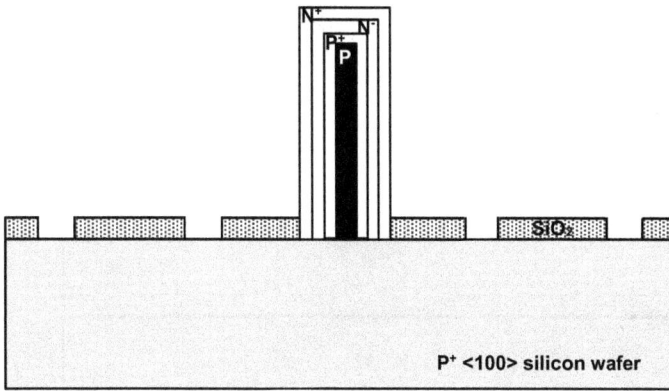

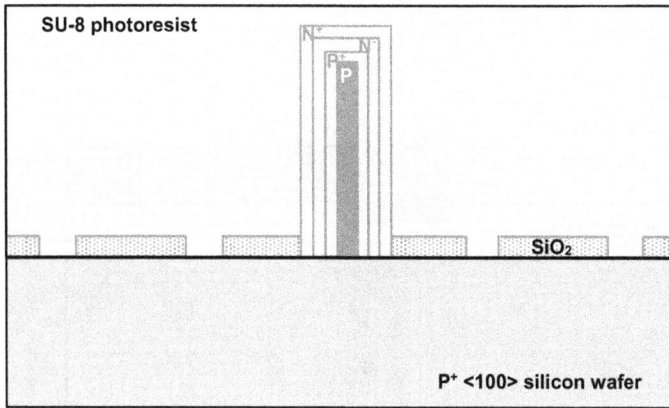

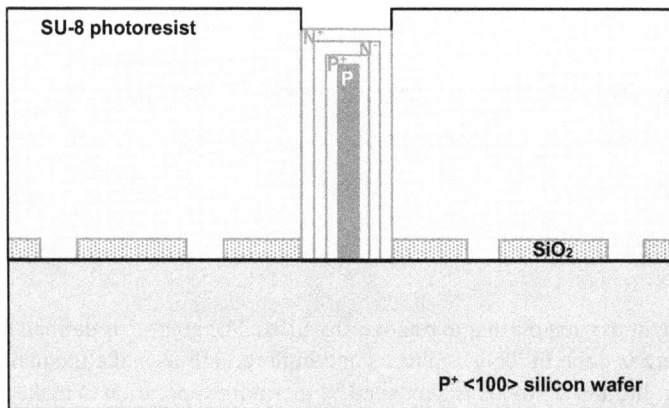

FIGURE 13.3 Fabrication of a single vertically oriented GaAs nanowire solar cell: (a) single GaAs P-core/P+shell/N−shell/N+shell nanowire on Si in patterned SiO₂/Si substrate, (b) spin-coating SU-8 photoresist, (c) O₂ plasma etching to release nanowire tip, (d) ITO top electrode deposition, and (e) applying silver glue on the underside of SiO₂/Si substrate for bottom connection.

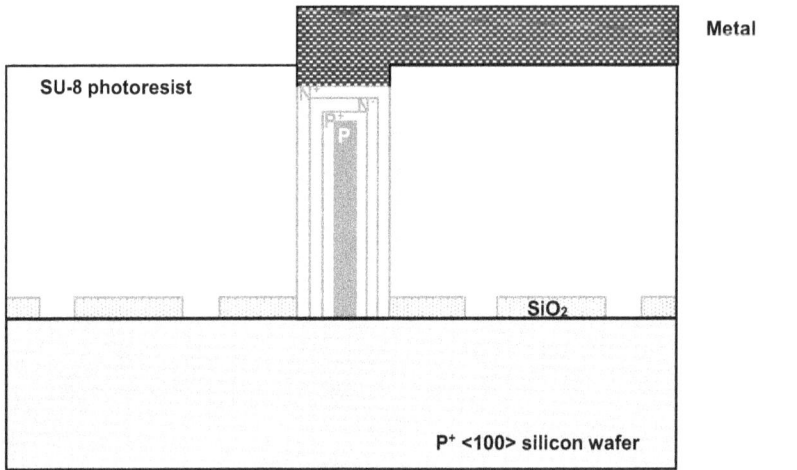

(d)

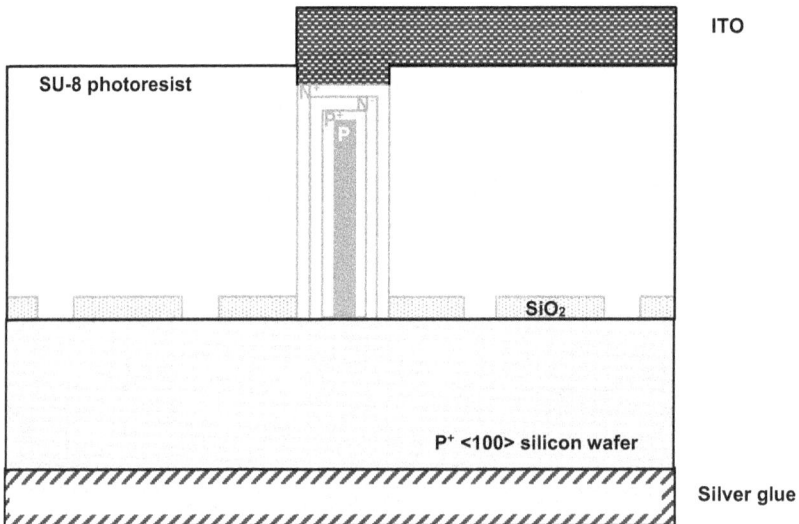

(e)

FIGURE 13.3 (Continued)

released by etching in oxygen plasma to remove any filler. The contact is defined by electron-beam lithography. Etching is done in oxygen plasma and buffered HF to make the nanowire top free of any silicon dioxide. Indium tin oxide is deposited by thermal evaporation to make the top electrode. The ITO is subjected to thermal treatment at 185°C for 3 min. to improve transparency of coating. Silver is glued to the underside of the substrate to establish the bottom contact with the nanowire.

13.1.1.7 Solar Cell Parameters
The short-circuit current density is measured as $180 \, mAcm^{-2}$, the open-circuit voltage is 0.43V, the fill factor is 52%, and the efficiency is 40% under 1 sun illumination. The extraordinarily large

value of short-circuit density is an order of magnitude higher than prophesied by the Lambert-Beer law (Krogstrup et al 2013).

13.1.2 Surface-Passivated Single GaAsP Nanowire Solar Cell in Horizontal Configuration ($\eta = 10.2\%$)

This device (Figure 13.4) uses a $Ga_{0.8}As_{0.2}P$ nanowire of controlled bandgap grown by direct epitaxy on silicon (Holm et al 2013). The criticality of passivation of nanowire surface is emphasized.

13.1.2.1 Fabrication Plan Outline

P-type core/N-type shell GaAsP vertical nanowires are grown on silicon substrate with native oxide, and surface passivated with N^+ InGaP (Figure 13.5). A nanowire removed from the parent substrate is laid down horizontally on another SiO_2/Si substrate. Contacts to P-core and N-shell are defined by e-beam lithography and metallized with appropriate metals for establishing ohmic contacts (Figure 13.6).

The main steps are growth of nanowires, their doping and surface passivation, detachment of a single nanowire from growth substrate, its alignment on a new substrate, and making contacts.

13.1.2.2 Growth of P-I-N Radial Junction Core-Shell $GaAs_{0.8}P_{0.2}$ Nanowires

Solid-source molecular beam epitaxy (SS-MBE) system for growth of III-V compound semiconductors is used. The substrate is silicon wafer, orientation < 111 >, P-type containing boron as

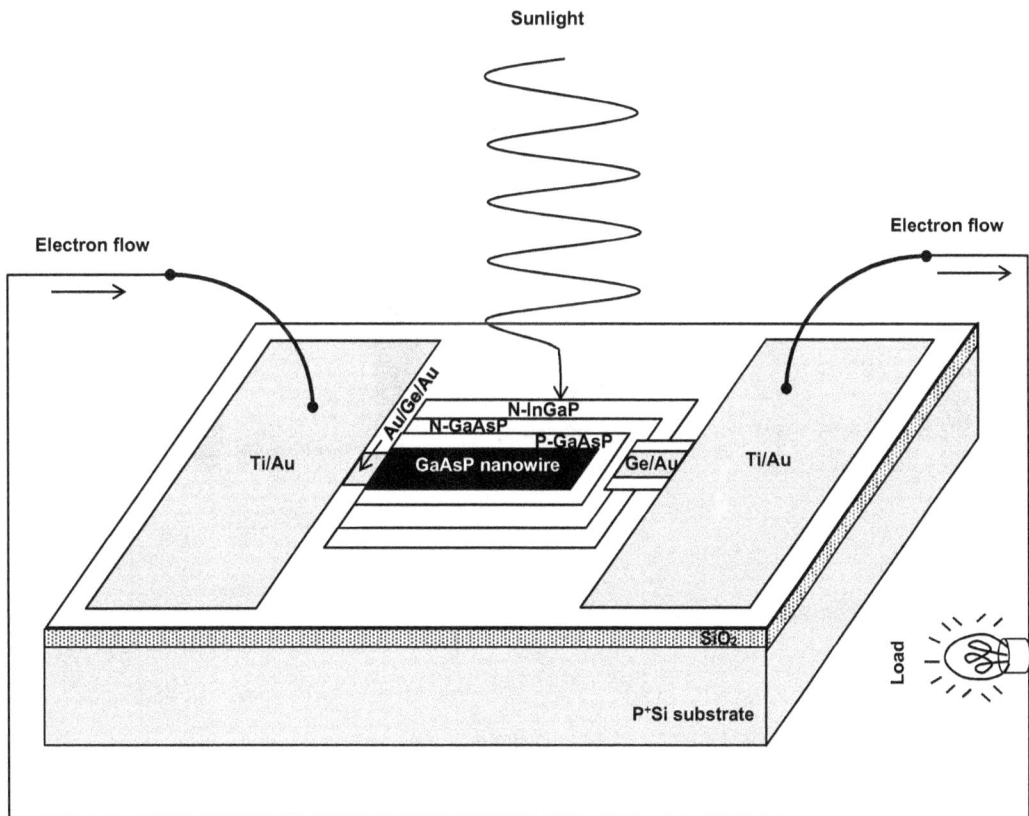

FIGURE 13.4 A single GaAsP nanowire solar cell in horizontal configuration showing the single P-core/N-shell nanowire with N-InGaP passivation layer laid down horizontally on an SiO_2/P^+Si substrate, and the P- and N-contact pads. Sunlight falling on the nanowire produces current flow between the contact pads.

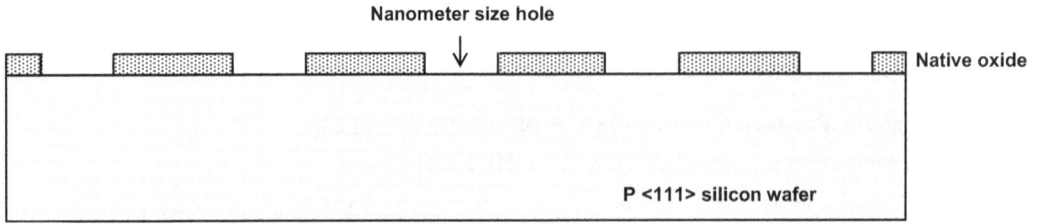

(a)

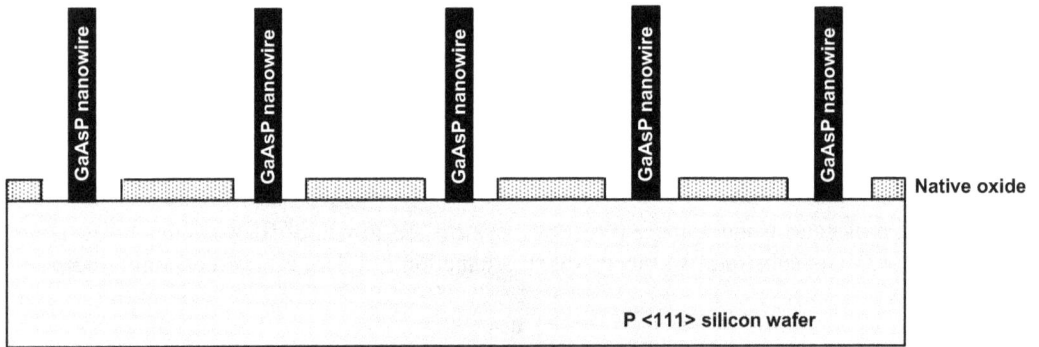

(b)

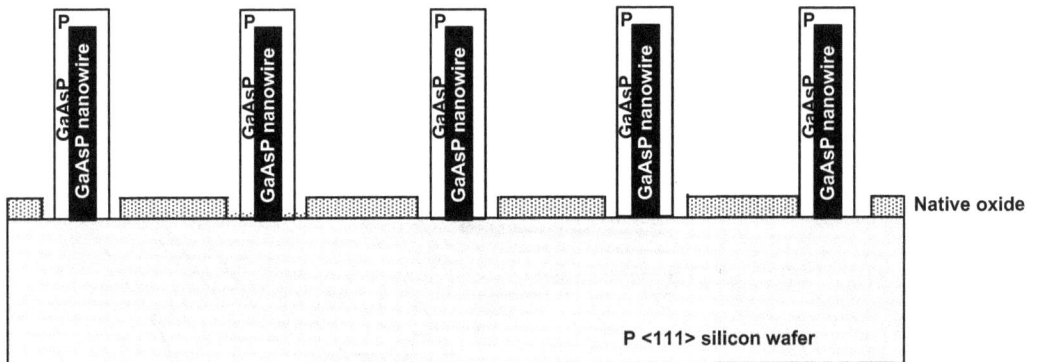

(c)

FIGURE 13.5 Growth, doping, and surface passivation of GaAsP nanowires for single horizontally laid out GaAsP nanowire solar cell: (a) P-type silicon wafer with orientation < 111 > and having native silicon dioxide, (b) growth of GaAsP nanowire, (c) growth of P-type shell, (d) growth of N-type shell, and (e) growth of N-InGaP passivation layer. The GaAsP nanowire with P- and N-type shells will be called GaAsP P-core/N-shell nanowire.

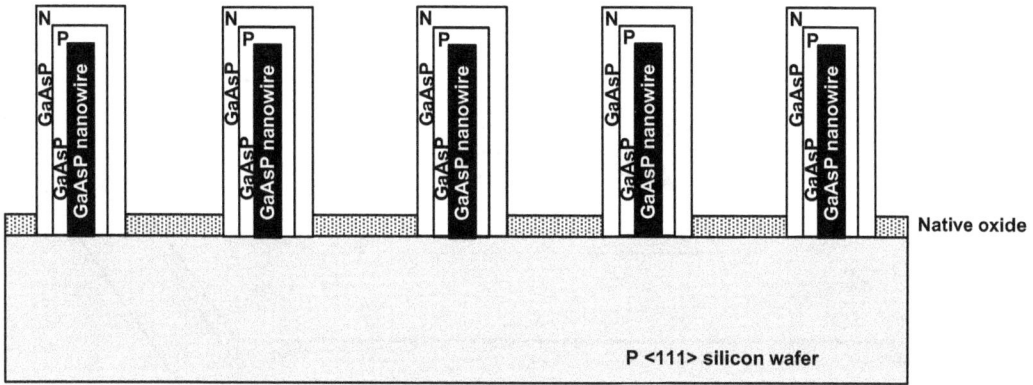

(d)

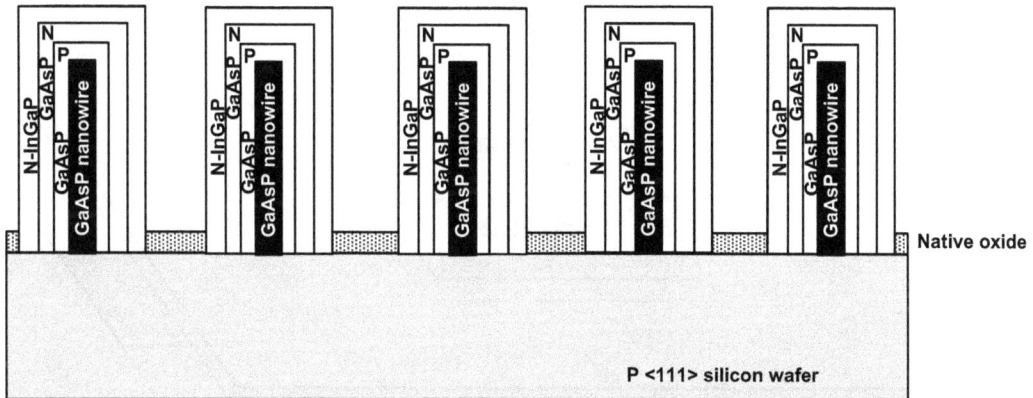

(e)

FIGURE 13.5 (Continued)

dopant, and having native silicon dioxide. Vertically aligned GaAsP nanowires with bandgap 1.7eV and structure close to zinc-blende crystal structure are grown by Ga-assisted vapor-liquid-solid mechanism. Nanowire core growth temperature is 620–650°C, Ga flux 4×10^{-8}–9×10^{-8} torr, group V flux 2.8×10^{-6} to 7×10^{-6} torr, time of growth 1 h. Axial nanowire growth is stopped for 10–20 min. by closing the Ga shutter. Temperature is decreased to 465°C for growing the shell. For the P-type shell, the doping is done with Be keeping doping concentration equivalent to 8×10^{17} cm^{-3} for planar growth. For the N-type shell, the doping is done with silicon keeping doping concentration equivalent to 0.8–1.3×10^{18}cm^{-3} for planar growth. Typical nanowire diameters are 300–400nm and lengths 6–8μm.

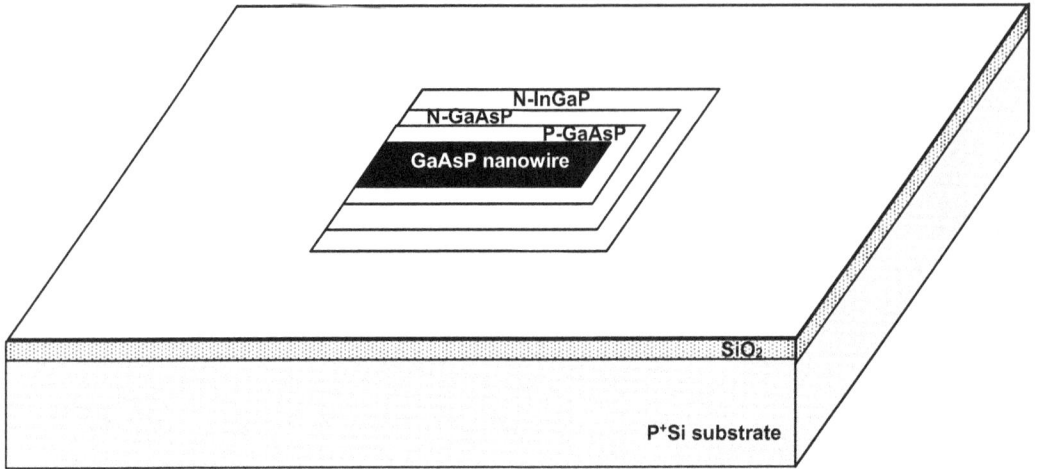

(a)

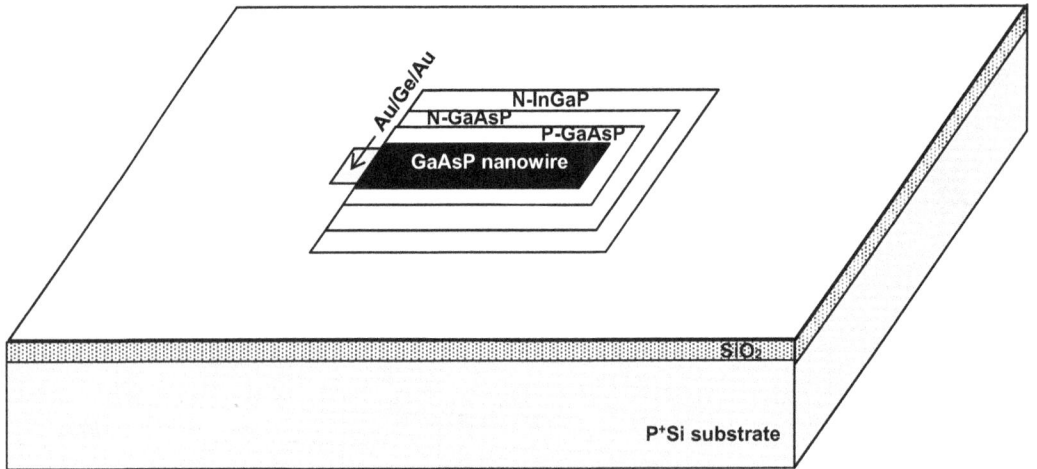

(b)

FIGURE 13.6 Fabrication of single GaAsP nanowire solar cell in horizontal configuration: (a) laying down a single P-core/N-shell nanowire with N-InGaAs passivation layer horizontally on an SiO_2/P^+Si substrate, (b) making P-contact, (c) making N-contact, and (d) defining contact pads.

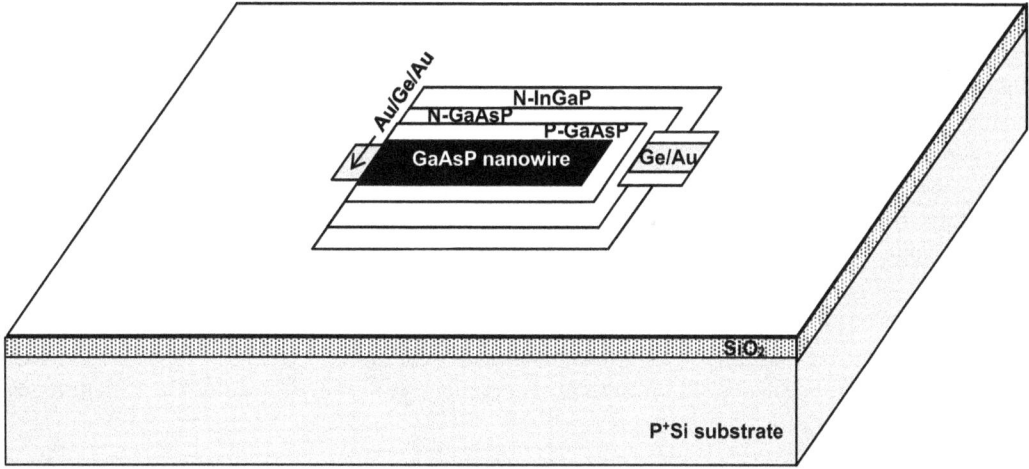

(c)

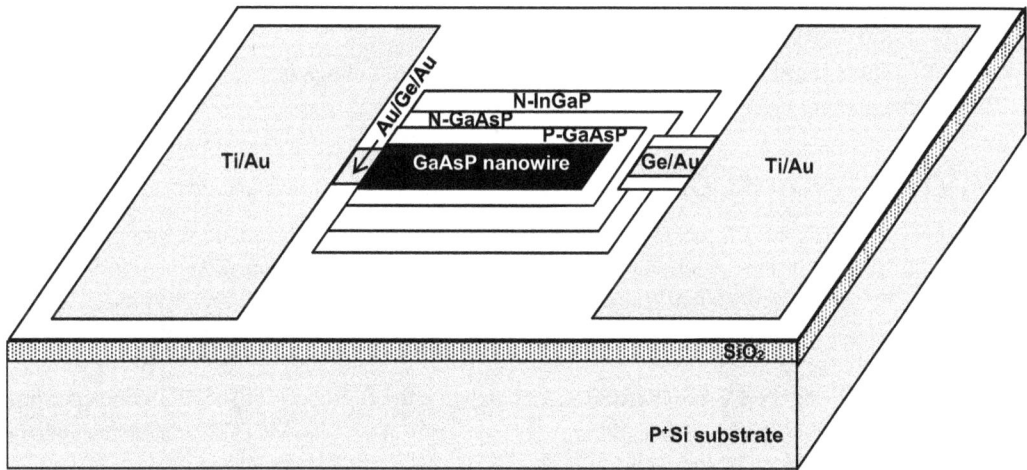

(d)

FIGURE 13.6 (Continued)

13.1.2.3   Surface Passivation

An extra shell of heavily N-doped InGaP is deposited for passivating the surface. Thickness of InGaP film is 10nm.

13.1.2.4   Nanowire Removal from Growth Substrate and Alignment on P⁺ Substrate

The nanowire removal is done by sonication in isopropanol. A P^+ substrate covered with 500nm-thick SiO_2 film is taken. Alignment marks are already defined on this substrate. Drip-drying method is applied to transfer the nanowire to the new substrate.

13.1.2.5   P-Contact to the Nanowire

The nanowire core is exposed by electron-beam lithography, followed by etching in $H_3PO_4:H_2O_2:H_2O$. Then, Au/Ge/Au film is deposited by thermal evaporation. Annealing is done at 420°C in nitrogen ambience for 2 min.

13.1.2.6   N-Contact to the Nanowire

Electron-beam lithography is applied for defining contact to the N-doped shell surface. Etching in buffered HF removes oxide. Thermally evaporated Ge/Au or In/Au-Ge are deposited and annealed at 300°C in nitrogen atmosphere for 2 min.

13.1.2.7   Contact Pads

These are defined by UV lithography, and Ti/Au metallization scheme is implemented.

13.1.2.8   Solar Cell Parameters

The short-circuit current density is 14.7mAcm⁻², open-circuit voltage is 0.9V, and the fill factor is 0.77. The efficiency is 10.2% (Holm et al 2013).

13.2   GₐAₛ NANOWIRE-ON-Sₗ TANDEM SOLAR CELL (η = 11.4%)

Simulations are performed to design the GaAs nanowire solar cell structure. A square array of nanowires (diameter 300nm, pitch 600nm) is taken (Yao et al 2015). A nanowire height of 900nm is optimized because it gives equal short-circuit current densities for the top and bottom cells. The current matching is done to ensure that the bottom cell does not receive insufficient light.

The bottom silicon cell is fabricated on a float zone N-type substrate (Figure 13.7). Boron implantation (dose 30keV, energy 1.5×10^{16} cm⁻²) is done to make the P^+/N emitter junction. Phosphorous implantation (dose 40keV, energy 8×10^{15} cm⁻²) is carried out to form the N/N^+ back surface field junction. Rapid thermal annealing at 1,050°C for 1 min. activates the dopants.

The top GaAs nanowire solar cell is fabricated by forming a silicon nitride layer over the silicon wafer by PECVD. Lithography is done, followed by reactive ion etching to define an array of holes for nanowire growth. Selective area growth (SAG) method is applied for nanowire growth by MOCVD. The nanowires are grown at 700–790°C using trimethyl gallium as gallium precursor and AsH_3 as arsenic precursor. A small segment of GaAs nanowire is N^+ doped to form an N^+ GaAs/P^+Si heterojunction with underlying silicon. This tunnel junction connects the bottom and top cells. Moving upwards on the nanowire, N-type, undoped, and P-type segments are grown to form an N-I-P diode. Transparent benzocyclobutene (BCB) polymer is applied to cover the nanowires. From the tips of the nanowires, the BCB is etched and RF sputtering is done to deposit indium tin oxide for top contact. The back contact is aluminum film deposited by electron-beam evaporation.

The tandem solar cell has an open-circuit voltage of 0.956V, indicating addition of voltages of the two solar cells. The short-circuit current density and fill factor of the cell are 20.64 mAcm⁻² and 0.578, respectively. Its efficiency is 11.4% (Yao et al 2015).

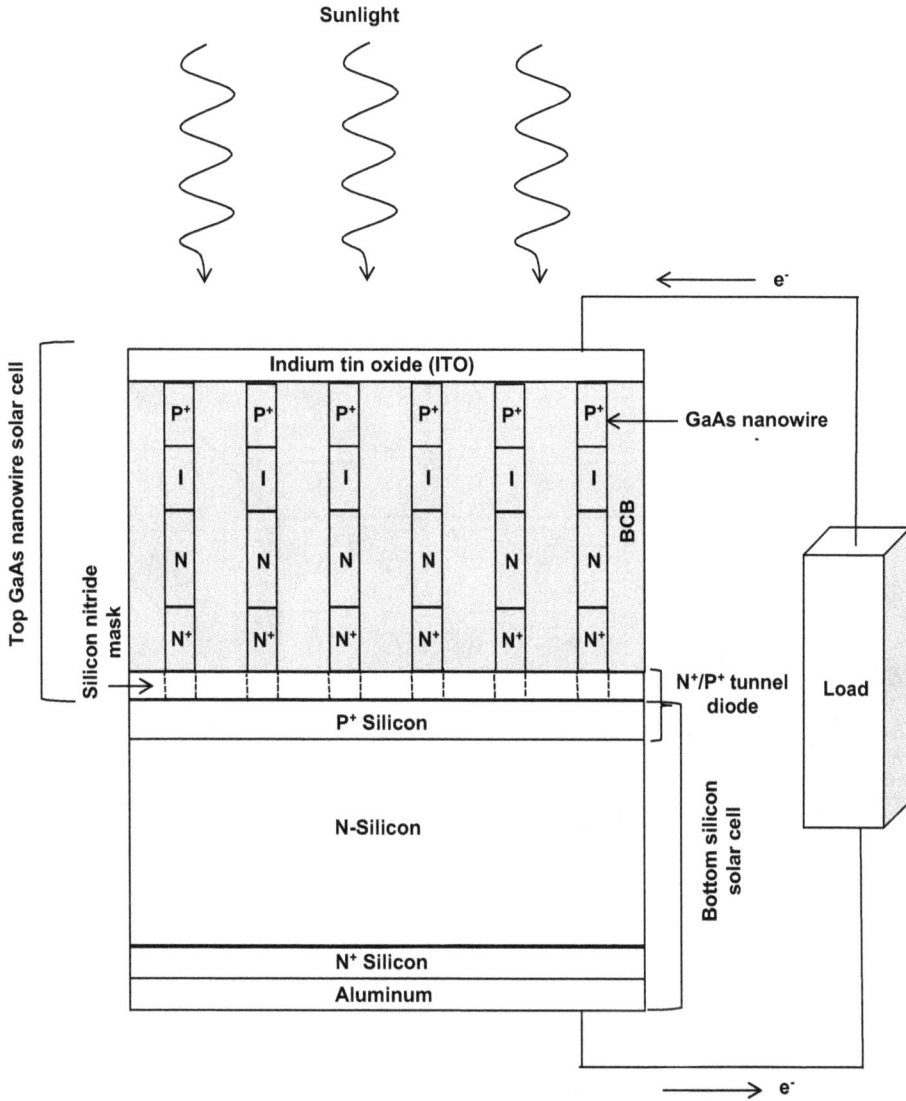

FIGURE 13.7 GaAs nanowire plus silicon tandem solar cell made of, (i) top cell, ITO glass, P-I-N GaAs nanowire diode, N^+ portion of nanowire to form a tunnel diode with underlying P^+ layer of (ii) bottom cell in which the subsequent layers are N-Si, N^+ Si BSF, and Al back contact. Also shown is a silicon nitride layer used as masking layer for nanowire growth and passivation layer for the bottom cell. Sunlight falls on ITO glass, and current flows between ITO and Al electrodes.

13.3 GaAs NANOWIRE ARRAY SOLAR CELL ($\eta = 15.3\%$)

Large-area solar cells are made from uniform arrays of nanowires. Åberg et al (2015, 2016) made an axial P^+-P-N^+ nanowire array solar cell on GaAs substrate of orientation (111)B. Figure 13.8 shows this solar cell, and Figure 13.9 the process for its fabrication.

13.3.1 MAKING Au DISK PATTERN

A pattern of Au disks is made on the substrate at a pitch of 400nm by conformal imprint lithography.

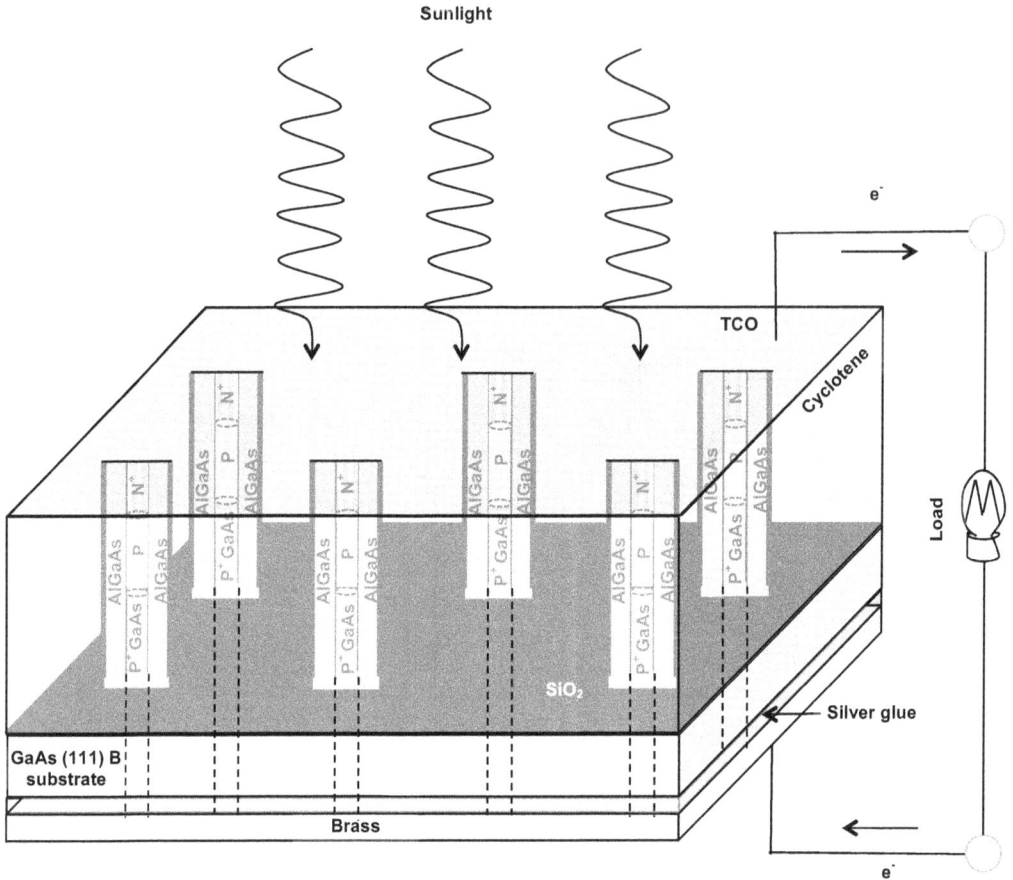

FIGURE 13.8   GaAs nanowire array solar cell showing the VLS-grown, P⁺PN⁺-doped, and AlGaAs passivated GaAs nanowires on GaAs substrate, cyclotene, TCO front contact, and brass back contact attached with silver glue underneath the substrate.

13.3.2   VLS Method of Nanowire Growth

VLS method is applied for nanowire growth using MOCVD. The nanowire has a core made of GaAs, and the diameter of the core is 165nm. The nanowire length is 3μm. The nanowire cores occupy 13% of the surface area. The growth of GaAs nanowire is done along the (111)B direction by the reaction between trimethyl gallium [TMG:Ga(CH₃)₃] and arsine (AsH₃) at 400°C:

$$Ga\left(CH_3\right)_3 + AsH_3 \rightarrow GaAs + 3CH_4 \tag{13.1}$$

13.3.3   P- and N-Type Doping

To make the P⁺-P-N diode, one end of the GaAs nanowire core is doped with P-type dopant diethyl zinc. The opposite end of the nanowire core is doped with N-type dopant tetraethyl tin.

13.3.4   Passivation

AlGaAs passivation layer of thickness 25–40nm is grown at 715°C over the GaAs nanowire core.

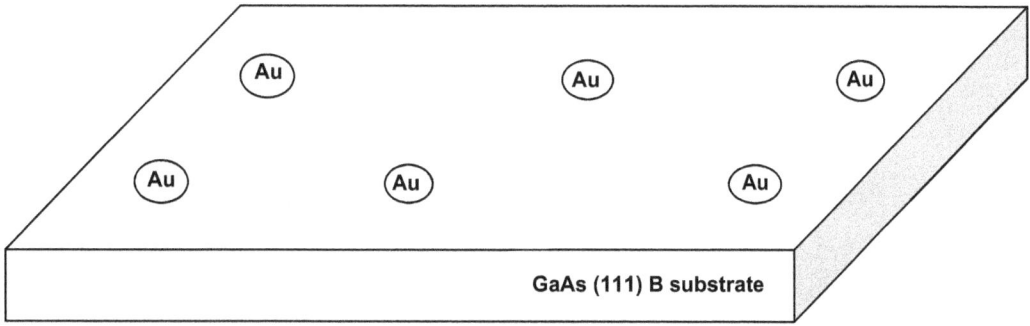

(a)

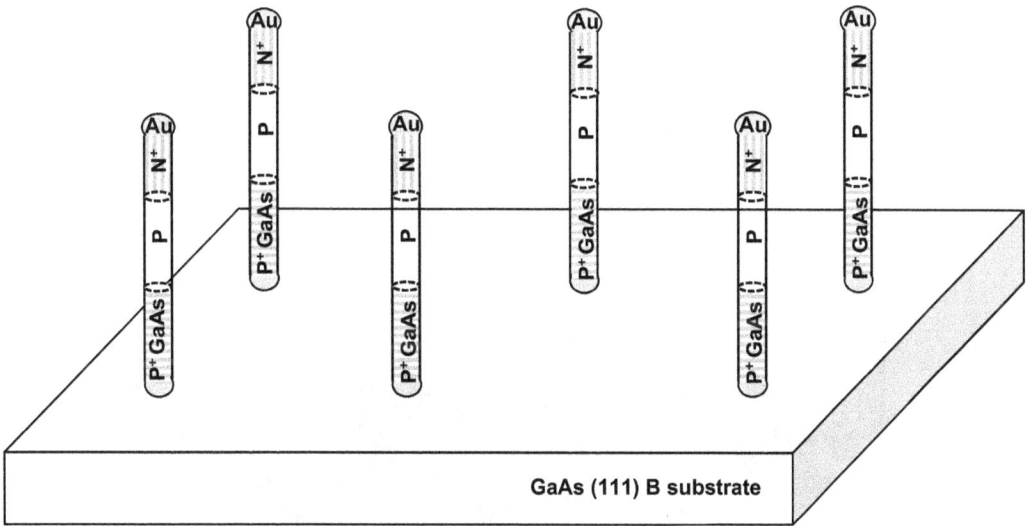

(b)

FIGURE 13.9 Fabrication of GaAs nanowire array solar cell: (a) defining Au disk pattern on the substrate, (b) growth of GaAs nanowires by VLS method and their P-, N-type doping, (c) passivation of nanowire surfaces with AlGaAs, (d) SiO_2 deposition, (e) cyclotene deposition, (f) removing the insulation from nanowire tips, (g) TCO deposition, and (h) attachment of brass base with silver glue underneath the substrate.

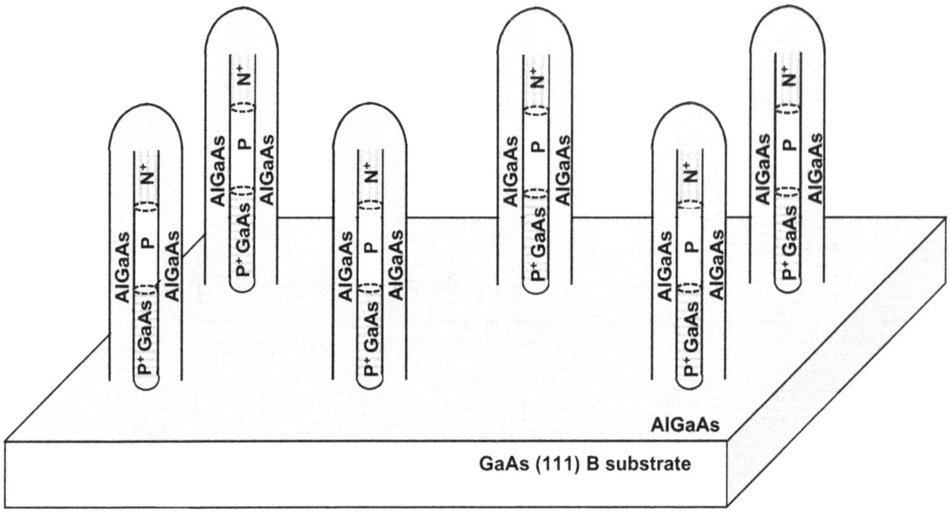

(c)

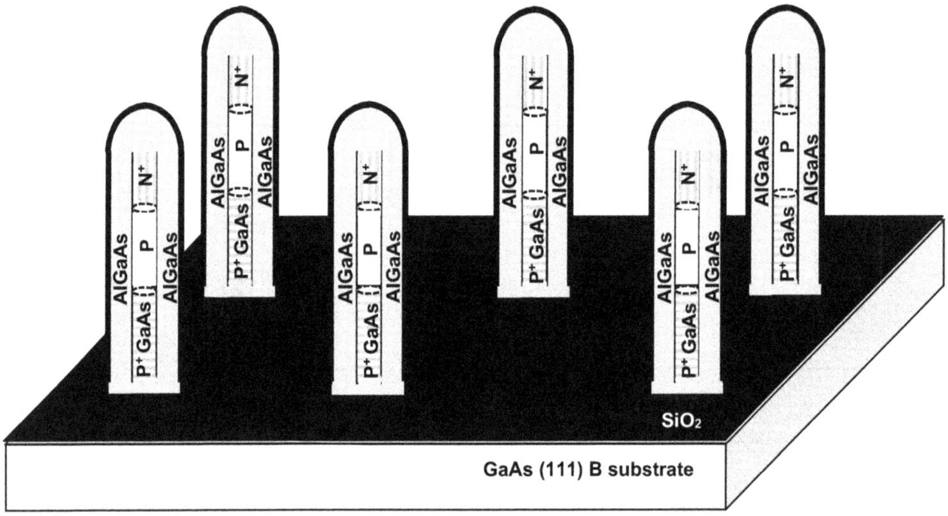

(d)

FIGURE 13.9 (Continued)

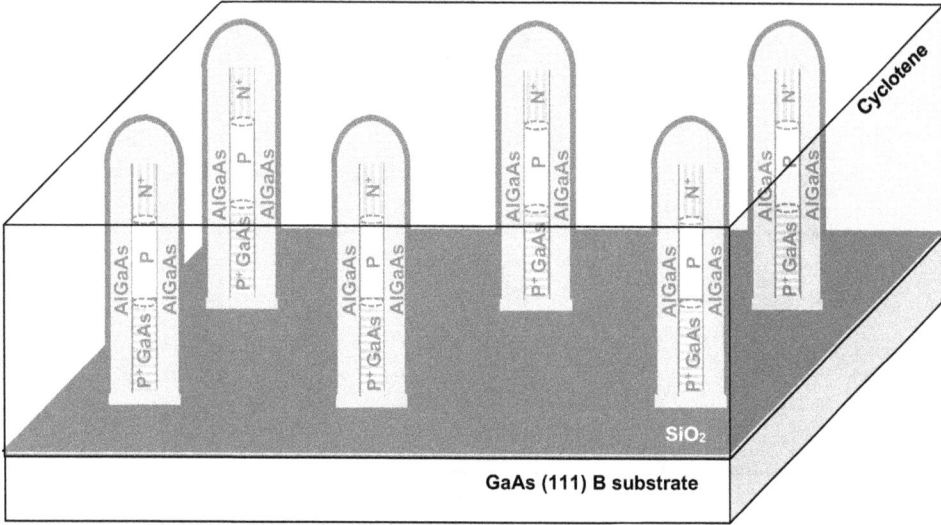

(e)

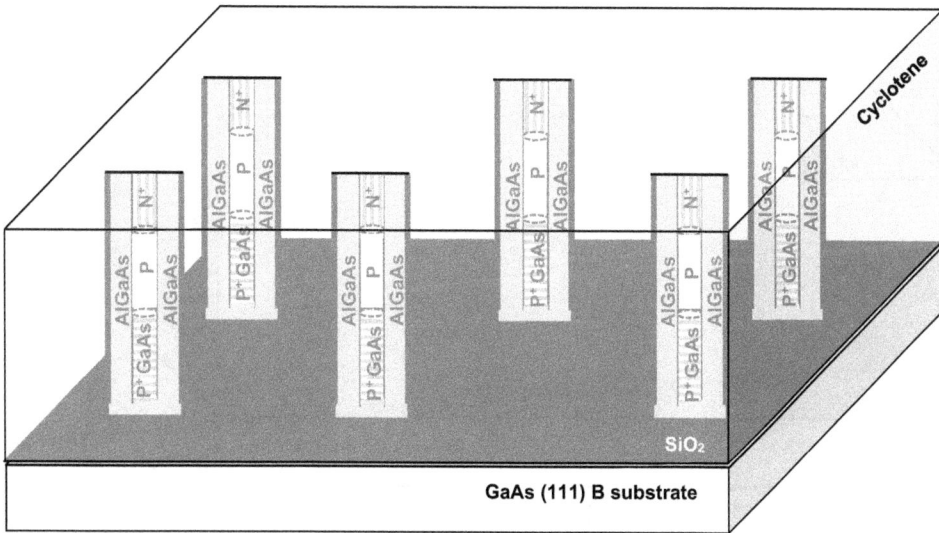

(f)

FIGURE 13.9 (Continued)

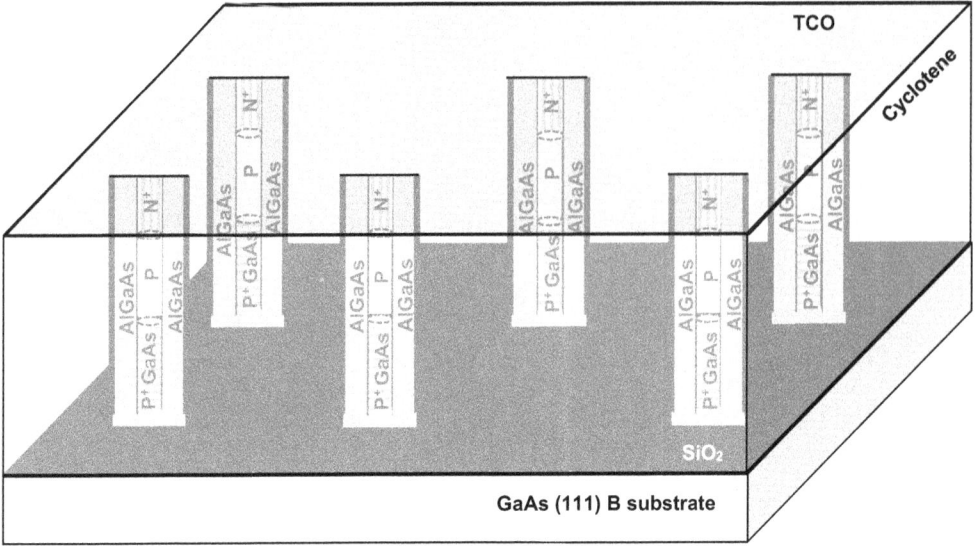

(g)

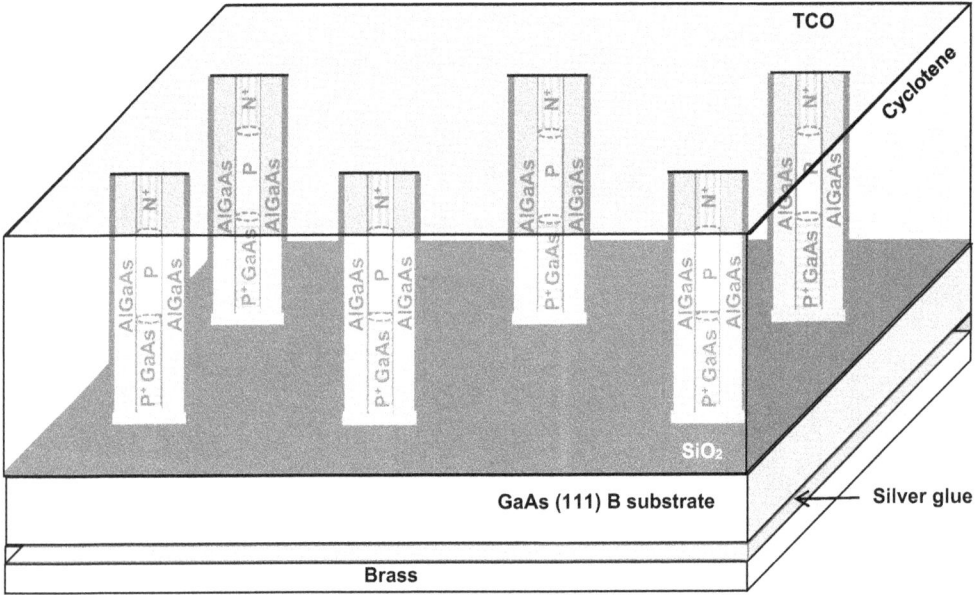

(h)

FIGURE 13.9 (Continued)

13.3.5 Nanowire Diameter, Length, and Segments

The total diameter of the AlGaAs covering + GaAs nanowire core + AlGaAs covering = 25nm + 165nm + 25nm = 215nm. In the 3μm-long nanowire, there are three segments: P$^+$ region (length 1.8μm, doping 10^{18}cm^{-3}), P-region (length 1μm, doping 10^{17}cm^{-3}), and N$^+$ region (length 0.2μm, doping 3×10^{18}cm^{-3}).

13.3.6 SiO$_2$ Deposition and Surface Planarization

SiO$_2$ is deposited over the nanowires. Spin-coating of cyclotene dielectric resin is done on the SiO$_2$-deposited nanowires for surface planarization.

13.3.7 Electrical Contacts

SiO$_2$ and cyclotene layers are removed from the tips of the nanowires by etching. Selective wet-etching of Au catalyst particles is also performed. A transparent conducting oxide (TCO) is deposited over the entire surface. Cells of area 1mm × 1mm are defined over the TCO. For probing, a Ti/Au bus bar is patterned. A silver glue is applied to the backside of the substrate, and the substrate is pasted on a brass coin. This coin serves as the back contact.

13.3.8 GaAs Cell Parameters

The solar cell performance parameters are short-circuit current density = 21.3mAcm^{-2}, open-circuit voltage = 0.906V, fill factor = 79.2%, and power conversion efficiency = 15.3% under AM1.5G illumination (Åberg et at 2015, 2016).

13.4 InP NANOWIRE ARRAY SOLAR CELL FABRICATION BY BOTTOM-UP APPROACHES

13.4.1 Solar Cell (η = 11.1%) with InP Nanowires Grown via Vapor-Liquid-Solid Mechanism and Surface Cleaning

In this process, the backside of a Zn-doped (111) B InP P-type substrate is coated with P$^+$ InGaAs film to make an ohmic contact (Cui et al 2013). Thickness of this film is 200nm, and the concentration of Zn dopant is 1.2×10^{19}cm^{-3}.

13.4.1.1 Nanowire Growth, Doping, and Passivation

Nanoimprint lithography is employed to arrange gold particles on the substrate. Diameter of these Au particles is 136nm. Spacing between any two particles is 513nm. The nanowires are grown in a metal-organic vapor phase epitaxy (MOVPE) system. The reactants are trimethyl indium and phosphine. The nanowire growth temperature is 450°C, and the growth mechanism is vapor-liquid-solid (VLS). Radial growth is prevented by *in situ* etching with HCl introduced at a molar fraction of 2.83×10^{-5}. The nanowires are 2.3μm long. Growth time of nanowires is 19 min. Using dimethyl zinc as P-type impurity and H$_2$S as N-type impurity, an axial P-N structure is formed. Etching in piranha solution cleans the sidewalls of the grown nanowires. The cleaned InP nanowires are covered with 40nm-thick SiO$_2$ film. SiO$_2$ covering helps to improve adhesion of nanowires with the benzocyclobutene (BCB) layer to be filled in the subsequent step. BCB is used for passivating the nanowires, isolating them and making the surface of nanowire-grown substrate planar. Extra BCB is removed by reactive ion etching. O_2/CHF_3 plasma is used for etching. Buffered HF is used to remove the SiO_2 from the tips of the nanowires.

13.4.1.2 Top and Bottom Contacts

Indium tin oxide film is deposited by sputtering over the nanowire tips to make the top contact of the solar cell. ITO film thickness is 300nm. Ti/Pt/Au film is deposited on the backside of the substrate for bottom contact of the cell. Patterning is done to define the solar cell size as 0.5mm × 0.5mm.

13.4.1.3 Solar Cell Parameters

The solar cell shows a short-circuit current density of $21mAcm^{-2}$, open-circuit voltage of 0.73V, a fill factor of 0.73, and efficiency of 11.1% at 1 sun. The dark current is 1pA/nanowire.

13.4.1.4 Role of Nanowire Surface Cleaning

Cleaning of the nanowires is crucial for fabricating high-efficiency solar cell. Piranha cleaning of grown InP nanowires provides a high rectification ratio of 10^7 at ± 1V. Etching of the sidewalls of the nanowires in piranha solution removes the short-circuiting paths for current flow and the laterally grown parts containing stacking faults (Cui et al 2013).

13.4.2 SOLAR CELL (η = 13.8%) WITH EPITAXIALLY GROWN InP NANOWIRES

Figure 13.10 shows the solar cell. Its fabrication is portrayed in Figure 13.11.

13.4.2.1 InP Nanowire Growth and Covering Its Sidewalls with SiO_2

Gold seed particles of various sizes are arranged in arrays of different pitches on the InP substrate by nanoimprinting (Wallentin et al 2013). InP nanowires having the P-I-N structure are grown by epitaxy. The nanowires are 1,500nm long. Their diameters lie in the range 130–190nm. The pitch of the nanowires in the array varies from 470 to 500nm. After the nanowires have been grown, the gold seed particles are removed by wet-etching. Their removal is essential to avoid reflection losses. An SiO_2 film formed by atomic layer deposition (ALD) covers the sidewall of each nanowire. Its thickness is 50.5nm. The top of the InP substrate, too, is covered with the SiO_2 film. Thickness of this film is 50nm.

13.4.2.2 Making Contacts

The silicon dioxide insulating film is removed from nanowire tip to define the top contact. A transparent conducting oxide (TCO) film is used to make the top contact. The oxide used is indium tin oxide. The TCO dome at the top of the nanowire is 30nm thick. The thicknesses of the TCO films over the SiO_2 film at the bottom and on the sidewalls of the nanowires are 50nm and 38nm, respectively. Contact pads are made with Ti/Au metallization.

13.4.2.3 Solar Cell Parameters

The size of the axial P-I-N solar cell is 1mm × 1mm. The nanowires cover 12% of the active area. The short-circuit current density is $24.6mAcm^{-2}$, open-circuit voltage is 0.779V, fill factor is 72.4%, and efficiency is 13.8% (Wallentin et al 2013).

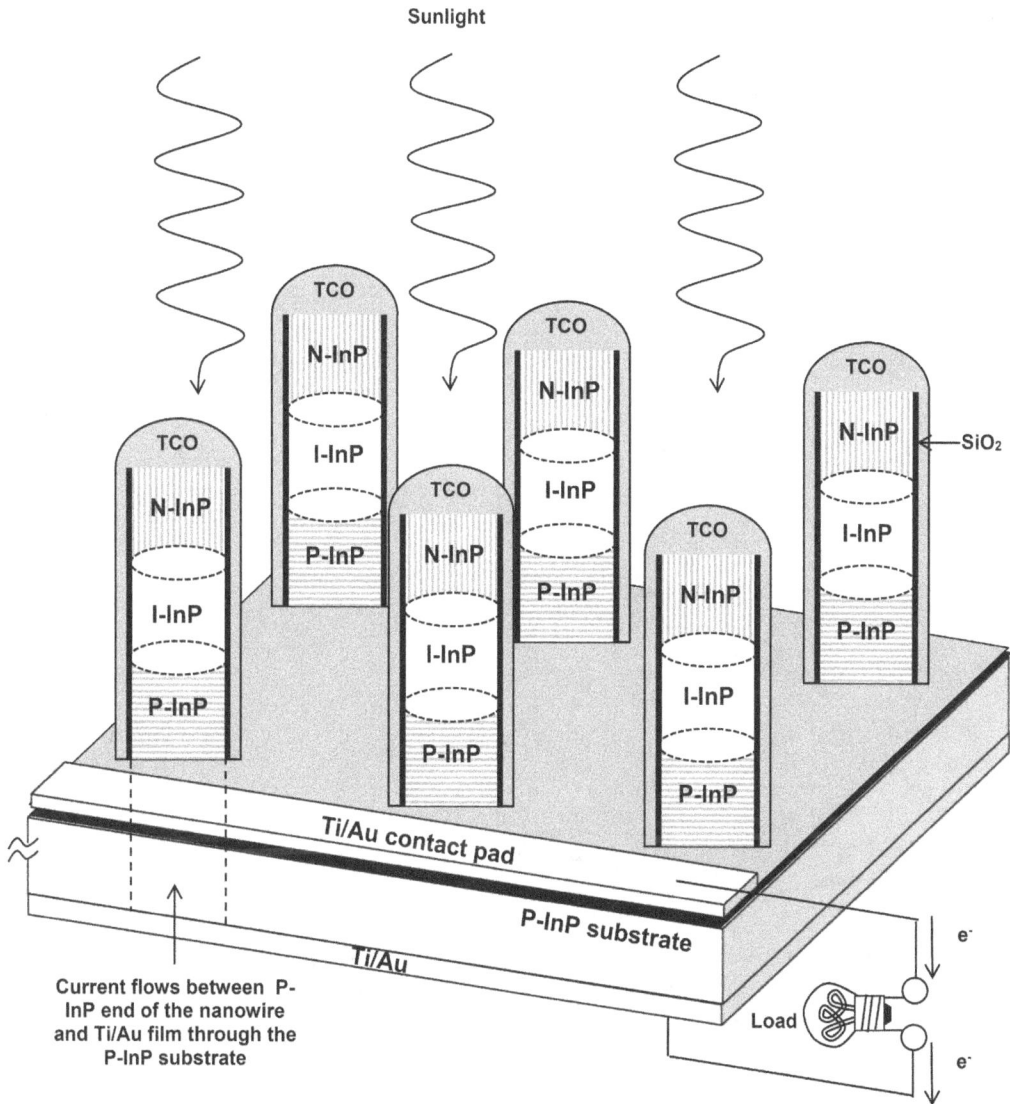

FIGURE 13.10 InP nanowire array solar cell showing the P-type InP substrate, VLS-grown InP nanowires doped to form axial P-I-N diodes, transparent conducting oxide film over the nanowires and the front surface of the InP substrate, Ti/Au contact pad on front side, and Ti/Au film on backside of the InP substrate. For performing measurements under sunlight, the axial InP nanowire diodes are illuminated with sunlight, and electric current flowing between Ti/Au contact pad and Ti/Au film is recorded.

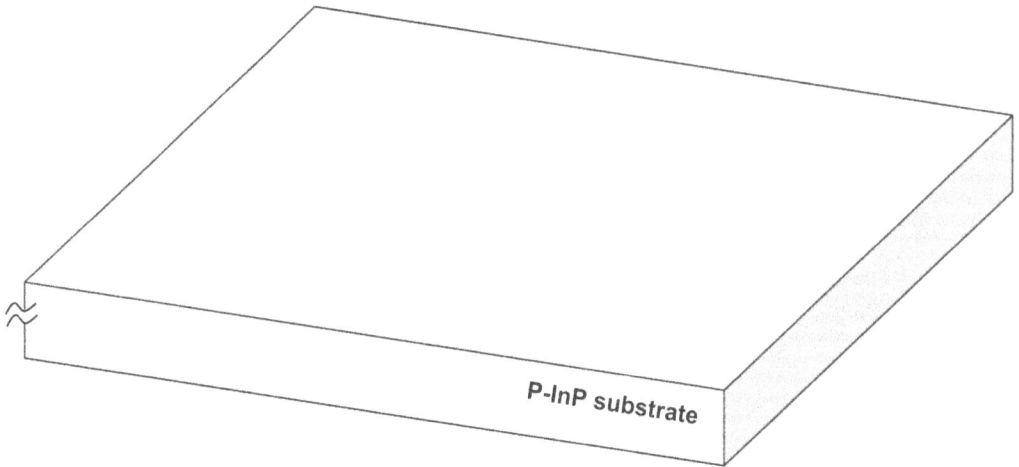

(a)

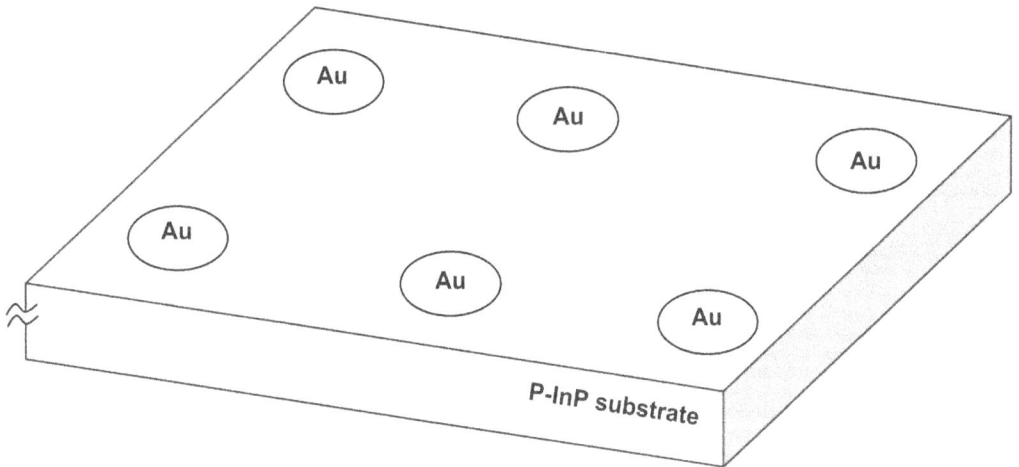

(b)

FIGURE 13.11 Moving stepwise to fabricate InP nanowire array solar cell: (a) taking P-type InP substrate, (b) arranging gold seed particles on the substrate, (c) growing InP nanowires by VLS method and doping them to form an axial P-I-N diode, (d) removing gold seed particles, (e) depositing SiO_2 film to cover the InP nanowires and the front surface of the InP substrate, (f) removing SiO_2 from the tips of the InP nanowires, (g) depositing transparent conducting oxide over the nanowires and the front surface of the InP substrate for electrical contact with the nanowires, and (h) metallizing the front and backsides of the InP substrate with Ti/Au, and making contact pad.

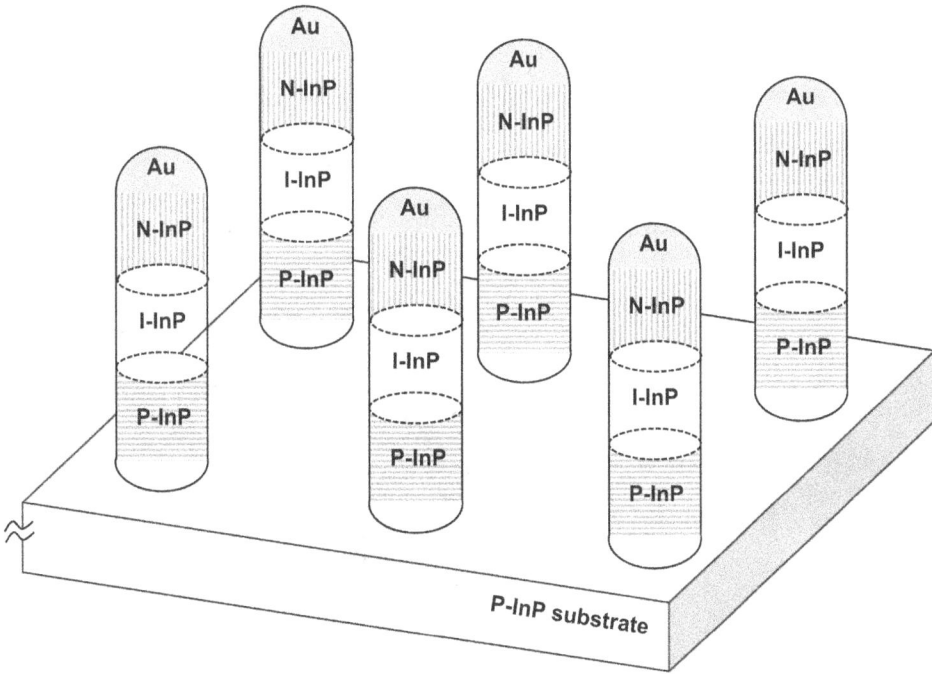

(c)

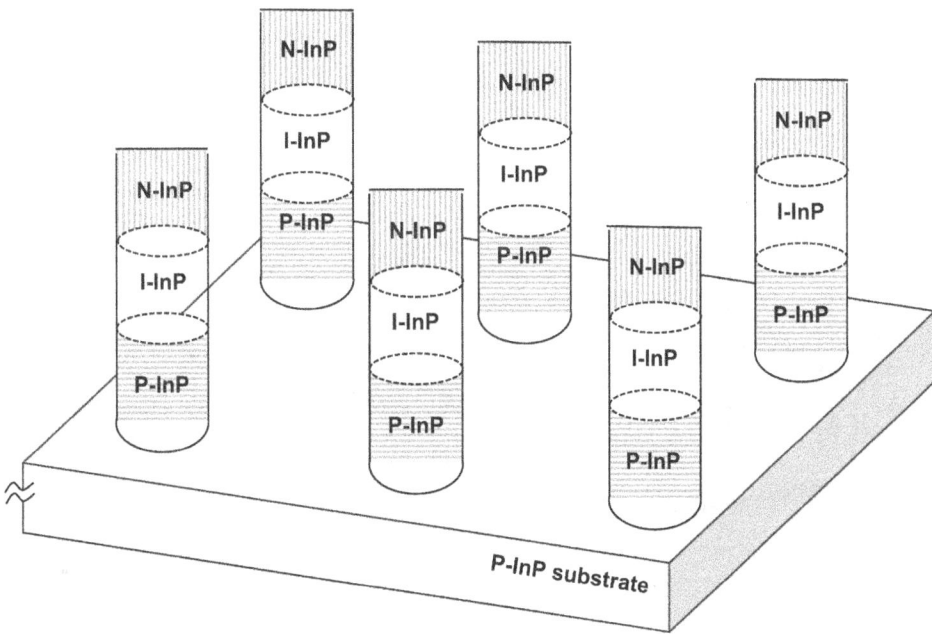

(d)

FIGURE 13.11 (Continued)

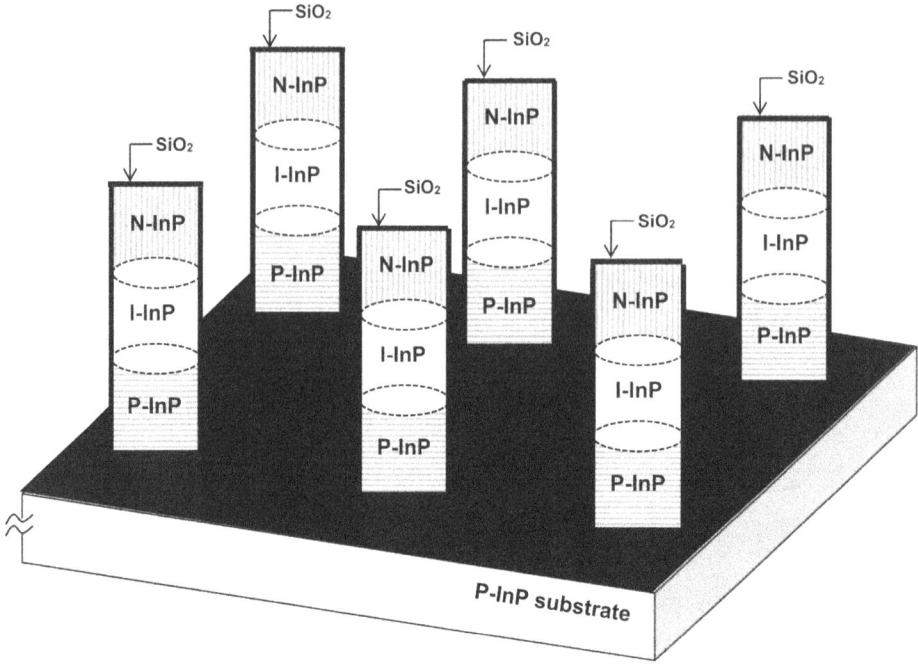

(e)

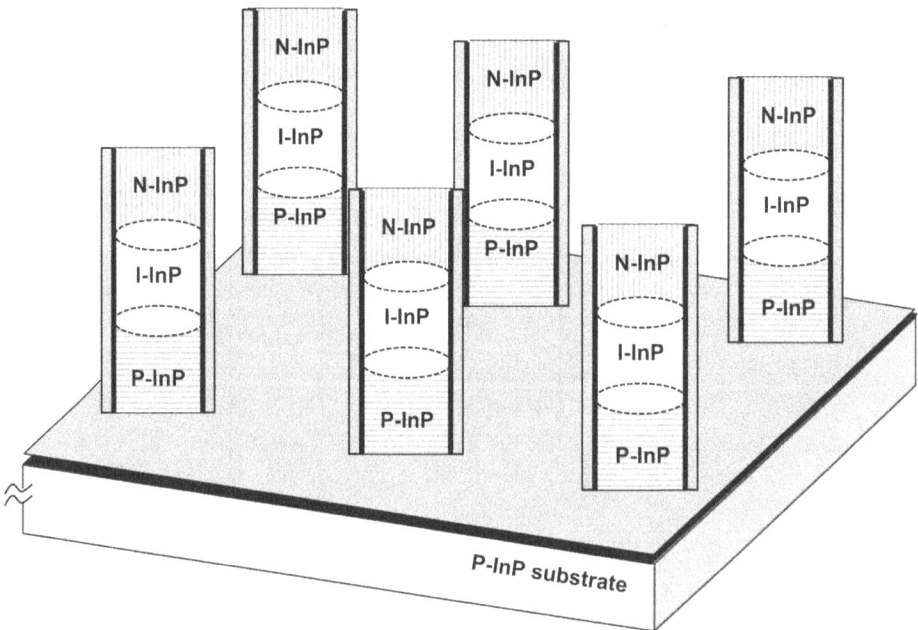

(f)

FIGURE 13.11 (Continued)

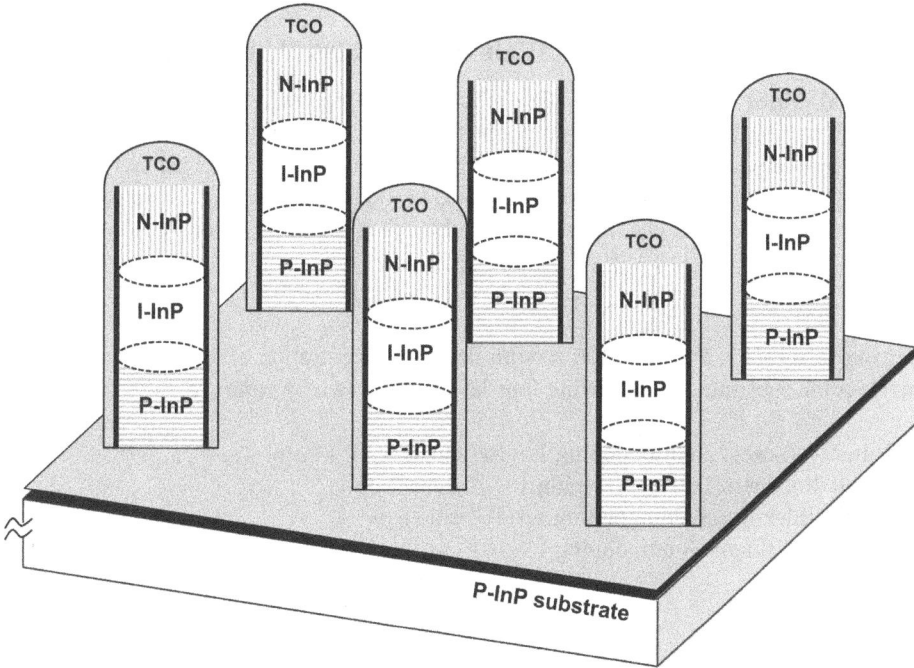

(g)

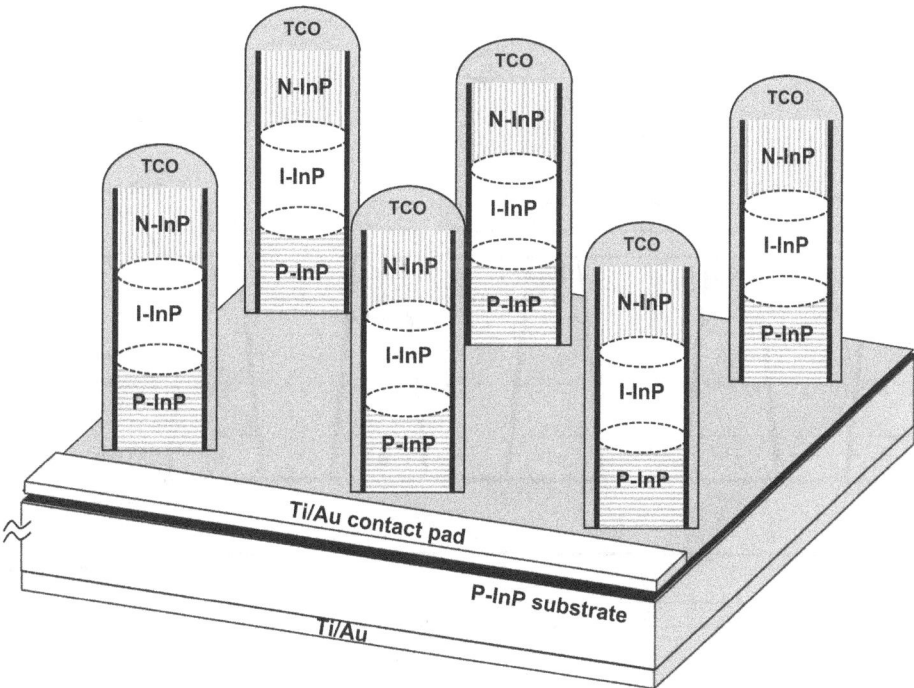

(h)

FIGURE 13.11 (Continued)

13.5   InP NANOWIRE ARRAY SOLAR CELL (η = 17.8%) FABRICATION BY TOP-DOWN APPROACH: DRY-ETCHING FROM EPITAXIALLY GROWN STACK

van Dam et al (2016) demonstrated a high-efficiency InP nanowire solar cell (Figure 13.12) in which Mie scattering by nanostructured contact layer considerably increases the absorption of sunlight. As explained in Section 5.3.2, Mie scattering is the elastic scattering caused by particles of size $\geq$ the wavelength of incident light as opposed to Rayleigh scattering due to particles of sizes smaller than the wavelength. The fabrication of solar cell is shown in Figure 13.13.

13.5.1   EPITAXY

The starting substrate is P$^+$-InP < 100 >, with doping concentration of 1.2×10^{18} cm^{-3}, thickness 300µm. Over the substrate, the following four layers are grown by epitaxy:

   (i) P-InP (thickness 200nm, doping 1×10^{18} cm^{-3})
   (ii) P-InP (thickness 1,600nm, doping 1×10^{17} cm^{-3})
   (iii) N-InP (thickness 170nm, doping 1×10^{18} cm^{-3})
   (iv) N$^+$-InP (thickness 30nm, doping 1×10^{19} cm^{-3})

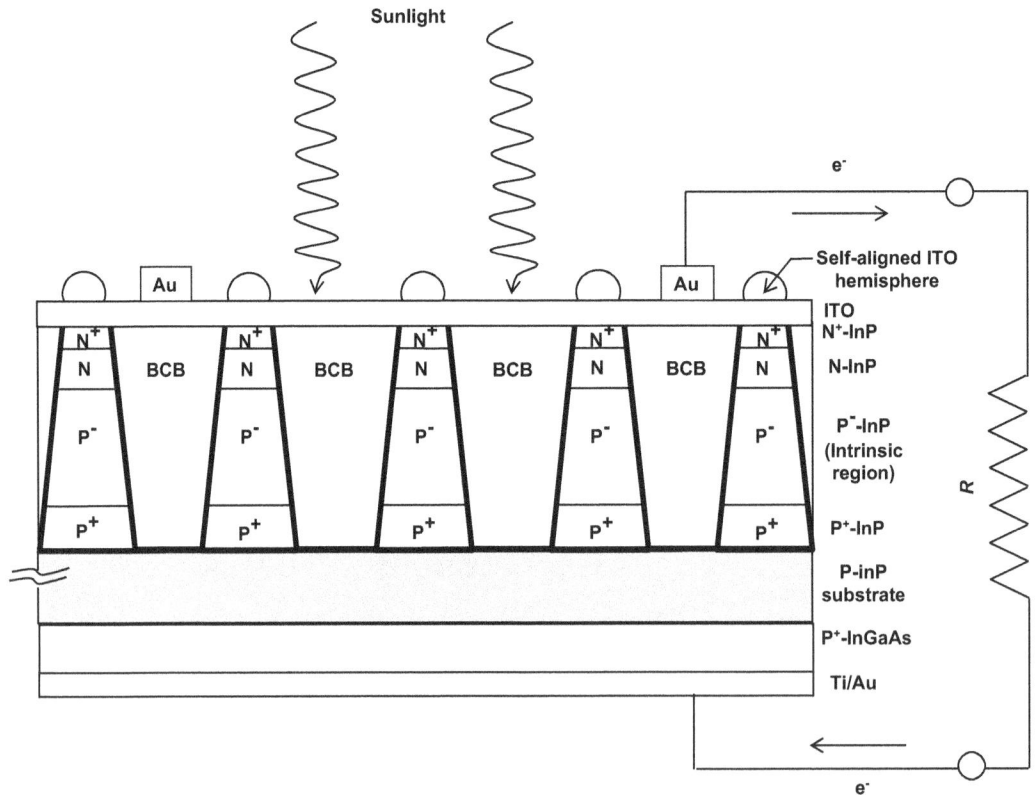

FIGURE 13.12   InP nanowire array solar cell fabricated by dry-etching. Diagram shows the P$^+$-InP substrate, epitaxially deposited P-InP, P-InP, N-InP, and N$^+$-InP layers above the substrate and P$^+$-InGaAs layer below the substrate, BCB filled for planarization, ITO arranged into self-aligned hemispheres, Ti/Au film deposited for backside contact, and gold film border acting as a bus bar. Sunlight strikes the solar cell on ITO surface, and current is drawn between the ITO and Ti/Au electrodes.

(a)

(b)

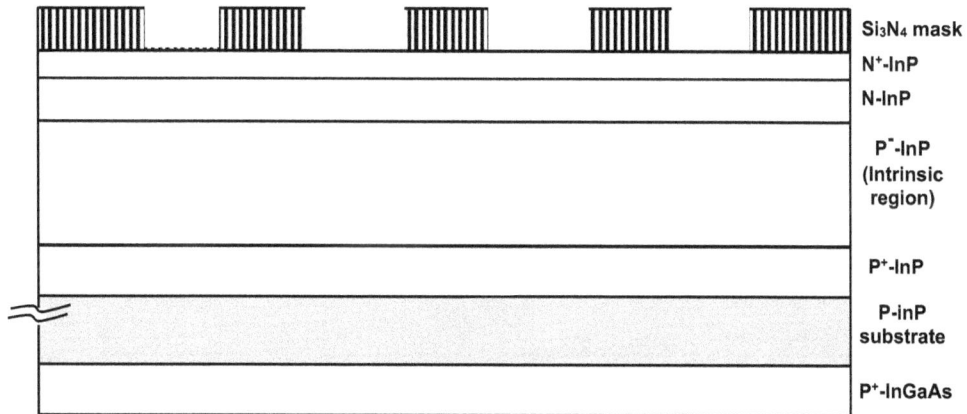

(c)

FIGURE 13.13 Fabrication of InP nanowire array solar cell by dry-etching: (a) the P⁺-InP substrate; (b) successive deposition of four layers epitaxially, P-InP, P-InP, N-InP, and N⁺-InP above the substrate and one layer (N⁺-InGaAs) below the substrate; (c) creating silicon nitride hard mask; (d) etching in ICP-RIE system to form nanowires; (e) Si_3N_4 mask removal; (f) coating the surfaces of nanowires with silicon dioxide; (g) filling the voids with BCB to planarize the surface; (h) removing SiO_2 from the tops of the nanowires for contacting; (i) ITO deposition and its arrangement into self-aligned hemispheres; (j) Ti/Au film deposition for backside contact; and (k) patterning the structure into squares of area 0.5 mm × 0.5 mm and making gold film border to act as a bus bar.

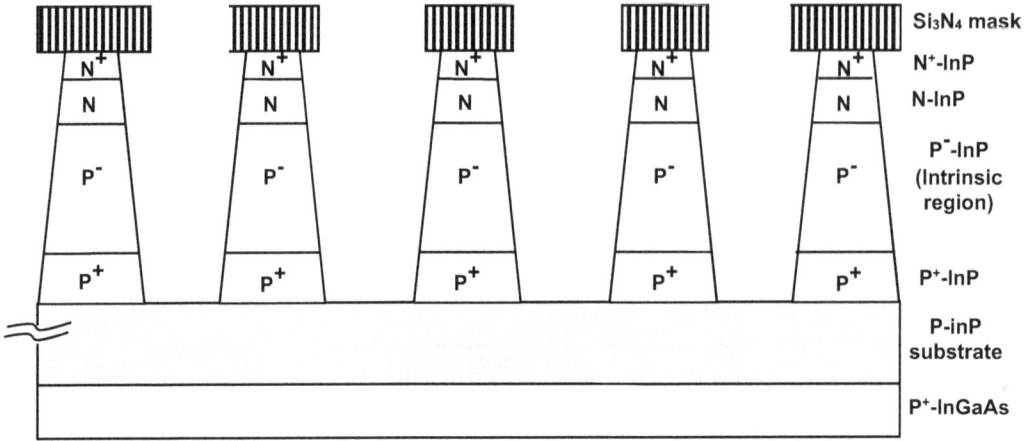

(d)

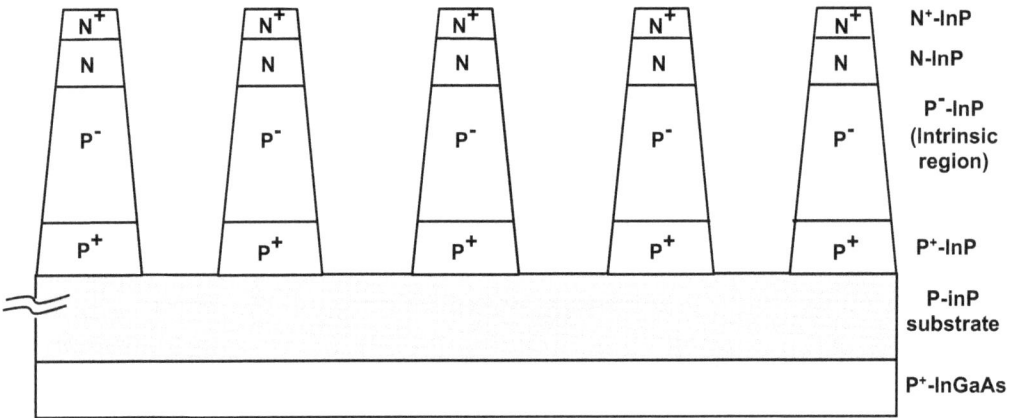

(e)

FIGURE 13.13 (Continued)

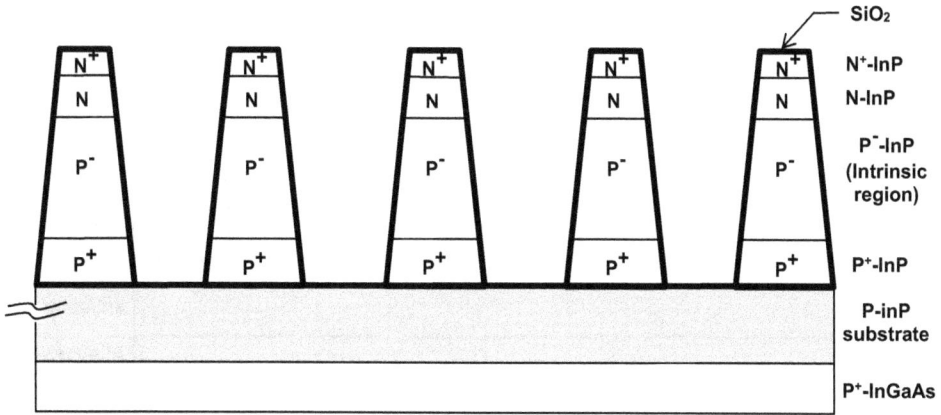

(f)

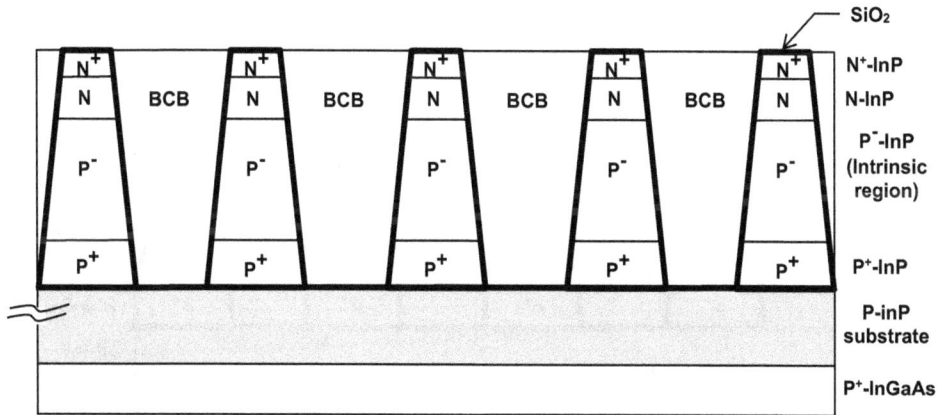

(g)

FIGURE 13.13 (Continued)

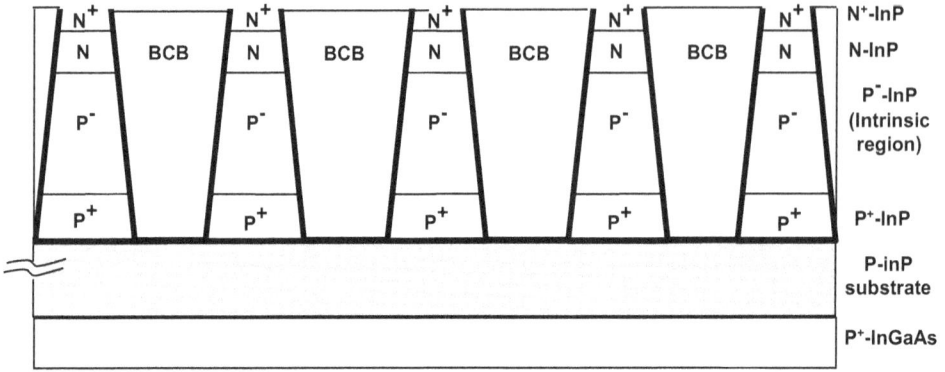

(h)

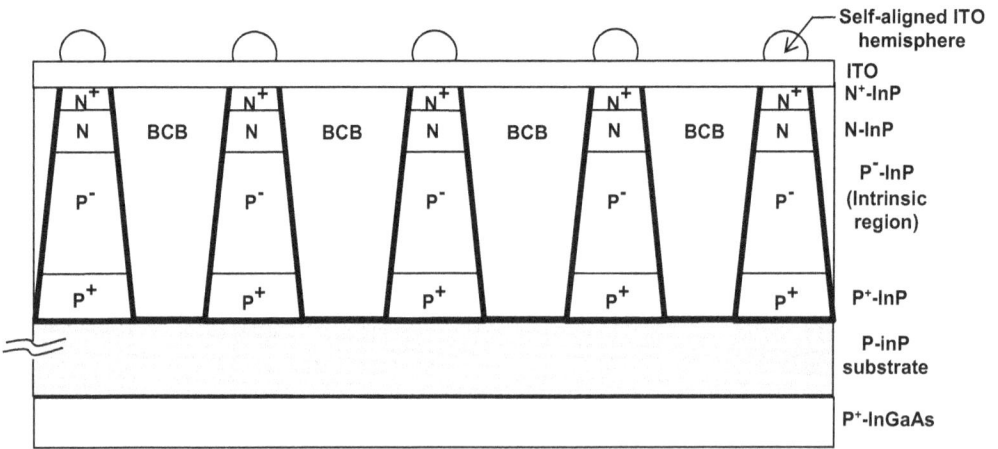

(i)

FIGURE 13.13   (Continued)

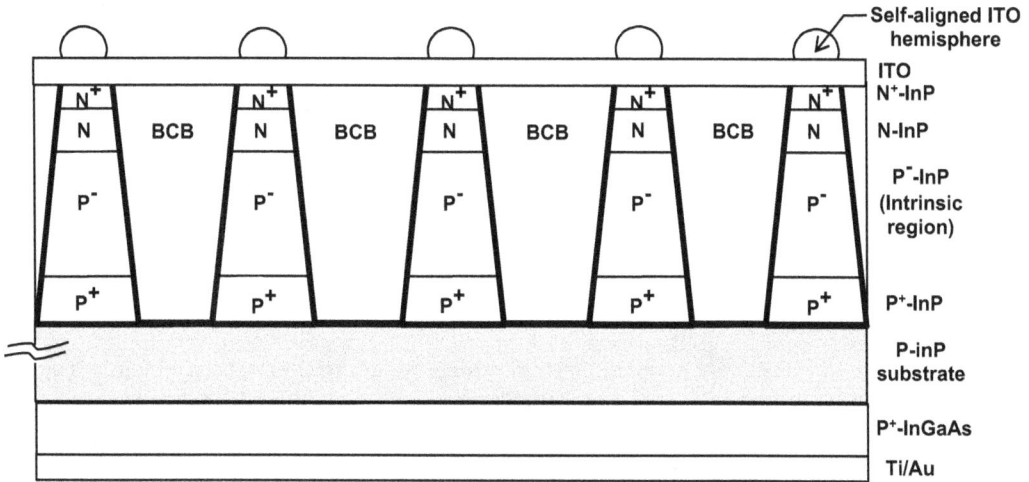

(j)

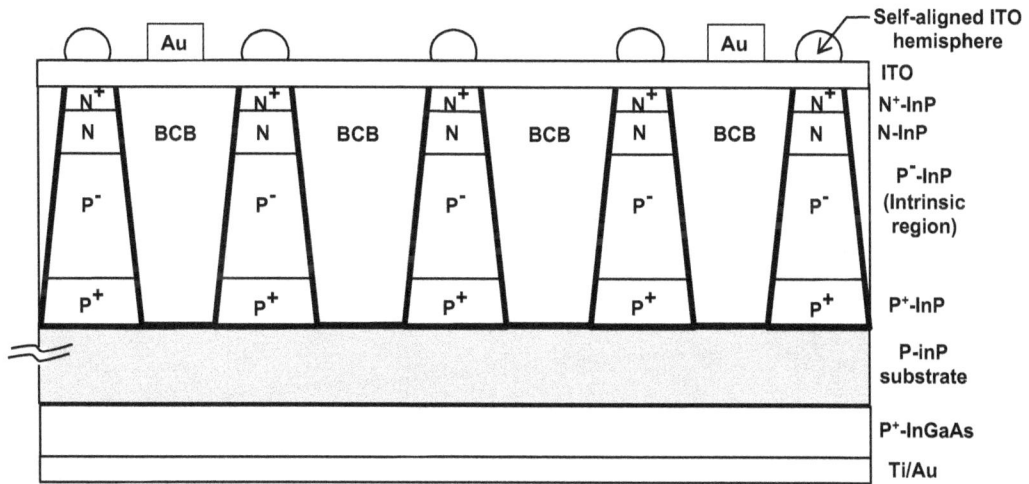

(k)

FIGURE 13.13 (Continued)

As a part of the bottom electrode, a $P^{|}$ InGaAs film (thickness 200nm, doping $1.2 \times 10^{19} cm^{-3}$) is deposited underneath the substrate. The solar cell stack consists of four epitaxial layers above the substrate and one below it.

13.5.2   LITHOGRAPHY

Nanoimprint lithography is used to make an etching mask from a silicon nitride layer. In this lithography, patterning is done by mechanical deformation of imprint resist.

13.5.3   DRY-ETCHING

InP is etched in an inductively coupled plasma etching system to make the nanowires. Digital etching of the sidewalls is carried out by oxidation of the surfaces. In digital etching, the surface is alternately bombarded with the etchant and an energetic beam causing chemical sputtering of the material. After digital etching, the oxide is removed chemically.

13.5.4   NANOWIRE DIMENSIONS

The nanowires of height 1,600nm have the appearance of pyramidal pillars, broad at the bottom ends with diameter of 350nm, and narrow at the top ends with diameter of 150nm.

13.5.5   SiO$_2$ DEPOSITION AND BCB FILLING

The nanowires are coated with a silicon dioxide layer. This coating provides better adhesion with any surrounding material. For planarization of the surface, the intervening empty spaces between the nanowires are filled with benzocyclobutene (BCB). The BCB is a transparent polymeric dielectric.

13.5.6   TOP ELECTRODE

The tips of the nanowires are exposed, and a transparent conducting film of indium tin oxide (ITO) is deposited as the top electrode.

13.5.7   ITO SPREADING AND REARRANGEMENT BY SELF-ALIGNMENT OVER THE InP AND BCB

The ITO arranges itself over the InP and BCB surface in the form of self-aligned nanoparticles. Let us examine its spreading closely. ITO is more adherent to InP than BCB. This causes preferential diffusion of ITO towards InP and away from BCB. As a result, hemispherical ITO nanoparticles are formed on InP nanowire tips. Thus, the top surface of the cell is covered with a nanostructured ITO layer.

13.5.8   ROLE OF NANOSTRUCTURED ITO

The ITO layer comprising self-aligned ITO hemispheres plays a special role. The hemispheres act as Mie scatterers. Broadband scattering of sunlight by ITO nanoparticles spread over the top contact layer increases the coupling of light with the absorber layer of the solar cell. Absorption is increased for the entire range of wavelengths and angles of incidence that are of interest for energy conversion. Such increase of absorption in all directions can be referred to as an omnidirectional enhancement. It is highly beneficial for cell performance.

13.5.9 Bottom Electrode

To make the bottom electrode, a Ti/Au layer is deposited on the underside of the substrate over the N$^+$-InGaAs film. The solar cells are defined as squares of area 0.5mm × 0.5mm.

13.5.10 Gold Border Film

A gold film of width 0.2mm makes a border surrounding the central 0.3mm × 0.3mm portion of the cell exposed to sunlight.

13.5.11 InP Cell Parameters

The InP nanowire solar cell gives a short-circuit current density of 29.3 mAcm^{-2} and efficiency of 17.8% under an illumination of 1 sun based on the exposed area of the solar cell receiving light; the short-circuit current density is 10.1 mAcm^{-2}, open-circuit voltage is 0.765V, and fill factor is 0.794 for the total cell area (van Dam et al 2016).

13.6 WET-ETCHING PROCESSES OF SILICON NANOWIRE ARRAY SOLAR CELL FABRICATION

13.6.1 Radial Junction Solar Cell (η = 13.7%) Fabrication on P-type Wafers with Si NWs Made by Wet-Etching

13.6.1.1 Fabrication Plan Outline

Silicon nanowires are formed on a P-type substrate by metal-assisted chemical etching in AgNO$_3$/HF/water solution (Kumar et al 2011). Regions where bulk silicon is to be retained are carefully protected from the etchant by an etch-resistant coating. Silver dendrites and residues are removed. N-type shell layer is formed by phosphorous diffusion using spin-on dopant. The front contact metal pattern is laid out by alignment on the protected bulk silicon areas. The backside contact is made with the wafer. Figure 13.14 shows the solar cell. Its fabrication process is given in Figure 13.15.

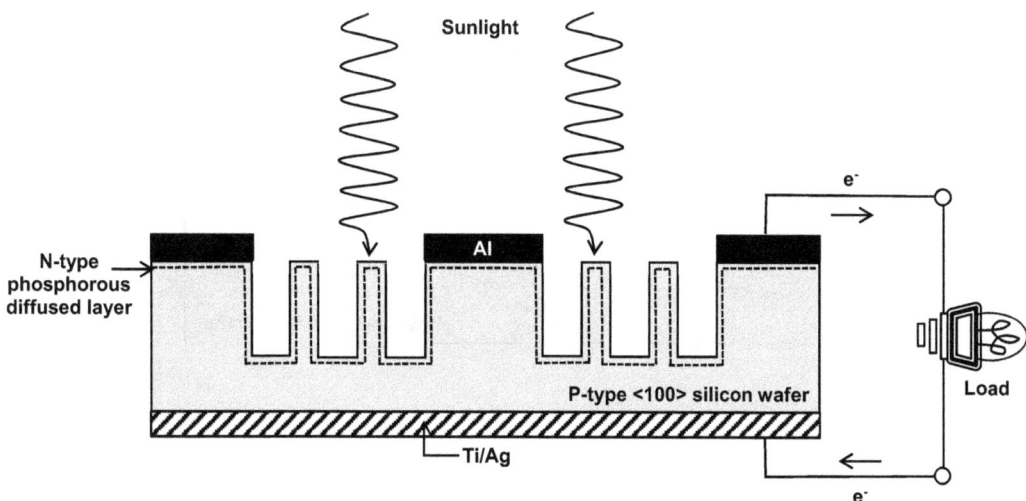

FIGURE 13.14 Silicon nanowire solar cell made by wet-etching. The layers are, from bottom upwards: back metallization, the P-type silicon wafer, phosphorous diffused layer (dotted line), and front metal grid. Sunlight falls on the front metal grid. The electric current is supplied through the metal electrodes.

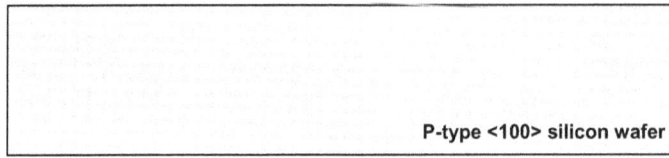

(a)

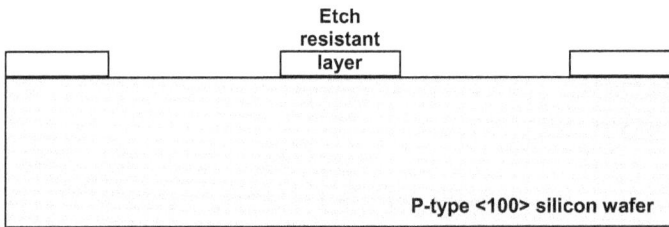

(b)

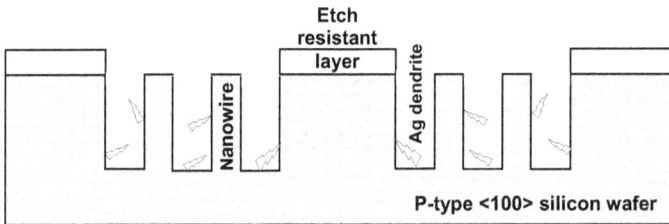

(c)

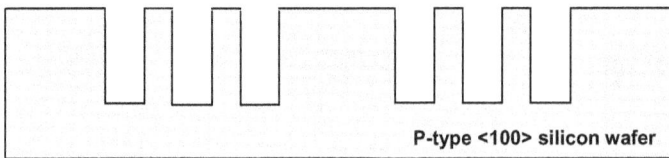

(d)

FIGURE 13.15 Process stages in the fabrication of silicon nanowire solar cell by wet-etching: (a) the silicon wafer, (b) deposition of etch-resistant layer on selected areas, (c) nanowire etching showing Ag dendrites, (d) removal of etch-resistant layer and Ag dendrites, (e) phosphorous diffusion, (f) front metal deposition, and (g) back metallization.

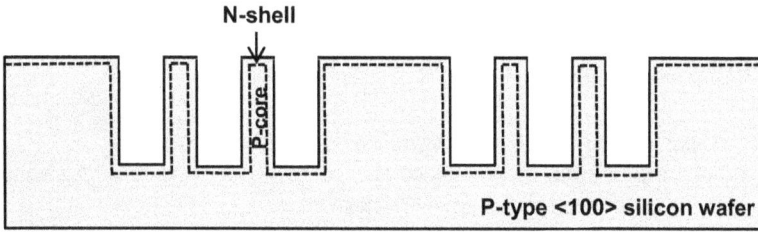

(e)

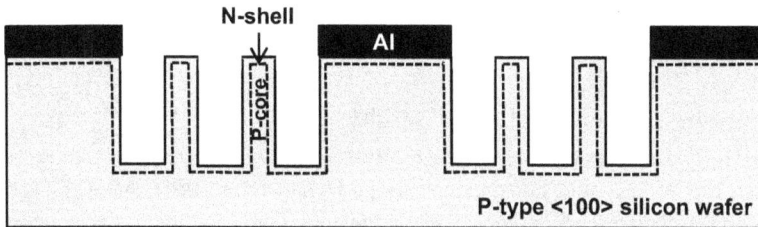

(f)

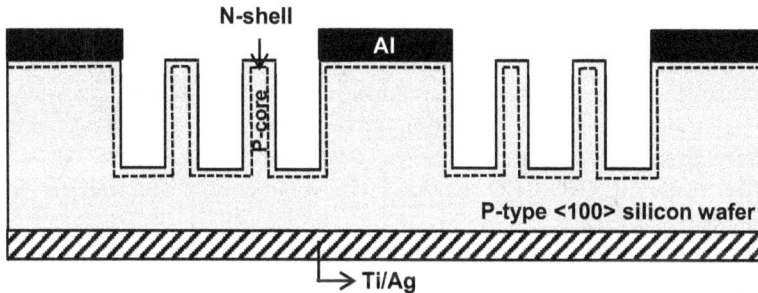

(g)

FIGURE 13.15   (Continued)

13.6.1.2   Metal-Assisted Chemical Etching (MACE or MacEtch) of Silicon

Boron-doped P-type < 100 > silicon wafer of resistivity 1–5Ω-cm is degreased by ultrasonification in acetone and rinsed with methanol and deionized water. The wafer is kept in piranha solution (1: 3 mixture of H_2O_2: H_2SO_4) for ½ h. The wafer is again rinsed in DI water. Then it is dipped in 10% HF to remove any native oxide, rinsed with water.

A layer of etch-resistant material is applied to protect the areas where contacts are to be made. This enables deposition of metal grid pattern for front contact of solar cell on a flat surface. Otherwise, the nanowires will be formed over the entire surface of silicon wafer, and the metal grid portions will be hanging on/filling into the nanowire gaps.

The protected wafer is transferred to 0.02M $AgNO_3$ in 5M aqueous HF solution taken in a Teflon vessel. Two simultaneously occurring processes are involved in the formation of vertically aligned silicon nanowires at room temperature:

  (i) Silver nanoparticles are deposited on the surface of the wafer.
 (ii) Catalytic etching of silicon takes place at the sites of deposition of Ag nanoparticles. The etching is catalyzed by the nanoparticles.

Thus, in this single-step process, deposition of Ag nanoparticles is combined with silicon etching induced by these nanoparticles.

As the regions where contacts are to be made had been carefully protected, silicon is etched selectively only in areas where nanowires are to be made. So the nanowires are formed in defined areas separated by a mesh of bulk silicon areas. The length of the nanowires is determined by the etching time. The diameter of the nanowires is 40–200nm.

13.6.1.3   Nanowire Diameter and Areal Density

The diameter of silicon nanowires and the number of nanowires per unit area (areal density) are functions of the $AgNO_3$ and HF concentrations (Khan et al 2014). As $AgNO_3$ concentration increases, the nanowire diameter decreases while the areal density of nanowires increases. Smaller nanowire diameter is explained as caused by reduction of nanoparticle size. The explanation for the formation of a larger number of nanowires is that a greater number of nanoparticles are produced. With increase in HF concentration, the nanowire diameter as well as the area density of nanowires decreases.

13.6.1.4   Removal of Ag Residues

When the etching is completed, the treelike Ag dendrites deposited on the wafer surface are removed by dipping in silver etchant consisting of NH_4OH and H_2O_2 in the ratio 3:1. The etching is done for 2 min. at room temperature. The wafer is rinsed with DI water. Any remnant Ag is removed by treatment with HNO_3/H_2O solution in 1:1 volume ratio. The wafer is rinsed in water and stored in methanol.

13.6.1.5   Formation of N-Type Shell Layer

The starting wafer being P-type, the nanowires are P-doped. To make the P-N diode, N-type layer is formed as a shell surrounding the P-type nanowire core. Phosphorous is diffused on the nanowire surface for creating the N-type layer. The wafer is rinsed in water, cleaned in piranha solution for ½ h., and rinsed again in DI water. It is dipped in 2% HF for surface oxide removal and rinsed in DI water. Phosphorous doping solution is spin-coated on the wafer. The wafer is kept on a hot plate for 10 min. and in 200°C oven for 15 min. and then loaded into a diffusion furnace at 900°C. Phosphorous diffusion is done for the required time decided by junction depth. Thus, a radial junction nanowire structure is obtained in which the core is P-type and the surrounding shell has N-type polarity. The wafer is unloaded, and phosphosilicate glass is removed in dilute HF.

13.6.1.6   Metallization

After water-rinsing and drying with methanol, the wafer is blown dry in nitrogen stream and loaded in a thermal evaporation system. Al film (thickness 1µm) is deposited in a vacuum of 1×10^{-6} torr for making the backside contact, followed by alloying at 500°C in forming gas for ½ h. On the front side of the wafer, Ti/Ag film is deposited through a metal grid mask. Sintering of the contact metals is done in forming gas.

13.6.1.7 Solar Cell Testing

Average reflectivity of Si nanowires is < 5% in the spectral range 300–1,000nm. An efficiency of 13.7% is achieved. The short-circuit current density is 37mAcm^{-2}, the open-circuit voltage is 0.544V, and the curve factor is 0.68 (Kumar et al 2011).

Note: Reflectivity vs. Reflectance Debate

To describe optical properties of nanowires, two terms, "reflectivity" and "reflectance," are frequently used. Reflectivity ρ of a surface is a property that measures its ability to reflect radiation. It is a parameter applicable to thick objects. When considering thin-layered materials, effects of internal reflections within the material can cause variations in reflection capability with thickness. Then the term "reflectance" is commonly used.

Suppose $\phi_{Reflected}(\lambda)$ is the flux of optical radiation of a given wavelength λ reflected from the surface, and $\phi_{Incident}(\lambda)$ is the flux received on the surface. Then reflectance R of the surface is defined as the ratio of reflected and incident fluxes, as

$$R = \frac{\phi_{Reflected}(\lambda)}{\phi_{Incident}(\lambda)} \qquad (13.2)$$

It is expressed as the ratio of powers as the ratio of reflected optical power $P_{Reflected}(\lambda)$ of radiation of wavelength (λ) to incident optical power $P_{Incident}(\lambda)$:

$$R = \frac{P_{Reflected}(\lambda)}{P_{Incident}(\lambda)} \qquad (13.3)$$

The flux or power ratio varies with the distribution of wavelengths in the incident radiation, e.g., the reflectance of silicon nanowires of diameter 20–200nm prepared by electroless etching is 3.5–19.2% in the UV region (200–400nm), 12.1–15.1% in the visible region (400–750nm), and 9.8–12.1% in the infrared region (750–1000nm); the corresponding values for bulk silicon substrate are 38.8–65.1%, 25.5–38.8%, and 22.1–25.5% (Hutagalung et al 2017).

As already mentioned, reflectivity equals reflectance when the sample is adequately thick so that its reflectance is independent of thickness. Hence, reflectivity = reflectance for a thick specimen. To reiterate, reflectivity is the limiting value of reflectance when the sample thickness is large. Reflectivity is distinguished from reflectance by asserting that reflectivity is a property of thick objects, whereas reflectance is applicable to thinner specimens. However, both reflectance and reflectivity are concerned with electromagnetic power. It may be noted that *reflection coefficient* is a term used for the ratio of electric fields, not powers. Reflectivity or reflectance is the square of the modulus of reflection coefficient.

13.6.2 SOLAR CELL (η = 17.11%) WITH DIELECTRIC PASSIVATION OF SI NANOWIRES

13.6.2.1 Fabrication Plan Outline

P-type core/N-type shell nanowires are formed by etching in two steps, first in AgNO$_3$/HF/ water, and then in HF/H$_2$O$_2$ (Lin et al 2013). After phosphorous diffusion, the nanowire surfaces are covered with silicon dioxide and silicon nitride layers, followed by annealing in forming gas (Figure 13.16).

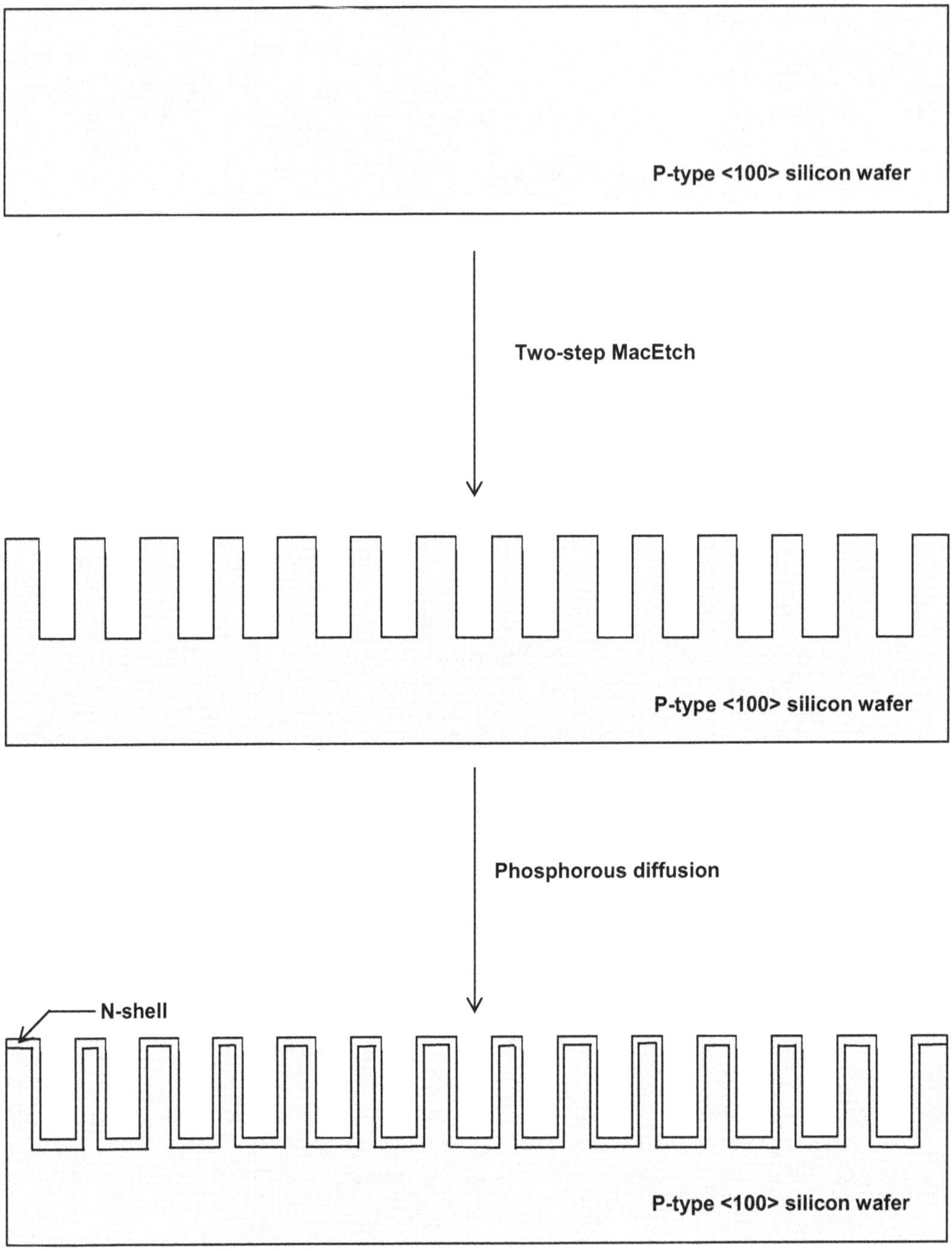

FIGURE 13.16 Process modification to include passivation of silicon nanowire surface by thermal silicon dioxide growth, silicon nitride deposition, and annealing: Beginning with P-type silicon wafer, silicon is etched in $AgNO_3$/HF/water, and then H_2O_2/HF, to make silicon nanowires. Phosphorous is diffused to make N-type shell forming P-N diode. On the nanowire surface, silicon dioxide is thermally grown and silicon nitride is deposited by PECVD. After forming gas annealing, further process is continued.

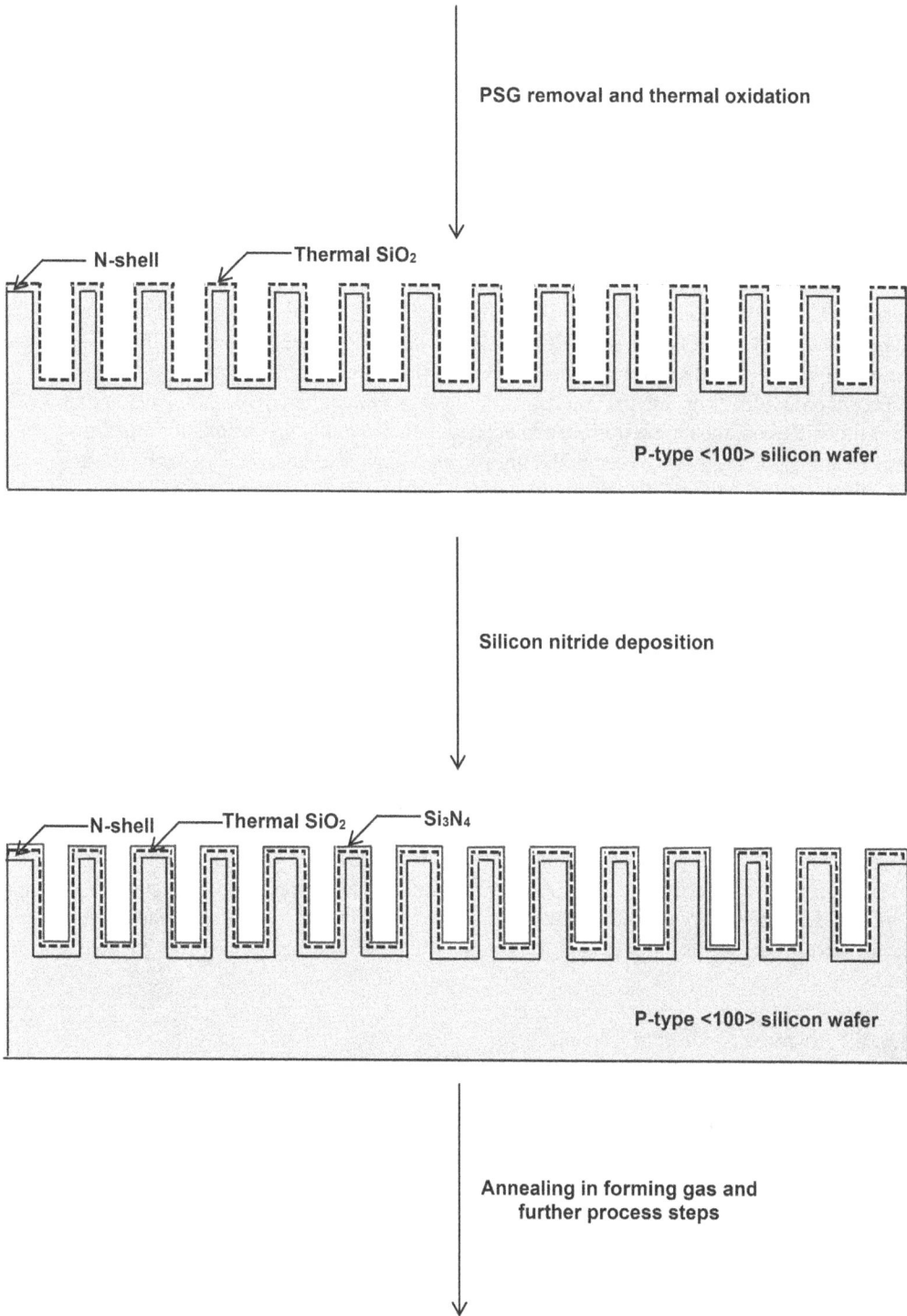

FIGURE 13.16 (Continued)

13.6.2.2 Nanowire Creation by Etching P-type Si Wafer, N-Type Shell Formation, and Surface Passivation

In a process incorporating passivation of nanowires with dielectric films, a two-step MacEtch technique is followed:

(i) After RCA cleaning, 2Ω-cm resistivity, boron-doped (100) P-type silicon wafer is immersed in (5M aqueous HF + 0.02M $AgNO_3$) solution for 90 s for silver nanoparticle deposition.

(ii) Then it is dipped in (aqueous buffered 5M HF + 0.01M H_2O_2) for required etching time to get silicon nanowires of stipulated length.

The average diameter of nanowires is 100nm, with variation between 60 and 200nm. The nanowires are 1,000nm long. The average period between the nanowires is 250nm.

Phosphorous diffusion for the N-type shell layer is carried out with $POCl_3$ source at 850°C. After this diffusion, the nanowires are protected with a thermally grown oxide of thickness 10nm. The thermal oxide layer is covered with 60nm-thick silicon nitride layer. The wafer is subjected to annealing at 450°C in forming gas.

13.6.2.3 Optical Reflectance, Carrier Recombination Properties, and Efficiency

This modified process scheme does not affect the light-trapping property of the nanowires. Reflectance of Si nanowires is ~ 4.6% in the wavelength range 300–1,100nm. But the passivation is very effective in reducing Shockley-Read-Hall and Auger recombination in regions adjoining the surfaces of nanowires. With this ($SiO_2 + Si_3N_4$) stack, a large-area solar cell (125mm × 125mm) of efficiency 17.11% is realized (Lin et al 2013).

13.6.3 Solar Cell ($\eta = 13.4\%$) Fabrication on N-Type Si Wafers

13.6.3.1 Nanowire Formation

N-type Czochralski wafer of orientation < 100 > and resistivity 1–2 Ω-cm is used (Leontis et al 2018). Si nanowires are formed by MACE process. The nanowires have diameters in the range 50–70nm. Distances between nanowires vary from 50 to 100nm. Among the nanowires of three heights investigated (500, 1,000, and 1,500nm), the 1,000nm-high nanowires yielded the best results.

13.6.3.2 Nanowire Doping

As the wafer is N-type, P-type diffusion for emitter is done using spin-on boron dopant. High-temperature annealing is carried out at 1,050°C for ½ h. The borosilicate glass is removed in buffered HF.

13.6.3.3 Contacts

The backside contact is aluminum. For front-side contact, aluminum is deposited by electron gun evaporation and patterned by photolithography in the shape of a grid. The Al metal grid coverage is 15%.

13.6.3.4 Comparison of Two Geometrical Designs

There are two geometrical designs (Figure 13.17):

(i) Nanowires are formed everywhere on the front surface of solar cell, and Al grid line is deposited over the nanowires, providing a conformal coating.

(ii) Nanowire formation is restricted to areas of size 200μm × 200μm separated by a tetragonal mesh of regions of bulk silicon, and Al grid line is aligned with bulk silicon areas without any nanowires.

Option ii was found to be superior to i. This solar cell showed a short-circuit current density of 32.5mAcm^{-2}, an open-circuit voltage of 0.587V, a fill factor of 70.3%, and efficiency of 13.4% (Leontis et al 2018).

13.6.3.5 Reflectance Dependence on Nanowire Length

Reflectance of nanowires varies with their lengths. From reflectance spectra spanning between 300 and 1,000nm wavelength, the maximum reflectance of 1,500nm-long silicon nanowires on

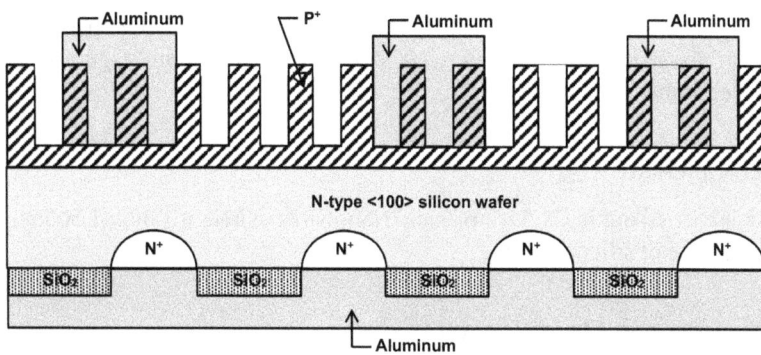

(a)

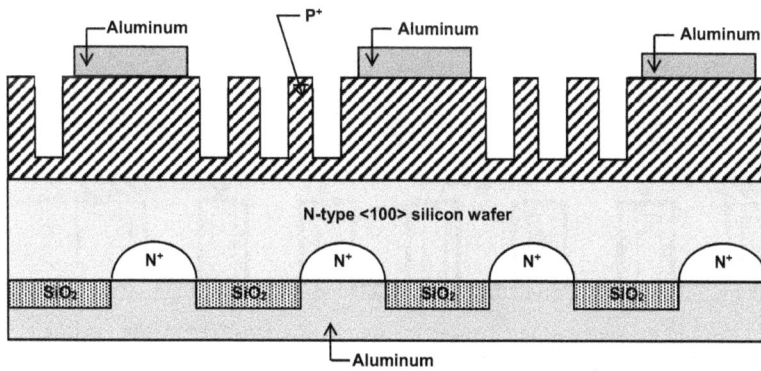

(b)

FIGURE 13.17 Comparison of the two geometrical designs of solar cells made on N-type silicon wafers: (a) nanowire growth takes place unrestrictedly all over the silicon surface, and (b) nanowire growth is allowed only in limited regions where aluminum grid line is not to be deposited for front contact; the remaining regions are reserved for contacting.

confined silicon areas after spin-on doping (option ii) is ~ 0.15. It is lower than the maximum value of 1,000nm-long nanowires ~ 0.2, and in both cases, the reflectance values are significantly lower than the reflectance of planar silicon ~ 0.55. The dependence of reflectance on nanowire length is explained on the basis of optical path length. Prolonging of the optical path in longer nanowires reduces the reflectance (Leontis et al 2018).

13.7   DRY-ETCHING PROCESS OF SILICON NANOWIRE ARRAY SOLAR CELL ($\eta = 11.7\%$) FABRICATION

13.7.1   SiO$_2$ Hard Mask Creation for Silicon Etching

Thermal silicon dioxide (150nm) is grown on P-type silicon wafer of resistivity 1–10 Ω-cm (Shieh et al 2015). Photolithography is carried out using the mask having the pattern for predeciced diameters of the nanowires and spacing between them, e.g., 400nm each. The pattern of nanowire diameters and spacing is transferred on the oxide layer. The oxide is etched in CF$_4$ plasma to produce the hard oxide mask for silicon etching. Figure 13.18 shows the solar cell, and Figure 13.19 presents the process sequence for making the cell.

13.7.2   Silicon Etching

Silicon etching is carried out in Cl$_2$/HBr plasma. Nanowires of height 1,000–1,500nm are obtained depending on the time of silicon etching.

13.7.3   Photoresist and Oxide Removal

A plasma asher is used to remove the photoresist. The oxide mask is removed in buffered HF. The nanowire of P-type polarity is ready.

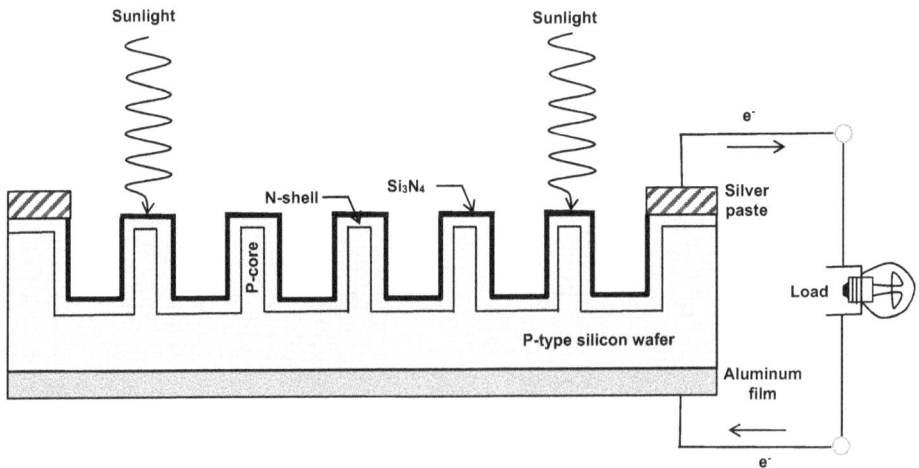

FIGURE 13.18   Core/shell silicon nanopillar solar cell made by dry-etching showing the P-type silicon wafer, P-type core, and phosphorous-implanted N-shell of silicon nanopillar, silicon nitride on the surfaces of nanopillars, aluminum film on the underside of the substrate for backside contact, and silver paste on upper side for front-side contact. Electric current produced by sunlight striking the front surface of the solar cell is withdrawn from Ag and Al electrodes.

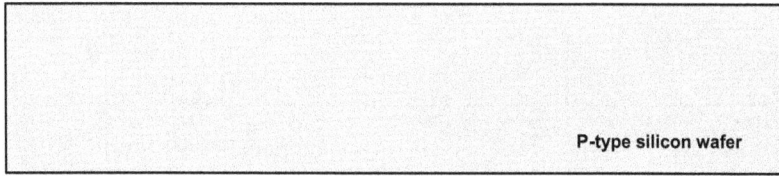

(a)

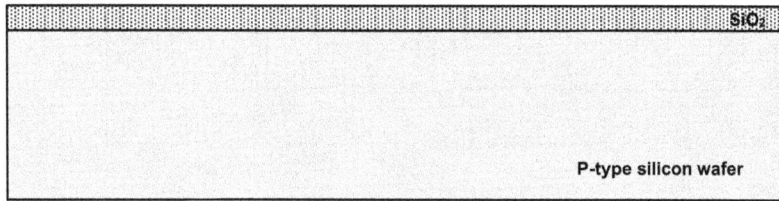

(b)

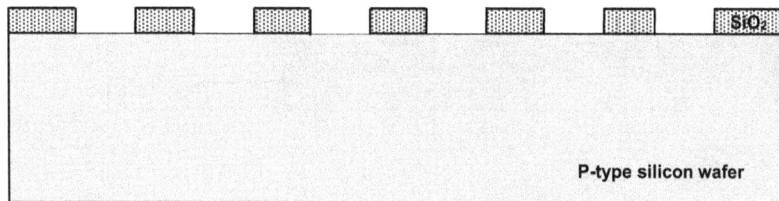

(c)

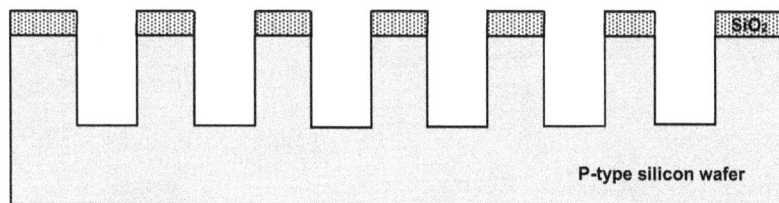

(d)

FIGURE 13.19 Fabrication of core/shell silicon nanopillar solar cell by dry-etching: (a) silicon wafer, (b) thermal oxidation, (c) patterning of the silicon dioxide layer and etching of the oxide layer in CF_4 plasma, (d) silicon etching in Cl/HBr plasma to create silicon nanopillars, (e) removal of oxide mask in plasma asher and buffered oxide etch, (f) phosphorous implantation and rapid thermal annealing, (g) silicon nitride formation on the surfaces of nanopillars by PECVD taking care to protect the corner regions for contact making, (h) electron-beam deposition of aluminum film on the underside of the substrate for backside contact, and (i) application and baking of silver paste on upper side for front side contact.

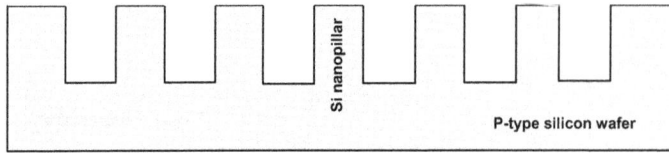

(e)

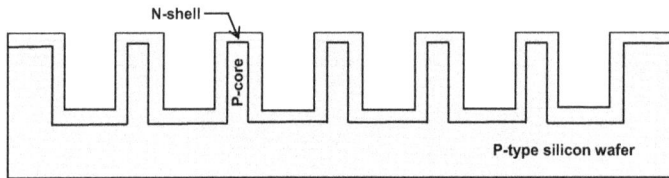

(f)

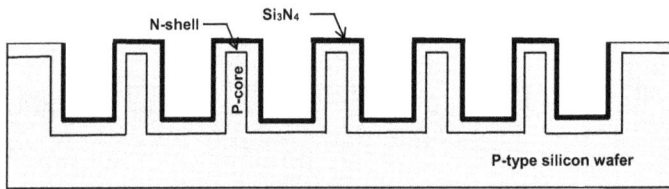

(g)

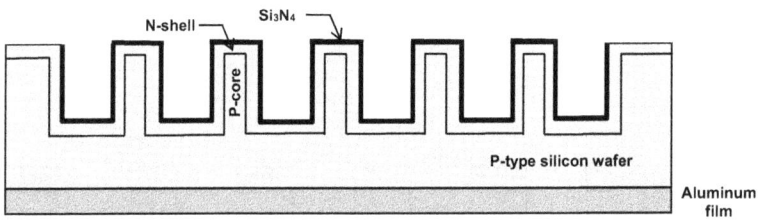

(h)

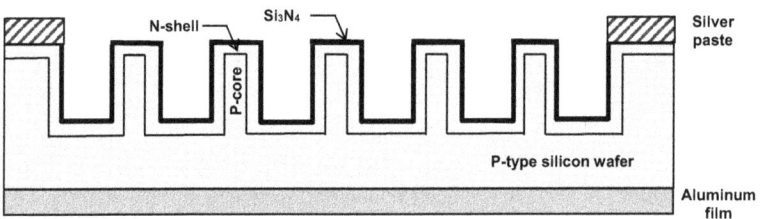

(i)

FIGURE 13.19   (Continued)

13.7.4 N-Type Layer Formation by Ion Implantation

Since a P-N junction diode is needed, the N-type layer must be formed as a shell surrounding the P-type nanowire core. This is done by ion implantation; the energy used is 36keV, and the dose is 5×10^{14}cm^{-2}. During ion implantation, the wafer is kept in a tilted position at an angle of $12°$ off the major plane. Tilting is necessary in order that doping occurs on the sidewalls of the nanowires. The uniformity of doping is achieved by using four rotational ion implantation through angles of $0°$, $90°$, $180°$, and $270°$.

13.7.5 Dopant Activation

Activation of dopant ions is required so that they occupy substitutional sites in the lattice. For activation, rapid thermal annealing (RTA) is done up to $1,000°$C. The RTA process also provides recovery of the damages inflicted on the lattice by the high-energy ion beam used in implantation.

13.7.6 Nanowire Surface Passivation

A silicon nitride film formed by plasma-enhanced chemical vapor deposition (PECVD) is formed for passivation of the surfaces of the nanowires.

13.7.7 Backside Contact

For the backside contact of the solar cell, an aluminum film is deposited by electron-beam evaporation. The aluminum is sintered at $600°$C in hydrogen ambience for 10 min.

13.7.8 Top Contact

Ag paste is applied to make the top contact of the solar cell. Baking of the paste is done for hardening it. The position of this front-side electrode is outside the nanowire array region.

13.7.9 Photovoltaic Properties of the Cell

Short-circuit current density of the solar cell is 32.931 mAcm^{-2}, open-circuit voltage is 0.448V, fill factor is 79.41, and efficiency 11.7% (Shieh et al 2015).

13.8 DISCUSSION AND CONCLUSIONS

The fabrication of III-V nanowire solar cells has extensively utilized vapor-liquid-solid mechanism of nanowire growth (Table 13.1). Alternatively, epitaxially grown stack of layers has been etched to form nanowires. For silicon nanowires, metal-catalyzed electroless etching of silicon requiring simple chemical laboratory ware has been largely used. Reactive ion etching has also been applied.

TABLE 13.1

Summary of Nanowire Solar Cells

Sl. No.	Material	Solar Cell	Structure	Type of Cell	Efficiency (%)	Reference
1	GaAs	Single nanowire: vertical	Ag glue/P$^+$ Si substrate/P-GaAs/P$^+$ GaAs shell/N-GaAs shell/N$^+$GaAs shell/ITO with SU-8 support	Radial	40	Krogstrup et al (2013)
2	GaAsP	Single nanowire: horizontal	Ti/Au/Ge/Au/GaAsP NW/P-GaAsP shell/N-GaAsP shell/Ge/Au/Ti/Au, with N-InGaAs passivation on P$^+$ Si/SiO$_2$ substrate	Radial	10.2	Holm et al (2013)

(Continued)

TABLE 13.1
(Continued)

Sl. No.	Material	Solar Cell	Structure	Type of Cell	Efficiency (%)	Reference
3	GaAs/Si	GaAs nanowire-on-Si	Al/N$^+$Si/N-Si/P$^+$Si/N$^+$GaAs/N-GaAs/I-GaAs/P$^+$GaAs/ITO	Axial, tandem	11.4	Yao et al (2015)
4	GaAs	Nanowire array	Brass/Ag glue/GaAs(111) B substrate/P$^+$GaAs/P-GaAs/N$^+$GaAs/TCO, with AlGaAs passivation	Axial	15.3	Åberg et at (2015, 2016)
5	InP	Nanowire array by bottom-up approach (VLS)	Ti/Pt/Au/P$^+$InGaAs/Zn-doped (111) B InP P-type substrate/P-InP/N-InP/ITO, with BCB passivation	Axial	11.1	Cui et al (2013)
6	InP	Nanowire array by bottom-up approach (epitaxial growth)	Ti/Au/P-InP substrate/P-InP/I-InP/N-InP/TCO/Ti/Au	Axial	13.8	Wallentin et al (2013)
7	InP	Nanowire array by top-down approach from epitaxially grown stack	Ti/Au/P$^+$InGaAs/P-InP substrate/P$^+$InP/P$^-$InP/N-InP/N$^+$InP/ITO/Au	Axial	17.8	van Dam et al (2016)
8	Si	On P-type by wet-etching	Ti/Ag/P-type < 100 > Si wafer/P core/N-type shell/Al	Radial	13.7	Kumar et al (2011)
9	Si	On P-type wafer with dielectric passivation	. . . P-type < 100 > Si wafer/P core/N-type shell/SiO$_2$/Si$_3$N$_4$ . . .	Radial	17.11	Lin et al (2013)
10	Si	On N-type wafer by wet-etching	Al/N$^+$/N-type < 100 > Si wafer/ P$^+$/Al	Axial	13.4	Leontis et al (2018)
11	Si	On P-type wafer by dry-etching	Al/P-type Si wafer/P-core/N-shell/Ag paste, with Si$_3$N$_4$ passivation	Radial	11.7	Shieh et al (2015)

Questions and Answers

13.1   Why can a nanowire of III-V semiconductor be grown on a low-cost silicon substrate without any defect generation in the nanowire? Answer: Because the nanowire base contacts the substrate over a very small area, the interfacial strain is relieved within a few monolayers of nanowire growth, enabling growth of defect-free nanowire.

Comment: This capability impacts the cost of nanowire solar cells. Cheaper nanowire solar cells can be made using common substrate materials as the growth substrate material need not of the same crystalline structure as the nanowire material.

13.2   What phenomenon determines the absorption of light by a vertically standing nanowire solar cell? Answer: Waveguide modes. The waveguide modes represent the unique distribution of transverse and longitudinal components of electric and magnetic fields in light. Two types of modes propagate in waveguides viz transverse electric or TE (in which only transverse electric field exists with the magnetic field in all directions) and transverse magnetic or TM (in which only transverse magnetic field exists with the electric field in all directions).

13.3   How is light absorbed in a horizontally laid nanowire solar cell? Answer: Through leaky (or tunneling modes) or Mie resonances. A leaky mode is a mode in which the electric field undergoes monotonic decay up to a finite distance in the transverse direction.

After that distance, it acquires an oscillatory nature. Mie resonances are mechanisms for generation of displacement current-based electric or magnetic resonances.

13.4 What special optical property does vertically standing nanowire solar cell provide? Answer: A very high degree of light concentration. This is achieved through a resonant increase in optical absorption cross section of the nanowire.

13.5 Comment on pairing GaAs nanowire solar cell with silicon solar cell in a tandem arrangement. Answer: GaAs has a bandgap of 1.424eV, and Si has a bandgap of 1.1eV. So the GaAs nanowire + Si tandem cell is able to use photons in a broader energy range for solar energy conversion into electricity in place of either a GaAs cell or a Si cell alone.

13.6 In what order should the two solar cells be stacked in a GaAs + Si tandem cell? Answer: GaAs has a larger bandgap than silicon. So sunlight should fall on the GaAs cell. The GaAs cell will absorb the high-energy photons-producing electron-hole pairs and allow the low-energy photons to pass through it. Emerging from the GaAs cell, the low-energy photons enter the Si cell and are absorbed by it, leading to photogeneration. This means that the silicon cell should be placed behind the GaAs cell.

13.7 What is vapor-liquid-solid mechanism of nanowire growth? Answer: This is a mechanism for growth of nanowires through chemical vapor deposition by directly absorbing the vapor on a solid surface with the help of a catalytic liquid phase, which can absorb the vapor to reach supersaturation so that crystalline nanowire growth can take place from seeds nucleated at the liquid-solid interface.

13.8 What is the process for liquid Au-Si droplet catalyzed VLS nanowire growth on a silicon substrate? Answer: For nanowire growth on a silicon substrate, a thin gold film is thermally evaporated or sputter-deposited on Si, nanometer-size Au areas are defined lithographically at designated places, depending on the required nanowire diameter and annealed to form liquid Au-Si droplets. A process catalyzed by the liquid Au-Si droplets is used to grow the nanowires.

13.9 What do you mean by self-catalyzed or Ga-assisted growth of III-V nanowires? Answer: Generally, a catalyst such as an Au nanoparticle initiates the growth of the nanowire. The catalyst aids in definition of nanowire diameter and stabilization of growth. Self-catalyzed growth proceeds without the presence of any foreign element used as a catalytic agent. The growth begins from the vapor phase through a droplet of native group III liquid. For a given temperature and pressures of source gases, there is a distinct diameter of stable nanowire growth and volume of the droplet. Both these parameters decrease as the III/V ratio increases (Colombo et al 2008; Ambrosini et al 2011; Tersoff 2015).

13.10 Why should the use of foreign element nanoparticle be avoided during nanowire growth? Answer: To avoid contamination of nanowire material by incorporation of a foreign element.

13.11 How are nanowires doped with impurities? Answer: Doping is done by introducing dopants during nanowire growth process. This is called *in situ* doping.

13.12 How is mechanical support provided to a single nanowire standing vertically on a substrate? Answer: By spinning SU-8 photoresist over the nanowire and hardening it by curing it. The photoresist is removed from the tip of the nanowire for making the top electrical contact.

13.13 What is the difference between MOVPE and MOCVD? Answer: Metal-organic vapor phase epitaxy (MOVPE) is the same as metal-organic chemical vapor deposition (MOCVD).

13.14 What is BCB used for? Answer: Benzocyclobutene (BCB), C_8H_8 is a low dielectric constant (2.65) material used to make photosensitive polymers for microelectronics and MEMS processing. It is also used as a passivating dielectric.

13.15 What is ALD? Answer: Atomic layer deposition is a vapor-phase thin-film deposition technique in which the substrate is alternately exposed to precursors in a sequence in a nonoverlapping manner.

13.16 How is nanostructured ITO formed over InP nanowire tips in InP nanowire array solar cell? Answer: Adhesion of ITO to InP differs from that on BCB. Consequently, preferential diffusion of ITO takes place towards InP and away from BCB. As ITO hemisphere-shaped nanoparticles are formed on InP nanowire tips, the top surface of the solar cell is covered with a nanostructured ITO film.

13.17 What role does nanostructured ITO play in the operation of the InP nanowire array solar cell? Answer: The hemispherical ITO nanoparticles produce Mie scattering, helping in absorption of light by the solar cell.

13.18 What is one-step MACE process? Silicon is immersed in an etching solution composed of $AgNO_3$ and HF.

13.19 What is two-step MACE process? Answer: In the first step, silver nanoparticles are formed on the silicon surface by thermally evaporating a 20–30nm thin film of silver over it; the silver thin film may also be deposited chemically. In the second step, the silver thin-film-coated silicon is immersed in aqueous HF and H_2O_2 solution.

13.20 What role is played by silver in one-step and two-step MACE processes? Answer: In one-step MACE process, silver catalyzes silver etching, whereas in two-step MACE process, silver nanoparticles protect the silicon underneath (Abouda-Lachiheb et al 2012).

13.21 Write down the chemical equations for silicon etching by one-step MACE process using $AgNO_3$ and HF. Answer: Mechanism I: The reaction involves the steps as under (Abouda-Lachiheb et al 2012):

(i) Silver nitrate dissociates as:

$$AgNO_3 \rightarrow Ag^+ + NO_3^-$$ (13.4)

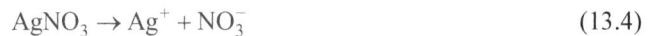

(ii) Hydrofluoric acid dissociation takes place as:

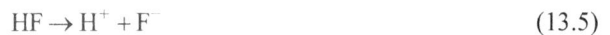

$$HF \rightarrow H^+ + F^-$$ (13.5)

(iii) H^+ and NO_3^- ions combine together as:

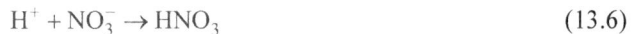

$$H^+ + NO_3^- \rightarrow HNO_3$$ (13.6)

(iv) Nitric acid oxidizes silicon as:

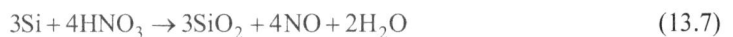

$$3Si + 4HNO_3 \rightarrow 3SiO_2 + 4NO + 2H_2O$$ (13.7)

(v) Silicon dioxide is etched by hydrofluoric acid as:

$$SiO_2 + 6HF \rightarrow H_2SiF_6 + 2H_2O$$ (13.8)

forming hexafluorosilicic acid and water.

Thus, H^+ and NO_3^- ions released respectively by HF and $AgNO_3$ react to form HNO_3, which oxidizes silicon to silicon dioxide. Then HF dissolves silicon dioxide. So the overall reaction consists of nucleation of metal catalyst (Ag) and anisotropic etching of silicon.

Mechanism II: The silver atoms form nuclei and, subsequently, nanoclusters. These silver nuclei and silver nanoclusters are distributed on the surface of silicon. So there are silicon areas that surround silver nuclei/silver nanoclusters. Silver nuclei/silver nanoclusters act as local cathodes and silicon areas as anodes, taking part in an electrochemical redox reaction consisting of two half-cell reactions (Ye et al 2009):

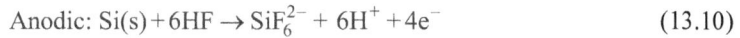

$$\text{Cathodic}: Ag^+ \rightarrow Ag(s) + e^- \tag{13.9}$$

$$\text{Anodic}: Si(s) + 6HF \rightarrow SiF_6^{2-} + 6H^+ + 4e^- \tag{13.10}$$

The half-reactions are combined into a concurrently occurring redox reaction, as:

$$4Ag^+ + Si(s) + 6F^- \rightarrow 4Ag(s) + SiF_6^{2-} \tag{13.11}$$

Silver nuclei/silver nanoclusters acting as cathodes are preserved, while silicon areas acting as anodes are etched away. The etching process is referred to as electroless because it does not require any external source of electrical energy.

13.22 How do the reactions for two-step MACE process differ from one-step MACE process? Answer: Here, the metal catalysts (Ag) are first deposited. Reduction of H_2O_2 is catalyzed at the Ag nanoparticle surface as (Li and Bohn 2000):

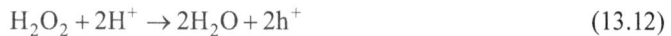

$$H_2O_2 + 2H^+ \rightarrow 2H_2O + 2h^+ \tag{13.12}$$

where h^+ represents a hole. The protons (H^+) are converted into hydrogen:

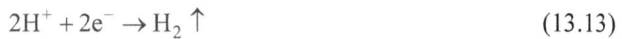

$$2H^+ + 2e^- \rightarrow H_2 \uparrow \tag{13.13}$$

The holes ($2h^+$) produced are injected into the valence band of silicon. These injected holes weaken the chemical bonds at the interface of Si/HF solution so that silicon is dissolved in HF through the reactions:

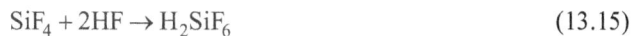

$$Si + 4HF + 4h^+ \rightarrow SiF_4 + 4H^+ \tag{13.14}$$

$$SiF_4 + 2HF \rightarrow H_2SiF_6 \tag{13.15}$$

The complete reaction is given by

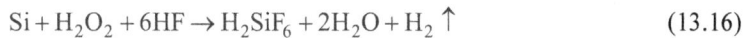

$$Si + H_2O_2 + 6HF \rightarrow H_2SiF_6 + 2H_2O + H_2 \uparrow \tag{13.16}$$

The holes themselves are consumed during dissolution of silicon.

13.23 Write the balanced chemical equation for silver etching using NH_4OH and H_2O_2. Answer:

$$2Ag + 2NH_4OH + 9H_2O_2 \rightarrow 2AgNO_3 + 14H_2O \tag{13.17}$$

13.24 Reflectivity represents the limiting value of —— when the sample thickness is large. Answer: Reflectance.

13.25 The term "reflectance" is used for thick films, while the term "reflectivity" is applied to thin films. Right or wrong? Answer: Wrong; reflectance is used for thin films, and reflectivity for thick films.

13.26 Why do thin and thick films differ in their reflection capabilities? Answer: Because internal reflections can affect reflection capability of a thin film when talking about thin layers of materials.

13.27 Why is passivation of silicon nanowires with dielectric beneficial for solar cell operation? Answer: Because it reduces carrier loss by surface recombination.

13.28 Arrange in order of decreasing reflectance: planar silicon, long silicon nanowires, short silicon nanowires. Answer: Reflectance is maximum for planar silicon, intermediate for short Si nanowires, and minimum for long Si nanowires. Hence, reflectance of long Si nanowire < reflectance of short Si nanowire < reflectance of planar silicon.

13.29 Of what material is the hard mask made for making silicon nanowire array by dry-etching of silicon? Answer: SiO_2.

13.30 Why is post-implantation thermal activation of dopant necessary? Answer: In order that any impurity dopant atoms in interstitial lattice sites move to substitutional sites, producing the desired electronic contribution to the host semiconductor.

REFERENCES

Åberg I., G. Vescovi, D. Asoli, U. Naseem, J. P. Gilboy, C. Sundvall, A. Dahlgren, K. Erik Svensson, N. Anttu, M. T. Björk and L. Samuelson 2015 A GaAs nanowire array solar cell with 15.3% efficiency at 1 sun, 2015 IEEE 42nd Photovoltaic Specialist Conference (PVSC), 14–19 June 2015, New Orleans, LA, IEEE, NY, pp. 1–3.

Åberg I., G. Vescovi, D. Asoli, U. Naseem, J. P. Gilboy, C. Sundvall, A. Dahlgren, K. Erik Svensson, N. Anttu, M. T. Björk and L. Samuelson 2016 A GaAs nanowire array solar cell with 15.3% efficiency at 1 sun, IEEE Journal of Photovoltaics, 6(1): 185–190.

Abouda-Lachiheb M., N. Nafie and M. Bouaicha 2012 The dual role of silver during silicon etching in HF solution, Nanoscale Research Letters, 7(1): 455, 1–5.

Ambrosini S., M. Fanetti, V. Grillo, A. Franciosi and S. Rubini 2011 Vapor-liquid-solid and vapor-solid growth of self-catalyzed GaAs nanowires, AIP Advances, 1: 042142–1 to 042142–12.

Cao L., J. S. White, J.-S. Park, J. A. Schuller, B. M. Clemens and M. L. Brongersma 2009 Engineering light absorption in semiconductor nanowire devices, Nature Materials, 8: 643–647.

Colombo C., D. Spirkoska, M. Frimmer, G. Abstreiter and Fontcuberta i Morral 2008 Ga-assisted catalyst-free growth mechanism of GaAs nanowires by molecular beam epitaxy, Physical Review B, 77: 155326–1 to 155326–5.

Cui Y., J. Wang, S. R. Plissard, A. Cavalli, T. T. T. Vu, R. P. J. van Veldhoven, L. Gao, M. Trainor, M. A. Verheijen, J. E. M. Haverkort and E. P. A. M. Bakkers 2013 Efficiency enhancement of InP nanowire solar cells by surface cleaning, Nano Letters, 13: 4113−411.

Holm J. V., H. I. Jørgensen, P. Krogstrup, J. Nygård, H. Liu and M. Aagesen 2013 Surface-passivated single-nanowire solar cells exceeding 10% efficiency grown on silicon, Nature Communications, 4(1498): 5 pages.

Hutagalung S. D., M. M. Fadhali, R. A. Areshi and F. D. Tan 2017 Optical and electrical characteristics of silicon nanowires prepared by electroless etching, Nanoscale Research Letters, 12(425): 1–11.

Khan F., S.-H. Baek and J. H. Kim 2014 Dependence of performance of Si nanowire solar cells on geometry of the nanowires, The Scientific World Journal, 2014(Article ID 358408): 7 pages.

Krogstrup P., H. I. Jørgensen, M. Heiss, O. Demichel, J. V. Holm, M. Aagesen, J. Nygard and A. F. i. Morral 2013 Single-nanowire solar cells beyond the Shockley—Queisser limit, Nature Photonics, 7: 306–310.

Kumar D., S. K. Srivastava, P. K. Singh, M. Husain and V. Kumar 2011 Fabrication of silicon nanowire arrays based solar cell with improved performance, Solar Energy Materials & Solar Cells, 95: 215–218.

Leontis I., M. A. Botzakaki, S. N. Georga and A. G. Nassiopoulou 2018 Study of Si nanowires produced by metal-assisted chemical etching as a light-trapping material in n-type c-Si solar cells, ACS Omega, 3: 10898−10906.

Lin X. X., X. Hua, Z. G. Huang and W. Z. Shen 2013 Realization of high performance silicon nanowire based solar cells with large size, Nanotechnology, 24(235402): 8 pages.

Li X. and P. Bohn 2000 Metal-assisted chemical etching in HF/H_2O_2 produces porous silicon, Applied Physics Letters, 77: 2572–2574.

Shieh J., Y. C. Li, C. Y. Ji, C. C. Chiu and H. Y. Lin 2015 Extracting high electrical currents with large fill factors from core/shell silicon nanopillar solar cells, Journal of Renewable and Sustainable Energy, 7: 033102–1 to 033102–8.

Tersoff J. 2015 Stable self-catalyzed growth of III—V nanowires, Nano Letters, 15(10): 6609–6613.

van Dam D., N. J. J. van Hoof, Y. Cui, P. J. van Veldhoven, E. P. A. M. Bakkers, J. G. Rivas and J. E. M. Haverkort 2016 High-efficiency nanowire solar cells with omnidirectionally enhanced absorption due to self-aligned indium–tin–oxide Mie scatterers, ACS Nano, 10: 11414–11419.

Wallentin J., N. Anttu, D. Asoli, M. Huffman, I. Åberg, M. H. Magnusson, G. Siefer, P. Fuss-Kailuweit, F. Dimroth, B. Witzigmann, H. Q. Xu, L. Samuelson, K. Deppert and M. T. Borgström 2013 InP nanowire array solar cells achieving 13.8% efficiency by exceeding the ray optics limit, Science, 339: 1057–1060.

Yao M., S. Cong, S. Arab, N. Huang, M. L. Povinelli, S. B. Cronin, P. D. Dapkus and C. Zhou 2015 Tandem solar cells using GaAs nanowires on Si: Design, fabrication, and observation of voltage addition, Nano Letters, 15,11: 7217–7224.

Ye W., C. Shen, J. Tian, C. Wang, C. Hui and H. Gao 2009 Controllable growth of silver nanostructures by a simple replacement reaction and their SERS studies, Solid State Sciences, 11: 1088–1093.

Zhang Y. and H. Liu 2019 Nanowires for high-efficiency, low-cost solar photovoltaics, Crystals, 9(87): 1–25.

14 Quantum Dot Solar Cells
Bandgap and Multicarrier Effects

A quantum dot solar cell is one that uses quantum dots as the light-absorbing medium in place of a bulk material. There are two distinct advantages of using quantum dots:

(i) Bandgap tuning. Quantum dots exhibit size-dependent bandgap. Changing the size of the quantum dot allows variation of its bandgap. As the solar spectrum contains energies in the range 0.5–3.5eV, quantum dots of the same material having different sizes will be able to exploit a greater fraction of photon energies. A bulk material needs to be replaced with a new material every time a new value of bandgap is required.

(ii) Multiple exciton generation. Quantum dots provide this opportunity which makes them capable of capturing the photon energies which are otherwise used in producing hot photogenerated carriers: hot electrons and hot holes having energies above the lattice temperature. These hot carriers are generated from photons having excess kinetic energy, i.e., energy greater than the bandgap. Excess kinetic energy = photon energy – bandgap energy. The hot carriers can be at such high temperatures as 3,000K, while the lattice temperature is the room temperature (300K) (Nozik 2002).

So in this chapter, both these phenomena, bandgap tuning and multiple exciton generation, will be explored. Furthermore, the operation of quantum dot solar cells is best understood by drawing their energy band diagrams. It is therefore expedient to explain the sketching of energy band diagrams of heterojunctions, in the context of quantum dots, and to apply these ideas to draw the band diagrams of some quantum dot solar cell structures.

14.1 BANDGAP TUNING OF QUANTUM DOTS

14.1.1 QUANTUM DOTS AS A PARTICLE-IN-A-BOX SYSTEM

The simple exercise of particle-in-a-box was performed in one dimension for the quantum well because the particle was free to move in the remaining two dimensions. Here, the particle is constrained from movement in all the three directions and is enclosed in a spherical box of radius R. The quantum dot is a box of spherical shape, and the electron/hole inside it is the particle being talked about. To extend the analysis to three dimensions, let us suppose that the electrons and holes are confined inside a three-dimensional potential well. Let the dimensions of this rectangular well be L_{wx}, L_{wy}, and L_{wz} along the X-, Y-, and Z-directions. Then the equation 10.32 of quantization of energy can be written for the three directions as:

$$E_n(x) = \frac{\pi^2 \hbar^2}{2m} \left(\frac{n_x^2}{L_{wx}^2} \right) \tag{14.1}$$

$$E_n(y) = \frac{\pi^2 \hbar^2}{2m} \left(\frac{n_y^2}{L_{wy}^2} \right) \tag{14.2}$$

and

DOI: 10.1201/9781003215158-19

$$E_n(z) = \frac{\pi^2 \hbar^2}{2m} \left(\frac{n_z^2}{L_{wz}^2} \right) \tag{14.3}$$

in terms of the quantum numbers n_x, n_y, n_z in the three directions and the lengths of the three sides of the well.

Combining together equations 14.1, 14.2, and 14.3, we get

$$E_n(x,y,z) = \frac{\pi^2 \hbar^2}{2m} \left(\frac{n_x^2}{L_{wx}^2} + \frac{n_y^2}{L_{wy}^2} + \frac{n_z^2}{L_{wz}^2} \right) \tag{14.4}$$

For a cubic box of side L,

$$L_{wx} = L_{wy} = L_{wz} = L \tag{14.5}$$

so that equation 14.4 reduces to

$$E_n(x,y,z) = \frac{\pi^2 \hbar^2}{2m} \left(\frac{n_x^2}{L^2} + \frac{n_y^2}{L^2} + \frac{n_z^2}{L^2} \right) = \frac{\pi^2 \hbar^2}{2mL^2} \left(n_x^2 + n_y^2 + n_z^2 \right) = \frac{\pi^2 \hbar^2}{2mL^2} n^2 \tag{14.6}$$

where n is the quantum number describing the system. Since the quantum dot is not a cubical box but a sphere of radius R, the energy equation for the quantum dot is written by replacing L by R in equation 14.6 as

$$E_n(x,y,z) = \frac{\pi^2 \hbar^2}{2mR^2} n^2 \tag{14.7}$$

For the ground state ($n = 1$) of the electron of effective mass $m_e{}^*$, equation 14.7 is written as

$$\left| E_n(x,y,z) \right|_{n=1,\text{ Electron}} = \frac{\pi^2 \hbar^2}{2m_e^* R^2} 1^2 = \frac{\pi^2 \hbar^2}{2m_e^* R^2} \tag{14.8}$$

An equation similar to equation 14.8 is written down for the holes of effective mass $m_h{}^*$:

$$\left| E_n(x,y,z) \right|_{n=1,\text{ Hole}} = \frac{\pi^2 \hbar^2}{2m_h^* R^2} \tag{14.9}$$

14.1.2 Effective Bandgap of the Quantum Dot

A detailed analysis was carried out by Brus (1986). The equation for the effective bandgap of the quantum dot is known as the Brus equation and is written as

$$E_G^{QD} = E_G^{Bulk} + \frac{\hbar^2 \pi^2}{2R^2} \left(\frac{1}{m_e^*} + \frac{1}{m_h^*} \right) \tag{14.10}$$

where the symbol E_G^{Bulk} stands for the bandgap energy of the bulk semiconductor. Note that the sum of the electron and hole ground state energies in a spherical box is added to the bulk energy gap in the same manner as we derived the equation for the quantum well in example 10.5, equation 10.66.

By virtue of electrostatic attraction, electrons and holes behave as mutually correlated particles called excitons. Therefore, the equation is presented in a modified form to include the effect of Coulomb interaction between the electron and the hole. The modified equation contains three terms on the right-hand side. The third term is the Coulomb attraction term. Thus, the bandgap energy E_G^{QD} of a quantum dot of radius R is (Brus 1986)

$$E_G^{QD} = E_G^{Bulk} + \frac{\hbar^2 \pi^2}{2R^2}\left(\frac{1}{m_e^*} + \frac{1}{m_h^*}\right) - \frac{1.8e^2}{4\pi\varepsilon_0\varepsilon_S R} \tag{14.11}$$

The symbols ε_0, ε_S denote respectively the free-space permittivity and dielectric constant of the semiconductor material of the quantum dot.

Example 14.1

For a given semiconductor, the exciton Bohr radius is expressed as

$$r_B = \frac{4\pi\hbar^2\varepsilon_0\varepsilon_S}{e^2}\left(\frac{1}{m_e^*} + \frac{1}{m_h^*}\right) \tag{14.12}$$

In this equation, the symbols have the usual meanings: $\hbar$ = reduced Planck's constant = $h/2\pi$, ε_0 = permittivity of free space, ε_S = the relative permittivity of the semiconductor, e = the electronic charge, m_e^* = the effective mass of electrons in the semiconductor from which the quantum dot is made, m_h^* = the effective mass of holes in the semiconductor.
Calculate the exciton Bohr radius for the following semiconductors:

(a) PbS: $\varepsilon_S = 17.2$, $m_e^* = 0.085m_0$, $m_h^* = 0.085\ m_0$ (Mukherjee et al 1994; Patil and Datta 2011).
(b) CdTe: $\varepsilon_S = 10.2$, $m_e^* = 0.11m_0$, $m_h^* = 0.4\ m_0$ (CRM 2010)
(c) CdSe: $\varepsilon_S = 9.56$, $m_e^* = 0.13m_0$, $m_h^* = 0.3m_0$ (ESI 2013)
(d) ZnO: $\varepsilon_S = 8.656$, $m_e^* = 0.24m_0$, $m_h^* = 0.59\ m_0$ (Norton et al 2004)
(e) ZnS: $\varepsilon_S = 8.9$, $m_e^* = 0.25m_0$, $m_h^* = 0.59m_0$ (ESI 2012)

Solution:

From equation 14.12, the unit of r_B is

$$\frac{(J.s)^2 \times Fm^{-1}}{(C)^2\ kg} = \frac{\left(kg.m^2s^{-2}.s\right)^2 \times kg^{-1}m^{-2}s^4A^2m^{-1}}{(C)^2\ kg} = \frac{kg^2.m^4s^{-2} \times kg^{-1}m^{-2}s^4C^2m^{-1}}{C^2.kg.s^2} = m \tag{14.13}$$

Considering equation 14.12, the common factor in problems a, b, c, and d is

$$CF = \frac{4\pi\hbar^2\varepsilon_0}{e^2 m_0} = \frac{4 \times 3.14159 \times \left(1.05457 \times 10^{-34}\right)^2 \times 8.854 \times 10^{-12}}{\left(1.602 \times 10^{-19}\right)^2 \times 9.109 \times 10^{-31}}$$

$$= \frac{4 \times 3.14159 \times \left(1.05457\right)^2 \times 10^{-68} \times 8.854 \times 10^{-12}}{\left(1.602\right)^2 \times 10^{-38} \times 9.109 \times 10^{-31}} \tag{14.14}$$

$$= \frac{4 \times 3.14159 \times \left(1.05457\right)^2 \times 8.854 \times 10^{-68-12}}{\left(1.602\right)^2 \times 9.109 \times 10^{-38-31}}$$

$$= \frac{123.73707}{23.3774} \times 10^{-68-12+38+31} = 5.293 \times 10^{-11}$$

Equation 14.12 will be applied in parts *a*, *b*, *c*, *d*, and *e*, and the common factor *CF* in equation 14.14 will be inputted. Also, the dielectric constants and effective masses of electrons and holes for the respective materials will be put to calculate the exciton Bohr radii.

(a)

$$|r_B|_{PbS} = 5.293 \times 10^{-11} \times 17.2 \left(\frac{1}{0.085} + \frac{1}{0.085} \right) = \frac{5.293 \times 10^{-11} \times 17.2 \times 2}{0.085}$$

$$= 2142.108 \times 10^{-11} = 21.4 \times 10^{-9} \text{m} = 21.4 \text{nm} \tag{14.15}$$

(b)

$$|r_B|_{CdTe} = 5.293 \times 10^{-11} \times 10.2 \left(\frac{1}{0.11} + \frac{1}{0.4} \right) = \frac{5.293 \times 10^{-11} \times 10.2 \times 0.51}{0.044}$$

$$= 625.777 \times 10^{-11} = 6.25777 \times 10^{-9} \text{m} \approx 6.26 \text{nm} \tag{14.16}$$

(c)

$$|r_B|_{CdSe} = 5.293 \times 10^{-11} \times 9.56 \left(\frac{1}{0.13} + \frac{1}{0.3} \right) = \frac{5.293 \times 10^{-11} \times 9.56 \times 0.43}{0.039}$$

$$= 557.909 \times 10^{-11} = 5.57909 \times 10^{-9} \text{m} \approx 5.58 \text{nm} \tag{14.17}$$

(d)

$$|r_B|_{ZnO} = 5.293 \times 10^{-11} \times 8.656 \left(\frac{1}{0.24} + \frac{1}{0.59} \right) = \frac{5.293 \times 10^{-11} \times 8.656 \times 0.83}{0.1416}$$

$$= 268.555 \times 10^{-11} = 2.68555 \times 10^{-9} \text{m} \approx 2.69 \text{nm} \tag{14.18}$$

(e)

$$|r_B|_{ZnS} = 5.293 \times 10^{-11} \times 8.9 \left(\frac{1}{0.25} + \frac{1}{0.59} \right) = \frac{5.293 \times 10^{-11} \times 8.9 \times 0.84}{0.1475}$$

$$= 268.274 \times 10^{-11} = 2.68274 \times 10^{-9} \text{m} \approx 2.68 \text{nm} \tag{14.19}$$

Example 14.2

Find the thermal de Broglie wavelength for electrons in the following semiconductors using parameters given in example 14.1 at *T* = 300K: (a) PbS, (b) CdTe, (c) ZnO, and (d) ZnS.

Solution: Recall equation 10.3. The thermal de Broglie wavelength is

$$\lambda_{Thermal} = \frac{h}{\sqrt{2m_e^* k_B T}} \tag{14.20}$$

From equation 14.20, we find that the unit of de Broglie wavelength is

$$\frac{h}{\sqrt{2m_e^* k_B T}} = \frac{J.s}{\sqrt{kgJK^{-1}K}} = \frac{J.s}{\sqrt{kgJ}} = \frac{\sqrt{J}}{\sqrt{kg}}.s = \frac{\sqrt{kg.m^2 s^{-2}}}{\sqrt{kg}}.s = m \tag{14.21}$$

When applying equation 14.20, we notice that the common factor in problems a, b, c, and d is

$$CF = \frac{h}{\sqrt{2m_0 k_B T}} = \frac{6.626 \times 10^{-34}}{\sqrt{2 \times 9.109 \times 10^{-31} \times 1.3806 \times 10^{-23} \times 300}}$$

$$= \frac{6.626 \times 10^{-34}}{\sqrt{2 \times 9.109 \times 1.3806 \times 10^{-31-23+2} \times 3}} \qquad (14.22)$$

$$= \frac{6.626 \times 10^{-34}}{\sqrt{2 \times 9.109 \times 1.3806 \times 10^{-52} \times 3}} = \frac{6.626 \times 10^{-34}}{8.6865} \times 10^{26}$$

$$= 7.6279 \times 10^{-9}$$

Equation 14.20 will be applied to calculate the thermal de Broglie wavelengths $\lambda_{Thermal}$ in parts a, b, c, and d by inputting the common factor CF in equation 14.22 and the electron effective masses in respective materials from example 14.1.

(a)

$$\left|\lambda_{Thermal}\right|_{PbS} = \frac{7.6279 \times 10^{-9}}{\sqrt{0.085}} = \frac{7.6279 \times 10^{-9}}{0.2915} = 26.1677 \times 10^{-9} \, m \approx 26.2 nm \qquad (14.23)$$

(b)

$$\left|\lambda_{Thermal}\right|_{CdTe} = \frac{7.6279 \times 10^{-9}}{\sqrt{0.11}} = \frac{7.6279 \times 10^{-9}}{0.3317} = 22.996 \times 10^{-9} \, m \approx 23 nm \qquad (14.24)$$

(c)

$$\left|\lambda_{Thermal}\right|_{ZnO} = \frac{7.6279 \times 10^{-9}}{\sqrt{0.24}} = \frac{7.6279 \times 10^{-9}}{0.489898} = 15.57 \times 10^{-9} \, m \approx 15.6 nm \qquad (14.25)$$

(d)

$$\left|\lambda_{Thermal}\right|_{ZnS} = \frac{7.6279 \times 10^{-9}}{\sqrt{0.25}} = \frac{7.6279 \times 10^{-9}}{0.5} = 15.2558 \times 10^{-9} \, m \approx 15.26 nm \qquad (14.26) \cdot$$

Example 14.3

Calculate the binding energy of the exciton in (a) PbS, (b) CdTe, (c) ZnO, and (d) ZnS using parameters of example 14.1.

Solution: The binding energy of the exciton in a material of dielectric constant ε_r is given by

$$E_{Binding} = \frac{\mu}{m_0 \varepsilon_r^2} R_y (H) \qquad (14.27)$$

where the Rydberg unit of energy $R_y(H)$ is evaluated as

$$R_y(\mathrm{H}) = \frac{m_0 e^4}{8\varepsilon_0^2 h^2} = \frac{9.109 \times 10^{-31} \times \left(1.602 \times 10^{-19}\right)^4}{8\left(8.854 \times 10^{-12}\right)^2 \times \left(6.626 \times 10^{-34}\right)^2}$$

$$= \frac{9.109 \times (1.602)^4 \times 10^{-31-76}}{8(8.854)^2 \times (6.626)^2 \times 10^{-24-68}} = \frac{59.9958}{27534.16} \times 10^{-31-76+24+68} \qquad (14.28)$$

$$= 2.17896 \times 10^{-3} \times 10^{-15} \mathrm{J} = \frac{2.17896 \times 10^{-3} \times 10^{-15}}{1.602 \times 10^{-19}} \mathrm{eV}$$

$$= 1.36 \times 10^{-3-15+19} \mathrm{eV} = 13.6 \mathrm{eV}$$

When applying equation 14.27, the common factor in problems a, b, c, and d is

$$C.F. = \frac{R_y(\mathrm{H})}{m_0} = \frac{13.6}{m_0} \mathrm{eV} \qquad (14.29)$$

Equation 14.27 will be applied for finding the exciton binding energies in parts a, b, c, and d by inputting the CF from equation 14.29, the reduced masses μ for the specific materials obtained from electron and hole effective masses given in example 14.1, and the dielectric constant ε_r values for respective materials.

(a)

$$\left| E_{\mathrm{Binding}} \right|_{\mathrm{PbS}} = \frac{\mu}{\varepsilon_r^2} \times C.F. = \frac{\mu}{(17.2)^2} \times C.F. = \frac{0.0425 m_0}{295.84} \times \frac{13.6}{m_0} \mathrm{eV} \qquad (14.30)$$

since

$$\left| \mu \right|_{\mathrm{PbS}} = \frac{m_e^* m_h^*}{m_e^* + m_h^*} = m_0 \left(\frac{0.085 \times 0.085}{0.085 + 0.085} \right) = m_0 \frac{0.007225}{0.17} = 0.0425 m_0 \qquad (14.31)$$

Equation 14.30 gives

$$\left| E_{\mathrm{Binding}} \right|_{\mathrm{PbS}} = \frac{0.0425 \times 13.6}{295.84} \mathrm{eV} = 0.001954 \mathrm{eV} = 1.95 \mathrm{meV} \qquad (14.32)$$

(b)

$$\left| E_{\mathrm{Binding}} \right|_{\mathrm{CdTe}} = \frac{\mu}{\varepsilon_r^2} \times C.F. = \frac{\mu}{(10.2)^2} \times C.F. = \frac{0.08627 m_0}{104.04} \times \frac{13.6}{m_0} \mathrm{eV} \qquad (14.33)$$

since

$$\left| \mu \right|_{\mathrm{CdTe}} = \frac{m_e^* m_h^*}{m_e^* + m_h^*} = m_0 \left(\frac{0.11 \times 0.4}{0.11 + 0.4} \right) = m_0 \frac{0.044}{0.51} = 0.08627 m_0 \qquad (14.34)$$

Equation 14.33 yields

$$\left| E_{\mathrm{Binding}} \right|_{\mathrm{CdTe}} = \frac{0.08627 \times 13.6}{104.04} \mathrm{eV} = 0.011277 \mathrm{eV} = 11.277 \mathrm{meV} \approx 11.3 \mathrm{meV} \qquad (14.35)$$

(c)

$$\left|E_{\text{Binding}}\right|_{\text{ZnO}} = \frac{\mu}{\varepsilon_r^2} \times C.F. = \frac{\mu}{(8.656)^2} \times C.F. = \frac{0.1706m_0}{74.926} \times \frac{13.6}{m_0} \text{eV} \tag{14.36}$$

since

$$\left|\mu\right|_{\text{ZnO}} = \frac{m_e^* m_h^*}{m_e^* + m_h^*} = m_0 \left(\frac{0.24 \times 0.59}{0.24 + 0.59}\right) = m_0 \frac{0.1416}{0.83} = 0.1706m_0 \tag{14.37}$$

From equation 14.36,

$$\left|E_{\text{Binding}}\right|_{\text{ZnO}} = \frac{0.1706 \times 13.6}{74.926} \text{eV} = 0.030966\text{eV} = 30.966\text{meV} \approx 31\text{meV} \tag{14.38}$$

(d)

$$\left|E_{\text{Binding}}\right|_{\text{ZnS}} = \frac{\mu}{\varepsilon_r^2} \times C.F. = \frac{\mu}{(8.9)^2} \times C.F. = \frac{0.1756m_0}{79.21} \times \frac{13.6}{m_0} \text{eV} \tag{14.39}$$

since

$$\left|\mu\right|_{\text{ZnS}} = \frac{m_e^* m_h^*}{m_e^* + m_h^*} = m_0 \left(\frac{0.25 \times 0.59}{0.25 + 0.59}\right) = m_0 \frac{0.1475}{0.84} = 0.1756m_0 \tag{14.40}$$

Equation 14.39 becomes

$$\left|E_{\text{Binding}}\right|_{\text{ZnS}} = \frac{0.1756 \times 13.6}{79.21} \text{eV} = 0.0301497\text{eV} = 30.1497\text{meV} \approx 30.15\text{meV} \tag{14.41}$$

Example 14.4

Quantum size effect becomes noticeable when the difference between energy levels of electrons in a quantum dot of a given material is greater than or equal to the thermal energy $k_B T$ at room temperature = 27°C. To ascertain whether the effect is realized or not, we can equate the thermal energy with energy level separation to write the equation for electrons as

$$k_B T = \frac{\hbar^2 \pi^2}{2R^2 m_e^*} \tag{14.42}$$

From this criterion, determine the radii of quantum dots for which the quantum size effect is realized in (a) PbS ($m_e^* = 0.085m_0$), (b) CdTe ($m_e^* = 0.11m_0$), (c) ZnO ($m_e^* = 0.24m_0$), and (d) ZnS ($m_e^* = 0.25m_0$).

Solution: Equation 14.42 is re-arranged as

$$R^2 = \frac{\hbar^2 \pi^2}{2k_B T m_e^*} \tag{14.43}$$

$$\therefore R = \frac{\hbar\pi}{\sqrt{2k_B T m_e^*}} = \frac{\hbar\pi}{\sqrt{2k_B T}} \times \frac{1}{\sqrt{m_e^*}} = C.F. \times \frac{1}{\sqrt{m_e^*}} \tag{14.44}$$

where CF is the common factor in problems a, b, c, and d. Its value is

$$C.F = \frac{\hbar\pi}{\sqrt{2k_BT}} = \frac{1.05457\times10^{-34}\times3.14159}{\sqrt{2\times1.38065\times10^{-23}\times300}}\frac{J.s}{\sqrt{JT^{-1}T}} = \frac{3.313\times10^{-34}}{9.1016\times10^{-11}}\frac{\sqrt{J}}{s^{-1}}$$

$$= 0.364\times10^{-23}\frac{\sqrt{kgm^2s^{-2}}}{s^{-1}} = 3.64\times10^{-24}\sqrt{kg}\times m \tag{14.45}$$

Equation 14.44 will be applied by putting the value of CF from equation 14.45, and the effective mass of electron in the particular material under consideration from example 14.1.

(a)

$$|R|_{PbS} = C.F.\times\frac{1}{\sqrt{m_e^*}} = 3.64\times10^{-24}\sqrt{kg}\times m\times\frac{1}{\sqrt{0.085\times9.109\times10^{-31}kg}}$$

$$= \frac{3.64\times10^{-24}}{2.7826\times10^{-16}} = 1.308\times10^{-8}m = \frac{1.308\times10^{-8}}{1\times10^{-9}}m = 13.08nm \tag{14.46}$$

(b)

$$|R|_{CdTe} = C.F.\times\frac{1}{\sqrt{m_e^*}} = 3.64\times10^{-24}\sqrt{kg}\times m\times\frac{1}{\sqrt{0.11\times9.109\times10^{-31}kg}}$$

$$= \frac{3.64\times10^{-24}}{3.1654\times10^{-16}} = 1.1499\times10^{-8}m = \frac{1.1499\times10^{-8}}{1\times10^{-9}}m = 11.499nm \tag{14.47}$$

$$\approx 11.5nm$$

(c)

$$|R|_{ZnO} = C.F.\times\frac{1}{\sqrt{m_e^*}} = 3.64\times10^{-24}\sqrt{kg}\times m\times\frac{1}{\sqrt{0.24\times9.109\times10^{-31}kg}}$$

$$= \frac{3.64\times10^{-24}}{4.6756\times10^{-16}} = 0.7785\times10^{-8}m = \frac{0.7785\times10^{-8}}{1\times10^{-9}}m = 7.785nm \tag{14.48}$$

$$\approx 7.8nm$$

(d)

$$|R|_{ZnS} = C.F.\times\frac{1}{\sqrt{m_e^*}} = 3.64\times10^{-24}\sqrt{kg}\times m\times\frac{1}{\sqrt{0.25\times9.109\times10^{-31}kg}}$$

$$= \frac{3.64\times10^{-24}}{4.772054\times10^{-16}} = 0.76277\times10^{-8}m = \frac{0.76277\times10^{-8}}{1\times10^{-9}}m = 7.6277nm \tag{14.49}$$

$$\approx 7.6nm$$

Example 14.5

Find the bandgap energies of quantum dots of diameters 7nm and 2nm made from the following materials: (a) CdS and (b) CdSe. Bulk bandgaps of CdS and CdSe are 2.42 eV and 1.84eV. For CdS, $m_e^* = 0.21$ and $m_h^* = 0.8$; for CdSe, $m_e^* = 0.13$ and $m_h^* = 0.3m_0$. Dielectric constants of CdS and CdSe are 5.7 and 9.56 (Rodríguez-Mas et al 2020; ESI 2013; Elements of Advanced Theory).

Solution:

(a) For CdS,

$$m_e^* = 0.21m_0 = 0.21 \times 9.109 \times 10^{-31}\,\text{kg} = 1.9129 \times 10^{-31}\,\text{kg} \tag{14.50}$$

$$m_h^* = 0.8m_0 = 0.8 \times 9.109 \times 10^{-31}\,\text{kg} = 7.2872 \times 10^{-31}\,\text{kg} \tag{14.51}$$

The unit of second term in equation 14.11 is

$$\frac{(\text{J.s})^2}{\text{m}^2\text{kg}} = \frac{(\text{J})^2}{\text{kgm}^2\text{s}^{-2}} = \frac{(\text{J})^2}{\text{J}} = \text{J} = \frac{1}{1.602 \times 10^{-19}}\,\text{eV} = 6.242 \times 10^{18}\,\text{eV} \tag{14.52}$$

The unit of third term in equation 14.11 is

$$\frac{(\text{C})^2}{\text{Fm}^{-1} \times \text{m}} = \frac{(\text{C})^2}{\text{s}^4\text{A}^2\text{m}^{-2}\text{kg}^{-1} \times \text{m}^{-1} \times \text{m}} = \frac{(\text{C})^2}{\text{s}^4\dfrac{\text{C}^2}{\text{s}^2}\text{m}^{-2}\text{kg}^{-1} \times \text{m}^{-1} \times \text{m}} = \frac{1}{\text{s}^2\text{m}^{-2}\text{kg}^{-1}} \tag{14.53}$$

$$= \text{kgm}^2\text{s}^{-2} = \text{J} = 6.242 \times 10^{18}\,\text{eV}$$

Equation 14.11 will be applied successively for radii = 3.5nm, 1nm corresponding to quantum dot diameters of 7nm, 2nm. For the 3.5nm-radius quantum dot,

$$\left| E_G^{QD,\,CdS} \right|_{R=3.5\text{nm}}$$

$$= 2.42\text{eV}$$

$$+ \frac{\left(1.05457 \times 10^{-34}\right)^2 (3.14159)^2}{2\left(3.5 \times 10^{-9}\right)^2} \left(\frac{1}{1.9129 \times 10^{-31}} + \frac{1}{7.2872 \times 10^{-31}} \right) \times 6.242 \times 10^{18}\,\text{eV}$$

$$- \frac{1.8\left(1.602 \times 10^{-19}\right)^2}{4 \times 3.14159 \times 8.854 \times 10^{-12} \times 5.7 \times 3.5 \times 10^{-9}} \times 6.242 \times 10^{18}\,\text{eV} \tag{14.54}$$

$$= 2.42 + \frac{1.112 \times 9.8696 \times 6.242 \times 10^{-68+18+31+18}}{2 \times 12.25}(0.5228 + 0.1372)$$

$$- \frac{1.8 \times 2.566 \times 6.242 \times 10^{-38+12+9+18}}{4 \times 3.14159 \times 8.854 \times 5.7 \times 3.5}\,\text{eV}$$

$$= 2.42 + \frac{45.2139}{2 \times 12.25} \times 10^{-68+67} - \frac{28.83}{2219.6879} \times 10^{-38+39}$$

$$= 2.42 + 0.1845 - 0.12988 = 2.47462 \approx 2.47\text{eV}$$

Equations 14.50 and 14.51 have been used.
Similarly, for the 1nm radius quantum dot,

$$\left| E_G^{QD,\,CdS} \right|_{R=1\text{nm}} = 2.42 + \frac{45.2139}{2} \times 10^{-68+67} - \frac{1.8 \times 2.566 \times 6.242}{4 \times 3.14159 \times 8.854 \times 5.7 \times 1} \times 10^{-38+39}$$

$$= 2.42 + 2.260695 - \frac{28.83 \times 10}{634.1965} = 2.42 + 2.260695 - 0.4546 = 4.226095 \tag{14.55}$$

$$\approx 4.23\text{eV}$$

(b) For CdSe,

$$m_e^* = 0.13m_0 = 0.13 \times 9.109 \times 10^{-31}\,\text{kg} = 1.1842 \times 10^{-31}\,\text{kg} \tag{14.56}$$

$$m_h^* = 0.3m_0 = 0.3 \times 9.109 \times 10^{-31}\,\text{kg} = 2.7327 \times 10^{-31}\,\text{kg} \tag{14.57}$$

By equation 14.11, for the 3.5nm-radius quantum dot,

$$\left| E_G^{QD,\,CdSe} \right|_{R=3.5nm}$$
$$= 1.84\text{eV}$$
$$+ \frac{\left(1.05457 \times 10^{-34}\right)^2 (3.14159)^2}{2\left(3.5 \times 10^{-9}\right)^2} \left(\frac{1}{1.1842 \times 10^{-31}} + \frac{1}{2.7327 \times 10^{-31}}\right) \times 6.242 \times 10^{18}\,\text{eV}$$
$$- \frac{1.8\left(1.602 \times 10^{-19}\right)^2}{4 \times 3.14159 \times 8.854 \times 10^{-12} \times 9.56 \times 3.5 \times 10^{-9}} \times 6.242 \times 10^{18}\,\text{eV} \tag{14.58}$$
$$= 1.84 + \frac{1.112 \times 9.8696 \times 6.242 \times 10^{-68+18+31+18}}{2 \times 12.25}\left(0.84445 + 0.36594\right)$$
$$- \frac{1.8 \times 2.566 \times 6.242 \times 10^{-38+12+9+18}}{4 \times 3.14159 \times 8.854 \times 9.56 \times 3.5}\,\text{eV}$$
$$= 1.84 + \frac{82.919}{24.5} \times 10^{-68+67} - \frac{28.83}{3722.845} \times 10^{-38+39}$$
$$= 1.84 + 0.3384 - 0.07744 = 2.10096 \approx 2.1\text{eV}$$

Equations 14.56 and 14.57 have been used.
   Similarly, for the 1nm-radius quantum dot, equation 14.11 gives

$$\left| E_G^{QD,\,CdSe} \right|_{R=1nm} = 1.84 + \frac{82.919}{2} \times 10^{-68+67} - \frac{28.83}{4 \times 3.14159 \times 8.854 \times 9.56 \times 1} \times 10^{-38+39}$$
$$= 1.84 + 4.14595 - \frac{28.83}{1063.67} \times 10^{-38+39} = 1.84 + 4.14595 - 0.2710 \tag{14.59}$$
$$= 5.71495 \approx 5.7\text{eV}$$

Example 14.6

What radii of the following quantum dots will give a bandgap of 3eV: (a) CdS quantum dot and (b) CdSe quantum dot?

Solution:

(a) Equation 14.55 for CdS quantum dot has been obtained from equation 14.11, which has R^2 in the denominator of the second term and R in the denominator of third term. Since equation 14.55 is written for a quantum dot of radium 1nm with $R = 1$, $R^2 = 1$, denominators of second and third terms are both unity. For a quantum dot of radius R nm, denominator of second term will be R^2, and that of the third term will be R. So equation 14.55 is written for a quantum dot of radius R nm in the form

$$\left| E_G^{QD,\,CdS} \right|_{R=?} = 2.42 + \frac{2.260695}{R^2} - \frac{0.4546}{R} \tag{14.60}$$

Since the required bandgap is 3eV, we have

$$3 = 2.42 + \frac{2.260695}{R^2} - \frac{0.4546}{R} \tag{14.61}$$

or

$$3R^2 = 2.42R^2 + 2.260695 - 0.4546R \tag{14.62}$$

or

$$3R^2 - 2.42R^2 - 2.260695 + 0.4546R = 0 \tag{14.63}$$

or

$$0.58R^2 + 0.4546R - 2.260695 = 0 \tag{14.64}$$

$$\therefore R = \frac{-0.4546 \pm \sqrt{(0.4546)^2 - 4 \times 0.58 \times -2.260695}}{2 \times 0.58} = \frac{-0.4546 + \sqrt{0.20666 + 5.2448}}{1.16} \tag{14.65}$$

$$= \frac{-0.4546 + \sqrt{5.45146}}{1.16} = 1.6209 \text{nm}$$

rejecting negative sign before square root as inadmissible being the value of a length (radius).

(b) From equation 14.59, with the radius R expressed in m,

$$\left| E_G^{\text{QD, CdSe}} \right|_{R=?} = 1.84 + \frac{4.14595}{R^2} - \frac{0.2710}{R} \tag{14.66}$$

where we have applied the same arguments to equation 14.55 as we had done for CdS quantum dot in part a.

$$3 = 1.84 + \frac{4.14595}{R^2} - \frac{0.2710}{R} \tag{14.67}$$

or

$$3R^2 = 1.84R^2 + 4.14595 - 0.2710R \tag{14.68}$$

or

$$3R^2 - 1.84R^2 - 4.14595 + 0.2710R = 0 \tag{14.69}$$

or

$$1.16R^2 + 0.2710R - 4.14595 = 0 \tag{14.70}$$

$$\therefore R = \frac{-0.271 \pm \sqrt{(0.271)^2 - 4 \times 1.16 \times -4.14595}}{2 \times 1.16} = \frac{-0.271 + \sqrt{0.073441 + 19.2372}}{2.32} \tag{14.71}$$

$$= \frac{-0.271 + \sqrt{19.3106}}{2.32} = 1.777 \text{nm}$$

14.2 MULTIPLE EXCITON GENERATION (MEG)

14.2.1 Difference between Bulk Solar Cell and Quantum Dot Solar Cell

In a conventional bulk solar cell, a single photon produces a single electron-hole pair. It will be highly beneficial if a single photon could yield more than one electron-hole pair, because the short-circuit current of the solar cell will increase. But contrary to need, when a photon has larger energy than the bandgap of the semiconductor, it is absorbed to form photoexcited carriers carrying additional energy. These photoexcited carriers equilibrate with other carriers in the material to reach a carrier distribution whose temperature is above the lattice temperature of the material. Hence, these carriers are said to be "hot carriers." The excess energy corresponding to the elevated carrier temperature is mainly carried by electrons, which have a low effective mass. The hot electrons attain equilibrium with the lattice by transferring the excess energy to it. This is known as the thermalization decay of carriers to the conduction band edge. During this process, optical phonons are produced. These phonons interact with other phonons. Through these steps, the energy contained in sunlight photons above the bandgap energy is wasted as heat and not converted into electricity by the solar cell. In a nutshell, high-energy photons give rise to hot carriers whose energy is wasted as heat. No extra electron-hole pair is produced. Encouragingly, the same is not true for the quantum dot.

Quantum dots exhibit a remarkable property, which is likely to exert a profound influence on solar cell performance. In a quantum dot, a photon having energy higher than a threshold level, which could be as low as twice the bandgap energy of the quantum dot, gives rise to two electron-hole pairs in the form of two excitons. This phenomenon is called multiple exciton generation.

Example 14.7

Calculate the minimum energy of photons required to generate two excitons in the 3nm-radius quantum dots of the following semiconductors: (a) CdSe and (b) PbS. Take the bandgap of PbS = 0.41eV. Other parameters may be taken from the preceding examples.

Solution:

The minimum energy is twice the bandgap energy. So we calculate the bandgaps of the quantum dots of the two semiconductors having radius 3nm and multiply by 2 to get the minimum energy.

(a) Application of equations 14.11, 14.52, 14.53, 14.56, 14.57 gives

$$\left| E_G^{QD,\,CdSe} \right|_{R=3nm}$$

$$= 1.84 eV$$

$$+ \frac{\left(1.05457 \times 10^{-34} \right)^2 \left(3.14159 \right)^2}{2 \left(3 \times 10^{-9} \right)^2} \left(\frac{1}{1.1842 \times 10^{-31}} + \frac{1}{2.7327 \times 10^{-31}} \right) \times 6.242 \times 10^{18} eV$$

$$- \frac{1.8 \left(1.602 \times 10^{-19} \right)^2}{4 \times 3.14159 \times 8.854 \times 10^{-12} \times 9.56 \times 3 \times 10^{-9}} \times 6.242 \times 10^{18} eV \tag{14.72}$$

$$= 1.84 + \frac{1.112 \times 9.8696 \times 6.242 \times 10^{-68+18+31+18}}{2 \times 9} \left(0.84445 + 0.36594 \right)$$

$$- \frac{1.8 \times 2.566 \times 6.242 \times 10^{-38+12+9+18}}{4 \times 3.14159 \times 8.854 \times 9.56 \times 3} eV$$

$$= 1.84 + \frac{82.919}{18} \times 10^{-68+67} - \frac{28.83}{3191.0099} \times 10^{-38+39}$$

$$= 1.84 + 0.46066 - 0.0903475 = 2.2103 \approx 2.2 eV$$

Hence, the threshold energy for CdSe quantum dot is

$$E_{\text{Threshold}} = 2 \times 2.2 = 4.4\text{eV} \tag{14.73}$$

(b) For PbS

$$m_e^* = m_h^* = 0.085m_0 = 0.085 \times 9.109 \times 10^{-31}\text{kg} = 0.774265 \times 10^{-31}\text{kg} \tag{14.74}$$

With the help of equations 14.11, 14.52, 14.53, 14.74, we get

$$\left| E_G^{\text{QD, PbS}} \right|_{R=3\text{nm}}$$
$$= 0.41\text{eV}$$
$$+ \frac{\left(1.05457 \times 10^{-34}\right)^2 (3.14159)^2}{2\left(3 \times 10^{-9}\right)^2} \left(\frac{1}{0.774265 \times 10^{-31}} + \frac{1}{0.774265 \times 10^{-31}} \right) \times 6.242 \times 10^{18}\text{eV}$$
$$- \frac{1.8\left(1.602 \times 10^{-19}\right)^2}{4 \times 3.14159 \times 8.854 \times 10^{-12} \times 17.2 \times 3 \times 10^{-9}} \times 6.242 \times 10^{18}\text{eV} \tag{14.75}$$
$$= 0.41 + \frac{1.112 \times 9.8696 \times 6.242 \times 10^{-68+18+31+18}}{2 \times 9}(1.2915 + 1.2915)$$
$$- \frac{1.8 \times 2.566 \times 6.242 \times 10^{-38+12+9+18}}{4 \times 3.14159 \times 8.854 \times 17.2 \times 3}\text{eV}$$
$$= 0.41 + \frac{176.95}{18} \times 10^{-68+67} - \frac{28.83}{5741.148} \times 10^{-38+39}$$
$$= 0.41 + 0.983 - 0.05022 = 1.34278 \approx 1.34\text{eV}$$

Hence, the threshold energy for PbS quantum dot is

$$E_{\text{Threshold}} = 2 \times 1.34 = 2.68\text{eV} \tag{14.76}$$

14.2.2 REASON FOR GREATER LIKELIHOOD OF MEG IN A QUANTUM DOT

What makes the occurrence of MEG in a QD more likely than in a bulk semiconductor? This happens because the threshold energy for MEG in a bulk semiconductor is much higher than in a QD. So chances of any photon in solar radiation exceeding this value are remote. Plausibly, the condition for MEG is difficult to be achieved in a bulk material. As an example, the threshold energy for MEG is 3.4 × the bandgap energy in a quantum dot of PbSe, while the same is 6.4 × the bandgap energy in bulk PbSe material (Tian and Cao 2013).

Why is the threshold energy for MEG lower in a QD than in a bulk semiconductor? Several reasons are offered:

(i) Electrons and holes are not free carriers in a QD as in bulk material. Instead, they exist as excitons.

(ii) The discreteness of energy levels occupied by electrons and holes in a quantum dot slows down the electron and hole cooling rates as compared to a bulk semiconductor so that MEG is more preferred in a QD than energy loss through cooling. To explain more clearly, the time taken by hot carriers to cool down is found to be longer in quantum-confined structures, such as quantum dots, as compared to bulk semiconductors. Typically, the cooling times in these nanostructures are in the range of tens of nanoseconds. The delayed cooling arises from a phonon bottleneck effect in these structures. This effect is understood by recalling that the carrier cooling takes place by interaction of hot carriers with optical phonons. The different response of quantum-confined structures is a consequence of the fact that the phonons are unable to attain equilibrium with the lattice rapidly in these structures. Hence, the cooling of hot carriers is slackened. On these grounds, the quantum dot structures can be utilized for hot carrier extraction.

(iii) Auger mechanisms are enhanced in a QD over a bulk semiconductor owing to the Coulomb interaction between electrons and holes. This enhancement favors the generation of multiple excitons.

14.2.3 Corresponding Terms for a Bulk Semiconductor and a Quantum Dot

In a bulk semiconductor, we are accustomed to calling such multiple-carrier generation as impact ionization. Therefore, MEG in a quantum dot is the counterpart of impact ionization in a bulk semiconductor. On the same line, MEG in a quantum dot is the inverse of Auger recombination in a bulk semiconductor. In Auger process, an electron recombines with a hole, and the excess energy of the electron is delivered to another electron or hole. The electron or hole receiving this energy is promoted to a higher energy state in the respective band. In MEG, a bound electron-hole pair is created by a photon having energy in excess of the bandgap energy. This is usual and expected. But the photon impinging on the semiconductor has excess energy above the bandgap energy. This extra energy is gainfully utilized to produce another bound electron-hole pair if the energy of the photon is greater than the threshold value for MEG.

Despite the fact that MEG provides a larger number of charge carriers, it must be noted that photon energies lying between the bandgap energy of the semiconductor and the threshold energy for MEG are still not tapped because this phenomenon occurs only when the photon energy exceeds the threshold energy.

14.3   DRAWING ENERGY BAND DIAGRAMS OF HETEROJUNCTIONS

After perusal of bandgap tuning and multiple electron generation, let us divert our attention to drawing energy band diagrams of heterojunctions in preparation for the ensuing Chapter 15.

14.3.1 Rules and Considerations in the Construction of Energy Band Diagrams of Heterojunctions

Anderson's rule, or electron affinity rule, is applied to draw the energy band diagrams. According to this rule, the vacuum energy level is the reference level. This reference level should be aligned for the two materials. Then, the electron affinity of each material, the distance of the conduction band edge from the vacuum energy level, is used to place the conduction band edge of each material on the band diagram. Using the electron affinity values of the two materials, the lines representing the conduction band edges of the two materials are drawn. Next, the bandgap of each material, the distance between the edges of conduction and valence bands, allows determination of locations of their valence band edges on the energy band diagram. Using the bandgaps of the two materials, the lines representing the valence band edges of the two materials are drawn. After the conduction and valence band edges of both materials have been drawn, the discontinuities of conduction band (ΔE_C) and valence band (ΔE_V) are easily calculated; indeed, they are obvious by looking at the diagram.

The positions of Fermi energy levels for the two materials are also indicated on the band diagram. The Fermi level position is given by the work function of the material, the energy required to remove an electron from the chemical bond, and move it to infinity. On the energy band diagram, the work function is the distance between the vacuum energy level and Fermi level. So the Fermi levels are placed from knowledge of work functions of the materials.

Once the band alignment diagram has been constructed by Anderson's rule, the Fermi energy levels for the two materials must be aligned. For making this alignment, one side of the diagram with all its energy levels (the vacuum level, the band edges, and the Fermi level) is moved, and this complete side is placed at a position where the Fermi levels on the two sides are in one straight line. During Fermi level alignment, the bands undergo bending, upward or downward. What does an upward bending of the conduction band edge of a material show? It tells us that electrons are transferred from this material to the other material. Consequently, these states become vacant near the interface, but not in the remaining bulk material. A downward bending of the conduction band edge implies gain of electrons so that more energy states are filled adjoining the interface than in the bulk of the material. Consequent upon the transference of electrons across the interface, there is a depletion of electrons on one side and an accumulation of electrons on the other side. The depleted region is positively charged, and the accumulated region is negatively charged. The electron transference takes place as long as further electron transfer is opposed. The repelling force of already-congregated electrons on opposite side opposes and thus prevents further electron transfer. This is the equilibrium condition. Now, oppositely charged regions exist on the two sides of the interface. The existence of these regions is tantamount to the establishment of an interfacial potential or built-in electric field.

14.3.2 DRIVING ENERGY FOR CHARGE TRANSFER ACROSS A HETEROJUNCTION

The band offset provides the driving energy. The band offset between two materials comprising a heterojunction consists of contributions from two factors:

(i) The discontinuities between conduction band edges and valence band edges of the materials, arising from the difference in bandgaps of the materials.
(ii) The built-in potential or electric field originating from charge imbalance across the interface.

For factor *i*, let us look at the energy difference between the band edges of the two materials. A point located at a lower energy on the energy band diagram is associated with a higher potential because more work is necessary to take a unit positive charge from this point and place it at infinity. Conversely, a point located at a higher energy on the band diagram is associated with a lower potential. Thus, an electric field always exists across the interface. It is directed from the lower-energy, higher-potential point to the higher-energy, lower-potential point. Under its influence, electrons move from the higher-energy, lower-potential point to the lower-energy, higher-potential point, and holes move in the opposite direction. This happens as long as an equilibrium state is reached. However, whenever additional carriers are generated, e.g., by illumination with sunlight, equilibrium is disturbed and transfer of relevant carriers is promoted in the aforesaid directions. It must also be noted that the band discontinuities also serve as energy barriers or potential barriers opposing the motion of charge carriers in a manner decided by their polarities, e.g., electrons from lower-energy, higher-potential to higher-energy, lower-potential point, and similarly for holes.

For factor *ii*, the underlying cause is already discussed in the context of band bending. To elucidate the reason of electron transfer, we note that electrons are transferred from the lower-work-function material to the higher-work-function material. This happens because electrons require less energy to be liberated from the chemical bonds of lower-work-function material than from those of higher-work-function material. As a result, the lower-work-function material becomes positively charged, and the higher-work-function material acquires a negative charge. Thus, an electric field is set up. It is directed from the positively charged, lower-work-function material to the negatively charged, higher-work-function material. When the heterojunction is illuminated with sunlight and electron-hole pairs are produced, the electrons are pulled towards the positively charged, lower-work-function material, while holes migrate to the negatively charged, higher-work-function material.

Thus, in a heterojunction solar cell, the combined effect of factors *i* and *ii* is responsible for separation of electrons and holes and transporting them to the two terminals of the device.

Example 14.8

Draw the energy band diagram of the ITO/PbS QD interface when (a) far apart and (b) in contact by applying Anderson's rule, and (c) in contact and after alignment of Fermi levels in the two materials. Indicate the values of the various energy level parameters taken for drawing the diagrams.

Solution: The energy band diagrams are shown in Figure 14.1.

Note: For solving examples 14.8–14.16, a literature survey may be performed and necessary values of parameters in the band structures of materials may be searched.

Example 14.9

Draw the energy band diagrams of PbS QD/Al interface (b) before contact between PbS QD and Al, (c) after contact by applying Anderson's rule, and (c) after contact and alignment of Fermi levels between PbS QD and Al.

Solution: The energy band diagrams are shown in Figure 14.2.

Example 14.10

Draw the energy band diagrams of ITO/PbS QD/Al solar cell based on Anderson's rule.

Solution: The energy band diagrams are shown in Figure 14.3. They are obtained by combining the diagrams of Figures 14.1 and 14.2.

Example 14.11

Draw the energy band diagram of ITO/PbS QD/Al by combining the energy band diagrams of ITO/PbS QD and PbS QD/Al from Figures 14.1 and 14.2 and aligning the Fermi levels.

Solution: The energy band diagrams are shown in Figure 14.4.

Example 14.12

Draw the energy band diagram of the FTO/TiO$_2$ interface when (a) far apart and (b) in contact by applying Anderson's rule.

Solution: The energy band diagrams are shown in Figure 14.5.

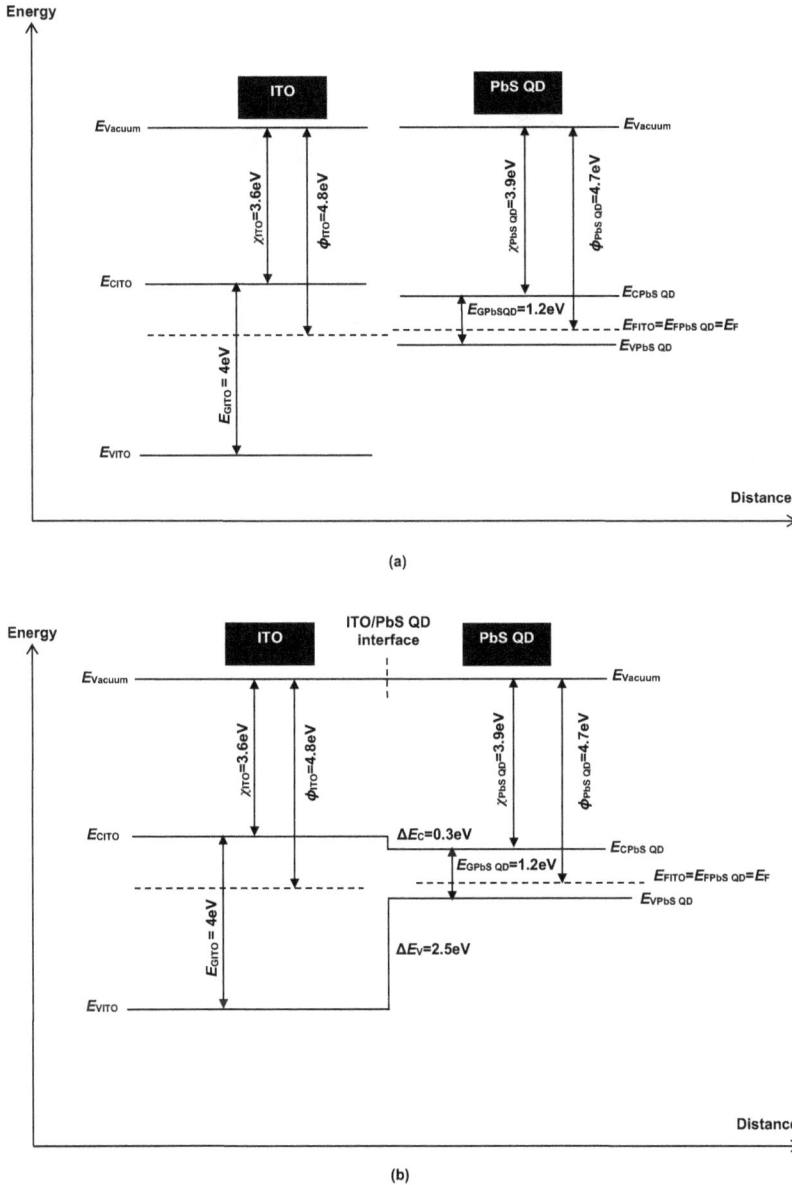

FIGURE 14.1 Energy band diagram of the ITO/PbS QD interface when (a) far apart and (b) in contact, and by applying Anderson's rule, and (c) in contact and after alignment of Fermi levels in the two materials. For ITO, work function $\phi_{ITO} = 4.8eV$, the electron affinity $\chi_{ITO} = 3.6eV$, and the energy gap $E_{GITO} = 4eV$. Various parameters such as energy gap and electron affinity of PbS QDs taken in drawing the diagram differ from bulk values, e.g., the bandgap E_G of PbS QD is taken as $E_{GPbSQD} = 1.2eV$ in place of $0.4eV$ for bulk PbS, the electron affinity χ_{PbSQD} for PbS QD is taken as $-3.9eV$ against $-4.6eV$ for bulk PbS. The work function ϕ_{PbSQD} of PbS QD is chosen to show P-type conduction. E_{Vacuum} is the vacuum energy level. For ITO, E_{CITO} is the energy of conduction band edge in ITO, E_{VITO} is the energy of valence band edge in ITO, and E_{FITO} is the Fermi level energy in ITO. In case of PbS QD, $E_{CPbS\,QD}$ is the energy of conduction band edge in PbS QD, $E_{VPbS\,QD}$ is the energy of valence band edge in PbS QD, $E_{FPbS\,QD}$ is the Fermi level energy in PbS QD. When the two materials come in contact, the Fermi energy levels are aligned. The energy bands bend downwards in ITO and upwards in PbS QD, indicating that electrons are transferred from PbS QD to ITO. PbS QD loses electrons and becomes positively charged, whereas ITO gains electrons and is negatively charged. A space charge region is formed. It is associated with a built-in electric field directed from PbS QD to ITO. The equilibrium energy level alignment in Figure *c* is predicted assuming a hypothetical flat-vacuum alignment in Figures *a* and *b*.

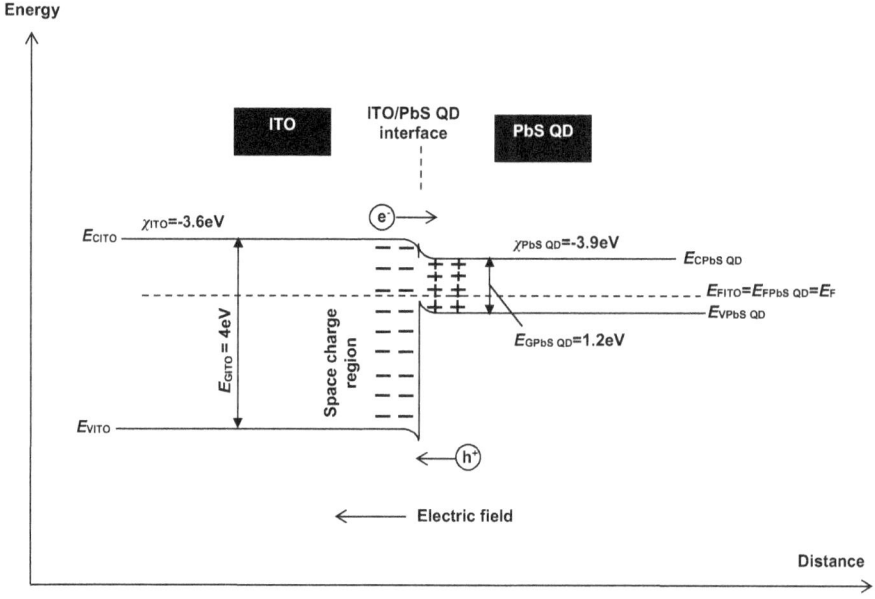

(c)

FIGURE 14.1 (Continued)

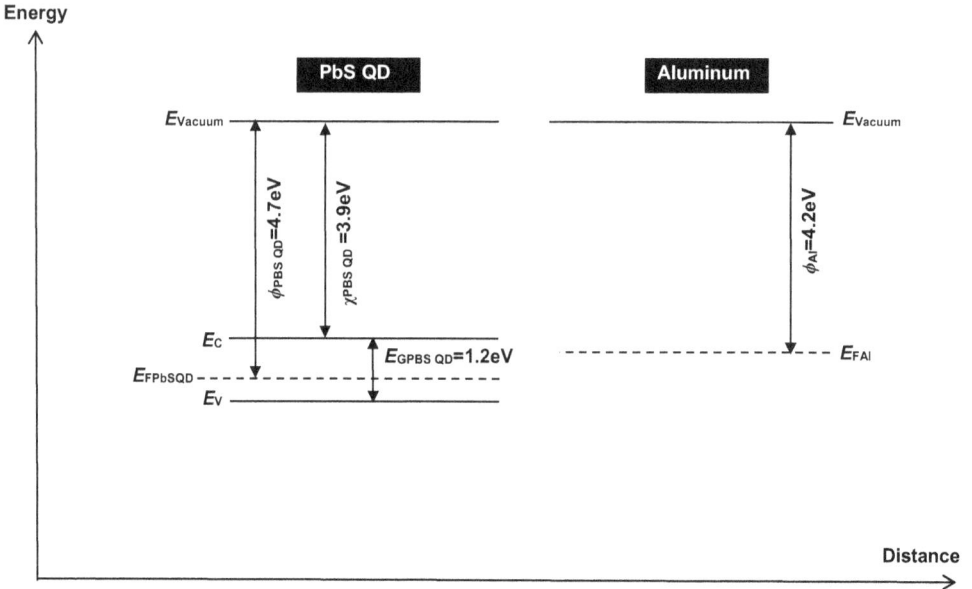

(a)

FIGURE14.2 Energy band diagrams of PbS QD/Al interface: (a) before contact between PbS QD and Al, (b) after contact by applying Anderson's rule, and (c) after contact and alignment of Fermi levels between PbS QD and Al. The values of parameters used for band structure of PbS QD are work function $\phi_{PbS\,QD} = 4.7\text{eV}$, the electron affinity $\chi_{PbS\,QD} = 3.9\text{eV}$, and the energy gap $E_{GPbS\,QD} = 1.2\text{eV}$. It is seen that the electric field due to band bending is acting from aluminum to PbS QD. It works on the charge carriers, providing effective separation.

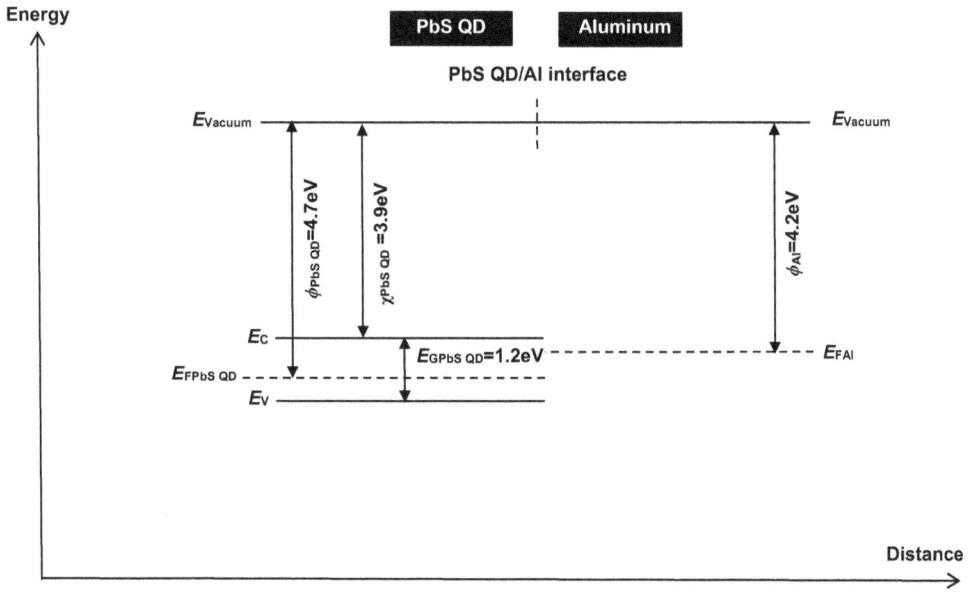

(b)

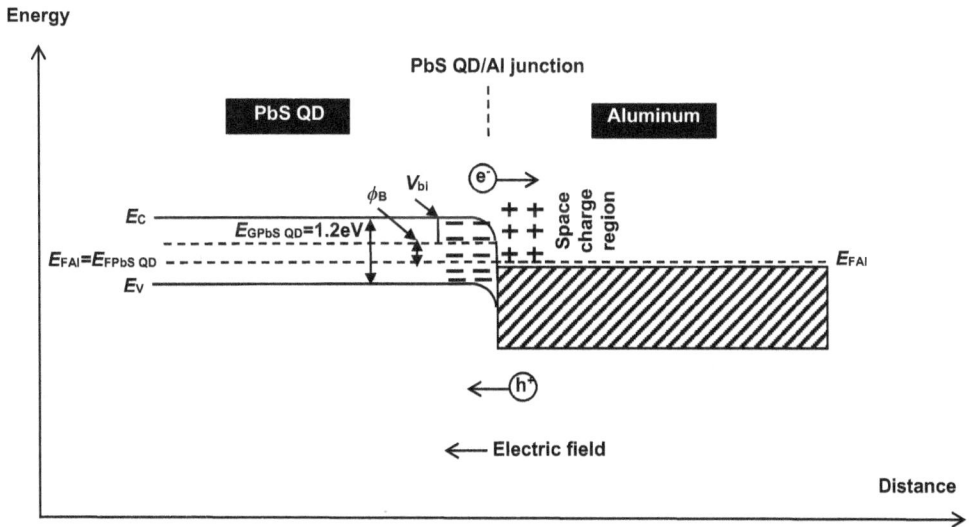

(c)

FIGURE 14.2 (Continued)

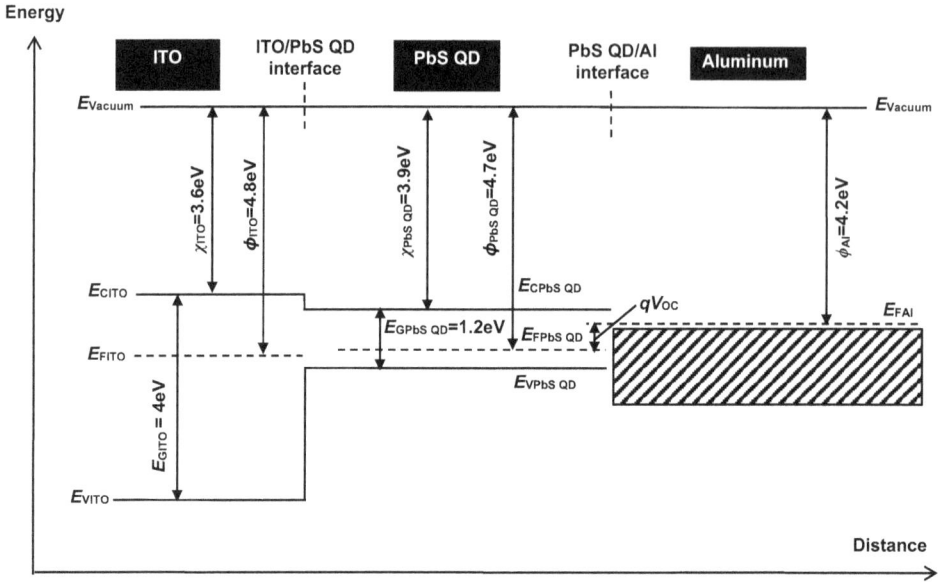

(a)

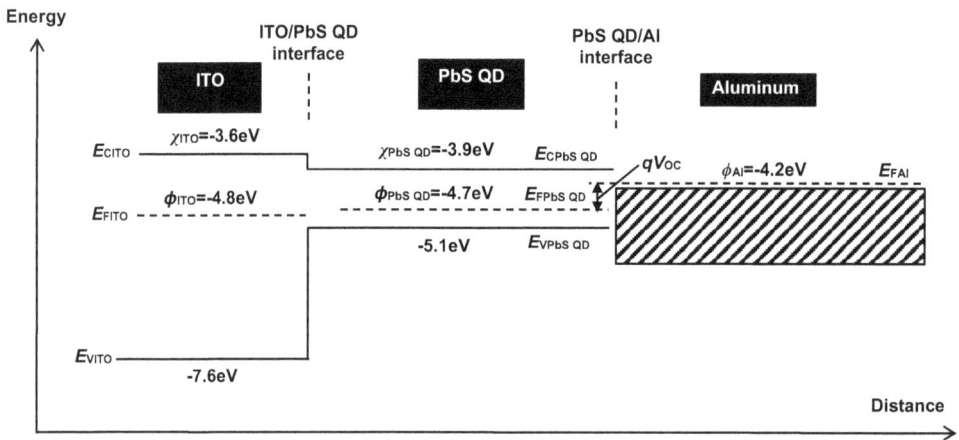

(b)

FIGURE 14.3 Energy band diagrams of ITO/PbS QD/Al solar cell based on Anderson's rule. They are obtained by combining the diagrams of Figures 14.1 and 14.2. Fermi levels are kept unaligned. V_{OC} is the open-circuit voltage of the cell. In a, the energy levels are shown with respect to the vacuum level. In b, the vacuum level is not shown, but the values of energies are prefixed with a negative sign to indicate the depths of the energy levels with reference to the vacuum energy level. In c, symbols for various energy levels are removed by considering them as implied, and boxes are drawn to represent the forbidden gaps of the materials.

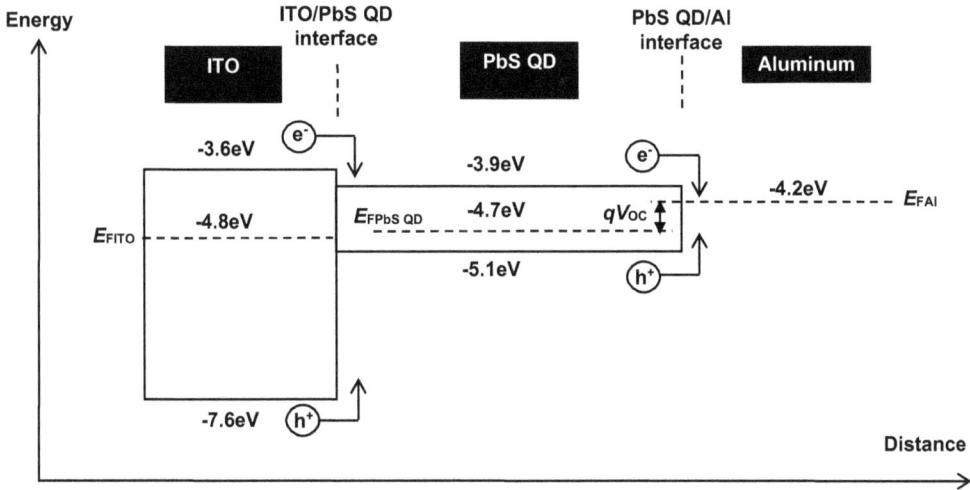

(c)

FIGURE 14.3   (Continued)

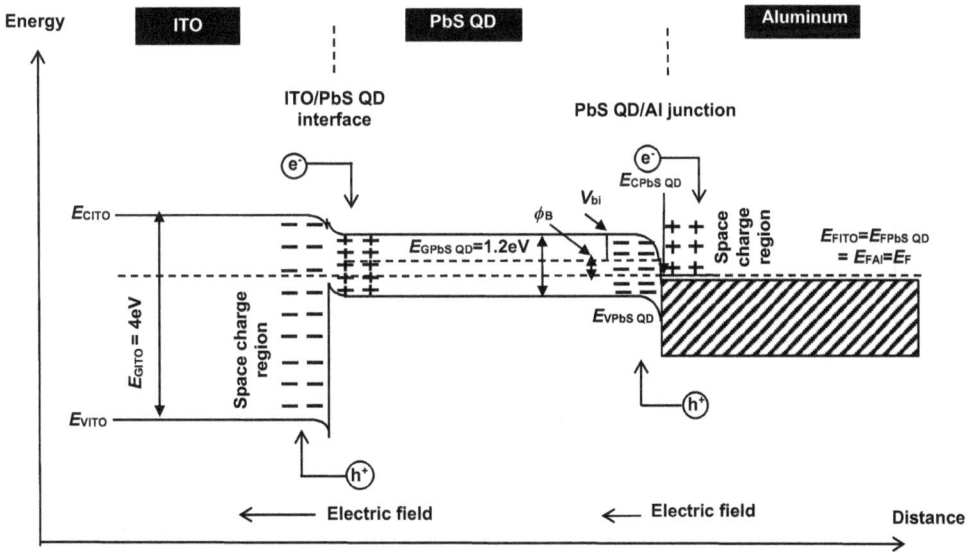

FIGURE 14.4   Energy band diagram of ITO/PbS QD/Al obtained by combining the energy band diagrams of ITO/PbS QD and PbS QD/Al from Figures 14.1 and 14.2 and aligning the Fermi levels. A unidirectional electric field exists across ITO/PbS QD/Al. Its direction is from Al to ITO.

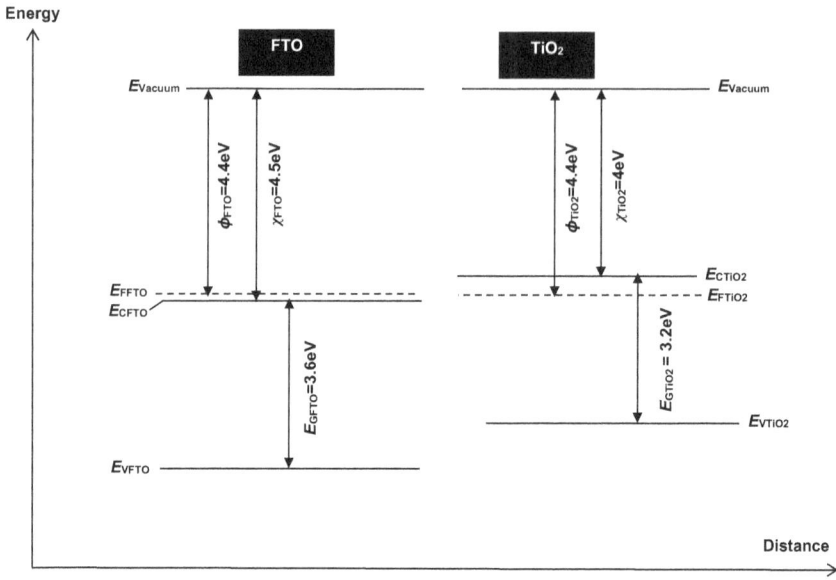

(a)

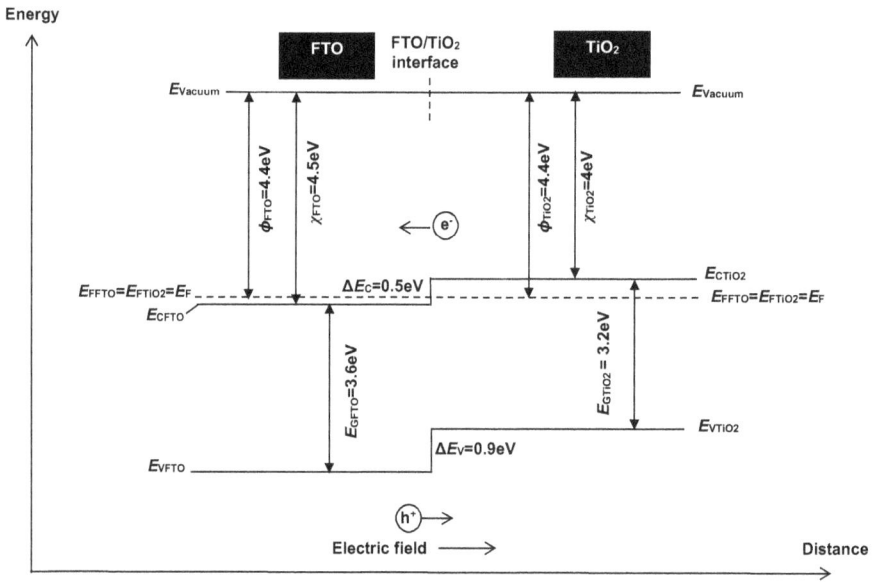

(b)

FIGURE 14.5   Energy band diagram of the FTO/TiO$_2$ interface when (a) far apart and (b) in contact by applying Anderson's rule. For FTO, the work function ϕ_{FTO} = 4.4eV, the electron affinity χ_{FTO} = 4.5eV, and the energy gap E_{GFTO} = 3.6eV. The values for TiO$_2$ are work function ϕ_{TiO2} = 4.4eV, the electron affinity χ_{TiO2} = 4eV, and the energy gap E_{GTiO2} = 3.2eV. E_{Vacuum} is the vacuum energy level. For FTO, E_{CFTO} is the energy of conduction band edge in FTO, E_{VFTO} is the energy of valence band edge in FTO, and E_{FFTO} is the Fermi level energy in FTO. In case of TiO$_2$, E_{CTiO2} is the energy of conduction band edge in TiO$_2$, E_{VTiO2} is the energy of valence band edge in TiO$_2$, and E_{FTiO2} is the Fermi level energy in TiO$_2$. When the two materials come in contact, the Fermi energy levels are already aligned. So there is no bending of bands at the interface.

Example 14.13

Draw the energy band diagram of the TiO$_2$/PbS QD interface when (a) far apart and (b) in contact, and by applying Anderson's rule, and (c) in contact and after alignment of Fermi levels in the two materials.
Solution: The energy band diagrams are shown in Figure 14.6.

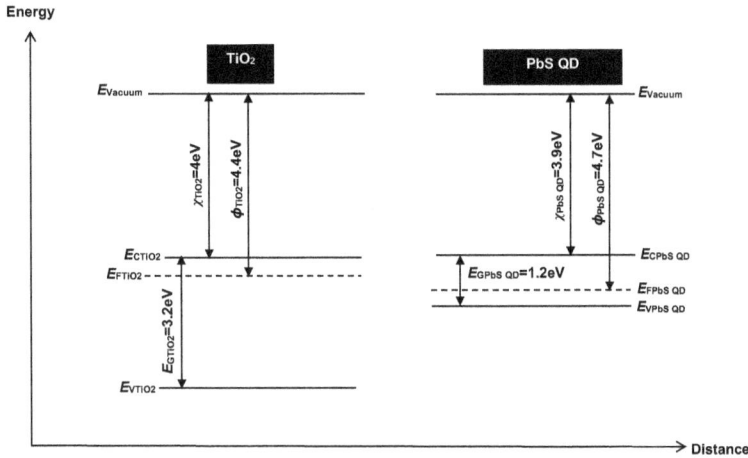

(a)

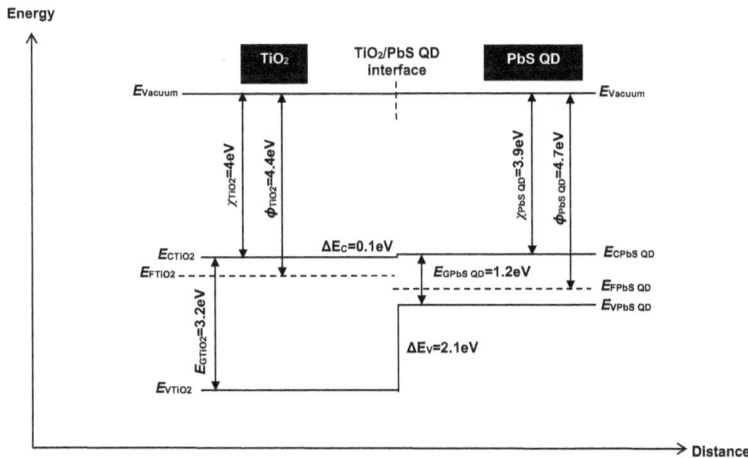

(b)

FIGURE 14.6 Energy band diagram of the TiO$_2$/PbS QD interface when (a) far apart and (b) in contact, and by applying Anderson's rule, and (c) in contact and after alignment of Fermi levels in the two materials. For TiO$_2$, work function $\phi_{TiO2} = 4.4$eV, the electron affinity $\chi_{TiO2} = 4$eV, and the energy gap $E_{GTiO2} = 3.2$eV. The values for PbS QD are work function $\phi_{PbS\ QD} = 4.7$eV, the electron affinity $\chi_{PbS\ QD} = 3.9$eV, and the energy gap $E_{GPbS\ QD} = 1.2$eV. E_{Vacuum} is the vacuum energy level. For TiO$_2$, E_{CTiO2} is the energy of conduction band edge in TiO$_2$, E_{VTiO2} is the energy of valence band edge in TiO$_2$, and E_{FTiO2} is the Fermi level energy in TiO$_2$. In case of PbS QD, $E_{CPbS\ QD}$ is the energy of conduction band edge in PbS QD, $E_{VPbS\ QD}$ is the energy of valence band edge in PbS QD, and $E_{FPbS\ QD}$ is the Fermi level energy in PbS QD. When the two materials come in contact, the Fermi energy levels are aligned. The energy bands bend upwards in TiO$_2$ and downwards in PbS QD, indicating that electrons are transferred from TiO$_2$ to PbS QD. TiO$_2$ is positively charged by losing electrons, whereas PbS QD is negatively charged by gaining electrons. A space charge region is formed. The space-charge region is associated with a built-in electric field directed from TiO$_2$ to PbS QD.

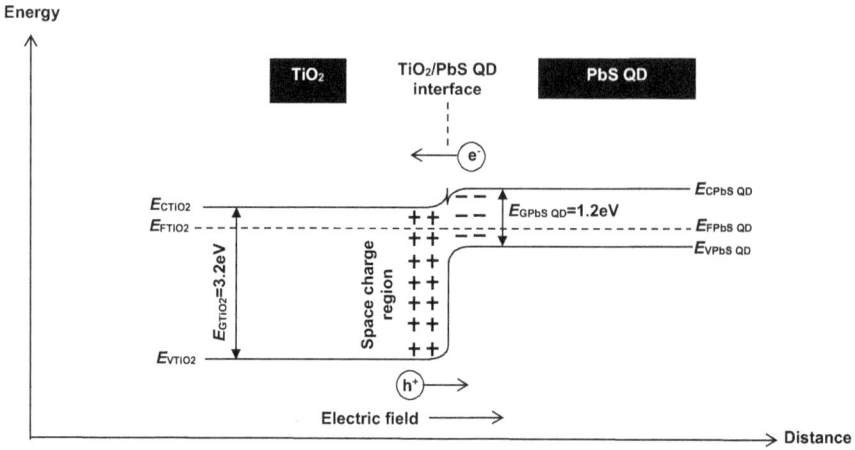

(c)

FIGURE 14.6   (Continued)

Example 14.14

Draw the energy band diagram of the PbS QD/Au interface when (a) far apart and (c) in contact, and after Fermi level alignment.

Solution: The energy band diagrams are shown in Figure 14.7.

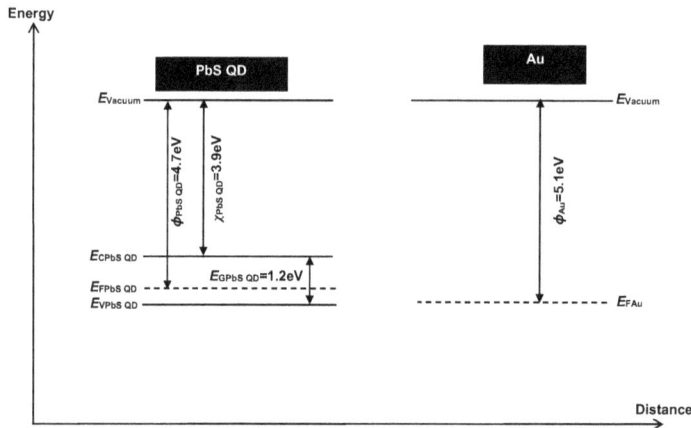

(a)

FIGURE 14.7   Energy band diagram of the PbS QD/Au interface when (a) far apart and (b) in contact, and after Fermi level alignment. For PbS QD, work function $\phi_{PbS\,QD} = 4.7eV$, the electron affinity $\chi_{PbS\,QD} = 3.9eV$, and the energy gap $E_{GPbS\,QD} = 1.2eV$. In case of Au, the values are work function $\phi_{Au} = 5.1eV$, and the energy gap $E_{GAu} = 0$ eV. E_{Vacuum} is the vacuum energy level. For PbS QD, $E_{CPbS\,QD}$ is the energy of conduction band edge in PbS QD, $E_{VPbS\,QD}$ is the energy of valence band edge in PbS QD, and $E_{FPbS\,QD}$ is the Fermi level energy in PbS QD. For Au, E_{CAu} is the energy of conduction band edge in Au, and E_{FAu} is the Fermi level energy in Au. When the two materials come in contact, the Fermi energy levels are aligned. The energy bands bend upwards in PbS QD and downwards in Au, indicating that electrons are transferred from PbS QD to Au. PbS QD loses electrons and becomes positively charged, whereas Au gains electrons and is negatively charged. A space charge region is formed. The space charge region is associated with a built-in electric field directed from PbS QD to Au

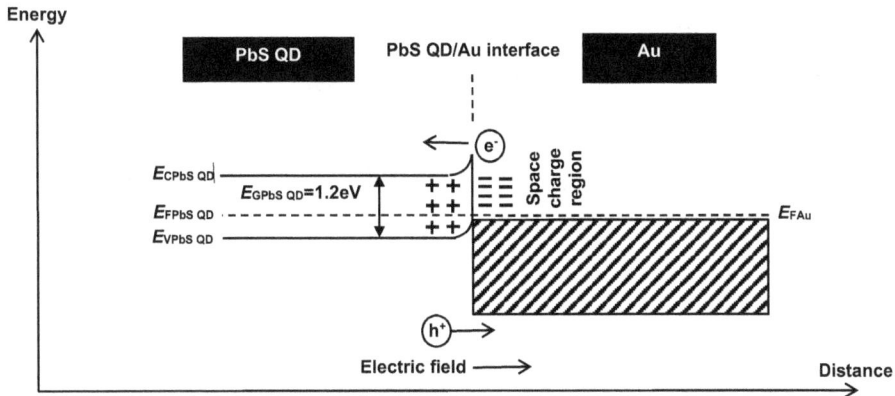

(b)

FIGURE 14.7 (Continued)

Example 14.15

Draw the energy band diagrams of FTO/TiO₂/PbS QD/Au solar cell based on Anderson's rule.

Solution: The energy band diagrams are shown in Figure 14.8. These diagrams are obtained by combining the diagrams of Figures 14.5–14.7.

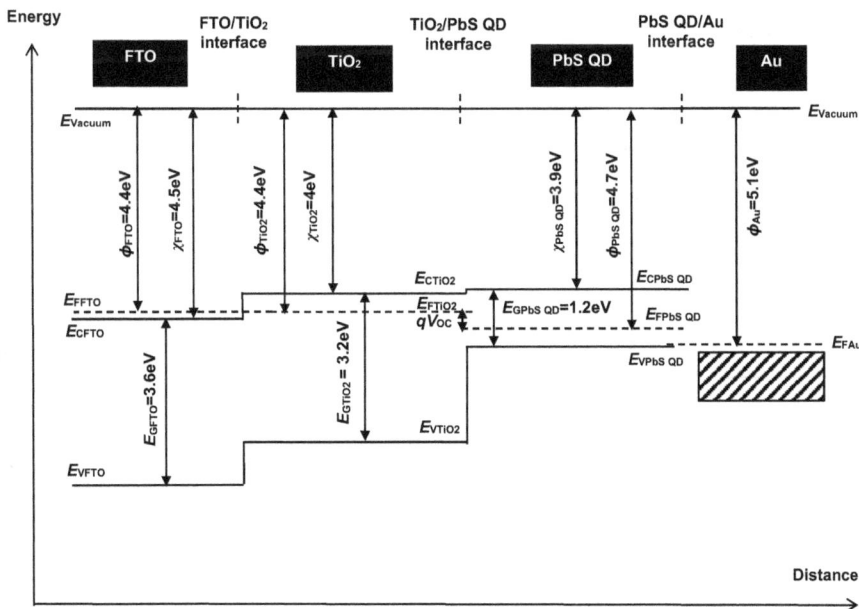

(a)

FIGURE 14.8 Energy band diagrams of FTO/TiO₂/PbS QD/Au solar cell based on Anderson's rule. These diagrams are obtained by combining the diagrams of Figures 14.5–14.7. The qV_{OC} is the open-circuit voltage of the cell; q is the electronic charge. In a, the vacuum energy level is marked. In b, the vacuum energy level is not shown. The energy levels are shown as negative, indicating their energy difference with respect to the vacuum energy level. In c, rectangular boxes are drawn to show the energy gaps of the materials. Symbols for work functions are also removed, considering these parameters as implied.

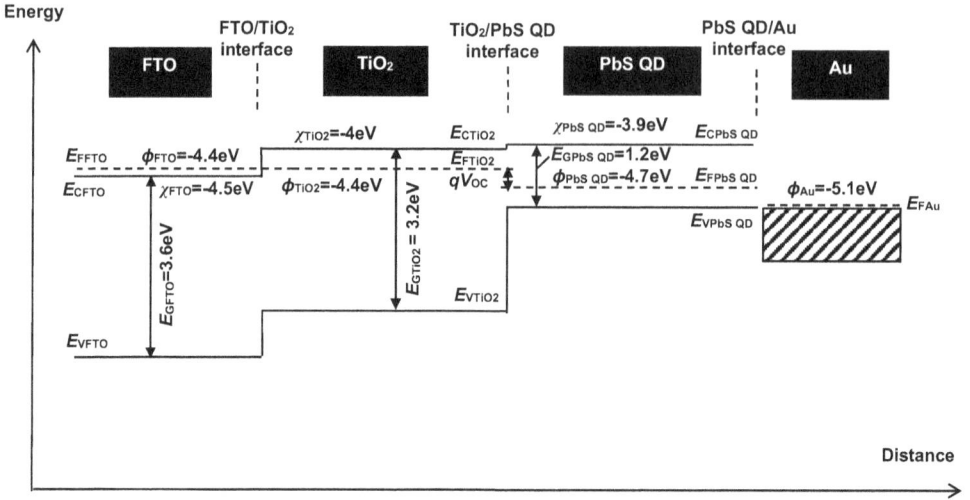

(b)

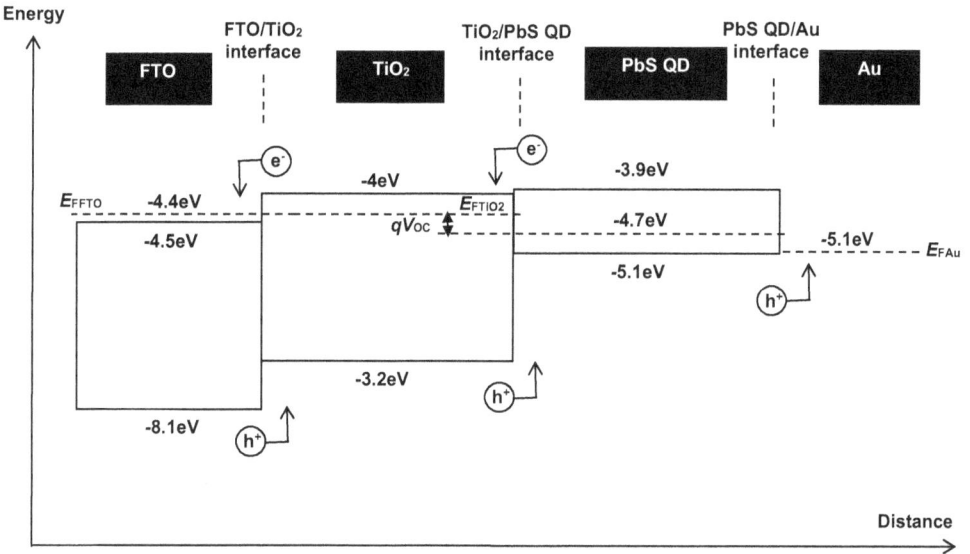

(c)

FIGURE 14.8   (Continued)

Example 14.16

Draw the energy band diagram of the FTO/TiO$_2$/PbS QD/Au obtained by combining the energy band diagrams of Figures 14.5–14.7, and aligning the Fermi levels.

Solution: The energy band diagrams are shown in Figure 14.9.

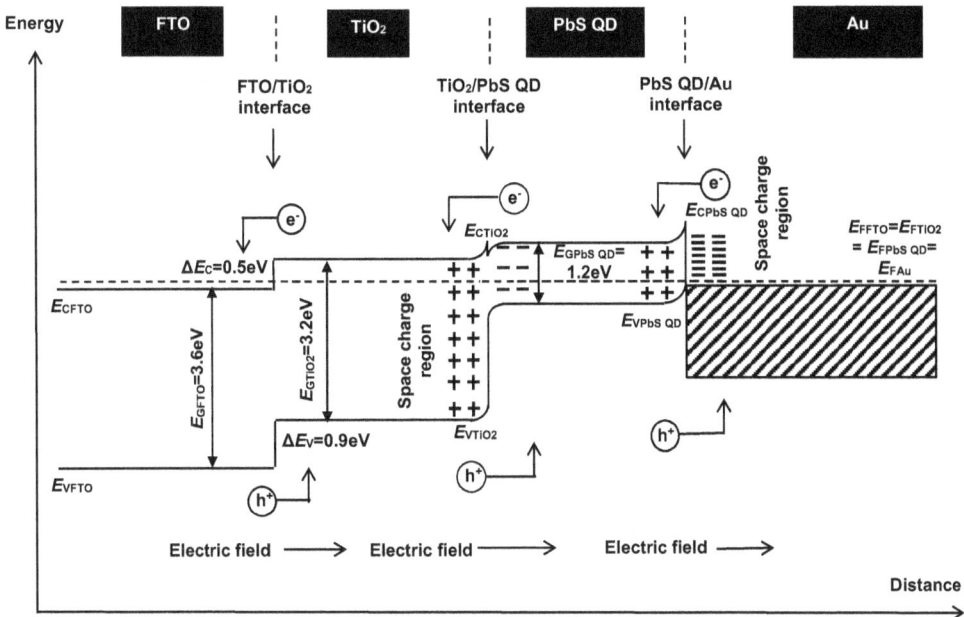

FIGURE 14.9 Energy band diagram of the FTO/TiO$_2$/PbS QD/Au obtained by combining the energy band diagrams of Figures 14.5–14.7, and aligning the Fermi levels. The diagrams show that electric fields are created at each interface: FTO/TiO$_2$, TiO$_2$/PbS QD, and PbS QD/Au, favoring conventional current flow from FTO to Au film, and electronic current in the opposite direction. Illumination of the solar cell with sunlight results in production of electron-hole pairs. The negative- and positive-charge carriers are separated, with electrons moving towards the FTO and holes towards Au.

14.4 DISCUSSION AND CONCLUSIONS

The special properties of quantum dots used in solar cells, namely, bandgap tuning and multicarrier generation have enabled the development of solar cells with improved characteristics. Owing to these special features, a variety of solar cells have been developed. The forthcoming chapter will describe the fabrication of solar cells using quantum dots.

Questions and Answers

14.1 What is bandgap tuning? Answer: It is a method to produce materials with different electronic and optical properties. For compound semiconductors, the composition of the semiconductor is changed through variations in percentages of its constitutional elements to alter the bandgap. For nanomaterials such as quantum dots, the size of the nanomaterial being talked about is varied, e.g., in the case of a quantum dot, the smaller the quantum dot, the larger is its bandgap.

14.2 What is multiple exciton generation? Answer: It is the phenomenon in which illumination of a semiconductor with a single high-energy photon leads to the production of multiple electron-hole pairs. This happens when the photon energy is much larger than the bandgap energy of the semiconductor material used in the solar cell.

14.3 What is meant by electron temperature? Answer: Electron temperature is a statistical temperature accounting for the observed energy of electrons assuming it to be purely kinetic. In absence of thermal equilibrium of electrons with their surroundings, the electron temperature may be higher than the surrounding temperature.

14.4   What is lattice temperature? Answer: Temperature of bulk crystal.

14.5   What is a hot carrier? Answer: An electron or hole at a temperature higher than lattice temperature. For an electron to be hot, the electron temperature is > lattice temperature.

14.6   Name two nanoscale properties which make quantum dots attractive for solar cell applications. Answer: (i) bandgap tunability, (b) multiple exciton generation.

14.7   In what way does multiple exciton generation influence the operation of solar cells? Answer: By utilizing the high-energy photons in producing additional electron-hole pairs, which increase the photocurrent of the solar cell. In conventional solar cells, these photons produce hot carriers, i.e., carriers at temperatures above the lattice temperature.

14.8   How is the particle-in-a-box model applied to a quantum dot? Answer: The model is applied by considering the quantum dot as a cubical box of dimensions L_x, L_y, L_z, writing the energy equations in X-, Y-, and Z-directions, and finally, putting $L_x = L_y = L_z = R$ = radius of a sphere with the understanding that the cubical box is replaced by a spherical box of radius R.

14.9   How is effective bandgap of a quantum dot related to its bulk bandgap? Answer: Effective bandgap is larger than the bulk bandgap of quantum dot material, and the increment in bandgap equals the sum total of electron and hole energies in the ground state of the quantum dot.

14.10 What is the equation for effective bandgap of a quantum dot called? Answer: Brus equation.

14.11 Why is an additional subtractive term for Coulomb attraction introduced in the effective bandgap equation? Answer: (i) To account for the fact that electrons and holes are charged particles and therefore experience an attractive force towards each other according to Coulomb's law, and (ii) to take into account the fact that they exist as mutually correlated particles in the form of excitons.

14.12 How is the Coulomb attraction term in Brus equation affected by the semiconductor material properties? Answer: The Coulomb attraction term decreases as the dielectric constant of the material increases.

14.13 How does electron confinement in an exciton differ from that in a hydrogen atom? Answer: In a hydrogen atom, the electron moves in vacuum of unity dielectric constant. In an exciton, the electron and hole move in a screening semiconductor medium. As the attractive force between opposite charge carriers in an exciton is weakened by a factor equal to the dielectric constant of the semiconductor, the spatial confinement of electron-hole pair in an exciton is less severe than in a hydrogen atom. The lesser the confinement, the larger the space occupied by the exciton.

14.14 What is Bohr radius? Answer: The distance between the proton in the nucleus of the hydrogen atom and the electron in the ground state:

$$a_0 = \frac{4\pi\hbar^2\varepsilon_0}{m_e e^2}$$

(14.77)

where m_e stands for the mass of electron and e is electronic charge.

14.15 What is exciton Bohr radius? Answer: The distance between electron and hole of the electron-hole pair constituting the exciton obtained by applying Bohr's model of hydrogen atom to exciton, noting that the electron and hole are present in a semiconductor of finite dielectric constant ε_r and so behave as a single particle of reduced mass μ given by the product of their masses m_e, m_h divided by the sum of their masses.

$$r_B = \frac{4\pi\hbar^2\varepsilon_0\varepsilon_r}{e^2}\left(\frac{1}{m_e} + \frac{1}{m_h}\right) = \frac{4\pi\hbar^2\varepsilon_0\varepsilon_r}{\mu e^2}$$

(14.78)

14.16 What is meant by strong quantum confinement in a quantum dot? What effects does it lead to? Answer: Strong confinement regime occurs when radius of the quantum dot < exciton Bohr radius. Then the bandgap of the quantum dot increases and the energy levels of electrons are discretized.

14.17 What is exciton binding energy? Answer: The interaction between electron and hole in an exciton is considered to be of the same type as in a hydrogen atom. Based on this hydrogenic type interaction, the binding energy E_B of exciton is

$$E_B = -\frac{\mu e^4}{32\pi^2\hbar^2\varepsilon_0^2\varepsilon_S^2} \tag{14.79}$$

where μ is the reduced mass of the electron-hole pair, e is the elementary charge, $\hbar$ is reduced Planck's constant, ε_0 is the permittivity of free space, and ε_S the relative permittivity of the semiconductor.

14.18 What is threshold energy for multiple exciton generation? Answer: It is the minimum energy of a photon at which it can produce more than one electron-hole pair in a semiconductor.

14.19 Why can MEG happen in a quantum dot while it does not occur in a bulk semiconductor? Answer: This is because the threshold energy for MEG is lower for a quantum dot than for a bulk semiconductor.

14.20 Why is threshold energy of MEG lower for a quantum dot than in a bulk semiconductor? Answer:

   (i) Bound electron-hole pairs or excitons in quantum dots take the place of free electrons and holes in a bulk semiconductor. The electrons and holes in a quantum dot take longer to cool down so that they retain their higher energies longer than in a bulk semiconductor.

   (ii) Auger recombination increases in a quantum dot by virtue of electrostatic interaction between electrons and holes. This increase is favorable for MEG.

14.21 MEG in a semiconductor is the same as —— in a bulk semiconductor. Answer: Impact ionization.

14.22 Can MEG utilize photons with energies intermediate between bandgap of semiconductor and threshold energy for MEG? Answer: No, because MEG will only occur when the photon energy exceeds threshold energy for MEG. Energies above bandgap energy and below threshold energy for MEG are wasted.

14.23 How will you indicate the following on an energy band diagram: (a) electron affinity of a material, (b) its work function, (c) its bandgap? Answer: (a) Electron affinity is shown as the distance of the bottom edge of the conduction band from the vacuum energy level, (b) work function is shown as the distance of the Fermi level from the vacuum level, and (c) bandgap is the distance between the bottom edge of conduction band and the top edge of valence band.

14.24 What does the vacuum energy level represent? Answer: The energy of a free electron at rest outside the material at an infinite distance away.

14.25 What is Anderson's rule for drawing energy band diagrams of heterojunctions? Answer: Also called electron affinity rule, it is an approximate rule for alignment of energy bands of two materials forming a heterojunction based on their electron affinity. It states that the vacuum energy levels on the two sides of the heterojunction should be in one straight line.

14.26 How do we find the band offsets after alignment of vacuum levels on the two sides of a heterojunction? Answer: Conduction band offset ΔE_C is the energy difference between the energies E_{C1}, E_{C2} of bottom edges of the conduction bands of the two semiconductors, while valence band offset ΔE_V is the energy difference between the energies E_{V1}, E_{V2} of the top edges of valence bands of the two semiconductors:

$$\Delta E_C = E_{C1} - E_{C2},\ \Delta E_V = E_{V1} - E_{V2} \qquad (14.80)$$

14.27 What does band bending in an energy band diagram indicate? Answer: A space charge region, i.e., a local imbalance in charge neutrality. An example is the depletion region in a P-N junction, where the free electrons and holes have diffused to opposite sides, leaving behind positively charged donor ions on the N-side and negatively charged acceptor ions on the P-side, between which an electric field is created at the junction. The depletion region becomes an insulator being devoid of free charge carriers.

14.28 If electrons leave an N-type material to move to P-doped side, how will the bands curve? Answer: The band bends upwards on the N-side and downwards on the P-side.

14.29 If holes leave a P-type material to move to N-doped side, how will the bands curve? Answer: The band bends downwards on the P-side and upwards on the N-side.

14.30 If the Fermi level on one side of a heterojunction is E_{F1} and that on the other side is E_{F2}, what is the built-in potential across the heterojunction? Answer:

$$V_{bi} = E_{F1} - E_{F2} \qquad (14.82)$$

14.31 When two semiconductors come into contact, why should Fermi levels in the two semiconductors align? Answer: Because Fermi level of a semiconductor represents its chemical potential. Upon joining the two semiconductors, the chemical potential of the system of two semiconductors should be equal in the same way as two materials at different temperatures acquire the same temperature when contacted together.

14.32 What is chemical potential of a substance? Answer: The chemical energy possessed by 1 mole of the substance. This is the same as Gibb's free energy, the maximum free energy available from a thermodynamic system to do useful work, i.e., maximum amount of work that can be extracted from a system under constant temperature and pressure:

Free energy of a system = Enthalpy of the system $-$ Temperature $\times$ Entropy $\qquad (14.83)$

where

Enthalpy of the system = Its internal enrgy + Its pressure $\times$ volume $\qquad (14.84)$

and

Entropy of the system = Amount of energy which is unavailable for performing useful work $\qquad (14.85)$

Entropy is a measure of the degree of randomness or disorder of the system.

14.33 The Fermi level in an N-type semiconductor is in the bandgap near the bottom edge of conduction band, while that in a P-type semiconductor is in the bandgap close to the top edge of the valence band. How will the Fermi levels align when the N-type semiconductor comes in contact with P-type semiconductor? Answer: The Fermi level in the N-type semiconductor will move downwards, while that in the P-type semiconductor will move upwards until the Fermi levels in the two semiconductors are aligned.

14.34 How do the Fermi levels of two dissimilar semiconductors come into alignment? Answer: Alignment is achieved by transference of charges across the interface between the two semiconductors.

REFERENCES

Brus, L. 1986 Electronic wave functions in semiconductor clusters: Experiment and theory, Journal of Physical Chemistry, 90: 2555–2560.

CRM2010, CopyRight ©2010 China Rare Metal Material Co., Ltd. http://www.china-raremetal.com/product/cadmium_telluride_cdte.htm

Elements of Advanced Theory, www.tf.uni-kiel.de/matwis/amat/semi_en/kap_2/backbone/r2_3_1.pdf.

ESI 2012: Electronic Supplementary Material (ESI) for Journal of Materials Chemistry. © The Royal Society of Chemistry 2012, www.rsc.org/suppdata/jm/c2/c2jm32982d/c2jm32982d.pdf.

ESI 2013: Electronic Supplementary Material (ESI) for Physical Chemistry Chemical Physics © The Owner Societies 2013, www.rsc.org/suppdata/cp/c3/c3cp52678j/c3cp52678j.pdf.

Mukherjee M., A. Datta and D. Chakravorty 1994 Electrical resistivity of nanocrystalline PbS grown in a polymer matrix, Applied Physics Letters, 64(9): 1159–1161.

Norton D. P., Y. W. Heo, M. P. Ivill, K. Ip, S. J. Pearton, M. F. Chisholm and T. Steiner 2004 ZnO: Growth, doping & processing, Materials Today, 7(6): 34–40.

Nozik A. J. 2002 Quantum dot solar cells, Physica E: Low-Dimensional Systems and Nanostructures, 14(1–2): 115–120.

Patil P. V. and S. Datta 2011 May Do we need to revisit the Bohr exciton radius of hot excitons?, arXiv: Mesoscale and Nanoscale Physics, arXiv:1105.2205 [cond-mat.mes-hall], 39 pages, https://arxiv.org/pdf/1105.2205.pdf.

Rodríguez-Mas F., J. C. Ferrer, J. L. Alonso, D. Valiente and S. F. de Ávila 2020 A comparative study of theoretical methods to estimate semiconductor nanoparticles size, Crystals, 10(3): 226, 17 pages.

Tian J. and G. Cao 2013 Semiconductor quantum dot-sensitized solar cells, Nano Reviews, 4(1): 22578, 8 pages.

15 Quantum Dot Solar Cells
Types of Cells and Their Fabrication

15.1 CLASSIFICATION OF QUANTUM DOT SOLAR CELLS

Quantum dot solar cells are classified into seven categories, depending on their principles of operation (Figure 15.1).

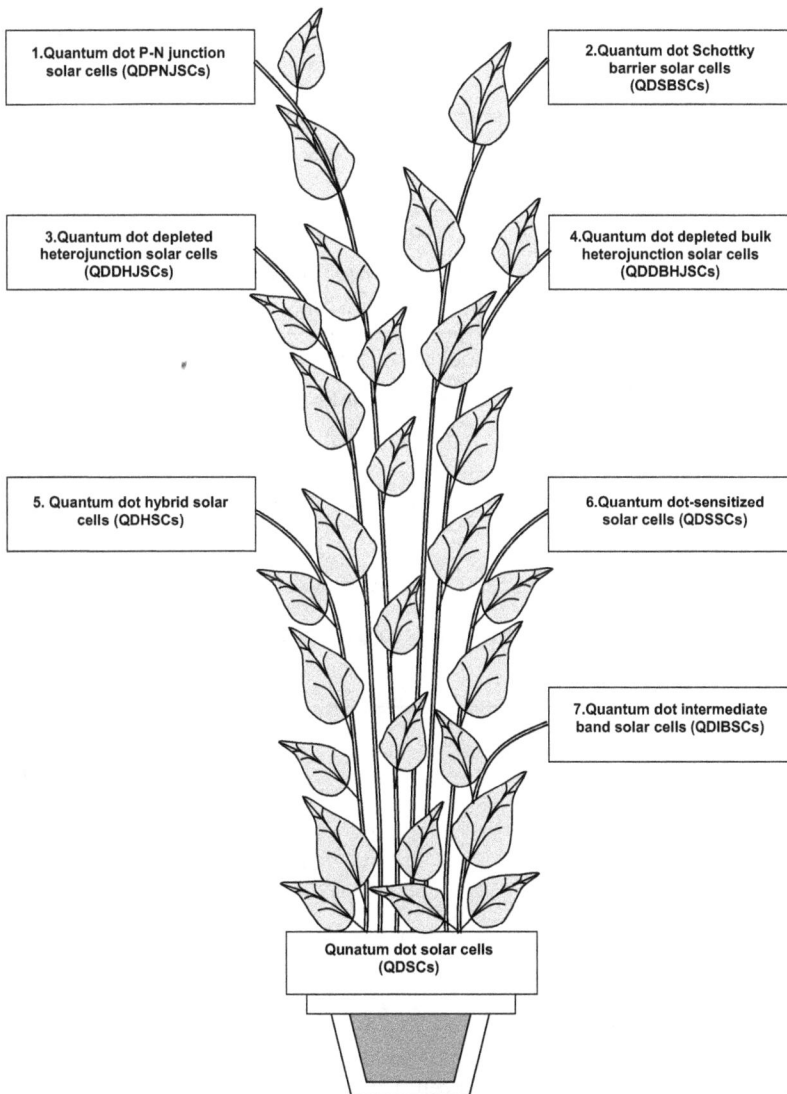

FIGURE 15.1 Family tree of the quantum dot solar cells. The branches are (i) P-N junction solar cells, (ii) Schottky barrier solar cells, (iii) depleted heterojunction solar cells, (iv) depleted bulk heterojunction solar cells, (v) sensitized solar cells, (vi) hybrid solar cells, and (vii) intermediate band solar cells.

DOI: 10.1201/9781003215158-20

15.2   QUANTUM DOT P-N JUNCTION SOLAR CELLS

The generic N-P structure of this solar cell consists of a transparent conducting substrate serving as one electrode; an N-type wide-bandgap metal oxide acting as an electron transport layer (ETL); N-type quantum dots, e.g., PbS QDs in the form PbS-X (X = halogen ions); P-type quantum dots, e.g., PbS-EDT, where EDT = 1,2-ethanedithiol; occasionally a hole transport layer (HTL); and a gold film as the second electrode (Goossens et al 2021).

15.2.1   PBS QD SOLAR CELL WITH NAHS-TREATED P-TYPE LAYER (η = 7.6%)

Figure 15.2 shows this solar cell (Speirs et al 2017). Postdeposition treatment with sodium hydrosulfide solution is developed as a simple and reproducible technique of P-type doping of PbS QDs. The fabrication process proceeds as follows:

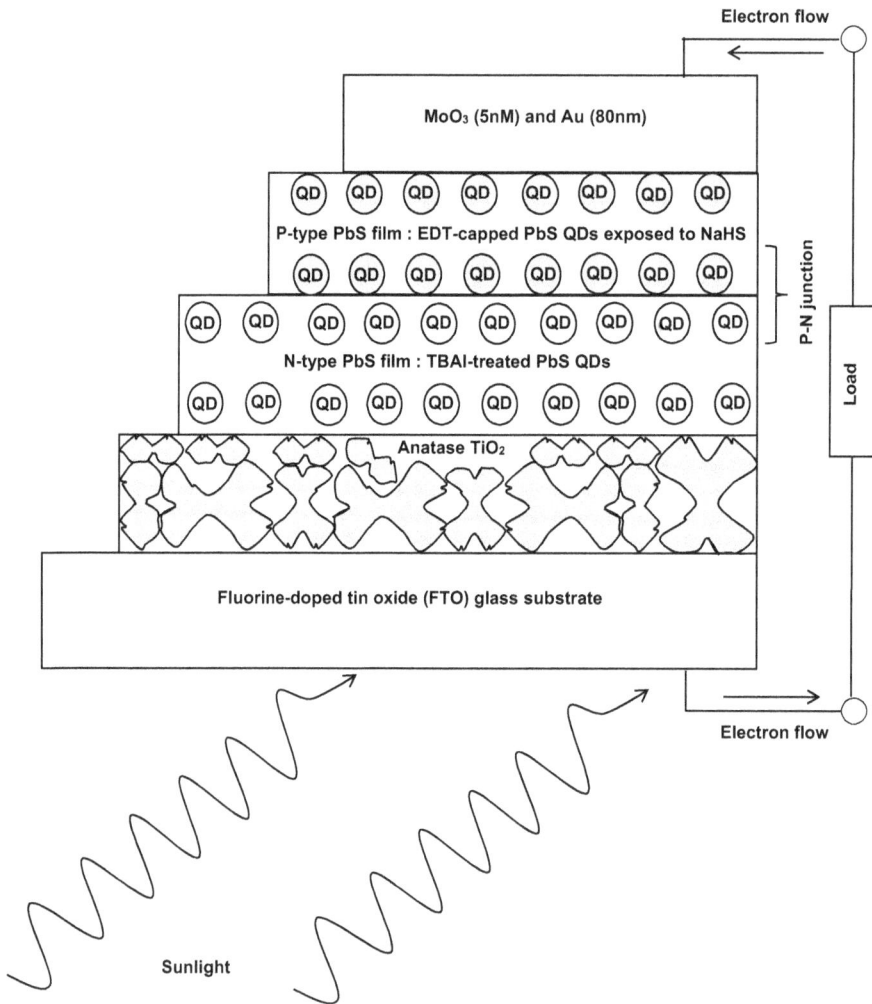

FIGURE 15.2   Quantum dot P-N junction solar cell showing the FTO glass substrate, upon which the anatase TiO_2 layer is formed, followed by deposition of an N-type layer consisting of oleic acid–capped PbS QDs treated with TBAI, a P-type layer containing oleic acid–capped PbS QDs covered with EDT and exposed to NaHS, and an MoO_3/Au film. Sunlight incident on FTO glass substrate produces current, which is collected between the MoO_3/Au and FTO glass electrodes.

15.2.1.1 Substrate

It is fluorine-doped tin oxide (FTO) glass with sheet resistance 13Ω/sq.

15.2.1.2 Oleic Acid–Capped PbS QD Synthesis

Hot injection method is applied (Lai et al 2014). Mixture of [lead (II) acetate trihydrate + octadecene + oleic acid] is dried under vacuum in a three-neck flask, and the temperature is set to 90°C. After removing the heating mantle, bis(trimethylsilyl)sulfide solution in octadecene is injected into the reaction flask at 82°C. Then oleic acid is injected and the solution is cooled down. The QDs are washed in toluene/ethanol and dissolved in anhydrous hexane. The size of QDs is 2.9nm.

15.2.1.3 Anatase TiO₂ Deposition

It is made by spin-casting a solution of ethanol/titanium(IV) butoxide/HCl in the ratio 20/2/1 on the substrate and annealing at 450°C for ½ h.

15.2.1.4 N-Type PbS Film (PbS QDs Treated with TBAI) Deposition

It is made by layer-by-layer spin-coating 10mg/mL solution of PbS QDs capped with oleic acid in hexane on anatase TiO₂ film, exposing the PbS film to TBAI (15mg/mL) solution in methanol, drying by spinning, and washing in methanol twice for removal of tetrabutyl ammonium cation.

15.2.1.5 P-Type PbS Film (PbS QDs Capped with EDT) Deposition

It is made by layer-by-layer spin-coating 10mg/mL solution of PbS QDs capped with oleic acid in hexane on PbS QDs-TBAI film, exposing the PbS film to EDT (0.01% v/v) solution in acetonitrile, exposing each layer to 0.1mM NaHS solution in MeOH for 15 s, and washing once in acetonitrile

15.2.1.6 MoO₃ (5nM) and Au (80nm) Deposition

It is deposited by thermal evaporation. MoO₃ acts as a hole-extracting layer. It also protects the active layer from deposition of electrode metal.

15.2.1.7 Enhancement of Power Conversion Efficiency by Increase in P-Type Doping with NaHS Treatment

Open-circuit voltage of the solar cell is somewhat depressed by the treatment. Both short-circuit current density and fill factor rise by small amounts. For the undoped solar cell, the short-circuit current density is 26mAcm^{-2}. The same for the doped device is 27mAcm^{-2}. The respective fill factor values are 0.49 and 0.51. The corresponding values of the open-circuit voltages are 0.56 and 0.55V. The overall effect is that the efficiency rises from 7.1 to 7.6% (Speirs et al 2017).

15.2.2 IMPROVED RELIABILITY PbS QD SOLAR CELL WITH ATOMIC-LAYER DEPOSITED TiO₂ ELECTRON TRANSPORT LAYER (η = 5.5–7.2%)

Sukharevska et al (2020) found that the solar cells made with sol-gel-synthesized, spin-coated TiO₂ films show a large spread in efficiencies in the broad range 0.5–6.9%. Also, the fabricated cells include a large fraction of short-circuited cells. Much better reproducibility is achieved with efficiency values restricted in a narrow range of 5.5–7.2% when atomic layer deposition (ALD) is used for TiO₂ film. The structure of this solar cell is fluorine-doped tin oxide (FTO) cathode/ALD-TiO₂ ETL/tetrabutylammonium iodide (TBAI)–treated PbS QDs for N-type layer/tetrabutylammonium chloride (TBAC)–treated PbS QDs for P-type layer/MoO₃ hole transport layer/Au back contact. The solar cell and its energy band are shown in Figure 15.3.

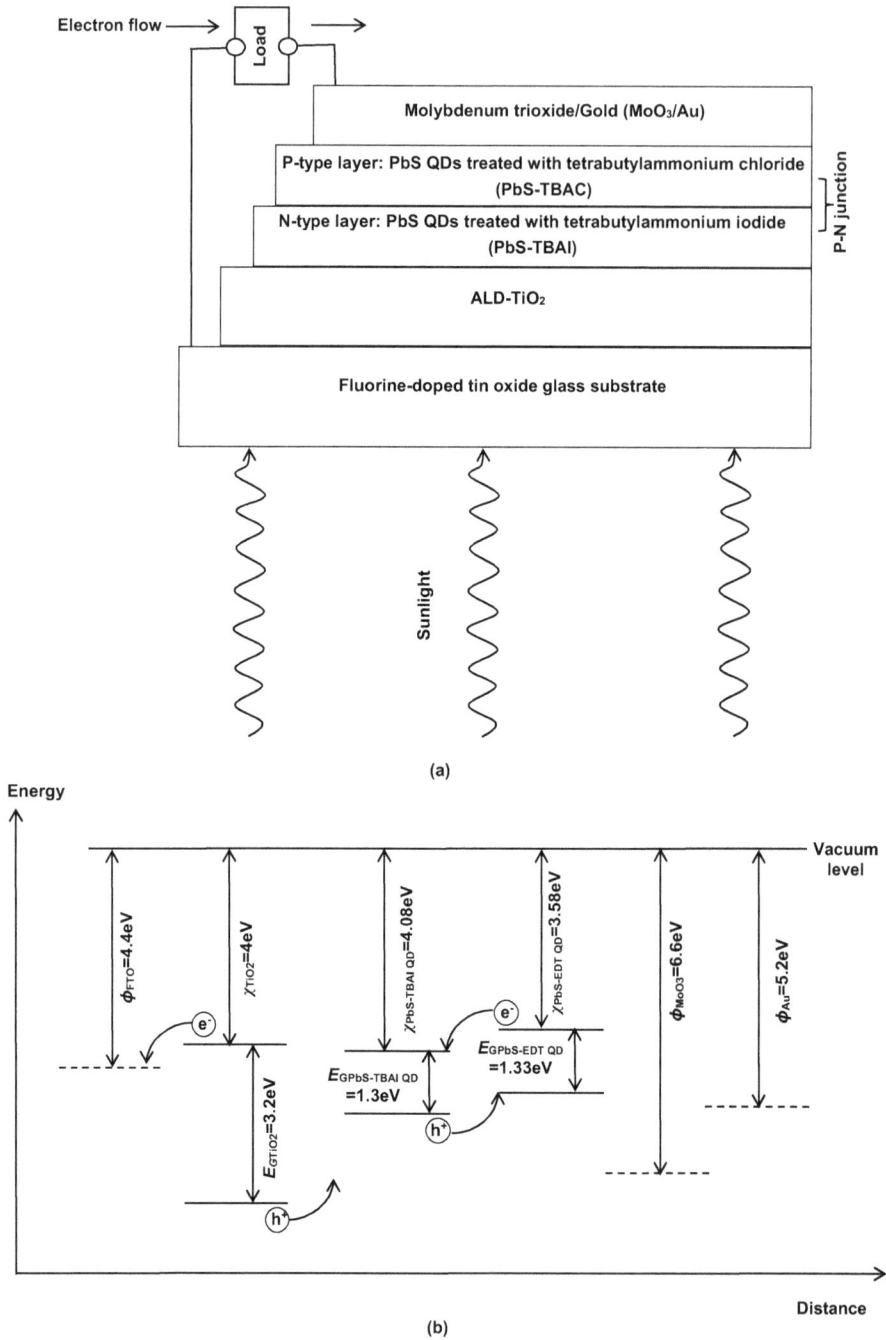

FIGURE 15.3 Quantum dot P-N junction solar cell: (a) cross-sectional diagram showing from bottom upwards: the FTO substrate, the TiO$_2$ layer, the N-type PbS-TBAI QD layer, the P-type PbS-EDT QD layer, and the MoO$_3$/Au electrode. Sunlight strikes the solar cell from the FTO substrate side. The electric current is drawn between the FTO substrate and the MoO$_3$/Au electrode. (b) Energy band diagram of the solar cell obtained by applying Anderson's rule. The work functions of FTO, MoO$_3$, and Au are $\phi_{FTO} = 4.4$eV, $\phi_{MoO3} = 6.6$eV, and $\phi_{Au} = 5.2$eV, respectively. The electron affinities of TiO$_2$, PbS-TBAI QD, and PbS-EDT QD are $\chi_{TiO2} = 4$eV, $\chi_{PbS\text{-}TBAI\,QD} = 4.08$eV, and $\chi_{PbS\text{-}EDT\,QD} = 3.58$eV. The bandgap of PbS-TBAI QD is $E_{GPbS\text{-}TBAI\,QD} = 1.3$eV, and that of PbS-EDT QD is $E_{GPbS\text{-}EDT\,QD} = 1.33$eV. Electrons move to the FTO side, and holes travel towards the Au electrode.

The TiO_2 layer is deposited on the FTO-coated glass substrate by ALD process with the help of $TiCl_4$ in nitrogen flow and H_2O in nitrogen flow at 260°C. The PbS QDs capped with oleate ligands are synthesized by the hot injection method. They are spin-cast from hexane solution on TiO_2 film. Ligand exchange is done by exposure to methanol solution of TBAI for N-type layer and TBAC for P-type layer. MoO_3 and Au are thermally evaporated.

15.2.3 LOW-COST PbS QD SOLAR CELL WITH ZnO ELECTRON TRANSPORT LAYER AND STABLE CR-AG ELECTRODES ($\eta = 6.5\%$)

Interfacing of chromium with silver yields an electrode which is highly stable in the atmosphere (Khanam et al 2019). Solar cell with spin-coated PbS QD and Cr-Ag electrode shows an efficiency of 6.5%. Its structure is ITO/ZnO/PbS-TBAI/PbS-EDT/Cr-Ag. The concentration of PbS QDs is restricted to 10mg/mL in toluene solvent as opposed to higher concentrations in the range 40–100mg/mL (Wang et al 2013; Xu et al 2017). Applying a higher concentration of PbS QDs raises the cost of the solar cell. The solar cell and its energy band diagram are shown in Figure 15.4.

Solution of ZnO nanoparticles is spun over ITO-coated glass substrate and heated at 110°C for 20 min. PbS QDs are spin-coated on ITO/ZnO substrate. Then TBAI (tetrabutylammonium iodide) solution is poured and left on the substrate for 40 s., followed by rinsing with ACN (acetonitrile). Again, PbS QDs are spin-coated. Then EDT (1,2-ethanedithiol) solution is poured, followed by same rinsing step. After depositing a few such layers, the substrate is heated at 80°C for 5 min. Cr-Ag electrode is deposited by thermal evaporation. Thus, we get the structure ITO-coated glass substrate/ZnO/PbS-TBAI/PbS-EDT/Cr-Ag. The solar cell made by spin-coating PbS QDs has efficiency of 6.5% (Khanam et al 2019).

15.2.4 PbS QD SOLAR CELL BY A SCALABLE INDUSTRIALLY SUITED DOCTOR BLADING PROCESS USING N- AND P-TYPE INKS ($\eta = 9\%$)

Making a P-type ink with EDT ligands is difficult (Goossens et al 2021). P-type PbS QD ink is prepared by capping the PbS QDs with 3-Mercapto-propionic acid (MPA). The N-type ink is made by capping Pbs QDS with lead halides. The ability to perform film casting and ligand exchanges in ambient conditions allows easy fabrication, achieving high throughput. Over the ITO substrate, PbS QDs-MPA P-type film, then PBs QDs-halogen N-type film, ZnO nanocrystals film as ETL and hole-blocking layer, and aluminum electrode are sequentially deposited. In the reverse sweep direction, the open-circuit voltage of the solar cell is 0.57V. Its short-circuit current density is $26.22 mAcm^{-2}$. The fill factor is 0.61, and efficiency is 9.04% (Goossens et al 2021).

15.2.5 PbS QD SOLAR CELL WITH PD2FCT-29DPP AS HTL ($\eta = 14\%$)

Desirable properties of the HTL in quantum dot solar cell are a high hole mobility, long carrier diffusion length, and lack of chemical interaction with other layers in the solar cell. The EDT-treated QD layer suffers from stability problems. Further, the density of trap states in this layer is high-yielding short diffusion lengths in the range of tens of nanometers. The layer also interacts with other materials in the solar cell, degrading the performance.

Here PD2FCT-29DPP (diketopyrrolopyrrole-based polymer with benzothiadiazole derivatives)–HTL is used as a replacement for the EDT-treated PbS QD layer (Kim et al 2020). It provides efficient transfer of charges together with absorption in NIR range. Stille-coupling reaction is applied for synthesis of PD2FCT-29DPP. This coupling reaction involves two organic groups, one partner taken as an organotin compound, and the other partner an electrophile, an electron-pair-deficient molecule. Purification is done by soxhlet extractions used for isolating a low-solubility compound from a solid specimen.

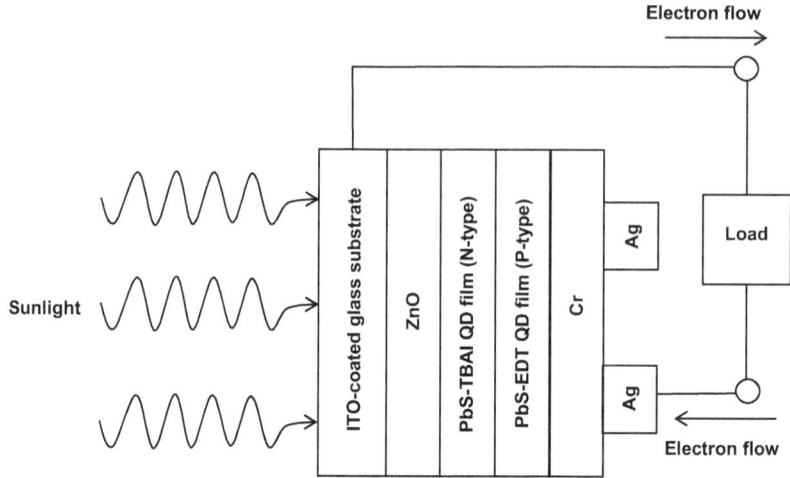

(a)

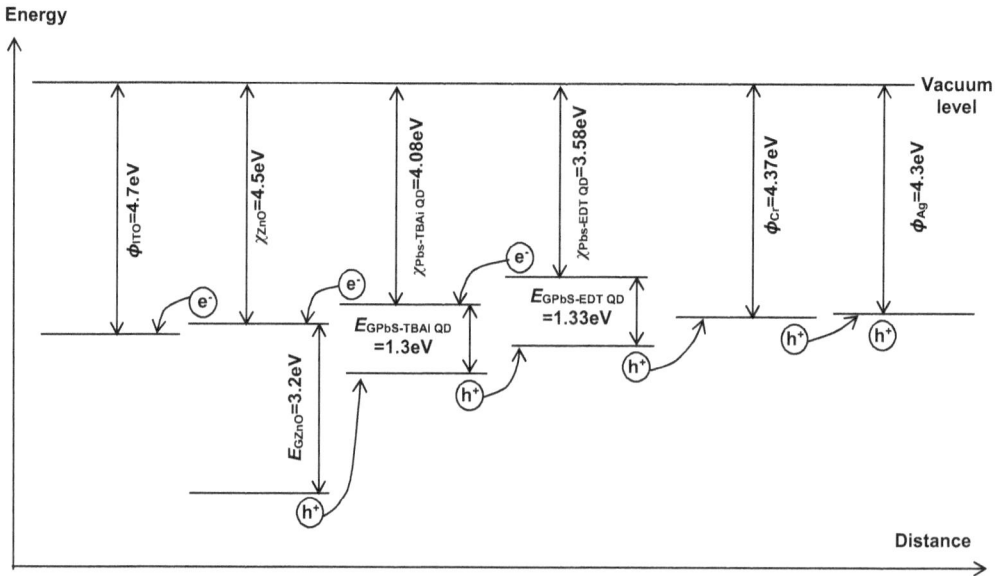

(b)

FIGURE 15.4 PbS quantum dot P-N junction solar cell: (a) schematic cross section showing the layers from left: the ITO substrate, ZnO film, PbS-TBAI QD film, PbS-EDT QD film, the Cr film, and Ag grid. Sunlight is incident on ITO substrate, and current output is taken between the ITO and Ag electrode. (b) Energy band diagram of the solar cell drawn by applying Anderson's rule. The work functions of ITO, Cr, and Ag are $\phi_{ITO} = 4.7$ eV, $\phi_{Cr} = 4.37$ eV, and $\phi_{Ag} = 4.3$ eV, respectively. The electron affinities of ZnO film, PbS-TBAI QD film, and PbS-EDT QD film are $\chi_{ZnO} = 4.5$ eV, $\chi_{PbS-TBAI\ QD} = 4.08$ eV, and $\chi_{PbS-EDT\ QD} = 3.5$ eV. The energy gap of ZnO film is $E_{GZnO} = 3.2$ eV, that of PbS-TBAI QD film is $E_{GPBS-TBAI\ QD} = 1.3$ eV, and that of PbS-EDT QD film is $E_{GPbS-EDT\ QD} = 1.33$ eV. Electrons move from higher to lower energies towards the ITO substrate side, while holes travel from lower to higher energies towards the Ag electrode side.

ITO substrate is used, and ZnO nanoparticles form the ETL. The PbS QDs in octane are ligand-exchanged using PbI_2, $PbBr_2$, and ammonium acetate dissolved in DMF. They are dispersed in butylamine, spin-coated in a glove box, and annealed at 70°C for 5 min. Then the HTL polymer is deposited by spin-coating. 10nm MoO_3 film + 120nm Ag film form the metal layer. The open-circuit voltage of the solar cell is 0.66V. Its short-circuit current density is 30.3 mAcm^{-2}. The fill factor is 70%, and efficiency is 14% (Kim et al 2020).

15.2.6 PbS QD Solar Cell (η = 10.06%) as the Back Cell in a Tandem Solar Cell (η = 18.9%)

A P-N junction quantum dot solar cell is used as the back cell of a four-terminal solar cell (Figure 15.5) in combination with a perovskite solar cell, exploiting the fact that the bandgap of perovskite is adjustable within the limits 1.2–2.3eV, making it suitable to wavelength < 1,000nm, whereas the bandgap of quantum dots (0.5–2eV) makes quantum dot solar cells capable of infrared absorption (Andruszkiewicz et al 2021).

15.2.6.1 Front Semitransparent Perovskite Solar Cell

The front solar cell is a methyl ammonium iodide perovskite solar cell. Its structure is glass/FTO/compact TiO_2/Mesoporous TiO_2/Perovskite/HTL/Au (10nm). On an FTO-coated glass substrate, a compact TiO_2 film is deposited by spray pyrolysis of titanium diisopropoxide bis (acetylacetonate) in isopropanol at 500°C and annealed for ½ h. Then a mesoporous TiO_2 film is spin-coated from a dispersion of TiO_2 paste in ethanol and annealed at 450°C for ½ h. Methylammonium iodide + PbI_2 dissolved in DMF: DMSO is spin-coated in a glove box with addition of chlorobenzene drops and annealed at 90°C for ½ h to form the perovskite layer. Then spiro-OMeTAD solution in chlorobenzene is deposited by spin-coating, followed by spin-coating of lithium bis(trifluoromethanesulfonyl) imide solution in acetonitrile and FK 209 Co(III)-TFSI salt solution in acetonitrile. 10nm-thick gold is thermally deposited. The power conversion efficiency of the solar cell is 17.37% (reverse scan) and 14.88% (forward scan).

15.2.6.2 Back Colloidal Quantum Dot Solar Cell

The back cell is a P-N junction solar cell. Its structure is glass/ITO/Al-doped zinc oxide (ETL)/PbS-PbI_2 N-type layer/PbS-EDT P-type layer/Au (80nm). The AlZnO (AZO) sol-gel is made by injecting ethanaloamine in heated zinc acetate dihydrate and aluminum nitrate nonahydrate solution in ethanol and allowing the resulting mixture to cool. The sol-gel is filtered and spin-coated on the ITO. Annealing is done at 200°C for ½ h and 300°C for ½ h. The PbS quantum dots are synthesized by injecting hexamethyldisilathiane (HMDST) solution in octadecane (ODE) into a heated solution of lead(II) oxide, ODE, and oleic acid (OA) and allowing cooling to take place. PbS QDs in octane are vortexed with a solution PbI_2, $PbBr_2$, and ammonium acetate in N,N-dimethylformamide and washed with octane, followed by addition of toluene, centrifuging, and dispersing in butylamine to get PbS-PbX_2 ink (X = halogen). The prepared ink is deposited by spin-coating over AZO film in two steps of increasing RPM. After annealing at 70°C for 10 min., the PbS-EDT film is formed by first spin-coating PbS QDs in octane, then spin-coating ethane-dithiol solution in acetonitrile (ACN) and double-rinsing with ACN. 80nm-thick gold film is formed by thermal deposition. The efficiency of this solar cell is 10.06% (reverse scan) and 7.4% (forward scan). For the filtered cell, with the front cell placed above the back cell, the efficiency values are 1.53% (reverse scan) and 1.05% (forward scan)

15.2.6.3 Stacking the Cells for Proper Light Coupling

Special attention is paid to ensure that there is no hindrance by the front cell to the light supply to the back cell. To this end, the gold electrode of the front cell is made only 10nm thick, with

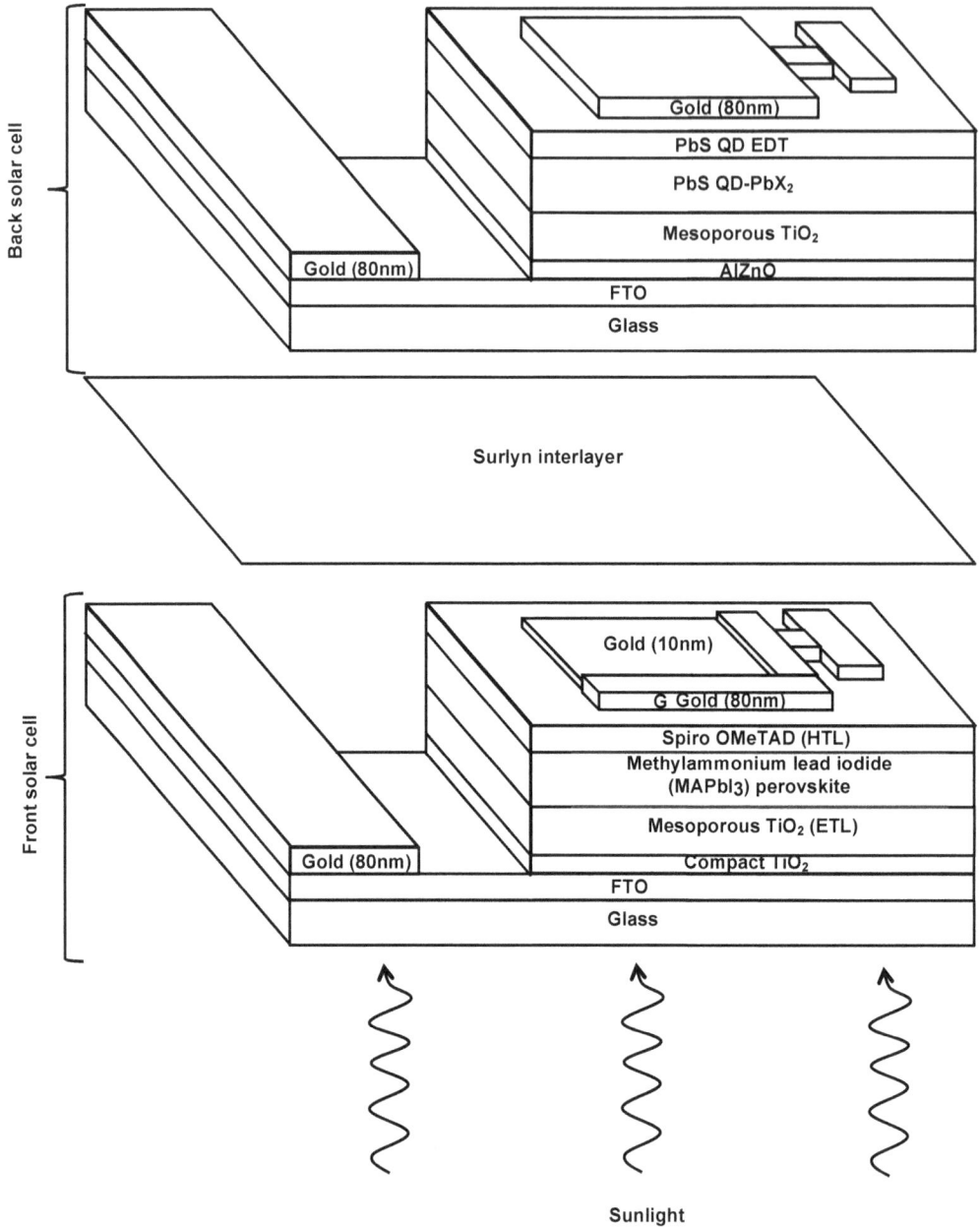

FIGURE 15.5 Four-terminal perovskite and PbS colloidal quantum dot tandem solar cell showing the various layers in (i) perovskite front cell: glass substrate, FTO, compact TiO_2, mesoporous TiO_2 ETL, $MAPbI_3$ perovskite, spiro-OMeTAD HTL, Au contact; and (ii) quantum dot solar cell: glass, ITO, N-type PbS-PbX_2 ink, P-type PbS-EDT, Au contact. Light is coupled between the two cells through a surlyn interlayer. Sunlight illuminates the combined solar cell from the glass substrate side of perovskite solar cell.

80nm thickness along a border strip. The 10nm thick gold film being semitransparent allows light to pass through it. Furthermore, an ionomer resin or molybdenum interlayer is fixed between the two cells. The ionomer resin taken is surlyn. Surlyn is a copolymer of ethylene and methacrylic acid. It is complexed with metal ions. Favorable qualities of surlyn include its high transparency, flexibility, chemical stability, and requirement of low temperature for fixation and sealing. Any air gaps between the two solar cells are avoided, reducing the reflection of light and thereby providing better coupling.

The two solar cells are stacked together, one above the other. The perovskite solar cell is placed on a hot plate, and 25μm-thick surlyn layer is deposited by heating to 70°C. After mounting the quantum dot solar cell over the sulyn layer on the perovskite cell, the two cells are manually pressed together.

The efficiency of the four-terminal perovskite–quantum dot tandem solar cell is 18.9% (reverse scan) and 15.93% (forward scan) (Andruszkiewicz et al 2021).

15.2.7 PbS QD Solar Cell (η = 11.6%) as the Back Cell in a Tandem Solar Cell (η = 20.2%)

1.63eV bandgap perovskite solar cell is stacked above the 1.15 eV bandgap quantum dot solar cell capable of IR trapping to fabricate a perovskite + quantum dot tandem solar cell which can collect photons with wavelength > 1000nm beyond the range of perovskites (Manekkathodi et al 2019). The salient feature of the front perovskite solar cell is a three-layer dielectric-metal-dielectric transparent conducting electrode (DMD-TCE). The structure of DMD-TCE is MoO_3 (5nm)/Au (1nm)/Ag (5nm)/MoO_3 (35nm). The films are deposited by thermal or electron-beam evaporation. The inclusion of dielectric helps to prevent oxidation of the metal film along with inhibition of surface plasmonic loss in the metal.

In the perovskite cell, the ETL used is SnO_2, owing to its higher electron mobility and low trap state density. The precursor for SnO_2 consists of SnO_2 nanoparticles solution which is diluted in water, spin-coated on ITO substrate, and annealed at 150°C for ½ h. The $Cs_{0.05}MA_{0.10}FA_{0.85}Pb_{0.85}Br_{0.15}$ perovskite is deposited in a nitrogen-filled glove box from a precursor made of PbI_2, $PbBr_2$, CsI, FAI, and MABr in DMF:DMSO, in two steps (200RPM and 6,000RPM), dropping chlorobenzene and annealing at 100°C for ½ h. Spiro-OMeTAD in chlorobenzene as HTL and DMD-TCE films are deposited to complete the fabrication.

In the QD cell, the ETL used is a ZnO film. The cell consists of a halide-exchanged PbS QD layer and an EDT-treated PbS QD layer. 120nm-thick gold film is used as the electrode.

The efficiency of perovskite solar cell is 17.6%, that of unfiltered PbS QD cell (without perovskite cell placed in front of PbS QD cell) is 11.6%, and that of filtered PbS QD cell (with perovskite cell placed in front of PbS QD cell) is 2.6%. The efficiency of perovskite + PbS QD cell is 20.2% (Manekkathodi et al 2019).

15.3 QUANTUM DOT SCHOTTKY BARRIER SOLAR CELL (η = 1.8%)

PbS quantum dot film is used as the semiconductor and a shallow work function metal such as aluminum or magnesium as the conductor to form a metal-semiconductor or Schottky diode (Johnston et al 2008). The ohmic contact applied to the opposite side of PbS QD film is indium tin oxide. All the processing work, including synthesis and deposition of PbS QD films, is carried out in an inert environment. Figure 15.6 (a) shows this solar cell.

15.3.1 Synthesis of PbS QD Film and Ligand Exchange for Improving Conductivity

Organometallic precursors are used for synthesizing the PbS QDs (Hines and Scholes 2003). The QDs are prevented from being closely packed by passivation with oleate ligands. But these ligands are excessively long at ~ 2.5nm in length. They are inconvenient to moving charges and hinder the transference of charge through the PbS film, rendering them insulating. Therefore, their removal is necessary to facilitate easy charge transport and increase carrier mobility. This is done

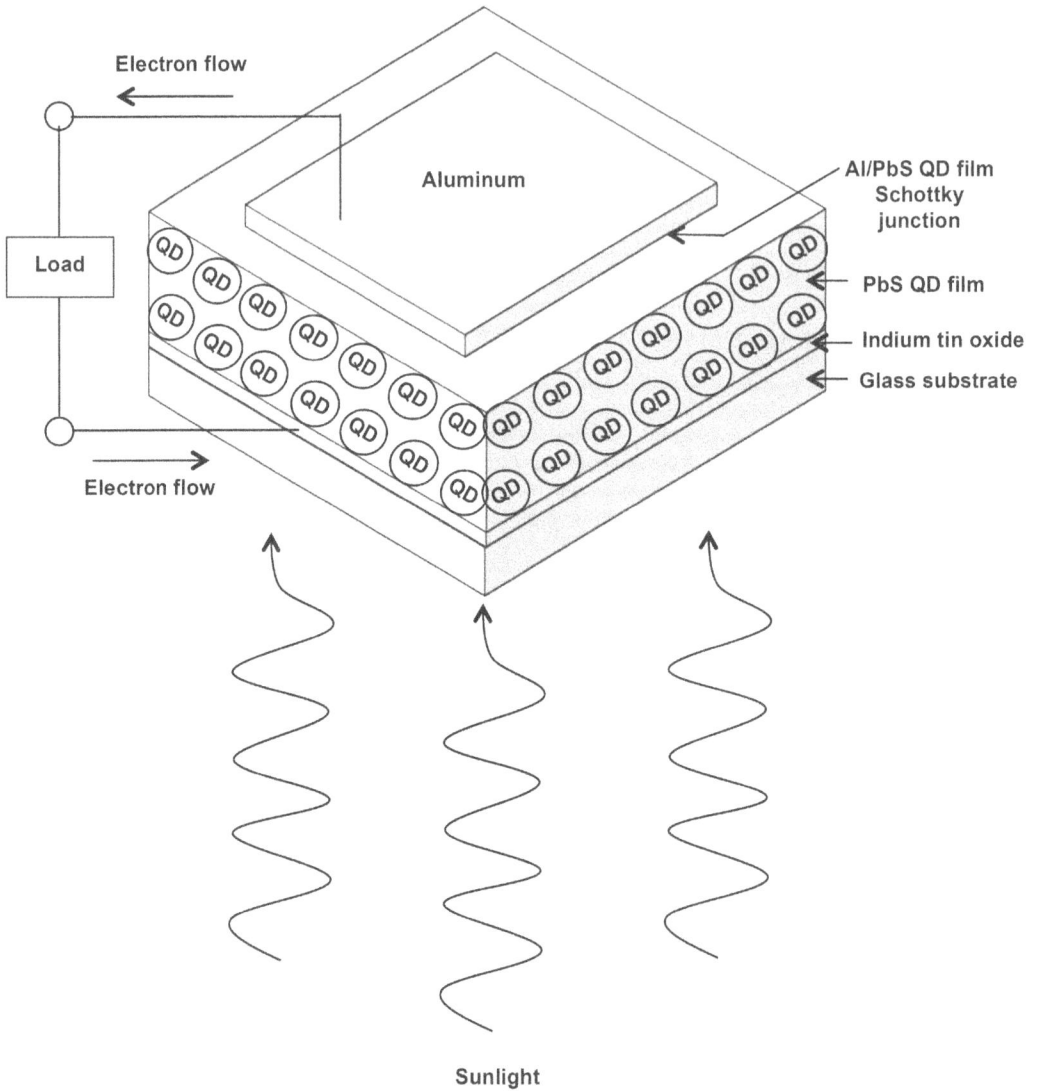

(a)

FIGURE 15.6 Quantum dot Schottky barrier solar cell and its energy band diagrams: (a) the construction of the solar cell showing the constituent layers from bottom upwards: glass substrate, ITO film, PbS nanocrystal film, and the top aluminum electrode. The Schottky barrier diode is formed at the interface of the aluminum electrode with the PbS nanocrystal film; *b* and *c* energy band diagrams of Al/PbS interface: (b) before contact between Al and PbS, (c) after contact and exposure to sunlight. Various parameters such as energy gap and electron affinity of PBS QDs taken in drawing the diagram differ from bulk values, e.g., the bandgap E_G of PbS QD is taken as 0.8eV in place of 0.4eV for bulk PbS, the electron affinity χ for PbS QD is taken as −4.4eV against −4.6eV for bulk PbS. The work function ϕ of PbS is chosen to show P-type conduction.

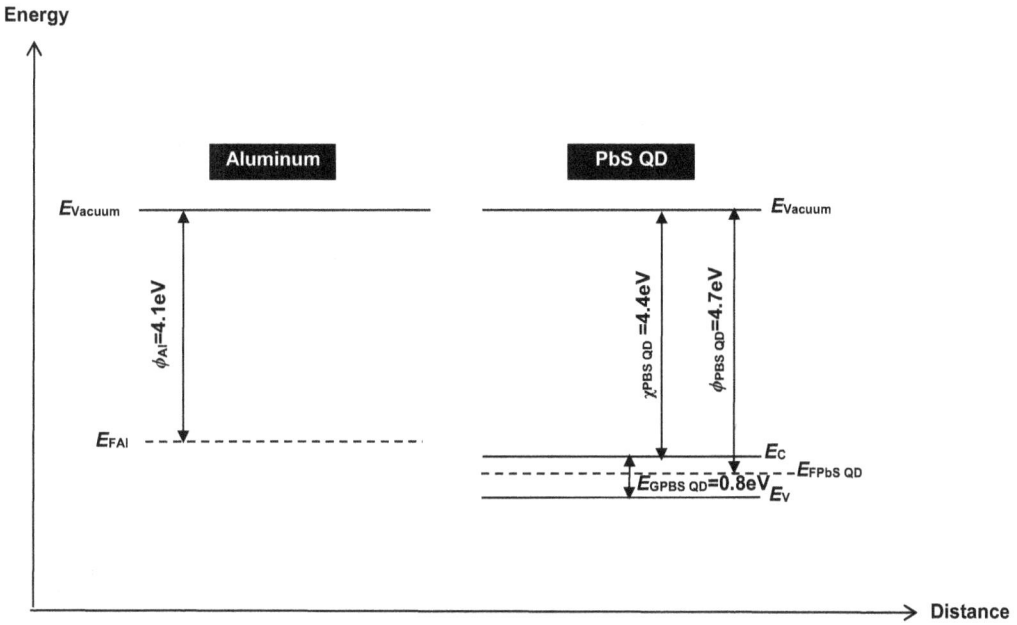

(b)

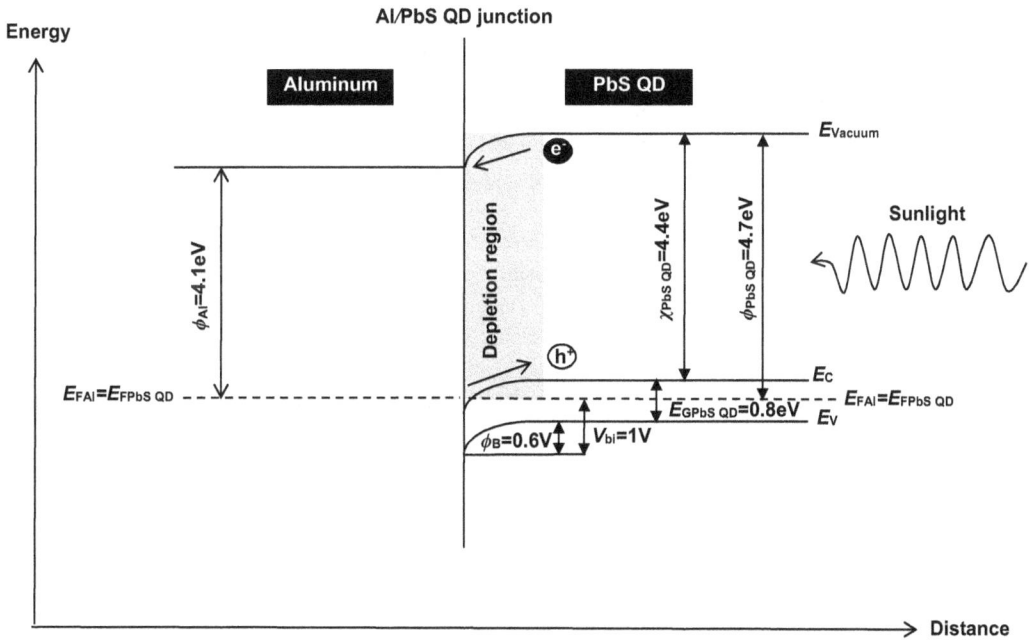

(c)

FIGURE 15.6 (Continued)

by repeatedly precipitating them with methanol. But this may not be sufficient. So additionally, the long ligands are exchanged with shorter *n*-butylamine ligands of 0.6nm length without sacrificing stability of colloidal solution. The process of ligand replacement is known as solution-phase ligand exchange. If adequate conductivity is not achieved even after ligand exchange, the butylamine ligands are also removed by immersing in methanol for 2 h. (Konstantatos et al 2006).

15.3.2   PbS QD Film Deposition and Making Contacts

The PbS QDs dispersed in an octane solution are spin-coated on an ITO film coated over a glass substrate. The concentration of octane solution is 150mgmL^{-1} (Johnston et al 2008). The PbS film contains smooth, densely packed arrays of PbS QDs. The Schottky contact is established with a three-layer stack consisting of LiF (0.7nm), Al (140nm), and Ag (190nm). These three layers are formed by successive thermal evaporation. LiF improves the quality of the contact. The Schottky junction is formed between the aluminum and P-type PbS QD film.

15.3.3   Operation

From the energy band diagram of the Al/PbS QD junction shown in Figures 15.6(b) and (c), it is seen that the vacuum energy level, the conduction, and the valence bands bend downwards at the interface (Clifford et al 2007). The curvature of the bands indicates the formation of a space charge region at the junction. Work function of aluminum is less than that of PbS QD, which means that less energy is required to liberate electrons from the chemical bonds in aluminum than in PbS QD. Hence, electrons are transferred from aluminum to PbS QD. Since PbS QD is a P-type material, it contains a high density of holes. The electrons transferred from aluminum to PbS QD fill the hole vacancies adjacent to the interface. As a result, the region of PbS QD film adjoining the junction becomes deficient of holes, forming a depletion region. Now we have acceptor ions in the depletion region, which are negatively charged. So the PbS QD side of the junction is negatively charged, whereas the aluminum side of the junction, which supplied electrons to the P-type PbS QD side, becomes positively charged. An electric field is produced across the junction due to the local charge imbalance created there. Obviously, the direction of electric field across the junction is from the positively charged aluminum side to the negatively charged PbS QD side because the direction of the electric field is taken from positive to negative charge. Owing to their mutually opposite charges, electrons and holes travel in opposite directions in the electric field. The direction of electron motion is opposite to that of the electric field. Holes move in the same direction as the electric field. Accordingly, electron motion takes place from P-type PbS QD side to N-type aluminum side, while hole motion occurs from N-type aluminum side to P-type PbS side. Thus, electrons are swept to the N-side and holes to the P-side. In this way, the field causes the separation of electron-hole pairs, which is responsible for the function of solar cell. To clarify, when the solar cell is illuminated with sunlight, electron-hole pairs are produced inside the depletion region or in its neighborhood. These charge carriers migrate under the influence of the built-in field across the junction, leading to current flow.

15.3.4   Solar Cell Performance Parameters

The solar cell shows an open-circuit voltage of 0.33V. It has a short-circuit current density of 12.3 mAcm^{-2}, and the efficiency is 1.8% under solar illumination of intensity 100mWcm^{-2} (Johnston et al 2008).

15.3.5   Shortcomings of Schottky Diode Quantum Dot Solar Cells

The drawbacks are:

(i) Although the light-absorbing film of a solar cell has small thickness, the thickness has a finite nonzero value. The finite thickness plays a crucial role in solar cell functioning. When light falls on the surface of the absorbing film, electron-hole pairs are generated at a region near the surface. As light goes deeper inside the light-absorbing film below the surface, it is weakened and becomes less energetic to create electron-hole pairs. Therefore, the surface regions or shallow depths of the light-absorbing film become the major source of electron-hole pairs and the deeper regions are less effective.

The junction where the electric field for separating the carriers is concentrated must be close to the surface of the light-absorbing region. If the junction is distant from the surface, the photogenerated carriers will have to travel a large distance to reach the junction. The larger this distance, the greater the chances of carriers recombining on the way before reaching the junction.

If we look at the diagram showing the different layers in the Schottky solar cell (Figure 15.6(a)), we notice that in the Schottky diode structure, this condition of placement of the junction near the surface of the illuminated side of solar cell is not satisfied. The sunlight incident from the side of the glass substrate falls on the QDs, which act as light absorbers. So absorption of light and consequent creation of electron-hole pairs is most pronounced at the glass-PbS QD interface and is comparatively feebler at the PbS QD-aluminum Schottky barrier. Electron-hole pairs are mostly created by light absorption near the glass substrate. These pairs have to traverse a large distance equal to the thickness of the PbS QD film to reach the Schottky barrier. Many of these charge carriers recombine during the journey and are therefore prevented from contributing to the photogenerated current. The loss of these charge carriers is detrimental to cell performance.

(ii) Schottky diode solar cells are prone to pinning of the Fermi level arising from defect surface states at the metal-semiconductor interface. Fermi level pinning is the phenomenon in which some point in the energy gap is pinned or locked to the Fermi level. As a consequence, the bend bending, and hence the Schottky barrier height, becomes independent of the work function of the metal. Fermi level pinning imposes restriction on the open-circuit voltage of the solar cell.

(iii) In a Schottky barrier cell, the barrier to hole injection at the electron extracting electrode is poor. Consequently, the open-circuit voltage and fill factor are low for a given short-circuit current density.

15.4 QUANTUM DOT–DEPLETED HETEROJUNCTION SOLAR CELL (η = 3.36%)

The depleted heterojunction solar cell overcomes the deficiencies of the Schottky barrier design. The cell shown in Figure 15.7 consists of a transparent titanium dioxide nanoparticle film deposited on an FTO substrate. Upon the TiO_2 film, the PbS QDs film is formed, and this film is coated with an Au- or Ni-based electrode (Debnath et al 2010). (Another commonly used electrode material for these solar cells is MoO_3.) TiO_2 serves as the electron-accepting contact with $\sim 1 \times 10^{16}$ cm^{-3} electron concentration. Its counterpart in the Schottky barrier cell is the Al electrode with electron density $\sim 10^{22}$ cm^{-3}. So the electron density in the electron acceptor close to the junction is minimal. Light falls on the cell from the FTO substrate side. The junction between TiO_2 and PbS QDs is placed on this side. So the difficulties faced regarding carrier recombination from distant locations of the illumination side and the junction are avoided. The problem of Fermi level pinning caused by the presence of defect states at the interface between the metal and semiconductor is circumvented because the interface of interest, the TiO_2/PbS QDs interface, is passivated when the oleic acid–capped PbS QDs are deposited from the solution phase. In the Schottky barrier solar cell, this interface was the PbS QDs/Al interface, which was not passivated during deposition of Al electrode. Further, when we draw the energy band diagram of the device, we shall observe that there exists a large discontinuity in the edge of the valence band of the heterojunction, which impedes hole injection into the electron extraction electrode. These accomplishments of the heterojunction cell will be revisited after describing its operation to gain further clarity. During fabrication of the heterojunction device, all the processing work is carried out in open atmosphere.

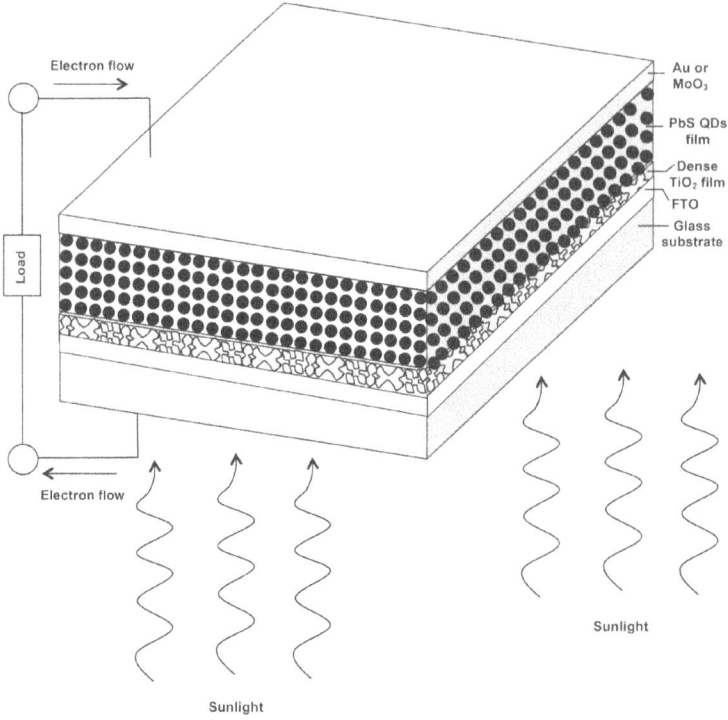

(a)

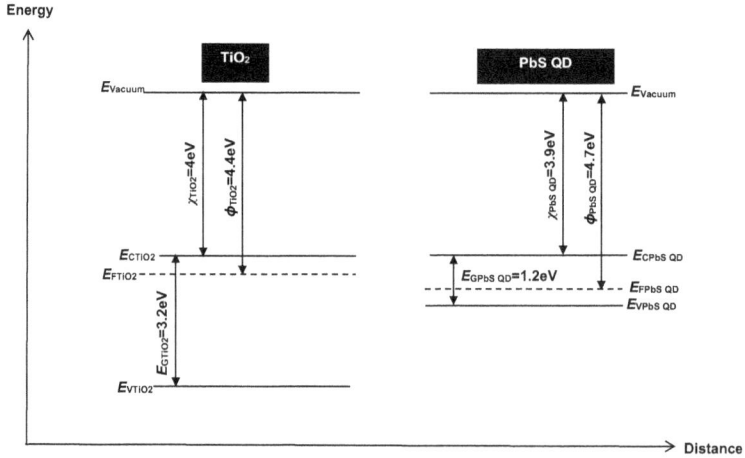

(b)

FIGURE 15.7 Depleted heterojunction solar cell: (a) construction of the cell showing from bottom upwards: the glass substrate coated with fluorine-doped tin oxide (FTO), the dense layer of TiO_2 nanoparticles, the PbS QDs, and the gold film. Light falls on the glass substrate of the solar cell and enters inside. Note the distinct boundary between TiO_2 and PbS QDs and mark that the QDs are placed above the TiO_2 film and do not infiltrate into the TiO_2 film apart from a few imperfections. *b* and *c* energy band diagrams of the TiO_2-PbS QDs film: (b) when they are far apart and (c) when they come in intimate contact. Note that the conduction band edge of PbS QDs lies slightly above that of TiO_2 and there is deep fall from the valence band edge of PbS QDs to that of TiO_2. The small difference in conduction band edges and the large difference in valence band edges play vital roles in the operation of the device.

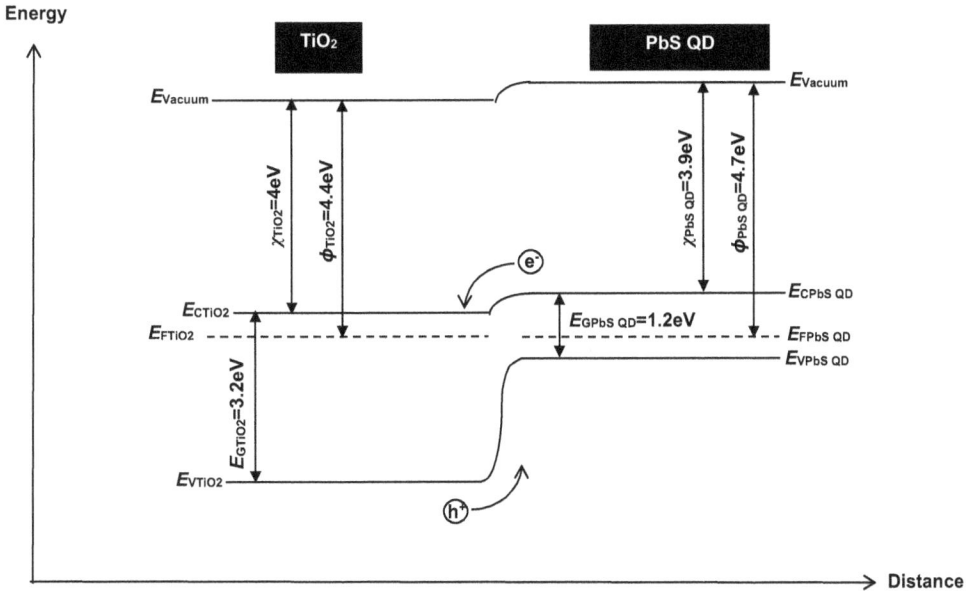

(c)

FIGURE 15.7 (Continued)

15.4.1 TiO₂ NANOPARTICLE FILM

The TiO_2 paste is diluted with ethanol, vortexed and spin-cast on FTO substrates at 1000RPM for 45 s. After 1 h sintering at 400°C, the substrates are dipped in aqueous $TiCl_4$ solution (120mM) for ½ h at 70°C, rinsed in DI water, dried, and heated at 520°C for 1 h.

15.4.2 PbS QD SYNTHESIS

A mixture of bis(trimethylsilyl)sulfide (TMS) with 1-octadecene (ODE) is dried and degassed to form the TMS/ODE mixture. Another mixture of lead oxide (PbO) with oleic acid and ODE is heated to 95°C in a flask under vacuum for 16 h and kept under Ar. Raising the temperature of the flask to 120°C, the TMS/ODE mixture is injected into the flask, the temperature decreased to 95°C, and the solution gradually cooled down to 36°C. Centrifuging is done in acetone to precipitate the PbS QDs. The QDs are dispersed in toluene, again precipitated with acetone and dried after centrifugation. After dispersal in toluene, storage in glove box, and a methanol wash, the QDs are ultimately dispersed in octane.

15.4.3 LAYER-BY-LAYER DEPOSITION OF PbS QD FILM ON POROUS TiO₂ FILM

The deposition is divided into eight steps of spinning at 2,800RPM for 10 s to achieve the desired film thickness. Each step involves spin-coating of PbS QD solution, addition of 3-MPA during spinning and rinsing with methanol, again during spinning.

15.4.4 TOP ELECTRODE DEPOSITION

LiF film of 1.2nm thickness is deposited by thermal evaporation. DC sputtering is done through a shadow mask for depositing Au and Ni films.

15.4.5 MEASUREMENTS

The solar cell with Au contact shows a short-circuit current density of 10.7mAcm^{-2}, open-circuit voltage of 0.51V, fill factor of 57.7%, and efficiency of 3.36% under 94mWcm^{-2} AM1.5 illumination. The device with LiF/Ni contact shows nearly equivalent performance with efficiency of 3.2% (Debnath et al 2010).

15.4.6 WORKING OF THE CELL

The focus of attention here is an N-p heterojunction, upper case "N" for wide-bandgap N-type material (TiO$_2$, E_G = 3.2eV), and lower case "p" for narrow-bandgap p-type material (PbS, E_G = 0.4eV for bulk and increasing with decrease in quantum dot size with typical value of 1.3eV for 3.7nm QD). Let us draw the energy band diagram for the TiO$_2$-PbS QD system. Inspection of the energy band diagram reveals that there is a band offset at the TiO$_2$/PbS QD interface. This band offset and the band bending indicates the formation of a space charge region. For annealed and stoichiometric material, the work function of TiO$_2$ is 4.4eV after exposure to water and annealing (Kashiwaya et al 2018). The value increases to ~ 4.9–5.1eV when the surface is oxidized. We have taken the value 4.4eV. The work function of P-type PbS QD is 4.7eV. Obviously, the work function of TiO$_2$ is less than that of PbS QD. Electrons will be released more easily from TiO$_2$ surface than from PbS QD surface. The electrons released in N-TiO$_2$ will be transferred to P-PbS QD. The N-TiO$_2$ region adjoining the interface will become positively charged by loss of electrons, while the P-PbS QD region near the interface will acquire negative charge. Thus, an electric field will be created at the interface. Its direction will be from N-TiO$_2$ to P-PbS QD. The electron transfer will continue unless the built-in electric field opposes further movement of electrons by repulsive force. A dynamic equilibrium condition will prevail.

15.4.7 SURMOUNTING THE DRAWBACKS OF SCHOTTKY DIODE CELL

Now let us look at the location of the electric field and the absorbing PbS QD film in the solar cell. We find that the electric field is established around the interface of TiO$_2$ and PbS QDs and the surface of the absorbing film is coincident with the interface. So any electron-hole pairs liberated inside the depletion region or within a diffusion length of this region will be easily captured by the electric field. The electrons will be transported towards the N-TiO$_2$ against the electric field direction, whereas holes will migrate towards the P-PbS QD in the direction of the electric field. Hence, the opposite charge carriers will be separated. Thus, in the depleted heterojunction solar cell, the disadvantage of Schottky diode structure about the distance between the surface of the absorbing film and the electric field region is avoided.

Note also that the edge of the conduction band of PbS QD lies slightly higher ~ 0.3eV than the conduction band edge of TiO$_2$. For quantum dots of diameter 3.7nm, the position of the 1S electron excited state, which is the conduction band edge for the QD, is ~ 0.3eV higher than TiO$_2$ conduction band edge (Hyun et al 2008). Electrons move down the hill from higher to lower energy, while holes climb up the hill from lower to higher energy. The interface between PbS QD and TiO$_2$, called a type II heterojunction, not only aids in the collection of electrons by TiO$_2$ but also obstructs their back flow. Hence, the driving stimuli for the charge separation are the built-in electric field of the N-p heterojunction and the band offset at the TiO$_2$/PbS interface.

Another difference from the Schottky diode structure becomes obvious from looking at the strikingly large descent of the valence band edge at the interface. The valence band discontinuity between PbS QD and wide-bandgap semiconductor TiO$_2$ shows that a high barrier > 1.5eV is faced by the majority-carrier holes moving from the P-type PbS QD film into the TiO$_2$ electron extraction electrode. So the unwanted hole injection in this direction is drastically reduced, stopping recombination of carriers at the PbS QD film/TiO$_2$ interface. The disadvantage of Schottky diode solar cell regarding this injection is alleviated. The holes are carried away by the ohmic contact to P-type PbS QD.

15.5 QUANTUM DOT–DEPLETED BULK HETEROJUNCTION SOLAR CELL ($\eta = 5.5\%$)

15.5.1 DISADVANTAGES OF SCHOTTKY DIODE AND DEPLETED HETEROJUNCTION STRUCTURES AND EVOLVING IMPROVED DESIGNS

On deeper examination of the structures of Schottky diode solar cell and the depleted heterojunction solar cells, it becomes evident that both these structures are disadvantaged by the presence of distinct, almost a straight line interface, apart from imperfections, between the light-absorbing PbS QD film and the electron-absorbing film, which is aluminum for Schottky diode and TiO$_2$ for depleted heterojunction case. For the schottky diode, the depletion region is formed in the PbS QD film. In the depleted heterojunction solar cell, too, a major part of the depletion region lies in the PbS QD film, with a small part in the TiO$_2$ film. The diffusion length of carriers in PbS QD film is small. So if we assume that photo-generated carriers within the depletion region or a diffusion length from the depletion region of the PbS QD film are apprehended by the electric field of the junction, we shall find that only a small thickness of PbS QD film is used for carrier collection. Carriers produced in the remaining PbS QD film are not useful. Thus, both these structures do not assure the best performance expected, and there is a need for a new architecture.

The carrier collection is considerably increased by employing a distributed interface between the PbS QD film and the TiO$_2$ in place of the sharp boundary interface (Figure 15.8). A nanostructured three-dimensional or bulk heterojunction is built to create this distributed interface. A thicker active layer consisting of quantum dots interspersed between TiO$_2$ nanoparticles is made. Its thickness is adequate enough to absorb all the parts of the solar spectrum, including the low-energy infrared photons. At the same time, care is taken to ensure that the PbS QD film is fully depleted so that charge collection efficiency is not compromised with.

Essentially, the volume of the PbS QD absorbing film is increased while simultaneously guaranteeing the depletion of the full QD film, unlike the depleted heterojunction device in which a large fraction of the volume of PbS QD film remained undepleted and hence unexploited. In the DBH cell, the photon interaction length is increased by using an infiltrated nanoporous TiO$_2$ electrode, but the length of electron exit route is kept small. By adopting this scheme, the efficiency limit on the heterojunction solar cell is uplifted.

Needless to say, the energy band diagrams of DH and DBH solar cells are identical. The only differentiating feature arises from the larger interfacial surface area between porous TiO$_2$ and PbS QDs. Owing to this surface area difference, there is a larger volume of PbS QDs in the depletion region of DBH solar cell than the DH cell.

15.5.2 DIFFERENCE BETWEEN FABRICATION PROCESSES OF DH AND DBH SOLAR CELLS

The fabrication process of the DH and DBH solar cells are different (Barkhouse et al 2011). In both cases, a dense layer of TiO$_2$ nanoparticles is spin-coated on the FTO substrate. The dense TiO$_2$ layer is necessary to prevent short-circuiting of current by direct flow between PbS QDs to be placed above the TiO$_2$ nanoparticles and the underlying FTO substrate. Next, the PbS QDs are deposited. Owing to the foundation of dense TiO$_2$ layer, they cannot penetrate this layer and contact the FTO substrate. The size of TiO$_2$ nanoparticles in this dense layer ranges between 10 and 30nm. For fabricating the DBH solar cell, another porous TiO$_2$ layer containing larger diameter particles of size range 150–250nm is coated by spinning a TiO$_2$ nanoparticle paste. On top of this porous layer, the PbS QDs are deposited. They infiltrate into the crevices of this porous layer and settle into a distributed TiO$_2$/PbS QD interface. As in the case of DH solar cell, the top electrode is

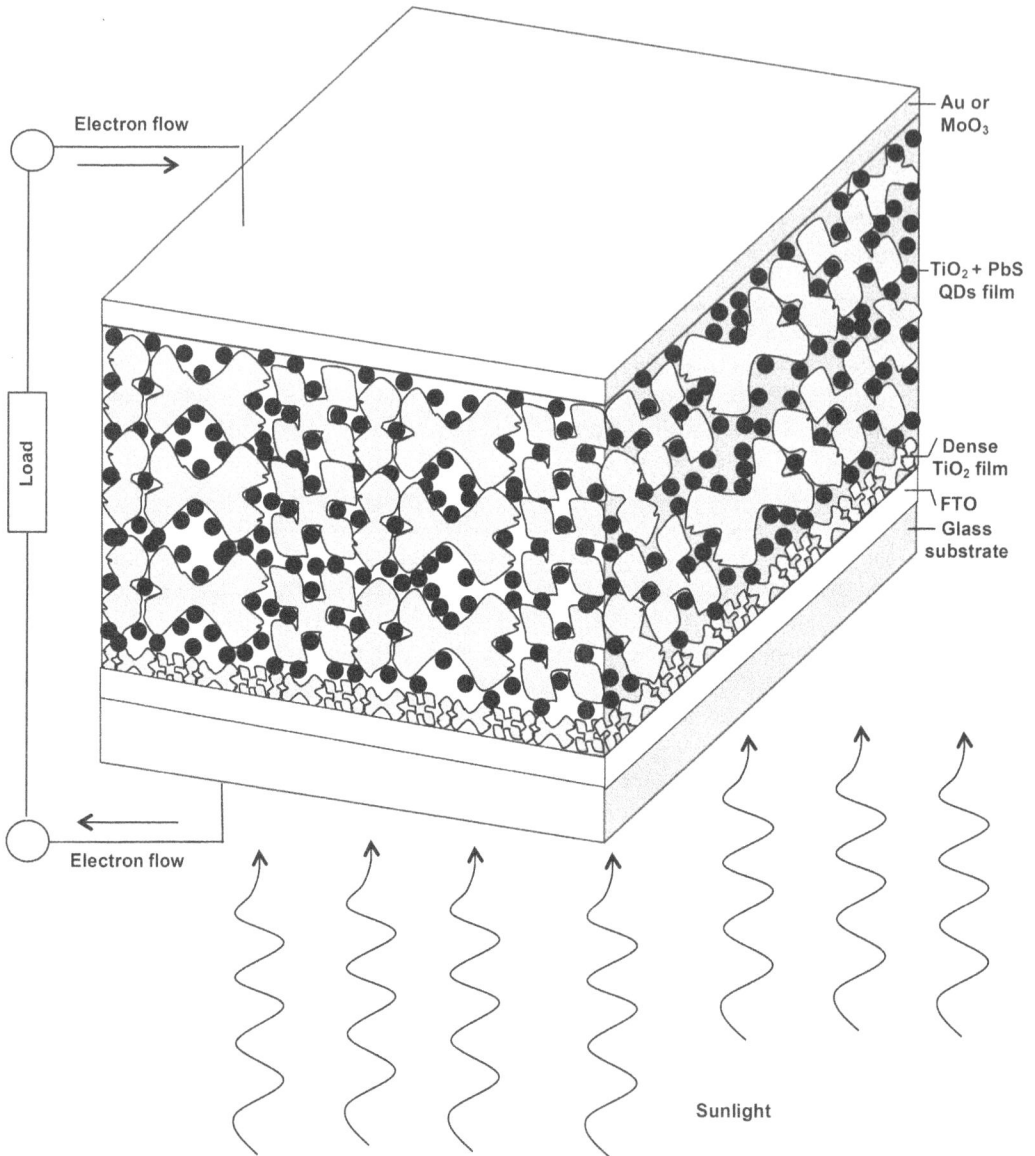

FIGURE 15.8 Depleted bulk heterojunction solar cell comprising, from bottom upwards, the FTO-coated glass substrate, the thin dense layer of smaller TiO$_2$ nanoparticles, the thick layer of larger-size TiO$_2$ nanoparticles intermixed with PbS QDs, and the top Au or MoO$_3$ electrode. The dense bottom layer of smaller TiO$_2$ nanoparticles is same as in the DH solar cell, but unlike the DH solar cell, there is another layer in which the TiO$_2$ nanoparticles and PbS QDs are not deposited as separate layers, but there is one (TiO$_2$ + PbS QDs) film. The QDs are distributed on the surfaces of TiO$_2$ nanoparticles. The thickness of (TiO$_2$ + PbS QDs) film is larger than the total thickness of separate TiO$_2$ and QDs layers in DH cell. Also, there is no separate layer consisting of QDs only. The top electrode is same, Au or MoO$_3$, as for DH solar cell.

made from either an Au or MoO$_3$ film. Scrutiny and comparison of the structures of DH and DBH solar cells shown in Figures 15.7 and 15.8 will bring out the differences between the two types of solar cells. With the help of TiCl$_4$ treatment, the DBH device exhibits a short-current density of 20.6mAcm^{-2}, open-circuit voltage of 0.48V, fill factor of 56%, and efficiency of 5.5% (Barkhouse et al 2011).

15.6 QUANTUM DOT HYBRID SOLAR CELL (η = 4.91%)

15.6.1 Necessity of Hybrid QD Solar Cell

One notable disadvantage of many solar cell categories discussed above is the complicated layer-by-layer deposition method followed in the preparation of PbS QD films. This obvious shortcoming restricts the applicability of the process to the realization of large-area cells, fostering the quest for simple one-step deposition methods. An easier way to avoid the lengthy process involves mixing a conjugated polymer with the quantum dots, and the device obtained by this method is named as the quantum dot hybrid solar cell (Nguyen et al 2021). To clarify further, in place of TiO$_2$ as the donor and PbS QD as the acceptor, we have now the polymer as the donor layer and PbS QD as the acceptor. An example of the polymer used is the block copolymer P3HT-b-PS [poly(3-hexylthiophene)-b-polystyrene].

By a hybrid solar cell, we mean a solar cell which works by transference of charge across the interface between an organic semiconductor and an inorganic semiconductor. Here, PbS QDs are the inorganic materials, whereas the polymer is an organic material. Hence, this is a hybrid solar cell made of organic donor and inorganic acceptor semiconductors. In the solar cell using TiO$_2$ and PbS, both were inorganic materials. So this was not a hybrid solar cell.

15.6.2 Hybrid QD Solar Cell Structure

The solar cell structure (Figure 15.9 (a)) is glass/ITO/PEDOT:PSS layer/(PbS QDs + polymer)/PbS QDs only/LiF/Al (Nguyen et al 2021). On the ITO-coated glass substrate is a hole transport layer made of PEDOT: PSS. Then there is a layer of PbS QDs blended with P3HT-b-PS polymer. Upon this layer, a pure layer of PbS QDs is deposited, and this is overlaid with LiF/Al cathode film. The energy band diagram for this solar cell is depicted in Figure 15.9(b). The fabrication process is described in the subsections below.

15.6.3 P3HT-Br Synthesis

Grignard metathesis (GRIM), a controlled polymerization method used for synthesizing semiconducting polymers, is applied to make P3HT-Br (P3HT with the bromo terminal group).

15.6.4 Formation of P3HT-b-PS

Suzuki coupling reaction, an organic cross-coupling reaction, is applied to combine the two blocks: P3HT-Br and PS-B(OR)$_2$ (pinacol boronic ester as functional group) to form the block copolymer P3HT-b-PS.

15.6.5 PbS QD Synthesis

Lead(II) acetate trihydrate (PbAc.3H$_2$O), oleic acid (OA), and 1-octadecene (ODE) are mixed in a three-neck flask to form lead oleate. After heating the flask to 100°C under nitrogen flow, bis(trimethylsilyl)sulfide (TMS) in ODE is injected into the flask at 100°C in nitrogen. Oleic acid–capped PbS QDs are formed, as indicated by the change in color of the solution to brown.

15.6.6 Substrate Cleaning

ITO-coated glass substrates are cleaned with detergent, DI water, acetone, and isopropyl alcohol and subjected to oxygen plasma.

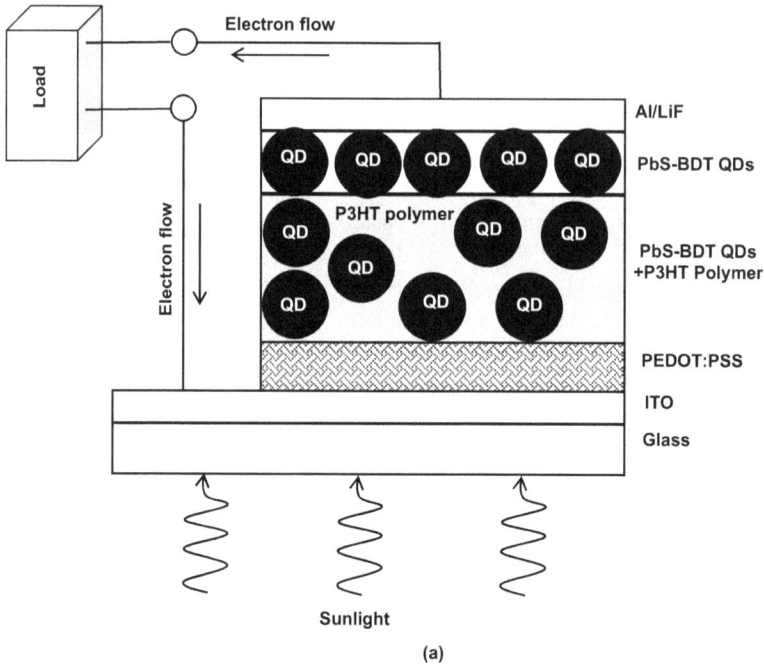

(a)

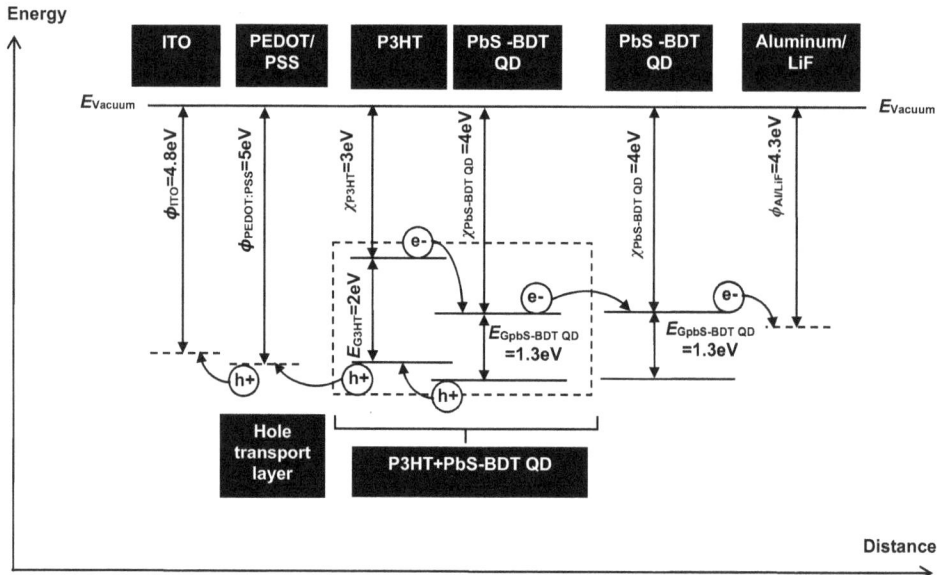

(b)

FIGURE 15.9 Quantum dot hybrid solar cell: (a) Schematic cross section showing the layers from bottom upwards: glass substrate, ITO, PEDOT:PSS, PbS-BDT QDs + P3HT, PbS-BDT QDs, and Al/LiF. Sunlight falls from the glass substrate side. (b) Energy band diagram of the solar cell. The work function of ITO is $\phi_{ITO} = 4.8\text{eV}$; its value for PEDOT:PSS is $\phi_{PEDOT:PSS} = 5\text{eV}$, and that for Al/LiF is $\phi_{Al/LiF} = 4.3\text{eV}$. The electron affinity of P3HT is $\chi_{P3HT} = 3\text{eV}$. The PbS-BDT QD has an electron affinity $\chi_{PbS-BDT} = 4\text{eV}$. The energy gap of P3HT is $E_{GP3HT} = 2\text{eV}$; the same for PbS-BDT QD is $E_{GPbS-BDT} = 1.3\text{eV}$. The electrons drift towards the Al side, whereas the holes migrate to the ITO side. The electronic current directions are indicated.

15.6.7 PEDOT:PSS Coating

A layer of poly(styrene sulfonic acid)-doped poly(3,4-ethylenedioxythiophene) is spin-coated on the substrate and dried at 140°C for 15 min. Thickness of PEDOT:PSS layer is 40nm.

15.6.8 P3HT-*b*-PS/PbS QDs Coating

A solution of P3HT-*b*-PS and oleic acid–capped PbS QDs in chloroform is spin-coated over the PEDOT:PSS layer.

15.6.9 Post-Ligand Exchange to BDT

Oleic acid is replaced with BDT (1,3-benzenedithiol) by spin-coating BDT solution in acetonitrile. Any excess BDT as well as oleic acid is removed by rinsing with acetonitrile two times. Thickness of polymer:BDT-PbS QD layer is 120nm.

15.6.10 Pure Layer of PbS QDs with Oleic Acid Ligands

This layer is deposited from an octane solution. Oleic acid ligand is exchanged with BDT. Cleaning is followed by 20 min. baking at 175°C in a glove box. Thickness of BDT-PbS QDs is 30nm.

15.6.11 Cathode Deposition

LiF film of 0.6nm thickness and 100nm-thick aluminum film are deposited by thermal evaporation through a shadow mask.

15.6.12 Solar Cell Testing

For P3HT/P3HT-*b*-PS/PbS QDs (0.7/0.3/20; polymer ratio is weight ratio), the open-circuit voltage is 0.57V, the short-circuit current density is 16.21mAcm^{-2}, fill factor is 53.2%, and power conversion efficiency is 4.91% (Nguyen et al 2021).

15.7 QUANTUM DOT–SENSITIZED SOLAR CELL

15.7.1 Similarities and Dissimilarities with DBH Solar Cell

This cell is similar to the DBH solar cell and also to the dye-sensitized solar cell from which it has evolved. Its difference from the DBH solar cell is that instead of hole extraction directly into an Au or MoO$_3$ electrode from the PbS QD film, it uses an electrolyte, either in liquid form or solid state, together with Au or Pt electrode for rapid hole extraction. The electrolyte, irrespective of being liquid or solid, acts as an efficient hole transporter material. In addition, it also blocks the back recombination of electrons and holes across the interface between TiO$_2$ and the electrolyte, preventing loss of carriers by this process.

15.7.2 Difference from Dye-Sensitized Solar Cell

It differs from the dye-sensitized solar cell in the respect that it uses quantum dots in place of ruthenium-based dyes. The advantage gained by the substitution of molecular dyes with QDs is that the quantum dots allow extension of the range of absorption of the solar spectrum beyond the cutoff wavelengths of the dyes. The additional flexibility provided by QDs has been widely exploited. On the downside, the absorption coefficient of TiO$_2$ coated with a monolayer of QDs is lower than that of a similar dye-coated layer. Therefore, higher short-circuit current densities are attainable only by sacrificing the fill factor, which is a deleterious effect.

15.7.3 CONSTRUCTION

Constructionally, the QD-sensitized solar cell shown in Figure 15.10 resembles the DBH cell displayed in Figure 15.8. The inclusion of electrolyte in the QD-sensitized cell distinguishes it from the DBH cell. Also, the top electrode is made of Au, Pt, or Cu_2S.

15.7.4 PRINCIPLE

This solar cell is a type II heterostructure-based solar cell. It is conceived as a device in which nanoparticles of a wide-bandgap semiconductor such as TiO_2, ZnO, or SnO_2 are sensitized with a

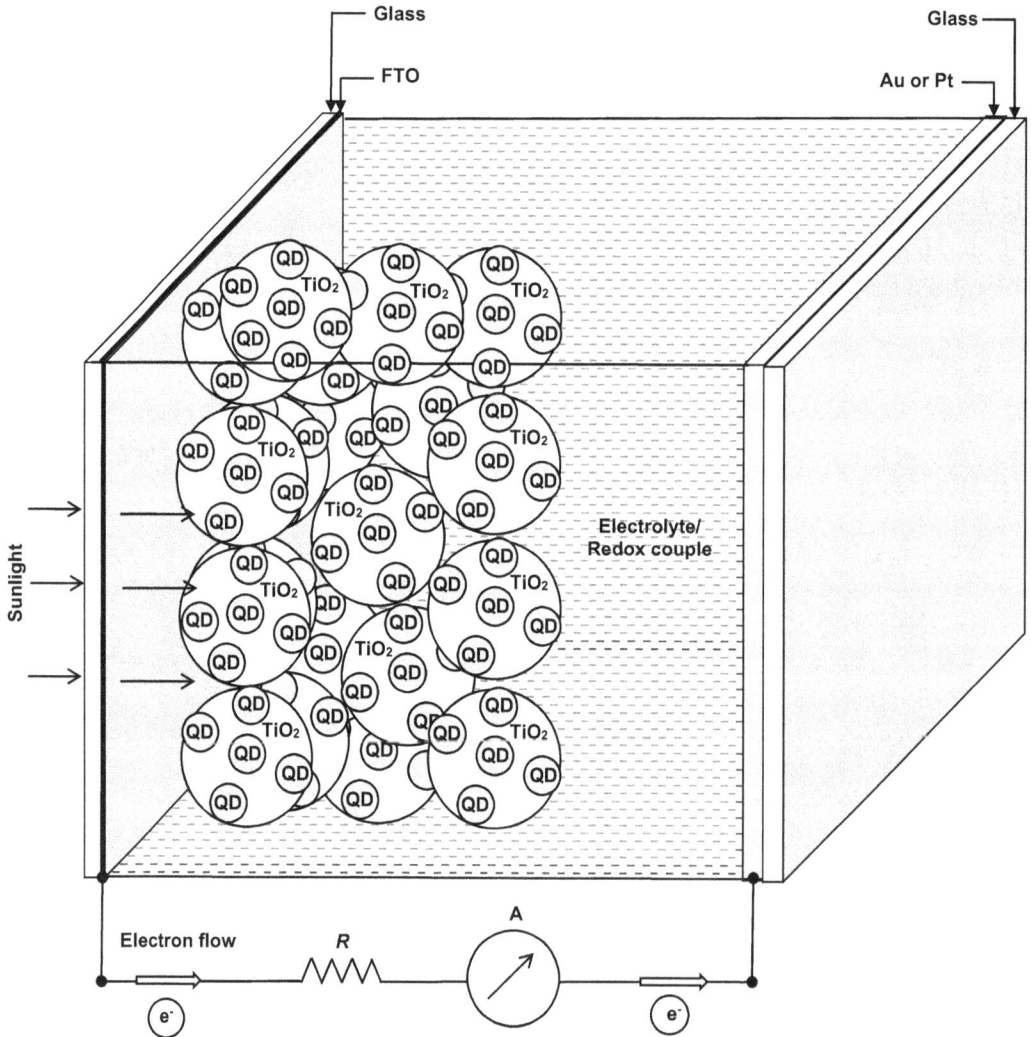

FIGURE 15.10 Generic construction of a quantum dot–sensitized solar cell showing the FTO-coated glass substrate, the TiO_2 nanoparticles, the quantum dots, the electrolyte, and the Au or Pt counter electrode. Quantum dots are used in place of dye molecules of a dye-sensitized solar cell.

quantum dot semiconductor, such as CdS, CdSe, PbS, ZnS, etc., having a lower bandgap responsive to visible or infrared radiations from the sun. The pair of semiconductors constituting the solar cell form a type II heterojunction. The smaller-bandgap semiconductor, e.g., CdS, absorbs the incident light. Electron-hole pairs are produced in this semiconductor upon irradiation. The large-bandgap semiconductor, e.g., TiO_2 is associated with the smaller-bandgap partner to separate the photogenerated charge carriers. The type II heterojunction between these two materials provides the charge separation. In this heterojunction, the lowest energies of the conduction and valence bands are not in the same semiconductor. Instead, they lie in the two semiconductors on opposite sides of the heterojunction. As the carriers move from higher to lower energies, both in the conduction and valence bands, the electrons are transported to one side of the heterojunction and holes on the other side and are thus mutually separated. Electrons are injected in the larger-bandgap semiconductor, while holes remain in the smaller-bandgap semiconductor. The electrons diffuse to the contact, and the holes interact with a redox couple from a liquid electrolyte or are regenerated from the valence band of a solid hole transporting medium. A commonly used redox couple is S_2^- / S_n^{2-}. This couple is obtained from a polysulfide electrolyte. The electrolyte is prepared from Na_2S, S, and KCl.

Taking the PbS semiconductor as an example, the processes taking place in the solar cell are:

(i) Production of electron-hole pairs in PbS upon receiving sunlight:

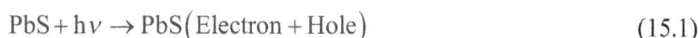

$$PbS + h\nu \rightarrow PbS(Electron + Hole) \tag{15.1}$$

(ii) Action of heterojunction. Injection of electron into TiO_2 occurs with hole residing in PbS. The necessary energy for electron transfer comes from the conduction band offset between the two semiconductors:

$$PbS(Electron + Hole) + TiO_2 \rightarrow PbS(Hole) + TiO_2(Electron) \tag{15.2}$$

(iii) Electron diffusion from TiO_2 to contact:

$$TiO_2(Electron) \rightarrow Electron(Contact) \tag{15.3}$$

(iv) Interaction of hole in PbS with redox system:

$$PbS(Hole) + S^{2-} \rightarrow PbS + S_n^{2-} \tag{15.4}$$

(v) Interaction of electrolyte with counter electrode

15.8 FABRICATION OF PbS QD-SENSITIZED SOLAR CELLS

15.8.1 PbS-ZnS QDs-Sensitized Solar Cells (η = 2.41, 4.01%)

Figure 15.11 shows a PbS QD-sensitized solar cell; the PbS QDs are passivated with ZnS. The layers in this solar cell are FTO/TiO$_2$/PbS-ZnS QDs/Electrolyte/Cu$_2$S counter electrode (Tian et al 2016). Look back at Figures 7.1–7.2 and note that the dye molecules in Figures 7.1–7.2 are replaced by quantum dots in Figures 15.10–15.11.

The performance of the solar cell was enhanced by optimizing the concentration of the precursor Pb(NO$_3$)$_2$ used in synthesis of PbS QDs and replacement of 30% DI water of polysulfide electrolyte with methanol. The energy band diagram is shown in Figure 15.12. Solar cell fabrication process is given in the following subsections:

15.8.1.1 Mesoporous TiO$_2$ Film Deposition on FTO Substrate

TiO$_2$ paste comprising a mixture of TiO$_2$ nanoparticles with ethyl cellulose and α-terpineol is deposited by doctor blading on the substrate and sintered at 500°C for ½ h.

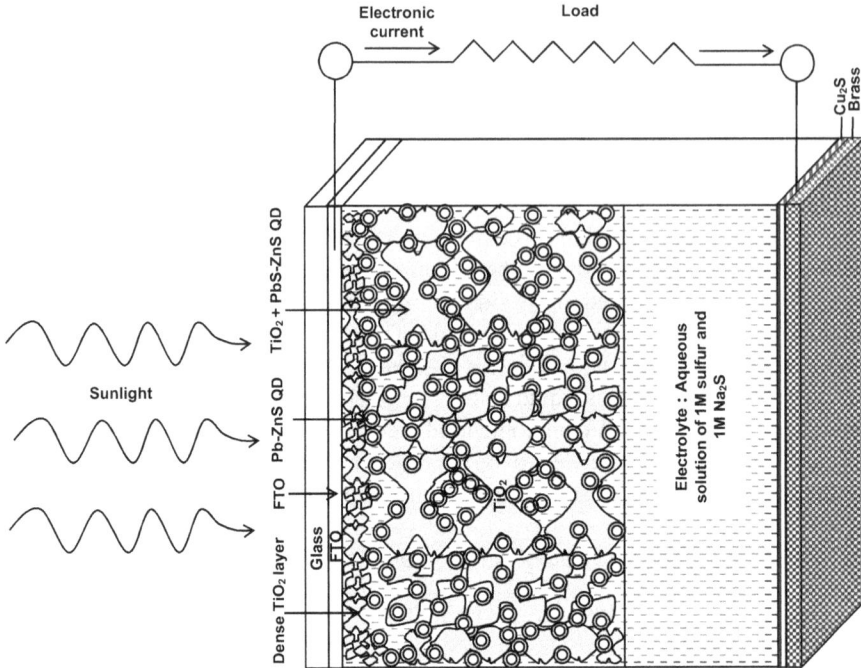

FIGURE 15.11 PbS-ZnS quantum dot–sensitized solar cell showing from left the glass substrate, the FTO layer, the dense TiO$_2$ layer comprising smaller particles, the mixture of PbS-ZnS QDs and bigger TiO$_2$ particles, the electrolyte, and the Cu$_2$S film deposited on brass foil as the counter electrode. Sunlight is incident from the glass-FTO side, and the electronic current produced flows from the glass-FTO side to the Cu$_2$S-brass foil electrode side.

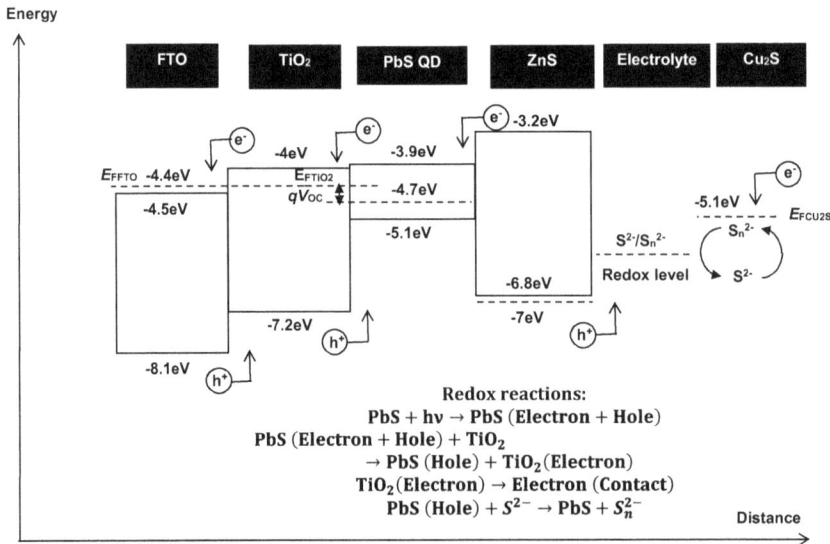

FIGURE 15.12 Energy band diagram of the FTO/TiO$_2$/PbS/ZnS/Electrolyte/Cu$_2$S solar cell. Electrons move towards the FTO side, and holes travel towards the Cu$_2$S side. Electron-hole pairs are produced in the PbS QDs under the action of sunlight. The electron migrates to TiO$_2$ and from TiO$_2$ to the FTO contact. Upon interaction with the electrolyte, the PbS (hole) becomes neutral PbS and the charge of the hole is delivered to the Cu$_2$S electrode.

15.8.1.2 Deposition of PbS QDs on TiO$_2$ Layer

The TiO$_2$ film is immersed in metallic precursor Pb(NO$_3$)$_2$ solution in DI water/methanol in the ratio 1/1 for 1 min., followed by washing with DI water and methanol. Then, the film is dipped in sulfur precursor Na$_2$S solution in DI water/methanol in the ratio 1/1 for 1 min., again followed by washing with DI water and methanol. These two steps constitute one cycle of successive ionic layer adsorption and reaction (SILAR) technique. The deposition is completed in three SILAR cycles. The concentrations of the precursors are altered to determine the optimal concentration (0.06M).

15.8.1.3 Deposition of ZnS Passivation Layer

Similar to PbS QDs, this deposition is done in two SILAR cycles using Zn(NO$_3$)$_2$ as the metallic precursor and Na$_2$S as the sulfur precursor.

15.8.1.4 Electrolyte

A mixture of 1M S and 1M Na$_2$S in DI water/methanol in the ratio 7/3 is used as the electrolyte.

15.8.1.5 Deposition of Cu$_2$S Film on Brass Foil to Make the Counter Electrode

The electrode is prepared by etching the brass foils in 37% HCl (70°C for 20 min.), washing, drying, and immersion in an aqueous solution of 1M sulfur and 1M Na$_2$S.

15.8.1.6 Photovoltaic Characterization of the Solar Cell

When the electrolyte contains 30% methanol, the open-circuit voltage of the solar cell is 0.43V, the short-circuit current density is 18.34mAcm^{-2}, the fill factor is 50.86%, and the efficiency is 4.01%. These values change to, respectively, 0.41V, 11.68mAcm^{-2}, 49.86%, and 2.41% without methanol. The difference is explained by the incomplete wetting of the TiO$_2$ nanoparticles in the absence of methanol. This happens due to adsorption of air molecules on their surfaces. The incomplete wetting inhibits the diffusion of redox couple through the electrolyte and thereby hinders the charge transport. Methanol helps improve the wettability of TiO$_2$ nanoparticles and therefore facilitates charge transfer (Tian et al 2016).

15.8.2 PbS-ZnS QDSSC (η = 5.82%)

This solar cell comprises an FTO substrate on which is deposited a compact TiO$_2$ layer, followed by a porous TiO$_2$ layer, then PbS QDs are deposited and passivated with ZnS (Bhalekar et al 2019). Polysulfide is used as the electrolyte. The cell has a molybdenum oxide counter electrode.

15.8.2.1 Compact TiO$_2$ layer

To TiCl$_3$ aqueous solution, 1M NaOH is added drop-wise until it becomes transparent. The FTO substrate is dipped in the solution and kept in incubator at 45°C for 15 h. The substrate is washed in water, dried, and annealed at 450°C for 1 h.

15.8.2.2 Porous TiO$_2$ Layer

TiO$_2$ paste is made by ultrasonifying TiO$_2$ powder in ethanol, ethyl cellulose, and terpanol and then adding acetyl acetone. Layer-by-layer deposition is done up to eight layers. The film is dried and, step-by-step, heated at 100°C, 200°C, and 300°C for ¼ h each and finally annealed at 450°C for 1 h.

15.8.2.3 Sensitization of Porous TiO$_2$ Layer with PbS QDs

0.02M Pb(NO$_3$)$_2$ solution is made in water. Separately, 0.02M Na$_2$S solution is made in ethanol. Six SILAR cycles are executed.

15.8.2.4 Passivation of QDs with ZnS, Electrolyte Injection, and Device Assembly

Two SILAR cycles are carried out for depositing ZnS over PbS QDs. Polysulfide electrolyte is made by grinding sodium sulfide and sulfur in ethanol and adding water and ethanol. FTO substrate is coated with MoO_3 by spray pyrolysis to make the counter electrode. The electrolyte is injected between the ZnS-passivated PbS QDs and the counter electrode to complete the assembly of solar cell.

The electrode with additional compact TiO_2 layer provides better charge recombination resistance and improved electron lifetime. Passivation of the surfaces of QDs with ZnS is also helpful, yielding an efficiency of 5.82% (Bhalekar et al 2019).

15.9 FABRICATION OF CdS QD-SENSITIZED SOLAR CELLS

15.9.1 CdS QDSSC WITH $\eta = 1.84\%$

Chang and Lee (2007) made this solar cell. The energy-band diagram of the solar cell is shown in Figure 15.13. It consists of two semiconductors: CdS quantum dots and TiO_2 nanoparticles. The two semiconductors CdS and TiO_2 are chosen because they together form a type II heterojunction. Due to its large bandgap, TiO_2 can at best respond to UV photons so that it has negligible role in carrier generation.

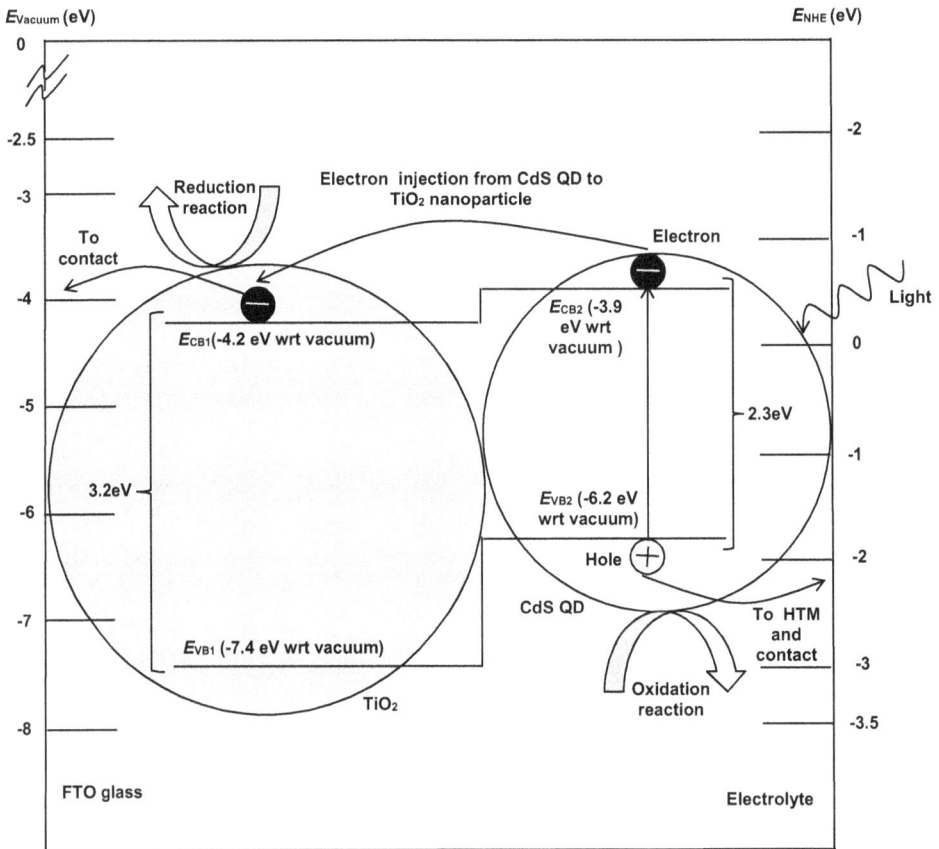

FIGURE 15.13 CdS quantum dot–sensitized solar cell. A CdS quantum dot is used as the photosensitive or light-absorbing material to sensitize the nanoparticle of a wide-bandgap semiconductor TiO_2. The conduction band edge and valence band edge energy levels E_{CB1} and E_{VB1} for TiO_2 and the corresponding energy levels E_{CB2} and E_{VB2} for CdS are marked relative to vacuum energy (E_{Vacuum}, eV) scale on the left and normal hydrogen electrode (E_{NHE}, eV) scale on the right. Electron injection and hole movement directions are shown. The reduction/oxidation reactions are also indicated.

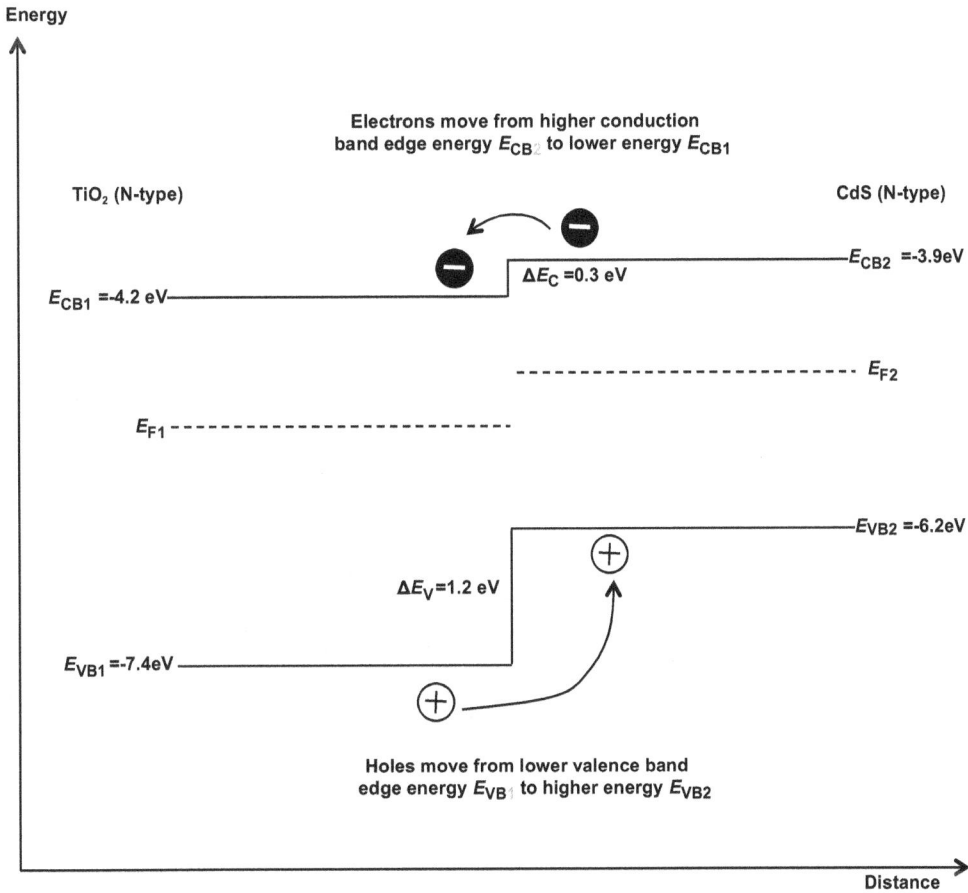

FIGURE 15.14 Type II heterojunction between TiO_2 and CdS semiconductors. E_{CB1} is the energy of conduction band edge of TiO_2, E_{VB1} is its valence band edge energy, E_{F1} is the Fermi level in TiO_2; E_{CB2}, E_{VB2}, E_{F2} are respectively the energies of conduction band edge, valence band edge, and Fermi level in CdS. The difference $\Delta E_C = 0.3$ eV is the conduction offset between TiO_2 and CdS, while $\Delta E_V = 1.2$ eV is the valence and offset between the two materials. Electron and hole movement directions are shown.

The behavior of type II heterojunction is explained in Figure 15.14. This heterojunction plays the vital role of separation of charge carriers produced by incident light in this solar cell. To appreciate its function, the peculiarity and special properties of type II heterojunction must be emphasized:

(i) The lowest energy of conduction band (E_{CB1}) is on the left-hand side of the heterojunction, while the highest energy of the valence band (E_{VB2}) is on the right-hand side of the heterojunction. This difference of location of conduction and valence band energy minima promotes electron transfer from right to left side and hole transfer in the reverse direction. Hence, this kind of heterojunction provides an efficient way of transference of charge carriers on opposite sides of the heterojunction, leading to their easy separation.

(ii) The conduction band edge of TiO_2 is at a lower energy than that of CdS so that electrons can migrate from CdS to TiO_2. But the valence band edge of CdS is at a higher energy than that of TiO_2. Hence, hole transfer from CdS to TiO_2 is not favored. Any hole in CdS will therefore not cross over to TiO_2 unlike electron, which immediately moves from CdS to TiO_2. Remember electrons move from higher to lower energy and holes from lower to higher energy.

From these behavioral aspects of CdS-TiO$_2$ heterojunction, we see that when sunlight shines on CdS and produces an electron-hole pair, the electron moves from CdS to TiO$_2$, while the hole is retained on the CdS side. The driving energy for electron transfer is the conduction band off-set energy which acts in the direction from higher- to lower-conduction band energy. The hole remains in the CdS side due to higher valence band level of this side. So the electron is in the TiO$_2$ side of the junction, whereas the hole is in the CdS side, and the two charge carriers have been separated. After separation, the electron moves towards the contact and a reduction reaction takes place. The hole moves through a hole transport medium to undergo an oxidation reaction near the contact.

The fabrication steps of this solar cell are given in the subsections that follow.

15.9.1.1   CdS QDs-Modified TiO$_2$ Electrode

Substrates: F-doped tin oxide (FTO) or indium tin oxide (ITO) substrates are used. These substrates are electrically conducting and optically transparent.

TiO$_2$ film: TiO$_2$ paste is spin-coated on the substrate and sintered at 450°C for ½ h, producing a 5.5μm-thick TiO$_2$ film.

QDs: For QD preparation, Cd(NO$_3$)$_2$ solution is made in ethanol, and Na$_2$S solution is made in methanol. A two-step chemical bath deposition (CBD) cycle is adopted. In the first step, the TiO$_2$ film is immersed in 0.5M Cd(NO$_3$)$_2$-ethanol solution for 5 min. and rinsed with ethanol. In the second step, it is dipped in 0.5M Na$_2$S-methanol solution for 5 min. and rinsed with methanol. The number of cycles repeated determines the amount of CdS QDs deposited. Thus, a CdS QDs-modified TiO$_2$ electrode is formed. This CdS deposition method is known as successive ionic layer adsorption and reaction (SILAR) as mentioned in Secs. 15.8.1.2, 15.8.2.3 and 15.8.2.4.

15.9.1.2   Counter Electrode

As a counter electrode, the platinum electrode is used.

15.9.1.3   Redox Electrolyte

3-methoxypropionitrile (MPN) solution is the redox electrolyte. It consists of lithium iodide (0.1M) + iodine (0.05M) + 1-propyl-2,3-dimethylimidazolium iodide (DMPII), 0.6M + 4-tert-butylpyridine (TBP), 0.5M.

15.9.1.4   Sealing

The CdS QDs-modified TiO$_2$ electrode and the platinum counter electrode are bound together with sealing material (60μm thickness).

15.9.1.5   Efficiency

The efficiency of the solar cell is 1.84% under 1 sun illumination (Chang and Lee 2007).

15.9.2   Cds QDSSC Using Graphene Oxide Powder (η = 2.02%)

To improve the efficiency of solar cell, Wageh et al (2013) reduced the electron-hole recombination because of the poor hole recovery rate across the counter electrode/electrolyte interface. In the method given by them, the CdS-modified TiO$_2$ film is covered with graphene oxide powder before fixing the Pt electrode.

TiO$_2$ powder is made by sol-gel method using the precursors titanium tetraisopropoxide (TTIP), ethyl alcohol (EtOH), water (H$_2$O), and hydrochloric acid (HCl) and converted into paste form. The TiO$_2$ paste is coated on FTO substrate by doctor blading, followed by sintering at 500°C for 1 h. CdS QDs are deposited by the same method as Chang and Lee (2007). Graphene powder made by modified Hummers method is distributed over the surface of the CdS-modified TiO$_2$ film. The electrolyte is polysulfide made from 0.5M Na_2S + 2M S + 0.2M KCl. Pt-deposited glass serves as the counter electrode. The efficiency of CdS QDs/TiO_2 solar cell with graphene oxide layer is 2.02%.

15.9.3 Increasing the QDSSC Efficiency by Modification of CdS with 2D g-C_3N_4 (η = 2.31%)

The modified solar cell is shown in Figure 15.15(a), and its energy band diagram is drawn in Figure 15.15(b).

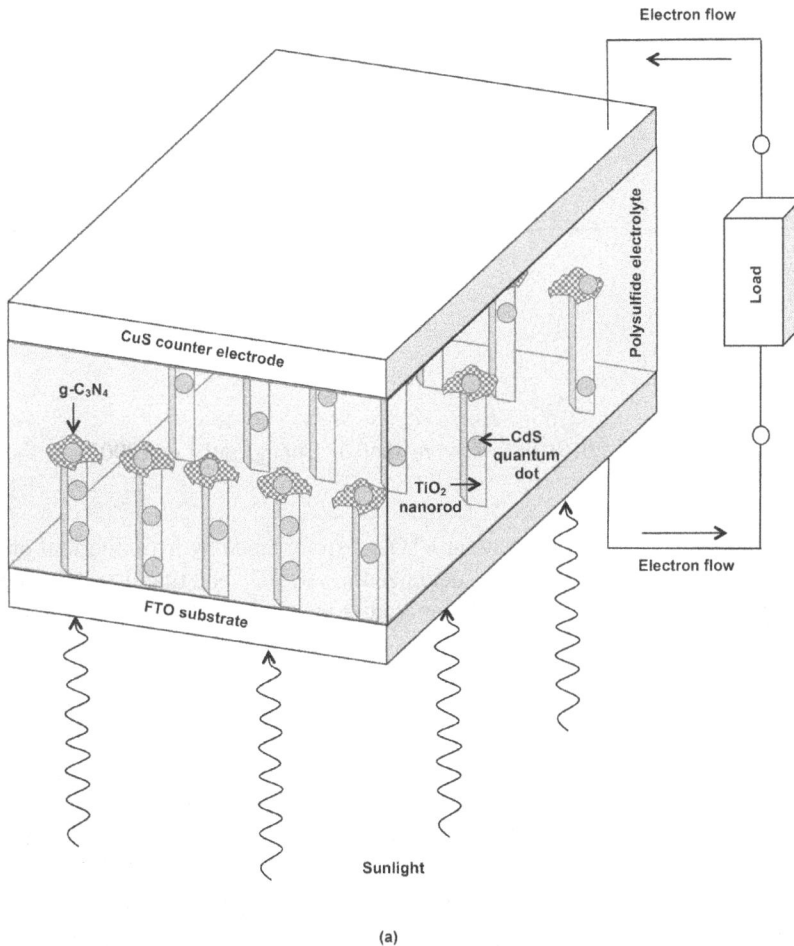

(a)

FIGURE 15.15 Quantum dot–sensitized solar cell enhanced with graphitic carbon nitride–modified TiO_2 nanorods: (a) cell construction showing the layers from bottom upwards: FTO substrate, TiO_2 nanorods decorated with g-C_3N_4, CdS quantum dots deposited over the decorated TiO_2 nanorods, the electrolyte, and the CuS counter electrode; (b) Anderson's rule-based energy band diagram showing the charge separation and transference mechanism in the solar cell. The electron affinity of g-C_3N_4 is χ_{C3N4} = 3eV, that of TiO_2 is χ_{TiO2} = 4.2eV, and that of CdS is χ_{CdS} = 3.9eV. The bandgaps of g-C_3N_4, TiO_2 and CdS are E_{GC3N4} = 2.7 eV, E_{GTiO2} = 3eV, and E_{GCdS} = 2.4eV, respectively. As electrons can move along downhill paths from higher to lower energies, the g-C_3N_4 layers on either side of TiO_2-CdS portion prevent electrons from travelling to the electrolyte, thereby opposing the backward electron recombination between TiO_2-CdS and electrolyte. The nonpermitted electron paths are indicated by dotted lines. Crosses are used to mark these paths, meaning, their impermissibility. Besides this electron-stalling action, the g-C_3N_4 layers contribute supplementary photogenerated electrons and holes contributing to current output from the cell.

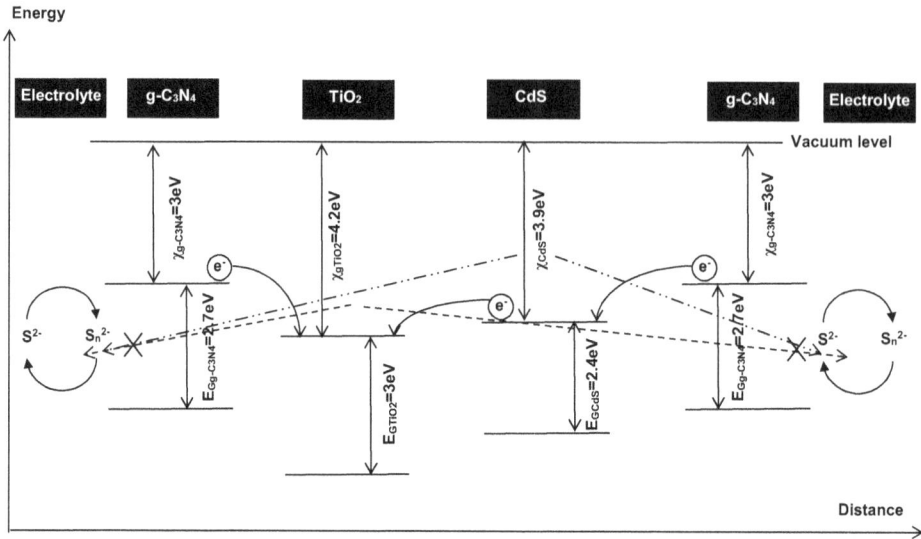

(b)

FIGURE 15.15 (Continued)

This solar cell (Gao et al 2016) differs from that of Chang and Lee (2007) in the following respects:

(i) Deposition of TiO_2 nanorod array on FTO substrate made by hydrothermal process at 150°C for 12 h from the mixture of deionized water, HCl, and titanium butoxide.

(ii) Decoration of the single crystal TiO_2 nanorod array with two-dimensional (2D) graphitic carbon nitride (g-C_3N_4). The 2D g-C_3N_4 is a metal-free conjugated polymer. Its layered structure is similar to that of graphite, and the basic structural unit is a layer of triazine ($C_3H_3N_3$) or heptazine ($C_6N_7H_3$) cores. It has a moderate bandgap of 2.7eV. It is highly stable, both chemically and thermally, and absorbs light in the visible spectrum range. It can be made by a facile synthesis procedure by heating a mixture of melamine and urea at 550°C for 2 h. The yellow crystals obtained are grinded into powder and converted into a paste by mixing with ethyl cellulose and α-terpinol in anhydrous ethanol and stirring.

(iii) CuS counter electrode is used in place of Pt electrode. It is made with the help of chemical bath deposition by immersing FTO substrates in 1M $CuSO_4$ aqueous solution + 1M $Na_2S_2O_3$ aqueous solution, with pH adjusted using acetic acid, heating to 70°C for 4 h, cooling, washing, and drying.

(iv) Polysulfide electrolyte is used: 1M Na_2S + 1M S in methanol + H_2O solution.

The solar cell fabrication process involves spin-coating of g-C_3N_4 paste on the TiO_2 nanorod array coated on FTO substrate. The CdS QDs are deposited on g-C_3N_4/TiO_2 by SILAR method. Finally, a sandwich solar cell is made by assembling together CdS QDs@ g-C_3N_4/TiO_2 with electrolyte and CuS electrode.

A champion efficiency of 2.31% is measured, which is attributed to the inclusion of g-C_3N_4. The g-C_3N_4 performs a dual function in the solar cell:

(a) As a light-absorption layer in the wavelength range 400–500nm, assisting the CdS QDs by supplying additional charge carriers.

(b) As an electron-blocking layer because the conduction and valence bands of g-C_3N_4 match well with those of TiO_2. So the bands of g-C_3N_4 and TiO_2 are properly aligned to provide charge transfer in the desired manner. Electron transport is permitted from g-C_3N_4 to TiO_2 and from g-C_3N_4 to CdS, but not in reverse directions, preventing the recombination of electrons, either from CdS or TiO_2 and the electrolyte. Therefore, the separation and transfer of photogenerated electrons and holes is considerably aided by including g-C_3N_4 through inhibition of the interfacial recombination processes (Gao et al 2016).

15.9.4 Raising the QDSSC Efficiency by Mn-Doping of CdS (η = 3.29%)

Mn doping affects the solar cell efficiency in a multiplicity of ways (Shen et al 2016):

(i) It increases the absorption of sunlight.
(ii) The range of wavelengths absorbed by the cell is extended in presence of Mn^{2+} so that more charge carriers are generated and higher current is obtained. Figure 15.16 shows the energy band diagrams of TiO_2-CdS without Mn^{2+} in CdS and with Mn^{2+} in CdS.
(iii) The rate of charge carrier transport is increased.
(iv) The electron diffusion length becomes longer; hence, carrier lifetime is improved. The electron lifetime in CdS cell is 0.17 s, while the same in Mn-CdS cell is 0.21 s. As a result, carrier recombination is lowered.

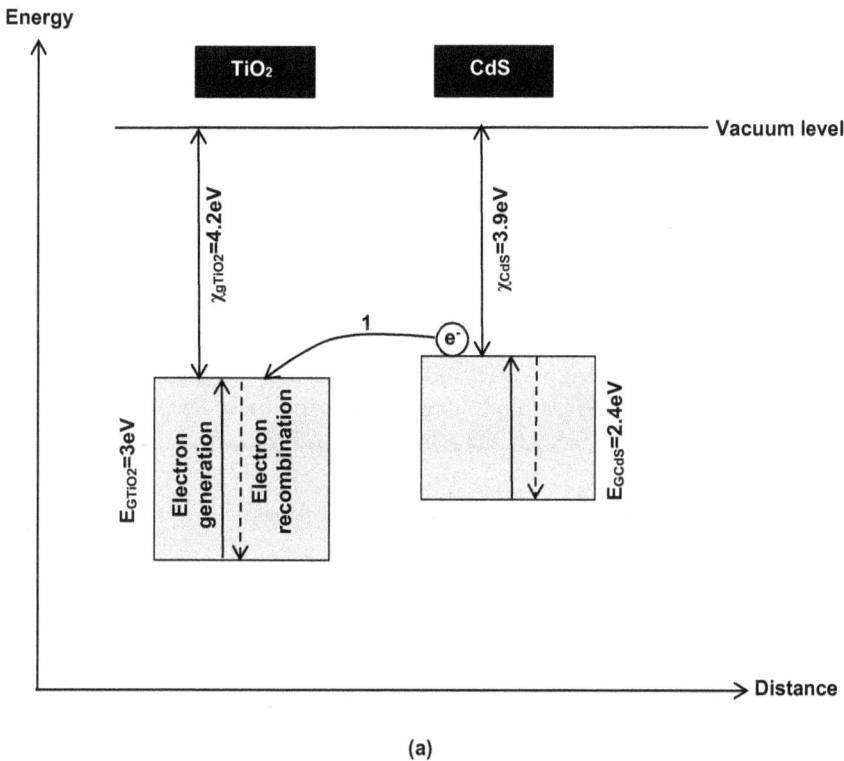

(a)

FIGURE 15.16 Anderson's rule-based energy band diagrams of TiO_2-CdS: (a) without Mn^{2+} in CdS and (b) with Mn^{2+} in CdS. Manganese inserts an energy level in CdS at the position shown. By virtue of this energy level, photons having energies lower than the bandgap of CdS can excite electrons to Mn^{2+} level from where they move to TiO_2. The range of frequencies absorbed by the cell increases. Electron injection from the conduction band of CdS to the conduction band of TiO_2 takes place along two routes 1, 2 when Mn^{2+} is present (Figure (b)) against only one path 1 in absence of Mn^{2+} (Figure (a)).

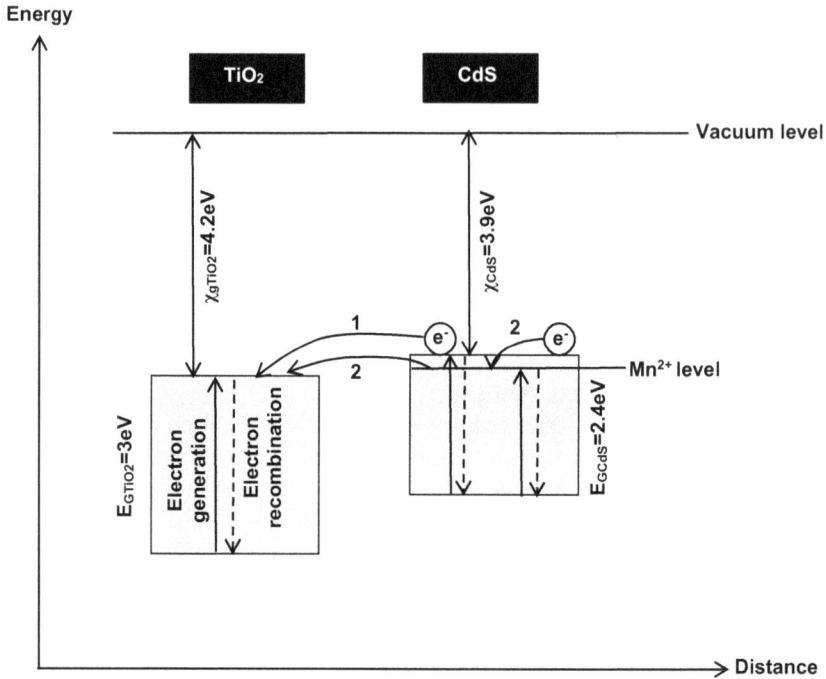

(b)

FIGURE 15.16   (Continued)

On the whole, Mn doping promotes the harvesting of sunlight by the solar cell and helps in the separation and collection of charge carriers.

This solar cell has a ZnS passivating coating over CdS QDs deposited by dipping in 0.1M Zn $(NO_3)_2$, $6H_2O$, and 0.1M Na_2S, $9H_2O$ solutions using SILAR process. The counter electrode is a Cu_2S film. The Cu_2S film is formed on brass foils by immersing in HCl and the polysulfide electrolyte is made from 1M S and 1M Na_2S solutions in methanol-water. For Mn^{2+} precursor solution concentration of 0.075M, the efficiency of the solar cell is 3.29% (Shen et al 2016).

15.9.5   GO/N-DOPED TiO₂/CDS/MN-DOPED ZNS/ZN-PORPHYRIN QDSSC (η = 4.62%)

Alavi et al (2020) presented a systematic analysis of solar cell efficiency amelioration as a function of process changes. Four types of photoelectrodes are made on FTO substrate (Figure 15.17):

(i) FTO/TiO₂/CdS/ZnS photoelectrode. This electrode is made by depositing CdS quantum dots on TiO₂ and then forming a ZnS passivation layer on CdS QDs. The TiO₂ paste is made by mixing nanocrystalline TiO₂ powder with water, acetylacetone, and polyethylene glycol. The paste is cast onto the substrate. Following CdS QDs deposition by SILAR method, the ZnS passivation film is deposited by successively dipping in 0.1M $Zn(NO_3)_2$ and 0.1M Na_2S solutions for 5 min. each.

(ii) FTO/GO/N-Doped TiO₂/CdS/ZnS phoelectrode. For making this electrode, N-doped TiO₂ nanoparticles are prepared by sol-gel method. Then GO/N-doped TiO₂ paste is formed with graphene oxide (GO). FTO substrates are coated with this paste by doctor blading, followed by calcining, dipping in $TiCl_4$ solution, and sintering. Remaining steps are the same as in *i*.

(iii) FTO/GO/N-Doped TiO_2/CdS/Mn-Doped ZnS photoelectrode: During ZnS deposition, manganese acetate is added.

(iv) FTO/GO/N-Doped TiO_2/CdS/Mn-Doped ZnS/Zn porphyrin photoelectrode: The photo-electrode is immersed in 0.5mM zinc porphyrin solution for 6 h and air-dried.

CuS counter electrode is used for all the four photoelectrodes. The CuS electrode is made by dipping FTO substrate in solution of copper sulfate pentahydrate (0.1M), sodium thiosulfate (0.4M), and acetic acid (0.7M) in water, keeping in oven at 60°C for 45 min., washing with water, and drying. Four versions of solar cells are assembled from the four types of photoelectrodes (named as versions i–iv) and the CuS counter electrode using the redox liquid electrolyte (1M Na_2S + 2M S + 0.2M NaOH in methanol + DI water).

Efficiency of version *i* is 1.8%. It increases to 2.52% for version *ii*, 3.47% for version *iii*, and 4.62% for version *iv*. GO and nitrogen increase the amount of light absorbed, decrease the carrier

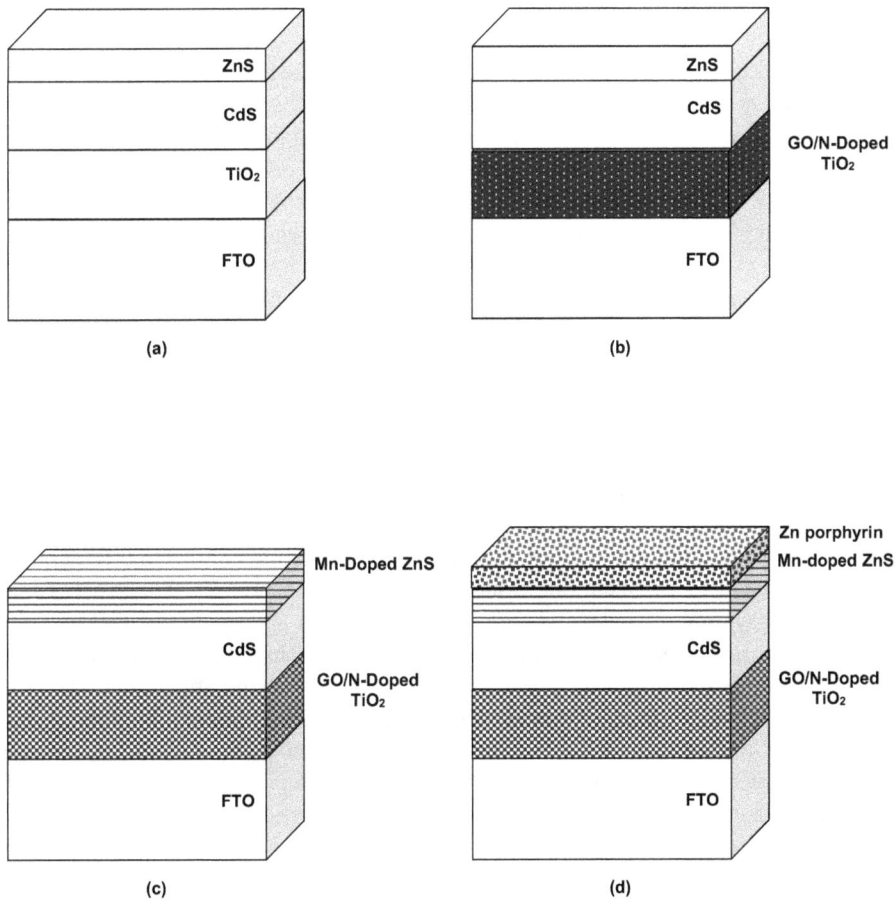

FIGURE 15.17 Four kinds of photoelectrodes: (a) FTO/TiO_2/CdS/ZnS photoelectrode, (b) FTO/GO/N-Doped TiO_2/CdS/ZnS photoelectrode, (c) FTO/GO/N-Doped TiO_2/CdS/Mn-Doped ZnS photoelectrode, and (d) FTO/GO/N-Doped TiO_2/CdS/Mn-Doped ZnS/Zn porphyrin photoelectrode. Starting from the basic FTO/TiO_2/CdS/ZnS structure in *a*, the TiO_2 layer is modified by GO/N-doping in *b*. The structure in *c* differs from that in *b* in the respect that the ZnS layer is doped with manganese in *c*. The structure in *d* is different from that in *c* because a Zn porphyrin layer is deposited in *d*.

recombination rate, and provide increased CdS/ZnS loading in all the versions, and also more dye loading in version *iv* by virtue of a larger surface area. Nitrogen raises the surface area of TiO_2 from 53.31 to 54.20 m^2/g, while GO sheets increment the same from 54.20 to 57.58 m^2/g. Additionally, the higher electrical conductivity of GO increases the charge carrier mobility from which the short-circuit current is benefitted. Inclusion of Mn^{2+} ions in ZnS lowers the recombination of photoelectrons. This recombination occurs with oxidized dye molecules or electrolyte at the TiO_2 surface. The Zn-porphyrin dye stretches the absorption wavelength range to near-infrared region (NIR), thereby increasing the short-circuit current.

15.9.6   Mixed-Joint CdS-ZnS QDSSC ($\eta = 6.37\%$) and ZnS QDSSC ($\eta = 2.72\%$)

CdS QDs are efficient light absorbers. ZnS QDs, too, absorb light and also provide surface passivation. The two types of QDs are combined to fabricate a mixed-joint CdS-ZnS QDSSC, giving a power conversion efficiency of 6.37% (Zheng et al 2019). A composite film of reduced graphene oxide is formed with TiO_2. This composite film is called the RGO/TiO_2 film. Over this film, layers of CdS QDs and ZnS QDs are alternately deposited by SILAR method to make the mixed-joint QDs.

The bandgap of ZnS is 3.2 eV, and that of CdS is 2.25 eV. The alignment of band structures of these semiconductors is such that the conduction band edge of ZnS lies at a higher energy than that of CdS. The advantage of this placement of conduction band edges of CdS and ZnS is that ZnS QDs impede back transport of photo-generated electrons from CdS to ZnS. This suppression of electron backflow improves charge collection. Furthermore, the valence bands of ZnS and CdS are placed in such a way that transference of holes from the valence band of CdS to electrolyte is prevented.

Since ZnS can act as both light-absorbing and recombination-blocking layer, QDSSC has also been made with only ZnS QDs, albeit giving lower efficiency (Mehrabian et al 2014). This solar cell and its energy band diagram are shown in Figure 15.18. Here, ZnO thin film is used in place of TiO_2. For ZnO thin film deposition, ITO-coated glass substrates are dipped in ZnO sol made by dissolution of zinc acetate dehydrate in absolute ethanol, adding diethanolamine (DEA) for stabilization and stirring. The ZnO film is annealed at 500°C for 1 h. Two-times repetition of the dip-coating and annealing procedures gave the desired thickness. SILAR technique is used for depositing the ZnS QDs. The ZnO-coated substrate is dipped in 25mM solution of $Zn(CH_3COO)_2$. $2H_2O$ in ethanol (1 min.), rinsed in ethanol, dipped in 25mM solution of Na_2S in methanol (1 min.), and rinsed in methanol to complete one cycle of the deposition process. The process is implemented for N cycles with N dependent on the required thickness.

Several problems are associated with liquid electrolytes, e.g., human safety issues, lack of stability, solidifying at low temperatures, and expanding at high temperatures, etc. Using a solid electrolyte is a viable remedy to solve these problems. P3HT and PCBM are mixed in 1,2-dichlorobenzene or orthodichlorobenzene (ODCB). This solution is coated on the ZnS QD by spinning in a glove box in N_2 ambience. Annealing is done at 150°C for ½ h. Ag film is vacuum-evaporated over this layer. The one-sun efficiency of the solar cell with mean particle diameter of 9nm is found to be 2.72% (Mehrabian et al 2014).

15.10   QUANTUM DOT INTERMEDIATE BAND SOLAR CELL ($\eta = 16.3\%$)

15.10.1   Intermediate Band Solar Cell Concept and Energy Band Diagram

Intermediate band is an energy band located in the forbidden energy gap between the conduction and valence bands of a semiconductor. In a solar cell (Figure 15.19a)), the purpose of this energy band is to utilize below-bandgap photons, i.e., photons having insufficient energy, to pump

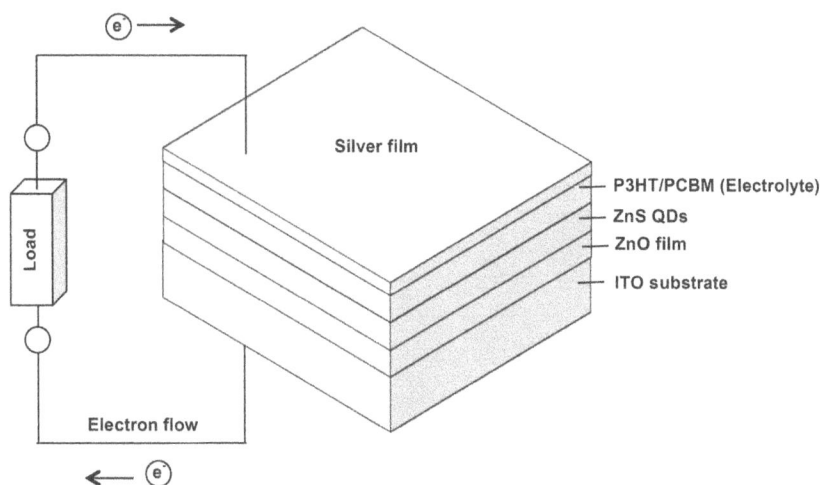

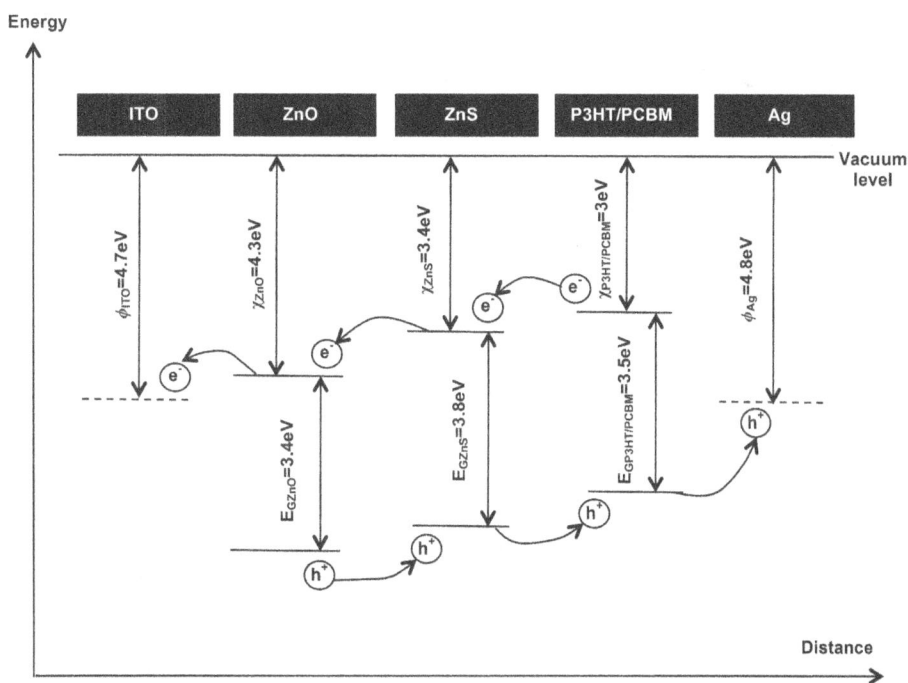

FIGURE 15.18 ZnS quantum dot–sensitized solar cell with P3HT/PCBM electrolyte: (a) structure of the cell showing the layers from bottom upwards: ITO substrate, ZnO film, ZnS film, P3HT/PCBM electrolyte, and silver film. (b) Energy band diagram of the solar cell drawn according to Anderson's rule. The work function of ITO is $\phi_{ITO} = 4.7eV$; that of silver is $\phi_{Ag} = 4.8eV$. The electron affinities of ZnO, ZnS, and P3HT/PCBM are $\chi_{ZnO} = 3.4eV$, $\chi_{ZnS} = 3.8eV$, and $\chi_{P3HT/PCBM} = 3.5eV$, respectively. The bandgaps of ZnO, ZnS, and P3HT/3PCBM are $E_{GZnO} = 3.4eV$, $E_{GZnS} = 3.8eV$, and $E_{GP3HT/PCBM} = 3.5eV$. Electrons move from higher to lower energies, i.e., P3HT/PCBM to ZnS to ZnO to ITO. Holes travel from lower to higher energies, i.e., ZnO to ZnS to P3HT/PCBM to silver. In the external circuit, electrons move from ITO to silver.

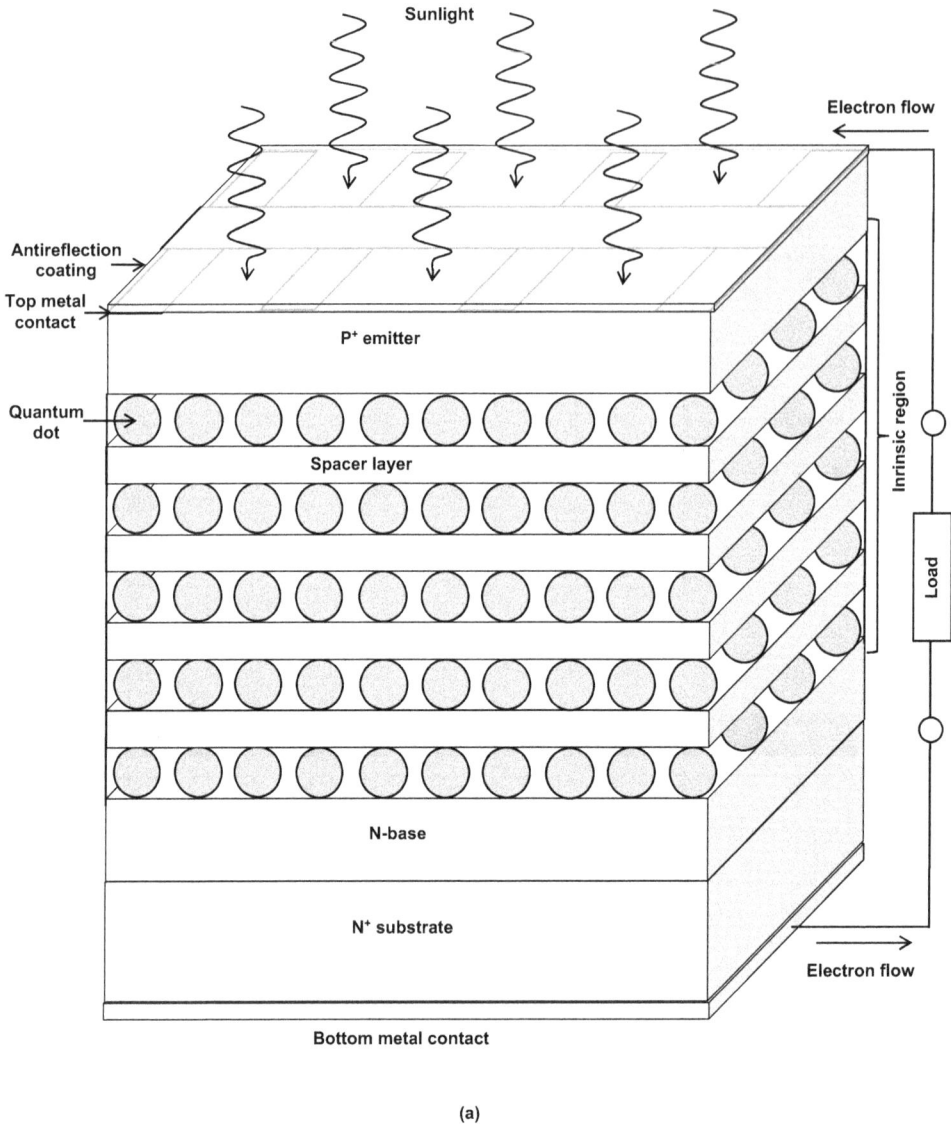

(a)

FIGURE 15.19 Quantum dot intermediate-band solar cell: (a) the P-I-N diode structure of the solar cell showing the various layers from bottom upwards: bottom metal contact, N⁺ substrate, N-base, the stacks of quantum dot arrays in the intrinsic region together with the spacer layers, the P⁺emitter, and the top metal contact with the antireflection layer. Sunlight strikes the solar cell from the top contact side, and the electricity generated is taken out from wires connected to the metal contacts. (b) Energy band diagram of the solar cell showing the conduction and valence bands and the intermediate-band produced by the insertion of stacks of quantum dot arrays in the intrinsic region of the P-I-N diode. Photon 1 of frequency v_1 and hence energy $E_1 = hv_1$ less than bandgap energy E_G is able to promote an electron from the conduction band to the intermediate band; symbol h stands for the Planck's constant. Photon 2 having frequency v_2 and energy $E_2 = hv_2$ less than bandgap energy E_G can uplift an electron from the intermediate band to the valence band. Photon 3 of higher frequency v_3 and greater energy $E_3 = hv_3$ exceeding the bandgap energy E_G can straightaway raise an electron from the conduction to the valence band. Thus, the solar cell is able to utilize electrons of energies smaller than the bandgap of the semiconductor along with those having energies greater than the bandgap. So it is able to cover a broader range of energies and frequencies across the solar spectrum than a solar cell without the intermediate band.

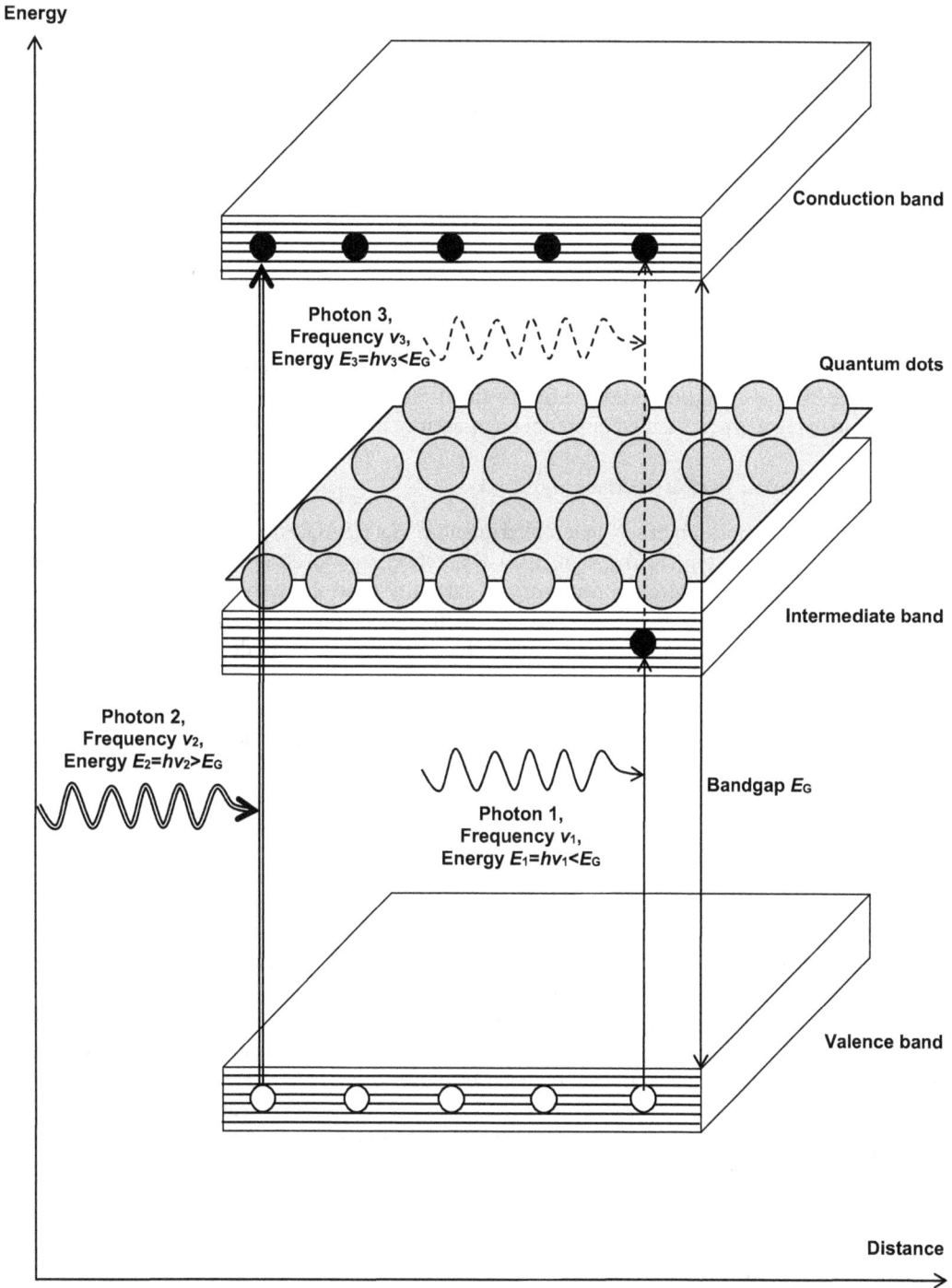

FIGURE 15.19 (Continued)

electrons directly across the forbidden gap for generation of electron-hole pairs. The absorption of sub-bandgap photons follows a two-step process in which the intermediate band acts as a stepping stone (Figure 15.19(b)). One electron-hole pair production is the outcome of absorption of two sub-bandgap photons. Sub-bandgap energy photon 1 promotes an electron from the valence band to the intermediate band, while sub-bandgap energy photon 2 pumps up the electron from the intermediate band to the conduction band. Greater-than-bandgap-energy photon 3 directly lifts the electron from the valence band to the conduction band. Thus, the solar cell contains electron-hole pairs from below-bandgap photons as well as from above-bandgap photons so that the total output current increases while a large open-circuit voltage is maintained.

The intermediate-band solar cell can be implemented with a quantum dot superlattice consisting of an array of quantum dots to provide the necessary intermediate band.

An array of self-assembled InAs quantum dots of precise shapes and sizes is incorporated in the intrinsic region of a GaAs P-I-N diode to open an energy gap between the conduction band and intermediate band of quantum states, leading to an intermediate-band solar cell (Beattie et al 2017). This is done using a process called quantum engineering.

15.10.2   QUANTUM ENGINEERING

Shape, size, and composition of self-assembled quantum dots (SAQDs) greatly influence their electronic and optical properties (Keizer et al 2010). Discrete energy levels of the quantum dots depend on these parameters. It is therefore necessary to carefully control the shape and size of quantum dots in the nanometer range, and also to maintain their composition. Temperature profile and depth of the spacer layer (Fig. 15.19(a)) critically affect the growth of quantum dots.

Consider vertically self-aligned, stacked, self-assembled InAs quantum dots with GaAs capping. The arrangement contains several layers of quantum dots. In such an arrangement, an intense correlation is observed in the lateral positions of quantum dots across the layers. The correlation arises principally from the segregation of indium in the overgrowth process and the impact of this process on the dot structure as well as its neighborhood. The process of growth of quantum dots is described in the next subsection.

15.10.3   GROWTH OF QUANTUM DOTS

A growth process ensuring uniformity of the shapes of the quantum dots is developed (Wasilewski et al 1999; Haffouz et al 2009). The process sequence involves an indium flush step to remove any traces of indium up to midway through the GaAs spacer layer to create similar growth conditions for each layer of InAs quantum dots. The sequence consists of steps such as annealing, deposition of self-assembled InAs quantum dot layer over 35 s using As_2 and In fluxes, capping the quantum dot layer with half the stipulated thickness of GaAs spacer layer, indium flush step at 510°C implemented by ramping the substrate temperature to 610°C over 30 s, annealing at 610°C for 70 s, returning the temperature to 510°C over 100 s, and depositing the remaining half of the GaAs spacer layer.

The indium flush step offers two main advantages:

(i)   Improvement in size uniformity of the dots within any given layer.
(ii) Prevention of any increase in size of the quantum dots from one layer to the next layer. The interdot spacing is also uniform.

Strain-induced competition for material between adjoining columns of quantum dots is the mechanism responsible for this process. This quantum dot growth process is known as quantum engineering because it is useful for tailoring the energy levels within the quantum dots.

In the incipient phase, an InAs wetting layer is formed. Afterwards, the formation of InAs quantum dots becomes energetically favorable. The quantum dots have the shape of a truncated pyramid. Typical dimensions of this truncated pyramid are base length = 15nm, length at truncation site = 6.7nm, and height = 0.9nm.

15.10.4 DEVICE STRUCTURE

The optimized quantum engineered device (Figure 15.20) is fabricated on an N⁺ GaAs substrate (Beattie et al 2017). The device is made in the form of circular mesas of diameters 2mm and 4mm. Molecular-beam epitaxy is applied to produce arrays of quantum dots that do not need compensation of strain. It is made of the following layers from bottom upwards: (i) heavily doped N-type

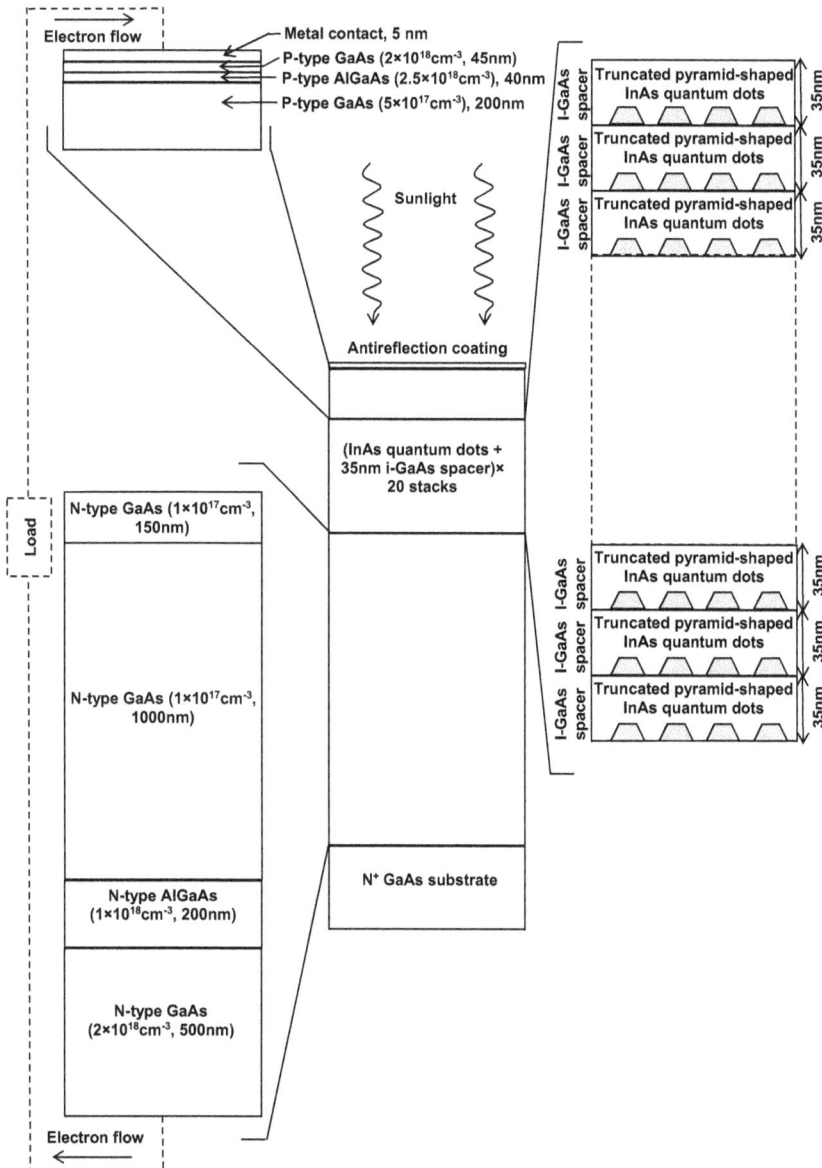

FIGURE 15.20 Quantum dot intermediate-band solar cell incorporating a sequence of 20 stacks of InAs quantum dots with GaAs spacers in the intrinsic region of a P-I-N diode. On the top left side of the diagram, the layers in the P-type region are shown on an enlarged scale. These include P-type GaAs layers of different concentrations together with a P-type AlGaAs layer; the carrier concentrations and thicknesses of the layers are indicated. In the bottom left side of the diagram, the layers in the N-type region are shown in magnified view. The N-type region is made of N-type GaAs layers of different concentrations along with an N-type AlGaAs layer. Carrier concentrations and thicknesses of the layers are given. On the right side of the diagram, the details of intrinsic region are shown, consisting of stacks of quantum dots with spacer layers.

GaAs layer (concentration $2 \times 10^{18} cm^{-3}$, thickness 500nm), (ii) heavily doped N-type AlGaAs layer ($1 \times 10^{18} cm^{-3}$, thickness 200nm), (iii) lightly doped N-type GaAs layers ($1 \times 10^{17} cm^{-3}$, 1000nm; $1 \times 10^{17} cm^{-3}$, 150nm), (iii) 20 stacks of InAs quantum dots with intrinsic GaAs spacer layer of thickness 35nm, (iv) lightly doped P-type GaAs ($5 \times 10^{17} cm^{-3}$, 200nm), (v) heavily doped P-type AlGaAs ($2.5 \times 10^{18} cm^{-3}$, 40nm), (vi) heavily doped P-type GaAs ($2 \times 10^{18} cm^{-3}$, 45nm), (viii) Ti/Au contact (5nm), and (ix) antireflection coating (ZnS/MgF_2). AlGaAs window layer is included to decrease surface recombination.

15.10.5 Signature of Intermediate Band

The absorption spectrum of the engineered quantum dot array solar cell reveals the existence of distinct regions of zero density of states (DoS) between the ground state e_0 and the first excited state e_1. This revelation indicates the formation of an energetically isolated intermediate band.

15.10.6 Photovoltaic Parameters of the QD-IBSC

Under 1 sun illumination, the open-circuit voltage V_{OC} of the solar cell is 0.822V, the short-circuit current density J_{SC} is 25.9 $mAcm^{-2}$, the fill factor FF is 76.3%, and the efficiency η is 16.3%. Under 5 suns, the efficiency is 18.4%. The active area efficiency obtained by taking the area of the contacts into consideration is 19.7% (Beattie et al 2018).

15.11 DISCUSSION AND CONCLUSIONS

Most of the experimental investigations on solar cells are conducted on PbS and CdS quantum dots (Table 15.1). High-quality quantum dots made with nontoxic, eco-friendly materials and passivated with suitable organic or inorganic capping ligands will lead to spectacular progress in the realm of quantum dot solar cells, presently largely hinged on quantum dots of lead-based or similar compounds. Nonetheless, the enormous potential and capabilities of quantum dot solar cells are abundantly obvious. Hydrophobic graphene quantum dots (HGQDs) have been used to ameliorate the efficiency of $CH_3NH_3PbI_3$ perovskite solar cells from 16% to 18.3% by passivating the defects and grain boundaries in the material (Khorshidi et al 2022).

TABLE 15.1
Summary of Quantum Dot Solar Cells

Sl. No.	QD Composition	Solar Cell Structure	Type of Solar Cell	Efficiency (%)	Reference
1	PbS	FTO glass/anatase TiO$_2$/PbS QD-TBAI/PbS QD-EDT/ MoO$_3$-Au	P-N junction with NaHS-treated P-type layer	7.6	Speirs et al (2017)
2	PbS	FTO glass substrate/ALD-TiO$_2$/ PbS QD-TBAI/PbS QD-TBAC/MoO$_3$-Au	P-N junction with atomic-layer deposited TiO$_2$ electron transport layer	5.5–7.2	Sukharevska et al (2020)
3	PbS	ITO glass/ZnO/PbS QD-TBAI/ PbS QD-EDT/Cr-Ag	P-N junction with Cr-Ag electrodes	6.5	Khanam et al (2019)
4	PbS	ITO glass/PbS QD-MPA/PbS QD-halogen/ZnO/Al	P-N junction by industrially suited process	9	Goossens et al (2021)
5	PbS	ITO glass/ZnO/PbS QD-halogen/PD2FCT-29DPP HTL/MoO$_3$-Ag	P-N junction with PD2FCT-29DPP as HTL	14	Kim et al (2020)

Sl. No.	QD Composition	Solar Cell Structure	Type of Solar Cell	Efficiency (%)	Reference
6	PbS	ITO glass/AlZnO/PbS QD -PbI$_2$/PbS QD-EDT/Au	P-N junction as the back cell in a (perovskite + PbS QD) tandem cell	10.06 for PbS-QD cell; 18.9 for tandem cell	Andruszkiewicz et al (2021)
7	PbS	ITO glass/ZnO/PbS QD-halogen/PbS QD-EDT/Au	P-N junction as the back cell in a (perovskite + PbS QD) tandem cell	11.6 for PbS-QD cell; 20.2 for tandem cell	Manekkathodi et al (2019)
8	PbS	ITO glass/PbS QD/Al	Schottky barrier diode	1.8	Johnston et al (2008)
9	PbS	FTO glass/dense TiO$_2$/PbS QD/ Au or MoO$_3$	Depleted heterojunction	3.36	Debnath et al (2010)
10	PbS	FTO glass/dense TiO$_2$/(TiO$_2$ + PbS QDs)/Au or MoO$_3$	Depleted bulk heterojunction	5.5	Barkhouse et al (2011)
11	PbS	ITO glass/PEDOT:PSS/ (PbS-BDT QDs + P3HT)/ PbS-BDT QDs/Al-LiF	Organic-inorganic hybrid	4.91	Nguyen et al (2021)
12	PbS	FTO glass/TiO$_2$/PbS-ZnS QDs/ electrolyte/Cu$_2$S counter electrode	Sensitized	2.41, 4.01	Tian et al (2016)
13	PbS	FTO glass/compact TiO$_2$/porous TiO$_2$/PbS-ZnS QDs/ Polysulfide electrolyte/MoO$_3$	Sensitized	5.82	Bhalekar et al (2019)
14	CdS	FTO glass/CdS QD/TiO$_2$/MPN electrolyte/Pt	Sensitized	1.84	Chang and Lee (2007)
15	CdS	FTO glass/CdS QD/TiO$_2$ with GO powder/polysulfide electrolyte/Pt/glass	Sensitized	2.02	Wageh et al (2013)
16	CdS	FT glass/CdS QD/TiO$_2$ nanorod with 2D g-C$_3$N$_4$/polysulfide electrolyte/CuS	Sensitized	2.31	Gao et al (2016)
17	CdS-ZnS	FTO glass/CdS-ZnS QD with Mn doping of CdS/TiO$_2$ CdS/ polysulfide electrolyte/Cu$_2$S	Sensitized	3.29	Shen et al (2016)
18	CdS-ZnS	FTO glass/GO/N-Doped TiO$_2$/ CdS QD/Mn-Doped ZnS/Zn porphyrin/electrolyte/CuS	Sensitized	4.62	Alavi et al (2020)
19	ZnS	ITO glass/ZnO/ZnS QD/solid electrolyte/Ag	Sensitized	2.72	Mehrabian et al (2014)
20	CdS and ZnS mixed QDs	CdS and ZnS mixed joint QDs in RGO/TiO$_2$	Sensitized	6.37	Zheng et al (2019)
21	InAs	Stacks of InAs quantum dots with intrinsic GaAs spacer layer in the intrinsic region of a P-I-N diode	Intermediate band	16.3	Beattie et al (2017)

"Cipher" in dimensions but splendid in action
Quantum dots are zero-dimensional
With size infinitesimal
But real champions, admirable and adorable
With properties exceptional

And activities fantastical
Unsurpassably influential
Reliable, indispensable, worthwhile, and invaluable!

Questions and Answers

15.1 Write the generic structure of a quantum dot P-N junction solar cell. Answer: Glass/ TCO/ETL/N-type QDs/P-type QDs/HTL/Au.

15.2 What is hot injection method of QD synthesis? Answer: A method of obtaining monodisperse QDs in which QD nucleation is triggered by swiftly injecting a cool solution of precursor into a hot solution of mixed ingredients. The temperature of the mixture decreases with injection of cool precursor, preventing formation of new nuclei. The temperature is raised to the required growth temperature but kept below that of the hot solution used previously. The existing nuclei grow slowly to form quantum dots without any new nucleation.

15.3 How are PbS QDs made to exhibit N-type behavior? Answer: By treatment with TBAI.

15.4 How to make PbS QDs show P-type nature? Answer: By treatment with EDT.

15.5 Write the chemical equation for ALD of TiO_2 using $TiCl_4$ and water. Answer:

$$TiCl_4 + 2H_2O \rightarrow TiO_2 + 4HCl \tag{15.5}$$

15.6 What is meant by ligand exchange? Answer: A chemical reaction during which one ligand in a compound is replaced with another ligand.

15.7 MoO_3 is an N-type transparent semiconducting oxide. Why is it used as a hole transport material? Answer: Because of its high work function, which makes it suitable as a hole-collection material. Its Fermi level is located a little above the edge of the valence band of the active layer in the solar cell. It thus fulfills the function of a metallic anode in the device.

15.8 How are the following materials used in solar cells: (a) TiO_2, (b) ZnO, (c) SnO_2? Answer: As electron transport layers (ETLs).

15.9 What conductivity do these materials show, N-type or P-type, and what is the cause: (a) TiO_2, (b) ZnO, (c) SnO_2? Answer: N-type conductivity. They are unintentionally N-doped. The N-type nature is ascribed to oxygen vacancies.

15.10 What is the polarity of PbS QDs capped with MPA: (a) N-type or (b) P-type? Answer: P-type.

15.11 What is the specialty of the industrial process for making PbS QD solar cells? Answer: P-type and N-type inks and doctor blading technique are used for making the layers of the P-N diode instead of lengthy film-coating processes.

15.12 What are the difficulties faced with P-type PbS-EDT layer? Answer: Instability, interaction with other layers of the solar cell causing cell degradation, and high defect density leading to short carrier lifetime.

15.13 Should HTL in a quantum dot solar cell have a (a) short carrier diffusion length or (b) long carrier diffusion length? Answer: (b) Long carrier diffusion length.

15.14 Name a P-type layer used to fabricate a P-N junction diode quantum dot solar cell with better efficiency than PbS QD-EDT layer. Answer: PD2FCT-29DPP layer.

15.15 List the favorable qualities of surlyn, which encourage its use as an interlayer between perovskite and quantum dot solar cells. Answer: Transparency, flexibility, chemical stability, and low-temperature processing.

15.16 Why is the MoO_3 dielectric used as a part of a conducting electrode? Answer: MoO_3 is used in DMD-TCE to prevent oxidation of metal and suppress plasmonic loss in metal.

15.17 What is the thickness of the MoO_3 dielectric layer in a DMD-TCE? What will happen if the thickness is increased? Answer: 5nm; a thicker MoO_3 layer will have a larger resistance. Such a DMD structure cannot be used as an electrode.

15.18 What is a glove box? Answer: It is a sealed container, either partly or wholly transparent. Inside the box, an inert atmosphere filled with nitrogen or argon is maintained. The worker can put hands in gloves, manipulate objects inside the container, and perform operations on them from outside without disturbing the environment. It is used for handling materials, which interact with oxygen and humidity in the outside atmosphere or for working with toxic/hazardous chemicals.

15.19 Why are perovskite and quantum dot solar cells paired together to make a tandem solar cell? Because perovskite has a larger bandgap (1.63eV) than PbS (1.15eV). High energy photons cause photogeneration in the perovskite light absorber layer. Low-energy IR photons pass through the perovskite cell and fall on the quantum dot solar cell, where they produce electrons and holes. Therefore, the combined cell supplies a larger photocurrent than either cell alone can do.

15.20 Discuss the use of a thin LiF layer as a component of the metal stack. Answer: LiF is a wide-bandgap semiconductor ($E_G = 13.6$eV) and is an excellent insulator. Due to its insulating property, it is used as an ultrathin buffer layer ~ 1nm. This ultrathin layer lowers the effective work function of the electrode, thereby enhancing injection of carriers. Further, a submonolayer of LiF slows down the oxidation of aluminum. Thus, LiF improves the electrical performance and lifetime of Al electrodes. However, after a critical thickness, the ability of LiF to improve charge transfer diminishes, deteriorating the solar cell performance (Tarak 2021).

15.21 What will happen if the electric field for charge separation is located on the opposite side of the absorber layer to that on which light falls on the solar cell? Answer: Light falling on the absorber layer has to pass through the full thickness of the absorber layer to reach the opposite electric field side. During propagation, light is continuously debilitated in intensity and therefore in its capability to create electron-hole pairs. The feeble-intensity light reaching near the electric field region produces only a small number of electron-hole pairs. Further, only a few of the large number of electron-hole pairs produced by the stronger-intensity light on the side of absorber on which light strikes are able to reach the electric field side. Many of them are lost by recombination in their long travelling distance to the electric field side. Neither is a large number of electron-hole pairs generated on the electric field side, nor is the large number of electron-hole pairs produced on the light receiving side able to reach the electric field side, resulting in a smaller photocurrent and hence lower solar cell performance.

15.22 What is the main drawback of a Schottky junction quantum dot solar cell? Answer: The junction is located on the opposite side of the absorber layer of solar cell to the light-receiving side.

15.23 How does the depleted heterojunction quantum dot solar cell meet the requirement of proximity of the junction to the light-receiving side of the solar cell? Because in this solar cell, the N-type TiO_2/P-type PbS QDs interface lies on the side of solar cell on which light strikes it.

15.24 What constructional difference between depleted heterojunction quantum dot solar cell and depleted bulk heterojunction quantum dot solar cell makes the latter perform better than the former? Answer: The difference lies in the straight-line interface between N-type TiO_2 and P-type PbS QDs in depleted heterojunction solar cell and the irregular distributed interface between them in the depleted bulk heterojunction solar cell. The distributed arrangement provides a thorough dispersion and mixing of quantum dots with TiO_2 nanoparticles, leading to a larger interfacial contact area and hence more efficient current production and collection than the sharp straight-line boundary between TiO_2 and PbS QDs.

15.25 What is the benefit of $TiCl_4$ treatment of TiO_2? Answer: The cracks and pinholes in TiO_2 are filled, and the work function of TiO_2 surface is lowered to facilitate carrier transfer. Due to reduced defects and by surface work function depression, not only is carrier recombination suppressed, but the TiO_2 also acts an efficient electron collector and hole-blocking layer (Xu et al 2019).

15.26 What prompts us to replace TiO_2 as the donor with a polymer along with PbS QDs as acceptors to make a solar cell? Answer: To eliminate the lengthy process of deposition of layers of TiO_2 and PbS QDs and espouse an easier polymer-QDs mixing procedure adaptable to bulk production.

15.27 What type of solar cell results by mixing a donor-type polymer with acceptor-type QDs? Answer: A hybrid solar cell.

15.28 In the hybrid solar cell formed by mixing PbS QDs with P3HT-b-PS block copolymer, what are the characters (donors or acceptors) of the two constituents? Answer: PbS QDs are acceptors, and P3HT-b-PS molecules are donors.

15.29 What is the polarity and use of PEDOT:PSS? Answer: P-type polymer used as a hole transport material.

15.30 Why are PbS QDs capped with oleic acid? Answer: A capping agent is a stabilizing ligand. It stabilizes the interface between the nanoparticle and the surrounding medium of preparation. It prevents overgrowth of nanoparticles and thus controls the nanoparticle size. It also avoids agglomeration of the nanoparticles.

15.31 How does a quantum dot–sensitized solar cell differ from a depleted bulk heterojunction solar cell? Answer: In the manner of hole extraction from the QD which occurs into an electrolyte here as opposed to that in an Au or MoO_3 electrode in a DBH solar cell.

15.32 In what respect does a quantum dot–sensitized solar cell differ from a dye-sensitized solar cell? Answer: In the replacement of dye molecules with quantum dots.

15.33 What is the function of counter electrode in QDSSC? Answer: (i) To catalyze reduction of the electrolyte and (ii) to extract electrons from the external circuit.

15.34 Refer to Figure 15.21(a). A heterojunction consists of a wide-bandgap semiconductor X and a narrow-bandgap semiconductor Y. The bandgap of the semiconductor Y lies completely within the bandgap of X. (a) What is this heterojunction called? (b) Where will the electrons and holes move in this heterojunction? (c) What will happen to them? (d) For making what kind of device is this heterojunction useful? (e) Can it be used to make solar cells? Answer: (a) Type I, or straddled gap heterojunction. (b) Electrons and holes will accumulate in semiconductor Y of smaller bandgap. (c) They will undergo recombination. (d) This heterojunction is useful for making light-emitting diodes and lasers. (e) No.

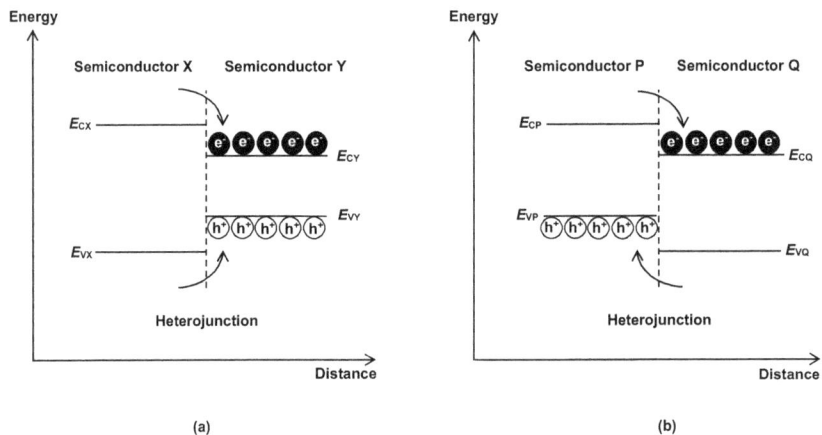

FIGURE 15.21 Energy band diagrams of heterojunctions: (a) type I heterojunction, in which both electrons and holes migrate to the semiconductor Y side, favoring carrier recombination, (b) type II heterojunction, in which electrons move to semiconductor Q side, whereas holes move to semiconductor P side, favoring charge carrier separation.

15.35 A heterojunction consists of a semiconductor P and semiconductor Q (Figure 15.21(b)). The bandgaps of the two semiconductors overlap, but the lowest energy position for electrons lies in semiconductor Q, while highest energy position for holest lies in semiconductor P. (a) What is this heterojunction known as? (b) In what directions will electrons and holes move in this heterojunction? (c) What will happen to them? (d) For making what kind of device is this heterojunction useful? Answer: (a) Type II or staggered gap heterojunction. (b) Electrons will move to semiconductor Q, while holes will move to semiconductor P. (c) They will be separated, with electrons moving to Q and holes to P. (d) This heterojunction is useful for making solar cells as it provides charge carrier separation.

15.36 What is meant by a redox couple? Answer: An oxidizing agent/reducing agent appearing on opposite sides of a half equation, an equation showing the phenomenon taking place at one of the electrodes in electrolysis.

15.37 Briefly write the five principal steps during operation of a quantum dot–sensitized solar cell. Answer: (i) Production of electron-hole pair in quantum dot by sunlight, (ii) electron movement to metal oxide semiconductor leaving hole behind, (iii) electron diffusion through metal oxide to reach contact, (iv) interaction of hole with electrolyte, and (v) electrolyte–counter electrode interaction.

15.38 What happens upon incorporation of methanol in polysulfide electrolyte? Answer: Wettability of TiO_2 nanoparticles with electrolyte is improved, whereby diffusion of redox couple through the electrolyte becomes easier, helping in transport of charge carriers and leading to better performance of solar cell.

15.39 What is the advantage of depositing a compact TiO_2 layer preceding porous TiO_2 together with passivating surfaces of PbS QDs with ZnS? Answer: It increases charge recombination resistance and electron lifetime.

15.40 How does covering $CdS@TiO_2$ film with graphene powder improve solar cell operation? Answer: By reducing electron-hole recombination.

15.41 How does modification of CdS QDs with g-C_3N_4 help in improving solar cell performance? Answer: By facilitating charge separation, by reducing interfacial recombination, and by providing extra electron-hole pairs by absorbing light in the 400nm- to 500nm-wavelength range.

15.42 What are the effects of doping CdS QDs with manganese? Answer: Increases light absorption by extending the range of absorption wavelengths and aids in separation/collection of charges.

15.43 Explain the effects of the following: (a) doping titanium oxide with nitrogen and graphene oxide, (b) doping zinc sulfide with manganese, and (c) modifying ZnS with Zn-porphyrin. Answer: (a) Increase surface area to provide better loading of CdS/ZnS QDs; higher mobility of graphene increases current output, (b) reduces carrier recombination, (c) enables IR photon absorption.

15.44 Why do we make a mixed-joint CdS-ZnS QDSSC? Answer: ZnS provides absorption of additional light and opposes carrier recombination.

15.45 Can we make a QDSSC with ZnS QDs only? Answer: Yes.

15.46 What is an intermediate band in a semiconductor, and what is its use in a solar cell? An artificially created band between conduction and valence bands used in helping produce electron-hole pairs from incident photons having energies less than the natural bandgap of the semiconductor.

15.47 How can an intermediate band be introduced in the bandgap of the semiconductor layer of a solar cell? Answer: By incorporating an array of quantum dots known as a superlattice in the intrinsic region of a P-I-N diode made from the given semiconductor.

15.48 Give an example of a solar cell made on intermediate-bandgap concept. Answer: A GaAs solar cell made in P-I-N diode configuration with an array of InAs quantum dots inserted in the intrinsic (I) region of the diode.

15.49 How are the quantum dots grown to make an intermediate-band solar cell? Answer: By molecular-beam epitaxy, in which a sequence of self-assembled quantum dot layers are deposited from As_2 and In fluxes by a several times' repetition of a process consisting of an anneal time, InAs quantum dot layer deposition, GaAs cap layer deposition, indium flush step, and again, GaAs deposition.

15.50 Why is the indium flush step needed? Answer: To ensure identical growth conditions for each InAs layer of self-assembled quantum dots by getting rid of indium traces during the second half of GaAs spacer deposition.

15.51 What will happen if indium flushing is not done? Answer: Indium flushing assures uniformity of size of quantum dots within the same layer as well as between successive layers, and also uniformity of spacing between any two dots, all of which will be spoiled if indium flush step is removed.

15.52 How do we confirm that an intermediate band has been formed? Answer: From the absorption spectrum of the solar cell, which clearly shows the existence of a zero density-of-states region between ground and first excited states.

REFERENCES

Alavi M., R. Rahimi, Z. Maleki and M. Hosseini-Kharat 2020 Improvement of power conversion efficiency of quantum dot-sensitized solar cells by doping of manganese into a ZnS passivation layer and cosensitization of zinc-porphyrin on a modified graphene oxide/nitrogen-doped TiO_2 photoanode, ACS Omega, 5: 11024–11034.

Andruszkiewicz A., X. Zhang, M. B. Johansson, L. Yuan and E. M. J. Johansson 2021 Perovskite and quantum dot tandem solar cells with interlayer modification for improved optical semitransparency and stability, Nanoscale, 13: 6234–6240.

Barkhouse D. A. R., R. Debnath, I. J. Kramer, D. Zhitomirsky, A. G. Pattantyus-Abraham, L. Levina, L. Etgar, M. Grätzel and E. H. Sargent 2011 Depleted bulk heterojunction colloidal quantum dot photovoltaics, Advanced Materials, 23: 3134–3138.

Beattie N. S., P. See, G. Zoppi, P. M. Ushasree, M. Duchamp, I. Farrer, D. A. Ritchie and S. Tomić 2017 Quantum engineering of InAs/GaAs quantum dot based intermediate band solar cells, ACS Photonics, 4(11): 2745–2750.

Beattie N. S., P. See, G. Zoppi, P. M. Ushasree, M. Duchamp, I. Farrer, V. Donchev, D. A. Ritchie and S. Tomić 2018 Design and fabrication of InAs/GaAs QD based intermediate band solar cells by quantum engineering, 2018 IEEE 7th World Conference on Photovoltaic Energy Conversion (WCPEC), 10–15 Jun 2018, Waikoloa Village, HI, USA, IEEE, NY, pp. 2747–2751.

Bhalekar V. P., P. K. Baviskar, R. Prasad M. B., B. M. Palve, V. S. Kadam and H. M. Patha 2019 PbS sensitized TiO_2 based quantum dot solar cells with efficiency greater than 5% under artificial light: Effect of compact layer and surface passivation, Engineered Science, 7: 38–42.

Chang C.-H. and Y.-L. Lee 2007 Chemical bath deposition of CdS quantum dots onto mesoscopic TiO_2 films for application in quantum-dot-sensitized solar cells, Applied Physics Letters, 91: 053503–1 to 053503–3.

Clifford J. P., K. W. Johnston, L. Levina and E. H. Sargent 2007 Schottky barriers to colloidal quantum dot films, Applied Physics Letters, 91: 253117–1 to 253117–3.

Debnath R., M. T. Greiner, I. J. Kramer, A. Fischer, J. Tang, D. A. R. Barkhouse, Xi. Wang, L. Levina, Z.-H. Lu and E. H. Sargent 2010 Depleted-heterojunction colloidal quantum dot photovoltaics employing low-cost electrical contacts, Applied Physics Letters, 97: 023109–1 to 023109–3.

Gao Q., S. Sun, X. Li, X. Zhang, L. Duan and W. Lü 2016 Enhancing performance of CdS quantum dot-sensitized solar cells by two-dimensional g-C_3N_4 modified TiO_2 nanorods, Nanoscale Research Letters, 11(463): 9 pages.

Goossens V. M., N. V. Sukharevska, D. N. Dirin, M. V. Kovalenko and M. A. Loi 2021 Scalable fabrication of efficient p-n junction lead sulfide quantum dot solar cells, Cell Reports Physical Science 2(12): 100655, 1–14.

Haffouz S., S. Raymond, Z. G. Lu, P. J. Barrios, D. Roy-Guay, X. Wu, J. R. Liu, D. Poitras and Z. R. Wasilewski 2009 Growth and fabrication of quantum dots superluminescent diodes using the indium-flush technique: A new approach in controlling the bandwidth, Journal of Crystal Growth, 311: 1803–1806.

Hines M. A. and G. D. Scholes 2003 Colloidal PbS nanocrystals with size-tunable near-infrared emission: Observation of post-synthesis self-narrowing of the particle size distribution, Advanced Materials, 15(21): 1844–1849.

Hyun B.-R., Y.-W. Zhong, A. C. Bartnik, L. Sun, H. D. Abruňa, F. W. Wise, J. D. Goodreau, J. R. Matthews, T. M. Leslie and N. F. Borrelli 2008 Electron injection from colloidal PbS quantum dots into titanium dioxide nanoparticles, ACS Nano, 2(11): 2206–2212.

Johnston K. W., A. G. Pattantyus-Abraham, J. P. Clifford, S. H. Myrskog, D. D. MacNeil, L. Levina and E. H. Sargent 2008 Schottky-quantum dot photovoltaics for efficient infrared power conversion, Applied Physics Letters, 92: 151115–1 to 151115–3.

Kashiwaya S., J. Morasch, V. Streibel, T. Toupance, W. Jaegermann and A. Klein 2018 The work function of TiO$_2$, Surfaces, 1: 73–89.

Keizer J. G., E. C. Clark, M. Bichler, G. Abstreiter, J. J. Finley and P. M. Koenraad 2010 An atomically resolved study of InGaAs quantum dot layers grown with an indium flush step, Nanotechnology, 21(215705): 4 pages.

Khanam J. J., S. Y. Foo, Z. Yu, T. Liu and P. Mao 2019 Efficient, stable, and low-cost PbS quantum dot solar cells with Cr—Ag electrodes, Nanomaterials, 9(1205): 1–12.

Khorshidi E., B. Rezaei, D. Blätte, A. Buyruk, M. A. Reus, J. Hanisch, B. Böller, P. Müller-Buschbaum and T. Ameri 2022 Hydrophobic graphene quantum dots for defect passivation and enhanced moisture stability of CH$_3$NH$_3$PbI$_3$ perovskite solar cells, Solar RRL, 2200023: 1–11.

Kim H. I., S.-W. Baek, H. J. Cheon, S. U. Ryu, S. Lee, M.-J. Choi, K. Choi et al 2020 Tuned alternating D—A copolymer hole-transport layer enables colloidal quantum dot solar cells with superior fill factor and efficiency, Advanced Materials, 32(2004985): 7 pages.

Konstantatos G., I. Howard, A. Fischer, S. Hoogland, J. Clifford, E. Klem, L. Levina and E. H. Sargent 2006 Ultrasensitive solution-cast quantum dot photodetectors, Nature, 442(13): 180–183.

Lai L.-H., L. Protesescu, M. V. Kovalenko and M. A. Loi 2014 Sensitized solar cells with colloidal PbS—CdS core—shell quantum dots, Physical Chemistry Chemical Physics, 16: 736–742.

Manekkathodi A., B. Chen, J. Kim, S.-W. Baek, B. Scheffel, Y. Hou et al 2019 Solution-processed perovskite-colloidal quantum dot tandem solar cells for photon collection beyond 1,000 nm, Journal of Materials Chemistry A, 7: 26020–26028.

Mehrabian M., K. Mirabbaszadeh and H. Afarideh 2014 Solid-state ZnS quantum dot-sensitized solar cell fabricated by the Dip-SILAR technique, Physica Scripta, 89(085801): 8 pages.

Nguyen D.-T., S. Sharma, S.-A. Chen, P. V. Komarov, V. A. Ivanov and A. R. Khokhlov 2021 Polymer—quantum dot composite hybrid solar cells with a bi-continuous network morphology using the block copolymer poly(3-hexylthiophene)-b-polystyrene or its blend with poly(3-hexylthiophene) as a donor, Material Advances, 2: 1016–1023.

Shen T., J. Tian, L. Lv, C. Fei, Y. Wang, T. Pullerits and G. Cao 2016 Investigation of the role of Mn dopant in CdS quantum dot sensitized solar cell, Electrochimica Acta, 191: 62–69.

Speirs M. J., D. M. Balazs, D. N. Dirin, M. V. Kovalenko, and M. A. Loi 2017 Increased efficiency in pn-junction PbS QD solar cells via NaHS treatment of the p-type layer, Applied Physics Letters, 110: 103904–1 to 103904–5.

Sukharevska N., D. Bederak, D. Dirin, M. Kovalenko and M. A. Loi 2020 Improved reproducibility of PbS colloidal quantum dots solar cells using atomic layer—deposited TiO$_2$, Energy Technology, 8(1900887): 1–9.

Tian J., T. Shen, X. Liu, C. Fei, L. Lv and G. Cao 2016 Enhanced performance of PbS-quantum-dot-sensitized solar cells via optimizing precursor solution and electrolytes, Scientific Reports, 6(23094): 9.

Turak A. 2021 On the role of LiF in organic optoelectronics, Electronic Materials, 2: 198–221.

Wageh S., A. A. Al-Ghamdi, M. Soylu, Y. Al-Turki, W. El Shirbeeny and F. Yakuphanoglu 2013 Improvement of efficiency in CdS quantum dots sensitized solar cells, Acta Physica Polonica A, 124(4): 750–754.

Wang H., T. Kubo, J. Nakazaki, T. Kinoshita and H. Segawa 2013 PbS-quantum-dot-based heterojunction solar cells utilizing ZnO nanowires for high external quantum efficiency in the near-infrared region, Journal of Physical Chemistry Letters, 4: 2455–2460.

Wasilewski Z. R., S. Fafard and J. P. McCaffrey 1999 Size and shape engineering of vertically stacked self-assembled quantum dots, Journal of Crystal Growth, 201/202: 1131–1135.

Xu W., F. Tan, Q. Liu, X. Liu, Q. Jiang, L. Wei, W. Zhang, Z. Wang, S. Qu and Z. Wang 2017 Efficient PbS QD solar cell with an inverted structure, Solar Energy Materials and Solar Cells, 159: 503–509.

Xu Y., C. Gao, S. Tang, J. Zhang, Y. Chen, Y. Zhu and Z. Hu 2019 Comprehensive understanding of $TiCl_4$ treatment on the compact TiO_2 layer in planar perovskite solar cells with efficiencies over 20%, Journal of Alloys and Compounds, 787: 1082–1088.

Zheng W., D. Wang, Q. Wang and H. Sun 2019 The high performance of quantum dot sensitized solar cells co-sensitized with mixed-joint CdS and ZnS quantum dots, ECS Journal of Solid State Science and Technology, 8(6): Q96–Q100.

Index A: Solar Cells

Note: Page numbers are in **boldface** font for text within tables and are *italicized* for text within figures/captions.

Index B: General

For Product Safety Concerns and Information please contact our EU
representative GPSR@taylorandfrancis.com
Taylor & Francis Verlag GmbH, Kaufingerstraße 24, 80331 München, Germany